Animal Behavior

ELSEVIER *science & technology books*

• *Companion Web Site:*

http://booksite.elsevier.com/9780128015322/

Animal Behavior, by Michael D. Breed and Janice Moore

Resources for Professors:

- **Answers to the end-of-chapter questions**
- **Figures and appendix material from the textbook**
- **Lab manual with fully developed and tested laboratory exercises available for courses that have labs (http://www.elsevierdirect.com/product.jsp?isbn=9780123725820).**

   Instructor Resources site can be found at http://textbooks.elsevier.com/web/Login.aspx, and it is password protected. Please contact your sales representative at textbooks@elsevier.com for access to the Instructor Resources site.

ACADEMIC PRESS

Animal Behavior

Second Edition

Michael D. Breed
Ecology and Evolutionary Biology
University of Colorado
Boulder, CO, USA

Janice Moore
Biology Department
Colorado State University
Fort Collins, CO, USA

AMSTERDAM • BOSTON • HEIDELBERG • LONDON
NEW YORK • OXFORD • PARIS • SAN DIEGO
SAN FRANCISCO • SINGAPORE • SYDNEY • TOKYO
Academic Press is an imprint of Elsevier

Academic Press is an imprint of Elsevier
125, London Wall, EC2Y 5AS.
525 B Street, Suite 1800, San Diego, CA 92101-4495, USA
225 Wyman Street, Waltham, MA 02451, USA
The Boulevard, Langford Lane, Kidlington, Oxford OX5 1GB, UK

Second edition 2016
First edition 2012

Notices
Knowledge and best practice in this field are constantly changing. As new research and experience
broaden our understanding, changes in research methods, professional practices, or medical treatment
may become necessary.

Practitioners and researchers must always rely on their own experience and knowledge in evaluating
and using any information, methods, compounds, or experiments described herein. In using such
information or methods they should be mindful of their own safety and the safety of others, including
parties for whom they have a professional responsibility.

To the fullest extent of the law, neither the Publisher nor the authors, contributors, or editors, assume
any liability for any injury and/or damage to persons or property as a matter of products liability,
negligence or otherwise, or from any use or operation of any methods, products, instructions, or ideas
contained in the material herein.

ISBN: 978-0-12-801532-2

British Library Cataloguing-in-Publication Data
A catalogue record for this book is available from the British Library.

Library of Congress Cataloging-in-Publication Data
A catalog record for this book is available from the Library of Congress.

For Information on all Academic Press publications
visit our website at http://store.elsevier.com/.

Typeset by MPS Limited, Chennai, India
www.adi-mps.com

Printed and bound in the USA
Transferred to Digital Printing, 2015

Working together
to grow libraries in
developing countries

www.elsevier.com • www.bookaid.org

To my wife Cheryl, with my thanks for all of her support, and my daughters, Ann and Elise, whose enthusiasm for learning has been a source of inspiration.

—Michael D. Breed

To my parents, Doyle and Tillie, with gratitude for decades of love, support, and common sense.

—Janice Moore

CONTENTS

Contents

The world of animal behavior changes rapidly, and in this second edition of *Animal Behavior* we reflect some of that change. We have remained true to the overall organization of the first edition; it works for us in our classrooms, and as we have heard from colleagues, it also works well for them. Throughout that organizational structure, we have added new material and deleted some material as well, updating references along the way. We have modified over 40 figures and added several new case studies. In particular, Chapter 2 (Neurobiology and Endocrinology for Animal Behaviorists) has been thoroughly updated. In Chapter 3 (Behavioral Genetics), the "toolbox" approach has given way to a more linear presentation after considerable rewriting. Behavioral syndromes and related material can now be found in Chapter 4 (Homeostasis and Time Budgets), and magnetoreception is addressed in greater detail (Chapters 2 and 8). We have enhanced coverage of parasite-induced behavioral changes, be they manipulative (Chapter 9) or defensive (Chapter 10). Throughout the book, we have increased examples demonstrating that small squishy and crunchy animals exhibit behaviors that are every bit as intriguing as the actions of the so-called charismatic megafauna. In almost every chapter, we have modified and added to our coverage of "Bringing Animal Behavior Home," topics that invariably capture student interest. In short, our aim has been that of every good revision—improve and update, keeping the best elements of the book and adding to them.

In so doing, we continue to offer students an accessible approach to the major principles, mechanisms and controversies in the study of animal behavior. Throughout the book, we use Tinbergen's four questions—causation, survival value, ontogeny, and evolution—to frame animal behavior and to lend coherence to a diverse and highly integrative field of scientific inquiry. We see that same inquiry at the heart of the discipline, so we emphasize how to test hypotheses about animal behavior, and we encourage students to think critically about experimental evidence.

xi

We take stands on controversial issues in this book and have clearly expressed our points of view in scientific interpretations. We do not expect faculty members who teach animal behavior courses to agree with all of our interpretations. (Indeed, the two of us have had some lively discussions about some of these topics as the work has progressed!) Instead, we see areas of disagreement as pedagogical tools to help students understand that not all scientific issues are settled. In fact, we have highlighted those unsettled areas, because a textbook is not a compendium of absolute knowledge; it is a snapshot of current scientific understanding. We encourage faculty and students who use this textbook to approach our statements critically, and to ask how further hypothesis testing will improve knowledge and understanding.

Every field has a history, but the history of animal behavior is particularly informative because it teaches us so much about how scientists sort through controversy and learn, collectively, how to think critically. It also tells us why we are only now asking questions that might have been off-limits to earlier workers. We therefore begin this book with a brief overview of that history. Then, because evolution is, famously, that thing without which nothing in biology makes sense, we follow history with a refresher on evolution. In our experience as teachers, we have realized that such a refresher is often desirable.

After this introductory chapter, our approach flows from the physiological and genetic underpinnings of behavior (Chapters 2 and 3) through behavioral concerns of individual animals to the complexities of social behavior (Chapters 13 and 14). Chapters 2 and 3 provide ample foundation for mechanisms-oriented animal behavior courses and offer background for students in courses that do not emphasize mechanisms. Building on mechanisms, we consider behavioral homeostasis (Chapter 4). Learning (Chapter 5) and cognition (Chapter 6) are aspects of animal behavior that link neural processes with the behavior of the animal itself. An understanding of communication (Chapter 7) involves not only the underlying mechanisms, but the behavior of at least two participants. Midway through the book a study of orientation and migration (Chapter 8) builds a bridge from mechanisms to behavioral ecology. Behavioral ecologists will find contemporary coverage of the key elements of behavioral ecology in chapters on foraging (Chapter 9), self-defense (Chapter 10), mating systems (Chapter 11), parenting (Chapter 12), and social behavior (Chapters 13 and 14). Chapter 15 introduces the exciting—and essential—field of conservation behavior.

In our coverage, we recognize two emerging topics in animal behavior, cognition (Chapter 6) and conservation (Chapter 15), with full chapters. Cognition presents engaging and difficult hypotheses that will challenge our students, perhaps more than any other topic in animal behavior. Given societal debates over our relationships with animals and the ethics of maintaining animals in farms, zoos, and human households, our chapter on cognition provides a timely overview of the scientific evidence in that field. Cognition is one of the Next Big Topics in animal behavior and reveals some dimly lit areas of the discipline; we invite students to step onto the ground floor of an area of study that in the future will likely revolutionize our concept of and relationship with animals.

As for conservation, there is no denying that many species are in peril. Effective animal conservation programs can exist only with a thorough understanding of the behavior of the species that the programs seek to save. In our experience, community, ecosystems, and landscape-level ecologists often give short shrift to the importance of animal behavior in making conservation choices, and in so doing, they run the risk of failure. Effective conservation programs require knowledge of foraging behavior, mating systems, dispersal, and migration. Successful release of captive-reared animals—an increasingly important strategy in saving endangered species—requires substantial knowledge of animal learning, ranging from imprinting to learned aspects of foraging and antipredator behavior. We offer the last chapter of this text as a platform for integrating behavior and conservation, and hope to inspire a generation of students to use behavior as a conservation tool.

The majority of animal behavior students will not go on to careers in animal behavior, but most of them will enjoy the companionship of animals. We have highlighted the behavior of companion animals in special features called "Bringing Animal Behavior Home." This dimension of animal behavior is often important to students but may be overlooked in textbooks intended for animal behavior courses. Teachers of animal behavior courses may opt to cover companion animals, but even if a lecturer does not mention dogs or cats, including them in the text is one more reason for a student to read this book. This material also appeals to preveterinary students, who often enroll in animal behavior courses as preparation for their careers.

Because of our emphasis on evolution, we have not strictly excluded humans from our textbook, but neither is human behavior emphasized. Instead, when we do refer to human behavior, it is often in the context of questioning the traditional distinction between nonhuman animals and humans—a distinction that may be artificial in places, lacking sufficient scientific scrutiny. We explicitly reject an overly simplistic "my-genes-made-me-do-it" approach to behavior in general and human behavior in particular. In asking students to consider the continuum of all living things, we aim to promote critical thinking and a new consideration of traits that we may share with other species as a result of descent from common ancestors.

The laboratory manual that accompanies this text (*Field and Laboratory Exercises in Animal Behavior*) contains proven exercises. We encourage instructors who have not incorporated a laboratory in their course to consider doing so. A successful laboratory experience in animal behavior need not be expensive to present. The laboratory manual puts a strong focus on inquiry-based student participation, developing student strengths in hypothesis testing, and on giving the students hands-on experiences with topics covered in animal behavior courses. The exercises in the manual include a mix of field studies and laboratory studies. The targeted species of the field studies are widespread, and if the species mentioned in the manual does not occur locally, substitutions should be easy. The laboratory studies rely on easily obtained study animals. The focus is slightly biased to experiments with invertebrate animals; this reduces the burden of obtaining IACUC approval for some of the laboratories. However, the most popular laboratory for our students is a multi-week exploration of *Betta* (Siamese fighting fish) behavior.

This book is an outgrowth of our decades of teaching animal behavior—what we have learned, what we have wished for, where we have found great joy. As we have worked through this immense undertaking, we have been impressed with the wide-ranging curiosity of our colleagues, friends, and students, and with their willingness to help in so many ways, surpassing our ability to thank them. We are particularly grateful to Phil DeVries, Scott Altenbach, Jeff Mitton, Randy Moore, and Ben Pless, who generously contributed their outstanding photography to this effort, and whose images have proven that the appeal of animal behavior knows no academic boundaries.

Michael D. Breed
Janice Moore

CHAPTER 1

Of Cockroaches and Wolves: Framing Animal Behavior

LEARNING OBJECTIVES

Studying this chapter should provide you with the knowledge to:

- Understand that behavior, broadly defined, includes movement, social interaction, cognition, and learning.
- See that adaptive mechanisms provided by behavior give animals tools for adjusting to their environments and for manipulating the world around them.
- Be able to illustrate that four central questions drive the study of behavior. These are mechanism, utility, development (ontogeny), and evolution. Use these questions to form testable hypotheses about behavior.
- Integrate the basic principles of evolution with an understanding of animal behavior.
- Discover that the roots of contemporary studies of animal behavior are in ethology, comparative psychology, sociobiology, and behavioral ecology.

Animal Behavior. DOI: http://dx.doi.org/10.1016/B978-0-12-801532-2.00001-5

1.1 INTRODUCTION: ANIMAL BEHAVIOR

Science is the outcome of human curiosity. We want to know the "why" and "how" of almost everything. In fact, all of biology can be addressed with two types of questions: proximate and ultimate. *Proximate* means "coming very soon" or "next"; *ultimate*, in contrast, means "coming at the end" of a process. In biological terms, proximate questions ask about mechanisms—how has something developed? how does it work?—and ultimate questions ask about how something has evolved—what is its selective advantage? what is its evolutionary history?

> **KEY TERM** Proximate questions ask about relatively imminent causes or mechanisms responsible for a trait.

> **KEY TERM** Ultimate questions ask about the evolution of a trait.

This curiosity and these types of questions have framed the study of animal behavior. All of us have watched animals, laughing at the antics of pets and marveling at the aerial acrobatics of birds. Shortly, we will discuss two animals in particular—the wolf and the cockroach—that invite us into the world of behavioral biology. That world differs from one in which we simply watch animals. Like the rest of science, it is grounded in questions—four of them, to be precise. Two are proximate and two are ultimate. These questions were put forward decades ago by Niko Tinbergen, one of the founders of the modern study of animal behavior. We present them here, before we discuss wolves or cockroaches. These questions will be our constant guides, and if we allow them to, they will transform our "watching" into scientific exploration.

1. *What is the mechanism that produces a behavior?* This question addresses the internal mechanism—nervous, hormonal, and physiological—that results in a specific behavior. When we ask this type of question, we test hypotheses about how nerves, muscles, hormones, and physiology in general interact to produce behavior. Tinbergen called this "causation," but today's scientists use the word "mechanism" more frequently.

2. *How does a behavior develop?* Development, or ontogeny, involves both a genetic component and an environmental, or learned, component. When we ask this type of question, we ask where a behavior "comes from." We test hypotheses about the relative contributions of genetics and environment—"innate" tendencies and experience—to behavior. Tinbergen used the word "ontogeny" in this question—a word that is still in current usage, but that inconveniently lacks a verb form.

3. *What is the survival value (utility) of a behavior?* Today, we take this question to encompass both survival and reproduction—that is, what is the utility, or usefulness, of the behavior in terms of fitness, or its adaptive significance? When we ask this type of question, we test hypotheses about how behavior contributes to the survival and reproduction of an animal.

4. *How did the behavior evolve from an ancestral state?* Evolution, or phylogeny, tells us about the origin of a behavior in distant time. When we ask this type of question, we test hypotheses about the beginnings of a behavior in ancestral organisms.

> **KEY TERM** Causation is the direct mechanism responsible for a behavior. Examples of causation are hormonal and neuromuscular events.

These questions were initially developed by one of the founders of the modern study of animal behavior, Niko Tinbergen, and have been updated to reflect conversations among biologists about how to best express the questions. Later in this chapter, we expand on these questions and how they might be applied. For now, we turn back to natural human curiosity about animal behavior.

Our Fascination with Animal Behavior

Long before humans envisioned science, animal behavior intrigued our distant ancestors. Understanding how and why animals behaved meant the difference between enjoying dinner and becoming dinner, between eating and being eaten (Figure 1.1). Within a social group, it meant the difference between understanding and even manipulating that group, and being marginalized (Figure 1.2). For these reasons and more, animal behavior is a course unlike any other. If scholarly interests have any genetic basis, we should all be intrigued by animal behavior!

3

FIGURE 1.1
Images from the Lascaux Caves in southwestern France, a home to Paleolithic humans some 16,000 years ago. The caves contain hundreds of images of animals, mostly horses and deer, and the paintings are so accurate that in many cases the animals they depict can be identified. The purpose of the paintings is lost to prehistory, but many people think that the humans who made these paintings might have used them in ceremonies that were meant to enhance hunting success. From left to right: an aurochs (extinct in the Middle Ages), running horses and bulls, and a narrative scene involving a human. *Images: http://www.sacred-destinations.com/france/lascaux-caves, public domain.*

FIGURE 1.2
Observation and analysis of behavior are deeply engrained in human evolutionary history. This Japanese temple carving suggests investigation of behavior, along with analysis, and interpretation. *Photo: Michael Breed.*

The idea of innate human fascination when it comes to animal behavior is not as far-fetched as it sounds—although, as will become apparent, questions about intellect, predisposition, and genetics are difficult to answer in a definitive manner, especially with human subjects. Nonetheless, evolution, the foundation of biological thought, makes the concept of an innate interest in behavior a reasonable supposition.

DISCUSSION POINT: DEFINING BEHAVIOR

We define behavior as movement, social interaction, cognition, and learning. Scientists have struggled with the term *behavior* and it is surprisingly tricky to define. We think our broad definition works better as a discussion point than as a factoid to be captured in a multiple-choice test. Some other scientists' definitions of behavior focus on movement: they view behavior as the result of the activity of cilia, flagellae, or muscles. Interactions with other animals—in pairs or groups—are key to many complex behaviors. Learning and memory are important aspects of animal behavior, and it is possible to learn without moving a muscle. Thought and understanding—cognition—are also very much a part of animal behavior. Given the importance of learning, memory and cognition in studies of behavior, a definition that focuses only on movement captures a very small portion of the richness of behavioral studies.

What value does behavior have for an animal? Behavior yields immediate flexibility. Using behavior, animals can change their locations, maximize their chances of survival by taking advantage of knowledge from previous experiences, and in some cases, gain the advantages of living in a group. Unlike a typical plant, rooted in the ground and limited to whatever conditions the environment creates for it, behavior gives animals tools for ecological choice. Even sessile (nonmoving) adult animals can have motile offspring that exhibit preferences about where to settle in preparation for adult life. In sum, the richness of animal diversity yields an equally rich and enchanting variety of animal behavior.

4

The central precept of evolution is that organisms with traits that lead to increased survival and reproduction leave more offspring than do other such organisms without those characteristics. If the advantageous characteristics have a genetic basis, those genes and the traits they support will also increase in the population. In short, the ever-so-early human ancestor who was fascinated by animal behavior was much more likely to leave descendants than the one who was not terribly interested in where wolves were likely to hide or which watering hole was particularly appealing to deer.

Fast forward to the twenty-first century, and here you are, reading this book. Wolves are still with us, living at the margins of human habitat and in the wilderness beyond. It is no accident that wolves play a central role in legends, from the foundation of Rome by Romulus and Remus, human children reared by wolves, to the tale of Little Red Riding Hood, a girl with a red cape and remarkably bad judgment (Figure 1.3). Like us, wolves are social animals. Like us, they seem clever at times; they may even solve problems and use past events to predict future ones (this is part of what we call *cognition*; see Chapter 6). In fact, domesticated wolves that we call dogs have done quite well for themselves. They have fared a lot better than their wild cousins, who inspire so much fear and resentment that they have been forced to near-extinction.

FIGURE 1.3
Little Red Riding Hood has come to symbolize human fear of predators or, in this case, blissful ignorance. Fear plays a powerful role in human interactions with animals, and one of the goals of studying animal behavior should be overcoming that fear through knowledge.

DISCUSSION POINT: GENES AND ENVIRONMENT IN BEHAVIOR

How might we test the hypothesis that some of our behavioral preferences may be largely inherited, whereas others are strongly influenced by the environment? Chapter 3 offers some good hints about how to do this for a given preference, but testing this hypothesis is not a trivial exercise. Also, note that we have called the idea that humans may have a genetic predisposition to be interested in animal behavior a "supposition" and "not … far-fetched." In other words, we are speculating. Is our speculation warranted? Why or why not?

1.2 WOLVES: LESSONS IN SOCIAL BEHAVIOR

Human fear of wolves notwithstanding, wolves reveal much about social behavior and how it functions in ecological associations, such as those between predators and prey. Evolutionarily, predators are under strong natural selection to eat, but they place their prey under strong selection to escape or defend themselves (see Chapter 10). This leads to a kind of predator–prey arms race.

For prey species, the evolution of large body size can be an extremely effective defense. Large size, while not in itself a behavioral trait, has major behavioral implications. We should not be surprised to discover that the largest herbivores in most ecosystems (elephants, baleen whales, and bison come to mind) are many times the size of the largest carnivores. A large, strong animal can be difficult or impossible for a predator, no matter how well armed, to take down.

What is the evolutionary response of predator species? Predators that cooperate as a group can often overcome prey that would be impossible for an individual to handle. Add social cunning—the ability of predators to work together against prey—and the result is truly formidable carnivores such as killer whales, African lions, and, of course, wolves. Chapters 13 and 14 reveal that the evolution of social behavior is a complex and interesting topic, as sociality has many costs and many benefits, and these must be balanced, both now (ecological time) and over evolutionary time, for the social behavior to persist.

How does cooperation help predators when they hunt animals three to five times their size? One highly effective strategy employed by wolves is "relay running." Tired prey, gasping for oxygen and their muscles weakened, are easier to kill than rested prey. If one wolf were to chase an elk for miles, the two animals would tire more or less equally. But what if the wolf can trade the lead in the chase with another wolf? The elk gets no respite, forced to run at full speed, while each wolf can slow its pace when another takes its turn. Over time, the elk can no longer outrun the wolves, which then group for the kill (Figure 1.4).

5

FIGURE 1.4
Wolves are cooperative hunters and work in packs. These photographs offer just a glimpse into the complex communication that allows successful cooperation. *Photo: Frank Wendland, W. O. L. F. Sanctuary, www.wolfsanctuary.net, www.facebook.com/wolf.sanctuary.*

FIGURE 1.5
This photograph of a wolf dining on "prey" is from a wolf sanctuary, but nonetheless allows the viewer to understand how "eye" might be an effective tool to use against potential prey. It is never a good idea to run from a canid. *Photo: Frank Wendland, W. O. L. F. Sanctuary, www.wolfsanctuary.net, www.facebook.com/wolf.sanctuary.*

Wolves also are clever at ambushing prey; watching the direction that a herd of prey is moving, they can hide near that possible pathway (Figure 1.5). This means that when the wolves spring out of hiding, they will be near to striking distance of the prey and can catch the prey before prey defenses are organized. Indeed, it is this inclination that is thought to be the basis for what has become, through artificial selection, the highly skilled herding ability seen in herding dogs. They work with the shepherd, circling around the "prey," giving them the "eye" (a stare that immobilizes sheep), and generally cooperating with the shepherd to move and control the sheep. It is the final sequence—the kill—that is thought not to be a desirable attribute of a herding dog. This general sequence can be discerned in many wild cooperative canid hunters.

It is worth mentioning that prey defenses reach well beyond running to escape (see Chapter 10 on self-defense). In the coevolutionary race between wolves and their prey, social behavior is as powerful a weapon for elk and bison as it is for the wolves. Social hunters such as wolves prey on ungulates (hooved mammals, like deer), and this fact may be one of the driving evolutionary forces leading ungulates to live in groups. Given the opportunity, herds of some ungulates will circle like a wagon train in a Western movie, with the more vulnerable young members of the herd in the center of the herd where they can be protected. (Later in this book, you will discover how some herds act in the best interests of the young members, whereas other herds behave more selfishly.)

Wolves can take smaller prey, such as rabbits and hares, as well, but this is more of an individual proposition than social foraging is. The ability of wolves to switch prey preferences, depending on the availability of each prey species and competition with other predators, gives them the flexibility to respond to changing environmental conditions. Wolves therefore serve as one of the classic examples of how predators can control prey populations, particularly when a given prey species population increases. To survive, the average Minnesota wolf needs to eat the nutritional equivalent of 15–20 deer a year. Deer are most vulnerable in the winter when the snow is deep, as they may be weakened from lack of food and the snow makes it more difficult for them to escape. This means that in a winter with little snow the wolves may have a hard time, and their population levels can be affected. Despite the claims of nature programs on television, old and sick deer are not sacrificed for the "good of the species." Instead, these are the easiest deer for wolves to take. For that reason, deer populations where wolves are present are likely to contain mostly healthy animals; in contrast, when wolves are absent, sick or malnourished deer may survive.

Wolf behavior came to the forefront in the conflict over reintroduction of wolves into the Greater Yellowstone Ecosystem of Wyoming and Montana. Like most animals, wolves do not recognize human boundaries, whether those are the limits of protected areas, state lines, or international borders. Radio collars reveal that wolves can move hundreds, or even thousands, of kilometers in startlingly short periods of time. These long-distance movements have resulted in wolf populations far from their protected reserves and have created strong conflict among conservationists, ranchers, and developers.

Wolves capture the imagination, and because of the devotion of scientists who study them, we understand something of their behavior. Unfortunately, they do not lend themselves to the close-in scrutiny required to answer some behavioral questions, and most of us do not have a lot of personal experience with wolves. We may be more familiar with, say, crickets or cockroaches. In fact, why is it almost impossible to step on a cockroach? No matter how fast we think we are stomping, the cockroach will probably be faster. Although this animal behavior question is not quite as gripping as how to avoid being a wolf's dinner, for most people in the modern world it may be more familiar and, in some respects, more puzzling. After all, humans are "smarter" than cockroaches (an assumption we may have to challenge later in this book)— why this lack of talent when it comes to smashing them? Surprisingly, exploring a bit about cockroach behavior will reveal some fundamental concepts that apply to all of animal behavior.

OF SPECIAL INTEREST: JUST-SO STORIES AND THE INTERPRETATION OF BEHAVIOR

Throughout human history, people have tried to explain the natural world. For most of that history, all that was needed was an observation and a vivid imagination. For instance, the following African folk tale explains the habits of the hippopotamus:

Once upon a time, the hippopotamus had beautiful, long hair—hair that, understandably, made the hippopotamus proud, and even vain. The other animals were jealous, and after some muttering and resentment, they decided to bring the hippopotamus down a few notches. They set fire to the beautiful hair. The hippopotamus, with hair aflame, rushed into the water and extinguished the fire. Alas, the lovely hair was gone. From that day forward, the hippopotamus, nearly hairless and ashamed of its looks, has remained in the water during the day and comes out only at night (Figure 1.6).

This engaging story is an excellent example of how we can look at a trait and imagine how it came to be, without any proof at all. Of course, we know that the hippopotamus's hairlessness and habits did not evolve as a result of jealousy and pyromania, but thinking critically about such stories is a good exercise in scientific reasoning. Scientifically, we rely on comparative or experimental tests of hypotheses, and because of that, we should always ask if the controls in the experiment were appropriate, if all variables were accounted for, and if additional hypothesis testing is needed to strengthen our conclusions. In studying animal behavior, it is extremely important not to imagine an interpretation for a behavior. It can be difficult to avoid telling just-so stories about behavior, but learning to recognize hypotheses that have actually been tested is key to practicing the science of animal behavior. In sum, behavioral hypotheses need to be accurately stated and tested with sound scientific methods.

FIGURE 1.6
This hippo, hair on fire, is a lesson in the dangers of pride and jealousy. The folk tale may teach good behavior and wise choices, but it does not provide a scientific explanation for the appearance and behavior of the hippopotamus. *Illustration: Rafael Rivero, a former student.*

DISCUSSION POINT: THINKING BACK TO THE FOUR QUESTIONS

Review Tinbergen's four questions at the beginning of this chapter. How would you investigate them using wolves? What would your hypotheses be and how would you test them? What difficulties might arise?

1.3 COCKROACHES: MODELS FOR ANIMAL BEHAVIOR

We can argue that natural selection has favored cockroaches that do not wait around to be obliterated when a foot is descending on them—that such cockroaches leave more offspring with that valuable trait than more sluggish conspecifics do—and we would be correct. That still does not tell us exactly how the cockroach manages to be so perceptive and fast; that is, it does not address causation. For that answer, we turn to the world of neuroethology—the study of the interaction of the nervous system and behavior.

Cockroaches are wonderful research animals. *Periplaneta americana*, the American Cockroach (a misnomer—the insect was introduced to the rest of the world from Africa), can be raised easily and in large numbers. Indeed, there are thousands of species of cockroaches (most do not live in houses!), allowing for fascinating interspecific comparisons. They are hardy and large enough to allow significant experimental tinkering; although determined entomologists can and have performed surgery on mosquitoes, cockroaches are more convenient candidates for the operating table.

Given the luxury of simply staring at a cockroach before it skitters away, we would see that there are two small appendages on the posterior end of the animal. These are called *cerci* (the singular is *cercus*, from a Greek word meaning "tail") and they are covered with little "hairs" (Figure 1.7). The hairs move when there is a sudden shift in the air currents around them. This does not have to be a breeze; it can be as subtle as a change in air pressure in front of a descending foot or predator's flicking tongue.

The direction of movement of the hairs tells the nervous system of cockroach about the direction of air current, and thus the source of the threat, in the following manner: An individual hair can move most easily in one direction only, so given an air current from a specific direction, only certain hairs will move. Each hair is connected to a nerve; the direction of the air current will determine which hairs move and, in turn, which nerves fire. The nerves are connected to a ganglion at the posterior end of the cockroach; the ganglion contains large nerve cells called *giant interneurons* (GIs). These GIs extend from the posterior of the cockroach, through the thorax, up to the head of the insect. In the thorax, they are connected to neurons that control leg motion and running. Action potentials (neuronal messages; see Chapter 2) can race up these GIs and communicate with ganglia in the thorax, which then cause the legs to move (Figures 1.8 and 1.9).

There are two behavioral consequences of all the wiring just described: (1) A few milliseconds after the hairs move, the cockroach turns away from the perceived threat, and (2) after turning, it runs. The turn-away direction is informed by which hairs are moving, that is, the direction of the air current; the impulse to run then overrides everything else that the cockroach might have been doing. After all, that is the nature of escape—it is a life-or-death matter.

FIGURE 1.7

American cockroaches, *Periplaneta americana*. Note the pair of small posterior appendages on the cockroach in the lower right-hand corner of the photograph. These are the cerci, important predatory-avoidance sensors. *Photo: Jeff Mitton.*

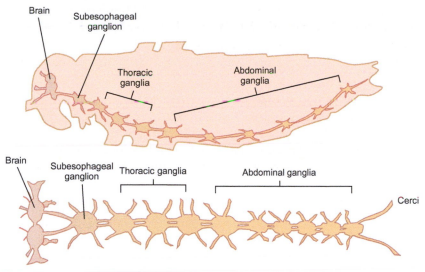

FIGURE 1.8
The nervous system of an insect. The nerve cord is ventral (true of arthropods and their relatives), and the spacing of the ganglia reflects the segmentation of the organism.

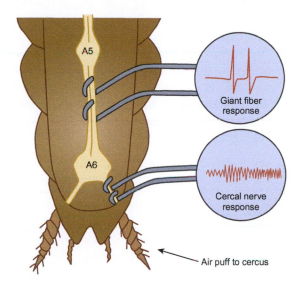

FIGURE 1.9
The cerci with neural connections. Changes in air pressure move the "hairs" on the cerci, initiating action potentials (nervous impulses) that travel up the GIs. These impulses override all other nervous activity and cause the animal to jump, orienting away from the air pressure source, and run away.

Because of this behavior, cockroaches have a remarkable story to tell us. Not only do we understand the mechanism that causes escape, but we understand how natural selection favors this behavior. We might then ask if very young cockroaches are as skilled as older ones at escape—in other words, how does the escape response develop? As expected, given the importance of the escape response, the hatchling cockroach escapes just as effectively as an adult does. Even though the newly hatched cockroach has far fewer hairs on its cerci, the ones that are there do their job perfectly. Moreover, molting does not seem to alter the escape response very much. Recall that cockroaches are hemimetabolous insects; that is, an immature cockroach, called a *nymph*, looks very much like an adult, with the exception of wings and sexual morphology. In the process of becoming an adult cockroach, the nymph grows and molts its exoskeleton several times. It is common for the process of molting to affect at least some other processes in insects, but the cockroach escape response is not affected in any significant way by this profound event in the life cycle. (There are some changes in the response, but they are on the order of a few milliseconds and may not be ecologically significant.)

The fact that the escape response remains the same throughout cockroach development probably reflects the fact that effective escape is advantageous for cockroaches of any age. This contrasts with a cockroach's ability to produce sex attractant, which is an adult feature. Although immature cockroaches look very similar to adults, they are not sexually mature; in cockroaches, as in vertebrates, hormones influence morphological and behavioral development. The ability to produce sex attractant, therefore, does not develop until adulthood and, in at least some species, is timed to coincide with egg production.

Finally, if we look at relatives of cockroaches, we can put the escape response in a context framed by evolutionary history, or phylogeny. Cockroaches are not the only animals with GIs; many arthropods have them, as do other organisms that are considered relatives of arthropods by some scientists. In most of the animals that have been investigated, one of the functions of the GIs is escape; for instance, the flick of a crayfish tail and the scamper of a cockroach are both mediated by the GIs and they both accomplish the same thing: predator avoidance.

Although the phylogeny of arthropods and other invertebrates is being reconsidered—a common occurrence in science as we increase our understanding—we might reasonably hypothesize that a ventral nerve cord, with GIs that assist in escape, is a shared trait among arthropods and their kin, inherited from a common ancestor. In light of this evolutionary context, the absence of such structures in, say, spiny lobsters becomes much more intriguing than if it were simply a nugget of information dropped out of someone's isolated laboratory. How do spiny lobsters cope without GIs? What led to their loss? We leave this question about evolution for contemplation, but it may well have everything to do with the interaction of behavior, the nervous system, and evolution.

10

FIGURE 1.10
Niko Tinbergen, a key figure in the establishment of ethology. Tinbergen's four questions—causation, survival value, development, and evolution—still form the foundation for studies of animal behavior.

1.4 THE FOUR QUESTIONS REVISITED

At the beginning of this chapter, we introduced four attributes—mechanism, development, utility, and evolution—that frame animal behavior. These are called "Tinbergen's questions," in honor of their author (Figure 1.10), and if we can answer Tinbergen's questions about a given behavior, we will understand that behavior at every level is addressed by science. Because of the core importance of these questions, first introduced at the beginning of this chapter, we will now consider them in more depth. In a 2013 paper (see the Further Reading list at the end of this chapter for the citation) contemporary biologists Kevin Laland and Patrick Bateson point out that even now there are few systems in which all four questions have been studied in detail. Figure 1.11 shows Bateson and Laland's illustration of how the four questions apply to the study of bird song.

The wolf and the cockroach offer good examples of how different organisms lend themselves to different questions. The wolf invites us to think about behavior from the perspective of ultimate questions—how social behavior evolved—which in turn causes us to ask about the survival value of social behavior. Cockroaches, because of their size, their compatibility with laboratory settings, and their overall versatility, allow a more detailed approach to development and causation; given their membership in that greatest of all phyla, Arthropoda, they tempt us with phylogenetic considerations. Both species offer windows into how Tinbergen's four questions can be applied in analyses of animal behavior.

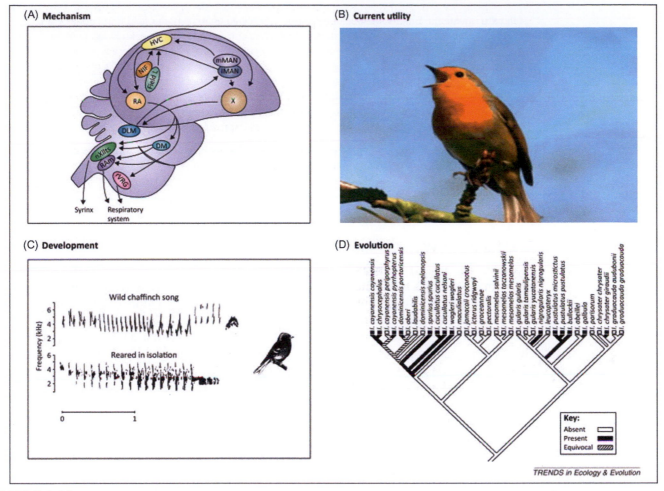

FIGURE 1.11

Tinbergen's four questions, applied to bird song. (A) The neural *mechanism* underlying song production. This is a cross section of the bird's brain, showing the centers involved in coordinating song production. Parts of the brain thought to integrate motivation (labeled the MAN regions, for magnocellular nucleus of the anterior neostriatum) give input to the higher vocal center (HVC), which then integrates signals to the respiratory system and the vocal box, or syrinx. The result is coordination via the arcopalladium (RA) and basal ganglion (X) of exhalation across the vocal cords to produce sound. (B) Bird songs often are used (have *current utility*) in territoriality and mate choice. This male chaffinch is singing in this type of context. (C) The *development* of birdsong can depend on learning. Wild chaffinches produce appropriate songs because they have had the opportunity to hear and learn songs of other chaffinches. When reared in isolation, chaffinches do not produce the appropriate song. (D) The ability to learn and produce the songs varies among species of bird, and this phylogeny shows that song has *evolved* differently in some groups of species than others. *Reprinted from Trends in Ecology and Evolution, Vol 28, Bateson, P. and K. N. Laland, Tinbergen's four questions: an appreciation and an update, Pages 712–718, Copyright 2013, with permission from Elsevier.*

OF SPECIAL INTEREST: PROXIMATE AND ULTIMATE QUESTIONS IN ANIMAL BEHAVIOR

In the beginning of this chapter, we briefly introduced the idea of proximate and ultimate questions. Proximate questions are concerned with the day-to-day (or millisecond-to-millisecond) understanding of how individual organisms develop and work. Ultimate questions address the products and process of evolution, including the shift from ancestral states to what we observe today, and the contribution of natural selection and other evolutionary forces to that observation.

These two approaches to discovery have had different influences on animal behavior, and they have different strengths and weaknesses. During the history of this fairly young field, comparative psychologists tended to focus on proximate questions. Their work was usually conducted in the laboratory and involved controlled experiments that revealed, for instance, how animals learned. In contrast, classical ethology research was usually conducted under natural conditions and posed ultimate questions. Behavioral research conducted on wild animals under field conditions tells us a lot about how animals behave in nature but also has some disadvantages: knowledge of the animal's previous experience—perhaps even its sex or age—is unavailable. Moreover, the observer has no control over weather, predators, and other influences that can interfere with comparisons. Comparative psychology solved these problems; the subjects—including their diets, rearing conditions, age, and history—were well known and the conditions of observation could be controlled and replicated. However, the laboratory conditions may bear small resemblance to the field conditions under which the behavior might actually be expressed.

Each type of research has strengths and weaknesses. Neither is "right" or "wrong"; the choice of methods depends on the organism and the question. A combination of field and laboratory work often provides the most convincing evidence with which to test a hypothesis.

Mechanism: What Is the Cause of a Behavior?

To understand what Tinbergen called "causation," or mechanism, we focus on the animal's behavioral responses to stimuli from the external environment. Does the animal handle stimuli the same way every time they are presented? Or does the nature of the response depend on the animal's internal state? Consider the difference in behavior between hungry and satiated (well-fed) animals. The sight of a potential prey animal, such as a deer, may elicit stalking, chasing, and pouncing by a hungry mountain lion, but a well-fed lion may have little interest in a passing deer. Studying causation allows us to delve into how behavior is regulated and how animals make choices about their activities. In the case of the wolf, this might mean understanding the relationship of hunger signals (e.g., nutrients in the blood), inclination to hunt, and effort expended in hunting. In the case of the cockroach, the exquisite connection between the cerci, the GIs, and the escape behavior, activated by a small change in air pressure, is a classic example of causation.

Ontogeny: How Does a Behavior Develop?

How does a behavior change during the lifetime of the animal? To survive and reproduce, an animal must change its behavior throughout its life. Animals that undergo metamorphosis are extreme examples of this: larval stages are devoted to foraging and growth, whereas adults disperse and reproduce. Many birds and mammals acquire skills while still under parental protection; young animals learn through experiences that shape their adult behavior. Sexual maturation plays an important role in development as well. Physiological and morphological changes, coordinated by hormones, prepare animals for the behavioral challenges of adulthood. Hypotheses about ontogeny (development) postulate learning and physiological development as driving forces in shaping the behavior of adult animals.

Utility: What Is the Utility of a Behavior?

In other words, how does the behavior help an animal to survive and reproduce—what is the adaptive significance of the behavior? Tinbergen called this "survival value," but the usefulness of a behavior, in evolutionary terms, extends far beyond survival; offspring must be produced and they must survive. Of course, much of behavior is clearly related to the day-to-day survival of animals, which is a prerequisite to any other consideration. Finding food, water, and shelter is a continuing behavioral challenge for animals. Because of this, it is all the more puzzling to observe a behavior and find, as we often do, that its survival value is not immediately clear. For instance, why does a dog circle three or four times before lying down? The immediate survival value of this behavior may not be obvious, but we would hypothesize that the behavior is rooted in an ancestral behavior that did improve survival. (See the next question: evolution!) Testing hypotheses about survival value plays an important role in developing an accurate understanding of behavior. As Tinbergen said, "… the part played by natural selection in evolution cannot be assessed without proper study of survival value."

Evolution: How Did the Behavior Evolve from an Ancestral State?

Behaviors do not arise spontaneously when animals need them. As with body structures, behaviors evolve through modification of previously existing, ancestral actions. Sometimes an action that is used in one context is "co-opted" for use in a different behavioral context; for example, the predatory behavior of a venomous snake may be co-opted and modified to serve as a defensive threat. Displays, such as those used in courtship, are often assembled over the course of evolution from a seemingly unrelated set of movements that the animal performs in other contexts. Testing hypotheses concerning behavioral evolution often requires constructing a "family tree," or cladogram, of related species and tracing how a behavior has changed in either its form (the way in which it is produced) or its function over evolutionary time.

1.5 EVOLUTION: A REVIEW

Behavior, like any other biological trait, evolves. Evolution is the foundation of biological thought in all fields, and understanding evolution paves the way to understanding animal behavior. Because of the explanatory power of the theory of evolution by natural selection, this book is grounded in evolutionary principles; they are the backbone of our understanding of animal behavior and of biology in general. Indeed, without those principles, biology shrinks to a collection of facts about the living world, much like a collection of cute stuffed animals—nice enough to look at, but not particularly coherent, and certainly no source of predictions about other collectible items.

13

Given the importance of evolution, then, it would be a mistake not to review that theory. In fact, we need to begin by visiting the word *theory*. To many nonscientists, the word *theory* can be used interchangeably with the word *hunch*. Scientists, however, have a clear definition for *theory* that goes well beyond the notion of a hunch. For scientists, a theory is an overarching concept that explains a large number of facts and observations about the natural world and that can be used to make predictions about future observations. A theory has such weight and scope and explains so many facts that it is unlikely to be refuted, although it may be refined as one tests the predictions it generates.

> **KEY TERM** A theory is an overarching concept that explains a body of facts about the natural world and that generates testable predictions.

Darwin's idea—the theory of evolution by natural selection—is one such theory, supported by evidence from geology, paleontology, agriculture, embryology, biogeography, anatomy, and molecular biology. It is based on the following fairly simple observations about populations of organisms: (1) Organisms can often produce far more offspring than those required to replace the parents—more than available resources can support. (2) In some species, this excess results in competition for those limited resources, and the outcome of that competition determines which individuals survive and reproduce. (3) These competing individuals are not alike; they have a variety of different traits. (4) The traits that belong to the successful competitors will be overrepresented in the next generation compared to the traits that conferred fewer benefits. This process resulting in differential survival and reproduction is called *natural selection*. (5) Some of this variation among traits can be inherited; it has a genetic basis.

> **KEY TERM** Natural selection is a process that results in increased survival and reproduction compared to that of competing organisms.

When the representation of the traits changes in the population—usually because of the action of natural selection—gene frequencies change. This change in gene frequency is called *evolution.* (Darwin did not know about genes, so he called evolution *descent with modification from a common ancestor.* We now know that gene frequency is the thing that is modified or changed.)

DISCUSSION POINT: NATURAL SELECTION

Note that there are two components of the process of natural selection: survival and reproduction. Although natural selection favors traits that improve survival and reproduction relative to those that do so to a lesser extent, sometimes traits that improve survival do not improve reproduction, or vice versa. How might increased reproduction, for instance, affect survival? Could increased survival have negative consequences for reproduction? What happens then? What kinds of experiments or observations might tease these apart?

Genetic Variation: A Necessary Ingredient of Genetic Change

In thinking about these observations, we can see that genetic variation is needed for evolution, or genetic change across generations. After all, if animals are genetically identical, there are no genetic advantages or disadvantages to be had, no distinctions to be made, and evolution cannot occur. What are some causes of genetic variation?

Two of the most important sources of new genetic variation in populations are mutation and gene flow:

1. *Mutation* is a change in an organism's DNA. If it occurs in the DNA of eggs and sperm, it can be inherited. The effects of mutations are random. Some will be beneficial, some will be detrimental, and some will be of no consequence.

2. *Gene flow* occurs when organisms move into a population, bringing their genes with them. In that way, new genes can be introduced and genetic variation increases.

In addition, genetic mixing, or recombination, can be very important in evolution. It does not change the amount of genetic variation in a population, but it does allow new, more advantageous, combinations of genes to emerge. Recombination occurs when chromosomes from parents line up during gamete production. Chunks of information from one parent may switch places with homologous (similar) chunks from the other parent during meiosis, thus creating new combinations of genetic traits. (See Chapter 3 for a more detailed discussion of these genetic mechanisms.)

Changes in Gene Frequencies

Given these causes of genetic variation, how do gene frequencies change; that is, how does evolution itself happen? Scientists are still exploring this question, and new discoveries are not in short supply, but four mechanisms are particularly important: mutation, genetic drift, migration, and natural selection.

Mutation not only serves as a source of genetic variation, but when it occurs, it also alters gene frequencies. Because of this, it is one cause of evolution, albeit a limited one. Mutations occur at very low rates, and because they are random, we do not expect them to produce directional change in a characteristic of a population in response to the environment.

Genetic drift is another accidental shift in gene frequencies, based on the fact that some individuals fail to survive simply because they are unlucky. They build their nest in the path of a tornado or in low-lying areas just in time for a particularly bad hurricane season. They are eliminated not because they are poorly adapted to the environment, but because of unpredictable misfortune. Nonetheless, the genes that these unfortunate organisms carry are removed from the population and gene frequencies change. As might be expected, this is especially influential in small populations, where a random event can eliminate some genotypes completely. Thus, genetic drift can actually reduce genetic variation. Once genetic variation is reduced, processes such as mutation and migration may restore variation, but these work very slowly.

Migration is another source of genetic variation that can also change gene frequencies and result in evolution. When organisms join a population and interbreed with residents, the subsequent generation will exhibit gene frequencies that differ from those in the population prior to the arrival of the migrants.

Mutation, genetic drift, and migration can all result in evolution, but such changes in gene frequencies do not necessarily proceed in any particular direction. Mutation and genetic drift have random consequences, and migrants are not necessarily better suited for the environment than the host population is. In short, there is no reason to expect increased survival and reproduction associated with traits that change as a result of this kind of evolution.

Natural selection, the fourth mechanism of evolution, is different. It is the only mechanism that causes an increase in the frequency of adaptations—the inherited traits that promote survival and reproduction. For natural selection to work, there must be genetic variation. Because of that variation, some individuals are better at surviving and reproducing than others. They leave more offspring, and those offspring carry the genes that are associated with the beneficial traits. In the process, gene frequencies change.

At this point, please note that although genetic variation emerges in an unpredictable fashion, and although mutation, genetic drift, and migration can lack any recognizable direction, the products of natural selection are anything but random. This is why evolution may be based on random events (e.g., mutation, genetic drift), but it does not often generate random results. Natural selection, a powerful engine of evolution, increases the frequency of adaptations in populations from generation to generation; it does not result in a random mix of traits, even though the material on which it acts may be randomly generated.

This, then, is *fitness*—the ability to contribute genes to the next generation. The popular media would have us believe that fitness means winning reality shows, getting rich, winning races, and doing other spectacular things. In contrast, biologists say that fitness has two main components: survival and reproduction.

> **KEY TERM** Fitness is the relative ability of an organism to contribute genes to the next generation.

Adaptation and Behavior

Although natural selection results in adaptation, not every trait is an adaptation. Some folks are eager to explain almost everything they see in terms of adaptation, but a bit of skepticism is the hallmark of a scientist. Recall that not every change in gene frequency (i.e., evolution) results from natural selection; changes that result from genetic drift or mutation are probably not going to be adaptive. In addition, some traits might have been adaptive at an earlier time but are no longer adaptive; the pelvic bones of whales are examples of traits that have lost their utility. We refer to these as *vestigial* traits. These bones served the terrestrial ancestors of whales, but are of little use to whales now. Yet other traits might not be adaptive in a perfect world, but may exist as a result of constraints imposed by different traits that are adaptive. Thus, although it might seem to be invariably adaptive to produce a large number of supremely healthy offspring, resource limitation or other environmental challenges may thwart this simple expectation. Instead, they may favor organisms that limit the number of offspring they produce, investing heavily in each one, or organisms that produce many offspring, each of which receives little investment.

15

Is every trait that increases survival and reproduction an adaptation? No, because some beneficial traits are not the products of natural selection. The fact that we do not wander aimlessly across busy highways certainly increases our probability of survival and reproduction, but that trait is learned, not inherited; in addition to the fact that an adaptation increases survival and reproduction, an adaptation is inherited.

Finally, an adaptation may be a beneficial trait for which the ancestral function has been modified. An original trait is often co-opted to result in something more beneficial. Some interesting terminology has grown up around this concept, but the core idea is that original traits are often co-opted. For instance, in animal behavior studies, signals used in communication are often co-opted from noises, movements, or odors that the animal already produces (see Chapter 7). Finally, a trait or structure that is evolutionarily malleable and can, over the course of natural selection, be modified for a new purpose is often termed a *preadaptation*.

> **KEY TERM** A co-opted trait, in evolutionary terms, is one that served a different function in an ancestor than the one it does today. These are also called *exaptations*.

> **KEY TERM** A preadaptation is a trait that is subject to modification for a new function by natural selection.

This reinforces an important evolutionary principle: natural selection acts on available variation, not the best variation imaginable. Thus, organisms do not get the adaptations they "need" in some evolutionary version of online shopping; their adaptations are a result of natural selection acting on existing traits.

Optimality and Behavior

A consideration of the limitations of adaptation and natural selection leads to the idea of optimality. An optimum, from a Latin word meaning "best," is a fairly unequivocal notion: it is not almost-best or sort of good; it is unsurpassed—the best. This idea will be covered in more detail in Chapter 9, but for now, given the discussion so far, it is probably apparent that there are many explanations for why adaptations are not always optimal. For a variety of reasons, ranging from developmental constraints to ecological trade-offs, the genetic variation available to natural selection may not include the ideal trait; in that case, the product of natural selection will not be optimal. In addition, the world is not a homogeneous place, either in time or in space. What might be optimal today, or in this location, may be suboptimal tomorrow, in another location. In fact, it is precisely the heterogeneous nature of the world at large that favors a particular and powerful adaption: sexual reproduction.

The thought of sexual reproduction as an adaptation might sound strange at first, but consider the fact that many organisms do not depend completely on sexual reproduction. Instead, they often clone themselves when reproducing. Plants frequently do this; strawberries, aspen trees, and daylilies all generate other plants from underground roots. In the world of animals, pieces of some flatworms can grow into entire individuals, some fish eggs can develop and hatch without being fertilized, and coral reefs grow when individuals bud and create new individuals. Initially, it seems that such organisms achieve high fitness; after all, 100% of their genes are passed on to each offspring compared to the 50% that sexually reproducing parents donate, often called *the cost of meiosis*. Given those differences, how can sexual reproduction be an adaptation?

Sexual reproduction is a major source of genetic variation. In addition to the genetic mixing that can occur between chromosomes when gametes are produced, fertilization also generates variation among offspring of animals such that each offspring carries a unique combination of genetic material bequeathed by its parents. Thus, while sexually reproducing animals must produce twice the number of offspring found in asexually reproducing animals to be able to convey the equivalent amount of genetic information to the next generation, the variety among sexually produced offspring means that at least some of them may be able to survive and reproduce in a changing world. Of course, if conditions are stable, then asexually reproducing animals have an advantage: Their own combination of traits succeeded, and as long as nothing changes, that combination should work well in subsequent generations.

In a world threatened by plagues, blistering droughts, and unrelenting floods, it is difficult to argue that the earth is a particularly stable environment. Indeed, the advantage of sexual reproduction is so significant that even asexually reproducing animals frequently participate in sexual reproduction as well. Chapter 12 reveals how all sexual reproduction is not equal and how mates are evaluated and selected. As will become apparent, the value of producing variable offspring has fueled much animal behavior, from the flash of peacock tails to the chorusing of frogs.

Speciation and Behavior

Finally, the formation of species (called *speciation*) provides a powerful tool for asking questions about the history of traits, including behavioral traits. So far, this review of evolution has addressed changes in gene frequency and the manner in which those changes happen. How can such genetic shifts within a population of animals result in the biological diversity we see today? Something else has to happen: the flow of genes within that population must be interrupted. This can occur when a population is subdivided and parts of that population are isolated from each other. Eventually, if environments of the subdivided groups differ, natural selection will favor different traits in the two new populations. As time passes, differences accumulate, and the two populations will no longer be able to interbreed were they to have that opportunity. For instance, foxes, coyotes, wolves, and domestic dogs all evolved from the same ancestor, with foxes diverging earliest in evolutionary time. Differences have accumulated so that now foxes cannot successfully interbreed with any of the other species. However, coyotes, wolves, and domestic dogs split more recently and remain so similar that hybrids are common. By the way, complete isolation is not necessary for species formation; a small amount of gene flow does not counteract the accumulation of differences.

Such reproductive isolation can result from a variety of causes:

1. *Geographic barriers.* If a population is subdivided by the emergence of a mountain range, river, or other inhospitable habitat, animals on one side of the barrier will be unable to breed with animals on the other side. The same effect occurs if part of the population moves away.
2. *Resource shifts.* For animals that live and reproduce on a resource, the ability to colonize new resources decreases the likelihood that they will encounter or mate with individuals in the parent population.
3. *Mate choice.* If females diverge in their preferences for male characteristics, for instance, and if that divergence has a genetic basis, then eventually there will be two distinct gene pools, each sporting one or the other preferred male trait.
4. *Genetic change.* Mutations that prevent proper meiosis can produce individuals that cannot mate with other members of the population. This is thought to be the origin of about 4% of plant species.

Phylogeny and Behavior

Such isolating events produce splits in lineages; if these divergences persist, two species will result from one. Tracing this history reveals a branching pattern—a "tree of life." (Actually,

17

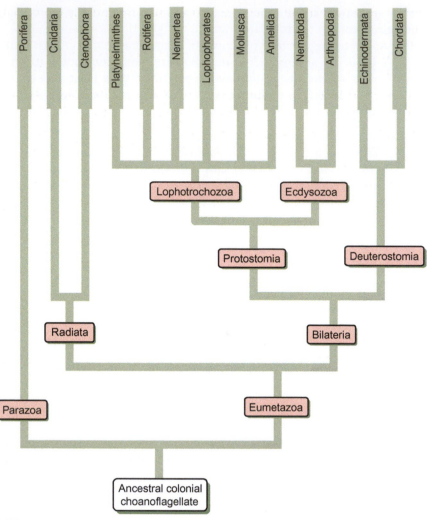

FIGURE 1.12

A simplified phylogeny of Animalia, showing the profound effect of behavior (movement) on the evolution of all animals. The Radiata have poorly developed musculature and nervous systems; many are sessile, and the ones that do move are often at the mercy of currents and tides. Because of this, the radiate animals must be equipped to confront the world as it approaches them from any direction. Bilateral symmetry (found in the Bilateria), along with a suite of other traits, is favored when locomotion is a more important part of the behavioral repertoire, allowing more controlled movement through the environment. Radially symmetrical members of the Bilateria (e.g., sea stars) are either sessile or thought to be descended from sessile ancestors. Note that "deep phylogeny"—the hypothesized relationships among phyla—is not a settled field; for instance, based on genetic studies, some scientists suggest that the Ctenophora diverged from the animal lineage before the sponges did. Likewise, relationships among the Bilateria may shift. What is clear is that the ability to move—that is, the fundamental behavioral state of most animals—set the stage for natural selection to favor a more streamlined shape, with bilateral symmetry and ultimately cephalization and sophisticated sense organs.

KEY TERM A phylogeny is a hypothesis of evolutionary relationship among types of animals (often species).

it is more like a disorganized bush.) The relationships among the branches—which branch came from what stock—are hypotheses to be tested, and the resulting bush-like representation is called a *phylogeny*, a combination of two Greek words meaning "phylum" or "race" and "origin." (The phylogeny shown in Figure 1.12 does not show all the branches, but it does show the relationships clearly.) Because these relationships are hypotheses, phylogenies will be refined as we understand more about them. Keep in mind that every phylogeny is a working hypothesis.

Phylogenies yield a variety of important information. Three nuggets to value in this course are the following:

1. Phylogenies are radiations over evolutionary time, not ladders of evolutionary progress. There is no "high" or "low," and the fact that humans can do calculus, and a sea star cannot, does not make us any better at living. If anything, it probably means that humans are a good deal worse than a sea star when it comes to life under water. (The notion of a ladderlike progression of life forms has been with us since Aristotle proposed the Great Chain of Being. During medieval times, angels were inserted between humans and God, so humans were almost at the pinnacle, but not quite.)

2. Because phylogenies are radiations, and not ladders, anything alive today represents a line that has been subject to natural selection for as long as the lineage of any other living organism. Sea stars are not "older" than humans, nor are they more "primitive," and they are in no way our ancestors. Instead, all living things share common ancestors to greater or lesser extents. A phylogeny reveals when that shared ancestry ceased, relatively speaking.

3. The places in a phylogeny where branching occurs are called *nodes*. These nodes represent the last common ancestor that the organisms in the branches shared. If organisms share a common ancestor, then they share characteristics (*characters*, in scientific terms) that they have inherited from that ancestor. The more characters that organisms share as a result of descent from a common ancestor, the more closely they are related. Based on the hypotheses of relationships portrayed in phylogenies, we can ask if traits, including behavioral traits, are influenced by evolutionary history and shared ancestry. If they are, we may be able to predict their occurrence in other animals.

Because of this, it is not surprising that the GIs of arthropods work in similar ways and that they often have similar functions. The ragged bush of evolutionary relationships will make sense of discoveries about animal behavior. It will also generate more questions. This is how science thrives.

1.6 THE STUDY OF ANIMAL BEHAVIOR: WHERE DID IT COME FROM?

In a world that is alive with the flashes of color, the buzzing and the splashing that come from animal behavior, why do we bother with history? When compared to other aspects of biology, animal behavior comes from unusual, and sometimes seemingly incompatible, places. Those origins influence what we study today and how we study it. They offer important considerations: Are field studies better than ones in the laboratory? How do we ask questions about development, causation, evolution, or survival value in each setting? Who are the people who wrested our knowledge away from the "just-so" storytellers and placed it into the realm of science? History answers these questions, and many more, and helps us understand our work today and our aims for the future.

Ancient Greece

Given the hypothesized hard-wired fascination with behavior, it will not be surprising that scholars and philosophers have written about behavior since antiquity. Aristotle (384–322 BCE) (Figure 1.13) probably thought most deeply about animal emotions and intelligence, writing, "Many animals have memory, and are capable of instruction; but no other

> **Note**
> From ancient times through the Enlightenment, scientists were called *philosophers*! This ancient tradition continues today with the advanced degree "Doctor of Philosophy" conferred upon scientists as well as scholars of other disciplines.

FIGURE 1.13
Aristotle contributed much to understanding the natural world and to forming a scientific basis for studying animal behavior. His works laid the foundation for accurately recorded natural history and for scientific testing of hypotheses about behavior.

19

creature except man can recall the past at will" (*The History of Animals*, Book I). These words, centuries old, capture a current debate in animal behavior. Is Aristotle correct, or are at least some animal species capable of "mental time travel"; that is, can they call on past experiences to shape future plans?

Darwin and the Victorians

Another highly notable figure in the science of animal behavior, Charles Darwin (1809–1882; Figure 1.14), lived centuries later. Darwin's thinking about evolution has had a tremendous effect on studies of animal behavior. His 1862 book on orchid pollination showed great insight into insect behavior and the coevolution of pollinator behavior with orchids. Later, he turned his full attention to behavior in his 1872 book *The Expression of Emotions in Man and Animals*. Articulating the fascination with social behavior that is shared by many people, Darwin explored how humans and other animals convey their emotions and act impulsively based on emotion. He argued,

That the chief expressive actions, exhibited by man and by the lower animals, are now innate or inherited,—that is, have not been learnt by the individual,—is admitted by every one. So little has learning or imitation to do with several of them that they are from the earliest days and throughout life quite beyond our control; for instance, the relaxation of the arteries of the skin in blushing, and the increased action of the heart in anger.

FIGURE 1.14
Charles Darwin had a keen interest in animal behavior and wrote extensively about his observations and hypotheses concerning behavior.

This gives great weight to the role of evolutionary history in shaping behavior, and it also poses some profound questions, for public debate over the relative contributions of genes and the environment to behavior has raged from Darwin's time to the present. The issue of nature versus nurture and how these shape behavior will be examined in more detail in Chapter 3.

The late 1800s and the early 1900s were the heyday of natural history exploration of the world. Earlier explorations, such as Lewis and Clark's of the American West, Darwin's voyage on the Beagle, and Alfred Russel Wallace's discoveries in the Amazon and the islands of the southwestern Pacific Ocean laid the groundwork for public fascination with natural history in Victorian England, the United States, and Europe. Published accounts of explorations found large and eager audiences of readers; the writings of natural historians fed the public imagination and had much the same effect on Victorian society that modern-day videos of animal behavior have on current public perceptions of wildlife and animal behavior. This era also saw the public display of exotic animals in zoos develop as entertainment and education; in zoos the public could directly see some aspects of the behavior of animals that they would never encounter in nature (Figure 1.15).

20

FIGURE 1.15
Zoos gained popularity as public entertainment in Victorian times. Live animal exhibits allowed the public to observe the behavior of exotic animals.

The Transition to Modern Science

Georges Romanes (born in 1848 in Canada, studied and worked in England, died 1894) was the academic successor to Darwin, bridging Darwin's work and contemporary evolutionary biology. Romanes is credited with founding comparative psychology as a field of study and contributed to our early understanding of cognition (thought processes) in pigeons. He was a creature of his time; he used anecdotal information and attributed attitudes such as "unselfishness" and "courage" to his study animals. The attribution of human traits to animal behavior began long before Romanes, of course, but Romanes was explicitly anthropomorphic in his work, claiming that he could infer the mental state of an animal from its behavior, based on what his own mental state would be if his behavior were similar. This was, as one writer claims, "poetic," but it did not serve the scientific study of animal behavior well.

OF SPECIAL INTEREST: CLEVER HANS

Clever Hans was indeed a clever horse (Figure 1.16). According to the popular press of the early twentieth century, he was able spell, read, and do arithmetic. He answered questions posed by strangers by tapping his hoof and was always correct, even when his trainer was absent. How could this be?

That was the question asked by the director of the Psychological Institute of the University of Berlin, who in 1907 sent a student, Oskar Pfungst, to find out what Hans could and could not do. Clever Oskar devised a mixture of trials; in some, the questioner knew the answer, and in others, the questioner did not. Pfungst showed that Hans was able to answer correctly only when the questioner knew the answer. It seems Hans had learned that questioners would make very small, involuntary movements as his taps neared the correct answer, and he used that to gauge his response. Hans was clever indeed, but he did not read and do arithmetic. This revelation did considerable damage to the idea that animals shared human mental abilities.

FIGURE 1.16
This drawing represents an old photograph of Clever Hans, the horse that was clever enough to learn a lot about human behavior. He learned, for instance, when to stop tapping his hoof—that is, when human handlers unconsciously exhibited signs as his foot tapping neared the right answer to an arithmetic question. Hans never learned arithmetic, but he knew a lot about human behavior.

C. L. Morgan (1852–1936), Romanes' successor, began the process of setting things aright. He is recognized as the first biologist to question the anthropomorphic approach that was so popular in the nineteenth century. In 1894, he wrote what Galef has called "possibly the most important single sentence in the history of the study of animal behavior" in a book called *An Introduction to Comparative Psychology*:

> In no case may we interpret an action as the outcome of the exercise of a higher psychical faculty, if it can be interpreted as the outcome of one which stands lower in the psychological scale.

FIGURE 1.17
Konrad Lorenz, followed by some of the geese that helped secure his position as one of the founders of animal behavior. Lorenz was deeply interested in instinctive behavior; among other things, he studied imprinting in goslings. Inset: Karl von Frisch, who decoded the dance language of bees.

FIGURE 1.18
A Skinner box, named after B. F. Skinner, who designed a similar apparatus. The classic box contains rewards (the food), signals (lights and speaker), and negative stimuli (the electric grid, in this case). Using rewards and punishments, Skinner studied animal learning.

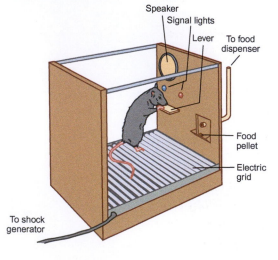

This statement is called *Morgan's Canon*. A canon is a rule or a law, and this rule has had both beneficial and regrettable effects on the study of animal behavior. It was, of course, essential to the development of animal behavior as a science rather than a storytelling device. However, it was sadly misinterpreted in some cases, leading to profound underestimation of the cognitive abilities of many animals. Such was the backlash against attributing human qualities to animals on the basis of behavior that even the study of cognition was frowned upon for decades. There is a difference between arguing for the most parsimonious explanation (as Morgan did) and denying the possibility of any other explanation.

Parsimony is difficult to implement in a world of anecdotal data. We have Edward Thorndike to thank for directing animal behavior away from that world. Thorndike studied learning in cats under laboratory conditions at Columbia University; he was also a major contributor to early work in human learning. He insisted that the study of animal behavior be quantitative and subject to rigorous scientific standards. He standardized his testing conditions and took care to expand sample sizes and do systematic comparisons.

European and American Traditions

In Europe, public fascination with natural history evolved into scientific inquiry about the mechanisms underlying animal behavior. Karl von Frisch, an Austrian biologist born in 1886 (died 1982; Figure 1.17 inset), studied the sensory biology of honey bees. Among his discoveries were the existence of ultraviolet perception, the ability of bees to perceive the polarization of light, and the use of color vision by bees. His observations of bees ultimately led to his decoding the dance language of honey bees.

Konrad Lorenz (born in Austria in 1903, died 1989; see Figure 1.17) and Niko Tinbergen (born in Holland in 1907, died 1988; see Figure 1.10) joined von Frisch in being leaders in the study of animal behavior in natural environments, and this developing scientific field became known as *ethology*. Von Frisch's, Lorenz's, and Tinbergen's contributions come up repeatedly in this book, and the three were awarded the 1973 Nobel Prize in Physiology or Medicine for their roles in developing important theories and conceptual approaches for the study of behavior. The focus of ethology was on the study of animals in natural, or minimally modified, environments. A critical common thread in the work of ethologists was a focus on innate, or instinctual, factors that shape behavior. This is an important point that we will revisit.

In America the development of animal behavior studies as a scientific endeavor took a very different path. J. B. Watson (1878–1958) founded *behaviorism* as a discipline of study; behaviorism focused on external manifestations of behavior, rather than the internal emotional or innate lives of humans and animals, and viewed modification of behavior by reinforcement as the predominant mechanism underlying behavior. B. F. Skinner (1904–1990) was a Harvard professor whose approach emphasized the ability of animals to learn and modify their behavior based on experience. Using animals such as rats and pigeons, which were easy to maintain in the laboratory, he developed paradigms for how animals could learn skills based on rewards and consequences. He is famous for developing the "Skinner box" (Figure 1.18), an isolated environment in which the animal could be presented with stimuli, rewards, and punishments, and the resulting behavior could be studied. Skinner's work was in the field of comparative psychology. In contrast to the ethologists, Skinner emphasized learning and developmental flexibility in shaping behavior. Interestingly, the work of the Russian behaviorist Ivan Pavlov (born 1849, winner of the 1904 Nobel Prize in Physiology or Medicine for his research on digestion) on conditioning in dogs had more influence on Skinner than on the European ethologists.

Daniel Lehrman (1919–1972, a professor at Rutgers University in New Jersey) became the focal point, on the American side of the Atlantic, of the split between the approaches of the ethologists and comparative psychologists. The nature versus nurture argument (see Chapter 3) became highly personal, as well as academic. The two sides of the debate finally came together to form a consensus at a meeting in Palo Alto, California, in 1957. However, the nature–nurture debate resurfaces from time to time, as it did in the bitter conflict between E. O. Wilson, a Harvard biologist who championed sociobiology as integrating neurobiology, physiology, behavior, ecology, and evolution, and Richard Lewontin (also at Harvard) and others who attacked sociobiology as social determinism, or a disguised version of social Darwinism.

Modern Science

Contemporary studies of animal behavior have outgrown much of this debate, as nearly all scientists recognize the intertwining of genes and environment in shaping an organism, including its behavior. The scientific field of animal behavior occupies much of the ground envisioned by Wilson for sociobiology, encompassing studies that range from the genetic and physiological mechanisms underlying behavior to those that place behavior in its ecological and evolutionary contexts. In the past three decades, *behavioral ecology* has succeeded *ethology* as the term used to characterize studies of behavior in natural environments. *Comparative psychology* has fallen into disuse, as animal research in many psychology departments has turned to the neurobiological bases of behavior.

DISCUSSION POINT: WHAT'S IN A NAME?

What do we—the folks who study animal behavior—call ourselves? It turns out that the answer varies. In this book, for convenience, we often use the term *behaviorist*. Is a person who studies wolves in the field a behaviorist? B. F. Skinner might disagree. What about *behavioral biologist*? Psychologists who study animal behavior might disagree. Much like the definition of animal behavior, the name of a person who studies that topic can be hard to pin down! The difficulty in defining the animal behavior and the difficulty in naming those who study it may well be a reflection of the spotted history of the field, gathered in from different traditions in field and lab. You will see a variety of identifiers applied to people who study animal behavior. The important point is what they do and the scientific rigor with which they do it.

In 1976, Donald Griffin courageously established the field of cognitive ethology, with a book titled *The Question of Animal Awareness: Evolutionary Continuity of Mental Experience*. This was no small task; scientists had embraced Morgan's Canon with fierce enthusiasm and were not inclined to deal with things they could not measure directly. Basing his argument in solid evolutionary theory, Griffin questioned the prevailing assumptions about the cognitive superiority/uniqueness of humans and opened the door for further scientific investigation. Griffin was not new to the field of animal behavior. He made other significant contributions to the study of behavior; almost 40 years before his first book about cognition, he and a fellow student at Harvard discovered that bats use sonar when navigating in the dark.

The world of animal behavior continues to change, reflecting novel scientific discoveries as well as emerging social concerns about animals. For instance, conservation behavior is a new and rapidly developing field; as the name implies, *conservation behaviorists* bring their expertise to important issues such as preservation of endangered species, design of natural reserves, and management of human–animal interactions in urban and suburban settings. From its early beginnings with clever horses and dancing bees, the study of animal behavior is helping to redefine the way we look at how behavior works, how it benefits animals, how it develops, and how it has evolved.

1.7 *UMWELT*: THE WORLD IN WHICH ANIMALS BEHAVE

This unlikely history, from Clever Hans to the contrasting worlds of the ethologists and the comparative psychologists, leads us to the central and continuing challenge in the study of animal behavior. Surprising as it may sound, this challenge is not about observing secretive organisms, not about discovering predator–prey linkages or the real meaning of some chirp or bellow. Those are indeed challenges, but our foremost obstacle is understanding the sensory-perceptual world of the study organism—that is, its *Umwelt*—in a way that allows us to ask meaningful questions. This German word meaning "environment" was used by Jakob von Uexkull (1864–1944), sometimes called *the father of ethology*. He explained this word by referring to how questing ticks locate hosts. Ticks need three cues—butyric acid (secreted from mammalian sebaceous glands), warmth, and then hair—in order to succeed in locating their next blood meal. For a tick, then, those three stimuli are its entire world (Figure 1.19). Never mind the green leaf it may be sitting on, never mind the color of the sky—butyric acid, body temperature, and hair inform the behavior of the questing tick. Thus, if we try to determine if a questing tick prefers a long, thin leaf or a lobed leaf, we will probably be wasting our time. To ask sensible questions in animal behavior and to sensibly interpret the results, we must understand the *Umwelt* of the animal we study.

FIGURE 1.19
The idea of *Umwelt* articulates one of the central ideas in animal behavior: each animal has a sensory-perceptual world. The most important task of the behavioral biologist is also the most difficult—to penetrate that world. A study of ticks inspired Jakob von Uexkull to think about *Umwelt*. Many animals have contributed to our knowledge of animal behavior, but few have contributed more than the tick! *Photo: Charles Schurch Lewallen.*

We stress that this is not as easy as it seems. Our main impediment is the fact that we are trapped in our own *Umwelt*. If there is any doubt, try counting how many times words like *see* or *look* are used in conversation—"I'm not sure I see the difference between the political philosophies of Smith and Jones" or "How does that vacation schedule look to you?" Is the philosophical difference really visible, or is the speaker using *see* to mean something like *comprehend*? Is the question really about the appearance of the schedule, or is it about the appeal of the proposed schedule? We, as humans, use *look* and *see* because vision is a large part of our sensory-perceptual world. It would be just as reasonable to sniff the difference between two philosophies, but no one ever says that; sniffing is not a huge part of the human *Umwelt*.

> **KEY TERM** *Umwelt* is the overall context in which an animal behaves, including its sensory environment and its behavioral capabilities.

This is perhaps the fuel for the excitement that has surrounded animal behavior since before it was named—the challenge, never quite met, of getting out of our own skins and perceiving the world in a completely different way. The animals around us are fellow travelers; that much has been clear for ages. Why do they do what they do? How do they "see" the world?

SUMMARY

The science of animal behavior is young, and Tinbergen's questions remain unanswered for many interesting behaviors. That is expected in the world of science. What is important is learning a bit more about how to ask the questions, and what questions are appropriate. The goal is to place observations of behavior in a framework of concepts. Using the four central questions of animal behavior—causation, development, survival value, and evolution—we can form testable hypotheses and come to a much deeper understanding of behavior.

Behavior includes movement, social interaction, cognition, and learning. Behavior is nonetheless difficult to define, and there may be definitions better than the one you encounter here. At its core, behavior provides animals with adaptive mechanisms for adjusting to changes in their environment and for manipulating the world around them.

We have provided you with two intriguing examples of animals that have contributed greatly to our understanding of animal behavior, and we have offered a quick refresher tour of evolution. This should help you to apply the basic principles of evolution to your understanding of animal behavior. You will probably want to refer to this section as you read the rest of this book.

This chapter concludes with a brief overview of the history of the study of animal behavior. Contemporary studies of animal behavior are rooted in ethology, comparative psychology, sociobiology, and behavioral ecology. As historians agree, the value of history is not in memorizing names and dates, but in understanding our collective path to the present day. This is as true of science as any other realm of history, and grasping how thought about animal behavior has developed over the past two and a half millennia will illuminate our current and future perspectives on how animals behave.

STUDY QUESTIONS

1. Make your own attempt to define behavior. What are the important elements? Do you see why movement alone does not capture all of behavior?
2. Without looking back in the chapter, list the four central questions of animal behavior (Tinbergen's questions).
3. Explain the importance of each of the four questions.
4. Wolves exemplify complex social behavior. Can you see how hunting in groups makes wolves more efficient predators and increases the range of prey they can exploit?
5. Cockroaches help us understand causation and adaptive value. Can you explain how cockroaches can so easily escape when you try to squash them?
6. Read some scientific papers about animal behavior that were written in the early and middle parts of the twentieth century. Would you view them as ethological or psychological? Why? Do you recognize influences from Lorenz, Skinner, and others?

Further Reading

Bateson, P., Laland, K.N., 2013. Tinbergen's four questions: an appreciation and an update. Trends Ecol. Evol. 28, 712–718.

Burghardt, G.M., 2010. Comparative animal behavior—1920–1973. In: Breed, M.D., Moore, J. (Eds.), Encyclopedia of Animal Behavior, vol. 1. Academic Press, Oxford, pp. 340–344. http://www.sciencedirect.com/science/referenceworks/9780080453378.

Darwin, C., 1872. The Expression of Emotions in Man and Animals. John Murray, London.

Drickamer, L.C., 2010. Animal behavior: antiquity to the sixteenth century. In: Breed, M.D., Moore, J. (Eds.), Encyclopedia of Animal Behavior. Academic Press, Oxford, pp. 63–67. ISBN 978-0-08-045337-8, http://dx.doi.org/10.1016/B978-0-08-045337-8.00204-7. http://www.sciencedirect.com/science/article/B6MV5-50G4S80-6B/2/8dc30829a6175ac6490be06ddf6aa7a8.

Drickamer, L.C., 2010. Animal behavior: the seventeenth to the twentieth centuries. In: Breed, M.D., Moore, J. (Eds.), Encyclopedia of Animal Behavior. Academic Press, Oxford, pp. 68–72. ISBN 978-0-08-045337-8, http://dx.doi.org/10.1016/B978-0-08-045337-8.00198-4. http://www.sciencedirect.com/science/article/B6MV5-50G4S80-66/2/15d6769a4160781c02e8015f8da6bd1a.

Espinoza, S.Y., et al., 2006. Loss of escape-related neurons in a spiny lobster, *Panulirus argus*. Biol. Bull. 211, 223–231.

Galef Jr., B.G., 1996. Historical origins. In: Houck, L.D., Drickamer, L.C. (Eds.), Foundations of Animal Behavior. University of Chicago Press, Chicago, IL, pp. 5–12.

Houck, L.D., Drickamer, L.C. (Eds.), 1996. Foundations of Animal Behavior: Classic Papers with Commentaries. University of Chicago Press, Chicago, IL.

Kays, R, Curtis, A, Kirchman, J.J., 2010. Rapid adaptive evolution of northeastern coyotes via hybridization with wolves. Biol. Lett. 6, 89–93.

Kellert, S.R., Wilson, E.O. (Eds.), 1993. The Biophilia Hypothesis. Island Press, Washington, DC, 484pp.

Kojola, I., Aspib, J., Hakalaa, A., Heikkinena, S., Ilmonic, C., Ronkainend, S., 2006. Dispersal in an expanding wolf population in Finland. J. Mammal. 87 (2), 281–286.

Kruuk, H., 2003. Niko's Nature: A Life of Niko Tinbergen and His Science of Animal Behaviour. Oxford University Press, Oxford, 391pp.

Libersat, F., Leung, V., Mizrahi, A., Mathenia, N., Comer, C., 2005. Maturation of escape circuit function during the early adulthood of cockroaches *Periplaneta americana*. J. Neurobiol. 62, 62–71.

Mech, L.D., 1981. The Wolf: The Ecology and Behavior of an Endangered Species. University of Minnesota Press, Minneapolis, MN, 384pp.

Mech, L.D., 1993. The Way of the Wolf. Voyageur Press, Stillwater, MN, 120pp.

Mech, L.D., Boitani, L. (Eds.), 2007. Wolves: Behavior, Ecology, and Conservation. University of Chicago Press, Chicago, IL, 472 pp.

Moore, R., Moore, J., 2006. Evolution 101. Greenwood Press, Westport, CT.

Ritzmann, R., 2010. Visuomotor control: Not so simple insect locomotion. Curr. Biol. 20 (1), R18–R19.

Ryan, J.F., Pang, K., Schnitzler, C.E., Nguyen, A.-D., Moreland, R.T., Simmons, D.K., et al., 2013. The genome of the ctenophore *Mnemiopsis leidyi* and its implications for cell type evolution. Science 342, 1327–1329.

Sorabji, R., 1993. Animal Minds and Human Morals: The Origins of the Western Debate. Cornell University Press, Ithaca, NY, 267pp.

Taborsky, M., 2010. Ethology in Europe. In: Breed, M.D., Moore, J. (Eds.), Encyclopedia of Animal Behavior, vol. 1. Academic Press, Oxford, pp. 649–651. http://www.sciencedirect.com/science/referenceworks/9780080453378.

Tinbergen, N., 1963. On aims and methods of ethology. Z. Tierpsychol. 20, 410–433.

Verhulst, S., Bolhuis, J., 2009. Tinbergen's Legacy: Function and Mechanism in Behavioral Biology. Cambridge University Press, Cambridge, 262pp.

Wilson, E.O., 1984. Biophilia. Harvard University Press, Cambridge, MA, 157pp.

Neurobiology and Endocrinology for Animal Behaviorists

LEARNING OBJECTIVES

Studying this chapter should provide you with the knowledge to:

- Analyze how axons carry electrical signals from point to point within the nervous system and the role of neurotransmitters in carrying information among neurons.
- Assess the relative roles of the central and peripheral parts of the nervous system in coordinating behavior and see how reflex loops provide a basic model for coordination of behavioral responses.
- Summarize how hormones are produced by glands and transported to target organs or tissues, where they have immediate or long-term effects.

Animal Behavior. DOI: http://dx.doi.org/10.1016/B978-0-12-801532-2.00002-7

- Diagram the impact of hormones on behavior in vertebrates, with a focus on hormonal effects on development and behavior, and gain a similar comprehension of hormones and behavior in invertebrates, with a focus on understanding the roles of juvenile hormone (JH) and ecdysone in invertebrate behavior.
- Generalize that perception of the external world is essential for all forms of life, including bacteria, and that information gained from perception is critical in behavioral decisions.
- Weigh the fact that each species' sensory world may be unique and understand that assumptions about perception based on human senses can dramatically mislead studies of animal behavior.
- Organize how animals may perceive a constellation of stimuli that includes light, polarization of light, temperature, sound, molecular energy (smell, taste), magnetic fields, electric fields, and inertia (touch) and understand how each stimulus is sensed.

2.1 NEUROBIOLOGY, ENDOCRINOLOGY, AND SENSORY SYSTEMS: AN OVERVIEW

This chapter is devoted, in large part, to Tinbergen's first question: causation (mechanisms underlying behavior). It covers the basic background about neural, endocrine, and sensory systems that is necessary for understanding animal behavior. An animal's neural system is composed of neurons, which collect information about the animal's internal and external environments and organize responses to the environment. An endocrine system relies on transmitters, hormones, that carry signals from glands via the body fluids or blood to target organs. Sensory systems, which are a subset of the neural system, are the interface between an animal and its external environment; sensory organs gather information about the world outside an animal. Animals are also capable of sensing their internal world, through perceptions of states such as pain and hunger (Figure 2.1).

These systems provide the foundation on which animal behavior is built; a basic understanding of these systems is essential for an animal behaviorist. For instance, consider sensory systems: What is their ecological role? True, they are extensions of the nervous system, but they are also gateways through which animals collect essential information about their surroundings. In turn, the abiotic nature of those surroundings can have profound effects on the way sensory systems function. A failure to comprehend the sensory world of

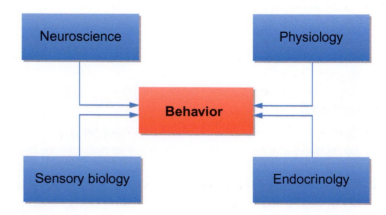

FIGURE 2.1
These four areas of study each add much to the understanding of the mechanisms underlying behavior. Adaptation drives these contributions, and although it may be tempting to consider mechanism without thinking about evolution, many of the most striking studies of animal behavior show how evolution shapes mechanism, and how the available mechanisms dictate evolutionary solutions for ecological problems that animals face.

animals sadly limits any understanding of their behavior. This is the message of *umwelt*, the concept of understanding an animal's behavior within its environment introduced in Chapter 1. Tinbergen was right when he included causation in his list of essential questions for those who want to understand animal behavior.

Likewise, even when one studies behavioral ecology or conservation, a basic knowledge of how the nervous and endocrine systems come together to shape animal behavior is imperative. The conservation of migratory birds, for example, is intricately linked to the hormones that shape their migratory behavior and how climate change or environmental pollutants might affect the endocrine triggers for migration. Some knowledge of how neurotransmitters shape personality and interact with social roles is fundamental to understanding social behavior. Fortunately, as a general rule, neural, endocrine, and sensory mechanisms use the same underlying framework across all animals, so mechanisms found in mammals are also likely to exist in some form in insects.

Much of the material in this chapter reviews what is normally covered in a general biology course, so referring to a general biology textbook, or to a beginners' physiology textbook, for additional information may be helpful. What follows is a brief refresher, a look at that "old" material with a behavioral twist.

Timescales and Behavioral Change

Behavior is *modulated*, or adjusted to fit conditions, on two very different timescales within the life of an animal: short-term, moment-to-moment change in activity and longer term change, perhaps over seasons.

Hormones and neurotransmitters are important agents of modulation that act at both timescales. A hormone is a molecule produced by a gland that affects a target organ within an animal. Epinephrine (adrenaline) is a hormone that has an immediate effect on an animal's physiology when it is forced to defend itself. Estrogen and testosterone, however, affect development over an animal's lifetime and regulate reproductive behavior from season to season. Age-related or seasonal changes in behavioral state are associated with changes in the underlying hormonal state of the animal. A good example of longer term behavioral changes induced by hormones is the female bird or mammal preparing a nest for her young; she likely has elevated levels of prolactin, a hormone that is generally associated with parental behaviors.

Short-term shifts in behavior, such as the transition from sleeping to searching for food, or feeding to mating, are usually mediated by changes in neurotransmitter levels. Neurotransmitters are small molecules used to convey signals within the nervous system. Neurotransmitters can be viewed as hormones that tend to act quickly. Thus, when a bird or a mammal enters into its daily inactive period (in other words, when it sleeps), the level of gamma-aminobutyric acid (GABA), a neurotransmitter, goes up in its brain. Drugs used by humans to treat insomnia, such as zolpidem (Ambien®), increase the levels of this neurotransmitter.

> **KEY TERM** Modulation is the ability to vary a behavior along a range of possible responses. For example, aggressiveness might be modulated depending on the strength of an opponent.

The following sections begin with a discussion of minute-to-minute regulation of behavior and continue with long-term seasonal and lifelong developmental changes.

2.2 WHAT DOES AN ANIMAL BEHAVIORIST NEED TO KNOW ABOUT NEUROBIOLOGY?

The nervous system takes information from cells and tissues, organizes it, and prioritizes the animal's behavioral responses. It is the physiological key to animal behavior; all behavior is organized by the nervous system. Sensory cells, discussed in Section 2.3, gather information from the environment and send it to the central nervous system (CNS) via axons. The CNS, which in most animals includes the brain and a central nerve cord (the spinal cord in vertebrates), organizes incoming information. It sets priorities for the animal's use of its time and energy, and sends signals to effector cells, such as muscles, which translate the decisions made in the CNS into behavioral and physiological actions. Learning and genetics meet in the nervous system; it is the storehouse for learned information, although genetic information in nervous system cells plays key roles in organizing the development and expression of behavior. Interconnecting cells in the nervous system house memory, decision making, and other integrative functions.

The neuron is the basic unit of the nervous system. A neuron is a type of nerve cell, consisting of an axon, a cell body, and dendrites. Neurons carry electrical impulses between locations via axons. They form networks of varying complexity, with the brain, a dense cluster of neurons, being the most complex. This section begins with insights into neurons and how they work, moves on to how neurons communicate with each other and with tissues and organs within an animal, and concludes with how the CNS functions as a coordination center for behavior.

CASE STUDY: A SIMPLE NERVOUS SYSTEM

Take a look at an organism that is, at once, seemingly simple and complex. The roundworm, *Caenorhabditis elegans*, has 302 neurons (compare this with the billions of neurons possessed by humans) and 81 muscle cells (Figure 2.2). Because of the anatomical simplicity of *C. elegans* and extensive knowledge of its genetics and development, it has become a model for understanding the basics of behavior. The role of each neuron can be investigated by looking at the behavior of mutants in which the activity of a single neuron is changed, by staining worms for specific neurotransmitters and observing which neurons contain those transmitters, and by measuring the metabolic activity in neurons. Genetic techniques can also be used to modify the production or reception of neurotransmitters, to disable sensory input, or to eliminate specific behavioral responses. Neurons can be removed or disabled using lasers. The combination of this vast array of techniques, the ability of investigators to rear many generations of worms in short periods of time, and the large number of laboratories focusing on *C. elegans* mean that this worm joins fruitflies and mice in the pantheon of well-known research organisms.

The nervous systems of roundworms are more complex than those of cnidarians and echinoderms in many ways, but not as complex as those of arthropods (crustacea, spiders, insects, and their relatives) and vertebrates. Roundworm behavior includes mating, egg-laying, foraging, and movement; in other words, they do everything an animal would be expected to do, and it is all organized by those 302 neurons. *Caenorhabditis elegans* senses the environment in at least two ways: chemosensation (smell or taste) and mechanosensation (touch or feel). "The worm" (as its students like to call it) also responds to temperature and light, even though there are no known sensory structures in this animal for these two inputs. This suggests that much is yet to be learned about the sensory world of *C. elegans*. It may not be as simple as it looks!

Mating behavior in *C. elegans* is complicated by an arrangement of sexes that is unusual from a human perspective. The worms are either male or hermaphroditic. The hermaphrodites pass through a male phase during their larval development and later function as females, but they have the possibility of self-fertilization. Males are attracted to females by chemical secretions and exhibit a stereotyped sequence of mating

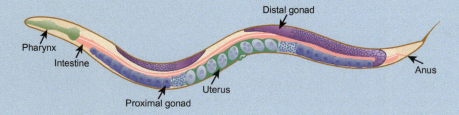

FIGURE 2.2

Caenorhabditis elegans, the "lab rat" of the nematode world. It is a typical nematode in many regards—typical shape, typical features, and typical anatomy. Note the tubular gonads in this female, tubular intestine, and tubular body wall. The tube-within-a-tube body plan of nematodes, together with an incredibly resistant cuticular covering, means that they are well adapted for a wide range of environments, famously ranging from the beer-soaked mats in European pubs to polar permafrost. Beginning with the work of Butschli in the late 1800s, nematodes have been in the forefront of developmental studies. It is therefore no surprise that "the worm" continues to answer our questions about function—and stimulate more!

CASE STUDY (CONTINUED)

behavior, culminating in ejaculation of sperm into the female's vulva and deposition of a copulatory plug (which may prevent subsequent matings by other males). About 300 progeny are produced during the worm's roughly 2-week lifespan. Like many other nematodes, *C. elegans* exhibits an interesting twist when unfavorable environmental conditions persist; larvae can enter the "Dauer" ("continuing" in German) stage, in which they cannot eat, do not age, and are protected from environmental stresses. This can prolong the worm's life for weeks or months.

One surprise to emerge from studies of *C. elegans* is that much the same suite of biogenic amines—octopamine, tyramine, dopamine, and serotonin—serves as neurotransmitters in *C. elegans* as in insects and vertebrates. This suggests that the basics of communication among neurons and between neurons and other types of cells appeared very early in animal evolution and that the fundamental framework of communication in the nervous system has been maintained through the development and diversification of neural systems. Dopamine is involved in mechanosensation (touch or feeling the environment), locomotion, and learning. Octopamine seems to be involved in

regulating egg-laying behavior. Another neurotransmitter, GABA, provides a good example of how the intimate details of nervous system function can be teased from such a simple system. Of the 302 neurons found within any *C. elegans* worm, 26 have GABA. GABA inhibits locomotion, movements of the head used in foraging, and stimulates defecation.

If so much can be accomplished with so few neurons, why do animals such as octopi and honeybees have millions of times as many neurons? As new sensory modes, more muscles, and more possibilities for learning and cognition have been added over the course of evolution, neurons for these functions have been built on top of the basic scaffold similar to that found in *C. elegans*. This principle, that evolution builds on existing structures, means that the neural systems can exhibit a fair amount of complex layering. Although no existing animal is an ancestor of any other existing animal, features—in this case, the nervous system—may have been subjected to different selection pressures in different evolutionary lineages. Thus, the nervous system of *C. elegans* may resemble that of an ancestor more closely than the nervous system of a grasshopper does.

Axons Carry Information to and from the CNS

NEURONS AND THE TRANSMISSION OF INFORMATION

Neurons come in many sizes and shapes, but they all have the same essential structure. A neuron consists of a cell body, or soma, with a nucleus. The axon is a signal-carrying extension that targets other neurons, muscles, or organs. The dendrites are branch-like structures of neurons that interface with other nerve cells or with organs and tissues, and an axon is the elongated portion of the neuron that carries the nervous signal.

Axons transmit electrical or chemical signals from one location in an animal's body to another. When the nerve cell is at rest—that is, when it is not transmitting information—there is an electrical difference between the inside of the cell and the surrounding fluid; this is called the *resting potential*. The nervous impulse, or *action potential*, starts when the cell membrane is stimulated by a neurotransmitter or by a sensory input. This stimulation causes the cell membrane to depolarize; the electrical difference between the outside and the inside of the cell reverses because of a temporary change in the permeability of the membrane. The depolarization can be visualized as a wave of electrical charge moving along the axon as shown in Figure 2.3. As one point on the membrane depolarizes, the depolarization spreads to adjacent parts of the membrane. In this way, the action potential travels the length of the axon and then is transmitted to another nerve, or to a muscle. The action potential can only travel in one direction, from the soma to the end of the axon. The connections with other nerves or with muscles are called *synapses*.

> **KEY TERMS** An action potential is the spike of electrical activity when a nervous signal is propagated. It results from depolarization of the axon membranes. The resting potential is the electrical state of a neuron when no signal is being carried.

Neurons are not the only nervous system cells. Given the focused role of neurons in transmitting and storing information in the nervous system, it is no surprise that they need support to be able to survive. This support is provided by glial cells. Glial cells are the helper

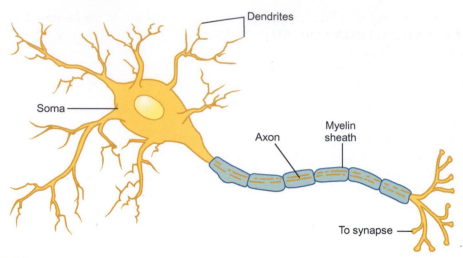

FIGURE 2.3

A myelinated neuron, with its basic elements labeled. In this diagrammatic representation of a nerve cell, the myelin sheath is blue. The myelin sheath greatly enhances the speed of transmission; this allows for faster reaction times. Some invertebrates lack myelin on their axons, although they do have giant neurons that can transmit impulses quickly. The action potential, a wave of depolarization, travels from the cell body to the dendrites at the end of the axon.

cells of the nervous system. They ensure that the metabolic needs of the neurons are met and provide structural support for the neuronal tissues. Two types of glial cells merit special mention: Astrocytes are important in the metabolic support of neurons in the CNS, and in vertebrates, Schwann cells provide metabolic and mechanical support to the axons that extend from the CNS to the muscles.

What determines the speed of a nervous impulse moving down an axon? Why does speed matter? This topic appeared in the discussion of cockroach escape responses in Chapter 1. When the action potential moves along the axon, as shown in Figure 2.3, its speed is determined by the diameter of the axon. This is similar to an electrical wire; a skinny wire has more resistance to electricity than a fat wire. Speed matters because a millisecond can make a life-or-death difference if a predator threatens. Thus, axons are expected to be large in more speed critical parts of the nervous system—where axons need to carry information over long distances to organize escape from predators. In invertebrates, this is exactly the case; axons that have critical functions in fast escape responses have large diameters and are called *giant axons*.

KEY TERM Giant axons are found in invertebrates, particularly in escape systems. Their large size speeds the transmission of action potentials.

Chapter 1 introduced cockroach giant interneurons (GIs) and the key role they play in a cockroach's ability to avoid being stomped on. The strategy of using GIs reaches a pinnacle in squid; those neurons are up to 1.0 mm in diameter and carry signals at about 20 m/s. Large diameter gives GIs less electrical resistance so that action potentials move faster in GIs than in other invertebrate nerves. Giant axons are used in nervous circuits in which speed of response is critical, such as those involved in organizing escape behavior (refer to Figure 2.3).

Another solution to the problem of speed is to "insulate" the neuron. Some animals—all vertebrates and a variety of invertebrates—do this with myelin sheaths. The myelin sheath consists largely of fatty materials and is maintained by Schwann cells. It is segmented by the nodes of Ranvier (see Figure 2.3); action potentials jump from node to node, speeding transmission. This is called saltatory conduction. Because the myelin sheath depends on fat metabolism in the glial cells, damage to the sheath, as happens in the human disease multiple sclerosis, impairs transmission of the signals along the axon. A myelinated vertebrate axon 20 µm in diameter can carry signals up to 100 m/s.

NEUROTRANSMITTERS SHUTTLE INFORMATION FROM NEURON TO NEURON

Neurotransmitters are the messengers of the nervous system. They are relatively small molecules that carry information across synapses from a nerve cell to its neighboring cells and are a critical part of the internal machinery controlling animal behavior. Generally speaking, the neurotransmitter is held in membrane-bound vesicles near the synapse. The nerve cell with these vesicles is the presynaptic cell. When stimulated, the vesicles merge with the cell membrane of the presynaptic cell, and the neurotransmitter is released into the synapse, or "synaptic space." The neurotransmitter molecules cross the synaptic space and match with receptor molecules in the membrane of the postsynaptic cell, causing depolarization in that membrane and continuing the transmission of the impulse. These receptors are critically important; for each neurotransmitter, there are several receptor molecule types, guaranteeing a transmitter-specific message. Each neurotransmitter has many different functions in the nervous system, and the receptor type involved in regulating a behavior often tells us more about the behavior than the identity of the neurotransmitter might tell us. The most commonly discussed neurotransmitter is *acetylcholine*, which often is the messenger between axons and muscles as well. Other common neurotransmitters are octopamine, serotonin, and dopamine; they usually function in the CNS. All of these neurotransmitters are found in both vertebrates and invertebrates.

> **KEY TERM** Neurotransmitters are small molecules that carry messages among axons and between the nervous system and other tissues and organs.

For the neuron-to-neuron signaling system to work, the neurotransmitter must be removed from the synapse after the signal is no longer needed. This happens by either cleaving the neurotransmitter to inactivate it or by re-uptake of the neurotransmitter into the presynaptic cell. For instance, a specialized enzyme called *acetylcholine esterase* breaks down acetylcholine in the synapse. The components can then be recycled. In contrast, serotonin is taken up directly by the presynaptic cell (Figure 2.4).

> **KEY TERM** Acetylcholine is a neurotransmitter that acts, in many animals, at synapses between nerves and muscles.

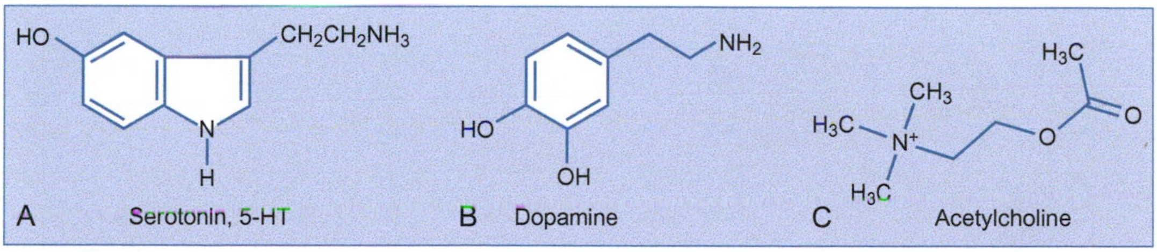

FIGURE 2.4
Chemical structures of three common neurotransmitters: (A) serotonin, sometimes called 5-hydroxy tryptamine, (B) dopamine, and (C) acetylcholine. These molecules share small size, the presence of a nitrogen molecule, and polarity, that is, having a chemical charge difference across the compound. Collectively, such compounds are sometimes referred to as biogenic amines. Their small size allows them to be easily transported across cell membranes, but their polarity reduces unintended diffusion across nonpolar cellular membranes. Insecticides such as malathion, which is commonly used in mosquito control, are acetylcholine esterase inhibitors. This means the insecticide prevents the enzyme that breaks down acetylcholine in the synapses from acting. The resulting accumulation of acetylcholine results in uncoordinated firing of nerves.

Acetylcholine also has a major function in the CNS, where its levels are critical in learning and memory. Drugs that inhibit the activity of acetylcholine esterase can enhance memory functions in humans and animals. Drugs for Alzheimer's disease, such as Aricept®, are acetylcholin esterase inhibitors; they interfere with the breakdown of acetylcholine in the brain, causing levels to rise so that normal learning and memory functions can operate.

OF SPECIAL INTEREST: WHAT DO FUNGI "KNOW" ABOUT NERVOUS SYSTEMS? THE CASE OF THE ZOMBIE ANT

Animals with parasites behave differently from animals that do not have those parasites. At times, this alteration results from the parasite manipulating the host in a way that improves parasite transmission or survival, and in other circumstances, it results from the host shifting away from its normal behavior in order to avoid parasites or combat them with fever, medicinal plants, or other strategies. (See Chapters 9 and 10.) The mechanisms that underlie behavior can be tricky to sort out in one animal, much less when considering two animals fighting, physiologically, what may be a life-and-death battle—certainly a battle over fitness outcomes. Nonetheless, with his questions, Niko Tinbergen encouraged us to think of behavior from the level of the neuron to adaptive significance and beyond, and studies of some parasite-induced behavioral alterations are beginning to achieve that breadth. The carpenter ant with its lethal fungus is one example.

In the rainforests of Thailand, there are "graveyards" of carpenter ants that exceed densities of 25 ants/m^2, ants that have descended from nests high in the canopy to die in a specific location

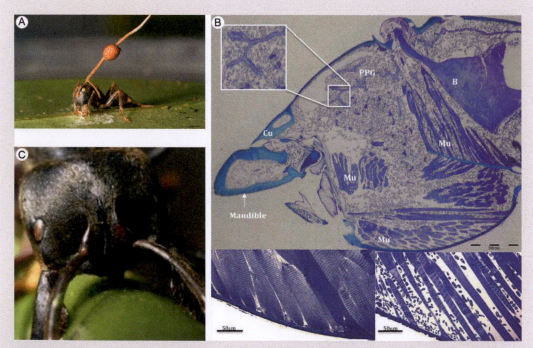

FIGURE 2.5

(A) The fungus *Ophiocordyceps unilateralis* manipulates its *Camponotus* ant host in a specific way, causing it to leave the nest, crawl out onto this epiphyte, and bite into it. A fungal stalk grows from the head and releases spores from this elevated location, which is good for fungal dispersal to other ants. (B) This is a section of the head of a *Ophiocordyceps*-infected ant, preserved just as it was biting a leaf. The muscle (Mu) is atrophied, and the head and mandibles are full of fungal bodies, which in this photograph look like small gray spots. The inset shows the fungus in the area of the postpharyngeal gland (PPG), a gland that functions in the chemistry of nestmate recognition in ants. Cu, cuticle, B, brain. The two tissue sections at the bottom, side by side, are muscle. Healthy muscle is on the left; on the right is muscle from an ant that is in the process of biting a leaf. The small bodies between the muscle fibers in this section are fungal cells. (C) The death bite of an ant infected with *Ophiocordyceps*. *Photos: David Hughes.*

approximately 25 cm above the ground; this is an area of understory that has a stable microclimate and high humidity. After an ant is infected by the fungus *Ophiocordyceps unilateralis*, the fungus begins to control the behavior of the ant. The ant leaves the colony the morning of its death. A series of convulsions and other pathological behaviors cause it to fall lower in the canopy to the graveyard area; around noon, it bites into the bottom of a leaf vein. (Uninfected ants are never observed to bite leaves.) The ant does not release the bite; its jaws lock it onto the bottom of the leaf, where it dies. The fungus grows from the intersegmental membranes of the ants, and a fungal stalk grows from the head and releases spores from this location, which favors fungal dispersal to other ants. Inside the ant's head, the fungus grows as a yeast-like stage, called a blastospores, releasing neuromodulators that affect both motor neurons and the CNS of the ant, resulting in behavior that takes the ant to an excellent location for fungal transmission. The mandibular muscles atrophy, making release of the jaws impossible (Figure 2.5).

Wouldn't an infected ant that remained in the colony be an even better source of infective fungal spores? Perhaps, but ants are diligent in removing dead and dying nestmates. The fungus would not have time to sprout a stalk and release spores before the ant was removed.[1,2]

The ability of *Ophiocordyceps* to alter ant behavior has been known since Alfred Russel Wallace, the person who shares credit with Darwin for proposing the theory of evolution by natural selection, first documented it in 1859. *How* it alters ant behavior—through neuromodulators and muscle atrophy—is only now coming to light. A variety of taxa, ranging from trematodes to fungi, cause ants to crawl onto plants and bite or otherwise suspend themselves. Have they converged on a single mechanism? Has every parasite managed to evolve a different way to create a "zombie" ant? This is part of the wonder of Tinbergen's first question.

THE CNS, WHEN PRESENT, ORGANIZES BEHAVIOR

The complexity and size of the CNS varies from simple *nerve nets* in organisms such as jellyfish (Cnidaria) to highly centralized systems (Figure 2.6). Cephalization is the concentration of elements of the nervous system, especially those involved in sensory activities and coordination of responses, into an anterior head. It is strongly linked to behavior, because it is usually found in mobile animals, and characterizes the part of the animal (anterior) that encounters the environment first. Sessile (nonmoving) animals are less cephalized; they encounter the environment from all directions, which favors a distributed nervous system.

Cephalized animals have brains of varying complexity. Complex brains are typically compartmentalized, with separate but interconnected structures for functions such as olfaction, vision, and integration. The structures of highly sophisticated brains that evolved independently, such as those found in octopi, insects, and mammals, share this compartmentalization.

> **KEY TERM** A nerve net is a simple, decentralized, multipolar nervous system found in animals such as jellyfish.

Does brain size matter? This is an intriguing but surprisingly complex question. In humans, people over a broad range of brain sizes seem to have roughly equal intelligence and cognitive abilities, a point that was vigorously argued by Stephen Jay Gould, one of the deepest evolutionary thinkers of the past century. Gould, in his brilliant book, *The Mismeasure of Man*, argued that the temptation to assess people's intelligence by looking at brain size was based in racism and deeply held assumptions about the superiorities of some human cultures over others.[3] Gould's argument is reasonable, yet within a larger (multispecies) taxonomic group such as birds or mammals, it does appear that cognitive abilities correlate with brain size across a wide range of species; in mammals, cognition and cerebral cortex volume have a particularly strong relationship. (Note that such broad interspecific comparisons differ from intraspecific comparisons.)

FIGURE 2.6

The anemones in the two upper photos are under water, with their tentacles and mouths exposed. In the lower photo, the mass of anemones sits above the tideline, exposed to the sun and air. They have retracted their tentacles, thus minimizing their surface area and reducing the risk of dessication and other threats. *Photos: Michael Breed and Tricia Soares (top left).*

The highly intelligent representatives of the molluscan phylum, squid and octopi, have very large brains relative to their more sedentary cousins, such as clams, oysters, and scallops. But the "smartest" insects, at least as measured by navigational and communication abilities, like the honeybees, have tiny brains compared to any squid, octopus, bird, or mammal. Adult human brains weigh 1300–1400 g and have over 100 billion neurons. Octopus brains have about 300 million neurons and weigh up to a gram. Honeybees have slightly less than a million neurons packed into less than a cubic millimeter (about 1 mg in weight). Some insects are so tiny as to be barely visible to the naked eye, yet they carry on all the basics of insect life—feeding, flight, and mating—with a brain that is 1000-fold miniaturized compared to a honeybee's brain. How basic functions are maintained in miniscule brains is a fascinating unanswered question.

DISCUSSION POINT: INTELLIGENCE

The idea of honeybees being the "smartest" insects, based on navigation and communication abilities, raises a question that has challenged biologists and psychologists for decades: How do we measure intelligence? We will return to this question in later chapters, but now would be a good time to begin thinking over this puzzle. What is intelligence? Is it the same for all animals—is dog intelligence the same as human intelligence? What aspects of intelligence are favored by natural selection in different habitats or lifestyles? After pondering this question (for which there is probably no single correct answer), think about how intelligence can be measured and whether it can be defined and measured in a way that would allow comparison among species. How does *umwelt* affect "intelligence"?

The brain receives information about the animal's physiological state and serves as a command center for translating physiological needs into behavioral responses. Cnidarians and echinoderms have neurons but no centralized brain. They still manage to organize fairly complex behaviors, as do protists. The echinoderms' lack of cephalization is taken to reflect sessile ancestry—will mobile echinoderms be cephalized millions of years from now? Cnidarians are anatomically different from most other animals, and their nervous system is no exception. First of all, although some cnidarians do move (e.g., jellyfish), that movement is largely at the mercy of currents; they are not powerful swimmers. In addition, many other cnidarians are mostly sessile (e.g., corals, anemones). In cnidarians, a net of neurons lies between the inner and outer body walls. Unlike nerves in most other animals, these nerves can transmit impulses in any direction. Sensory nerves project to the surfaces of the animal, and nerves also attach to contractile cells, which function similarly (if weakly) to muscles in other animals. This nerve net allows transmission of information from one part of the animal to another and coordination of simple movements. The stinging organs of cnidarians, nematocysts, discharge without nervous control. Well-fed cnidarians are less likely to discharge their nematocysts. Nervous connections to the nematocysts may affect the likelihood of discharge, depending on the need for food (Figure 2.6).

FIGURE 2.7
A freshwater mussel, *Lampsilis* sp. Mussels are an example of bivalves, molluscs that have a sessile, filter-feeding lifestyle. Their nervous system lacks any cephalization. They have ganglia, connected by neurons, that receive sensory input and control basic functions such as movement of the foot and gill. Some bivalves, including many scallops, can swim by flapping their shells, a motion that can be coordinated by a very simple nervous system. *Photo: Phillip Westcott, US Fish and Wildlife Service, NCTC Image Library, Public Domain.*

Comparing Brains

Molluscs are a particularly interesting case of brain evolution and how natural selection has acted on locomotion and the nervous system to influence that evolution. They are one of three major groups of animals to have evolved a highly sophisticated brain, the other two being arthropods and vertebrates. Molluscs (slugs, snails, clams, squid, and octopi) have ganglia—collections of neurons—associated with the mouth, foot, and gut. Cephalopods (squid and octopi) are active predators; they have complex brains and are capable of sophisticated movements and color changes, along with learning. Mussels and other bivalves such as clams have limited mobility, and consequently natural selection has not favored a complex nervous system (Figure 2.7).

The nervous system of insects reflects the segmented body plan shared by all arthropods. Each of the insect's segments has its own nerve center, called a *ganglion*, which is connected by a pair of nerves to the ganglia of the adjacent segments. The insect's head consists of a set of fused segments; scrutiny of the head capsule of a grasshopper reveals some of the sutures between the segments. Inside the head, the segmental ganglia have fused to form a brain.

> **KEY TERM** A ganglion (pl. *ganglia*) is a concentration of axons that integrate sensory input and locomotory output, and which also report the status of a portion of the animal to the brain.

OF SPECIAL INTEREST: INSECT BRAINS

The insect brain performs nearly all of the same functions as the brain of a bird or mammal but is much more compact. It is also organized much differently. The major parts of an insect's brain (shown in Figure 2.8) are as follows:

- Optic lobes, which provide a link between the receptor surfaces of the eyes and the brain. The optic lobes perform the first steps in interpreting visual information for the insect.
- Ocelli (sing. ocellus), simple, accessory eyes that are important in perceiving the intensity of illumination. Many insects have three, although some have only two or even none.
- Antennal lobes. The antennal lobes receive input from millions of olfactory receptors and reduce it to manageable taste and smell information.

- Mushroom bodies. If an insect "thinks," this is where it happens. These are the "higher" centers in the insect brain, where learning, memory, and integration occur.
- Neurosecretory cells. Peptide hormones produced by the neurosecretory cells regulate endocrine and homeostatic functions in insects. These cells are analogous to the hypothalamus of the vertebrate brain.
- Corpora cardiaca and corpora allata (CA). Analogous to the pituitary in vertebrates, these organs secrete hormones into the circulation. The CA produce JH, which plays a critical role in the regulation of development and behavior in many insects.
- Subesophageal ganglion. The input for this structure comes from the insect's mouthparts; this ganglion coordinates the action of the mouthparts when the insect feeds.

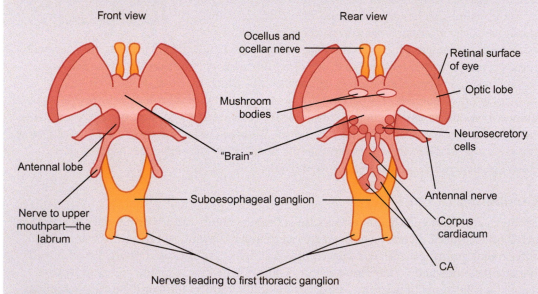

FIGURE 2.8

The insect brain is a fusion of anterior ganglia, with distinct areas devoted to vision (e.g., ocellar nerves, optic lobes), antennae, and even memory (mushroom bodies). Note the subesophageal ganglion—the first separate ganglion leading to the ganglionated ventral nerve cord typical of arthropods. It forms the ventral part of the nerve ring that circles the esophagus—hence subesophageal—and attaches the dorsal brain to the nerve cord. The CA (sing. corpus allatum) synthesize and release JH. The corpus cardiacum (which is actually a fusion of two corpora cardiaca) synthesizes some other hormones and stores other neurohormones that are synthesized in the brain.

Vertebrate brains appear to have been built in an evolutionary sequence starting with fish, which have olfactory and optic lobes, whose basic functions are for smell and vision. The cerebral hemisphere, which becomes the center of tasks such as learning and cognition in birds and mammals, is less well developed in fish. The cerebellum coordinates physical movements. The hypothalamus directs physiological functions by secreting neurohormones, whereas the medulla oblongata has axonal connections that control the cardiac and respiratory systems (Figure 2.9).

Comparison of reptilian, avian, and mammalian brains (Figure 2.10) reveals that the neocortex of the cerebrum becomes progressively larger and more dominant in function.

DISCUSSION POINT: BRAINY AND BRAINLESS ANIMALS

How do the differences between the nervous systems of the "brainy" predatory octopus and the brainless filter-feeding clam reflect the differences in their behavior? Is it surprising to find such different nervous systems in animals that are both molluscs?

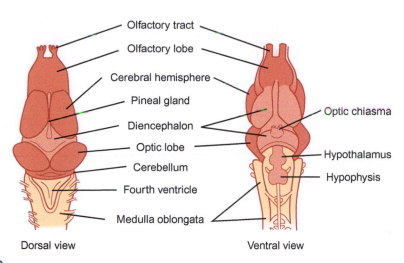

FIGURE 2.9

Many of the basic elements of the fish brain can be recognized in the brains of other vertebrates. The fish brain is organized in sections that are correlated with function. The olfactory lobes receive chemosensory input, the optic lobes receive visual input, the cerebellum coordinates motor activity, and the brain stem regulates the basic physiology of the organs.

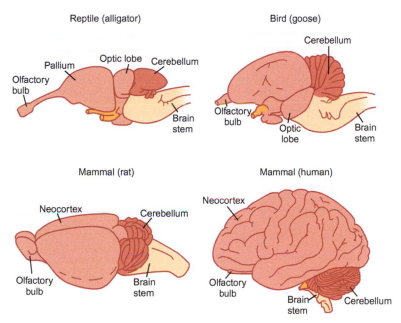

FIGURE 2.10

Diagrams providing a comparative look at vertebrate brains, which in turn can reveal much about differences in behavior. For instance, the rat's olfactory bulb is larger than that of humans relative to the overall size of the brain; rats have a much keener sense of smell than humans do.

How Behavior Is Generated

Quick responses are necessary in critical situations. *Reflex loops* are a simple integrating mechanism for organizing quick responses. Reflex loops require no "thought" or decision making on the part of the animal. Sensory input is transmitted to the CNS, it is linked to the appropriate output, and a signal to act is then sent

KEY TERM A reflex loop is a simple connection from a sensory cell to the spinal cord and back to a muscle. An example is the human patellar reflex (see Figure 2.11).

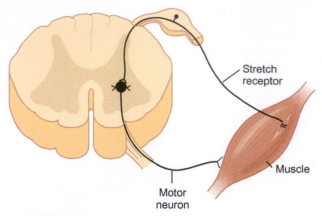

FIGURE 2.11
A diagram of a simple reflex loop in a vertebrate. Stretching the muscle sends a signal via an axon to the spinal cord. This sensory input then connects with a motor neuron that signals the muscle to move. When a physician taps a knee tendon with a mallet, the tendon stretches, and just such a signal is sent to the patient's spinal cord. The motor neuron then fires, causing the leg to jerk. In general, the function of this reflex is to reduce the stress on the tendon; in the physician's office, it serves as a basic test of the nervous system.

to a muscle (Figure 2.11). Sometimes, animal behavior, particularly the behavior of insects, is characterized as being entirely instinctive, or *reflexive*. Another term for this view of behavior is *hard-wired*. When behavior is characterized as being hard-wired, the animal is not credited with possessing decision-making processes to shape that behavior. The relationship between hard-wired behavior and thought, or cognition, is interesting and complex; we will discuss this issue in detail in Chapter 6 (Figure 2.12).

Connecting the nervous system to the muscular system is critical to making all this work. Motor neurons have axons that carry signals from more central locations in the nervous system to the muscles that will execute the intended actions. In most animals, acetylcholine, discussed earlier in this section, released from the end of the axon stimulates the muscular contraction.

2.3 WHAT DOES AN ANIMAL BEHAVIORIST NEED TO KNOW ABOUT ENDOCRINOLOGY?

Hormones and Regulation: The Basics

The study of hormones and behavior offers an opportunity to examine all of Tinbergen's questions from a single scientific viewpoint: endocrinology. By way of review, hormones are the messengers that help integrate physiological functions and behavior, communicating from glands to organs and tissues; they are chemical signals between tissues/organs and other tissues/organs. Typically, hormones regulate longer term changes in behavior, such as seasonal differences or development from juvenile to adult. There are exceptions, such as the fast-acting short-term effects of adrenaline on fight-or-flight responses.

Hormones mold development, and in doing so illuminate behavioral ontogeny. Many hormones are conserved evolutionarily so that the same molecules appear across a wide range of taxa. For example, insulin, a regulator of sugar metabolism in humans, is present in many insects and has similar functions! This conservation of molecules across vast

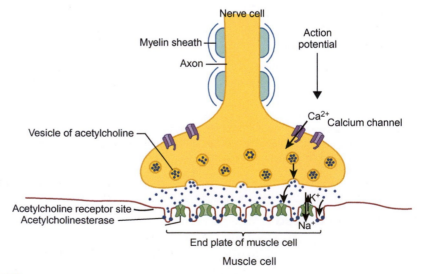

FIGURE 2.12
The electrical impulse (action potential) travels through the axon to the neuromuscular junction. Acetylcholine is released from vesicles in the nerve and initiates depolarization of the muscle membrane, which in turn causes the release of calcium, triggering muscle contraction.

40

evolutionary distances is yet one more statement of how evolutionary history and phylogenic relationships shape the survival mechanisms of animals.

Before beginning this review, consider a general principle, true across all of science—perhaps all of life—but particularly relevant to the study of hormones and behavior: correlation should never be confused with causation. Simply said, the fact that events are correlated does not mean that one has caused the other.

A straightforward approach to studying behavioral neurobiology or behavioral endocrinology is to sample hormone levels and associate those levels with behavior. For example, blood samples can be collected from male birds during the mating season and compared with samples collected from the same birds in the fall. Does the testosterone level differ between the two samples? Likely, the answer is yes. Does this mean that testosterone caused the difference? Probably not. The existing information does not indicate if the behavioral change caused the hormone level to change, if the hormone caused the behavior to change, or if some other variable(s) caused either or both to change. This experiment demonstrates a correlation between testosterone levels and behavior, but offers no information about causation.

What are some ways to test hypotheses about causation when considering hormones and behavior? Usually the approaches involve some combination of the following manipulations: (1) removal of the organ or tissue that produces the hormone, (2) supplementation/addition of the hormone in experimental animals' bloodstreams, (3) blockage of the receptors for the hormone or neurotransmitter using drugs, and (4) use of RNA interference techniques to block the synthesis of the neurotransmitter or hormone. These types of direct experimental manipulation provide stronger evidence for causative effects than the observation of a correlation.

41

OF SPECIAL INTEREST: THE RING DOVE

Consider the ring dove (Figure 2.13). It is a domesticated bird with wild relatives—a wonderful choice for a study that addresses causation with an eye to ultimate questions. The courtship and reproduction cycle proceeds as follows: If a male and a female are together, the male will display and court the female. After a day or two, the couple begins to build a nest. They copulate and continue to work on the nest, and eggs are laid. Both parents incubate the eggs. Upon hatching, the young are fed with crop milk, a special substance that both parents produce in the linings of their crops (a pouch in the upper part of the digestive tract in birds used for food storage and production of food for nestlings). This wanes as the young reach fledging age, and the parents return to courtship.

All this may seem unremarkable, and in one sense, that is correct. Mating and producing offspring are not particularly exceptional occurrences. But how do the two doves decide to build a nest at the same time? How do they coordinate incubation and crop milk production? Hormones coordinate these events; the production of the hormones is influenced by the environment, including the behavior of the pair of doves, along with endocrine feedback. What is exceptional about this behavioral sequence is not the fact that it happens, but the endocrine–behavior–environment synthesis that allows it to happen—in synchrony—in the first place.

Now revisit that behavioral sequence and mark where these influences are important. The preceding text is repeated here in bold, with the hormonal influences inserted in italics.

If a male and a female are together, the male will display. *Androgens are necessary for male display.* **After a day or two, the couple begins to build a nest.** *Male courtship display causes the release of follicle-stimulating hormone (FSH) from the pituitary. FSH initiates a cascade of development in the female reproductive tract, including estrogen release. Ovaries and uterus grow and develop.* **They copulate and continue to work on the nest, and eggs are laid.** *The nest's presence stimulates progesterone secretion; progesterone in both sexes promotes incubation behavior. (Other hormones are involved in egg production as well.)*

Both parents incubate the eggs. *Incubation stimulates the release of prolactin from the pituitary. Prolactin inhibits the release of previous hormones, so that sexual behavior wanes. It also stimulates crop milk production.* Upon hatching, the young are fed with crop milk, a special substance that both parents produce in the linings of their crops. *Prolactin levels fall during the course of nestling rearing. Hormones that promote courtship and reproduction once again increase, no longer inhibited by prolactin.*

This wanes as the young reach fledging age, and the parents return to courtship.

Hormones are not the only influences on this cycle, however. For instance, males that are deaf or males that are with an ovariectomized female do not produce as much testosterone as males that can hear or males that are with a female in reproductive condition. Indeed, it seems that both males and females need to hear vocalizations during courtship for necessary endocrine changes to ensue and result in egg development. Visual cues are also important. If males and females are separated after nest construction and held behind a glass partition, males that can see the female incubate eggs will proceed to produce crop milk, whereas males that do not see mates incubate will not produce crop milk.

Ring doves have been carefully studied for decades, and this summary is a fraction of what is known. The purpose here is not to provide an exhaustive summary, but rather present a hint of the riches that are just being discovered in the biology surrounding the interaction of behavior, hormones, and the environment.

FIGURE 2.13
Ring doves (*Streptopelia risoria*) have played a central role in our understanding of the relationships among hormones, behavior, and environment.

Hormones, Physiology, and Behavior

Returning to the review of endocrinology, organs and tissues cannot operate effectively without information from other parts of the body. For instance, in birds and mammals, reproduction is usually seasonal, so the ovaries need to be told the appropriate time of year for egg production. Males of the same species may compete intensely for mates when females are ovulating; male muscles and nervous systems need to be primed for this competition. Longer term behavioral changes, of the sort that happen as an animal matures or between seasons of the year, are usually coordinated by the endocrine system. The endocrine system produces hormones, messenger molecules that communicate among organs and tissues. Glands are frequently the source of hormones; the adrenal glands of vertebrates are a good example. Hormones have effects over short time periods (these are called *releasers* because they have an immediate effect in releasing a response) and over long ones (these are called *primers*, because they prepare the body for a response, even if that response is not immediate). An immediately obvious advantage of hormonal control over neural control is that changes can be easily maintained over long periods of time by continued release of a hormone.

KEY TERM A receptor is a molecule, typically a protein, that receives the hormone and passes the information carried by the hormone onto metabolic or genetic mechanisms within a cell. Hormone receptors can be embedded in the cell membrane or may be found in the nucleus.

Hormones are usually relatively small molecules; this allows them to be carried as signals from tissue to tissue in body fluids. Every hormone has unique *receptor* molecules. These receptor molecules transduce (change) the hormonal signal into an intracellular (within cell) signal that triggers a physiological or genetic response to the hormone.

Some hormones act on receptors on the surfaces of cells; others can pass through cell membranes and act directly on receptors in the cell nucleus. This ability to penetrate cell membranes can be critical, as the "blood–brain barrier" separates the nervous system from some types of circulating hormones, but not others. Steroids, a class of hormones that include testosterone and estrogen, generally enter cells and act directly to affect gene expression. Peptide hormones, such as oxytocin and vasopressin, rely on receptors on the surfaces of cells, which then cause a cell to produce secondary messengers (such as cyclic adenosine monophosphate) that act within the cell to affect its function.

Hormones and Behavior in Vertebrates

In this overview of some of the major ways in which hormones affect behavior in vertebrate animals, the focus is on development, a topic that exemplifies how hormones and behavior interact. Two additional major topics will appear in later chapters. Hormonal priming for sexual behavior in vertebrates and the relationships among steroid hormones, aggression, and territoriality are covered in Chapter 11 on mating. Chapter 12, on nesting and parental behavior, includes the hormonal bases of parenting in vertebrates.

Among the vertebrate hormones, two *steroids* are most familiar: testosterone, best known for its role in male reproduction and behavior, and estrogen, which plays similar roles in females. Steroids are structurally related to cholesterol; because they dissolve in fats and oils, steroids easily pass through cell membranes and can work in the nucleus of cells by turning gene expression off and on. These molecules are not easily degraded, so steroid and steroid-like compounds persist in the environment and build up in the fatty tissues of animals; this is a problem from a conservation standpoint and is covered in more detail later in this chapter.

> **KEY TERM** Steroids are a class of hormones that include testosterone, estrogen, and ecdysone. Chemically, they are related to cholesterol.

In addition to steroids, there are other kinds of hormones, including peptides, relatively short chains of amino acids. These peptides act by triggering receptors on cell membranes. They are more fragile than steroids; the most familiar one is insulin, which helps to regulate the energy metabolism of animals. Other major vertebrate hormones, along with their sources and their action in regulating behavior, are summarized in Table 2.1 and Figure 2.14.

Nonapeptides, peptide chains consisting of nine amino acids, are a class of hormones found in vertebrates. Generally speaking, these hormones are associated with bonding, grouping, affiliation, parental behavior, and positive reinforcement of social behavior. In addition to behavior, nonapeptides regulate salt concentrations in the blood. The most familiar of the nonapeptides is oxytocin, commonly touted in the popular press as the "hormone of love" in mammals. Vasopressin is also found in mammals and has behavioral effects related to those of oxytocin. In birds, there are also two nonapeptide hormones: arginine vasotocin and mesotocin. In zebra finches, a highly social species, arginine vasotocin levels are low in isolated birds and high in birds after exposure to their mate or to a flock of zebra finches. Less social and territorial species of finches show no response or even a decrease of arginine vasotocin when they encounter another bird.[4]

The hormones involved in sexual development and behavior in vertebrates are summarized in Figure 2.14. The environment and an animal's internal condition, including its age, come together to determine when the brain (specifically, the hypothalamus) should produce gonadotropin-releasing hormones. These hormones act on the pituitary, which in turn produces two hormones, FSH and luteinizing hormone (LH), which act on the gonads. The primary hormone of the female gonads (ovaries) is estrogen, which is a steroid. The primary hormone of the male gonads (testes) is testosterone, also a steroid. These steroid hormones then regulate, among other things, physical and behavioral readiness to mate (Figure 2.15).

TABLE 2.1 Vertebrate Hormones and Behavior: An Overview

	Produced by	Target Organs/ Tissues	Steroid, Biogenic Amine, or Peptide?	Behavioral Effects
LH	Pituitary gland	Gonads	Peptide (glycoprotein)	Indirect, via stimulation of estrogen or testosterone production by gonads
FSH	Pituitary gland	Gonads	Peptide (glycoprotein)	Indirect, via stimulation of estrogen or testosterone production by gonads
Estrogen	Gonads, primarily the ovaries	Brain, tissues, and organs involved in reproduction	Steroid	Female sexual receptivity, maturation, and female secondary sexual characteristics
Testosterone	Gonads, primarily the testes	Brain, tissues, and organs involved in reproduction and dominance/ aggression	Steroid	Dominance, aggression, male sexual receptivity, maturation, and male secondary sexual characteristics
Progesterone	Gonads (the ovaries) and the placenta	Brain and mammary glands	Steroid	Reproduction (particularly postfertilization) and parental behavior
Oxytocin	Hypothalamus, released from the pituitary	Uterus, mammary glands, and brain	Peptide	Birth, lactation, and bonding
Vasopressin	Hypothalamus, released from the pituitary	Brain and kidneys (for regulation of hydration)	Peptide	Bonding
Prolactin	Pituitary	Mammary glands and brain	Peptide	Parental behavior
Insulin	Pancreas	Metabolic tissues, possibly the brain	Peptide	Possibly hunger
Epinephrine (adrenaline)	Adrenal	Brain, vascular, and respiratory tissues	Biogenic amine	Fight or flight— vascular and respiratory control and sense of alertness/ awareness

44

During development, the endocrine system plays a critical role in the expression of age-appropriate behavioral patterns. For vertebrates, estrogen and testosterone have critical roles, whereas JH is important in insects. These hormones affect both physical and behavioral development; because they affect both of these developmental areas, physical and behavioral development are appropriately coordinated.

Behavioral changes during growth and development are key to an animal's survival. Very obviously, immature animals do not behave like adults do. Steroid hormones, such as

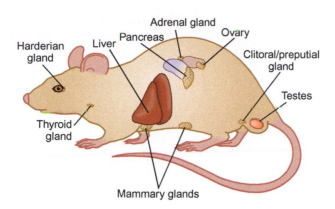

FIGURE 2.14

Distribution of vertebrate endocrine glands in hypothetical (hermaphroditic) rat; the hermaphroditic rat, which does not exist in nature, is drawn to show both male and female endocrine organs. The hypothalamus is the main neuroendocrine interface in vertebrates. It coordinates the production of many hormones and can have a central role in the control of behavior. Neuropeptides from the hypothalamus induce the pituitary gland, which is suspended from the hypothalamus to produce a number of hormones.

estrogen and testosterone, help to regulate behavioral development in vertebrates. Hormonal changes drive striking seasonal patterns in sexual receptivity, conflict and aggression, and parental behavior. In vertebrates, behavioral changes of this type are strongly associated with the levels of steroid hormones. Seasonal changes are equally marked in some invertebrates, but their physiological bases are less well understood.

After an animal is hatched or born, its nervous system often is incompletely formed. Nevertheless, it coordinates the activities needed to sustain the young animal's life, such as seeking food or its mother, hiding from potential predators, and finding the appropriate environmental conditions. Vertebrate species are termed either *altricial*—needing intensive parental care at hatching or birth—or *precocial*—being relatively independent at birth (Figure 2.16). Mouse pups are altricial; they have incompletely developed eyes and lack the ability to feed themselves; they must be carefully nurtured by their mother. Altricial young cannot move about or thermoregulate, and their visual systems may not be fully developed. They require intensive parental care, usually centered around a nest or den where the young are protected from the environment. Marsupials, songbirds, bats, rodents, carnivores, and humans are other examples of animals with altricial young.

Species in which the female lays eggs and then leaves or dies have precocial development. For example, baby sea turtles are precocial. Upon hatching on a beach, they are programmed to seek the water, where they then can maintain themselves without parental assistance.

What selective forces lead the young of some species to be semi-independent, whereas the young of others are

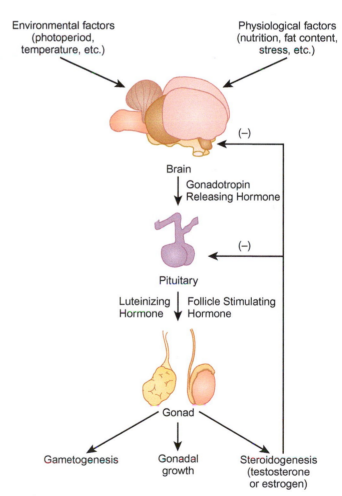

FIGURE 2.15

The brain, pituitary, and gonads form a unified system that regulates sexual maturation and reproduction. Steroid hormones (estrogen and testosterone) provide negative feedback, indicated by the arrows, to the brain, signaling that the gonads have been activated.

FIGURE 2.16
Canada goose young are able to swim and feed themselves soon after hatching. Although they rely on their mother for protection and for information about feeding, their mother does not pass food to them. Contrast these precocial goslings with the intensive parental care needed by altricial young. *Photo: Michael Breed.*

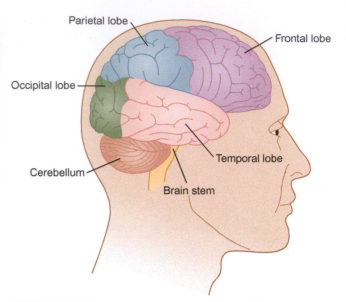

FIGURE 2.17
Major areas of the human brain.

so poorly developed that they need intensive parental care? One hypothesis is that having semi-independent offspring allows the mother to have a larger clutch. Alternatively, some habitats or feeding preferences may favor offspring independence. In mammals, the young of herbivores are more fully developed at birth than the young of carnivores. This difference may reflect a number of selective forces, such as the fact that many herbivores must continually be on the move to find suitable forage. Herbivore young may also have fewer challenges in terms of learning how to select and process food than do carnivores, which must learn complicated strategies of prey capture.

Vertebrates begin gender differentiation early in development. Sex steroids (estrogen, testosterone, and progesterone) play critical roles in the embryonic development of males and females (Figure 2.17). In the earlier stages of development, all embryos are female-like in their morphology. In male embryos, testosterone production from the testes induces male external genitalia. Testosterone may also induce the development of male-specific details of brain structures and sets the stage for male behavior later in life. Embryonic cells in the hypothalamus and telencephalon have working estrogen receptors; the function of these receptors is not well understood, but exposure of these cells to estrogen probably influences later behavioral development.

One of the most interesting aspects of endocrine influences on behavior is the effect that developing embryos have on each other in mammals. Field studies of mammals often yield observations of behavioral variability among family members; one of the factors that may cause these differences in personality (see Chapter 6) is uterine position. Most mammalian species produce litters of two or more young; these young shared their mother's uterus during development. In the uterus, the embryos are lined up so that except on the ends of the rows, each embryo has two neighbors. Because steroid hormones easily move through biological membranes, hormone produced by one embryo may pass in the uterine fluid to adjacent embryos. Testosterone from male embryos is particularly potent; in addition to controlling the development of the male genitalia/reproductive system, it also apparently induces the development of testosterone receptors in the developing male brain so that later in life high numbers of receptors make the male brain more susceptible to the behavioral influences of testosterone. Female fetuses produce estradiol, but less is known about the effects of fetal estradiol on neighboring fetuses.

The effects of uterine position are best known in rodents, but these principles seem to apply broadly across the mammals. An embryo can have two male neighbors (2 M), two female neighbors (0 M), or one of each (1 M). The 2 M young, whether genetically male or female, are more masculinized, and 0 M young are more feminized. The 2 M embryos (male or female) are much more sensitive as adults to the effects of testosterone; for example, they hold larger territories, and when injected with testosterone, they are more aggressive. 2 M mice urine mark more frequently, and male mice urine mark more often when exposed to a 2 M female than to a 0 M female (presumably because the 2 M female smells more masculine and induces a greater territorial response from the male mice). When treated with testosterone, female mice that are 2 M are more likely than 0 M females to exhibit male sexual behavior (mounting). However, 0 M females are more likely to exhibit female sexual behavior (lordosis) and to copulate successfully. In addition to these hormonal influences, embryos in some positions in the uterus have better access to nutrients, and are born larger and stronger than other embryos. These size and strength differences carry over into juvenile and adult development, yielding animals that are more likely to be dominant.

BRINGING ANIMAL BEHAVIOR HOME: TESTOSTERONE, CASTRATION, AND MALE BEHAVIOR

Male animals can be strong and behaviorally difficult to manage; male strength and behavior are largely under the control of testosterone, which is produced in the testicles. Removal of the testicles from male birds (caponization) and mammals (castration) is a technique that was established early in human history, many years before scientists demonstrated the existence of hormones, defined their chemical structures, or discovered their physiological effects.

Castration results in lowered levels of aggression, decreased muscle mass, and an overall fattier body composition. Castrated animals have more trainable, malleable personalities. In agriculture, the most commonly castrated animals are horses (the castrated stallion is a gelding), cattle (a castrated bull is a steer), sheep, and chickens (the rooster becomes a capon).

Companion animals, particularly dogs and cats, are often castrated, with the goals of making male animals more manageable and reducing the potential for both population growth and health issues, including hormone-sensitive cancer. Testosterone primes the expression of male behavior by affecting development of the brain; males castrated after male behavior has developed often retain male behavioral patterns.

A simple example of this is leg-lifting (territorial marking) during urination by male dogs. Lifting the leg is a common sign of maleness; young male dogs develop this behavior as they mature. Males castrated before the behavior develops rarely express it, whereas males castrated after the behavior develops retain the behavior even though the hormonal signal has been removed. Contemporary studies of the effects of testosterone on male behavior often employ chemical castration or supplemental testosterone delivered by implants; typically, territoriality and aggression can be manipulated with these treatments.

As indicated earlier, establishing linkages between hormones and behavior often involves, as a first step, measuring hormone levels and looking for correlations between shifts in hormones in the blood and behavior. This approach has been powerful in suggesting a variety of behavioral roles for steroid hormones in vertebrates, and in fact some of the relationships, such as that between testosterone and aggression in males, seem to hold up under further scrutiny (Figure 2.18). Other relationships, such as the hypothesized role for hormones in regulating paternal behavior, are not so well supported. This points out the danger of leaping from a demonstrated correlation to an assumption of causation.

47

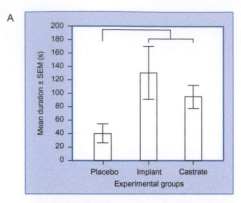

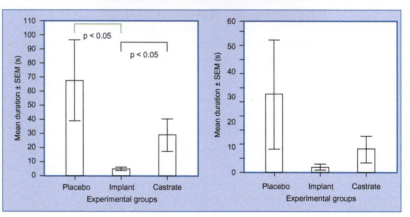

48

FIGURE 2.18

The effects of castration and the surgical implantation of a chemical castrator, deslorelin, on the behavior of male ferrets are shown in these graphs. The "placebo" males in this experiment are intact—they have their testes—and are the experimental controls. In (A), play behavior is measured; note that both types of castrated males are more likely to play than intact males. The upper graph in (B) shows an overall count of aggressive and defensive behaviors, and the lower graph shows biting. These behaviors are lower in the castrated males. From studies like this, we can conclude that testosterone plays a key role in priming animals for aggression and that affiliative behaviors like play are reduced in mature intact males. *Reprinted from Applied Animal Behaviour Science, volume 115, Claudia M. Vinke, Remko van Deijk, Bart B. Houx, Nico J. Schoemaker, The effects of surgical and chemical castration on intermale aggression, sexual behaviour and play behaviour in the male ferret (Mustela putorius furo), pages 104–121, copyright 2008, with permission of Elsevier. Photo (inset): A similar species, the black footed ferret, Mustela nigripes, Ryan Hagerty, US Fish and Wildlife Service, Public Domain.[5] Photo (inset): A similar species, the black footed ferret, Mustela nigripes, Ryan Hagerty, US Fish and Wildlife Service, Public Domain.*

OF SPECIAL INTEREST: FIGHT-OR-FLIGHT RESPONSES

When confronted with a life-threatening event, an animal needs to mobilize many systems at the same time to save itself. The hormone epinephrine (adrenaline) does exactly this. It is produced by the adrenal glands, which are associated with the kidneys, in response to startling stimuli or fearful reactions. Notable effects of epinephrine include increased heart and respiration rates; vasocontraction, particularly in the peripheral circulation; and heightened sensory and cognitive awareness (Figure 2.19).

FIGURE 2.19
The external appearance of an animal clearly indicates whether its mode is "fight" or "flight." Inside the skin, endocrine changes are just as dramatic.

In animals with seasonal mating/migratory cycles, the testes typically enlarge as mating season approaches, corresponding to increasing intermale competition, courtship displays, and willingness of males to copulate. Migration seems, at first glance, to be a strong candidate for hormonal control. Spring migrations precede mating seasons, and winter migrations culminate the reproductive cycle for many animals. Yet if there is a single hormone that regulates migration, it has so far eluded discovery; the effects of treatments such as gonadal ablations and hormone supplements do not support a consistent model for hormonal regulation of migration itself, although events surrounding it are under hormonal control. For instance, animals must store substantial fat in preparation for migration, and hormonal regulation of food intake and fat deposition plays a role in readiness for the migratory journey.

Territoriality is more clear-cut. Testosterone injections or implants result in males that hold larger territories or have higher dominance status. However, hyperaggressive behavior does not always translate into more mating success because males may shift their time budget to expend too much time and energy on territoriality and dominance, and consequently have too little time and energy to mate or feed young successfully. Testosterone levels probably represent the result of an evolutionary balance that maximizes mating potential, and not hyperaggression. This topic receives more scrutiny in Chapter 11.

OF SPECIAL INTEREST: HUMANS AND GONADAL TRANSPLANTS

The manipulation of human hormones has a long, if somewhat undignified, history. Some of the more extravagant efforts were fashionable for several decades during the late nineteenth and early twentieth centuries (Figure 2.20). These efforts included radiation of *in situ* human gonads, injection of various concoctions (e.g., seminal fluid, blood, minced testes), and transplantation of testes from a variety of species. (Some practitioners were fastidious in their choice of donors, restricting them, for instance, to one variety of goat; others were more catholic in their approach and included, say, a broad range of ungulates or great apes.) Patients/customers were the rich and the famous of the day, including one of the greatest poets in the English language, W. B. Yeats, and Sigmund Freud, the founder of modern psychoanalysis. The best of these attempts, carefully experimental by the standards of the time, paved the way for the modern understanding of reproductive endocrinology, a field that, as you might imagine, was nonexistent in the first part of the twentieth century. Possibly, the worst was typified by the "goat gland doctor," "Dr." John Brinkley. Brinkley was a radio commentator, evangelist, and would-be governor of Kansas. (After 16,000 goat transplants into humans, he went bankrupt and fled to Mexico, where he continued his practice.)

FIGURE 2.20
The Gland Stealers satirizes the work of Serge Voronoff, a Russian physician working in France who reported success with testicular transplants. In this book, wealthy men of a certain age head into Africa in search of 100 gorillas—or more precisely, their testicles. Although this may not qualify as the Great American Novel, it is evidence that testicular transplants were a significant part of the culture in the early 1900s.

Hormones and Behavior in Invertebrates

Among invertebrates, the regulation of behavior by hormones is best understood in insects. The endocrine systems of other arthropods, such as spiders, millipedes, centipedes, and crustacea, are similar to those of insects, and many of the principles presented here apply to arthropods in general. Beyond arthropods, invertebrates are a heterogeneous lot, and the neuroendocrine functions of bag cells in molluscs stand out as being well-enough understood to merit some discussion.

JH AND ECDYSONE

For invertebrates, the best-studied hormone with behavioral effects is JH, a nonsteroidal terpenoid and a product of the CA in insects. JH affects mating behavior, pheromone secretion, and parental behavior in adult insects, as well as worker behavior in eusocial insects, such as honeybees. Although JH is the best known of insect hormones, there are other important hormones as well. For instance, the other primary hormone of arthropod growth and development is *ecdysone*, a steroid. Ecdysone is primarily known for its role in regulating the cycle of exoskeleton molting (ecdysis) in growing arthropods, but it also regulates egg production by stimulating yolk synthesis in adult females. Ecdysis is a phase in molting; this is the time when arthropods shed their exosekeleton in order to molt. A neurohormone, *prothoracicotropic hormone* (PTTH), released from the brain stimulates the prothoracic glands to secrete ecdysone. Interestingly, PTTH is a peptide that is evolutionarily related to insulin. The insect neuroendocrine system and the hormones of behavioral interest are shown in Figure 2.21.

> **KEY TERM** JH is a hormone that regulates development, reproduction, and behavior in arthropods. Chemically, it is a terpenoid. It is produced by the CA, glands behind the brain that are connected to the brain by neurosecretory axons.

> **KEY TERM** Ecdysone is a hormone that regulates molting and reproduction in arthropods. It is a steroid and is produced by several glands, including the prothoracic glands and the ovaries.

PTTH is secreted at a certain time of day, in conjunction with the circadian rhythm of the insect (see Chapter 4). Transplanting the brain of an insect reared on one light–dark cycle into the body cavity of an insect on a different cycle will cause the recipient to follow the molting rhythm of the donor.

In addition to its many behavioral functions, JH regulates two important developmental and physiological processes in insects: maturation and reproduction. In insects that undergo complete metamorphosis, JH determines whether larval, pupal, or adult characteristics develop during the molt cycle. As ecdysis approaches in a larva, if JH levels are high, the larva will molt to the next larval stage. If JH levels are diminishing, the pupal transformation will ensue. Absence of JH results in the development of the adult insect. In fact, experimental removal of the CA results in premature development of adult characteristics, and application of JH to final-stage larva will inhibit pupation and result in "supernumerary" instars, that is, unusually large larvae that have undergone more molts than the normal number for that species.

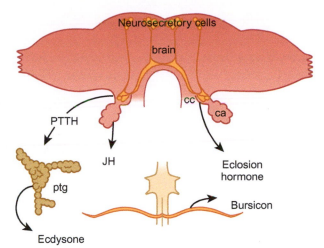

FIGURE 2.21
A dorsal view of the insect brain and endocrine glands. Neurosecretory cells in the brain have axonal connections with the corpora cardiaca (CC; sing. corpus cardiacum), small paired glands behind the brain and also release PTTH, prothoracicotropic hormone, which stimulates the prothoracic gland (ptg) to release ecdysone. The CC, in turn, have axonal connections with the CA (sing. corpus allatum). JH is secreted from the CA. A releasing hormone, PTTH, stimulates ecdysone production by the prothoracic glands. Ecdysone is also produced by the ovaries.

OF SPECIAL INTEREST: JH, LABOR, AND THE HONEYBEE

In an intensively studied species, the honeybee, many scientists think JH regulates the behavioral activities of workers throughout their lives. Aggressiveness of guard bees is correlated with their blood JH levels, for example. Even though guards have high JH levels, their ovaries are relatively undeveloped. Guarding behavior prevents robbing of honey from the nest by bees from other colonies. It is not an easy trick for the guards to tell which incoming bees to attack and which to tolerate. They do this by smelling odors on the surface of the incoming bees; the bees that do not smell like nestmates are attacked. Nestmate recognition is similar to kin recognition and is expressed by most eusocial insects—termites, ants, bees, and wasps. A guard bee may attack a bee from another colony, preventing her from entering the guard's nest. JH titers in worker honeybees progressively increase through the first 15 or so days of the worker's life. During this period, workers perform tasks inside the hive, such as nursing larvae, constructing comb, and cleaning cells. JH titers peak around day 15; workers this age guard, remove dead bees from the colony, and fan at the colony entrance to cool the nest. Older workers forage for pollen and nectar.

JH has numerous behavioral roles, including the induction of courtship behavior, sex pheromone production, diapause, and migration, in many insects, along with influencing caste differences in eusocial insects. It is particularly interesting, in contrast with vertebrate hormones, because it regulates so many systems (Figure 2.22).

JH also regulates reproductive cycles in adults. Because mating behavior is often synchronized with the ovarian cycle, it makes sense for mating behavior and pheromone production to be linked with JH (Figure 2.23). In some insects, including some species of cockroach, this is exactly the case. In other species, the role of JH has evolved one step further so that the linkage with ovarian activity is lost.

MOLLUSCAN HORMONES

Bag cells (Figure 2.24) provide an interesting noninsect endocrine example from among the invertebrates. The bag cells are specialized neurons found in the abdominal ganglion of molluscs such as the sea slug, *Aplysia*. These cells produce an egg-laying hormone that coordinates reproductive behavior in these molluscs.

KEY TERM Bag cells are hormone-producing secretory cells in molluscs.

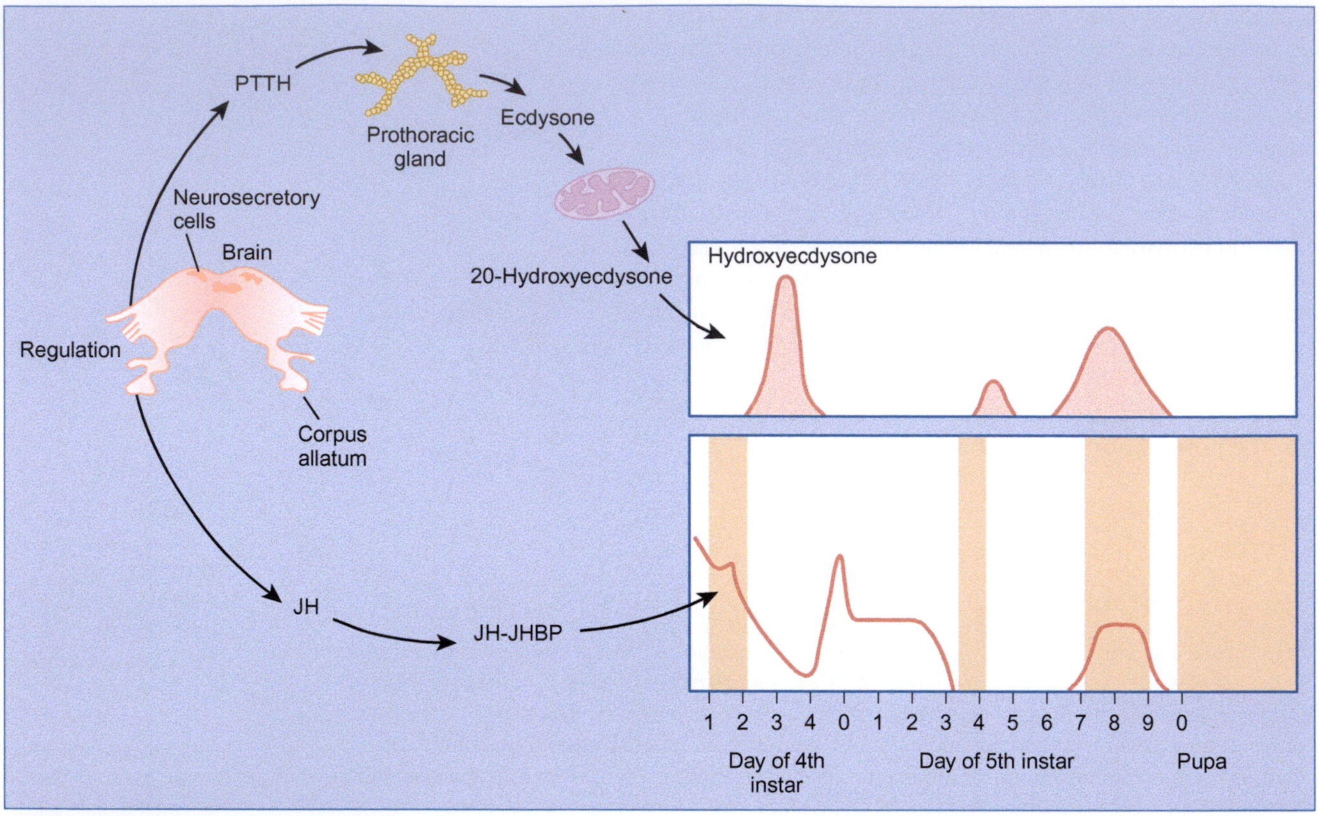

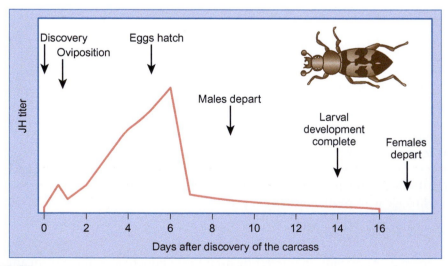

FIGURE 2.22

JH is produced by paired glands, the CA, that are attached by neurosecretory axons to the insect brain (refer to Figure 2.21). This figure illustrates the interaction of JH and a steroid hormone, ecdysone, in the development and molting of insects. Ecdysone peaks in correspondence with the molt, and JH peaks correspond to developmental changes from juvenile to adult. Behavioral changes also occur during development and may also be linked to these hormones. Removal of the CA from an adult disrupts mating behavior, whereas often treatment of an insect with JH stimulates mating behavior and sexual receptivity. *Adapted from: Nijhout, H. F., 1994. Insect Hormones. Princeton University Press, Princeton (1994).*

FIGURE 2.23

Changes in JH titers through a reproductive cycle of a burying beetle. These beetles find a small carcass, bury it underground, and then lay their eggs on the carcass. Both parents remain with the carcass and their offspring until after the eggs hatch, and the females remain until their larvae finish developing. JH increases in the period between egg-laying and egg-hatching, and experiments suggest that JH plays a role in regulating parental behavior in this species. *Adapted from Scott and Panaitof, 2004, Hormones and Behavior 45: 159–167.*[6]

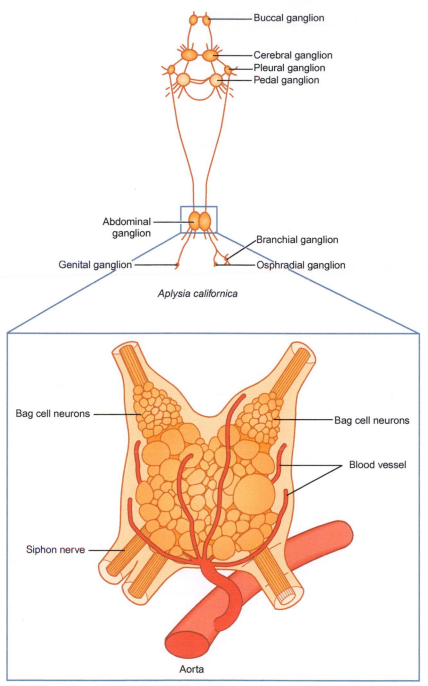

Aplysia californica

FIGURE 2.24

The nervous system of the sea slug *Aplysia californica* (top) and a detailed drawing of the abdominal ganglion (bottom). Signals from the cerebral ganglion cause release of egg-laying hormone from the bag cells in the abdominal ganglion. This hormone coordinates egg-laying behavior, causing the sea slug to stop feeding and moving, to initiate contractions for egg-laying, and to deposit the eggs on the substrate.

Conservation: The Behavioral Effects of Environmental Endocrines

Before leaving this review of endocrinology, consider that the widespread dissemination of estrogen-like compounds into ecosystems (particularly, aquatic environments) has had important behavioral effects. The hormones in birth control pills, for example, are excreted in urine and move, largely unaltered, into wastewater systems and ultimately into streams

FIGURE 2.25

Estrogen-contaminated sewage affects the reproductive behavior of many aquatic vertebrates. In this example, the ability of fathead minnows (*Pimephales promelas*) to hold nests and reproduce was measured. The pink bars represent males that were not exposed to environmental estrogens; the light pink areas on those bars show how many not only held nests but had eggs in the nest. Males exposed to the environmental estrogens (blue bars) rarely nested and never succeed in fostering eggs. Asterisks indicate significant ($p < 0.05$) differences between treatments and controls. *After Martinovic et al.*[7]

and rivers. These estrogens affect morphological development, so intersex (partially female, partially male) fish and amphibia have become common in some locations. Exposure to estrogenic compounds during development also affects mating behavior and has the potential to alter parental behaviors. This is an important emerging issue in applied animal behavior and conservation (see Figure 2.25).

2.4 WHAT DOES AN ANIMAL BEHAVIORIST NEED TO KNOW ABOUT SENSORY SYSTEMS?

Sensing the Environment

Perception of the external world is essential for all forms of life, including bacteria. Animals gather information about many aspects of the environment. To do this, they use organs that act as transducers, converting various kinds of energy from the environment into action potentials. Because there are indeed various kinds of energy, depending on the type of information being received, there are different kinds of transducers, each specialized to convert a specific type of energy into action potentials. For instance, the rods and cones in eyes transduce photons into nervous impulses. This information then moves along the nerves in the form of action potentials until it reaches the CNS, where it is processed and used in determining the animal's next behaviors. In contrast, the sensory mechanisms used with sound differ depending on the sound's wavelength. The following sections contribute to an overall discussion of the sensory systems used by animals to take advantage of these different forms of environmental information.

Sensory Biology: Dangerous Assumptions

Humans, scientists included, have had difficulty accepting that animals can live in sensory worlds that differ remarkably from that of humans. Only in recent decades have people generally accepted the fact that animals can use sensory information impossible for humans to perceive. Some of the first breakthroughs came from studies of honeybees and established that bees can see ultraviolet wavelengths of light and perceive the plane of polarization of light.

OF SPECIAL INTEREST: ULTRASOUND AND BATS

Sound is a particularly interesting communication medium because it can be used in such varied ways. Infrasound—very low-pitched sound—can travel great distances. On the other end of the sound spectrum, bats use ultrasound—very high-pitched calls—for communication and echolocation. Because bats prey on insects, many insect species are attuned to bat echolocation calls and take evasive measures if they hear a bat call. Males produce a calling song to attract females in greater wax moths (*Galleria mellonella*) and lesser wax moths (*Achroia grisella*); they stop calling if a calling bat approaches; presumably the bat can orient to the mating call of the moth (Figure 2.26).

FIGURE 2.26
Bats not only capture flying insects using echolocation but also can sense ground-dwelling prey, like this large centipede. In fact, they can even use echolocation in fishing. (More about that in Chapter 7.)
Photo: J. Scott Altenbach.

As may be apparent by now, it takes quite a leap of the imagination to comprehend that bees see this vastly different world and that, in fact, the colors of flowers may include colors invisible to humans. These colors evolved under natural selection for visibility to pollinators; whether humans could see them or not had no evolutionary consequence. Sensory biologists and animal behaviorists now recognize that the colors, odors, and sounds of an animal's world can extend far beyond what humans perceive, and entire sensory realms, such as electrical fields, are apparently closed to humans. Or are they? What are humans perceiving unconsciously, and are there ways to investigate that?

OF SPECIAL INTEREST: *UMWELT* AND SENSORY BIOLOGY

We introduced the central concept of *umwelt* in Chapter 1. *Umwelt* is a core concept in animal behavior because the price of ignoring it is extremely high; the price that we pay for ignoring it is accuracy and understanding. In studying animal behavior, we must never assume that an animal can only perceive what we think we can perceive. An animal may have no perception of a sound, odor, or wavelength of light that seems obvious to the student or scientist but may have keen perception of information in its environment that is outside the human sensory realm. Seemingly good experiments have been confounded and misinterpreted because an investigator failed to realize that the study animals could sense something in the environment that was unknown and imperceptible to the scientist.

In the words of the poet Mary Oliver, "A dog can never tell you what she knows from the smells of the world, but you know, watching her, that you know almost nothing."[8]

Transduction

Animals perceive energy, in its many forms, by transducing energy from the environment into signals within the nervous system. Transduction is an extremely important process in biology as well as in engineering. Transduction is the conversion of energy from one mode to another, such as light to electricity, or sound to vibration. When a laser "reads" the information on a CD or DVD, the information recorded on the disk is transduced from minute bumps on the surface of the disk to electrical signals that are then amplified into sound or pictures. The same process takes place at the sensory interface between an animal's external and internal worlds. Information comes to the animal in the form of energy—light, sound, magnetic, and so on. The external energy causes changes in specially designed cells, which send the energy along in the form of action potentials in the nervous system. Thus, the receptive cells serve as transducers, converting one form of energy into another.

Generally, the first physiological response to perception is in the realm of specialized molecular receptor molecules on cells. Examples of types of energy that animals perceive include light, polarization of light, temperature, sound, molecular energy (chemoreception), magnetic fields, electric fields, and inertia (touch).

CHEMORECEPTION

Assessing chemical information in the environment was probably one of the earliest sensory elements to evolve. Animals that lack complex sensory organs are nonetheless able to respond to chemicals in their environment. It is thought that what chemosensory structures lack in anatomical complexity—all that is really needed is a fluid-covered membrane—they may make up for in receptor variety. In essence, in the course of chemosensation (the perception of chemicals in the environment), a chemical dissolves in the fluid that covers a membrane, and eventually that chemical or an intermediate messenger system encounters an appropriate receptor. Transduction occurs when the chemical and the receptor produce an action potential. In many invertebrates, chemosensory tissue can be found in specialized depressions on the animal's surface.

The importance of chemosensation to even the earliest life forms becomes more clear upon considering how all living things "leak" chemicals into the environment. Imagine a halo of chemicals surrounding each organism, the unavoidable result of ongoing metabolic activity. To other organisms—everything from potential mates to potential predators—this halo is an advertisement to those with chemoreceptors to perceive it (Figure 2.27).

Taste is a form of chemoreception dedicated to chemicals associated with food. The microvilli at the distal end of the taste cells are exposed to saliva through a pore in the taste bud. There are five types of taste recognized for humans: meat (umami), salt, sweet, bitter, and sour. Meat and sweet guide us to nutrition, bitter and sour steer us away from toxins, and salt helps us maintain osmotic *homeostasis*. These types of taste are then modulated considerably by the sense of smell. Likely other mammals have similar arrays of taste sensors, although the number and responsiveness of each type of sensor evolves to fit the dietary needs of any given species.

In invertebrates, chemosensors can occur anywhere on the surface of the animal; natural selection favors their development where they are most useful. Thus, many insects have chemosensillae on their front legs; they

56

FIGURE 2.27
A chemoreceptive sensor—in this case, a taste receptor.

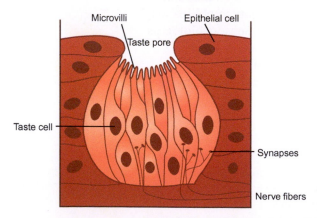

Microvilli

Epithelial cell

Taste pore

Taste cell

Synapses

Nerve fibers

KEY TERM Homeostasis is the maintenance of constant internal conditions in an animal. This is usually done by using feedback loops, in which the internal state is monitored and physiological or behavioral adjustments are made.

can "taste" with their feet (Figure 2.28). Vertebrate chemosensory structures are generally concentrated in the upper part of the respiratory system and are divided among the tongue (taste buds), the vomeronasal organ (on the roof of the mouth in vertebrates), and the olfactory epithelium, which is somewhat anterior in the respiratory system. Sometimes, the vomeronasal organ is called *Jacobson's organ*. The vomeronasal organ may be specifically sensitive to pheromones, and in snakes, tongue-flicking brings molecules sampled from the air into the vomeronasal organ. Until recently, birds have been thought to have limited chemosensory abilities, but at least some birds, such as homing pigeons, are now known to have well-developed chemosensory abilities (Figure 2.28)

DISCUSSION POINT

If flies taste with their feet, why don't we taste with our fingers?

LIGHT AND VISUAL PERCEPTION

Vision begins with receptive cells that transduce light (generally considered to range from longwave infrared radiation to shortwave ultraviolet radiation) into action potentials. Structurally, visual receptors may appear complex, but the underlying principles are simple. When light is perceived, photons are absorbed, and this causes a pigment called *retinal* to change its energetic state. The energized pigment combines with a protein, *opsin*, the primary light-sensitive pigment in the eyes, which leads to depolarization of the membrane of a nerve cell. One photon/retinal/opsin chain is not enough to stimulate an action potential to the brain, but the combined action of many pigment molecules in the cell can generate an action potential.

Light receptors are widespread in metazoan animals. Even cnidarians and flatworms have light-sensitive pigment cells and collections of cells that allow them to discriminate light and dark. In animals with vision that allows some image formation, light-receptive cells are usually concentrated in a light-receptive surface, the retina, and light is focused on the retinal surface by a lens or lenses (Figures 2.29 and 2.30A and B). The processing of light information is complex and involves filtering and enhancement of the image in several stages, beginning at the level of individual receptor cells and proceeding next to interactions among adjacent receptors. Image formation and pattern recognition are functions of the CNS.

MONOCHROMATIC VERSUS COLOR VISION

One of the basic ways in which most animals differ from humans is in how their visual receptors respond to "color." What is color? "Color" is the name for the internal representation of the spectrum of wavelengths reflected by objects. Light itself seems colorless to humans, who perceive color when some of the wavelengths or spectra that comprise light are reflected from a surface. Thus, a surface that appears to be green is in reality reflecting light that is mostly in the green part of the spectrum (Figure 2.31).

There is a wealth of information available in ultraviolet and infrared spectra, and although humans cannot see these wavelengths, many animals can. Bees see ultraviolet, but not

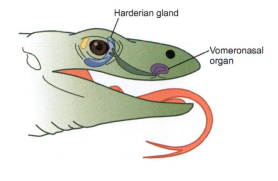
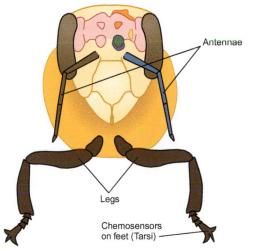

57

FIGURE 2.28
Locations of chemosensors in a mammal (top); a lizard and a honeybee (bottom).

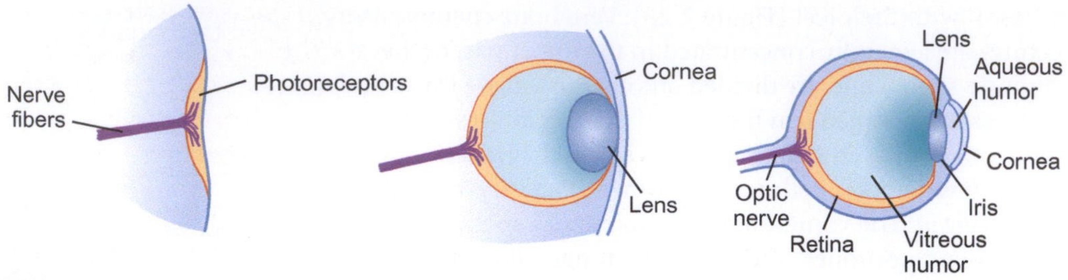

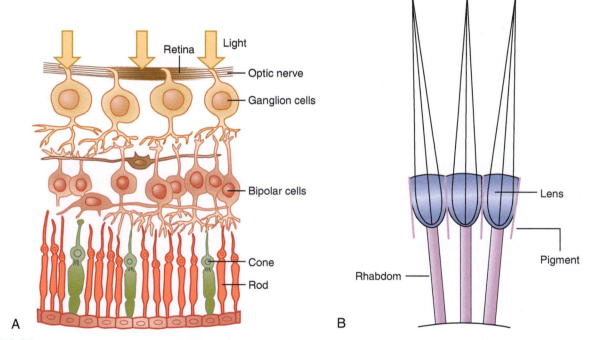

FIGURE 2.29
A variety of eyes, or photoreceptors. Some invertebrates (e.g., flatworms) have photosensitive pigment cups that do not form images. Others have eyes that form images (e.g., octopus) or that are sensitive to the smallest movement (e.g., many insects). The eye on the right is typical of a vertebrate. Animals that live in dark places such as caves or intestines usually have no eyes at all, leading to the inference that the production and maintenance of an eye might be energetically expensive.

58

FIGURE 2.30
(A) The general organization of the vertebrate retina. The photoreceptive cells form a layer behind the nerve (ganglion) cells. In vertebrates with color vision, cone cells provide wavelength-specific responses. (The cone cells illustrated here follow the human pattern of red-, blue- and green-sensitivity.) The rod cells are sensitive across a broad range of the visual spectrum, providing monochromatic (grayscale) vision. Visual pigments are stacked in layers in the rods and cones. (B) In the insect eye, hundreds or thousands of photoreceptive structures called ommatidia are packed together making a compound eye. Three are pictured here. Each ommatidium corresponds, more or less, to a rod or a cone in a vertebrate eye. Although the array of ommatidia in a compound eye can be impressively beautiful, the number of photoreceptors is much smaller than what is found in a typical vertebrate retina, resulting in a lower resolution image. Because each ommatidium essentially captures a point of light, representations of insect vision as an octagonal grid overlaid on a picture are incorrect.

red. Not surprisingly, flowers pollinated by bees are rarely red, but many have ultraviolet patterning that is invisible to the unaided human eye. Most birds can see near ultraviolet wavelengths, as well as the entire human visual spectrum.

Some primates, including humans, gain color vision from having three types of cone receptors. These receptors respond to different wavelengths and have broadly overlapping response curves, so some primate color perception results from the combined effects of two receptors responding to an intermediate wavelength. "Purple" is the name assigned to the perception of the combination of red and blue receptors. No one wavelength produces purple, and in that sense, such a combination can be called *nonspectral*.

FIGURE 2.31
(Left) An image shown in grayscale, as if an organism perceiving this image has only one type of visual receptor that is sensitive across a broad range of wavelengths. (Right) An image shown in color, representing the visual world for an organism with several types of light receptors, each most sensitive at a different wavelength.

OF SPECIAL INTEREST: THE DISCOVERY OF COLOR VISION

Thomas Young (1773–1829) was the first person to hypothesize that humans could see colors with only three types of receptors (Figure 2.32). When he was not busy figuring out the nature of light, discovering the cause of astigmatism, learning how the lens of the eye worked, or practicing medicine in London, he was deciphering the Rosetta Stone, a task that was no doubt enhanced by his knowledge of at least six languages.

FIGURE 2.32
Thomas Young, a London physician who also deciphered the Rosetta Stone, was the first person to develop a fruitful hypothesis about color vision in humans.

Color perception by nonprimate animals can differ from primates in several important ways: different numbers of color receptors, more broadly or narrowly tuned receptors, use of different parts of the wavelength spectrum, or absence of perception in parts of the primate visual range.

For instance, most nonprimate mammals have only two color receptor types: one in the blue–violet (or even in the ultraviolet) range and the other in the green. (So much, by the way, for red being a meaningful warning coloration for the majority of the mammals.) Most birds and some fish (e.g., goldfish) have four color receptor types. Some insects have five or even more receptor types. Having multiple color receptors creates more possibilities for nonspectral colors, as each receptor type can pair with each other receptor type to yield a nonspectral color. Having four receptor types translates into a difficult to imagine minimum of four nonspectral colors, illustrating the difficulty of imagining the visual world of other species.

DISCUSSION POINT: COSTS AND BENEFITS OF COLOR VISION

Can you come up with evolutionary hypotheses about the costs versus the benefits of having a certain number of color receptor types? What might impose an upper limit on the number of receptor types? (*Hint:* Animals that live in very dim light usually have only one receptor type, giving them grayscale vision. This allows them to take advantage of minimal light energy. Each time a different receptor type is added, the light energy is split among the receptors.)

Broadly overlapping receptor ranges, such as those found in humans, result in the perception of intermediate colors, and intermediacy reduces contrast. More narrowly defined receptor ranges, such as the honeybee's, increase the perceived contrast between colors. Contrast may be critical for picking small items, such as flowers, out of monotonous green-colored backgrounds. The answer to the question "What is color?" is difficult. The most important point is that no one should assume that any animal species perceives the world in the same way that humans do (Figure 2.33).

Most crepuscular (active at dawn or dusk) or nocturnal animals have monochromatic vision. This means that they have only one type of visual receptor in their retina, which responds to the presence or absence of light energy. Under low levels of light, human vision is monochromatic, using receptors called *rods*. Because objects reflect different wavelengths of light and different amounts of light, objects that reflect a variety of wavelengths—that is, that are different colors—will produce "pictures" that the animal can discriminate, even if an animal has only one type of visual receptor. The example of a seahorse photographed against a colored background (Figure 2.31) shows that in a monochromatic rendition, contrast between the light and dark areas preserves an interpretable image.

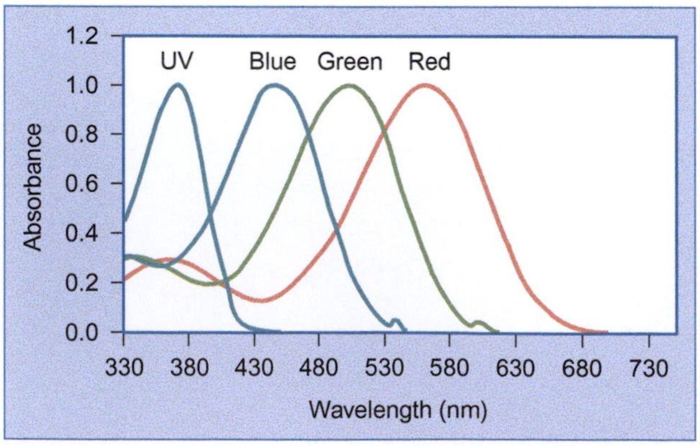

FIGURE 2.33
The spectral sensitivity of the European Starling, showing this bird's four color receptors and its visual responsiveness to light.

SHAPES AND IMAGES

Despite the difficulty of imagining how nonhuman animals perceive and remember images, there are some clear general principles. First, most animals respond to shapes within their visual fields and to the spatial relationships among those shapes. This observation suggests that "snapshot" images of the environment may be formed and stored. Second, interactions between adjacent visual receptors (ommatidia in the compound eyes of arthropods, or rods and cones in vertebrate retinas) emphasize the edges at the margins of shapes. (This will become ecologically relevant in the study of defenses against predators.) The perimeter of a square, for example, is more strongly perceived than the center part. The acuity of vision (the detail of the image) depends on the number of receptors in a given area of receptive surface. An insect eye has a few thousand ommatidia in a few square millimeters; the center of the human eye has 160,000 cones per square millimeter; and the hawk's eye has over 300,000 cones per square millimeter. The hawk has by far the highest visual acuity.

BRINGING ANIMAL BEHAVIOR HOME: WHAT DOES A CAT HEAR? SEE?

Cats' ability to move stealthily through the dark is the stuff of legends and the envy of so-called "cat burglars," but how do cats do this?[9] The most obvious explanation would be those huge eyes, eyes that are almost the size of human eyes in a much smaller head, with pupils that can be much larger. The fact that cats' eyes glow in the dark is part of this enhanced light-gathering efficiency; there is a reflective layer behind the retina called the tapetum, so light can hit the retina when it enters the eye, or when it is reflected from behind the retina. Light that manages to miss the retina (both coming and going) exits the eye and creates that eerie glow. When that light-gathering ability is combined with the very large population of rods in a cat's eye, the result is a predator that can see exceptionally well in the dark. Cats "pay" for this nocturnal accuracy with less accurate diurnal vision and an inability to focus on close objects (<1 foot away). This may seem counterproductive; what is the point of seeing a mouse in the dark if, in that final, close moment, the cat can't focus on it? Tactile information comes into play at this time. Cats can move their whiskers forward and use them to get information about objects within the grasp of their jaws. Meanwhile, the next time you see a cat seeming to nap in the bright sunlight, pupils nothing more than a slit and eyes half-closed, remember that it may simply be shielding its retina from a surplus of light.

As fascinating as cats' eyes are—and we have not even mentioned the flicker fusion rates that allow them to see tiny movements and the muscles that move their lenses—their other senses also contribute to their grace and skill in the dark. For example, cats are amazingly tactile animals. In addition to whiskers, cats have stiff hairs near their ankles and on the sides of their heads, all the better to move through small spaces. Even their canine teeth are sensitive to touch in a way that enables them to perform their killing bite between prey vertebrae. That bite is stimulated by receptors in their lips. Their paw pads and claws, often the first part of the cat to sink into prey, are also rich with nerve endings. Cats are fascinating and complex predators with rich sensory lives.

Flicker fusion is about how rapidly an animal can process motion as separate images. For humans, flicker fusion occurs at something less than 60 images per second; images flashed at a rate faster than the flicker fusion rate blur together, whereas slower rates result in the perception of flicker. In the United States, alternating current electricity is 60 cycles per second, and generally no flicker is perceptible. A strobe light flashing at a lower rate yields a flickering image. Film or video images are shown at slower rates—16–25 images per second—but the similarity of image from frame to frame eliminates the perception of flicker. Animals that rely on seeing small movements by prey, such as cats and hawks, have much higher flicker fusion rates, over 300 cycles per second. This allows them to pick out minute prey movements and perhaps accounts for the fascination of cats with motion.

SOUND AND SOUND PERCEPTION

Humans usually experience sound as the result of vibrations in air or water. Although sound that humans can sense is usually carried through these media, vibrations can also travel through soil, including rocks. Thus, sound can travel through a variety of substances with different densities, and the physical characteristics of the medium through which the sound travels have a major influence on how the sound can be used. For instance, it requires more energy to make water vibrate than to vibrate air, and it requires a great deal of energy to make soil vibrate. Thus, the use of vibrations in communication depends on the ability of the sender to make a substance vibrate. Because of this, large animals such as elephants are more likely than small animals to use vibrations in the soil for communication. In addition, the speed at which sound travels depends on the density of the medium through which it is traveling. This, in turn, varies with physical conditions such as temperature and chemical composition.

OF SPECIAL INTEREST: COMMUNICATION IN THE OCEAN

The speed of sound in the surface waters of oceans is approximately 1520 m/s, almost five times faster than its speed in air (334 m/s at 20°C). In some ocean depths, cold temperatures and moderate pressures yield ideal conditions for transmission; these ideal conditions are called the *sound fixing and ranging* (SOFAR) *channel*, and sounds under these conditions can travel thousands of kilometers. Marine mammals can communicate across oceans via the SOFAR channel.

Vibrations in soil move even faster—up to 13,000 m/s. The frighteningly rapid travel of tidal waves (tsunamis) and the even faster travel of earthquake tremors through the earth's crust reflect the speed at which any vibration if strong enough will travel in these media.

Analytically, sound is usually treated as waves, analogous to ocean waves or ripples in water caused by dropping a stone into a pond or puddle. The reason is that sound, in essence, is vibration; vibrations that are sensed with ear-like structures are called *sound*.

CHARACTERISTICS OF SOUND

Sound has a number of characteristics and properties that affect its perception. Wavelength is the period between waves of sound, also known as *pitch* or *frequency* (Figure 2.34). Expressed in *hertz* (Hz), cycles per second, the human ear perceives frequencies ranging from 20 to 20,000 Hz, although as humans age they tend to lose their ability to hear high-frequency sounds. Hearing ranges for some animals are shown in Table 2.2. A high-pitched sound results from sound waves that have a high frequency; a low-pitched sound results from sound waves that have a low frequency, or relatively few waves per unit time.

> **KEY TERM** A hertz is a unit of frequency that is approximately one cycle per second. It is named after Heinrich Hertz, who discovered electromagnetic radiation in 1888. Electromagnetic radiation is characterized by very long wavelengths.

Intensity is the energy of the sound, measured in watts per square meter. The greater the amount of energy in the sound wave, the higher the amplitude, and the louder the sound. Intensity is a function of the square of the amplitude, corrected for the density of the medium and the speed at which the sound travels in the medium. Another way of expressing this is that the amplitude is the volume, or loudness, of a sound (refer to Figure 2.34). The more dense the medium, the faster the speed at which sound travels.

Sound is vulnerable to distortion and dissipation; these can result from loss of energy over distance, reflection upon bouncing off denser media, or refraction upon moving into a different density medium. In the first instance, sound dissipates as it travels. One important factor in loss is the effect of spreading: as the sound spreads out over a larger area, less energy is present at any one point. Mathematically, this dissipation is described by the inverse-square law, which states that sound energy decreases exponentially over distance.

Reflection happens when sound energy encounters and bounces, or echoes, off the surface of the denser medium. Such echoes can be useful, as in the echolocation system of bats, or can result in confusing noise. Finally, refraction occurs when sound moves into a different density medium and the wavelength frequency changes. This results in distortions of the

TABLE 2.2 Hearing Ranges for a Number of Animals	
Domestic cats	100–32,000 Hz
Domestic dogs	40–46,000 Hz
African elephants	16–12,000 Hz
Bats	1000–150,000 Hz
Rodents	70–150,000 Hz

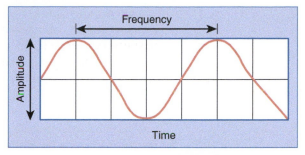

FIGURE 2.34
In this graphical representation of a sound, the *x*-axis is time and the *y*-axis is amplitude, or the strength of the signal. A louder sound produces higher waves, but does not change the distance between the waves. The wave makes repeated full cycles; the number of cycles completed per second is the frequency in hertz. When a recording is played into an oscilloscope, sound is represented in this way.

sound. This is another cause of loss of intensity, because sound dissipates due to absorption and scattering by the medium in which the sound is traveling.

Auditory organs often include an external structure that gathers, focuses, and perhaps adds resonance to sound. Sound may then pass through a tube, channel, or chambers, which also can function as resonators. In general, a body part, such as a hair or a membrane, vibrates in sympathy with the movement of molecules in the surrounding medium (air, water, or soil). This starts the chain of events that lead to transduction.

Just as human visual perception does not encompass the entire visual spectrum, neither do humans perceive the entire range of environmental sound. Although the principles underlying perception of sound are basically the same for different frequencies, sounds above the normal human range of hearing (ultrasound) or below it (infrasound) sometimes differ in their perception.

ULTRASOUND AND INFRASOUND

Ultrasound is an extremely high-frequency sound. Its waves are reflected at the interfaces between different types of tissue. The first attempt at a practical application of ultrasound was in the (unsuccessful) search for the *Titanic* in 1912. It was first used in medicine in 1942, when it helped localize brain tumors. Ultrasound presents two challenges for the animal trying to hear it. First, high frequency translates to short wavelengths; the hearing organ must be miniaturized to match the wavelength. Second, high-frequency sounds tend to be supported by little energy. Not only do they dissipate rapidly as the sound travels, making them relatively faint even close to the source, but they are also subject to absorption by the hearing organ without being transduced into a signal to the CNS.

To accommodate the lower energy of ultrasound, the hearing membrane, or tympanum, is typically thinner in animals that rely on ultrasound for communication or navigation. The outer ears (pinnae) of mammals that perceive high-frequency sound may be quite complex; bat ears are characterized by grooves and channels that help to carry sounds to the tympanum and maintain small differences in frequency (pitch) and amplitude (volume), which can be used to localize sound sources.

Ultrasonic signals are produced in two contexts. First, in echolocation, an animal (generally, bats come to mind) produces high-pitched sounds that are reflected off objects in the bat's flight path. The use of high-pitched sounds (ultrasound) has several advantages in echolocation. (1) The short wavelength of these sounds makes them more likely to bounce back to the bat, rather than bend around the object. This, of course, is essential if echoes are to be used in orientation. (2) It takes relatively little energy to produce these sounds, and (3) they dissipate rapidly, reducing confusion from "old" sounds that could still be bouncing around an area.

Second, ultrasounds are used by several kinds of animals in social contexts. Bats use ultrasounds to communicate with mates, as do murid rodents (rats and mice) and various sorts of moths. In addition, rodent pups use ultrasound to call their mothers if they become isolated from her.

Note
An interesting side note is that humans have less sensitivity at lower frequencies. This is the reason that sound quality appears to diminish when the volume is turned low on music; sophisticated music systems augment (amplify) the lower frequency sounds when set to low volumes.

Infrasound has the special characteristic of traveling well in the ground or water; in fact, the waves of an earthquake can be thought of as a form of infrasound. Because sound travels much faster in the ground than in air (or water), ground-borne vibrations, if perceived, can serve as an early warning system, arriving well before airborne sound from the same source arrives. Infrasound also dissipates less rapidly in air, making it ideal for long-distance communication. Perception of infrasound, however, presents some specific problems. An object smaller than the distance between waves is a poor receiver for those waves. Thus, infrasonic receivers need to be large. This is probably the reason that infrasonic communication is used by only a few animals, and the best understood infrasonic communication system is the African elephants'.

OF SPECIAL INTEREST: INFRASONIC COMMUNICATION IN ELEPHANTS

Infrasonic calls, because they travel great distances, provide an important mechanism for communication and affiliation in elephants (Figure 2.35). By remembering the identity of different callers, elephants can maintain contact with specific individuals, although the social use of individual

FIGURE 2.35
African elephants move in social groups that band and disband over time. Contact among members of social groups is maintained, in part, by infrasonic communication. *Photo: Pen-Yuan Hsing.*

FIGURE 2.36
The large ear of the African elephant serves several functions, including infrasonic hearing and thermoregulation. *Photo: Pen-Yuan Hsing.*

identifications is not known. Female African elephants use "contact calls" to communicate with other elephants in their bands (usually a family group). These infrasonic calls, with a frequency of about 21 Hz and a normal duration of 4–5 s, carry for long distances (several kilometers) and help elephants determine the location of other individuals. Calls vary among individual elephants, so that they respond differently to familiar calls than to unfamiliar calls. Perhaps elephants can recognize the identity of the caller. African elephants have a social structure best described as fluid; animals move freely over wide areas, sometimes affiliating with other animals. Female members of a family tend to stay together, and of course, their juveniles travel with them. These female-centered groups may merge with other such groups periodically. Adult males are less likely to join groups (Figure 2.36).

THERMAL PERCEPTION

Heat and cold receptors play critical roles in habitat choice, avoidance of dangerous conditions (such as overheating), and behavioral thermoregulation. Cold receptors respond to cool temperatures by release of calcium and sodium ions into a signaling pathway when the receptor cell is cooled. These are familiar to us as the receptors that respond to menthol as if it were cold.

Heat perception can also rely on transient receptor potential (TRP) channel receptors. TRP channel receptors are the primary transducers for heat and cold. Some heat-sensitive TRP receptors respond to capsaicin, the chemical that gives hot chili peppers their heat. This is not the case for all heat-sensitive TRP receptors, however, and birds, for example, seem not to suffer from eating capsaicin-laced seeds.

In the most dramatic example of thermal detection, pit viper snakes, heat is perceived as infrared radiation. The pit of a pit viper is lined with infrared-sensitive receptors that are protected by a membrane. Lore among biologists and snake hobbyists falsely states that detailed images, as in the vertebrae eye, are produced by the receptors in the pits. The images are much less detailed, and more subject to interference, than this belief asserts. Pit vipers need a cool background to provide contrast for perception of their prey; this accounts for the fact that most pit vipers hunt at night, when surfaces are cooler, and choose hunting locations with fairly uniform surfaces.

MECHANORECEPTION

Mechanoreception, also termed *tactile sense*, or *touch*, is an important source of information for animals. Touch provides information when light is unavailable, when noise interference obscures sound, or when vocalizations might attract predators. In mammals, touch is an important aspect of affiliative behaviors in social groups. Touch also provides important information about proximity of food, predators, and other environmental features. The tactile sense in vertebrates is supported by pressure-responsive nerve cells near the surface of the skin. These nerve cells act directly as transducers.

Tactile receptors are not evenly distributed over the animal's surface. They are in higher concentration, and therefore closer together, on surfaces that are used for exploration, manipulation, or expression, or that are critical to protect. A standard neurological test in humans involves touching the skin with two needles. On the hands or face, the two stimuli are felt as separate points when only a few millimeters apart. On the upper arm or back, the needles will be felt as a single point when close together and will be felt as separate stimuli only when they have been moved 1 or 2 cm apart. The upper arm and back are less critical areas and are not involved so much in exploration, manipulation, or expression; consequently there are not as many tactile receptors in these areas. In insects, hairlike sensillae play the same role. Again, there are many touch receptors in critical areas, such as the antennae, and many fewer on the thorax and abdomen.

> ## BRINGING ANIMAL BEHAVIOR HOME: EXTRAORDINARY SENSORY ABILITIES
>
> Dogs and cats are startlingly able to sense some things about their environment that humans sense poorly or not at all. Dogs perform much better than humans in olfaction and detecting ultrasound. Cats also hear ultrasound and are much more sensitive to motion than humans (or dogs), and in their play respond very strongly to moving items.
>
> Selective breeding in particular has heightened olfactory abilities in "scent hounds"—dog breeds that include beagles, bloodhounds, and coonhounds. Scent hounds can reliably track the odor of a specific animal or human, and can detect odors left when an animal passed hours or even days earlier. The olfactory epithelium of scent hounds is 6–50 times as sensitive as that of humans. In addition to being employed in hunting, scent hounds work in criminal investigations, including tracking suspects, and in searches for drugs or other contrabands. Scent hounds are often used at disaster scenes as "cadaver dogs," in searches for the remains of fatally injured humans.
>
> Ultrasonic sensitivity in dogs is most commonly exploited in training whistles. Sound from a high-pitched, or ultrasonic, whistle carries long distances and is easy for the dog to pick out from background noise. A dog's hearing may extend to 60,000 Hz, well above the human range; the ability to hear at high frequencies may have assisted the ancestors of domestic dogs in locating prey or in communicating.
>
> Motion sensitivity in cats is likely an adaptation for seeing the movements of small prey, such as mice, when the cat is hidden some distance from the potential prey. The optical processing regions of cat's brains are highly sensitive to movement across the visual field. Motion also seems to feed into a cat's sense of alertness or perhaps pleasure; cats seek play with objects that move rapidly (such as a string that is made to "dance" by a human).

ELECTRORECEPTION

Electroreception is common in fish and amphibia, which live in water, a good electrical conductor. It is absent (or at least poorly understood) in terrestrial vertebrates and invertebrates. Animals cause local distortions of electrical fields and produce electrical currents whenever a nerve cell depolarizes (sends a signal) or a muscle contracts. Usually, there are enough ions in the surrounding water to make the medium a good electrical conductor, and electrical currents and fields are excellent sources of information. Additionally, the interaction of water currents and the earth's magnetic field produces electrical fields that can be used in navigation.

Electroreceptors of fish include ampullary receptors, which are canals opening from the surface of the fish into cavities lined with nerve cells, and tuberous receptors, found on weakly electric fish (mormyriforms and gymnotoforms). The tuberous receptors have a similar structure to the ampullary receptors but are designed to reduce the loss of the incoming signal.

MAGNETORECEPTION—THE ELUSIVE SIXTH SENSE

Wouldn't it be useful to carry an inborn compass so that there would never be a doubt about north and south? Magnetic "signposts"—areas on the surface of the earth or in the oceans with unique magnetic signatures—would be obvious. The world would seem different.

Humans cannot do this, but many animals possess the ability to sense magnetic fields and to extract information about the strength, direction, and polarity of the earth's magnetic field. The list of animals known to use magnetic fields to orient their movements includes sea turtles, salmon, migratory birds, ants, bees, and even bacteria.

Many bacteria are able to perceive magnetic fields and follow lines of strength of those fields when they move. Magneto-orientation may provide bacteria a way of achieving straight-line movements, which otherwise would be difficult or impossible. Bacteria also may combine magnetic field information with other environmental factors, such as oxygen concentration, to find favorable habitats. Magnetoreceptors, which are iron-rich, membrane-bound structures within the bacterial cells, apparently evolved more than once in bacteria.

In multicellular animals, the first suggestion of magnetic field perception came from work on homing pigeons.[10,11] Pigeons seemed to retain their navigation abilities under adverse conditions for compass orientation, such as cloudy skies. William Keeton, an influential figure in the study of pigeon homing, found that, at least in some experiments, equipping pigeons with magnets on their heads disrupted their orientation.[12] He took this to implicate perception of the earth's magnetic field in pigeon orientation. Magnetite crystals were later found at the base of pigeon skulls, suggesting a magnetic perception organ (Figure 2.37). The realization that fluctuations of the earth's magnetic field are correlated with systematic errors in honeybee dances (*Missweisung*) led investigators to explore whether honeybees, as well, might have the ability to perceive the earth's magnetic field. This prompted the discovery of magnetite crystals in the abdomens of honeybees.

The distribution and importance of magnetoreception organs among animals has yet to be appreciated. The typical experimental demonstration of such organs includes placing magnets in or on an animal and monitoring the animal for changes in orientation. Recent experiments with salmon show that they have magnetite crystals in their olfactory organs, but implanting magnets in salmon did not affect their orientation. Similarly, sea turtle migration was not disrupted by magnets placed on the animals. In field orientation experiments, animals have access to a wide range of information and may discard magnetic influences in favor of other stimuli when the magnetic field is disrupted.

Beyond knowing that many animals can sense magnetic fields, little is known about how they actually do this. One possibility is that they use sensory cells containing magnetite, an iron oxide mineral that is highly magnetic. The discovery of magnetite in the beaks or snouts of a variety of animal species suggested this area as the sensory organ for magnetism, and further research showed that in at least some birds and mammals animal species magnetite is found in cells in the olfactory epithelium. Magnetite crystals are also found in otoliths, small particles found in the inner ear of vertebrates that are essential for balance and perception of movement. Additional evidence for the magnetite hypothesis comes from a concentration of magnetite in the abdomen of honeybees.

Another possibility, called the paired radical hypothesis, postulates that the electrical charges of eye pigment molecules called cryptochromes might cause them to couple, and that then the two molecules respond to magnetic fields. The presence of light in the ultraviolet-A/blue range of the spectrum is necessary for cryptochrome-based magnetic orientation. Paired radicals that could function like this have been found in the eyes of birds such as the European robin and the domestic chicken. Cryptochromes are also important in the magnetic orientation behavior of fruitflies, *Drosophila melanogaster*. Genetically engineered flies that lack a functional gene for cryptochrome do not orient to magnetic fields.

There is no strong evidence about what part of the brain information about magnetic fields is analyzed. Obviously, the olfactory epithelium gives information to the olfactory part of the brain, otoliths to the auditory part of the brain, and cryptochromes to the visual part. Can this information be integrated

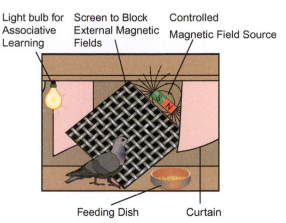

Light bulb for Associative Learning Screen to Block External Magnetic Fields Controlled Magnetic Field Source

Feeding Dish Curtain

FIGURE 2.37
In this clever experiment, pigeons were trained to respond to a light in one way when the earth's magnetic field was turned off and in another when it was turned on. They were then tested with and without the magnetic field and responded correctly, showing that pigeons can perceive the magnetic field. This experimental apparatus is derived from the Skinner Box, which is discussed in more detail in Chapters 1 and 5. Mora, C. V., Davison, M., Wild, M. J. and Walker, M. M. 2004. *Magnetoreception and its trigeminal mediation in the homing pigeon. Nature 432, 508–511.*

67

KEY TERM *Missweisung* are systematic errors in bee dances due to fluctuations in the earth's magnetic field.

if an animal is using more than one of these inputs? The evidence for both the magnetite and the cryptochrome hypotheses seems strong, and it is important to understand that the hypotheses are not mutually exclusive. Both sensory mechanisms could function in the same animal. The results from the genetically engineered fruitflies suggest that in flies the system must have the cryptochrome input to function, but this still does not exclude the possibility of the flies using magnetite. Discovering how the magnetoreception system works in animals is one of the most intriguing unsolved problems in sensory biology.[10,11,13]

SUMMARY

The nervous system receives sensory inputs, coordinates movements, controls physiology, and stores information. The basic unit of the nervous system, the neuron, carries electrical signals from point to point within the nervous system and works in a similar manner across all species. Neurotransmitters carry information among neurons. There are a variety of neurotransmitter molecules, including acetylcholine, dopamine, serotonin, and GABA. Each neurotransmitter has several distinct functions; acetylcholine, for example, carries signals from nerves to muscles but is also involved in learning and memory. Neurotransmitters can set an animal's behavioral responsiveness and prime an animal to behave appropriately in a given context. They are responsible for states that resemble mood in animals.

In some animals, the nervous system amounts to a net of interconnections that allows communication among locations in the body and basic coordination of movements. Most animals, though, have a brain, an anterior mass of neurons. Concentration of neural function in a brain allows for more control over complex functions, and brain complexity is correlated with behavioral complexity. The central and peripheral parts of the nervous system coordinate behavior; reflex loops provide a basic model for coordination of behavioral responses.

The interaction between hormones and behavior is intricate. Hormones are produced by glands and transported to target organs or tissues. They can either prime the tissue or organ for activity or stimulate it into action. Many hormones are steroids, whereas others are peptides (short amino acid chains). For vertebrates, numerous hormones, including estrogen, testosterone, progesterone, prolactin, oxytocin, and vasopressin, affect behaviors as well as physiology. JH and ecdysone are especially important in the behavior of some invertebrates. JH promotes many behaviors, including mating and labor in social insects. Human use and disposal of endocrines and endocrine-like compounds in the environment have important behavioral effects. Steroid hormones and chemicals that mimic steroid hormones persist in the environment and affect development, reproduction, and reproductive behavior in vertebrates. Aquatic organisms, such as fish and amphibia, have been particularly hard hit by these effects.

All living things must perceive the external world. Even bacteria have chemoreceptors and use that information. Information from the surrounding world is critical to behavioral decisions as well as survival, and animals have evolved sensory mechanisms that transduce a wide range of energy from the environment. Animals may perceive the environment in a variety of modalities. As a result, each species' sensory world may be unique. Assumptions about perception based on human senses may be dramatically misleading in studies of animal behavior. Animals transduce energy from their environment into signals within their nervous systems. Sensory systems are derived, evolutionarily, from systems that enable unicellular organisms to perceive molecules in their environments.

How do neurobiology, endocrinology, and sensory biology come together in shaping behavior? These three topics form the core of studying causation in animal behavior. Neurobiology gives us insight into the inner workings of an animal's brain and into the interaction of the brain with the rest of the body. Endocrinology shows us how the development of organ systems is coordinated and how seasonal behavioral changes are orchestrated. Sensory biology tells us about an animal's windows into the external world; sensory systems provide a bridge between the external world and the internal neural and endocrine mechanisms.

Generally speaking, evolution is a relatively slow, multigeneration process that affects regulation of behavior over very long timescales. Hormones stand at an intermediate point in the timescale, operating within the animal's lifespan. Neurotransmitters support immediate change, as when a nerve signals a muscle to twitch, or when the brain needs to organize a response to a specific stimulus. This line is blurry, though, because some hormones, such as adrenaline and insulin, have almost instantaneous effects, although the overall levels of neurotransmitters and their receptors in an animal's brain determine much about the animal's behavioral potential.

Moving forward with study of animal behavior, you will probably need to occasionally refer to this chapter or to explore some of the books listed in the Further Reading section. Even the most field-based, ecologically oriented animal behaviorist needs to have a basic mastery of these topics as they provide the keys to full understanding of animal behavior.

STUDY QUESTIONS

1. What are the parts of a neuron, and how is an action potential transmitted by a neuron?
2. What are the roles of neurotransmitters in behavior?
3. What glands produce estrogen and testosterone? What are the behavioral effects of the removal of the male glands in a developing bird or mammal?
4. What kind of molecule is JH and what gland produces it? What behavioral effects does JH have?
5. How does the pit of a pit viper work? How is this different from heat and cold perception in most animals?
6. Can you explain how scientists can test for the presence of magnetoreception as a sensory mode?

Further Reading

Heiligenberg, W., 1990. Electrosensory systems in fish. Synapse 6 (2), 196–206.

Nijhout, H.F., 1998. Insect Hormones. Princeton University Press, Princeton, NJ, 280pp.

Norris, D.O., 2006. Vertebrate Endocrinology, fourth ed. Academic Press, San Diego, CA, 560pp.

Rudy, J.W., 2008. The Neurobiology of Learning and Memory. Sinauer, Sunderland, MA, 380pp.

Smagghe, G. (Ed.), 2009. Ecdysone: Structures and Functions. Springer, Berlin/Heidelberg.

Verhulst, S., Bolhuis, J., 2009. Tinbergen's Legacy: Function and Mechanism in Behavioral Biology. Cambridge University Press, Cambridge, 262pp.

Zupanc, G.K.H., 2010. Behavioural Neurobiology: An Integrative Approach. Oxford University Press, Oxford, 400pp.

Notes

1. Hughes, D.P., Andersen, S., Himaman, W., Hywel-Jones, N.E., Billen, J., Boomsma, J.J., 2011. Behavioral mechanisms and morphological symptoms of zombie ants dying from fungal infection. BMC Ecol. 11, 13. (10 pages).
2. Hughes, D.P., 2013. Pathways to understanding the extended phenotype of parasites in their hosts. J. Exp. Biol. 216, 142–147.
3. Gould, S.J., 1996. The Mismeasure of Man. W. W. Norton & Co., New York, NY, 448pp.
4. Goodson, J.L., Kelly, A.M., Kingsbury, M.A., 2012. Evolving nonapeptide mechanisms of gregariousness and social diversity in birds. Horm. Behav. 61, 239–250.
5. Vinke, C.M., van Deijk, R., Houx, B.B., Schoemaker, N.J., 2008. The effects of surgical and chemical castration on intermale aggression, sexual behaviour and play behaviour in the male ferret (Mustela putorius furo). Appl. Anim. Behav. Sci. 115, 104–121.

6. Scott, M.P., Panaitof, S.C., 2004. Social stimuli affect juvenile hormone during breeding in biparental burying beetles (Silphidae: Nicrophorus). Horm. Behav. 45 (3), 159–167.

7. Martinovic, D., Hogarth, W.T., Jones, R.E., Sorensen, P.W., 2007. Environmental estrogens suppress hormones, behavior, and reproductive fitness in male fathead minnows. Environ. Toxicol. Chem. 26 (2), 271–278.

8. Oliver, M., 2004. "Her Grave," new poems (1991–1992) New and Selected Poems, vol. 1. Beacon Press, Boston, MA.

9. Bradshaw, J., 2013. Cat Sense: How the New Feline Science Can Make You a Better Friend to Your Pet. Basic Books, New York, NY.

10. Gegear, R.J., Casselman, A., Waddell, S., Reppert, S.M., 2008. Cryptochrome mediates light-dependent magnetosensitivity in Drosophila. Nature 454, 1014–1018.

11. Nießner, C., Denzau, S., Stapput, K., Ahmad, M., Peichl, L., Wiltschko, W., 2013. Magnetoreception: Activated cryptochrome 1a concurs with magnetic orientation in birds. J. R. Soc. Interface 10 (88). http://dx.doi.org/10.1098/rsif.2013.0638.

12. Keeton, W.T., Larkin, T.S., Windsor, D.M., 1974. Normal fluctuations in earths magnetic-field influence pigeon orientation. J. Comp. Physiol. 95, 95–103.

13. Eder, S.H.K., Cadiou, H., Muhamad, A., McNaughton, P.A., Kirschvink, J.L., Winklhofer, M., 2012. Magnetic characterization of isolated candidate vertebrate magnetoreceptor cells. Proc. Natl. Acad. Sci. USA 109, 12022–12027.

Behavioral Genetics

71

LEARNING OBJECTIVES

Studying this chapter should provide you with the knowledge to:

- Realize that comprehending genetic underpinnings of behavior is essential to understanding how behavior evolves.
- Apply the difference between proximate and ultimate causes in behavioral genetics to solving biological problems.
- Use behavioral genetics tools to explore the physiological and neurobiological systems that control behavior.
- Know that most behavior is shaped by a combination of genetic and environmental factors.
- Be able to explain why single-gene effects on behavior should not be confused with the erroneous idea that complex behavior is "controlled" by those genes.
- Show why quantitative genetics provides better explanations than single-gene models for most animal behavior traits.
- Use molecular approaches to behavior genetics as tools that provide important techniques for exploring the regulation of behavior.
- Be able to approach any question about the relative importance of genetic and environmental effects on behavior with the impressive array genetic tools available for testing hypotheses.

Animal Behavior. DOI: http://dx.doi.org/10.1016/B978-0-12-801532-2.00003-9

3.1 INTRODUCTION: PRINCIPLES OF BEHAVIORAL GENETICS

Behavior, like all characteristics of animals, is shaped by a combination of genes and environment. This chapter presents techniques used by animal behaviorists to discover how genes and environment come together to determine behavior. Before Mendel saw his first pea plant, people knew that traits could be inherited. They also knew that the environment could affect inherited traits. Humans have manipulated the genetics of plant and animal species by selective breeding for at least 10,000 years, using the process of domestication to shape crops and livestock to meet our needs. The domestication of animals includes selection for behavioral traits that improve manageability and work characteristics of domestic animals; this selection can be strong, as animals that are unproductive or uncooperative are often culled.

The study of behavioral genetics is a surprisingly controversial enterprise. Behavioral genetics is at the core of the nature–nurture debate, and responsible scientists need to understand the political overtones of nature–nurture discussions. Thinkers at least as far back as Shakespeare understood that behavior is usually the outcome of intrinsic influences (*nature*) and environmental ones (*nurture*). Based on this understanding, in *The Tempest* (IV.i.188–189), Prospero bemoans his attempts to reform the intrinsically savage Caliban: "A devil, a born devil, on whose nature/Nurture can never stick." (In other words, Prospero cannot modify Caliban's nature.) Unlike Shakespeare, scientists have the tools to study the relative contributions of genetics and environment to behavior. What follows in this chapter will not help reform the Calibans of this world (it is to be hoped that there are few!), but it will introduce the genetic basis of behavior, which, in turn, is the key to understanding how behavior evolves.

> **KEY TERM** In the nature–nurture debate, nature is the extent to which genetics influence behavior.

> **KEY TERM** In the nature–nurture debate, nurture is the extent to which environment influences behavior.

The goal of this chapter is to provide an introduction to a broad spectrum of genetic approaches to behavior.[1,2] Genetic studies of behavior take a much stronger problem-solving approach than other areas of behavioral science. An important part of solving a problem is choosing the right tool, and progress through this chapter can be measured by the facility with which genetic tools are understood and used. Knowledge of genetics gives students of behavior powerful windows into the evolution of behavior and the physiological regulation of behavior. The problem-solving abilities gained from behavioral genetics can be applied in the study of almost any type of behavior and form the scientific basis for understanding the nature–nurture debate.

Genetics as a field of study derives from six distinct traditions and scientific cultures (Figure 3.1):

1. Domestication. Humans gained knowledge about artificial selection over centuries of experience and applied that knowledge in domestication of animals and plants. In particular, humans learned that offspring resemble their parents and that selectively breeding animals can be used to develop generations of animals with preferred behavioral traits.
2. Phylogeny. Phylogeny gives us tools to understand the evolutionary roots of behavior, to know the time frame for the evolution of innovations, and to understand evolutionary interactions at the genetic level. Phylogenies were traditionally developed by studying similarities in phenotypes among organisms, but current studies of phylogenies also rely on data gathered using molecular genetic techniques.
3. Classical (Mendelian) genetics. With the advent of more structured, scientific approaches to the natural world, scientists starting with Gregor Mendel tracked how phenotypes

were passed from generation to generation. This led to Mendel's laws of segregation and independent assortment. The concepts of genetic dominance and of epistasis also developed from breeding and selection experiments following Mendel's model. Simple behavioral traits, such as mutations affecting coordination in fruit flies, can be studied using Mendelian approaches.

4. Quantitative and biometrical genetics. Quantitative genetics applies to traits that are continuously distributed, like height or weight, which are not easily associated with the single-gene effects highlighted by Mendel's laws. Many genes in combination influence the expression of quantitative traits. Quantitative genetics gives us heritability, a valuable tool for understanding the relative importance of genes and environment in determining a phenotype. Such genetics are valuable in sorting out the genetic correlations of complex behavioral characteristics such as behavioral syndromes or personality. Biometrical genetics brings detailed measurements into genetic analyses, allowing correlations between the measurements and quantitative genetic influences.

5. Evolutionary and population genetics. These fields study how gene frequencies change over generations, often involving measurements of the effects of selection. Evolutionary and population genetic techniques allow scientists to make models that predict how the frequencies of genes in populations change due to selection, to measure the genetic effects of emigration and immigration on populations, and to assess how artificial selection affects behavioral traits. Heritabililty, a tool from quantitative genetics, comes into play in evolutionary genetics in the assessment of the magnitude of the impact of selection on a phenotype. Phylogeny is also a tool of evolutionary biology that gives insight into how behavior changes over evolutionary time, which can then lead to hypotheses in behavioral genetics.

6. Molecular genetics. Molecular genetics provides tools for determining the location of a gene on a chromosome, for studying how the expression of a gene is regulated, and for identifying genes that are candidates for having a role in regulating a particular behavioral phenotype. Genetic sequencing gives opportunities to explore animals' genomes to find genetic similarities that correlate with behavior.

73

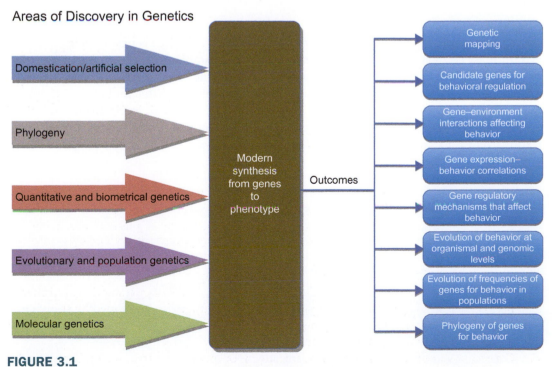

FIGURE 3.1
Contemporary genetics, as a field of study, is an integration of several very different intellectual pathways.

Knowing how and why behavior has evolved is a central goal in the study of animal behavior, one that is greatly informed by an understanding of the genetic influences on behavior. Studying genes that seem to influence behavior gives insight into the physiological and neurobiological underpinnings of behavior. In addition, separating genetic from environmental effects allows both identification of the selective forces that shape behavior and determination of how behavior responds to those forces across generations. In effect, behavioral genetics allows observation of evolution in action.

The identification of specific genes that regulate specific behaviors has been something of a holy grail in behavioral genetics. As more refinements come to the science of behavioral genetics, it has become apparent that for most behavioral traits this is an impossible goal. Candidate genes usually prove to play important roles in sensory systems, channels in membranes, neurotransmitter dynamics, or muscle physiology but are rarely linked only to one behavior. This reflects the fact that growth and development, physiology, reproduction, and behavior in a typical organism are the products of 15,000–20,000 genes. Most genes have multiple functions depending on the type of tissue where the gene expression occurs and the developmental context of the gene expression. Another way of making this point is to say that quantitative analyses, discussed later in this chapter, usually suggest a relatively small number of genes, typically three to eight, that have major effects on expression of a behavior, but when these genes are identified, they turn out to be genes that operate in many other contexts within the organism.

OF SPECIAL INTEREST: PHENOTYPE AND GENOTYPE

Quick refresher: *Phenotype* is the outward manifestation of a trait, that is, the actual behavior, morphology, or physiology. *Genotype* is the characterization of the genes associated with the phenotype. Phenotype does not always reflect genotype in the same way because of environmental influences on the phenotype. The interplay between genes and environment is often measured using *heritability*, a key concept in genetics. Heritability is the extent to which variation in phenotype in a population of animals is correlated with genetic variation. Calculating heritability allows scientists to investigate the genotypic and environmental roots of phenotypic variability.

The most fruitful aspect of behavioral genetics is the ability to place genes in their evolutionary context at many different levels. How, for example, are the sugar receptors of fruit flies different from the sugar receptors of honeybees, and how do these differences reflect the evolutionary results of divergence in diet and foraging behavior between these two branches of the insect phylogeny? This question can be studied at all levels, from genomic exploration of the DNA sequences of sugar receptors to comparative behavioral studies of responsiveness to sugars of the two species.

CASE STUDY: THE "KILLER" BEE

The "killer" bee story is a classic example of plans gone awry. In the early 1950s, Brazilian geneticist Warwick Kerr came up with a well-intentioned scheme to introduce honeybees into the tropical Americas. Honeybees provide an outstanding subsistence source of food and wax for farmers, and these products can also be cash crops that supplement income from products such as bananas and papayas. Kerr planned to keep tropically adapted honeybees, imported from southern and eastern Africa, isolated and to use controlled crosses to develop hybrid bees that would be manageable and productive in the Brazilian tropics. Unfortunately, an unintended release of the African bees in 1957 resulted in the spread of very dangerous, unmanageable insects across the Western hemisphere.

(Continued)

CASE STUDY (CONTINUED)

One of the most extreme examples of ecotypic differentiation is the Western honeybee, *Apis mellifera*. Because honeybees provide many of the classic examples in behavioral genetics, we'll give some background here on their *ecotypes*. The geographic distribution of this species, from southern Africa to northern Europe, through the Middle East and into central Asia, encompasses both tropical and temperate environments, as well as arid and mesic (moist) environments. More than two dozen subspecies of *A. mellifera*—ecotypes differentiated by color, body size, and behavior—have been recognized. This species has been introduced in the Americas and is the common honeybee seen in many North American habitats. One subspecies is also, famously, the "killer" bee, the highly defensive bees now found in much of Central and South America.

> **KEY TERM** An ecotype is a genetically differentiated population within a species that is adapted for a particular habitat.

Why introduce bees to the tropics? Honeybees were already present in the Americas, but these bees originated in Europe and are not well adapted to tropical environments. Honeybees in most of North America are the descendants of bees brought by settlers from Europe beginning in the mid-1600s. They are mostly derived from *A. mellifera ligustica*, the ecotype that is characteristic of the Italian peninsula, but they also have genetic influences from a variety of other European ecotypes. The "Africanized" or "killer" bee is *A. mellifera scutellata*, but it has hybridized with other ecotypes as it has spread through South and Central America and, more recently, into Texas, Arizona, and California.

The differences among honeybee ecotypes have a genetic basis; this was well established by beekeepers long before scientists approached the question, but *common garden* and controlled crosses of ecotypes by scientists support this conclusion.

> **KEY TERM** Common gardening is a technique in which animals (or plants) are maintained in the same environment while observed for differences. If the animals differ in behavior across their native habitats, then keeping them in a common garden helps to separate environmental from genetic influences by standardizing environmental influences. If they all behave the same when in the common garden, then the behavioral differences observed across their native habitats are likely due to environment. If the differences persist in a common garden, then a genetic hypothesis is supported.

The most striking difference between "Africanized" bees and the honeybees to which most North Americans are accustomed is the heightened defensive response of the "Africanized" bees. With bees of European origins, under most conditions an animal can walk to within a few meters of a hive without risk of being stung, and if approaching the hive from behind, can probably walk up to it and sit on it without risk. In contrast, "Africanized" bees often respond to an animal's movement 50–100 m from their hive by flying and stinging. Each "Africanized" bee sting is no more potent that a European bee sting, but mass stinging can be lethal to large animals such as humans, dogs, horses, and cattle. A conservative estimate is that several hundred people have died in South and Central America from mass stinging events since the release of the bees in 1957. "Africanized" bees exhibit extreme responses in all phases of nest defense when compared with their European counterparts; the differences between the ecotypes are largely due to genetics (Figure 3.2 and Table 3.1).[3]

FIGURE 3.2
Ecotypes of the Western honeybee, *A. mellifera*, showing how within a single-species external morphology can vary. (A) A dark honeybee, characteristic of some areas in Europe. (B) An Italian honeybee. *Photo: (A) Emmanuel Boutet, Creative Commons CC-BY-SA-2.5 license; (B) Ken Thomas, public domain.*

(Continued)

CASE STUDY (CONTINUED)

TABLE 3.1 Comparison of the Behavioral Attributes of the African Ecotype with the Western European Ecotype *A. mellifera mellifera*

	A. mellifera scutellata	*A. mellifera mellifera*
Geographic origin	South and Eastern Africa	Northwestern Europe
Reproduction	Many swarms, absconds from nests to migrate when climate turns unfavorable	One or possibly two swarms a year, rarely absconds
Honey storage	Low	Very high
Defensiveness	Very high	Moderate
Size and color	Small, black	Medium, black

These observations raise the question of why the root African stock from which the "Africanized" bees are derived exhibits such extreme responses. The best hypothesis is that honeybees in Africa have faced a combination of predators that includes humans and their evolutionary predecessors. This amounts to strong selection favoring extreme defense. The *scutellata* ecotype has other adaptations for its tropical environment, such as propensity to invest in reproduction rather than storing large quantities of honey, which makes it less attractive to beekeepers; its one strongly positive attribute from a human point of view is its ability to do well in tropical climates, where honeybees of European origin usually fail.

A key lesson from this story is that containment and isolation of potentially dangerous or ecologically damaging introduced animals need to be extremely secure. Knowledge of the behavior of species in their native habitats does not always predict how well they will do in a new habitat, and species as diverse as the "killer" bee, cane toads, Burmese pythons, European starlings, and English sparrows have demonstrated unexpectedly high behavioral abilities to reproduce and disperse in new (to them) habitats.

As noted in Chapter 1, when biologists talk about *ultimate causes*, they are thinking about evolution and the long-term selective pressures that shape an organism's phenotype. Studies of ultimate causes often boil down to understanding where, in the course of evolution, a trait first appears, or where a structure first is used for a particular function. All mammals have digits—fingers and toes—but few have opposable thumbs and the high manipulative ability that comes with having a thumb-like digit. Knowing when opposable digits evolved, and how they were adaptive, helps biologists to understand the ultimate questions associated with these structures, that is, how they evolved. Because genes evolve and help to determine behavioral phenotypes, understanding ultimate causes plays a key role in behavioral genetics.

Often, studies of ultimate causes suggest approaches to finding *proximate causes*, in this case, the present-day physiological or behavioral expressions of gene activity. Biologists who focus on how genes regulate the functions of organisms study the proximate causes of traits. They discover the sequence of events that start with the transcription of DNA and end with the expression of a behavior. The most elegant behavioral genetic studies link the evolutionary roots of behavior to its proximate causes.

In fact, virtually every behavior is shaped by genes acting in sequence or in a coordinated fashion to produce that behavior. Most investigators in behavioral and neural genetics now feel that genetic influence on any one behavioral trait is dispersed over a large number of neural locations, rather than being coordinated by "executive neurons" that integrate the behavior. If many neural elements are involved in the production of a behavior, it logically follows that many genes must underlie the production and regulation of those neural elements. Because behavior is typically shaped by a large number of genes acting in concert, one of the main goals of contemporary behavioral genetics is to understand how multiple interacting genes can shape specific behavioral patterns.

Although a fine-grained discussion of the genetics that underlie behavior may address Tinbergen's causation question, his questions about utility and evolution are not far away. As discussed in Chapter 1, evolution is a change in frequencies of genes in populations. Such a change normally takes place over the course of generations; the environment, however, may change within the course of a single lifetime. In the face of short-term environmental change, behavior often provides the route for an individual animal to adapt to that change. Migration, for example, is usually a behavioral adaptation to seasonal fluctuations that occur during an animal's lifetime. Similarly, learning the location and types of available food resources allows animals to accommodate shifts in food organisms—again, during a single lifetime. Thus, within-generation phenotypic flexibility is very important to animal survival.

Equally important to the survival of lineages is response to longer scale environmental change, such as a slowly shifting climate. This type of change invokes evolutionary responses by favoring the survival of some phenotypes over others, which in turn leads to shifts in gene frequencies that then produce physiological, morphological, and behavioral modifications. All of these modifications are phenotypic; that is, they are all outward manifestations of genetic traits.

If an animal's behavior allows flexibility, it can be a short-term tool with which an individual animal can respond to a variable environment. To the extent it has a genetic basis, it can also be molded over generations by natural selection, in response to long-term environmental variation.

3.2 THE NATURE VERSUS NURTURE DEBATE

What is the balance between genetics (instinct or nature) and learning (nurture) in shaping behavior?[4] Few topics have wasted more emotional energy and created more futile academic fury than the question of behavioral plasticity. An extreme view holds that all animal (including human) behavior is instinctive, with little room for learning and flexibility of response. At the other extreme, some scientists argue that behavior is plastic, modifiable over a great range of possibilities, and that individual experience determines behavior. As is typically the case in such debates, the truth lies between the extremes.

In the animal world, behavior can be envisioned on a continuum between learning and instinct. Depending on the behavior and in some cases, the individual, the explanation may (rarely) involve one of the ends of the continuum, or (more likely) any one of an infinite array of intermediate locations along that continuum. The appropriate scientific goal is that of understanding the balance between genetic constraints and phenotypic flexibility in shaping the behavior of animals. It is this understanding, and not a squabble of extremes, that gives great insight into how evolutionary forces shape behavior. What follows is a discussion of the history of ethology, animal behavior, and sociobiology as it specifically relates to the nature–nurture debate. Refer to Chapter 1 for a broader view of the history of the discipline.

The "nature" school of thought came to the forefront in the early to mid-twentieth century among European ethologists, such as Konrad Lorenz. Their studies emphasized the roles of instinct, fixed patterns of behavior, and the influence of evolution on behavior. Looking back at the work of the early ethologists, we can see the most striking feature of their work was its combination of analytical field studies and well-designed experiments that tested the hypotheses emerging from those field studies. The leading ethologists—Karl von Frisch, Niko Tinbergen, and Lorenz—were interested in observing behavior in a field, or naturalistic, setting and then in using experimental approaches to explore the neurophysiological basis for the behavior. Lorenz enjoyed self-promotion and capturing the public eye; he used his work on animal behavior to make larger arguments about human behavior in books such as *On Aggression, King Solomon's Ring, Beyond the Mirror,* and *Civilized Man's Eight Deadly Sins.* Lorenz's penchant for publicity helped to bring attention to his involvement with the Nazi party in prewar Germany, as well as some of his writings in which he appeared to use his scientific findings to support Nazi social theories. For many, this darker side of Lorenz's past casts a shadow on his work.

Ironically, Niko Tinbergen, the other leading ethologist of the era (von Frisch was a generation older than Lorenz and Tinbergen), was Dutch and lived in the Netherlands through the German occupation of that country. Tinbergen was held at a prison camp (Beekvliet) by the occupying forces for 2 years (1942–1944). Thus, in important ways, Tinbergen and Lorenz seem strange bedfellows, but both before and after the war, they were academic collaborators and personal friends.

In the United States, the study of animal behavior had a much stronger laboratory component than it did in Europe. As a result, psychologists contributed a great deal to its history. American psychologists championed the "nurture" school of thought. Because they worked in a laboratory setting, they could study the role of learning in behavior, something that is difficult to address under field conditions. They started with simple models of learning, such as conditioning, and argued that most behavior is learned, modifiable due to experience, and at least in humans, not constrained by evolutionary history.

A critical evaluation of ethology by the American psychologist Daniel Lehrman[5] provided a touchstone for American comparative psychologists who preferred to isolate animals from their natural environments and study behavioral plasticity (learning) in a laboratory context, where all stimuli could be controlled. (Lehrman and his laboratory discovered much of the behavior–environment–endocrine interaction in ring doves, introduced in Chapter 2.) Lehrman's harsh reaction to the ethologists and to the study of instinct also must have been shaped by his repugnance against the Nazis, although Lehrman's biographer, Rosenblatt, suggests that Lehrman de-emphasized Lorenz's Nazi sympathies in his 1953 critique of ethology. Lehrman died relatively young (in 1972, aged 53).[6] It is interesting to contemplate the role that he might have had in the synthesis of ethology, neurobiology, and genetics had he lived long enough to participate in these scientific revolutions. (See Chapter 1 for more on this aspect of the history of the field.)

Application of the nature–nurture question to human behavior nearly always generates trouble. Data interpreted to show genetic bases for differences among humans in intelligence, motor learning capabilities, criminality, and a broad range of other behaviors have, unfortunately, been used to support racism and other forms of bigotry. Advocates for human social reform are repelled by the thought that human social behavior might be predetermined and might not be subject to modification by social forces. Scientific discoveries should not be ignored if they are disturbing or if they fall outside cultural norms, but neither should they be overinterpreted. It is critical to ask about the costs and benefits of basing policy decisions on an understanding that may be a work in progress. If potential costs—social as well as financial—exceed benefits, then some caution is warranted. Remember that scientific knowledge is a progression based on improvements of methods and collection of more data, and conclusions are less than permanent; as new methods become available, scientific understanding changes. There is danger that much harm can be done by rigidly applying a scientific dogma to human social behavior because that dogma may prove to be flawed, based on incomplete understanding. In short, scientists often do well to avoid stating conclusions in a strongly affirmative manner, especially when those statements may have harmful sociological or political effects on groups of people, unless the benefits of such statements undeniably exceed the cost of being wrong. William Shockley, a Nobel prize-winning physicist who made arguments about racial differences in intelligence and social capabilities, stands as an extreme example of application of "scientific" findings to support negative social agendas.[7]

DISCUSSION POINT: SCIENTISTS' RESPONSIBILITIES TO SOCIETY

Scientists asking seemingly innocuous questions about intelligence and differences among animals and making hypotheses about the evolution of social behavior found their work at the center of a sociocultural and political debate. How responsible are scientists for the uses to which society puts their discoveries? Were strongly worded arguments based on incomplete or inconclusive data? Does this mean that scientists should retreat to ivory towers and address questions that have no consequences for the way humans live? What is the role of scientists in educating the public about the potentially transient nature of scientific conclusions?

Sociobiology provided a major arena for the nature–nurture debate in the 1970s and 1980s. This discipline, championed by E. O. Wilson, integrates thought from ethology, ecology, evolution, and genetics in an attempt to develop a deeper understanding of the evolution of behavior.[8] Although this approach attracts many behavioral biologists, its detractors, such as R. C. Lewontin,[9] suspect that sociobiology (genetic determinism) ultimately supports racist or class-based justifications for inequities in human societies. This vituperative conflict ranges far outside the boundaries of science and, again, becomes an argument that uses science in the pursuit of policy and, in some cases, power.

A more recent example of the nature–nurture debate is the reaction to the argument that rape is an adaptive reproductive strategy in humans.[10] This assertion not only attributes rape to genetic influences, thus raising questions about individual responsibility, but also challenges the notion that sex criminals can be rehabilitated. Some people think that it raises the possibility that potential sex criminals could be genetically identified and segregated from society. These assertions are repugnant to those who think that human behavior is shaped by experience and that all humans are capable of improvement and rehabilitation. No matter what one's stance, this is an understandably volatile question.

It seems that the scientists and popularizers who have attempted to construct an interpretation of human behavior around biological principles typically overreach because they do not recognize the limitations of the scientific base from which they draw their conclusions. The vicious responses of critics take the debate wholly out of the realm of science (and of acceptably polite discourse). The resulting standoff does not increase the understanding of behavior, nor does it help institutions and individuals who struggle to cope with the real outcomes of aggression and other disturbing conditions. Taking extreme stands is, however, a behavioral pattern that is frequently seen in human conflict. In the world of ideas, this often involves oversimplification that is appealing to those who seek a less complicated world.

What is clear is that culture stands between humans and their biology in many interesting ways; some aspects of human culture reflect the evolutionary history and biological constraints of humans, whereas other aspects of culture are counterpoised to biology, regulating or opposing biological forces so that human societies can function. Understanding the complex relationship of biology and culture is beyond current understanding of either topic. Intellectual surrender in the face of this complexity is wrongheaded—this interplay forms one of the great questions of the twenty-first century!—but caution is reasonable when making sweeping "scientific" statements about the biological bases of human social behavior.

79

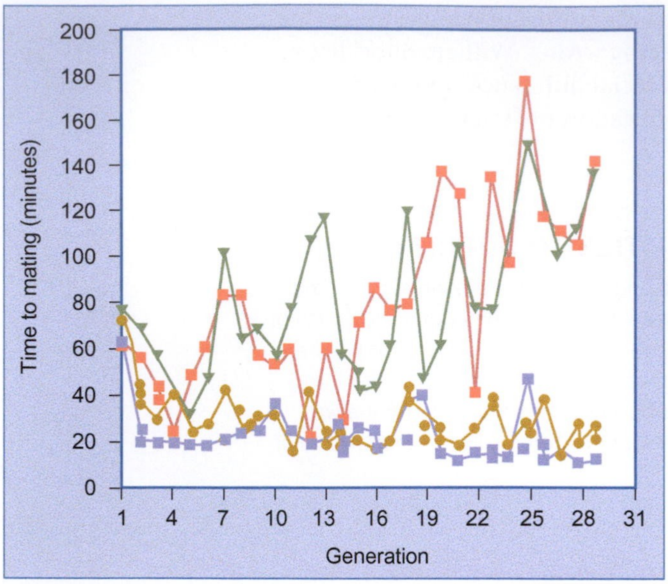

FIGURE 3.3

The response of fruit flies to artificial selection for mating speed. Each symbol represents a line of flies selected for high or low speed. Over generations, the lines diverge; after roughly 7–10 generations, the differences are apparent, and by the 30th generation, the differences are extreme. The end-of-chapter discussion of the application of microarrays to behavioral genetics will include this example. *Adapted from Mackay T.F., et al., 2005 Proc. Natl. Acad. Sci. USA 102 (Suppl. 1), 6622–6629.*[11]

3.3 DOMESTICATION

Artificial selection and *inbred lines* allow exploration of behavioral genetics by testing the responses of behavior to selection or to reduction of genetic variation.

Artificial selection, in scientific laboratories and in animal husbandry, has dramatic effects on behavior. Perhaps the broadest range of artificially selected behavior is seen in domestic dogs, which display a wide variety of behavioral attributes. These behavioral patterns are the result of selection for dogs that assist humans in work (e.g., retrievers, shepherds) or as companion animals. Most domestic livestock (such as chickens, horses, cattle, sheep, goats, and swine) reflect the results of artificial selection for manageability in confinement, ease of training, and docility (Figure 3.3). Strong artificial selection, such as that applied by animal breeders to domestic species (e.g., rabbits, chickens, dogs,[12,13] cats, and cattle), can have substantial effects over three to five generations. This suggests that populations of species in new environments (such as invasive species) or species that are experiencing rapidly changing environmental conditions could have the flexibility to exhibit rapid evolutionary responses if sufficient genetic variation is present.

3.4 PHYLOGENY

Evolution is often perceived as taking many generations to have a visible effect. This may be generally true, but there are many examples of rapid evolutionary change in response to strong selection. It is possible to use genetics to understand evolutionary history. The history, even the recent history, of sea birds is particularly difficult to infer because of their highly mobile lifestyle (Figure 3.4). Genetic studies of these animals in the Pacific Northwest of North America have revealed that there are two genetically distinct populations: one on the Aleutian Islands and one from the eastern Alaskan Peninsula. Although they nest in different locations, the genetic distinctions can be traced back to a single ancestral population that expanded in the early Pleistocene period and then was separated by later Pleistocene glaciations.[14]

FIGURE 3.4

Marbled murrelet (*Brachyramphus marmoratus*). These sea birds have two genetically distinct populations in the Pacific Northwest as a result of late Pleistocene glaciations.

Comparing families, subspecies, and species and looking at differences at those levels help to indicate the extent to which genes influence behavior.[15] Strains, sometimes called *ecotypes* or *subspecies* (see the Case Study), can be produced naturally or by artificial selection. When subspecies are formed naturally, they can be compared to see if the subspecies differ behaviorally; if so, there is a possibility of genetic influence on those behaviors. In the case of artificial selection, animals are bred to produce strains with differing characteristics, including behavior; comparisons among these strains are also important sources of behavioral genetic information.

Because an animal's behavioral phenotype is determined by a combination of environmental and genetic factors, scientists can estimate genetic influences by eliminating environmental variation in the animal's life; if they are successful in doing so, then any between-strain differences can be attributed to

KEY TERM A cladogram represents a hypothesis of evolution within a group of species with a treelike drawing.

FIGURE 3.5
A gray wolf (left), a Mexican wolf (center), and an Arctic fox (right). *Photos: (Left) Frank Wendland, W.O.L.F. Sanctuary, www.wolfsanctuary.net, www.facebook.com/wolf.sanctuary; (Center) US Fish and Wildlife Service; (Right) Keith Morehouse, US Fish and Wildlife Service.*

genetics. For instance, constructing a *cladogram*, or phylogeny, allows an investigator to follow changes in behavioral patterns in evolutionary time and to identify key behavioral innovations.[16]

The gray wolf (*Canis lupus*) is an excellent example of a species with population-level behavioral differences that may reflect selective effects of differing environments on those populations. Prior to the spread of humans from Europe into North America, wolves were widely distributed on the continent and could be easily separated into subspecies based on their habitats. The gray wolf ranged across the north-central part of the continent, whereas the Mexican wolf (*C. lupus baileyi*) was found in the arid southwest and was half the weight of the gray wolf (Figure 3.5). Thus, gray wolves are widespread, and we find much more behavioral variation among such widely distributed populations than among populations with more limited distribution and narrower habitat requirements (e.g., the Arctic fox, *Alopex lagopus*, which is specifically adapted to cold northern climates).

Larger patterns of behavioral evolution can sometimes be visualized by superimposing behavior on a phylogeny or cladogram of a taxonomic group. A phylogeny is a "tree" that expresses supposed relationships among taxa (see Chapter 1). Phylogenies can be constructed in a variety of ways and may follow the intuition of the scientist studying the evolution of the taxa in question. A cladogram is also an evolutionary "tree," but it is a well-defined hypothesis about relationships that is constructed following explicit rules about the analysis of evolutionary relationships; oftentimes, computers are required to implement the "cladistic analysis" of large data sets that contain multiple traits belonging to many species. Cladograms are typically constructed using a set of morphological characteristics, molecular characteristics, or both. Behavioral characteristics can be included in a cladistic analysis, but investigators interested in the evolution of behavior usually avoid circular reasoning by constructing a cladogram based on morphological or molecular (nonbehavioral) characteristics and then observing how behavioral patterns fit into the cladogram (Figure 3.6). Cladistic analysis is most important in helping to test whether a particular behavior or syndrome of behaviors evolved only once in evolutionary history or if it evolved multiple times (convergent evolution). It can also help piece together the evolutionary sequence of events that lead to a behavior; for instance, there may be a shift from a simple behavioral pattern in basal species in the tree to a complex behavioral pattern in more derived species, or the behavior may have become simplified over evolutionary time, with derived species exhibiting only the core attributes of a behavior.

81

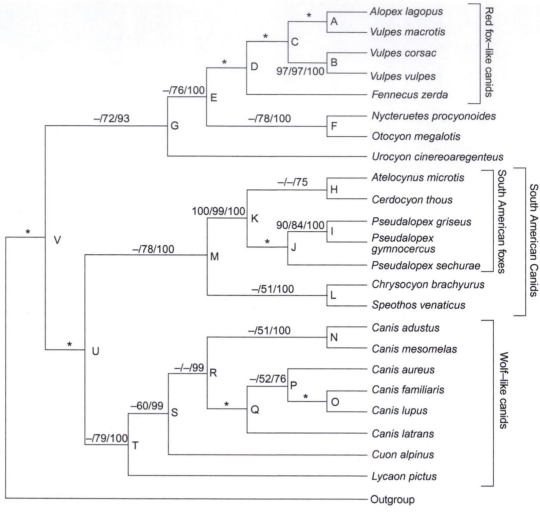

FIGURE 3.6

This phylogeny of the canid family, which includes foxes, jackals, wolves, and domestic dogs, was developed using data from DNA sequences of six genes. Phylogenies like this have value in helping to understand domestication and the evolution of behavior. Domestic dogs, *Canis domesticus*, are most closely allied to wolves, *C. lupus*. Coyotes are also clustered with dogs and wolves, but slightly further from wolves than are dogs. Asterisks indicate phylogeny branches that are statistically significant, p < 0.05. *Reprinted from Molecular Phylogenetics and Evolution, volume 37, Carolyne Bardeleben, Rachael L. Moore, Robert K. Wayne, A molecular phylogeny of the Canidae based on six nuclear loci, pages 815–831, copyright 2005, with permission of Elsevier.*[17]

OF SPECIAL INTEREST: BIRD'S NEST SOUP—THE EVOLUTION OF NESTING BEHAVIOR IN SWIFTLETS

Swiftlets are abundant in South and East Asia. The nests of some swiftlet species are used as the basis for bird's nest soup in Asian cooking. Many swiftlet species build nests that are glued to a tree or a rock face (Figure 3.7A). The glue is a salivary secretion that is alleged to have marvelous nutritional and medicinal properties, and is the key ingredient in the expensive soup. In a few species, the main nesting material is small twigs or feathers. In other species, the glue gains more prominence as a construction material, and in a few species, the glue is the predominant material. Lee et al.[18] chose four characteristics of swiftlet nests: whether a glue or a rock ledge serves as the primary support for the nest, whether feathers are used in construction, whether twigs or other vegetation is used, and proximity of nests to each other (whether the birds nest in colonies).

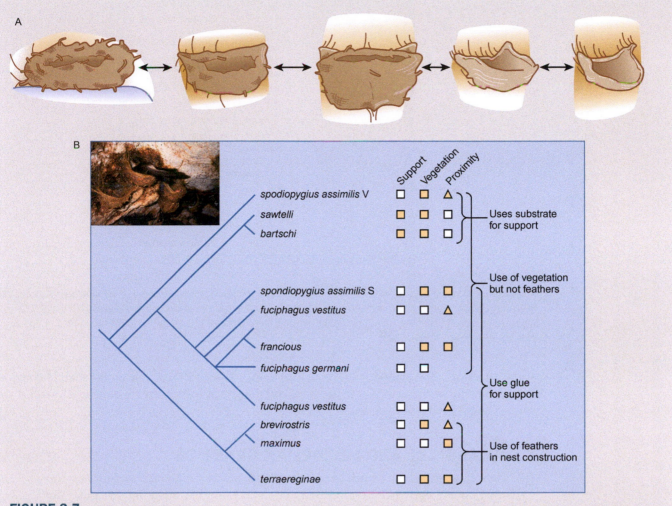

FIGURE 3.7
(A) Variation in swiftlet nests. Some are made almost entirely with salivary glue and adhere to a rock, whereas others sit on rock and are made of twigs or other plant material. (B) A cladogram of swiftlets, with the behavioral traits shown on the right. *Adapted from Lee, P.L., Clayton, D.H., Griffiths, R., Page, R.D., 1996. Does behavior reflect phylogeny in swiftlets (Aves: Apodidae)? A test using cytochrome b mitochondrial DNA sequences. Proc. Natl. Acad. Sci. USA 93, 7091–7096.*[18] *Photo: Dale Clayton.*

They then constructed a cladogram for the bird species and examined how well the behavioral variables fit with the pattern of swiftlet evolution (Figure 3.7B). If the use of salivary glue as the main support for the nest had evolved only once in swiftlets, the expected result would be for this behavior to appear on only one branch of the cladogram. In fact, the behavior appears on three different branches. This evidence shows how a behavior can evolve more than once within a single group of related species.

The same principles of cladistic analysis apply to the study of social evolution in sweat bees (family: Halictidae). Some species of sweat bees are solitary; females establish nests of their own and do all the work without assistance. Others are social; several females (usually a mother and her daughters or a group of sisters) occupy a nest and labor in the colony is divided among the colony members. Did this shift from solitary nesting to complex social behavior happen only once in the course of sweat bee evolution, or has it happened repeatedly? Is it possible for the reverse to happen—for a solitary species to be derived from

83

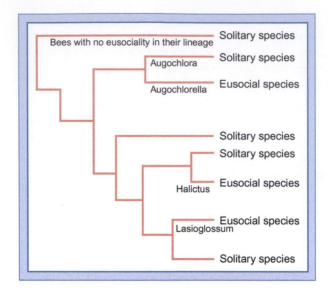

FIGURE 3.8

A cladogram of sweat bees, showing the points at which the worker caste has evolved. This cladogram shows that even very complex behavior can evolve multiple times within a larger taxonomic group. *Adapted from Brady, S.G., Sipes, S., Pearson, A., Danforth, B.N., 2006. Recent and simultaneous origins of eusociality in halictid bees. Proc. R. Soc. B Biol. Sci. 273 (1594), 1643–1649.*[19]

social ancestors? The cladogram (Figure 3.8) helps to answer these questions; social behavior has evolved more than once in sweat bees, and there is at least one case of reversal, with a solitary species derived from social ancestors.[19]

3.5 CLASSICAL AND MENDELIAN GENETICS

The science of genetics was stimulated by the work of Gregor Mendel and others who measured how phenotypes are expressed in parents and their offspring. For some traits, such as the floral colors studied by Mendel, simple parent–offspring patterns of inheritance give insight into the genetic control of the trait. In the early part of the twentieth-century geneticists developed the powerful tool of studying mutations and how mutations are transmitted from generation to generation. Among the important outcomes of studies of mutations was the ability to associate mutations with chromosomes and to establish linkage groups—genes that are close to each other—on chromosomes.

A single-gene mutation is a genetic change in just one gene. Such a mutation, spreading rapidly through a broad geographic area, has dramatically changed the social behavior of the black imported fire ant, *Solenopsis richteri*, and the red imported fire ant, *Solenopsis invicta* (both introduced via cargo ships to the United States in the twentieth century). Colonies of *S. invicta* are normally monogynous, meaning each colony has a single queen and the workers in the colony are daughters of that queen. Growth of monogynous colonies is limited by the number of eggs this single queen can lay and by how much food can be collected to feed to the developing larvae. For this ant in its native habitats in Argentina and southern Brazil, the reproductive capacity of a limited number of queens is adequate for colonies to be competitive. Once this species was introduced into the southeastern United States, it encountered a habitat devoid of its natural predators, parasites, and competitors. These conditions favor colonies with the capacity for explosive growth. This growth can be attained by having many queens per colony (polygyny), each laying large numbers of eggs.

The difference between monogynous and polygynous fire ants lies in the expression of a single gene, *Gp-9*, that codes for a pheromone receptor molecule. The mutation appears to rob the worker ants of the ability to discriminate among queens (Figure 3.9); consequently, they tolerate a large number of queens, including queens that are genetically unrelated to each other or the workers. This is an excellent example of how a change in a single genetic component that underlies a complex social system can have major effects on the function of the entire social structure. Recent discoveries of the polygynous form of *S. invicta* in Taiwan, Australia, and China underscore the effectiveness of the mutation in facilitating ant invasions.[20]

FIGURE 3.9
A fire ant queen with workers and eggs. *Photo courtesy of Sanford Porter/ USDA-ARS.*

Another invasive ant, the Argentine ant (*Linepithema humile*), has been highly successful, possibly because it exhibits reduced intraspecific aggression. A genetic study using microsatellite DNA markers showed that these ants experienced a population bottleneck that reduced genetic diversity during introduction to California. This is consistent with the hypothesis that reduced genetic variability may impair nestmate or colony recognition, permitting the invaders to behave like one large colony. This, in turn, allows them to rise to higher densities than they do in their native habitat.[21]

The *Solenopsis* story is significant in several ways. First of all, the ant is a serious, invasive pest whose arrival in an area is associated with undesirable ecological and economic consequences. That said, this ant does present a rare example of rapid evolutionary change. It is a rare example because the behavior of animals in their natural setting normally results from many generations of natural selection; the results of this natural selection are observed more often than is natural selection itself in action. In contrast, the introduction of exotic animals (and plants) into ecosystems creates new opportunities for natural selection to act. Animals in new (to them) environments face dramatically changed regimes of selection, which can lead to strikingly rapid evolutionary change.

85

The effect of single-gene mutations can also be readily seen in that workhorse of laboratory genetics, the fruit fly, *Drosophila melanogaster*.[22] It is raised easily in large numbers in the laboratory and has a fast generation time. This makes it a good "model system" (see Chapter 1). Many genetic mutations affect fruit fly mating behavior. Male behavior is inhibited by genes named *Nerd* and *fruitless* and is enhanced by *Voila*. Female receptiveness to mating is impaired by genes such as *dissatisfaction*, *spinster*, and *chaste*. Mutations that affect the visual system reduce fruit fly male mating success by making it difficult for males to find females (Figure 3.10). Auditory mutations make females less receptive to mating because they cannot hear the male courtship song. Chemosensory mutations can affect the courtship performance of sexes. In fact, *Voila* seems to affect chemosensory cells in the male's front legs. Fruit flies can be rendered less receptive to mating by modifying the gene (the *Icebox*, or *ibx*, mutation) that is involved in normal formation of brain structures. Unable to respond to a potential mate, the modified flies cannot mate. Does that mean that the *ibx* gene "controls" mating? Not at all. It merely controls one small but crucial step in the mating sequence.[23]

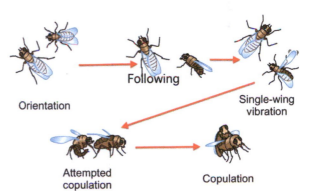

Orientation

Following

Single-wing vibration

Attempted copulation

Copulation

FIGURE 3.10
Drosophila courtship. These are examples of behaviors that mutations may affect.

Mutations that affect processes in the brain are even more interesting because they can give insight into the organization of the central nervous system and the generation of complex behaviors such as mating. Fruit flies offer insights into this level of behavioral control as well. The ways that these genes might affect courtship and mating become more apparent when the sequence of courtship behavior and related sensory modes are considered. Visual, auditory, and chemosensory systems may all be affected.

In addition to the study of mutations and their inheritance, a number of helpful techniques are used in classical studies of genetics and behavior. *Cross-fostering* is a simple and highly effective technique for separating the effect of rearing environment from genetic (instinctive) influences on behavior. Consider two species of swiftlet: species A, which grows up in nests that resemble weaving, and species B, which grows up in well-constructed mud cups. If members of species A are reared in nests of species B, then how they build their own nests as adults will depend on whether experience is important shaping nest-building behavior. If the behavior is entirely genetically programmed, then species A animals will weave nests, even if they are reared in a species B nest. If the behavior is shaped by learning and experience, then species A will learn and adapt to species B's mud construction.

The basic procedure of *cross-fostering* involves transferring some newly born or hatched young of species A from their parental nest to the nests of species B. Ideally, if there are multiple offspring from a given mating, the clutch or litter would be subdivided, with some going to B, and some staying home with A. The converse happens as well—that is, some young of species B are moved to homes with A, whereas others remain with species B. This results in four groups: species A raising A, species A raising B, species B raising A, and species B raising B, setting the stage for asking what happens to the behavior under these four conditions. Does the behavior reflect their genetic background (the behavior of their parents) or their social background (the behavior of the foster parents that reared them)? After a period of time, during which the behavior of the focal animals develops, the behavior of the transferred (fostered) animals and the nonfostered controls is documented, measured, and compared. If the transferred animals behave like the controls of their species, then the usual conclusion is that genetics dominates, and if they behave like their adopted "family" in the host nest, then environmental influences are predominant.

> **KEY TERM** Cross-fostering is a transfer of young between mothers. This technique is used to separate genetic from environmental influences on behavior.

A recent study highlighted the utility of cross-fostering experiments in separating genetic from environmental (learned) components in bird song (Figure 3.11).[24] Johannessen et al. wondered if great tits and blue tits, both common in Europe, learned songs from parental birds in their own nest. Because these species nest in the same habitat, it would be possible for young birds to pick up songs from birds in their own nest, from neighboring nests of the same species, or from neighboring nests of other species. The investigators located nests and transferred some eggs between nests of the two species. This resulted in a classical cross-fostering design, in which birds of other species rear treatment individuals and control animals are reared by their own species. Because the experiment involved two species, the test had the added dimension of allowing the scientists to ask if song learning in these birds is flexible enough to accommodate songs from a different species.

Blue tit songs were very similar between control and cross-fostered birds, whereas the songs of cross-fostered great tits took on many of the characteristics of blue tit foster parents. Elements of the song of the biological (same species) parent remained, but because the experiment was conducted in the field, the results do not indicate if these song elements are genetically derived or if they are learned from birds in neighboring nests. Further experimentation using a cross-fostering design in a more controlled setting would sort this out. After leaving the nest, cross-fostered birds in this experiment appeared to prefer the company of their foster species to that of their own species, suggesting that song learning is part of a larger syndrome of species-specific imprinting (see Chapter 5).[25]

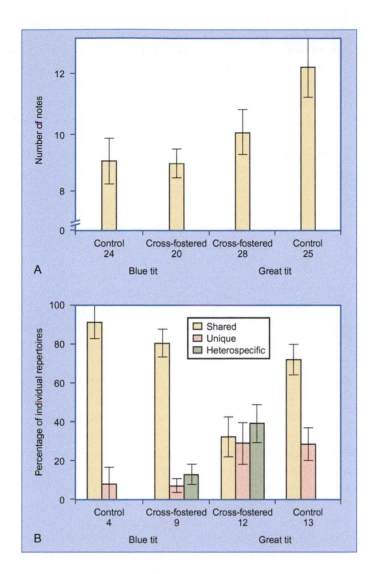

FIGURE 3.11
Results of cross-fostering experiments between blue and great tits in which song learning of the nestlings was measured. Cross-fostered birds use songs that are intermediate between the two species. (A) Cross-fostered great tits have fewer notes and shorter songs than control birds of the same species. (B) Heterospecific songs appear as part of the repertoire of cross-fostered birds of both species. *Adapted from Johannessen, L.E., Slagsvold, T., Hansen, B.T., 2006. Effects of social rearing conditions on song structure and repertoire size: experimental evidence from the field. Anim. Behav. 72, 83–95.*[24]

DISCUSSION POINT: CROSS-FOSTERING

What needs to be done to make sure a cross-fostering experiment is adequately controlled so that genetic and environmental components of a behavior can be separated?

Although intrauterine conditions are often not considered when studying environmental effects, they are some of the first environmental conditions that an animal encounters. Can the intrauterine environment affect the later behavior of animals? (See Chapter 2 for more discussion of this topic.) How can maternal influence and genetic effects be distinguished? Cross-fostering can play a role here as well. Embryos implanted into recipient uteri can serve as a control for prenatal effects of maternal physiology on the behavior of offspring. This technique has been used in mice, with the intriguing result that cross-fostered embryos behave like the recipient strain of mice, yet if cross-fostering is performed after birth, then

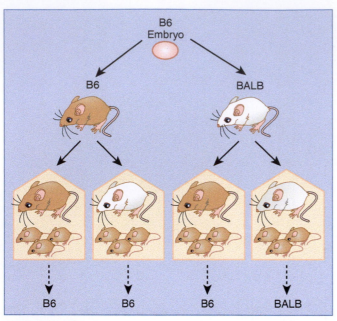

FIGURE 3.12
B6 mice embryos were implanted in mice of two strains (B6 or BALB). The offspring, which were all genetically identical, were raised by either B6 or BALB mothers. The only offspring that exhibited BALB behavioral traits were those that had been both prenatally and postnatally fostered by BALB mothers. *Adapted from Crabbe, J.C., Phillips, T.J., 2003. Mother nature meets mother nurture. Nat. Neurosci. 6, 440–442.*[4]

the mice behave like their biological parents (Figure 3.12). This result does not invalidate cross-fostering experiments done after birth or hatching, but it does raise the interesting caution that prenatal effects need to be considered when designing cross-fostering experiments. In humans, some investigators think that altered maternal well-being (e.g., stress) changes the uterine environment, resulting in lifelong effects on the behavior of the offspring.

In other mammals, position of the embryo in the placenta can also affect adult behavior; in mammals with three or more embryos in the uterus, offspring behavior after birth is affected by relative position (front to back) in the uterus and by the gender of adjacent embryos in the uterus. The influence of maternal effects and interactions between embryos on gene expression in offspring is one example of "epigenetic" effects, that is, variable functions of genes that do not involve actual changes in DNA. This variation can be caused by a broad array of environmental influences that are only beginning to be understood but that are clearly critical to attempts to parse the phenotypic contributions of "nature" and "nurture" (see Section 3.2 for more on this topic).

Twin studies are similar to cross-fostering tests of the relative importance of genes and environment in shaping behavior. These studies are most commonly performed in humans, for whom cross-fostering experiments are ethically questionable. They take advantage of the fact that sometimes human twins are monozygotic (MZ) and sometimes they are dizygotic (DZ). MZ twins come from the same fertilized egg and are genetically identical (Figure 3.13); DZ twins come from separate fertilized eggs and thus are no more related to each other than if they were siblings born at different times, except that they share the same uterine environment. Because both twins in a pair experience the same rearing environment, experimenters can compare similarities (concordances) and dissimilarities (discordances) in behavior between MZ and DZ twins; if MZ twins show greater degrees of behavioral concordance than DZ

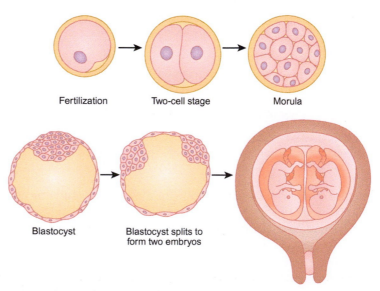

FIGURE 3.13
Embryogenesis of MZ twins.

twins do, those differences are probably due to genetics because rearing environments are controlled. For instance, Deater-Deckard et al. have recently found that behavioral attributes such as task persistence, anger/frustration, and conduct problems in children were much more similar in MZ twins than in DZ twins, although environmental influences also have substantial importance. Perfect concordance of behavioral traits is rare even in MZ twins, suggesting that genes are not the sole influence on behavior; environment and chance play large roles as well. MZ twins occur in at least some nonhuman primates[26] and in other animals, so the possibility of using twin techniques for behavioral studies extends beyond humans, although detection of MZ offspring in nonhumans requires genetic testing.

> **KEY TERM** Twin studies can segregate environmental from genetic influences because twins are genetically identical, but may be reared in different environments.

Cloned animals offer a major potential for application of the principles of twin analyses in unique ways: a clone offers the opportunity to observe the genetic equivalent of MZ twins born at different times! If a favorite pet is cloned, will the clone's behavior resemble the original pet's behavior in the desired ways? Human MZ twins are often quite similar behaviorally, sometimes eerily so in behavioral quirks and mannerisms, but also can differ in significant ways; if cloning of pets gains widespread acceptance, it will be interesting to see how well pet owner expectations are met.

> **KEY TERM** An inbred line is a population in which closely related animals, such as siblings or parents and offspring, have been repeatedly mated so that nearly all genetic variation is lost. This is similar in effect to cloning.

An other approach to separating genetic and environmental effects, called the common garden technique, merits mention in this section. (See the Case Study and Key Term on page 75 for more on this technique.) In a "common garden" experiment, animals with differing genetic backgrounds are all kept under the same conditions. The assumption is that if there are genetic differences among the animals in their behavior, these will become apparent, as all the animals are experiencing the same environment. This is an extension of the reasoning used in cross-fostering experiments.

> **KEY TERM** Cloned animals are genetically identical. Use of cloned animals extends the idea of twin studies by allowing larger sample sizes and more controlled conditions.

In common garden experiments, the environment is held constant in an attempt to assess genetic variation. Another approach aimed at exploring the relative contributions of genetics and environment involves moving an animal to a new environment and asking if a behavior persists. If behavior changes with the change in environment, then a strong genetic influence on behavior may be doubtful. However, conclusions are more difficult to come by if the behavior persists.

What if parental influences on offspring behavior might be important? The approach is basically the same, using translocated parents, and then focusing on offspring (F1 individuals) reared in a shared environment. If parent–offspring interactions are important in shaping the behavior of the F1 generation, then a more sophisticated approach, such as cross-fostering or planned matings between strains, may be required to completely test the hypothesis that behavioral differences among ecotypes have a genetic basis.

3.6 QUANTITATIVE AND BIOMETRICAL GENETICS
Heritability

Quantitative genetics focuses on the variability of phenotypes and the correlation between genotype and phenotype. *Genetic variation* is necessary for either natural or artificial selection to produce shifts in gene frequencies. The amount of genetic variation present in

89

a population is key to understanding how selection will affect the population. Measures of genetic variation also help to understand how past selection has affected a trait. The study of genetic variation is approached through heritability, which quantifies the portion of the phenotypic variation which is due to genetic variation. Genetic variation also links to the powerful method, *quantitative trait loci* (QTLs) analysis, for identifying genes that may regulate behavior through genetic mapping.[27]

The first focus of this section is on calculations of heritability. *First, when thinking about heritability, it is very important to remember that heritability is about measuring variance and is not the same as inheritance.* The concept of heritability was developed by quantitative geneticists and has powerful applications in behavioral genetics. Neither environment nor genes *determine* an animal's behavior, but both environment and genes contribute to the behavioral phenotype. A major goal of behavioral genetics is to understand the extent of these environmental and genetic correlations with the behavioral phenotype and this is accomplished by estimating heritability.

The second key to understanding heritability is to remember that it is a population, not an individual, measure. By knowing that evolution occurs within a population over generations, it can be seen that heritability assesses the potential for evolution in a population. The potential for modifying a trait through selection is assessed by measuring its heritability, which is estimated by measuring genetic and phenotypic variation.

> **KEY TERM** Heritability is the proportion of phenotypic variation that is explained by genetic variation.

> **KEY TERM** A QTL is a gene that contributes, with other genes, to a phenotype. Because multiple genes contribute to the phenotype, no one gene "determines" the phenotype.

Natural selection acts on genetic variation; if there is no genetic variation, then traits cannot change over time and evolution cannot occur (Figure 3.14). It follows from this that behavioral geneticists often focus on understanding how genetic variation affects a behavioral trait. Inevitably, studying the effects of genetic variation leads to considering the effects of environmental variation on the behavior as well. The behaviors an animal expresses are partly a reflection of the animal's environment and partly a reflection of the animal's genes. Not surprisingly, some behavioral phenotypes, such as signals involved in courtship,[28] often show little variance among environments, whereas others, such as foraging behavior, are typically quite responsive to environmental differences.

How do scientists measure the variation in a behavior and allocate that variation to correlations with genes or variation correlated with environment?[29] As has been pointed out already, heritability is the proportion of phenotypic variation in a population due to genetic variation. *Variation* is the operative word here: It is important to understand and remember that heritability is all about variation; studies of the heritability of a behavior are actually studies of the variability of that behavior within a population.

Understanding heritability and how it is used rests on the following key concepts:

1. Calculating heritability is done by estimating the genetic and environmental contributions to phenotypic variation at the population level.
2. Heritability is a population-level measure, not an individual measure; only traits in populations have heritability.
3. Heritability is NOT a measure of the degree of genetic control of a behavioral trait.
4. Heritability can differ among environments; the same population may show a different heritability for a trait if the environment is changed.
5. The behavior in question must vary among individuals in the population for it to have a measurable heritability; invariable behavior has no heritability.
6. Strong selection (natural, sexual, or artificial) on a behavioral trait reduces the heritability of that trait (because it reduces the variation).

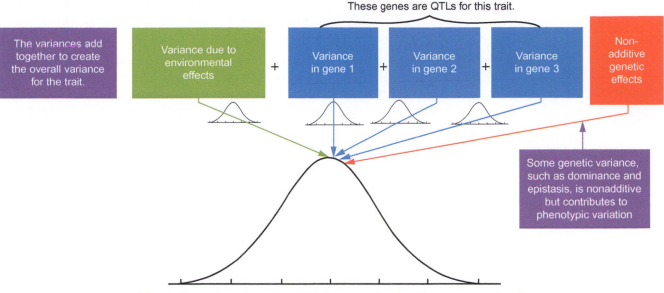

How phenotypic variation for a
quantitative trait is built

These genes are QTLs for this trait.

The variances add together to create the overall variance for the trait.

Variance due to environmental effects

$+$

Variance in gene 1

$+$

Variance in gene 2

$+$

Variance in gene 3

Non-additive genetic effects

Some genetic variance, such as dominance and epistasis, is nonadditive but contributes to phenotypic variation

The trait, such as length, weight, or a behavior like shyness, is often normally distributed in the population. A "bell curve" describes its variance in a population of animals.

FIGURE 3.14
Quantitative genetic traits are usually normally distributed. The overall variation of the trait is due to variation contributed by each gene that affects the trait, as well as by environmental effects. These sum to determine the trait variation.

Keep these principles in mind through the discussion of heritability that follows.

Consider a phenotype, such as shyness, in a population of animals, using the following notation:

V_p = variation of the phenotype
V_e = variation due to the environment
V_g = variation due to genetic effects

With these three expressions, a simple equation can be generated that describes the relationship among phenotype, environment, and genes:

$$V_p = V_g + V_e$$

This is a restatement of the central concept: phenotypic variation is the sum of environmental and genetic variation.

There are two types of *genetic variation* (V_g): additive and nonadditive. Recall that animals can differ genetically because they have differing alleles (forms of a gene) at loci (locations on the chromosome) that influence the trait in question. Usually, the effects of the different alleles that work together are the sum of their activity, so genetic variation due to allelic differences is called *additive genetic variation* (V_a). Sometimes one allele is dominant to another in determining the effect of a locus, or loci may interact in ways that change

KEY TERM Genetic variation is a measure of the variation in phenotype that is due to all variation in genotype.

KEY TERM Additive genetic variation is the proportion of genetic variation that is due to differences among alleles that sum together.

KEY TERM Nonadditive genetic variation results from interactions between genes and from gene dominance.

KEY TERM Epistasis occurs when genes interact.

KEY TERM Broad-sense heritability is the proportion of phenotypic variation that is explained by all genetic variation for the trait.

KEY TERM Narrow-sense heritability is the proportion of phenotypic variation that is explained by only the additive genetic variation for the trait.

the action of the genes (*epistasis*). Variation due to dominance, epistasis, and other types of interactions among alleles and loci is *nonadditive* (V_{na}).[30] Total genetic variation, V_g, is the sum of the additive and nonadditive components, $V_g = V_a + V_{na}$.

Heritability is the proportion—or percentage—of the phenotypic variation that can be attributed to genetic influences. To get this, like any other percentage, the genetic variation is divided by the phenotypic variation (V_g/V_p). This yields a measure called broad-sense heritability. Notice that *broad-sense heritability*, H^2, involves both additive and nonadditive genetic variation.

When only additive genetic variation is used in the calculation, the result is *narrow-sense heritability*, $h^2 = V_a/V_p$ (Figure 3.15).

Given this explanation, how might heritability be used in studies of animal behavior? What have studies of heritability of behavior found? Heritability is useful in two ways:

1. Broad-sense heritability is used as a measure of the magnitude of all of the genetic influences on a trait. Some phenotypic variation depends on the environment; if animals are measured in two different environments, the result will be different heritabilities for a trait, even if the animals in the two environments are genetically the same.

2. Narrow-sense heritability is particularly useful in predicting how animals will respond to artificial or natural selection. If a trait has a high heritability, selection or controlled breeding can change that trait. This is because high heritability is calculated from a high level of genetic variation that can be subject to selection. Low genetic variation does not give selection much "wiggle room."

92

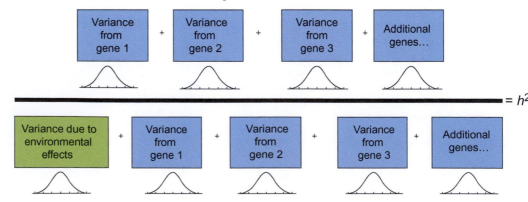

Narrow-sense heritability

Phenotypic variation due to additive genetic variation

Total phenotypic variation

FIGURE 3.15

Narrow-sense heritability is the proportion of phenotypic variance due to additive genetic variance. In this illustration, the numerator shows how effects of different genes sum to equal the additive genetic variance of the trait and the denominator shows the contributors to the overall variance of the trait.

BRINGING ANIMAL BEHAVIOR HOME: HERITABILITY AND THE BEHAVIORAL GENETICS OF DOGS

The fact that much of dog behavior has genetic underpinnings is patently obvious. Differences in temperament and ability among breeds are well known. These differences are generally associated with the purposes for which the breed was developed. This brief summary, which is typical of how dog breeds can be classified (Figure 3.16), suggests some of the traits that have been subject to selection:

1. Sheepdogs (shepherds, collies, and the like) are selected for their keen ability to focus on sheep or cattle, their ability to manipulate the behavior of these animals, and their ability to learn and follow their handler's commands. Some think that much of the herding behavior is derived from predatory behavior, but the actual killing behavior has been suppressed.

2. Terriers are energetic hunters, very attracted to small animals. They are quite willing to follow their prey down burrows, hence their name—derived from *terra*, or earth—not terror, as some owners insist. They dig.

3. Scent hounds (including beagles, bassets, fox- and coonhounds) are able to behaviorally exploit their keen sense of smell in tracking prey.

4. Retrievers (including Labrador and golden), also known as gun dogs, are selected for retrieving ability. A specific behavior that has been selected is a "soft mouth," the ability to handle prey without damaging the item or attempting to consume it.

5. Companions and Toys (miniature and toy poodles, Pekingese, Chihuahua) display behavioral traits that make them attractive household pets.

6. Sighthounds (Afghans, borzois, greyhounds) use their distance vision to track prey. They are also selected for high running speed and endurance.

FIGURE 3.16
The domestic dog, *Canis familiaris*, is strong testimony to the variation that can exist within a biological lineage.

93

In general, dog breeds that are recognized by groups such as the American Kennel Club "breed true." This means that pairing any male and any female in the breed will result in pups with the breed-specific characteristics. If you think about this, you will recognize that the only way to accomplish "breeding true" is through reduction of genetic variation related to those breed-specific traits. Any one breed of dogs will have less genetic variation within the breed than you find if you look across all dogs. At some point, breeding of this sort will eliminate most or all of the additive genetic variation in the breed for some traits. At the point at which the additive genetic variation for a trait has been exhausted, no further "improvement" of the breed (depending on the trait) is possible even through carefully designed pairings.

Thinking about how heritability is calculated, it follows that within any dog breed you would expect to find low heritabilities, particularly for the traits that are thought to characterize the breed. Does this mean that the traits do not have a genetic underpinning? No—not at all. It simply means that further attempts to select for the trait will be futile. Following this line of reasoning, you would expect to observe higher heritabilities for traits (including, of course, behavioral traits) if you include a variety of dog breeds in a study, and lower heritabilities if you focus on only one dog breed.

Most measures of heritabilities of dog behavioral traits are from studies of single breeds. A wide range of behaviors have been measured, such as "willingness," fighting the leash, hare tracking, and "obedience." Nearly all of the studies were performed within breeds (the alternative would be to do controlled matings between breeds). Generally, heritabilities for behavioral traits range from 0 to 0.25. For instance, heritabilities for personality traits of German Shepherds are 0.24 or less.[29]

Does this mean, then, that mixed-breed dogs have more hybrid vigor? A careful breeder of purebred dogs will have used genetic testing and careful pedigree scrutiny to minimize the occurrence of genetic diseases in her kennel's lineage. If the same scrutiny is not applied to the production of a mixed-breed litter, there is no reason to expect such a healthy outcome. A puppy mill breeder of purebred dogs will often use inbreeding and back-crossing, with no concern for genetics, to produce unhealthy puppies. In short, it is the exclusion (or inclusion) of undesirable genes, and not the heritability of other traits, that contributes to healthy or unhealthy puppies.

How are the estimates of genetic and environmental contributions to phenotype estimated so that heritability can be measured? Assessing heritability often boils down to how much parents resemble their offspring, or offspring resemble each other. Generally, heritability is calculated by looking at associations in the expression of the trait within families.

Mechanically, this is easy to do; for any given trait, the average value of the two parents (not surprisingly, this is called the *midparent value*) is graphed on the *x*-axis of a plot, and the value for the offspring is plotted on the *y*-axis (Figure 3.17). If enough parent–offspring pairs are analyzed, the slope of the resulting line is the heritability of that trait. In cases of traits that relate to sexual behavior, such as antlers in deer, the trait is expressed in only one gender; if single-parent values, rather than midparent values, are used, the heritability is twice the slope. Similar techniques can be used if data are available for groups of siblings. The result of these heritability analyses is a number between 0.0 and 1.0, with the extremes being no phenotypic variation due to additive genetic variation (slope ~0.0) and no phenotypic variation associated with environmental variation (slope ~1.0). The most common way of calculating heritability of a trait is by using linear regression analysis, a statistical technique that yields the equation for a line describing the effect of one variable on another. When the phenotypic values for offspring are regressed on the parental values, the slope of the resulting line is the heritability.

Another way of looking at heritability is to explore how a behavioral trait responds to either natural or artificial selection. If selection on the trait results in change from generation to generation, this suggests that the trait is heritable because selection can act only if there is genetic variation for the trait. However, if selection has no effect on a trait, then its heritability is probably low. Another way of saying this is that a heritable trait is a selectable trait.

OF SPECIAL INTEREST: CALCULATING AND INTERPRETING HERITABILITY

Figure 3.17 shows the heights of 50 male and 50 female students in centimeters. We also asked these students to give us their parents' heights. We averaged the father's and mother's height; this gives us the *midparent* value for each student.

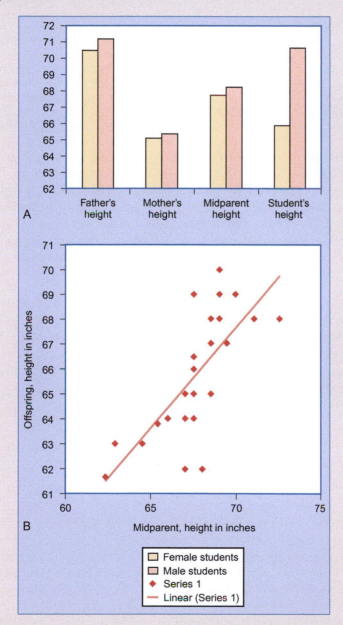

FIGURE 3.17
(A) Bar graph of male student heights and female student heights compared to the height of their parents. Midparent is the average of the two parents' heights. (B) Regression analyses of the male and female student heights against their midparent values. Both the slope of the regression line and the heritability are 0.82.

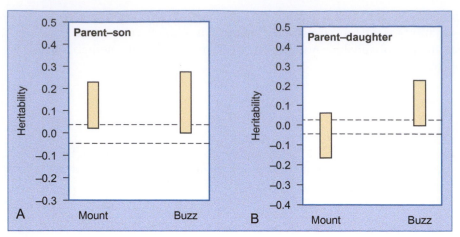

FIGURE 3.18
Heritability analysis of courtship in houseflies. For both behaviors, buzz and mount, the heritability is significant for the parent–son analysis, but only buzz is significant for the parent–daughter treatment. Thus, heritabilities may be influenced by sex. *Adapted from Meffert, L.M., Hicks, S.K., Regan, J.L., 2002. Nonadditive genetic effects in animal behavior. Am. Nat. 160 (Suppl. S DEC), S198–S213.*[30]

Heritability studies have many applications in animal behavior (Figure 3.18). Indeed, using artificial selection (ten generations) and house mice, Swallow et al.[31] showed that voluntary wheel-running could be increased by 75%, but that by the end of the experiment no additional effect of selection could be found. Interestingly, these increases were expressed as increased revolutions per minute, not increased minutes per day, meaning that the mice did not run longer, but ran faster. Following the information in this section about heritability and selection, it is reasonable to predict that strong selection would reduce heritability to zero.

Why? In theory, strong natural selection should eliminate all additive genetic variation because the alleles favored by selection will be the only ones remaining in the population. Without additive genetic variation, the narrow-sense heritability is zero, and no further evolutionary change in the trait would be expected. This theoretical prediction is supported by data in some cases, whereas in others there is substantial heritability of traits that nonetheless seem to have been under strong selection.

What does high or low heritability mean, in terms of history of selection? Low heritability results from strong selection on a trait, so key features for an animal's survival are expected to have low heritability. High heritability, correspondingly, reflects a more benign selective environment; genetic variation is tolerated if selection is not high.

OF SPECIAL INTEREST: DOMINANCE AND HERITABILITY IN JAPANESE QUAIL

The calculation of heritability of traits in Japanese quail is an outstanding example of the use of heritability in a behavioral study. It addresses the heritability of dominance, a trait that has important fitness consequences. Both male and female Japanese quail (*Coturnix japonica*) start pecking other birds soon after hatching. In adults, males and females establish dominance hierarchies (see Chapter 9). Heritability for a large number of behavioral traits, including dominance, was determined in Japanese quail (Figure 3.19). Dominant males have more mating success, supporting the hypothesis that strong selection has reduced additive genetic variation for dominance in males. Dominance rank has a higher heritability for males than for females. Heritability may be maintained in females because the trait has no fitness consequences or because there are balancing factors.[32,33]

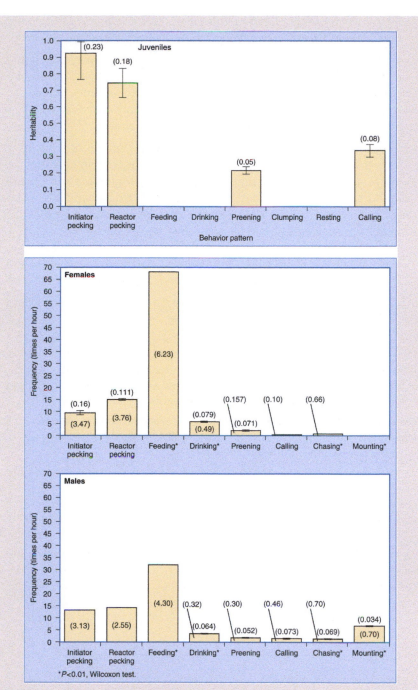

FIGURE 3.19

Heritabilities of behavioral traits of Japanese quail (*Coturnix japonica*). In young birds (upper panel), pecking is heritable in both sexes (0.74 overall). In adults, however, heritability for pecking is high in females (middle panel) (1.33) but 0 in males (lower panel). Other traits have moderate or low heritabilities. *Adapted from Nol, E., Cheng, K., Nichols, C., 1996. Heritability and phenotypic correlations of behaviour and dominance rank of Japanese quail. Anim. Behav. 52, 813–820.*[32]

Quantitative Trait Analysis

What is a quantitative trait? A quantitative trait varies continuously, in contrast to qualitative traits, which have discrete values. Another aspect of a quantitative trait is that it is determined by a number of genes acting together. A good example of a quantitative trait is height in humans; "normal" adult humans range in height over a span of more than half a meter (>2 feet). Human height is greatly influenced by genes, but no one gene is solely responsible for height. Usually, quantitative traits are normally distributed; a graph of the trait results in a bell-shaped curve.

In these cases, a normally distributed trait, whether it is physical, such as height, or behavioral, such as pollen collection, is usually the result of the contribution of a number of genes. Recent advances in genetics allow scientists to map the genes having the greatest influence on the trait. These genes, as described previously, are called QTLs. QTL analysis is now a standard format for studying genetic influences on behavior.[33,34]

It follows that a QTL is a location on a chromosome that is thought to regulate an organism's phenotype for quantitative trait. Other experimental approaches will give the same result, but all QTL analyses rely on a linkage map and good behavioral measures. Investigators find numerous markers such as generally single-nucleotide polymorphisms (SNPs), amplified fragment length polymorphisms, or random amplification of polymorphic DNA (RAPDs), and determine their linkage group map locations by cross-breeding.

OF SPECIAL INTEREST: HOW IS A QTL ANALYSIS DONE?

To perform a QTL analysis, a behavioral biologist needs three critical sets of information: a linkage map, a behavioral assessment, and a breeding experiment (Figure 3.20).

The first kind of information needed for QTL analysis is a linkage map of the genome of the study species. This is a significant limiting factor for QTL analyses. In construction of a map, the first step is to find physical (such as the bands on *Drosophila* chromosomes) or molecular markers scattered throughout the animal's genome. This map needs to be well saturated; that is, the markers must be distributed evenly and frequently enough to assort during recombination.

Many types of molecular markers are available; the linkage map for honeybees, which will be our main example in this section, was constructed using RAPDs; these are small DNA segments that can be identified using polymerase chain reaction techniques.

To visualize how the data are analyzed, think of a chromosome, remembering that in most animals chromosomes come in pairs. The scientist has genetic markers (short sequences of DNA) scattered along the chromosome; at a given location, the DNA sequences of these markers vary between the two copies of the chromosome. During meiosis, the two members of the chromosome may pair cross over and segments of the chromosomes are exchanged, resulting in genetic recombination.

Most importantly, the chances of recombination occurring are high if the markers are far apart and low if the markers are close together. Now what happens to genes that regulate behavior? If they are close to a marker, they likely stay linked with that marker during recombination. If they are far from a marker, then they are much less likely to stay linked. Hundreds, perhaps thousands, of markers are needed to produce a well-saturated map of an animal's genome. In honeybees, mapping has been facilitated by the fact that recombination rates are much higher than in most organisms.

The second type of information needed for QTL analysis is a good method of measuring how the behavior varies among animals.

The third type of information needed for QTL analysis is that derived from a breeding experiment. Typically, animals whose behavior has been measured and that have been genotyped are then mated. These crossed animals with different behavioral phenotypes permit observation of how the behavioral phenotype is correlated with the markers. The behavior and genotypes of their offspring are then determined as well.

The following questions are usually addressed in a QTL analysis:

- How many genes influence the expression of a quantitative trait?
- What is the level of influence of each gene on the trait?
- Where are the genes located on the chromosomes?
- What is the function of each gene?

Answers to these questions are discussed in the following section. The last question, about function, is the most difficult to answer.

Using a rather elaborate computer program, scientists can ask whether the variation in behavior is correlated with each marker. They do this by focusing on one "family" (a male, female, and their offspring) and asking if the behavior is always correlated with a certain marker and always low with other markers. In that case, the gene for the behavior is likely to be close to the marker on the chromosome. If the variation in the behavior is more or less random in relation to the marker, then there is little or no linkage between the behavior and that chromosomal location. Most quantitative traits, including behavioral traits, correlate well with from two to four locations in an organism's genome, each of these locations suggest a gene in the region that helps to regulate the behavior. There are, of course, exceptions, but this is a good rule of thumb. The higher the correlation, the more important a gene may be in regulating the behavior. These chromosomal locations probably do not have genes that code directly for a particular behavior, but rather ones that code for factors that shape the behavior. For example, a behavioral trait may be influenced by three genes, one of which affects the activity (or arousal), another that affects sensory perception, and a third that influences the latency to respond to a stimulus.

Many behaviors are quantitative traits. Aggressiveness, for example, varies among individuals in a wide range of animals, such as honeybees, rodents, horses, dogs,[35] and various primates. Another good example is expression of play behavior in mammals. Intelligence, to the extent that it can be measured, is also a quantitative trait. Activity levels are particularly good examples of quantitative traits.

Figure 3.20 shows a typical set of activity measurements—in this case, a QTL analysis of aggressive behavior (stinging) in honeybees. The purpose of this analysis was to identify

QTL Mapping

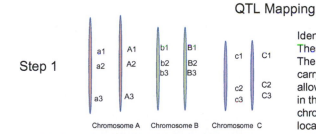

Step 1

Identify genetic markers on chromosomes. These can be RAPDs, SNPs, or microsatellites. The key is to be able to mate animals that carry different alleles for the markers. This allows the investigator to follow the markers in the animals' offspring. Hypothetical chromosomes are shown here with marker locations.

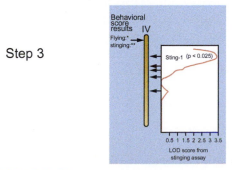

Step 2

Measure the animals' behavior and allow the animals to mate. Next the behavior and genetic markers of the offspring are assessed. In this example, a queen from a gentle honeybee colony, which is less likely to sting, is crossed with a male from a more defensive colony.

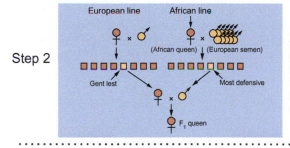

Step 3

Measure the offspring behavior and correlate the behavior with the markers on the chromosomes. If a gene controlling the behavior is close to the marker on the chromosome, the behavior and marker likely remain together after meiosis. This correlation is called a LOD (logarithm odds ratio) score. Here, the relationship between stinging markers on honeybee chromosome IV are shown.

FIGURE 3.20
The QTL technique for mapping genes associated with behaviors. This example is drawn from honeybees and focuses on their stinging behavior. *Adapted from Hunt, G.J., Guzmán-Novoa, E., Fondrk, M.K., Page Jr., R.E., 1998. Quantitative trait loci for honey bee stinging behavior and body size. Genetics 148, 1203–1213.*[36]

the locations in the genome of loci responsible for the large difference in stinging behavior between the "African" honeybees in Mexico and the gentler bees found farther north. In an attempt to further focus in on the regulation of aggressiveness in honeybees, a later study sequenced the genome around the sting-2 QTL; this is the *candidate gene* approach, which is discussed in more detail in the next section of this chapter.[37,38]

What does knowing the locations of QTLs accomplish? In working with domestic animals and wanting to select for certain behavioral traits, the knowledge of the QTLs would allow the design of an efficient breeding program to achieve the desired result. This has not actually been done in animals, but the principle is commonly applied in crop plants. By knowing which chromosomal locations are important, scientists can, conceivably, track how their genetic manipulations are affecting the regulation of the trait.

> **KEY TERM** A candidate gene is a gene that has been identified as having a strong possibility of playing a role in regulating a phenotype.[36]

3.7 EVOLUTIONARY AND POPULATION GENETICS

Population genetics relies on techniques that allow measuring of the effect of selection on gene frequencies within a population. Gene frequencies in populations are at the Hardy–Weinberg equilibrium when selective forces have not acted. This equilibrium is rarely achieved in nature, because selective forces are almost always in play, but it is a useful theoretical baseline. The extent to which gene frequencies deviate from the Hardy–Weinberg equilibrium reveals the existence and intensity of differential fitness and/or survival. The Hardy–Weinberg equilibrium is simple to express if one gene with two alleles, a and b, is considered. The frequencies of a and b must total one—all of the genes in the population must be one of the two alleles, using p as the probability for an allele and a subscript letter to indicate which allele is being considered:

$$p_a + p_b = 1$$

The possible genotypes in the population are the combinations of the a and b alleles: aa, ab, and bb. When the population is at equilibrium, the frequency of animals with the aa genotype is p_a^2 and the frequency of b homozygotes is p_b^2. Heterozygotes, the genotype ab, are present with the frequency of $2p_ap_b$. These are important measures for the study of behavior because if differential mortality affects an allele in a population, then the genotype frequencies in the population will not fit the Hardy–Weinberg expectation, which is often expressed as follows:

$$p_a^2 + 2p_ap_b + p_b^2 = 1$$

In addition to selection, other factors such as mutation and immigration affect gene frequencies, and calculation of Hardy–Weinberg frequencies allows investigators to assess whether a population has been under various kinds of pressure, such as strong selection or immigration of animals from other locations. Because selection is often associated with mortality and population reduction, a population that is not in Hardy–Weinberg equilibrium is often hypothesized to have suffered recent reductions in numbers. In a recent study of puff-throated bulbuls in a national park in Thailand, investigators made inferences about population and conservation status by measuring how well several genetic loci conformed to Hardy–Weinberg expectations.[39]

The study of population genetics informs many areas of biology, from animal husbandry to studies of fitness. Speciation is one major topic in population genetics that has a strong behavioral component. Because of behavioral choices (e.g., habitat choice, shift in activity regimes, migration, and the like), populations may become reproductively isolated, a condition that can lead to speciation. Generally speaking, new species arise when a population becomes reproductively isolated from other populations and then evolves (gene frequencies change) so that the reproductive isolation is reinforced by mating incompatibilities. Mating behavior is a major potential isolation mechanism among animal species, and selection can favor divergence in mating behavior between populations of newly split species. For example, dolphins encounter two very different types of environment, pelagic (open ocean) and coastal. Specialization by subpopulations in these habitats has led to differentiation between pelagic and coastal dolphins in foraging and social behavior. The dolphins are considered to be a single species, but further isolation could give rise to distinct species of dolphin in each habitat.[40] This has already occurred in the Amazon River dolphin, which has evolved characteristics for successful living in a freshwater environment that offers very different food than the oceanic environment that supports its close relatives.

OF SPECIAL INTEREST: EVOLUTIONARY MECHANISMS THAT IMPACT BEHAVIOR

Some evolutionary processes have particularly notable effects on behavior. Many scientists consider these outcomes of population genetics:

- *Correlated characteristics in males and females.* If a trait that is important in one sex, either for competition for mates (as in horns or antlers in male ungulates) or in mate choice, is produced by both sexes, then the deleterious effect on the other sex may outweigh selection for maximizing the trait. For example, in barn swallows, tail length varies in both males and females but is a sexually selected trait only in males (Figure 3.21).[41]

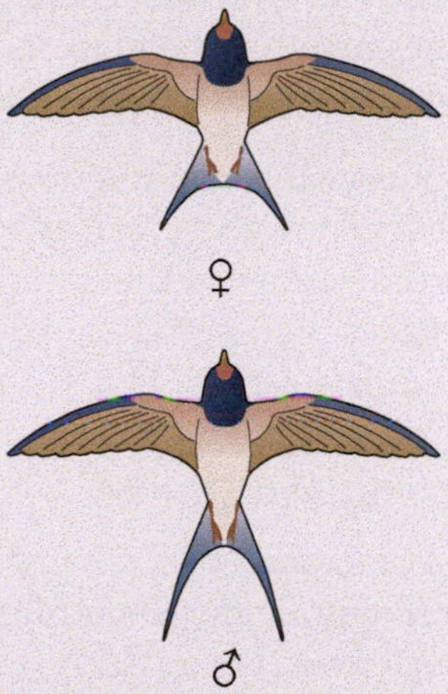

- *Phylogenetic inertia* is the effect of ancestral traits carried over to present-day organisms even when the trait is no longer particularly adaptive. Why, for example, does a dog turn around a few times before lying down (see Chapter 1, page 12)? The adaptive roots, or ultimate causes, of this behavior may lie in ancestral preparation for sleep by trampling vegetation or in checking for potential parasites; in a domestic dog, the behavior is charming but meaningless. It is always possible that an observed behavior or phenotype had a function in evolutionary history that is now lost; the behavior persists because it is engrained in the genetic instructions the animal receives from previous generations. Phylogenetic inertia is the carryover of traits that evolved in previous habitats, even if those traits have little relevance in the current habitat.

FIGURE 3.21
Male and female barn swallow tails.

- *Disruptive selection* is a type of selection that occurs when more than one phenotype is favored. It could occur among animals in competition (as in calling males and satellite males in a mating chorus) or between generations, as when environmental conditions shift, so that more than one phenotype is favored over time. Disruptive selection occurs when extreme phenotypes are at an advantage, so selection favors the maintenance of both phenotypes. For example, very large and very small animals may be at an advantage in mating, so selection would favor these phenotypes. In mating systems, sometimes males of two distinctive behavioral and morphological types are found. One type are dominant, territory-holding males and the other type are small sneaker males that attempt mating near the territories of the dominants. Disruptive selection maintains both types of behavior in the population.

- *The handicap principle* relates to maintaining traits that seem to be costly—which may appear to put an animal at a selective disadvantage. If producing a phenotype is costly enough, selection against extreme individuals may counterbalance selection favoring that extreme phenotype. The handicap principle states that production of a phenotype may be costly, but that can make the phenotype an honest signal of an animal's condition. If one sex chooses mates based on the costly signal, then the signal is maintained because having it confers fitness on the animal.

3.8 MOLECULAR GENETICS

One major goal of many genetic studies is to map the location of genes that affect behavior on the chromosomes of an animal. Mendelian genetics provides tools for mapping, as genes that are close to one another on a chromosome do not segregate independently. Quantitative genetics contributes through tools such as QTL analysis, which allows an investigator to focus in on a gene's location on a chromosome and to provide the bridge from quantitative to molecular genetics.

For organisms with known linkage maps (e.g., nematodes, fruit flies, honeybees, mice,[42] humans), QTL analyses have provided intriguing windows into how variation in a small set of genes can explain much of the phenotypic variation in behavioral traits.[43] QTL analyses then open the door to identification of specific genes that are involved with a trait, and move the discussion from whole-organism approaches to looking at proximate causes at the level of the gene. Molecular techniques can pinpoint gene candidates for control of behavior. The exact location on the chromosome of a QTL location is not precisely known; the data show only that it is between two markers. If the distance between the two markers is not too large, molecular geneticists can overcome this problem by sequencing the chromosome from the two markers ("upstream" from one, "downstream" from the other). By comparing the DNA sequences with sequences of known genes, the scientist may determine the identity of the genes between the markers. The scientist can then form a hypothesis about which of the genes in the QTL region has the behavioral effect; this is a candidate gene hypothesis.

Sequence data from molecular genetics can show the position of a gene relative to other genes, and ultimately, sequenced data can be correlated with known markers on chromosomes so that sequenced genes can be proposed as candidates for modulating the behavior being studied. Further experiments can selectively modify the expression of the candidate gene to test the hypothesis that it is important in regulating the behavior.

Candidate Genes

A recent study of chickens[44] illustrates how interesting candidate genes are identified. Chickens are the domesticated descendants of junglefowl, a southeast Asian bird species. Junglefowl (Figure 3.22) are wilier and less manageable than their domestic relatives, in part because they are more aggressive. In this study, the investigators hypothesized a relationship between genes for growth, identified through QTL analyses, and domestication.

The investigators measured dominance and inspection of strangers (birds new to the social group). Differences among birds in these behaviors suggest that the arginine vasopressin receptor 1a (*AVPR1a*) gene may be involved in these social behaviors. Thus, *AVPR1a* is a candidate gene for social behavior in chickens. This is a fascinating discovery because vasopressin is the same hormone discussed in Section 2.2 as being important in pair bonding and lifetime mating in voles. To verify the function of vasopressin in chicken social behavior, the investigators will now have to perform experiments directly addressing the effects of vasopressin on the behavior, but this finding is certainly highly suggestive. The candidate gene approach is only now coming into its own, in terms of being useful in behavioral investigations, but it holds extremely high promise in drawing links between genes and behavior and in allowing scientists to see similarities in behavioral regulation among species.

Microarrays and Assessments of Gene Expression

Are there different patterns of gene expression between cells of animals with different behaviors? This question can be answered by looking at expressed sequence tags (ESTs). An EST is a tag based on a known

102

FIGURE 3.22
Junglefowl, the wild ancestor of domestic chickens.

functional gene sequence from an animal. It can be amplified—reproduced many times—and then be placed as part of a microarray (Figure 3.23). A microarray holds each amplified EST in a small well, like an indentation on a microscope slide. A sample of tissue from an animal is then stimulated to produce proteins, which are labeled fluorescently. If a protein matches an EST, it binds, and the well becomes a fluorescent dot in the microarray, as shown in Figure 3.23. ESTs can then be compared between tissues, between animals, or between species. In the simplest analyses, the number of ESTs fluorescing is compared—for example, between sleeping and awake animals. But if details about the ESTs and which genes they derive from are known, this analysis can also provide clues about specific genes, leading to candidate gene hypotheses. ESTs provide a direct window into the relationship between gene expression and the functional context of the products of those genes.

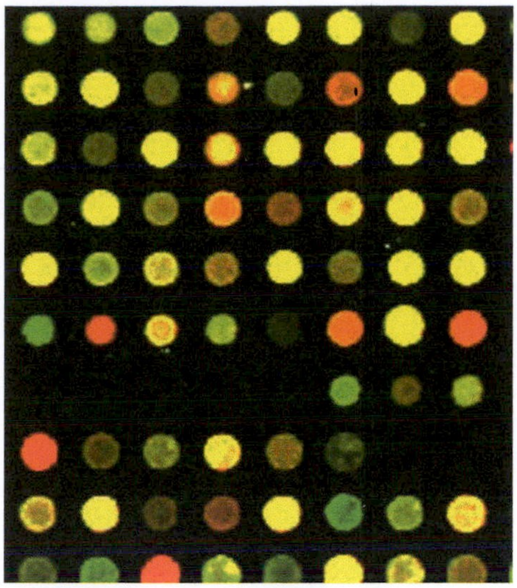

FIGURE 3.23
An EST microarray, showing strong expression of some ESTs and little or no expression of other ones.

RNA Knockouts

RNA knockouts are a relatively new tool in the exploration of genes and behavior. This technique involves synthesizing oligonucleotides that are complementary to the RNA products of genes thought to influence a behavior. When the synthetic oligonucleotide is introduced into the organism, it binds with the RNA from the target gene, effectively inactivating (or knocking out) that gene. Key to this approach is having a candidate gene with a hypothetical role in regulating a behavior.

Epigenetics: Genomic Imprinting

One of the most fascinating discoveries in genetics over the past few years has been that animals begin their embryonic development; with genetic regulatory effects that are carried from one generation to the next. Gene regulation that was established in the parental generation can carry over from the parents to the offspring. This is called genomic imprinting, and behavioral traits, such as shyness, boldness, anxiety, and parental care, can be affected. Genomic imprinting will certainly be an important area of future inquiry in behavioral genetics.

> **KEY TERM** An EST is a genetic marker that is linked to the gene being studied. When the gene of interest is expressed, the EST "reports" that activation. This allows an investigator to see how gene activation correlates with physiological and behavioral activity.

Epigenetics: DNA Methylation

The mechanism underlying genomic imprinting is often DNA methylation, which is the addition of methyl groups to the bases that form the backbone of DNA. Methylation silences genes, so that cells in which genes are methylated have a narrower potential range of function than cells in which genes are not methylated. Within the tissues of an animal, methylation is an important mechanism of cellular specialization. In some animals, such as honeybees and ants, methylation is associated with behavioral specialization within a social group.[45]

BRINGING ANIMAL BEHAVIOR HOME

Canine compulsive disorder (CCD), in which behaviors are executed repeatedly and with some distress, shows a high resemblance to human obsessive–compulsive disorder. Dogs may lick, tail chase, flank suck, engage in pica (consumption of indigestible substances), and pace or circle. Using SNPs from Dobermans diagnosed with this malady, scientists have recently determined that a gene on chromosome 7 may be responsible for some susceptibility to CCD.[46] Sixty percent of dogs with multiple compulsive behaviors have the allele thought to confer risk (CDH2) compared to 22% of control Dobermans. Such studies open the door for early intervention into compulsive disorders and perhaps even treatment.

DISCUSSION POINT: CAUSE AND CORRELATION IN BEHAVIORAL GENETICS

We pointed out in the beginning of this chapter that single-gene mutants can give misleading information about genetic regulation of a behavior. A standard technique for producing mutants involves treating animals with either radiation or a chemical mutagen and then screening their offspring for behavioral abnormalities. Genetic investigation can then pinpoint the mutant gene that correlates with the abnormality. Is this proof that the gene "controls" the behavior? Why or why not?

Molecular Studies: Overview and Future

Many surprises have emerged as our understanding of genes and genomes has improved. There are far fewer genes in the typical genome than was thought. Candidate genes are far less important than had been hoped, as there is no pattern of strict single-gene/behavior correspondence. The evolution of regulatory genes and regulatory interactions has major importance in understanding behavioral genetics, but these studies are in their infancy. Epigenetics adds another complex regulatory layer to understanding how genetics determines behavior. A new synthesis in behavioral genetics will develop around integrating regulatory genetics and epigenetics into explanations for behavioral expression.

SUMMARY

Behavioral genetics forms a critical component of nearly all behavioral studies. Understanding genetics establishes a thread that runs from the ultimate evolutionary causes of behavior to the behavior's proximate underpinnings, and knowing the genetic underpinnings of behavior is essential to understanding how behavior evolves. Behavioral genetics also helps in understanding the physiological and neurobiological systems that control behavior. Although the approach of this chapter is genetics, most behavior is shaped by a combination of genetic and environmental factors. This chapter and the following two chapters—Chapter 4 on behavioral homeostasis and Chapter 5 on learning—work together to shape a picture of how behavior is controlled.

To understand the evolutionary roots of behavior (ultimate causes), behavioral geneticists often employ phylogeny. Genetics brings a broad range of tools to behavioral investigations. These tools are used to establish the phylogeny of animal groups, which can then be used to understand patterns of evolution for specific behaviors, such as nest construction by birds or colony defense by bees.

Peeling away the layers of genetic and physiological regulation of a behavior using behavioral genetics—starting with differences among species and moving down to the regulation of gene expression—holds great promise for solving many of the mysteries of animal behavior. Single-gene effects on behavior are easily documented, but such effects are relatively rare and may produce the erroneous notion that complex behavior is "controlled" by those genes. In fact, behavior is most often the result of a combination of the environment and many genes acting together in a regulatory system. Quantitative genetics can be used to provide better explanations than single-gene models for most animal behavior traits. Studying heritability helps to unravel genetic and environmental influences on behavior. Molecular approaches to behavior genetics provide useful techniques in exploring the regulation of behavior.

Behavioral genetics is less a scientific discipline than a set of approaches to be applied to almost any behavioral question. Mastering the impressive array of behavioral genetic tools takes considerable work, but the reward is an ability to understand the relationship between ultimate and proximate causes, and to see the promise experimental genetics holds for unraveling complex scientific problems (Table 3.2).

TABLE 3.2 The Behavioral Genetics Toolbox

Tests of Ultimate Causes: The Evolution of Behavioral Phenotype	Proximate Causes: Whole-Organism Studies of the Genetic Bases of Behavior	Proximate Causes: Genetic Dissection of Mechanisms Underlying Behavior
1. Geographic variation, subspecies, and ecotypes	1. Mutational studies	1. Microarrays and other assessments of gene expression
2. Comparative phylogenetic studies	2. Heritability	2. Candidate gene approaches
3. Cross-fostering	3. QTL analysis	3. RNA knockout
4. Twin analyses		
5. Artificial selection and inbred lines		
6. Natural selection and behavior		

STUDY QUESTIONS

1. Food preference in a species of mouse has both innate (genetic) and experiential (learned) components. Design experiments that will test the following three hypotheses: (A) Learned information is used in preference to genetic information when both are available; (B) there is a critical period for learning food preferences; and (C) the ability to learn food preferences has a high heritability. Make sure that the experimental designs include specification of sample sizes and controls.

2. The heritability of dominance behavior among males of a species of monkey is high (>80%). Does this observation support a prediction that females use male dominance status in their choice of mates? Why or why not?

3. One of the central issues in behavioral genetics is how animals balance the use of genetically based and learned information. Under what circumstances might selection favor the use of inherited information? What circumstances favor the use of learned information? What are the general principles that determine the relative importance of learned and inherited information in shaping animal behavior? Remember to revisit this question after reading Chapter 5 on learning.

4. Without looking back in this chapter, define *heritability* and *additive genetic variation*. Why might additive genetic variation be more informative than total genetic variation? How does natural selection act on additive genetic variation?

5. If the heritability of a trait is high (close to 1.0), does this support a hypothesis that past selection on that trait has been high or low? Why?

6. Explain the candidate gene approach to discovering the regulatory pathways for a behavior. What are the advantages and drawbacks to making hypotheses about candidate genes?

7. How are microarray analyses useful in behavioral genetics?

Further Reading

Anholt, R.R.H., Mackay, T.F.C., 2009. Principles of Behavioral Genetics. Academic Press, San Diego, CA, p. 334.

Kyriacou, C.P., Tauber E., 2010. Genes and genomic searches, In: Breed, M.D., Moore, J., Editor(s)-in-Chief, Encyclopedia of Animal Behavior, vol. 1. Academic Press, Oxford, pp. 12–20, ISBN 978-0-08-045337-8, <http://dx.doi.org/10.1016/B978-0-08-045337-8.00166-2>. <http://www.sciencedirect.com/science/referenceworks/9780080453378>.

Morris, D., 1969. The Naked Ape. Dell Publishing Co., New York, NY.

Oldroyd, B.P., 2010. Social insects: behavioral genetics, In: Breed, M.D., Moore, J., Editor(s)-in-Chief, Encyclopedia of Animal Behavior, vol. 3. Academic Press, Oxford, pp. 251–259, <http://www.sciencedirect.com/science/referenceworks/9780080453378>.

Plomin, R., DeFries, J.C., McClearn, G.E., McGuffin, P., 2008. Behavioral Genetics. Worth. p. 560.

Wilson, E.O., 1975. Sociobiology: The New Synthesis. Belknap Press of Harvard University Press, Cambridge, MA.

Notes

1. Tschirren, B., Bensch, S., 2010. Genetics of personalities: no simple answer for complex traits. Mol. Ecol. 19, 624–626.
2. Boake, C.R.B., Arnold, S.J., Breden, F., Meffert, L.M., Ritchie, M.G., Taylor, B.J., et al., 2002. Genetic tools for studying adaptation and the evolution of behavior. Am. Nat. 160 (Suppl. S), S143–S159.
3. Ruttner, F., 1988. Biogeography and Taxonomy of Honeybees. Springer-Verlag, Berlin, Germany. Whitfield, C.W., Behura, S.K., Berlocher, S.H., Clark, A.G., Johnston, J.S., Sheppard, W.S., et al., 2006. Thrice out of Africa: ancient and recent expansions of the honey bee, *Apis mellifera*. Science 314 (5799), 642–645.
4. Crabbe, J.C., Phillips, T.J., 2003. Mother nature meets mother nurture. Nat. Neurosci. 6, 440–442.
5. Lehrman, D.S., 1953. A critique of Konrad Lorenz's theory of instinctive behavior. Q. Rev. Biol. 28, 337–363.
6. Rosenblatt, J.S., 1995. Daniel Sanford Lehrman: June 1, 1919–August 27, 1972. Biogr. Mem. Natl. Acad. Sci. 66, 227–245.
7. Shockley, W., 1992. In: Jensen, A.R., Pearson, R. (Eds.), Shockley on Eugenics and Race: The Application of Science to the Solution of Human Problems. Scott-Townsend Publishers, Washington, DC.
8. Wilson, E.O., 1975. Sociobiology: The New Synthesis. Belknap Press of Harvard University Press, Cambridge, MA.
9. Lewontin, R.C., Rose, S., Kamin, L.J., 1984. Not in Our Genes: Biology, Ideology, and Human Nature. Pantheon Books, New York, NY. Kitcher, P., 1985. Vaulting Ambition: Sociobiology and the Quest for Human Nature. MIT Press, Cambridge, MA.
   Hellman, H., 1998. Great Feuds in Science: Ten of the Liveliest Disputes Ever. Wiley, New York, NY.
   Gould, S.J., 1981. The Mismeasure of Man. W. W. Norton, New York, NY.
10. Thornhill, R., Palmer, C.T., 2000. A Natural History of Rape: Biological Bases of Sexual Coercion. MIT Press, Cambridge, MA.
    Thornhill, R., Palmer, C.T., Coyne, J.A., Berry, A., 2000. A natural history of rape: biological bases of sexual coercion. Nature 404, 121.
    Ward, T., Siegert, R., 2002. Rape and evolutionary psychology: a critique of Thornhill and Palmer's theory. Aggress. Violent Behav. 7, 145–168.
11. Mackay T.F., Heinsohn, S.L., Lyman, R.F., Moehring, A.J., Morgan, T.J., Rollmann, S.M., 2005. Genetics and genomics of *Drosophila* mating behavior. Proc. Natl. Acad. Sci. USA 102 (Suppl. 1), 6622–6629.
12. Wilsson, E., Sundgren, P.E., 1997. The use of a behaviour test for selection of dogs for service and breeding. Heritability for tested parameters and effect of selection based on service dog characteristics. Appl. Anim. Behav. Sci. 54 (2–3), 235–241.
13. Hare, B., Brown, M., Williamson, C., Tomasello, M., 2002. The domestication of social cognition in dogs. Science 298 (5598), 1634–1636.
    Koler-Matznick, J., 2002. The origin of the dog revisited. Anthrozoos 15 (2), 98–118.
14. Congdon, B., Platt, J.F., Martin, K., Friesen, V.L., 2000. Mechanisms of population differentiation in marbled murrelets: historical versus contemporary processes. Evolution 54, 974–986.
15. Deater-Deckard, K., Petrill, S.A., Thompson, L.A., 2007. Anger/frustration, task persistence, and conduct problems in childhood: a behavioral genetic analysis. J. Child Psychol. Psyc. 48 (1), 80–87.
16. Ossi, K., Kamilar, J.M., 2006. Environmental and phylogenetic correlates of Eulemur behavior and ecology (Primates: Lemuridae). Behav. Ecol. Sociobiol. 61 (1), 53–64.
17. Bardeleben, C., Moore, R.L., Wayne, R.K., 2005. A molecular phylogeny of the Canidae based on six nuclear loci. Mol. Phylogenet. Evol. 37, 815–831.
18. Lee, P.L., Clayton, D.H., Griffiths, R., Page, R.D., 1996. Does behavior reflect phylogeny in swiftlets (Aves: Apodidae)? A test using cytochrome b mitochondrial DNA sequences. Proc. Natl. Acad. Sci. USA 93, 7091–7096.
19. Brady, S.G., Sipes, S., Pearson, A., Danforth, B.N., 2006. Recent and simultaneous origins of eusociality in halictid bees. Proc. R. Soc. B Biol. Sci. 273 (1594), 1643–1649.
20. Ross, K.G., Krieger, M.J.B., Shoemaker, D.D., 2003. Alternative genetic foundations for a key social polymorphism in fire ants. Genetics 165 (4), 1853–1867.
    Ross, K.G., Vargo, E.L., Keller, L., 1996. Social evolution in a new environment: the case of introduced fire ants. Proc. Natl. Acad. Sci. USA 93, 3021–3025.
21. Suarez, A.V., Tsutsui, N.D., Holway, D.A., Case, T.J., 1999. Behavioral and genetic differentiation between native and introduced populations of the Argentine ant. Biol. Invasions 1, 43–53.
22. Yamamoto, D., Nakano, Y., 1999. Sexual behavior mutants revisited: molecular and cellular basis of *Drosophila* mating. Cell. Mol. Life Sci. 56, 634–646.

23. Carhan, A., Allen, F., Armstrong, J.D., Goodwin, S.F., O'Dell, K.M.C., 2005. Female receptivity phenotype of icebox mutants caused by a mutation in the L1-type cell adhesion molecule neuroglian. Genes Brain Behav. 4 (8), 449–465.

24. Johannessen, L.E., Slagsvold, T., Hansen, B.T., 2006. Effects of social rearing conditions on song structure and repertoire size: experimental evidence from the field. Anim. Behav. 72, 83–95.

25. Airey, D.C., Castillo-Juarez, H., Casella, G., Pollak, E.J., DeVoogd, T.J., 2000. Variation in the volume of zebra finch song control nuclei is heritable: developmental and evolutionary implications. Proc. R. Soc. Lon. B Biol. Sci. 267, 2099–2104.

26. Zhang, Y., Lawrance, S.K., Ryder, O.A., Zhang, Y., Isaza, R., 2000. Identification of monozygotic twin chimpanzees by microsatellite analysis. Am. J. Primatol. 52, 101–106.

27. Flint, J., 2003. Analysis of quantitative trait loci that influence animal behavior. J. Neurobiol. 54 (1), 46–77.

28. Boake, C.R.B., Konigsberg, L., 1998. Inheritance of male courtship behavior, aggressive success, and body size in *Drosophila silvestris*. Evolution 52 (5), 1487–1492.

29. Ruefenacht, S., Gebhardt-Henrich, S., Miyake, T., Gaillard, C., 2002. A behaviour test on German Shepherd dogs: heritability of seven different traits. Appl. Anim. Behav. Sci. 79 (2), 113–132.

30. Meffert, L.M., Hicks, S.K., Regan, J.L., 2002. Nonadditive genetic effects in animal behavior. Am. Nat. 160 (Suppl. S DEC), S198–S213.

31. Swallow, J.G., Carter, P.A., Garland Jr., T., 2004. Artificial selection for increased wheel-running behavior in house mice. Behav. Genet. 28, 227–237.

32. Nol, E., Cheng, K., Nichols, C., 1996. Heritability and phenotypic correlations of behaviour and dominance rank of Japanese quail. Anim. Behav. 52, 813–820.

33. Gleason, J.M., Nuzhdin, S.V., Ritchie, M.G., 2002. Quantitative trait loci affecting a courtship signal in *Drosophila melanogaster*. Heredity 89, 1–6.

34. Ruppell, O., Pankiw, T., Page, R.E., 2004. Pleiotropy, epistasis and new QTL: the genetic architecture of honey bee foraging behavior. J. Hered. 95 (6), 481–491.

35. Netto, W.J., Planta, D.J.U., 1997. Behavioural testing for aggression in the domestic dog. Appl. Anim. Behav. Sci. 52 (3–4), 243–263.

36. Hunt, G.J., Guzmán-Novoa, E., Fondrk, M.K., Page Jr., R.E., 1998. Quantitative trait loci for honey bee stinging behavior and body size. Genetics 148, 1203–1213.

37. Guzman-Novoa, E., Hunt, G.J., Uribe, J.L., Smith, C., Arechavaleta-Velasco, M.E., 2002. Confirmation of QTL effects and evidence of genetic dominance of honeybee defensive behavior: results of colony and individual behavioral assays. Behav. Genet. 32 (2), 95–102.

38. Hunt, G.J., Collins, A.M., Rivera, R., Page, R.E., Guzman-Novoa, E., 1999. Quantitative trait loci influencing honeybee alarm pheromone levels. J. Hered. 90 (5), 585–589.

39. Page, R.B., Sankamethawee, W., Pierce, A.J., Sterling, K.A., Reed, D.H., Noonan, B.P., et al., 2014. High throughput sequencing enables discovery of microsatellites from the puff-throated bulbul (*Alophoixus pallidus*) and assessment of genetic diversity in Khao Yai National Park, Thailand. Biochem. Syst. Ecol. 55, 176–183.

40. Louis, M., Fontaine, M.C., Spitz, J., Schlund, E., Dabin, W., Deaville, R., et al., 2014. Ecological opportunities and specializations shaped genetic divergence in a highly mobile marine top predator. Proc. R. Soc. B Biol. Sci. 281, 20141558.

41. Cuervo, J., de Lope, F., Moller, A., 1996. The function of long tails in female barn swallows (*Hirundo rustica*): an experimental study. Behav. Ecol. 7, 132–136.

42. Gershenfeld, H.K., Neumann, P.E., Mathis, C., Crawley, J.N., Li, X.H., Paul, S.M., 1997. Mapping quantitative trait loci for open-field behavior in mice. Behav. Genet. 27 (3), 201–210.

43. Jordan, K.W., Morgan, T.J., Mackay, T.F.C., 2006. Quantitative trait loci for locomotor behavior in *Drosophila melanogaster*. Genetics 174 (1), 271–284.

44. Wiren, A., Gunnarsson, U., Andersson, L., Jensen, P., 2009. Domestication-related genetic effects on social behavior in chickens—effects of genotype at a major growth quantitative trait locus. Poult. Sci. 88, 1162–1166.

45. Yan, H., Simola, D.F., Bonasio, R., Liebig, J., Berger, S.L., Reinberg, D., 2014. Eusocial insects as emerging models for behavioural epigenetics. Nat. Rev. Genet. 15, 677–688.

46. Dodman, N.H., Karlson, E.K., Moon-Fanelli, A., Galdzicka, M., Perloski, M., Shuster, L., et al., 2010. A canine chromosome 7 locus confers compulsive disorder susceptibility. Mol. Psychiatry 15, 8–10.

107

Homeostasis and Time Budgets

109

LEARNING OBJECTIVES

Studying this chapter should provide you with the knowledge to:

- Appreciate how homeostasis describes the behavioral and physiological processes that animals use to maintain appropriate internal and external environments.

- Examine the use of contemporary versions of motivation theory to understand how animals make behavioral decisions to help maintain their homeostasis.

- Employ the measures of displacement behavior and redirected behavior, in the form of self-directed behaviors (SDBs), as powerful tools for assessing animal welfare.

- Develop models for how biological clocks, pain, sleep, appetites for food or mating, and fear influence animals' choices among possible activities.

- Be able to explain and use time budget analyses to quantify and interpret how and why animals allocate their time in maintaining homeostasis.

Animal Behavior. DOI: http://dx.doi.org/10.1016/B978-0-12-801532-2.00004-0

4.1 INTRODUCTION

This chapter addresses a seemingly simple set of questions: How and why does an animal behave in a given way at a given time? What causes an animal to stop one activity and start another? What, for example, moves a sleeping dog to get up, stretch, and solicit play or food? What prompts bouts of sexual activity? Why do animals behave one way in the spring and another in the winter?

Beneath these simple questions—all facets of Tinbergen's core issue of causation (mechanism)—lies a wealth of complexity and unsolved puzzles in animal behavior. This chapter introduces concepts that are at the root of much of the contemporary study of animal behavior. Some of the answers to the questions reflect deep evolution—the biorhythms shared by every living thing, for instance.

There are compelling reasons for understanding behavioral homeostasis as a key foundation for animal behavior. The complex mechanisms that underlie behavioral regulation affect the entire range of an animal's potential behavior. Behavioral homeostasis gives us a key window into the adaptive value of important behavioral states such as fear and sleep (Figure 4.1).

Coverage begins with a concept from the early days of ethology: *drive theory*. Drive[1] is an animal's internal motivation to initiate or complete a task, and drive theory was one of the foundations of understanding causation in animal behavior. Another way of expressing this concept is motivation—hypotheses about animals' motivations for performing behaviors can be tested.

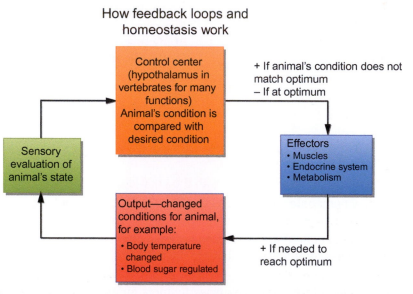

FIGURE 4.1

Feedback loops serve as regulators in behavioral homeostasis. Homeostasis is the maintenance of a constant state; animals use behavior to regulate their body heat, water content, nutrients, and virtually all other important microclimate or internal needs. A feedback loop requires an integrator (top orange square) which signals effectors to upregulate or downregulate their output. The animal's condition changes as a result of the action of the effectors (red square) and the changes are evaluated (green square) by sensors that then feed information back to the integrator. The most commonly used analogy for homeostatic feedback loops is a home heating system, in which the thermostat is the integrator and sensor, the furnace is the effector, and the heat produced by the furnace is the output. Rising temperature in the house feeds back to cause the thermostat to turn the furnace off. Because the effector is downregulated when it has produced enough heat, this is termed negative feedback.

Modern neuroscience has largely replaced drive theory in our understanding of in-the-moment behavioral choices by animals. We now see behavior as largely being modulated by neurotransmitters, and much of the focus of discovery about motivation is currently on measuring and correlating neurotransmitter levels with behavior. On a deeper level, neurotransmitter levels are the manifestation of a combination of gene expression and experience. Neuroscience provides the potential for contemporary models of *behavioral homeostasis*, but most neuroscience (excepting neuroethology) does not link mechanism with external environment, and establishing these links remains a goal for the future.

Pharmacology provides powerful tools for studying the correlations of neurotransmitter levels with behavior, as chemicals that act as agonists or antagonists for neurotransmitter action are readily available for experimental use. By using pharmacological techniques scientists studying animal behavior have been able to tap the mechanisms that underly behavioral choices made by animals.

Homeostatic mechanisms include appetite, pain, fear, and sleep. The use of these states in studying behavior is discussed in this chapter, as well as the physiological mechanisms that underlie the expression of each.

Despite the move forward from drive theory to neuroscience, some concepts from drive theory remain quite useful in developing ways of measuring *animal welfare* and assessing techniques for ameliorating the stresses of captivity for animals. In this chapter, SDBs, displacement behaviors, and repetitive behaviors are discussed in the contexts of drive theory and evaluation of animal welfare.

The chapter concludes with time budgets, an analytical tool in the study of behavior. On a practical level, *time budgets* are introduced as an analytic tool for understanding trade-offs between homeostatic demands. An animal's time budget is the culmination of its behavioral choices. As the behavioral phenotypic expression of the animal's internal state, time budgets give windows into documenting the interaction between internal state and behavioral expression, as well as assessing the effects of factors such as development, seasonality, and parenting on allocation of effort.

111

> **KEY TERM** Drive theory postulates that animals have a store of drive energy that is expended when a task is performed. This aims to explain how animals allocate their efforts among tasks.

CASE STUDY: HAVE NO FEAR

What happens when fear is unnecessary? Fear is useful but restricting. It keeps animals out of trouble but costs them opportunities. A fearful animal may survive but may not forage or mate as often as a fearless competitor. In this chapter, we discuss how fear affects animal activity.

Islands are places where fear seems to disappear. Islands tend not to have large predators, so herbivores and small predators live out their lives in peace. Fearlessness in island animals may result from generations of relaxed selection for avoiding predators. Visitors to the Galapagos Islands, including Charles Darwin, have noticed the remarkable "tameness" of the island's animal residents. Antipredator fear can also be a matter of learning. Geese and deer famously adjust their behavior during hunting season. Late in the hunting season, experienced animals assiduously avoid hunting blinds, decoys, and areas that they associate with gunshots.

What happens when much-needed fear is absent? This is a major conservation concern as humans introduce predators to islands. Marine iguanas on the Galapagos are remarkably placid, allowing visitors to walk nearby and even to touch them (Figure 4.2). The introduction of dogs and cats to the Galapagos creates a threat that the iguana populations have not experienced for millions of years. Galapagos marine iguanas show some flexibility in response, even though adaptation has largely taken away their antipredator behavior. When threatened by predators, they have higher levels of stress hormones than unthreatened animals and are more likely to attempt to flee, albeit inefficiently and largely unsuccessfully.[2,3]

CASE STUDY (CONTINUED)

FIGURE 4.2
Galapagos marine iguana. These organisms have lived in a predator-free environment during evolutionary history since colonizing these distant islands. As a result, they are fearless around potential predators (such as humans) that cause animals in most other environments to flee. Lack of fear occurs in many insular organisms, and as humans and their associates (cats, rats, dogs, etc.) have colonized the earth, fearlessness has been the cause of extinction in more than one instance. *Photos: Randy Moore.*

Behavioral regulation is part of the general concept of homeostasis, or maintenance of appropriate internal states, in animals. Behavior is the gateway between the external, or ecological, world of an animal, and its internal, or physiological, world. Homeostatic regulation generally refers to maintenance of constant internal state, but to achieve this, animals must continually modify their behavioral state to accommodate changes in the outside world. Behavior is the necessary homeostatic link that gives animals the means to obtain food, regulate their internal temperature, and locate themselves in an appropriate habitat. It is also the major dividing line between the homeostatic strategies of animals, which often depend on mobility, and the homeostatic strategies of the relatively immobile plants. Behavioral homeostasis can be as simple as behavioral thermoregulation, causing an animal to bask in the sun when it is cold or seek shade when it is hot, or as complex as migrating thousands of miles to find a seasonally appropriate habitat.

> **KEY TERM** Animal welfare is the well-being and health of animals. Welfare includes behavioral welfare, so that management of anxiety, fear, and pain is the major issue in animal welfare. Humans take responsibility for the welfare of companion, captive, and farm animals.

> **KEY TERM** Behavioral homeostasis is the maintenance of appropriate living conditions by using behavior to achieve those conditions by minimizing their fluctuations. It comes from two Greek words meaning "similar" and "stoppage."

> **KEY TERM** A time budget is a description and quantification of how an animal divides its available time among its activities.

Neurobiology and endocrinology provide the regulatory foundations on which much of homeostasis is built. Homeostatic regulation requires information relayed in *feedback loops* (see Figure 4.1). A feedback loop functions by first using a sensor to gather information about internal state. This information is then *fed back* to the regulatory system, much in the way you watch a car's speedometer and give the car more or less gas to keep your speed constant. In response to information from internal heat sensors, the basking lizard in Figure 4.3 moves into the sun to warm itself and out of the sun when it is too warm. Thus, simple feedback (skin and blood temperature) serves to regulate basking behavior.

Physiology texts often use the analogy of a home furnace regulated by a thermostat to explain homeostatic regulation by feedback. If the house temperature falls below a fixed point, the furnace turns on. Once the desired temperature is reached, negative feedback from the thermostat turns off the furnace. Generally, negative feedback loops provide the

FIGURE 4.3
A lizard basking in a sunfleck on the floor of a tropical rainforest. Basking behavior is a common mechanism for increasing body temperature and improving muscle performance and digestion. Basking can also generate "behavioral fever," elevated body temperature that helps the animal fight infections. Cold-blooded animals, like the lizard in (A), are typically assumed to benefit from behavioral thermoregulation, but basking is also common in warm-blooded birds (B) and mammals, such as these cormorants. *Photo: Michael Breed.*

most effective behavioral regulation. When an animal's stomach is full, negative feedback inhibits further feeding behavior. However, positive feedback, such as a food reward or praise during learning, plays a key role in shaping behavior and also fits the general model for feedback loops. (See Chapter 5 on learning for more details on this topic.)

> **KEY TERM** A feedback loop occurs when an animal continually assesses its condition and modifies its behavior or physiology to achieve a targeted condition.

113

4.2 DRIVE THEORY AND HOMEOSTASIS

Drive, or motivation, theory holds an interesting place in the history of animal behavior, and it is key to understanding much of how we assess and work to improve animal welfare. It is also among the most interesting concepts in animal behavior.[1,4-6] Ethologists developed the concept of drive to answer questions about behavioral choice and switching. The study of drive has gone out of style in animal behavior, in part because it addresses "deep motivation," those urges that even highly articulate human beings may not be able to describe or explain. Clearly, these impulses do not lend themselves to crisp, testable predictions, and scientists rightly shy away from ideas that are untestable.

DISCUSSION POINT

Animal behaviorists who trained over the past 40 years were strongly discouraged from discussing or considering the motivations of animals. Thinking about motivation was viewed as unscientific as it involves imagining what is going on inside the mind of an animal, which is essentially a black box to us. When Chapter 6 (Cognition) is reached, mental time travel—the ability of an animal to project forward in time using the imagination—is considered an important cognitive attribute. But mentally time travelling into the mind of another organism, human or animal, has its pitfalls, largely because humans tend to project their own feelings, thought processes, and social expectations onto the unknown mind of the other organism. Recognizing this tendency, behavioral biologists have shied away from such pitfalls, and in so doing, have created another problem; they built a firewall against considering motivation that resulted in the neglect of the study of emotions, which are now known to be real, palpable, aspects of the lives of many animals. How does a scientist set appropriate boundaries that allow the development of data-based hypothesis testing while acknowledging the real, if complex, consequences of motivation and emotion?

> **KEY TERM** Displacement behavior is the release of drive energy in an irrelevant task when the desired behavior cannot be completed. Animals often groom as displacement behavior.

> **KEY TERM** Redirected behavior is the direction of behavior to a third party or an inanimate object. If you kick the wall when you are angry at your roommate, you have redirected your aggression.

> **KEY TERM** SDB is a behavior, such as grooming, that an animal performs on itself. It can become pathological in anxious animals.

> **KEY TERM** Repetitive behaviors (sometimes called *stereotyped behaviors*) such as pacing are often pathologies resulting from the displacement or redirection of energy when an animal cannot perform a desired task.

114

Drive theory enjoys prominence here because the concepts it has yielded, such as *displacement behavior* and *redirected behavior*, are important in interpreting the welfare of animals. When taken to the extreme, these behaviors become pathological and manifest as *SDB* and *repetitive*, or *stereotypical*, *behavior*.[7,8] Using tools developed from drive theory, applied animal behaviorists can assess the condition of captive animals or companion animals and improve the welfare of those animals.

In 1972, McFarland and Sibly suggested analyzing drives as vectors, much as vectors of force might be analyzed by physicists.[4] This suggestion has not gained much attention, but it remains the most recent approach to reconciling drive theory with a modern understanding of homeostatic regulation. Drive models are at once crude and clever. The graphical representation of two hydraulic models of drive (Figure 4.4) shows the crudeness of drive models from physiological and neurobiological standpoints. In contrast, contemporary biology generally views homeostatic regulation as a complex interaction among control systems. The cleverness of drive models derives from the attempt by McFarland and Sibly to explain, with little knowledge of animals' inner workings, how competing needs are assigned behavioral priorities.

Drive theory asks us to imagine the source of an animal's internal energy for behavior (refer to Figure 4.4). In simple models, this energy flows into a reservoir, which then empties as the animal expends "motivational energy" in its activities. In the model, activities are ranked by positions of valves on the tank, so less-important activities receive energy only if higher priority behaviors have already been accomplished. But what happens if excess energy accumulates or if an animal is conflicted between expressing two behaviors? Approach–avoidance conflict

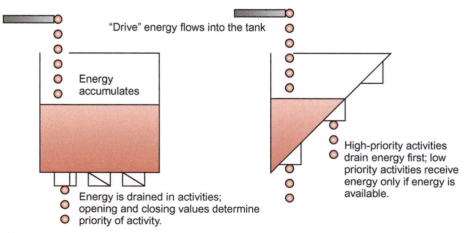

"Drive" energy flows into the tank

Energy accumulates

Energy is drained in activities; opening and closing values determine priority of activity.

High-priority activities drain energy first; low priority activities receive energy only if energy is available.

FIGURE 4.4
Hydraulic models of drive. Early ethologists thought of animals as having a supply of energy that could be allocated to tasks. Over time, this energy accumulates and ultimately is released through performance of a behavioral act. Energy models of drive gave ethologists tools for explaining how animals prioritized their behavior.

is particularly common: if a food item is attractive, but collecting it is risky, an animal may sneak forward and then jump back nervously as it attempts to snatch the item. Similarly, the relative risks and benefits of fighting a territorial intruder versus fleeing from it result in a fight-or-flight dilemma. In both instances, the animal is caught on the cusp of indecision between approach and avoidance, or fight and flight. This fits well with the model of drive, as energy for action builds up but is unexpended.

How might this energy be dissipated? This is the point at which displacement behavior and redirected behavior come into play. If a conflict occurs between allocating energy to one of two or more behaviors (such as occurs in approach–avoidance conflict), energy might be dissipated in a third irrelevant act, such as grooming.

Given this explanation, it becomes clear that animals engaged in activities such as displacement resulting from approach–avoidance conflict may be under a good deal of stress. Because of this, drive theory offers value for the assessment of animal welfare. It is the foundation for important tools—introduced in the following sections—that explain and evaluate some behaviors in both natural and confined habitats. These tools can provide a window into the internal state of animals in captivity. Given advances in contemporary science, many behavioral states (hunger, motivation to court and mate, migration, for example) are best understood on physiological and neurobiological levels, as functions of learning, hormones, and neurotransmitters. These states can lead to motivational conflict that can be addressed by drive theory. Although drive theory was developed decades ago, its utility in welfare assessments is strong justification for contemporary students to understand the theory.

In the sections that follow, drive theory will be incorporated into contemporary explanations for everyday behavior of animals as they resolve motivational conflicts. It is also the basis for understanding the expression of animal behavior at the extremes of these conflicts, along with resulting pathology.

115

Displacement and Redirection

Understanding how animals deal with internal conflicts contributes to animal welfare assessment. Studies in this area provide a larger picture of what occurs in abnormal circumstances, such as captivity. These responses stem from mechanisms animals exhibit in response to conflicting motivations. Such situations can become welfare issues at the extremes, but under more moderate circumstances, they are part of everyday life for every animal. Measurement of the frequency and intensity of displacement and redirected behaviors gives a window to the welfare of an animal, particularly how it is being affected by mental states that are homologous with anxiety and depression.

Drive theory attempts to explain how animals resolve conflicted situations. A young animal may be conflicted, torn between the desire to play and the desire to be near its mother's protection. The baby cannot resolve the conflict, so it vacillates between the two activities. Observations often reveal that conflicted animals groom themselves; this SDB may be a reflection of the anxiety caused by this conflict.[9]

DISPLACEMENT BEHAVIOR

Displacement behavior occurs when an animal performs an act that is irrelevant to the behavioral context. Animals caught in approach–avoidance conflict often groom; this is displacement behavior. Displacement behavior is obvious in humans under socially stressful situations. When attracted to a person at a party, shy humans often groom their hair with their fingers or touch their faces as a form of displacement behavior. Much of this behavior is hypothesized to be energy dissipation associated with conflicting drives.

Displacement behavior includes SDBs such as self-grooming, touching, or scratching, displayed when an animal has a conflict between two motivations, such as the desire to

FIGURE 4.5
Displacement behavior occurs when an animal experiences conflicting drives, which in some cases is called approach–avoidance conflict. The animal may do something irrelevant, such as scratch or preen, in these situations.

approach an object while at the same time being fearful of that object. Many, perhaps most, birds and mammals groom in similar ways when faced with a conflict between approaching and avoiding another animal (Figure 4.5). In social hierarchies, lower ranking animals groom more frequently than do higher ranking animals, possibly reflecting the conflict between attraction to the social group and avoidance of the higher ranking animals in the group.

REDIRECTED BEHAVIOR

Redirected behavior is usually aggressive; because a response cannot be directed at the cause of the animal's impulse to attack, the behavior is directed elsewhere. This often happens when a dominant animal would be the target of aggression, but social restraints or possible consequences of attacking the dominant animal in the social group prevent direction of behavior at that animal.[10] A person who kicks his dog or flings her pillow after an argument with a roommate has redirected the aggression that developed as a result of an event or behavior that occurred previously. Classically, most discussion of redirected behavior centers on aggression; redirection can also emerge in behaviors such as nest building or cleaning.[11]

An animal that has been attacked by the dominant individual in its social group may not retaliate (due to the possibility of stimulating a more severe attack by the dominant) but may instead redirect its aggression to other less-dominant members of the social group. Humans who have older siblings often experience their brother's or sister's redirected aggression when their sibling has been punished by a parent. Musth elephants redirect aggression to other species or to inanimate objects.

In animal welfare assessments, redirected behaviors are not always aggressive. They frequently involve behaviors such as *cribbing* (licking or chewing cage or stall bars) and repetitive pacing. Sometimes, these behaviors, particularly repetitive pacing, are termed *stereotypical* behaviors, but this is potentially confusing because *stereotypy* has a distinctly different meaning in the study of the evolution of communication (see Chapter 7).

OF SPECIAL INTEREST: MUSTH

Bull elephants annually cycle between a state of heightened aggressiveness, called *musth*, and nonmusth. Excess energy is dissipated in displacement behavior or redirected behavior by the elephants. In elephants, musth presents an extreme example of redirected aggression; often an elephant in this condition will attack any object or animal that happens to be in the way.

A musth elephant is primed to mate, fights other bull elephants, attacks other animals, and may destroy inanimate objects in its way. Musth bulls produce a distinctive low-frequency vocalization, the musth rumble; have thick secretions from their temporal glands (the duct from the temporal gland opens between the eye and the ear); and continuously dribble urine. Testosterone levels peak in musth males and probably regulate this extreme form of reproductive behavior. Legendary "rogue" elephants were probably musth bulls, redirecting their aggression at random objects, including villages and their inhabitants.

Young bulls do not go into musth and may be inhibited from doing so by older males. Asian elephant males start their musth cycles by the time they are 20, but African elephants do not reach this stage of maturity until they are 25 or so years old. Musth has implications for maintenance of captive populations of elephants in conservation programs. Musth males are unmanageable and extremely hazardous to elephant handlers (Figure 4.6).

Musth is an interesting reproductive strategy.[12] Presumably, males cannot maintain themselves in this physiological state for extended periods of time. In a sense, it might be a costly advertisement of mate quality, as only well-fed, healthy males may be able to physiologically support the energetic demands of being in musth. When more than one bull is in musth, fights result in risk of serious injury. Bulls may adopt a strategy of avoiding fights by coming into musth out of synchrony with stronger or more powerful bulls. Because female ability to conceive is highly seasonal, this strategy can be costly; an elephant whose musth is poorly timed may sire few offspring. Bulls that exhibit off-season musth are displaying a type of satellite strategy in which their mating efforts are peripheral to the main mating competition. This and other mating strategies will be covered in more detail in Chapter 11 on mating.

FIGURE 4.6
During the mating season, mature male elephants redirect their reproductive energy and enter a condition of musth, an extremely aggressive state in which they are likely to attack whatever is in their path, including inanimate objects.

SDBs and Repetitive Behaviors

SDB can be another result of dissipation of behavioral energy. They are behaviors that an animal performs on itself, without a social partner. Strictly speaking, it ranges from benign activities to potentially

> **KEY TERM** A focal animal is an animal within a group that an investigator targets for observation.

destructive behaviors such as feather plucking, masturbation, and obsessive licking. Self-grooming, or auto-grooming, is a normal part of the behavioral repertoire of a wide range of animals and usually serves to clean the surface of the animal of parasites. In behavioral evaluations of domestic or caged animals, SDBs often are grouped with redirected behaviors, sharing the umbrella term of SDB.

Social stress in natural habitats can also give rise to SDBs. SDBs increased in wild olive baboons when the animal nearest the *focal animal* was dominant.[13] The proximity of dominant animals was associated with a 40% increase in SDBs, indicating a higher level of social anxiety in animals near dominant individuals.

Assessment of Animal Welfare

As drive theory fell out of favor, studies of displacement behavior suffered collateral damage. Animal behaviorists paid little attention to displacement behavior until a study by University of Chicago primatologist Dario Maestripieri pointed out that it might be a good measure of anxiety levels.[14] Since that time, a growing body of literature has argued that self-directed and repetitive behaviors accurately indicate anxiety. Particular attention has been paid to primates, including humans. Measures of displacement behavior, for example, have been applied in psychiatric studies of human anxiety.

Measuring SDB gives animal behaviorists powerful tools for assessing the welfare of animals in captivity. For assessment purposes, both SDBs, such as grooming, and stereotypies, such as repetitive pacing, are evaluated. Quantification of SDBs allows an animal behaviorist to measure both the level of stress caused by a captive environment and the effectiveness of interventions intended to ameliorate that stress. If SDBs are reduced after the intervention, then it is judged to be effective.

Pathological SDBs most often are grooming to the point of self-damage, such as feather-plucking in birds or paw-licking in dogs. Domesticated ungulates sometimes display redirected behavior by cribbing, in which they lick or chew on the bars of their enclosure. Pacing, excessive grooming, and cribbing are common in zoo animals and can be interpreted as symptoms of suboptimal conditions for the animal. Carnivores and primates exhibit pacing and excessive grooming more than other animals, but these behaviors can be seen in a wide range of other birds and mammals.

Measuring SDBs in captive animals offers a window into the animals' "emotional" state. Animals with high rates of SDBs require changes in their conditions. The most common response in zoos to high rates of SDBs is habitat enrichment. Depending on the species, there may be opportunities to search, "hunt," or play with toys. Additional space (larger cage size) can help, but the key factor seems to be giving the animal something interesting to do (Figure 4.7). In general, habitat modifications that increase the complexity of the animals' environments result in reduced expression of SDBs; this reduction can be interpreted as a result of reduction of the stresses inflicted by the captive environment. Zookeepers may present primates with food that is hidden or disguised, requiring the captive primates to solve simple problems to be able to find their food. Similarly, the presence of objects to play with and manipulate reduces the stress of captivity for primates.[15] This modified understanding of the needs of captive animals has also influenced the companion animal world and is reflected in the increased number of dog toys that challenge dogs to find and dislodge a treat; most owners find this preferable to dislodging, say, pillow stuffing.

Chimpanzees are an excellent example of animals that display higher levels of SDBs in anxiety-inducing situations (Figure 4.8). In the case of responses to vocalizations from animals in neighboring cages, captive chimpanzees displayed SDBs when housed in groups but did not do so when socially isolated. If the vocalizations suggest to the chimpanzees that an attack is imminent, isolated animals may realize that no other chimpanzees are in their cage, and therefore may feel safe even when hearing the vocalizations.[9]

Problem-solving tasks can also induce SDBs in captive chimpanzees, depending on the difficulty of the problems and the way they were introduced. If the chimpanzees started with an easy problem and then progressed to more difficult problems, they displayed more SDBs when confronted with the difficult problems. Chimpanzees who received only difficult problems did not display more SDBs and positive auditory reinforcement during the problem-solving reduced SDBs.[15]

In addition to mental challenges, social environment is important. Animals reared in captivity require a social environment that mimics the richness and complexity of the

A   B   C

FIGURE 4.7
Modern zoos work to enrich habitats for captive animals by (A) offering animals their natural foods—in this case, nuts that the hooded capuchin monkey must crack; (B) giving carnivores like tigers meals in "puzzle" containers that require some effort prior to dinner; and (C) providing "toys," such as this elephant toy. *Photos: Michael Breed.*

species' normal social environment. Because of this, social animals reared in social isolation or in socially deprived environments typically express pathological SDBs. Harry Harlow's famous experiments on rhesus macaque monkey infants, mostly conducted in the 1950s and 1960s, showed that social isolation resulted in physiological and behavioral pathologies related to stress (Figure 4.9).[16,17] Today, scientists have an increased understanding of how nonhumans experience the world (in part, thanks to Harlow's work!), and Harlow's experiments would not be done. Nonetheless, they have given us important windows into how social deprivation during development and the stress associated with isolation affect behavior later in life. Moreover, Harlow's research itself is part of what informs current judgment about the effects of social deprivation on young macaques. (Note that rearing in social isolation differs from the condition of socially isolated chimpanzees in the previous study; the animals in the first study were not necessarily reared in isolation.) Take heart from the fact that Harlow also did rehabilitation experiments; in the right social conditions, deprived youngsters could regain normal behavior in relatively short order.

In canids, such as wolves, or felids, such as lions or tigers, larger enclosures and complexity in food discovery are key to reducing social stress (optimally, live prey that must be killed, although few zoos are willing to do this). The discussion point invites consideration of the importance and role of zoos and other captive environments for nondomesticated animals.

FIGURE 4.8
Grooming serves many functions, from parasite removal to emotional expression. In the case of social animals such as these monkeys, it can also reinforce social bonds. *Photo: Michael Breed.*

FIGURE 4.9
Harry Harlow asked how social deprivation and aspects of mothering such as warmth and food influenced the development of young rhesus monkeys. The effects were wide ranging and tended to reflect the severity of the treatment, providing important insights into the psychological and social requirements of young social mammals. Fortunately, the experimental subjects suffered no lasting effects after social rehabilitation.

120

DISCUSSION POINT: ZOO ANIMALS, ANIMAL WELFARE, AND THE EDUCATIONAL FUNCTION OF ZOOS

The continued existence of zoos is one of the most passionately debated issues in animal behavior and conservation. Here, we present both sides of the argument; a class discussion of this topic will probably reveal a wide range of opinion on this issue. Thus, drive theory, far from being out of date, provides critical tools in evaluating captive animal welfare.

Proponents of zoos argue that maintaining animals in captivity serves educational and conservation missions. Zoos remain the primary point of public exposure to exotic animals, and people who become engaged with what they see in the zoo are likely to favor conservation of natural habitats. Children, in particular, love zoos and may be drawn into a lifelong commitment to conservation by zoo visits. For species that are in serious trouble in their native habitats—many of which are being destroyed or diminished—zoos provide captive breeding programs, which may be the last defense against extinction and may provide animals for reintroduction into natural habitats.

Opponents of zoos argue that no positive benefit of zoos outweighs the harm done to animals by their conditions in captivity. Based on the observations of captivity-induced behavioral pathologies, opponents argue that no animal can be maintained humanely in a zoo. Conservation goals can be better met by preserving natural habitats and perhaps by establishing populations on large reserves outside the native habitat where visitation is limited. Modern media allow observation of a wide range of animals from around the globe in movies and television, making zoos unnecessary for public education. In the extreme, this argument extends to the idea that it would be more ethical to allow a species to go extinct than to maintain its only population in captivity.

Do the educational and conservation benefits outweigh the animal welfare issues? Is the ability to make behavioral assessments of welfare and to use these measures to evaluate the success of enrichments adequate to ensure the welfare of animals in captivity?[18]

The presentation of SDBs and animal welfare so far has mostly focused on wild animals kept in captive conditions. How do these principles extend to domesticated animals, and what is the neural and endocrine basis of these behaviors? When observing an animal expressing SDBs, most people would characterize the animal's state with words such as *anxious*, *nervous*, *tense*, or *distressed*. Collectively, these terms capture an animal's *affective state*. Although there are serious pitfalls associated with using terms that have a human emotional context to describe the internal state of a nonhuman animal, the evolutionary relationships among humans and other mammals establish a scientifically sound argument for homologies between the internal states (emotions) of humans and nonhuman mammals.

Neuroscience and endocrinology tell us that the same suite of neurotransmitters and hormones underpins behavior across a much broader range of animals than mammals. Given this line of thinking, it is reasonable to hypothesize that birds and mammals with SDB-type behavioral pathologies are "anxious." In addition, it is not difficult to see that repetitive pacing and similar behaviors resemble obsessive–compulsive disorder (OCD) in humans.

Is there a neuroscience-based argument for this homology between human and nonhuman anxiety? How can a hypothesis about the similarities between apparent animal anxiety and human anxiety be tested? Anxious mammals, such as domestic dogs or cats, may lick their paws obsessively, claw furniture, chew destructively, or drool excessively. These behaviors fit in the general classifications of SDBs and redirected behavior. Humans do not claw furniture, and drooling is frowned upon, but they do indeed feel anxious; if a human reports an overwhelming sense of anxiety to his or her physician, he or she would likely be prescribed a drug such as fluoxetine (Prozac). These drugs act as selective serotonin reuptake inhibitors (SSRIs), increasing the amount of serotonin in the space between neurons; behaviorally, taking such a drug has the effect of ameliorating symptoms of depression. The relationships between neurotransmitters, such as serotonin, and behavioral state are discussed below. If this type of treatment does not work, that person might be prescribed a medication such as Xanax, which enhances production of the neurotransmitter gamma-aminobutyric acid (GABA).

All of this should make it no surprise that anxious animals respond to therapy with drugs that increase the action of serotonin, a neurotransmitter implicated in depression and OCD in humans. This is consistent with the hypothesis that animal anxiety is similar to human anxiety. Fluoxetine and similar drugs are commonly used to treat dogs and cats for separation anxiety and similar behavioral issues. Interestingly, these drugs are effective in reducing certain kinds of aggression in dogs and cats, as well, supporting another hypothesis discussed later in this chapter: that much aggression in animals is fear based. Typically, medication of a dog or cat for anxiety-related behavioral problems is accompanied by behavioral training, and for long-term treatment success, it is usually as important, or more important, to teach the animal's owner to modify its behavior as it is to medicate the pet.

FIGURE 4.10
An American goldfinch captured in a mist net during a study of migratory birds. The behavioral effects of such captures on wild animals are largely unknown, but the corticosteroid response to events such as mist-netting or close encounters with predators is probably costly to the animal, especially during an energetically demanding season. *Photo: Michael Breed.*

Pathological behaviors arise in other domesticated animals, such as swine, horses, and cattle. Generally, these behaviors are regarded as symptomatic of welfare problems, such as keeping horses confined in stalls for long periods of time. The same solutions used with zoo animals—increasing habitat complexity and making the animals' lives more interesting—apply to treatment of anxiety-related problems in these species. The best-known nonmammalian behavioral pathology associated with captivity is feather-plucking in birds.[19] Again, reducing behavioral stressors in the environment is important, but birds also respond to drugs developed for the treatment of OCD in humans. Knowing that the neurobiological analogy with human anxiety extends to birds puts an exclamation point on the argument that many behavioral pathologies have the same neurobiological roots across a broad range of vertebrates.

Before leaving this topic, let us note an important point regarding stress and the endocrine system. Stress is a physiological response to difficult conditions, including hunger, fear, or pain. It may also involve the social stress of being low in a dominance hierarchy or the individual stress of being caged. Stress is usually accompanied by high levels of corticosteroids, hormones produced by the adrenal glands in response to stress. In humans, corticosteroid levels are associated with internal symptoms of anxiety, such as high blood pressure. Chronically high corticosteroid levels can result in a variety of physiological pathologies. For nonhuman animals, handling can dramatically raise corticoid steroid levels as part of a stress or fear response. In agricultural settings, repeated handling can reduce growth rates of poultry and other animals; this is probably due to the effects of corticosteroids in raising metabolic rates. Less is known about handling effects on wild animals (Figure 4.10), but it is a safe assumption that handling, via the action of elevated corticosteroids, causes the animal to use valuable energy resources and may disrupt the future behavior of the animal. Corticosteroids can be measured in fecal samples from animals, providing a tool for assessing stress on a physiological level. Biologist Shelley Adamo has pointed out that invertebrates have stress responses as well, and that stress responses are linked across all animals by their role in fight or flight situations and by their impact on the immune system.[20]

4.3 BEHAVIORAL SYNDROMES, PERSONALITY, EMOTION, AND MOOD

In this section, we introduce the concept that animals within a population can vary in predictable ways in how they interact with the external world. Thus, individuals can be predictable in their behavior, but they can also be different from other animals in the same population. This kind of predictable behavior is commonly called "personality," but that word has an overlay of meaning from human social interactions. To avoid anthropomorphism, biologist Andy Sih proposed the term *behavioral syndrome* as an alternative to the word *personality*. Essentially, these terms are interchangeable and both capture the concept that an individual behaves predictably through its lifespan. Behavioral syndrome or personality has strong genetic underpinnings; it is a set of correlated responses that are relatively stable over time for individual animals.[21] There is some question, in both animals and humans, how modifiable underlying personality is through experience (Figure 4.11).

> **KEY TERM** Behavioral syndrome and personality refer to consistently expressed behavioral tendencies of an individual.

Nonhuman animals can definitely have personality. Humans think of the animals with which they interact—dogs, cats, horses, and so on—as having personality. This is part of what makes the company of a pet so interesting. Scientists working on social behavior of birds or mammals are often struck by differences in personality among their study animals. Primates, canids, parrots and their relatives, crows and their relatives, and dolphins are known for the variety of personalities they present, but such variation can be found in a broad, and sometimes surprising, range of animals, including, for example, insects[22] and spiders.

Personality is the consistent expression of behavioral tendencies over time by an animal. Looking at a given dog, cat, mouse, pigeon, or honeybee, scientists might observe that it is always shy, bold, aggressive, passive, eager to bond with other animals, unfriendly, and so on. Gene expression, developmental influences, and memories come together to shape the neurochemical profile of an animal's brain. At any given moment, the levels of neurotransmitters such as dopamine and serotonin in the brain create a profile that

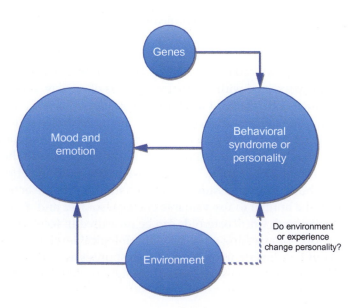

FIGURE 4.11
Genes and environment act together to shape an animal's behavioral syndrome or personality. Behavioral syndrome is a predictable pattern of behavior through an animal's lifetime, such as always being shy (not exploring or taking risks) or always being bold (engaged in exploration, taking risks). Mood and emotion are more "in the moment" expressions of mental state that reflect in part behavioral syndrome and in part current events. A dog, for example, may have a personality that causes it to be an enthusiastic greeter of familiar humans. This personality is always present, but the dog's emotional state changes when a familiar person comes into view.

has substantial influence on behavior; this is part of the chemical basis for personality. An animal's circulating levels of testosterone, oxytocin, vasopressin, cortisol, and other hormones also influence its personality.[23–26]

Personality is also an adaptive trait. Like other BPs, personality is subject to natural selection and personality can evolve in a population. As would be expected with a trait that is subject to selection, typically personality varies among animals in a population.

For nonhuman animals, the most commonly studied behavioral syndrome is the shy/bold continuum. In many species, some animals are predictably bold in behaviors such as environmental exploration, foraging, and soliciting mates, whereas other animals are predictably shy in these same behavioral contexts. The continuum of behavior persists in evolutionary time because animals at both extremes of the spectrum can succeed; they just have very different styles of achieving their goals.[27–30]

What shapes personality in animals? We can think of personality as a scaffold upon which behavior tendencies are built. At the base of the scaffold are genetically determined behavioral traits. Taking boldness as an example of personality, bold parents tend to have bold offspring. On the second level are developmental (ontogenetic) influences such as prenatal exposure to testosterone, which may also cause offspring to be more bold. On the third level are past experiences; an animal that has never confronted a potential predator may be bolder than one that has had near-death experiences. On the top level are current conditions such as hunger, which may also drive an animal to be more bold in its search for food. Thus, personality is the outcome of a mixture of genes, ontogeny, experience, and current events.

One of the commonly studied behavioral syndromes, the bold–shy spectrum, is an excellent example of how animals with personalities at either end of the spectrum experience both selective advantages and disadvantages. Bold animals may experience a selective edge in mating but also be more subject to predation. Shy animals may not do quite so well in mating but survive longer so that they get more chances at mating during their lifetime. Depending on the habitat, one end of the spectrum or the other might be favored by the selective balance at any given moment, but over evolutionary time fluctuations in features such as predator abundance probably allow both phenotypes to persist.

In European sparrows, the amount of testosterone in the egg yolk influences aggression, dominance, and sexual behavior in the young. This testosterone comes from the mother sparrow, whose testosterone levels in turn are influenced by her social environment. Prenatal effects of testosterone on the behavior of birds and mammals are well known and generally speaking play out in the same behavioral contexts—dominance and sex—as in the European sparrows.

Recent studies show that experiences prior to birth can also have a strong role in shaping personality. These include levels of maternal testosterone and maternal cortisol as well as maternal diet. In mammalian species that give birth to large litters, the position in the womb influences physical characteristics, such as body size, in addition to personality traits. Intrauterine position can also affect behavior via hormonal influences among fetuses. In rodents, a female from a fetus that developed next to a male fetus may have more masculinized behavior than a female fetus that was next to a female fetus. This effect is caused by testosterone from the male fetus.

In a recent study, aquatic spiders within a population were classified along a continuum of shyness–boldness.[31] These spiders skim on the surface of the water, supported by surface tension. The shy spiders tended to hide for long periods of time when disturbed, whereas the bold spiders quickly returned to the water surface. Shy spiders may be better protected from predation, but miss opportunities for feeding and mating that more bold spiders can exploit. Often, differences within populations in behavioral syndromes reflect behavioral options that are each successful, allowing animals to use different pathways to survival and reproduction. The behavioral syndrome concept puts the focus on how behavioral differences among animals affect strategies in a broader range of contexts than that of social behavior. It also emphasizes that behavioral syndromes can be constant within a population but differ among populations depending on evolutionary history and ecological conditions.

In contrast to personality, *mood* and *emotion* describe the state of an animal at any particular moment.[32,33] These terms are laden with anthropomorphic temptations and must be used carefully when characterizing animal behavior. Mood and emotion reflect the interaction of the animal's behavioral syndrome/personality with environmental events. A term that captures these concepts while avoiding anthropomorphic temptation is "affective state." The response of animals' affective states to a particular situation is determined by the event and by the animals' underlying behavioral syndrome/personality. Because behavioral syndromes have a strong genetic basis, the affective response of a specific individual may be predictably different from another individual of the same species, even though both are confronted with the same situation.

What does the concept of personality have to do with Tinbergen's questions? Personality can be subject to natural selection and evolve over time because it emerges, in part, from a genetic basis. It has utility and adds survival value in the way it enhances social interactions. For instance, knowing the personality of a friend or enemy helps shape interactions with that friend (or enemy) to be able to achieve goals. This brings the assessment of personality clearly within the scope of social cognition (see Chapter 6). One of the reasons that natural selection might favor social cognition is that in a group setting, it is beneficial to be able to discriminate among members of the group and predict what they might do.[34–36] Finally, the underlying mechanisms that help form personality involve the neurotransmitters that are central to many behavioral studies.

Genes play a role in the continuum of personalities. Personality may be genetically influenced for a given animal but may vary among individuals. Different personality phenotypes (and their underlying genotypes) continue to coexist because several strategies can be successful; for example, both aggressive and passive animals may succeed in finding mates, but the strategies they use in mate-finding will differ greatly. (These two sources of variation in personality—experience and genetics—are reminiscent of the nature–nurture controversy discussed in Chapter 3; it may be useful to review that material now.)

If personality varies among animals but is strongly influenced by genetic makeup for an individual, then the study of personality lies within the realm of behavioral genetics (refer to Chapter 3). The relevance of personality for this chapter, no matter what the genetic basis, is that personality creates the opportunity for multiple strategies within an animal population. This is critically important in social groups where, if all animals had exactly the same strategy, there could be no effective cooperation, the group might then descend into chaos, and the individual reproduction of group members would be affected.

Among nonhuman animals, personality is best known in chimpanzees and domestic dogs. In chimpanzees, personality is described by these variables:[35]

- Dominance
- Extraversion
- Dependability
- Emotional stability
- Agreeableness
- Openness

The last five of these dimensions are thought to describe human personality; their presence in both chimpanzees and humans can be thought of as representing shared evolutionary history. Interestingly enough, in populations of both humans and chimpanzees, these personality traits are strongly correlated between parents and their offspring and show virtually no effect of rearing environment. In a study of chimpanzees, there was a particularly strong heritability for social dominance and weak heritabilities for the other dimensions of chimpanzee personality. As in human studies of personality, the same study found little effect of environment (in this case, different zoos) on personality.[35] Human twins who are separated at birth and reared in very different environments show startling similarities in personality. These findings support the hypothesis that for both humans and chimpanzees, genes strongly influence personality, and variability of personality in a population may result from many different personalities, each of which tends to be fixed, rather than a population full of highly variable individuals.

What are some underlying mechanisms that influence personality? Neurotransmitters modulate general behavioral state at the level of the central nervous system and play key roles in behavioral syndrome as well as affective state. Behavioral expression is affected by neurotransmitter levels, the quantity of neurotransmitter receptors, and the presence of enzymes that activate or deactivate the transmitters. Figure 4.12 depicts how three major neurotransmitters interact in determining mood. This concept is applied most frequently to humans, but pharmacological studies and veterinary practice indicate that some of the

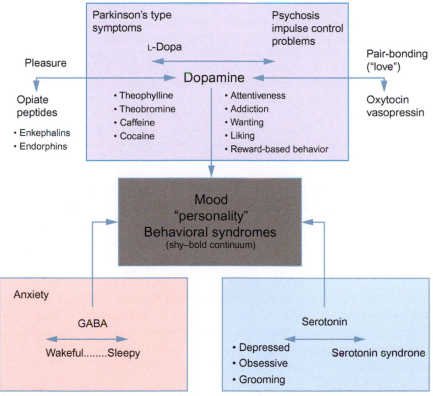

FIGURE 4.12
An animal's current behavioral state is regulated by the concentrations of neurotransmitters and their receptors in its brain. This figure highlights three important transmitters found in mammal and bird brains, dopamine, serotonin, and GABA. These contributed to current behavior that might be interpreted as "depressed," "angry," "addicted," and so on in humans. For each of these neurotransmitters, the figure indicates the specific behavioral states associated with the transmitter and common pharmacological agents that impact that transmitter. Neuropeptides, oxytocin and vasopressin, interact with the dopamine system in pair bonding. Internal opiates, endorphins and enkephalins, interact with the dopamine system in the perception of pleasure.

same principles apply to many, perhaps most, mammals, although the exact behavioral effects can vary dramatically. These same neurotransmitters are present in the brains of other animals, such as birds, but their role in regulating behavior is less well understood in nonmammals.[37,38]

Serotonin is the most commonly manipulated neurotransmitter in pharmacological interventions in animals. Serotonin levels in synapses can be regulated by drugs called SSRIs. Drugs such as Prozac®, Zoloft®, and Lexapro® are SSRIs; they alleviate depression in humans by increasing the levels of serotonin in the brain. The action of these drugs in regulating behavior is not, however, limited to humans; many mammals respond to SSRIs in ways that are reminiscent of human responses.[39]

BRINGING ANIMAL BEHAVIOR HOME: DOGS, CRUSTACEANS, AND ANTIDEPRESSANTS

SSRIs are effective in moderating aggression in dogs. Clomipramine, an older drug that also affects serotonin-based behavioral responses, is commonly used in dogs to treat anxiety. In humans, clomipramine has been used to treat depression and OCD, although it (and other antidepressant drugs in this general class, which are called *tricyclics*) has largely been replaced by SSRIs in the treatment of human depression. If canine aggression is sometimes induced by anxiety—and that is a reasonable hypothesis—this would explain the efficacy of serotonin-related treatments for aggression in dogs. In mice, the expression of "anxiety" is higher in strains that have been genetically selected to be nonaggressive. Although this anxiety–aggression relationship is the opposite of that suggested for dogs, the existence of a correlation between serotonin and behavior does imply a similarity in how these behaviors are controlled in the two species. Recently, serotonin has been shown to increase behaviors, such as hiding in the dark and reduced exploratory behavior, that are associated with anxiety in crayfish; crayfish subjected to electrical shock display these behaviors more than unstressed crayfish do. Injection with serotonin elicits these behaviors in unstressed crayfish, and injection with clordiazepoxide reduces stress behavior in these crustaceans. Clordiazepoxide is in a family of drugs used to manage stress in humans. This suggests that stress is an ancient response to adverse environmental conditions, one with potentially great survival value.[40]

GABA, discussed previously, is most strongly associated with wakefulness/sleepiness. Sleep aids, sometimes called *hypnotics* by physicians, such as Ambien®, activate GABA receptors, imitating the effect of increased GABA concentrations. Low GABA may also be associated with depression. In animals, low GABA levels are generally associated with aggression, and treatments that increase GABA lower aggression. Binding GABA receptors mimics the effect of increasing GABA levels.

Another neurotransmitter, octopamine, merits mention because it is often implicated in the regulation of behavior in invertebrates. Examples of behaviors associated with octopamine levels include foraging activity in honeybees, olfactory responsiveness in fruit flies, and firefly flashing. Octopamine appears to be associated with general arousal in insects; injecting or feeding octopamine to insects results in high levels of activity.

Dopamine is most commonly considered in the context of Parkinson's disease in humans. Parkinson's disease, a progressive loss of muscular control due to death of dopamine-producing cells in the brain, can be counteracted to a certain extent by treatment with compounds that supplement brain dopamine, such as L-dopa. Dopamine is known to modulate mood and behavior, especially pleasure. Endorphins, a class of neuropeptides that occur in the brains of many animals, regulate pleasurable sensations by stimulating dopamine release. They are found in many animal species, including insects, but their function outside of vertebrates is not well known.

> **BRINGING ANIMAL BEHAVIOR HOME**
>
> **Petting, Pet Therapy, and Oxytocin**
> People believe their dogs are good for them, and they are good for their dogs. Much has been written about the benefits of pet therapy and the value of a trusted relationship with an animal for emotionally at-risk humans. Is there physiological evidence for these values?
>
> It turns out that only a few minutes of interacting with a dog raises the oxytocin levels of both the human and the dog! These short-term bursts of the hormone associated with attraction and affiliation likely have long-term psychological benefits, increasing feelings of security and well-being for the human. In mammals, oxytocin is tied into the dopamine reward system in the brain so that the behavioral affiliation is rewarded neurochemically.
>
> Animal-assisted therapy, typically involving dogs or horses, is used in long-term care facilities and children's hospitals, and with autistic children. Strong evidence exists for positive effects of dog visits on residents of long-term care facilities, with reductions of feelings of loneliness and depression. Working with and riding horses may have positive effects on socialization and self-esteem for some children with autism spectrum disorders, but the evidence is weaker for this use of animal-assisted therapy, opening the door for further studies. Other applications of animal-assisted therapy include preschool children, dementia patients, and anxiety disorders; in these cases, much more data are needed to know if the interactions are beneficial.
>
> The downside risks of animal-assisted therapy are small, if human–animal interactions are carefully managed. Examples of risks include dog bites, being kicked or stepped on by a horse, or having allergic reactions to the animal. A variety of agencies offer therapy certification programs for dogs. Horses are chosen for therapy programs based on gentleness and ease of handling. Given the manageability of the risks and the possible magnitude of the benefits, animal-assisted therapy is a valuable addition to therapy plans for many people.[41–44]

127

Opioids, such as heroin, mimic the internal effects of endorphins and cause massive dopamine releases, resulting in similar, but stronger, pleasurable sensations. In dogs, selegiline hydrochloride, a drug developed as a Parkinson's treatment for humans, may enhance training. Generally, behavioral rewards stimulate dopamine receptor systems, so learning that is in itself rewarding, or is rewarded by a trainer, involves dopamine. Because of the link between reduced dopamine and neuromuscular control, veterinarians may be wary of lowering dopamine concentrations. Dopamine also plays a significant role in regulating obsessiveness and impulsive behavior in dogs.

In sum, personality is a set of attributes, such as sociability, aggressiveness, and willingness to please, that come together to form both the individual and the social behavior of animals. It is real and measurable, and seems to be strongly influenced by genes. Variability in personality reflects, in a sense, variability in genetic information and expression. Tinbergen's question of causation is difficult to address with personality, but evidence is mounting that supports the action of specific neurotransmitters in various behaviors that are hypothesized to reflect mood and emotion. At the level of evolution, personality variability suggests that, like other traits for which there exists variation, different personalities can be successful and persist in evolutionary time; if only one personality type were successful over time and space, natural selection would eliminate this variation.

4.4 BIOLOGICAL CLOCKS AND CIRCADIAN RHYTHMS

How does a pet seem to know what time it is? This behavior becomes especially apparent if, say, she gets a treat on a regular schedule. If someone has forgotten the treat, the pet assuredly has not, and there will almost always be a reminder in the form of a wet nose

Note
Even Darwin was intrigued by de Mairan's work and studied daily rhythms in plants. In *The Power of Movement in Plants*, he hypothesized that "sleeping" plants conserved energy, a hypothesis that has yet to be rejected as knowledge about rhythms in plants increases.

or an insistent purr. All animals—and even protists and fungi—can tell time, in a manner of speaking. In fact, the study of biological rhythms—activities that are linked to regular environmental cues and occur cyclically—may be the only topic in animal behavior that began with the study of plant behavior!

In the early 1700s, a French astronomer, Jean-Jacques d'Ortous de Mairan (Figure 4.13), realized that the leaves of the heliotrope plant opened and closed with a daily rhythm. By placing the plant in continuous darkness, he showed that the leaf action continued; it was not a response to sunlight. Instead, it was something now called *free-running behavior*, the maintenance of daily rhythm in the absence of any cues. Almost all organisms have *biological clocks*, that is, endogenous timing mechanisms (pacemakers) that allow an organism's activity to occur in a rhythmic manner, linked to external events. The external expression of a biological clock is called a *biological rhythm*.

Biological Clocks and Behavioral Rhythms

Because the earth rotates on its axis every 24 h, many environmental cues and significant environmental changes occur with regularity (Figure 4.14). The ability to predict these changes and to prepare for them, be they daily or annual, is clearly adaptive. Daily rhythms are called *circadian*, from a pair of Latin words, *circa* (about) and *dies* (day). A circadian rhythm runs on a 24-h cycle and is entrained to the day–night cycle of the natural world. Behaviors that happen at certain times of day—for instance, insect molting or rodent peak activity—are circadian. *Circannual rhythms* are also important and include seasonal behaviors such as hibernation or migration. Intertidal animals often exhibit *lunar rhythms* (28-day cycles); their activities coordinate with the presence or absence of water in their immediate environments. Biological clocks thought to be molecular timekeeper and clock mechanisms can be found in a variety of tissues in a typical animal.

> **KEY TERM** A biological rhythm is a behavioral or physiological attribute that changes over time on a predictable cycle.

128

Credit: National Library of Medicine

FIGURE 4.13
Astronomer Jean-Jacques d'Ortous de Mairan was among the first scientists to document biological clocks. *From http://www.hhmi.org/biointeractive/museum/exhibit00/index.html.*

Free-running rhythms are one type of evidence for biological clocks. An animal placed in constant darkness, without any environmental cues, will nonetheless exhibit an activity rhythm that approximates (but is not quite equal to) 24 h.

Even jet lag is a form of evidence for a biological clock. When an animal that feeds at a certain time every day is moved to a different time zone, it will continue to feed based on the original rhythm, at the time of day that corresponds to the feeding time in the previous time zone.

Of course, animals lose synchrony with their environment when their day/night cycle is shifted. Circadian rhythms in the real world track 24-h day/night cycles much more accurately than the free-running rhythm does. This happens because the endogenous pacemaker can be entrained by external cues. The external cue is called a *zeitgeber*, or time-giver; the most common *zeitgeber* in the terrestrial world is the day–night cycle. The light part of that cycle is called the *photophase* or *photoperiod*. Many environmental events have been shown to act as *zeitgebers*—even the regular sound of a school bell can serve that function—but the predominance of photoperiodic *zeitgebers* is thought to reflect the dependability of that cue.

The circadian rhythm is the building block for the other rhythms that overlie it. Thus, the circannual rhythm regulates migration (see Chapter 8), but it can do so only if the organism perceives seasonal photoperiodic changes and, indeed, uses its clock mechanism in conjunction with other cues to discern compass direction. Sleep/wake cycles, feeding bouts, and foraging compasses all depend on daily rhythms. Seasonal changes in reproductive status, migration, and hibernation are also tied to biological clocks (Figure 4.15).

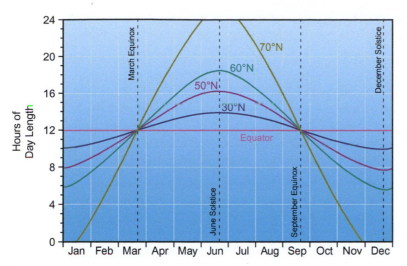

FIGURE 4.14

The tilt of the earth on its axis causes day length to vary seasonally depending on latitude (distance from the equator). At the equator, 0 degrees longitude, the 24 hour daily cycle is always equally divided between day and night, twelve hours each. One of the most surprising events when traveling to the tropics is the combination of summer-like weather associated with relatively early sunsets. As an animal travels from the equator towards one of the poles it encounters increasing seasonal differences between the light and dark parts of the 24 hour cycle, with long days associated with a warm or hot season. This translates into temperate and polar animals being able to employ information about changing day/night patterns into predictions about weather, and to time their migrations, mating, and feeding habits to fit expected weather. An animal's circadian clock helps it to track changing day length, as well as to time its activities within any 24 hour cycle.

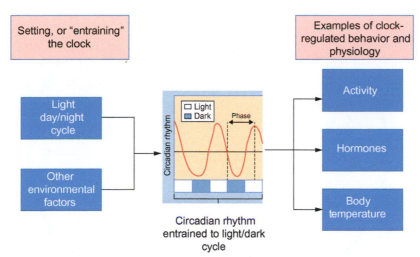

FIGURE 4.15

This diagram summarizes how a circadian clock regulates the activity of an animal. The *period* is the time it takes to complete an entire cycle— in this case, 24 h. When light and dark alternate in a 24-h cycle (left half of figure), activity peaks during the light part of the cycle (called the *photophase*) and decreases to a minimum during the dark phase (*scotophase*). When the photophase is removed and the animal is kept in constant darkness (right side of figure), cyclic activity continues but becomes free-running. Thus, the cyclical presence of light does not create the activity cycle, but serves as a *zeitgeber* to keep activity on a 24-h pattern. The activity cycle shown in this figure typifies a *diurnal* (day-active) animal; it would be reversed in a nocturnal animal.

The Molecular Basis for Biological Clocks

Genetics provides the key to understanding exactly how biological clocks function. In 1971, Ron Konopka and Seymour Benzer discovered a gene, one they called *period*, or *per*, that occurred in fruit flies (*Drosophila melanogaster*).[46] Variants of this gene were associated with variations in the circadian activity in the flies. Animals that have mutations in the genes associated with biological clocks will breed true for those mutations and pass the mutant rhythms on to their offspring. Thus, a mutant might free-run at a rhythm of 22 h, instead of 24 h. Genomics has brought a much deeper understanding to biological clocks.[47]

The genetic basis for clock function varies across animals, but the underlying mechanism appears to be universal.[48] To function, a biological clock depends on the presence of a molecular cycle that can be reset to stay in synchrony with the 24-h solar cycle. The basic principle underlying biological clocks is the daily buildup of a protein, which is phosphorylated as the day progresses. This buildup is reinforced by positive feedback at the molecular level. When an upper threshold of phosphorylation is reached, negative feedback then causes the protein to degrade until a lower threshold is reached and the cycle restarts. In birds, mammals, and fruit flies, genes such as *period*, *clock*, *timeless*, and *cryptochrome* play key roles in the clock (Figure 4.16).

In mammals, the clock that most affects behavior is found in the suprachiasmatic nuclei in the hypothalamus region of the brain. The daily oscillation of the molecular clock is linked with the endocrine system of the animal, which directly regulates behavioral and physiological changes over the 24-h cycle.

Although biological clocks have a general role in regulating animal activity cycles, they also play key roles in daily and seasonal movements (see Chapter 8).[49] For example, the sandhopper, a small crustacean found on European beaches, uses the moon to set its

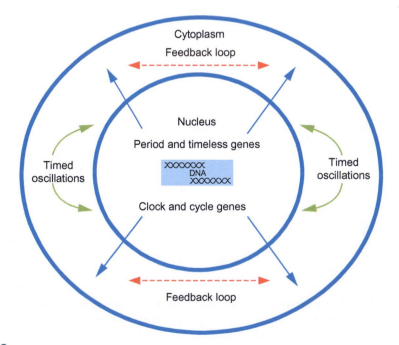

FIGURE 4.16

This schematic of the molecular workings of a biological clock shows how alternating cycles driven by the buildup of products of four genes creates a pendulum-like timekeeping mechanism in cells. The products of one pair of genes builds up until it reaches a threshold at which point feedback causes those products to diminish. This triggers the other pair of genes into activity, which continues until feedback curtails those genes. This alternates back and forth, creating a timed cycle within the cell. Environmental input of some type is required to synchronize this process with environmental events (e.g., solar cycle, lunar cycle). *Bell-Pederson et al., Nature Reviews Drug Discovery, http://dx.doi.org/10.1038/nrd1633.*[45]

direction when it swims from the open water to the wet part of the beach at night.[50] Moon compasses are common in nocturnal animals. On a moonlit light, the position of the moon allows the sandhopper to find the beach. But what about times of the month when the moon is not visible in the night sky? The sandhoppers use their biological clocks to calculate the expected position of the moon and are still able to find the beach. Similarly, honeybees can use their biological clocks to compensate for the apparent movement of the sun in the sky as they fly to and from food, or when they dance to demonstrate the location of food to other bees. The ability to use biological rhythm to compensate for the "movement" of the sun is called *clock-shifting*, and will be revisited in Chapter 8.

4.5 MODERN CONCEPTS OF HOMEOSTATIC REGULATION

Some areas of homeostatic behavior, such as sleep, feeding and appetite, and pain, lend themselves readily to scientific exploration. As scientific understanding of the hormonal and neurobiological regulation of behavior has increased, animal behaviorists have shifted from discussions of motivation and drive to examinations of those physiological regulators. After all, behavior is the result of physiological needs (e.g., thirst, hunger) or biological imperatives (e.g., reproduction) regulated via neurochemical and hormonal mechanisms. Consequently, many behaviors directly improve an animal's physiological condition and have direct effects on internal homeostasis.

Sleep

Sleep plays a key role in time budgets and behavioral decision making. A sleeping animal is immobilized (reversibly, not paralyzed), is not easily disturbed, and generally expresses a preference for a sheltered or protected location before entering sleep. Sleep, or sleeplike states, occur in all vertebrates, including fish,[51,52] and have also been observed in some insects.[53] Sleep probably has three adaptive functions: (1) A sleeping animal conserves energy; this is particularly true of endotherms (birds and mammals), which lower their metabolic rates while sleeping. (2) Sleep deactivates an animal during phases of the daily cycle when activity might put it at risk. (3) Sleep allows a time when brain repair and memory consolidation can take place.

DISCUSSION POINT: SLEEP

We have put forward three adaptive hypotheses for sleep: energy conservation, risk avoidance, and brain repair/memory consolidation. Are these hypotheses mutually exclusive? Are they testable, and if so, how would you go about testing them? Can you think of other possible adaptive advantages of sleep?

Although the details of how sleep is regulated are not fully worked out, the basic framework for sleep starts when an animal's circadian cycle primes it to sleep at appropriate times. This priming sets the stage for regulation of overall arousal/drowsiness. Deep sleep is also physiologically regulated. Melatonin, a hormone of the vertebrate pineal gland, plays a key role in regulating daily activity cycles; high levels of melatonin are associated with drowsiness.[54] A peptide hormone, called *orexin* or *hypocretin*, also has a major regulatory role in the sleep cycle. Orexin causes wakefulness and drives appetite; in its physiological absence (or in genetic mutants lacking the ability to produce orexin), an animal has a low state of arousal, or is *narcoleptic*. GABA plays in important role in the transition between light sleep and deep *rapid eye movement (REM) sleep*. Human sleep aids, such as zolpidem (Ambien®), promote the effect of GABA in suppressing brain activity and inducing sleep. The importance of sleep in the biology of most animals is emphasized by the pathological, impaired mental states of animals that have been sleep deprived. Sleep deprivation is a powerful torture device among humans; this fact demonstrates the importance of sleep in homeostasis.

> **KEY TERM** Sleep is a period of inactivity during which brain activity is depressed or modified.

> **KEY TERM** A narcoleptic is a hormone, neurotransmitter, or drug that induces sleep. It comes from narcolepsy, a nervous disease that is characterized by excessive sleep (Gr. *narco* for "numbness"; L. *epilepsia* for "to take hold of").

> **KEY TERM** REM sleep is a deep sleep state that is key to the neural repairs and memory consolidation that occur during sleep.

Sleep is, then, a key behavioral trade-off. The sleep state is beneficial to the organism, but the increased arousal threshold during sleep puts the animal at high predation risk. In addition to finding preferred locations for sleep, animals use a variety of behavioral strategies to reduce risk while sleeping. Birds such as wrens and warblers, which migrate to tropical climates during the nonbreeding season, tend to sleep far out on the ends of branches. This makes it harder for foraging snakes to find them, and the flexibility of the branch makes it likely that the snake will vibrate the branch, arousing the bird. Some birds are known to "sleep with one eye open," so that one hemisphere of the brain is in a sleep state, whereas the other side of the brain is aroused enough to perceive threats. Many animals become more physically tolerant of close contact with family, flock, or herd members during sleep. This close contact allows shared vigilance and ensures that if one member of the group is aroused by a disturbance, the others will be as well.

Feeding and the Regulation of Appetite

Some phases of an animal's life pose particular challenges for energy uptake.[55] These phases include growth periods, preparation for overwintering or migration, competition for mates, and, in the case of mammals, pregnancy and lactation for females. Control of food intake, and consequently of foraging behavior, is extremely complex and involves three interrelated control mechanisms: (1) the sense of satiety, or fullness; (2) the responsiveness of the body, and of behavioral reactions, to the insulin control system; and (3) seasonal and developmental changes that affect fat storage (and consequently feeding behavior) for use in egg production, overwintering, or migration. These three control mechanisms are detailed below.

SATIETY

Satiety is a sense of fullness after eating. The key neurotransmitters controlling appetite, at least in vertebrates, are serotonin (5-HT) and catecholamine. These neurotransmitters act to reduce feeding behavior and consequently food consumption. Carbohydrate intake can result in both increased levels of serotonin and catecholamine, and also changes in the number of their receptors. One effect of increased carbohydrate intake is to elevate raw materials available in the brain for neurotransmitter synthesis; this feeds back in the form of suppressed appetite. However, the relationship is complex and carbohydrate intake does not always result in satiation.

The picture concerning satiation in insects is much simpler. Stretch receptors detect the degree of stomach expansion that accompanies feeding. At least in the insects most studied for satiation (mosquitos, houseflies, and honeybees), feeding stops when the stomach is fully stretched. Making a small incision through the body wall and piercing the stomach so that ingested fluids drain out result in continuous unregulated feeding, as does severing the nerve that serves the stretch receptors.

INSULIN CONTROL SYSTEM

Sugars are the basic energy substrate of living things, so it is not surprising that sugar circulation and storage are critical to animal life. Perhaps more surprising is that virtually all animals seem to share a common metabolic pathway for regulating circulating sugars—the insulin signaling system (ISS), or *insulin control system*. (Insulin is a small peptide that, in vertebrates, is produced by the pancreas.)

Some systems are so fundamental to organismic function that they show little, if any, change over the course of evolutionary time. These systems regulate the basic components of life. The functions of the families of genes that code for these highly conserved processes are intricately linked to animal behavior. Here, the ISS illustrates the complexity of a highly conserved signaling system.

> **KEY TERM** The insulin control system regulates carbohydrate metabolism in nearly all animals. Insulin is a peptide hormone, secreted by the pancreas in vertebrates, that is released in response to rising blood sugars; it stimulates sugar storage or conversion to fats.

As indicated previously, the genetic code for insulin, in some form, is found in all animals. Correspondingly, all animals also have the genetic codes for insulin receptors—proteins that respond to the presence of insulin. Although it may not be surprising that insulin from pigs or cattle has strong effects on human physiology, it is startling to learn that insulin from cattle has physiological effects in very distantly related organisms, such as locusts, flies, and honeybees. The reason is that the genetic code for both insulin and the insulin receptor molecule is so important to survival—so closely linked to the function of cells and cell membranes—that they have remained relatively unchanged over millions of years of evolution.

In addition, the function of insulin has been conserved across a wide range of taxa. Recent studies have implicated the insulin signaling pathway (or insulin-like signaling) in regulation of feeding behavior in a range of animals, including roundworms, insects, and, of course, vertebrates. The sugars themselves may vary across taxa. For instance, in vertebrates, blood sugar (glucose) is regulated by the insulin signaling pathway. The insulin signaling pathway is also important in metabolic regulation in insects, even though the blood sugar in most insects is trehalose, which is composed of two glucoses.

How does blood sugar relate to behavior? Blood sugar regulation leads to a cascade of behavioral events, beginning with the internal perception of hunger or satiation and ultimately leading to food searches. The primary physiological role of insulin is to regulate blood sugar.

Insulin is the hormone of glucose storage; it is secreted in response to elevated blood sugar levels and triggers cellular storage of glucose in the form of complex carbohydrates or fats. Lowered blood sugar then triggers further feeding behavior. In vertebrates, another hormone, glucagon, mobilizes stored energy substrates. The insulin signaling pathway appears to affect feeding and related behaviors through the action of neuropeptide Y, a neurotransmitter produced as a result of activation of the insulin signaling pathway (Figure 4.17). Like many molecules with deep evolutionary roots, insulin has also evolved functions that differ from those of the original molecule. Insulin-like growth factors play key roles in the development and maintenance of the nervous system in vertebrates. Age-related loss of insulin-like growth factors negatively affects neural functions such as learning and memory. Mutations in the insulin signaling pathway interfere with heat-seeking behavior in roundworms, a behavior important in foraging for some of these animals. In vertebrates, the insulin signaling pathway is tied into learning and memory, playing a key role in behavioral integration.

FAT STORAGE

Fat storage is essential for egg production, migration, or prevention of starvation during lean times. It also carries potential costs, particularly in predation risk. (Fat animals may move more slowly and can be easier and more attractive targets for predators.) Regulation of appetite through satiety control and homeostatic control of blood sugar results in fat storage if the animal overeats relative to its current energy demands. Excess energy intake triggers the liver (in vertebrates) or the fat body (in insects) to metabolize sugars into fats, which

133

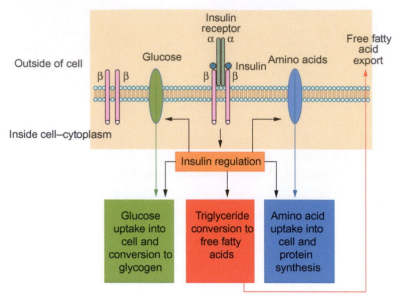

FIGURE 4.17

This figure shows the roles of insulin in regulating energy substrates. Insulin receptors and transport mechanisms for glucose and amino acids are embedded in the cell membrane. When an animal eats, the pancreas produces insulin, which when it reaches cellular insulin receptors stimulates glucose uptake by the cells (green box), conversion of triglycerides into fatty acids (red box) and amino acid uptake combined with protein synthesis (blue box).

can then be stored. Behaviorally, this means that feeding must be triggered by factors that do not completely depend on blood sugar regulation. The reason is that an animal must take in more food than is necessary to maintain blood sugar.

Pain

Pain is a homeostatic mechanism that allows animals to recognize and avoid potential injuries. A painful experience may cause an animal to withdraw from a risky location and may also cause it to avoid that location in the future. Pain is hard to study and quantify because pain is a subjective human characterization of a response to bodily damage. Some scientists and philosophers argue that pain is restricted to humans because the experience of pain suggests self-knowledge or self-awareness of a type that might be restricted to humans. Yet even casual observations of injured mammals and birds suggest a commonality of experience, convincing most animal behaviorists that at least some vertebrates feel pain in the same way as humans do. Going beyond vertebrates, it is also reasonable to suspect some avoidance response to harmful stimuli on the part of most animals.[56,57] Pain is emphasized here because of its link to the behavioral welfare of animals and the use of behavioral techniques for assessing pain in animals.

Most pain results from stimulation of *nociceptors*, specialized pain receptors found throughout the body. There are three types of nociceptors: thermal (sensitive to potentially damaging temperature differences), chemical, and mechanical. (Because the thermal receptors respond to temperature differences, extreme hot and extreme cold yield the same sensation.) When nociceptors are stimulated, the signals travel to the primary somatosensory cortex of the brain. Substance P, a peptide neurotransmitter, is important for the transmission of signals in nociceptive nerve fibers. The pain is represented in the brain, which assigns the pain to the appropriate location so the animal can, if possible, remove the part of its body experiencing pain from the cause of the pain (Figure 4.18).

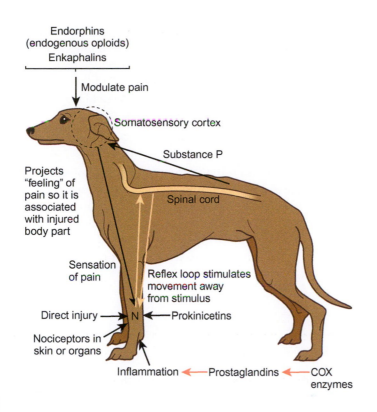

Endorphins
(endogenous oploids)
Enkaphalins

Modulate pain

Somatosensory cortex

Substance P

Projects
"feeling" of
pain so it is
associated
with injured
body part

Spinal cord

Sensation
of pain

Reflex loop stimulates
movement away
from stimulus

Direct injury ——→ N ←—— Prokinicetins

Nociceptors in
skin or organs

Inflammation ←—— Prostaglandins ←—— COX
enzymes

FIGURE 4.18
In this schematic diagram of pain perception pathways, an injury to the dog's ankle (lower left in figure) results in stimulation of nocioceptors, which relay the signal to the spinal cord. A reflex loop (labeled in yellow) causes quick movement of the leg away from the painful stimulus, and at the same time the impulse travels up the spinal cord to the somatosensory cortex of the brain, which projects the "feeling" of pain to the appropriate part of the body. An immune response to infection or continued irritation of the wounded area causes inflammation (pathway labeled in red).

Pain can be subdivided into two categories. The first type is sharp or immediate pain, which is the instant result of an injury such as a bite, sting and stepping on a sharp object. These stimuli directly activate the nociceptive system and can cause a "reflex" avoidance action that is processed in the spinal cord, well before the signal reaches the brain. The second type of pain is due to inflammation of tissues. Inflammation is a complex immunological response, including the dilation of small blood vessels that increases circulation to damaged tissue, but it has the additional adaptation of stimulating the perception of pain.

> **KEY TERM** COX enzymes play a key role in perception of pain by helping to synthesize prostaglandins.

Key elements in the perception of pain from inflammation are the *cyclooxygenase enzymes* (COX enzymes) that catalyze a critical step in the synthesis of prostaglandins. Prostaglandins have numerous functions; they are important in mediating inflammatory pain and sensitizing nerve cells to pain. The link from inflammation to pain via the action of COX enzymes and prostaglandins is present in at least birds and mammals, and perhaps other vertebrates.

Inflammatory pain in animals can be modulated pharmacologically. Steroidal anti-inflammatories, including cortisone, reduce inflammation and, consequently, decrease pain. Over time steroids have potentially serious side effects; they may best be used for short-term reactions, such as insect stings. Nonsteroidal anti-inflammatory drugs, such as aspirin, ibuprofen, and naproxen, inhibit prostaglandin synthesis and release. COX-2 inhibitors (brand names include Rimadyl®, Celebrex®, and Vioxx®) specifically inhibit the COX enzyme that catalyzes the production of the type of prostaglandin most directly involved in inflammation and pain. Controversy over potential cardiovascular side effects has greatly reduced the prescription of COX-2 inhibitors for humans.

In the vertebrate brain, opioid receptors modulate the overall feelings of pain and euphoria. Endorphins, enkephalins, and related compounds are internal messengers in the opioid system. Opioid receptors in the brain respond to opiates (opium-like compounds) and are responsible for feelings of pleasure. Endorphins are the body's internal opiates; production of endorphins results in pleasurable sensations. Drugs, such as morphine, serve as analgesics by binding with opioid receptors. Opioid-like receptors have also been discovered in invertebrates, but because opioid receptors and somatostatin (a growth hormone) receptors are similar, it is difficult to assign a function to these receptors.

OF SPECIAL INTEREST: DO ANIMALS FEEL PAIN?

Here are the lines of evidence that scientists generally consider when testing hypotheses for the perception of pain in nonhuman animals:

1. Possession of Neurotransmitters, Biochemical Pathways, or Neural Systems Homologous to Nociceptive Pathways in Mammals

Arguments for pain perception in animals can be built on genome searches for genes that code for products, such as substance P or prokineticins, or for receptors that may be involved in pain modulation, such as opioid receptors. A flaw in this argument is that DNA sequences may be deeply rooted in evolution, but the function of the products from similar DNA can differ substantially across organisms. In other words, the presence of matching DNA sequences does not guarantee matching function. Similarly, the presence of matching neurotransmitters does not prove identical signaling function. Prokinecitins and opioid receptors, in particular, occur in many diverse animals, and their presence has been used to argue for perception of pain as a general phenomenon among animals.

The anatomical version of this type of approach is to investigate the functional anatomy of the nervous system. In salmon, receptors on the skin map to the brain in much the same way that nociceptors do in mammals. Low-intensity electric shocks result in smaller brain responses than high-intensity shocks, suggesting that fish can perceive painful stimuli, based on structures that are similar in function to the mammalian somatosensory system.

2. Sensitivity/Response to the Painkilling Effects of Known Analgesics

One standard test for response to pain is the application of acetic acid, which is relatively mild, to an animal's skin. The animal typically responds by wiping the spot with a leg; water, applied as a control, does not induce wiping. In frogs treated with analgesics, tolerance of acid was increased. This result shows that amphibians respond behaviorally as if the acid causes pain. The fact that analgesics reduce the pain response in both frogs and mammals supports the hypothesis that the biochemical basis for pain response is similar in the two groups.

3. Behavioral Avoidance/Response to Situations Thought by the Investigator to Be Painful

Recent studies show that crustacea avoid circumstances under which they have previously experienced conditions that can be defined as painful. Application of mildly noxious chemical stimuli to the surface of crustacea cause grooming movements that suggest nociception.

In sum, birds and mice (and by extension, presumably all mammals) meet all three of these criteria for pain perception; the most commonly used nonhuman model system for the study of pain is mice. Other vertebrates, including fish and amphibia, also meet the criteria. Among the invertebrates, roundworms meet the first criterion and crustacea meet the third.

> **BRINGING ANIMAL BEHAVIOR HOME: ARTHRITIS IN DOGS**
>
> A commonly observed type of inflammation that causes pain in animals is arthritis. Anyone who has cared for an arthritic dog will have little doubt that the dog is in pain. Dogs with mild arthritis are often prescribed aspirin or naproxen under a veterinarian's supervision. A COX-2 inhibitor is commonly prescribed for dogs with more serious arthritis or with hip dysplasia (a painful congenital misalignment of the hip joint). Some reports suggest that COX-2 inhibitors may have unfortunate side effects for some dogs, but they remain in fairly widespread veterinary use.

The evidence for pain as a physiological response to allow an animal to avoid injury is widespread and convincing. Does it "feel" the same to nonhuman animals as it does to humans? We will probably never know, just as, in truth, we do not know if "sad" feels exactly the same to every human being. Nonetheless, attempting to answer this question, and to develop objective methods of assessing pain in animals, has great significance to the societal debate about ethics and the treatment of companion and farm animals. If well-intentioned avoidance of anthropomorphism has overstepped reasonable bounds in science, this is the area in which it has done so most frequently.

Fear

Fear, like pain, is a homeostatic safeguard that can keep an animal from lethal risk. But, also like pain, fear is a subjective characterization drawn from human emotions. What is the evidence for fear in animals? Fearful behavior in domestic animals is easily recognized; fearful dogs have a typical tail down, ears back, mouth and face neutral posture when moderately afraid. This can progress to a more defensive posture (teeth bared) or to a submissive posture (rolled on their back, stomach exposed) depending on the type of threat and the dog's likelihood of winning a direct encounter with the threat. Many other mammals, such as cats and horses, give signals that can be interpreted as reflecting fear.

Fear has a very real adaptive value in an animal's life.[58] Fear is what keeps animals from harm's way. Like much of what has been discussed in this chapter, it is easier to spot fearful-seeming behaviors in birds and mammals than in animals that share fewer communicatory channels with humans. In order to remove subjectivity from assessments of fearfulness, animal behaviorists sometimes refer to a certain kind of fear called *neophobia*—the avoidance of the new and unfamiliar in the environment. Neophobia is discussed in more detail in Chapter 15 (on conservation and behavior); as members of species such as coyotes, raccoons, and bears have overcome neophobia, they tolerate closer contact with humans. This close human–wildlife interaction in suburban environments presents difficult conservation challenges.

> **KEY TERM** Fear is more easily recognized by behavior—cowering, defensive threats, and flight—than it can be defined. We define fear as a set of adaptive behaviors and their corresponding internal state (emotion) that help an animal to avoid danger.

Fear can, of course, be a response to abuse (either at the hands of a human or by a dominant animal in a social group), and veterinarians and animal welfare agencies often use measures of fearfulness when assessing prior treatment of animals. Fearful animals can sometimes be dangerous to work with because the boundary between fearfulness and defensive aggression can be difficult for a human observer to predict. Some animal workers believe that fear is the cause of most aggression problems in domestic animals. For many pets, patience and a positive social environment can overcome fearfulness that results from an abusive history.

4.6 TIME BUDGETS AND TRADE-OFFS: BALANCING DEMANDS IN HOW ANIMALS BUDGET THEIR TIME

An animal's time budget is the nexus among all of these regulatory mechanisms. How much time is spent foraging, versus courting, versus sleeping? All are necessary activities for survival and reproduction, but knowing how appetite, sex drive, parental behavior, and foraging are each independently regulated does not help much with understanding how relative priorities among the behavioral possibilities are determined. This section focuses on the practical aspects of time budgets: How is a time budget constructed? How does it reflect behavioral trade-offs for animals? How can time budgets be used to ask interesting questions about animal behavior?

The time budget for an animal is a simple documentation of the proportion of time spent in each available activity by that animal. Time budgets are deceptively simple and low-tech; construction of a time budget gives investigators a powerful tool for testing hypotheses. For example, does disturbance affect an animal's time budget? How much time does it spend in each possible behavioral mode? How frequently does it switch from sleep to foraging? Time budgets provide a critical tool for analyzing an animal's internal state and for comparing animals across developmental stages, among seasons, or between sexes.

A good time budget relies first on the construction of an accurate ethogram of the animal's behavior. An ethogram is usually based on many hours of observation, including analysis of video recordings of a species' behavior. The finished ethogram will have captured the full range of behavioral possibilities for a species and provided a list, with definitions, of the behaviors (Figure 4.19). Photos or drawings are very helpful accompaniments. When coupled with records of the frequency of each behavior, an ethogram becomes a time budget.

Other interesting approaches to time budgets deal with developmental and seasonal changes in behavior, and comparisons of males and females during mating and nesting. Clearly, there will be differences between juvenile and adult organisms in time budgets. How does the frequency of play behavior change with maturation? Are there specific hormonal correlates

138

Playful observation — Play balancing

Play climbing
on partner

FIGURE 4.19

Behaviors that form the basis for a small portion of a sample ethogram. The ethogram was designed to compare the behavior of five species of monkey, including the Hanuman langur. Playful observation is defined as follows: "A playful monkey performs play intention movements … while watching its play partner." Play balancing involves jumping on top of another monkey and attempting to balance. Play climbing occurs when one monkey climbs on another. These simple definitions allow observers to easily differentiate among behaviors. The descriptions must be precise enough so that another investigator can replicate the analysis. *Adapted from Petru, M., Spinka, M., Charvatova, V., Lhota, S., 2009. Revisiting play elements and self-handicapping in play: a comparative ethogram of five old world monkey species. J. Comp. Psychol. 123, 250–263.*[59]

with changes in play? A time budget is a quantitative tool for investigating these questions. In many eusocial insects—bees, ants, wasps, and termites that live in colonies—the time budgets of workers change as they age; age-related specialization allows workers to be efficient in their use of time. Typical seasonal changes in time budgets in birds and mammals reflect shifts associated with mating, nesting, and preparing to migrate or overwinter. Finally, a question such as "Do males contribute equally in nest care in birds?" is easily answered when male and female time budgets are compared. Time budgets and the need to allocate time for tasks may even influence the geographic distribution of species.[60]

Occasionally, time budget analyses can seem a little obvious. For example, mallard ducklings (Figure 4.20) shift their time budgets depending on how resource rich or poor their foraging area is.[61] Ducklings in resource-poor lakes spend more time moving, presumably searching for food. It is not surprising that hungry animals look for food but consider this question from a different point of view: by observing the behavior of the ducklings, ecologists can infer much about the environmental conditions under which they live. The time budget of ducklings whose time budget is biased to feeding tells us that their environment is resource poor relative to populations that invest proportionally less time in feeding. Likewise, seals adjust their time budgets during the breeding season (Figure 4.21).[62]

Recall that ethograms represent the behavior of a single animal, or what that single animal does when it is interacting with other animals. Two related approaches focus on transitions, or sequences of behavior. When the objective is to study single animals, the probability of shifting from one behavior to other behavioral possibilities is recorded. An interaction diagram captures the sequence of events when two or more animals communicate (Figure 4.22).

Behavioral interactions can also be quantified in a *transition matrix*. In the matrix, the ethogram for the first animal in the interaction is listed in the descending column and the reaction of the second animal runs along the top of the matrix. The matrix shows how often each behavior of the first animal stimulates each possible response of the second animal. Through the use of transition matrices, comparisons can be made, for example, between sib and nonsib playmates, or between male–male encounters and male–female encounters. A more statistically formal way of analyzing the sequences captured in transition matrices uses *Markov chains*, which allow investigators to look at each behavioral transition in a long sequence.

139

> **KEY TERM** A transition matrix quantifies the actions of one animal in an interaction and the resulting responses of the second animal.

FIGURE 4.20
Mallard ducklings adjust their activity budgets to reflect, among other things, foraging success. *Photo: Michael Breed.*

FIGURE 4.21
Habitat characteristics influence the amount of time that gray seals spend in the water during breeding season. *Photo: Michael Breed.*

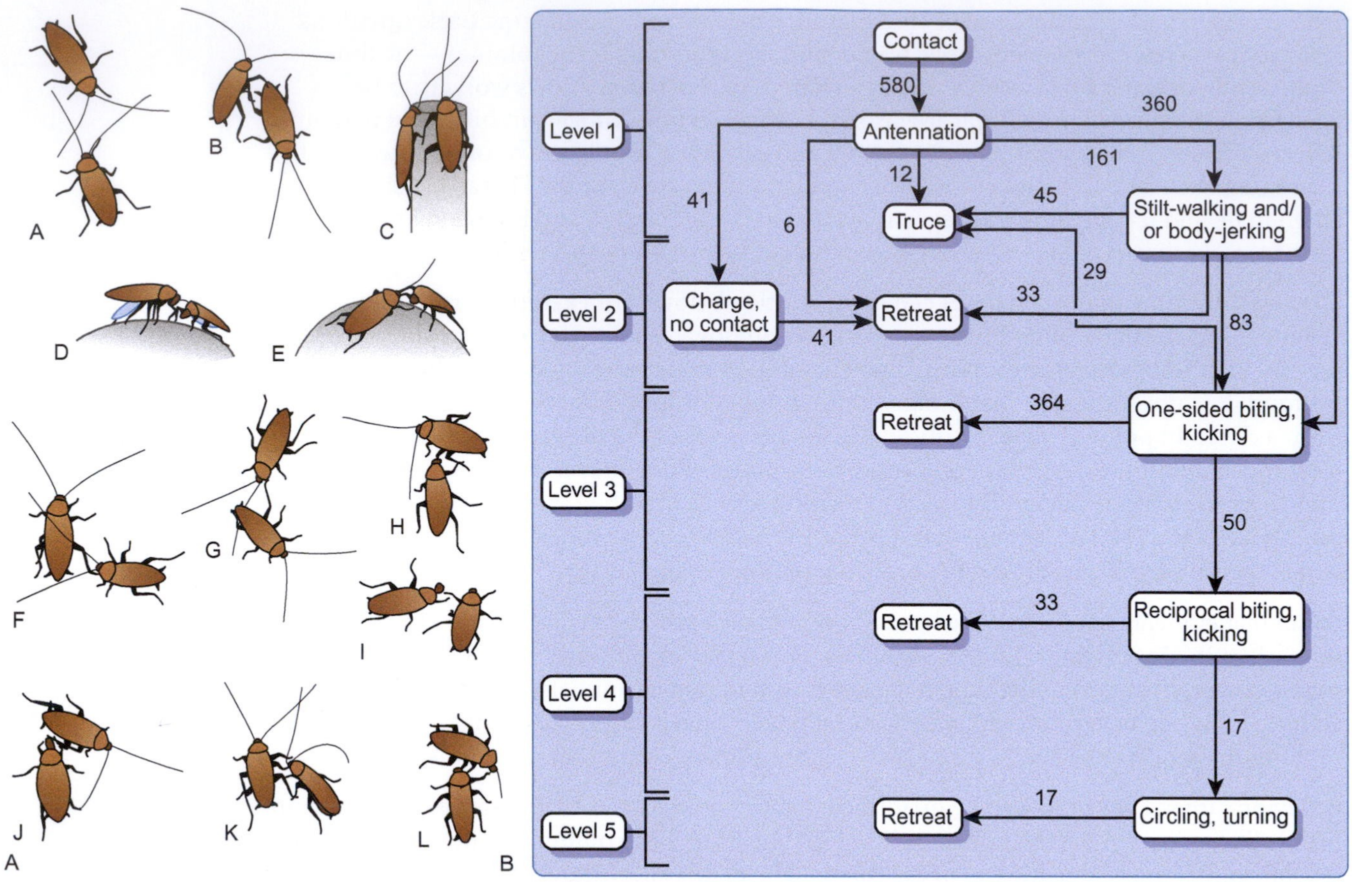

FIGURE 4.22

(A) These diagrams illustrate a variety of interactions that occur during aggressive interactions between cockroaches. The letters A through L indicate increasing levels of aggression and correspond to boxes in diagram B. A = Contact, B = Antennation, C = Truce, D = Charge, no contact, E = Stilt-walking and/or body jerking, F = One=sided biting, kicking, G = Retreat, H = Reciprocal biting, I = kicking, J, K, L = Circling, turning. Note that these interactions can be placed into a kind of flow chart (B) that shows the frequency with which one such interaction leads to another. *Adapted from Bell, W.J., Sams, G.R., 1973. Aggressiveness in the cockroach Periplaneta americana (Orthoptera, Blattidae). Behav. Biol. 9, 581–593.*[63]

A special type of matrix captures the result of dominant/subordinate interactions between animals (Figure 4.22). Animals A, B, C, and D are represented in the left column and also across the top of the matrix. The number of times each animal dominated another animal is then recorded. In a perfectly linear dominance hierarchy, A always dominates all of the others, B dominates C and D, and C dominates D. *Linear* means that the animals are lined up in a hierarchical order, with strict rules of priority access to a resource. (Different resources may produce different hierarchies; thus, an animal that always competes successfully for water may not be as successful for nesting sites.) In the real world of animal behavior, dominance interactions are not quite so predictable, and nonlinear relationships (A dominates B, B dominates C, but C dominates A) can occur. A somewhat different approach to this same type of question is to construct a sociogram, which is a diagram of how frequently animals are located near one another. Sociograms have recently come into the forefront in analyses of social networks in animals (see Chapter 13 for more information on social networks).

OF SPECIAL INTEREST: MAINTAINING SKIN, FUR, AND FEATHERS

Many animal species devote time and effort to keeping their body surfaces free of parasites, waterproof, or as good insulation. Simple behavior, such as licking or scratching, can be an efficient means of removing parasites, dead skin, and old fur or feathers (Figure 4.23A and B). Bathing is also

FIGURE 4.23
Maintaining skin, fur (A, B), and feathers (C, D). *Photos: (A and B) Ben Hart; (C and D) Michael Breed.*

a common means of skin, fur, and feather maintenance (see Figure 4.23C). Animals with disease conditions or which are suffering from old age often lose their ability to maintain themselves. This results in poor-appearing pelage or feathers, may increase external parasites, and may affect an animal's ability to court mates. Some animals, including many aquatic birds, groom oils to their feathers to improve their waterproofing and allow them to float on the water surface without absorbing water into their feathers (see Figure 4.23D). Oils on mammalian hair, such as lanolin on sheep, may have similar waterproofing functions.

Like humans, animals have 24 h in a day. Time budgeting by animals inevitably involves behavioral trade-offs. An animal that spends time feeding is doing that at the expense of another activity, such as courting potential mates. From an evolutionary perspective, the single critical result is having offspring and ensuring their survival—establishing genetic fitness. Natural selection directs all of an animal's activities to this ultimate goal. Self-maintenance and longevity are usually critical to reproduction, and keeping this point in mind helps us to understand the decisions that animals make as they choose among possible activities.

SUMMARY

This chapter covers an intellectual journey from asking how animals make decisions about their behavior at any one moment to important regulators of behavior such as hunger, sleep, fear, and pain. It finished with some practical tools for field and lab analyses of how animals spend their time.

Homeostasis is an umbrella term for the behavioral and physiological processes that animals use to maintain appropriate internal and external environments. Motivation and drive theory help us to understand homeostatic decisions by animals; although drive theory has been replaced by more sophisticated ways of looking at the physiology behind behavioral decisions, it remains a useful concept, especially in the area of animal welfare. Behavior plays a crucial role in the maintenance of animal homeostasis, and homeostatic needs often motivate behavioral choice.

Concepts of mood, emotion, affective state, behavioral syndrome, and personality describe variation in behavioral responses of individual animals over time and between animals within populations. Mood, emotion, and affective state are immediate responses to behavioral conditions and environmental experience, but they interact strongly with behavioral syndrome and personality, which are rooted in an animal's genetics. Until recently, animal behaviorists shied away from these concepts because of the dangers of anthropomorphism, but the importance of these mechanisms in behavioral regulation is now generally accepted.

Much of the discussion in this chapter, such as that about SDBs and measures of pain, gives students of animal behavior objective, quantifiable tools for assessing animal welfare. Displacement behavior and redirected behavior, in the form of SDBs, are tools from drive theory that provide powerful insights for assessing animal welfare. Time budget analyses allow quantification and interpretation of behavioral decisions that maintain homeostasis. Time budgets are among the simplest, yet most powerful, tools available to biologists studying animal behavior in field settings.

STUDY QUESTIONS

1. What is the value of fear in an animal's life? What are the costs of fear?
2. At the zoo, you observe a primate pacing repetitively, plucking its hair, and rubbing against the cage bars until it wounds itself. What is your interpretation of this behavior and why? What would you suggest to the zookeeper in terms of changes in the animal's habitat?
3. Spend an hour watching an animal—any animal you can find will do as long as it remains in view for a while—and construct a time budget for that animal. When the animal switches between activities, is the motivation for the shift always apparent? Does the animal's allocation of time seem well adapted to its needs?
4. What is pain? How can pain be measured in a nonhuman animal?
5. What is a *zeitgeber*? What happens to an animal's circadian clock when the animal is kept in continuous light or dark?
6. What is satiety and how is it regulated?

Further Reading

Bouchard, T.J., 1994. Genes, environment, and personality. Science 264 (5166), 1700–1701.

Braithwaite, V., 2010. Do Fish Feel Pain? Oxford University Press, Oxford.

Giebultowicz, J., 2010. Circadian window of opportunity: what have we learned from insects? J. Exp. Biol. 213, 185–186.

Harlow, H.F., Dodsworth, R.O., Harlow, M.K., 1965. Total social isolation in monkeys. Proc. Natl. Acad. Sci. USA 54, 90–97.

Harlow, H.F., Suomi, S.J., 1971. Social recovery by isolation-reared monkeys. Proc. Natl. Acad. Sci. USA 68, 1534–1538.

Hinde, R.A., 1959. Unitary drives. Anim. Behav. 7, 130–141.

Hughes, B.O., Duncan, I.J.H., 1988. The notion of ethological need, models of motivation and animal-welfare. Anim. Behav. 36, 1696–1707.

Huntingford, F.A., 1976. An investigation of the territorial behaviour of the three-spined stickleback (Gasterosteus aculeatus) using principal components analysis. Anim. Behav. 24, 822–834.

McFarland, D.J., Sibly, R.M., 1972. 'Unitary drives' revisited. Anim. Behav. 20, 548–563.
Sneddon, L.U., Elwood, R.W., Adamo, S.A., Leach, M.C., 2014. Defining and assessing animal pain. Anim. Behav. 97, 201–212.
Toates, F.M., 1986. Motivational Systems. Cambridge University Press, Cambridge.

Notes

1. Hinde, R.A., 1959. Unitary drives. Anim. Behav. 7, 130–141.
2. Blumstein, D.T., 2002. Moving to suburbia: ontogenetic and evolutionary consequences of life on predator-free islands. J. Biogeogr. 29, 685–692.
3. Berger, S., Wikelski, M., Romero, L.M., Kalko, E.K., Roedl, T., 2007. Behavioral and physiological adjustments to new predators in an endemic island species, the Galapagos marine iguana. Horm. Behav. 52, 653–663.
4. McFarland, D.J., Sibly, R.M., 1972. "Unitary drives" revisited. Anim. Behav. 20, 548–563.
5. Toates, F.M., 1986. Motivational Systems. Cambridge University Press, Cambridge.
6. Hughes, B.O., Duncan, I.J.H., 1988. The notion of ethological need, models of motivation and animal-welfare. Anim. Behav. 36, 1696–1707.
7. Troisi, A., Belsanti, S., Bucci, A.R., Mosco, C., Sinti, F., Verucci, M., 2000. Affect regulation in alexithymia—an ethological study of displacement behavior during psychiatric interviews. J. Nerv. Ment. Dis. 188 (1), 13–18.
8. Manson, J.H., Perry, S., 2000. Correlates of self-directed behaviour in wild white faced capuchins. Ethology 106 (4), 301–317.
9. Baker, K.C., Aureli, F., 1997. Behavioural indicators of anxiety: an empirical test in chimpanzees. Behaviour 134, 1031–1050.
10. Watts, D.P., 1995. Post-conflict social events in wild mountain gorillas 2. Redirection, side direction, and consolation. Ethology 100, 158–174.
11. McGinnis, M.Y., 2004. Anabolic androgenic steroids and aggression—studies using animal models. Youth violence: scientific approaches to prevention. Ann. N. Y. Acad. Sci. 1036, 399–415.
12. Chelliah, K., Sukumar, R., 2013. The role of tusks, musth and body size in male-male competition among Asian elephants, Elephas maximus. Anim. Behav. 86, 1207–1214.
13. Castles, D.L., Whiten, A., Aureli, F., 1999. Social anxiety, relationships and self-directed behaviour among wild female olive baboons. Anim. Behav. 58, 1207–1215.
14. Maestripieri, D., Schino, G., Aureli, F., Troisi, A., 1992. A modest proposal—displacement activities as an indicator of emotions in primates. Anim. Behav. 44 (5), 967–979.
15. Leavens, D.A., Aureli, F., Hopkins, W.D., Hyatt, C.W., 2001. Effects of cognitive challenge on self-directed behaviors by chimpanzees (Pan troglodytes). Am. J. Primatol. 55 (1), 1–14.
16. Harlow, H.F., Dodsworth, R.O., Harlow, M.K., 1965. Total social isolation in monkeys. Proc. Natl. Acad. Sci. USA 54, 90–97.
17. Harlow, H.F., Suomi, S.J., 1971. Social recovery by isolation-reared monkeys. Proc. Natl. Acad. Sci. USA 68, 1534–1538.
18. Mason, G.J., Veasey, J.S., 2010. How should the psychological well-being of zoo elephants be objectively investigated? Zoo Biol. 29, 237–255.
19. van Zeeland, Y.R.A., van der Aa, M.M.J.A., Vinke, C.M., Lumeij, J.T., Schoemaker, N.J., 2013. Behavioural testing to determine differences between coping styles in Grey parrots (Psittacus erithacus erithacus) with and without feather damaging behavior. Appl. Anim. Behav. Sci. 148, 218–231.
20. Adamo, S.A., 2012. The effects of the stress response on immune function in invertebrates: an evolutionary perspective on an ancient connection. Horm. Behav. 62, 324–330.
21. Sih, A., Bell, A., Johnson, C.J., 2004. Behavioral syndromes: an ecological and evolutionary overview. Trends Ecol. Evol. 19, 372–378.
22. Liang, Z.S., Nguyen, T., Mattila, H.R., Rodriguez-Zas, S.L., Seeley, T.D., Robinson, G.E., 2012. Molecular determinants of scouting behavior in honey bees. Science 335, 1225–1228.
23. Partecke, J., Schwabl, H., 2008. Organizational effects of maternal testosterone on reproductive behavior of adult house sparrows. Dev. Neurobiol. 68, 1538–1548. http://dx.doi.org/10.1002/dneu.20676.
24. Wells, D.L., Hepper, P.G., 2006. Prenatal olfactory learning in the domestic dog. Anim. Behav. 72, 681–686. http://dx.doi.org/10.1016/j.anbehav.2005.12.008.
25. Hudson, R., Bautista, A., Reyes-Meza, V., Montor, J.M., Rodel, H.G., 2011. The effect of siblings on early development: a potential contributor to personality differences in mammals. Dev. Psychobiol. 53, 564–574. http://dx.doi.org/10.1002/dev.20535.
26. Adkins-Regan, E., Banerjee, S.B., Correa, S.M., Schweitzer, C., 2013. Maternal effects in quail and zebra finches: behavior and hormones. Gen. Comp. Endocrinol. 190, 34–41. http://dx.doi.org/10.1016/j.ygcen.2013.03.002.
27. Jandt, J.M., Bengston, S., Pinter-Wollman, N., Pruitt, J.N., Raine, N.E., Dornhaus, A., et al., 2014. Behavioural syndromes and social insects: personality at multiple levels. Biol. Rev. 89, 48–67.
28. Carlson, B.E., Langkilde, T., 2013. Personality traits are expressed in bullfrog tadpoles during open-field trials. J. Herpetol. 47, 378–383.
29. Sih, A., Cote, J., Evans, M., Fogarty, S., Pruitt, J., 2012. Ecological implications of behavioural syndromes. Ecol. Lett. 15 (3), 278–289.
30. Sih, A., Bell, A.M., 2008. Insights from behavioral syndromes for behavioral ecology. Adv. Study Behav. 38, 277–281.

31. Johnson, J.C., Sih, A., 2007. Fear, food, sex and parental care: a syndrome of boldness in the fishing spider, Dolomedes triton. Anim. Behav. 74, 1131–1138.
32. Marzouki, Y., Gullstrand, J., Goujon, A., Fagot, J., 2014. Baboons' response speed is biased by their moods. PLoS One 9, e102562.
33. Mendl, M., Paul, E., Chittka, L., 2011. Animal behaviour: emotion in invertebrates? Curr. Biol. 21, R463–R465.
34. Bouchard, T.J., 1994. Genes, environment, and personality. Science 264 (5166), 1700–1701.
35. Weiss, A., King, J.E., Figueredo, A.J., 2000. The heritability of personality factors in chimpanzees (Pan troglodytes). Behav. Genet. 30 (3), 213–221.
36. Svartberg, K., Forkman, B., 2002. Personality traits in the domestic dog (Canis familiaris). Appl. Anim. Behav. Sci. 79 (2), 133–155.
37. Gubert, P., Aguiar, G.C., Mourao, T., Bridi, J.C., Barros, A.G., Soares, F.A., et al., 2013. Behavioral and metabolic effects of the atypical antipsychotic ziprasidone on the nematode Caenorhabditis elegans. PLoS One 8, e74780.
38. Komiyama, T., Iwama, H., Osada, N., Nakamura, Y., Kobayashi, H., Tateno, Y., et al., 2014. Dopamine receptor genes and evolutionary differentiation in the domestication of fighting cocks and long-crowing chickens. PLoS One 9, e101778.
39. Bubak, A.N., Renner, K.J., Swallow, J.G., 2014. Heightened serotonin influences contest outcome and enhances expression of high-intensity aggressive behaviors. Behav. Brain Res. 259, 137–142.
40. Fossat, P., Bacqué-Cazenave, J., De Deurwaerdere, P., Delbecque, J.-P., Cattaert, D., 2014. Anxiety-like behavior in crayfish is controlled by serotonin. Science 6189, 1293–1297.
41. Nagasawaa, M., Kikusuia, T., Ohtaa, M., 2009. Dog's gaze at its owner increases owner's urinary oxytocin during social interaction. Horm. Behav. 55, 434–441.
42. Marcus, D.A., 2013. The science behind animal-assisted therapy. Curr. Pain Headache Rep. 17, 322.
43. Beetz, A., Uvnas-Moberg, K., Julius, H., Kotrschal, K., 2012. Psychosocial and psychophysiological effects of human–animal interactions: the possible role of oxytocin. Front. Psychol. 3, 234. http://dx.doi.org/10.3389/fpsyg.2012.00234.
44. Handlin, L., Hydbring-Sandberg, E., Nilsson, A., Ejdeback, M., Jansson, A., Uvnas-Moberg, K., 2011. Short-term interaction between dogs and their owners: effects on oxytocin, cortisol, insulin and heart rate—an exploratory study. Anthrozoos 24, 301–315. http://dx.doi.org/10.2752/175303711X13045914865385.
45. Bell-Pederson et al., Nature Reviews Drug Discovery, http://dx.doi.org/10.1038/nrd1633.
46. Konopka, R.J., Benzer, S., 1971. Clock mutants of Drosophila melanogaster. Proc. Natl. Acad. Sci. USA 68, 2112–2116.
47. Giebultowicz, J., 2010. Circadian window of opportunity: what have we learned from insects? J. Exp. Biol. 213, 185–186.
48. Robinson, I., Reddy, A.B., 2014. Molecular mechanisms of the circadian clockwork in mammals. FEBS Lett. 588, 2477–2483.
49. Maerz, J.C., Panebianco, N.L., Madison, D.M., 2001. Effects of predator chemical cues and behavioral biorhythms on foraging, activity of terrestrial salamanders. J. Chem. Ecol. 27, 1333–1344.
50. Meschini, E.A., Gagliardo, F.P., 2008. Lunar orientation in sandhoppers is affected by shifting both the moon phase and the daily clock. Anim. Behav. 76, 25–35.
51. Sigurgeirsson, B., Porsteinsson, H., Sigmundsdottir, S., Lieder, R., Sveinsdottir, H.S., Sigurjonsson, O.E., et al., 2013. Sleep-wake dynamics under extended light and extended dark conditions in adult zebrafish. Behav. Brain Res. 256, 377–390.
52. Zhdanova, I.V., 2011. Sleep and its regulation in zebrafish. Rev. Neurosci. 22, 27–36.
53. Yokogawa, T., 2007. Characterization of sleep in zebrafish and insomnia in hypocretin receptor mutants. PLoS Biol. 5 (10), 2379–2397.
54. Chabot, C.C., Menaker, M., 2004. Effects of physiological cycles of infused melatonin on circadian rhythmicity in pigeons. J. Comp. Physiol. A 170, 615–622.
55. Cuthill, I.C., Maddocks, S.A., Weall, C.V., Jones, E.K.M., 2000. Body mass regulation in response to changes in feeding predictability and overnight energy expenditure. Behav. Ecol. 11 (2), 189–195.
56. Magee, B., Elwood, R.W., 2013. Shock avoidance by discrimination learning in the shore crab (Carcinus maenas) is consistent with a key criterion for pain. J. Exp. Biol. 216, 353–358.
57. Elwood, R.W., 2012. Evidence for pain in decapod crustaceans. Anim. Welf. 21, 23–27.
58. Laundre, J.W., Hernandez, L., Medina, P.L., Campanella, A., Lopez-Portillo, J., Gonzalez-Romero, A., et al., 2014. The landscape of fear: the missing link to understand top-down and bottom-up controls of prey abundance? Ecology 95, 1141–1152.
59. Petru, M., Spinka, M., Charvatova, V., Lhota, S., 2009. Revisiting play elements and self-handicapping in play: a comparative ethogram of five old world monkey species. J. Comp. Psychol. 123, 250–263.
60. Korstjens, A.H., Lehmann, J., Dunbar, R.I.M., 2010. Resting time as an ecological constraint on primate biogeography. Anim. Behav. 79, 361–374.
61. Nummi, P., Sjoberg, K., Poysa, H., Elmberg, J., 2000. Individual foraging behaviour indicates resource limitation: an experiment with mallard ducklings. Can. J. Zool. 78 (11), 1891–1895.
62. Caudron, A.K., Joiris, C.R., Ruwet, J.-C., 2001. Comparative activity budget among grey seal (Halichoerus grypus) breeding colonies—the importance of marginal populations. Mammalia 65, 373–382.
63. Bell, W.J., Sams, G.R., 1973. Aggressiveness in the cockroach Periplaneta americana (Orthoptera, Blattidae). Behav. Biol. 9, 581–593.

Learning

LEARNING OBJECTIVES

Studying this chapter should provide you with the knowledge to:

- Discover how learning allows animals to cope with unpredictable elements in their environment.

- Know that cellular changes, including protein synthesis and remodeling of synapses, underlie memory formation.

- Integrate the simple mechanisms of habituation and sensitization into a model that helps animals sort important from unimportant information.

- Observe how the concept of conditioning provides important models for understanding how animals respond to repeated experiences.

- Predict that in social learning, animals observe one another and gain critical information from the experiences of other animals.

- Be able to apply the concept that play is important in learning and development for many animals, and understand how play provides experiences that enhance the development of both physical and social skills.

Animal Behavior. DOI: http://dx.doi.org/10.1016/B978-0-12-801532-2.00005-2

5.1 INTRODUCTION

Learning is a change in behavior as a result of experience. Genetic information is molded across generations by evolution; learned information is molded during each animal's lifetime. Learning is particularly useful for animals with relatively long lives, because experience takes time. Recalling a previous experience and using that experience to efficiently solve a current problem provide animals with key tools for survival and reproductive success (Figure 5.1).

> **KEY TERM** Learning is the modification of behavior due to stored information from previous experience.

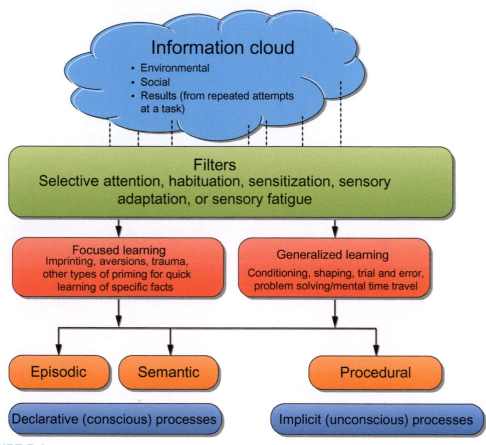

FIGURE 5.1

An animal's environment is filled with information, some of which needs to be learned for future use. The first task is to filter the information so that useful facts are separated from unessential clutter. Some filters, such as habituation and sensitization, are learning processes. After filtering, certain types of information are rapidly incorporated into an animal's memory through focused learning. Information learned in this pathway has critical immediate survival value, such as aversion to poisonous foods. This is particularly true of information associated with immediate threats to an animal's welfare. Generalized learning conditions future behavior, shapes movements, or yields information that stored for problem solving. Depending on the type of information acquired, it can be stored as part of conscious processes (episodic and semantic memory) or unconscious processes (procedural memory).

CASE STUDY: LEARNING IN ANTS

We may think that an animal such as an ant has little to teach us about learning, but we would be wrong. Some ants engage in a behavior called "tandem running." This behavior is briefly discussed in Chapter 7 ("communication") and, like many behaviors, could easily be covered as part of several other topics as well. We address it here because of the surprising way that the rock ant *Temnothorax albipennis* integrates individual learning with what it experiences as part of a tandem team when its colony seeks better nesting conditions.

(Continued)

CASE STUDY (CONTINUED)

The search for better nesting conditions is a fairly common occurrence in *T. albipennis*, which lives in warm crevices among rocks, stumps, branches, and the like. For the type of animal that has become a synonym for industrious domesticity, this ant is surprising, preferring easily and quickly excavated nest sites with weak structure to those that are stronger, but take more time to excavate. Because of this, *T. albipennis* moves readily; it lives in small colonies, and because it has few scouts, it is probably much less likely to use pheromone trails to communicate location (see Chapter 7) than other species of ants with larger colonies. Tandem running seems to take the place of those pheromone trails and can be investigated in the context of teaching and learning.

In tandem running, two ants move to a destination. The first ant knows the destination and leads the way; the second ant follows, staying in antennal contact with the first. The progress is often slow, and it is thought that this may assist the second ant in learning landmarks and the like, so that it can find the goal on its own later (Figure 5.2).

Elizabeth Franklin and Nigel Franks set out to understand how tandem running contributes to what ants know about the location of the new nest. For instance, do ants learn landmarks along the way, or do they use other mechanisms, such as path integration, in which an animal comprehends the locations and distances of the beginning and end of a journey and bases the route on that information? (These aspects of navigation and more are covered in Chapter 8.)

In order to ask this question, Franklin and Franks collected 15 colonies of *T. albipennis* and put them into laboratory nests that were considered "poor"—that is, the nests were well lit and the ants did not have much space. From the previous work, they knew that such ants would be motivated to seek a new nest. They placed the colony (they used each colony only once) in a small arena that was connected to a larger arena with a better (darker and larger) nest by two tunnels. They then set out to do some serious ant-watching.

The ants were allowed to explore the area until one ant found the new nest. At that time, other ants were removed from the large arena, and the remaining ant was watched and her movements recorded until she had led a tandem run to the new nest. At the end of that run, the leader of the tandem was removed, and the follower was allowed to move around the large arena until she, too, recruited a partner and completed a tandem run to the new nest. Franklin and Franks paid special attention to two aspects of this ant behavior: (1) the ability of a follower ant to lead a tandem run and (2) the extent to which an ant explored the arenas prior to leading the next tandem run.

The scientists found that follower ants that explored the arena prior to leading their own tandem run made improvements to the run that they led. If they explored the area on their own before leading a tandem run, that run took less time and followed a more direct route than their previous tandem run had taken before their exploration. In other words, not only did they learn the desired goal from their first tandem run, but by exploring, they then learned enough about the environment to improve on the route to that goal. In fact, exploration resulted in improved travel to the new nest whether the location of the nest was first discovered by an ant's own exploration or by the ant having been led there in tandem. Even so, there is a point of diminishing returns; more than one bout of exploration did not yield additional improvements in a route.

Franklin and Franks showed that individual learning— the learning that an individual ant accomplishes through exploration—contributes significantly to increasing the efficiency of a tandem run. The more efficient the tandem run is, the less likely that the partners will be separated and the more likely that an increasing number of ants will know the location of the new nest. When a critical number of ants (called a "quorum") know about the new nest, they can then shift to yet a third mode of moving the colony—"social carrying." Social carrying is exactly what it sounds like—the ants physically carry other ants to the new site. This is a remarkably efficient way to move ants—three times as many ants can be socially carried in the same amount of time that tandem running requires.

Given this threefold difference, you may reasonably wonder why ants bother with any other form of moving. Why not simply engage in social carrying? Again, the answer rests in how the ants learn. Ants that are carried to a new location stay there; they do not attempt to leave and recruit other ants. It is impossible to know exactly what an ant knows, but given the fact that both exploration and tandem running involve ants that move around under their own volition, probably noticing landmarks and other geographic cues, and given the fact that carried ants are not moving themselves through the environment, nor do they have a particularly normal view of that landscape (they are carried upside down, facing backward!), it may well be that the carried ant does not know where it is in relation to the old nest nor much of anything else.

Indeed, there is evidence that these ants not only learn but actually engage in *teaching* their tandem partner. The tandem leader's behavior fits the criteria for teaching in that it is costly for the leader, who could move much faster without the tandem

FIGURE 5.2
These two ants are engaged in tandem running. Both contact (tactile communication) and pheromones are used to lead nestmates to food sources. Tandem running can result in long chains of moving ants.

147

CASE STUDY (CONTINUED)

partner, and who stops periodically although the partner seems to explore the route. In fact, the leader will slow down or stop if it loses touch with the partner. Franks and Richardson have argued that in ant species with small colonies, in which information loss is a serious risk, individualized teaching may be more important than in species with larger colonies, where pheromone trails are common.

For these ants, learning about their environment and spatial relationships within that environment is at the heart of a successful colony move, with individual learning through exploration making a measurable difference in that success. In this regard, they join the food-caching birds and squirrels discussed later in this chapter, animals whose spatial learning and memory are near legendary.[1–4]

The optimal environments for behavior based on learned information and that based on genetic information are different. Chapter 3 focused on the behavioral value of inherited information, which is greatest when an animal's environment is predictable from generation to generation. In contrast, one advantage of learned information is that it allows more tolerance for unpredictability across generations. Predictability still plays a role, because within any one animal's lifetime, predictability is what makes learning advantageous; there is no point in learning if the situation arises only once per lifetime. Predictability makes learned behaviors applicable to future events within a lifetime.

Chapter 1 discussed the uneasy relationship between comparative psychology and ethology in the study of learning and memory. Some students may have been taught that there is a fundamental gulf between laboratory studies on learning in animals such as mice, rats, and pigeons, and field-based studies of learning in animals in natural environments. In fact, field studies of learning should be informed by the laboratory findings, and the bridges between the two areas of inquiry are often not too difficult to find.

The important questions that scientists ask about learning and memory can all be explored in detail, under controlled laboratory conditions, and they all have applications to field studies of behavior. The following questions are frequently asked: (1) How long is memory retained? (2) Is developmental stage important, for example, do nestlings show special proclivities for learning certain information? (3) Do sex differences in learning reflect the differences in male and female activities in a given species? (4) Does context affect what is learned or how well it is learned? The last question in particular reveals the critical relationship between studies of learning and *umwelt*. Learning ability in most animals is quite context-specific. For example, a bird species may be a champion at learning and remembering where it has cached food, but be rather dull at other tasks. This difference in abilities is reasonable, from an evolutionary standpoint, but reflects one complexity addressed in Chapter 2.

This chapter first discusses some standard topics in the study of learning. Once the basic structure of memory is covered, we address habituation and sensitization, followed by imprinting and conditioning. These topics provide an important background for studying the fascinating area of cognition, which is covered in Chapter 6.

148

FIGURE 5.3
A sample learning curve. Learning curves are usually a graphical expression of change in performance as time/experience increases. Thus, a "steep learning curve" means that performance changes rapidly.

5.2 LEARNING AND MEMORY

Steep learning curve is a term with ominous overtones. No one seems to want to experience a steep learning curve. What is this curve? *Learning curves* are an important tool in studies of learning because they allow learning, especially its rate, to be quantified. They are the visual representation of learning as defined here: a change in behavior as a result of experience, which takes time. Figure 5.3 shows a sample learning curve. The *x*-axis represents the number of trials or time; a measure of learning performance, such as percentage of correct attempts, is on the *y*-axis. Thus, a steep learning curve means that behavior changes rapidly over time; in other words, tasks with steep learning curves can be challenging because they offer a lot to learn! What is undesirable, however, is a flat learning curve, which indicates an inability to modify behavior with experience.

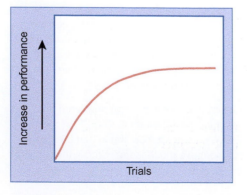

KEY TERM A learning curve is a graph of the time, or the number of trials, an animal performs (*x*-axis) versus the likelihood it will perform the task correctly (*y*-axis).

The reverse of the learning process is captured in *extinction* and *forgetting curves*.[5-8] The concepts of extinction and forgetting will be considered in greater detail later in this section.

Humans can create memories through imagination, that is, forming a mental image. Imagination is certainly linked to sensations—touch, smell, and vision—experienced in the past, but it also represents new and sometimes startling combinations of what is already known. Whether nonhuman animals can imagine is an interesting question, one that is revisited when we explore cognition in Chapter 6.

Memory generally starts with a sensory input. This results in sensory or "electrical" memory, which is transient; it lasts just a few milliseconds as the sensation is being formed and transmitted. Sensory input that has immediate importance is stored in short-term memory. *Short-term*, or working, *memory* is stored and available for relatively short periods of time thereafter—seconds or minutes.[9] It has limited capacity; in most animals, six or seven elements, or chunks, can be remembered at a time. Short-term memory is useful in remembering the sequence of a series of tasks. For example, a bee probably does not need to revisit flowers on the same plant within a short span of time. It can store recently visited locations in short-term memory and consequently avoid wasted effort.

At a cellular level, cyclic adenosine monophosphate (cAMP), an important messenger within cells, and protein kinase, an enzyme involved in synthesizing proteins, are essential elements of the formation of short-term memories (Figure 5.4). If protein synthesis is biochemically prevented, short-term memory fails. As expected, the proteins associated with short-term memory do not remain intact for long.

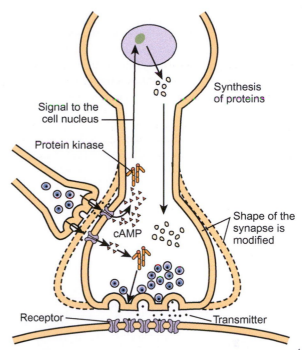

FIGURE 5.4
The cellular pathways of memory. As animals learn, synapses are activated, and within the target cell, protein kinase and cAMP synthesis is increased. As new proteins are synthesized and the synapse grows, the synapse performs more efficiently. The key elements in memory at the cellular level are the activities of protein kinase and changes in the shape of the synapse. *Adapted from http://www.nobelprize.org/nobel_prizes/medicine/laureates/2000/illpres/kandel.html.*

Of course, many memories last longer than short-term memories and are used over extended time, up to weeks or months. These are called *long-term memories*. The gateway between short- and long-term memory is *selective attention*. Based on prior strong experiences or genetic programming, the brain recognizes certain types of information as being important in the longer term.[10,11] An interesting example of auditory selective attention is the *cocktail party effect*, which is the ability to sort sounds of interest, such as your name, from a very noisy background (see Chapter 7, "Auditory Noise"); more generally, it is the ability to sort the important from the unimportant and store the former information for future use.

> **KEY TERM** Selective attention is the ability to focus on the most relevant stimuli while excluding irrelevant information.

Long-term memory also relies on kinases, but in addition involves transcription (RNA synthesis) and the growth of synaptic connections among neurons. The changes in the synapses appear to be particularly important for the formation of long-term memories. Maintenance of synaptic connections allows continued ability to access long-term memories.

Why Have Short- and Long-Term Storage of Information?

Assume that learning is a costly act for an animal. The costs include the time it takes to learn something, the storage space for information in the brain, and the neurochemicals required to store the information. If this assumption is true, then animals should learn only those parts of their experience that may be relevant and useful in the future.

KEY TERM The engram is the encoded storage of memory in the brain. Discovering the exact nature of the engram has presented a deep scientific challenge.

KEY TERM The neocortex is the outer layer (gray matter) of the vertebrate brain.

KEY TERM The amygdalae are small nuclei (concentrations of neurons) within the temporal lobe of the vertebrate brain.

KEY TERM The hippocampus is a nucleus (concentration of neurons) within the temporal lobe of the vertebrate brain.

KEY TERM The mushroom bodies are concentrations of neurons in the upper part of the arthropod brain. So named because of their mushroom-like shape when viewed in cross section, the mushroom bodies are thought to be the center of higher functions in arthropod brains.

150

Hippocampus

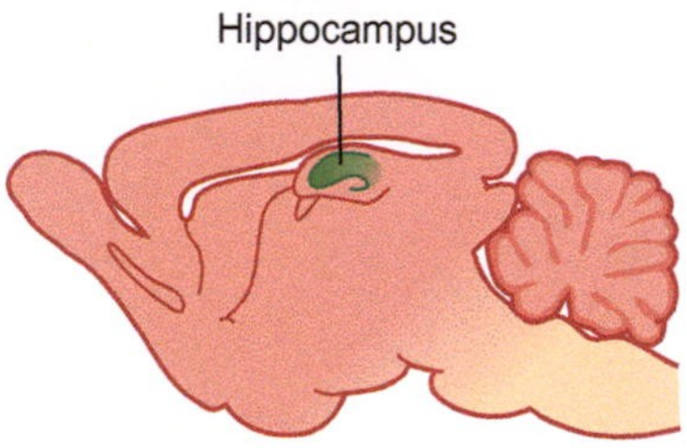

FIGURE 5.5
The vertebrate brain, highlighting the hippocampus, which has been strongly implicated in memory.

In general, animals divide learned information into two types. The first is information that may be useful in the next few minutes, but which, if remembered later, just clutters thinking. A coyote might spy a mouse running under a specific bush. This information is useful although the coyote hunts for the mouse, but later (considering the large number of bushes in the coyote's world), it is probably not helpful. The same coyote may remember the location of a stream where it drinks; the water remains in the same location and a long-term memory of that location could be quite useful.

This logic predicts that short- and long-term memory should be handled somewhat differently in animals' brains and that the persistence of a memory should be tied to the length of time over which it might be useful. An excellent example of the importance of long-term memory in animals is the retrieval of cached food, a topic explored in greater detail later in this chapter.[12]

Where Is Memory?

If there is a holy grail of neurobiology and behavior, it is the *engram*—the internal record of a remembered event.[13] The basic principle underlying long-term memory is that synapses are "plastic," that their responses to the synapses of neighboring cells can change, and that this change becomes the record of a remembered event. The big question is where this occurs in the vertebrate brain. The most likely locations are the *neocortex*, the *amygdala*, and the *hippocampus* (Figure 5.5).[14-16] Substantial debate over the past 40 years has failed to completely resolve this question. In addition, not all memory is formed in the same way. Human long-term memory is thought to be divisible into several kinds of memory (episodic, working, conditioning, and skill); it is entirely possible that different pathways and storage locales are invoked for each kind of memory.[17] In insects, long-term memory resides in the Kenyon cells of the *mushroom bodies*, which exhibit the same sort of synaptic plasticity that has been postulated for the parts of the vertebrate brain that are involved in memory.[18-21]

OF SPECIAL INTEREST

Have you ever wondered if moths and butterflies can remember being lowly caterpillars? There has long been good evidence that larval food strongly influences later adult Lepidoptera choice of oviposition sites and even mates,[22] but what of other experiences? Tobacco hornworm caterpillars (*Manduca sexta*) can learn to associate a shock with an odor and will avoid that odor both as caterpillars and later, after complete metamorphosis, as adult moths. Apparently associative learning that persists through metamorphosis must happen later in larval development, for although third instar caterpillars could learn this aversion and retain it throughout their remaining two larval instars, caterpillars had to learn the aversion as fifth instars if they were to remember it as moths.[23]

Reinforcement, Consolidation, Strength of Memory, and Forgetting

Some investigators have hypothesized that in addition to short- and long-term memory, there is a third type of memory called *long-lasting memory*. Although long-term memory lasts from hours to months, long-lasting memory stays with the animal for time spans in the range of months to an entire lifetime.[24] Long-lasting memory seems to differ from long-term memory in the strength of the associations formed between the remembered item and other memories.

Understanding these strong associations involves a process called *reinforcement*, a term that occurs repeatedly in the study of learning and memory. Reinforcement is anything that increases the probability of a behavior. At the structural level, one hypothesis is that periodic reinforcement of an item in long-term memory causes the synaptic connections associated with that item to persist in time, thus strengthening the memory. *Periodic reinforcement* is a critical component of the process of *memory consolidation*, in which long-term memory is converted into long-lasting memory.[25]

As with selective attention, consolidation is an important process in separating valuable from expendable information. Access to experiences that are not reinforced and consolidated is, after a time, lost. It may in fact be that the memory itself is not lost, but that its synaptic connections disappear, so there is no route within the brain to find the information. In humans, this would explain why memories can be "dredged up" when a person is presented with an appropriate set of associations.

> **KEY TERM** Periodic reinforcement occurs when a behavior is sometimes rewarded, usually at unpredictable intervals. At other times, the behavior is unrewarded. Periodic reinforcement can be a powerful shaper of behavior. Humans playing a slot machine are periodically reinforced with jackpots; they continue to play even though the inevitable outcome is the loss of all their money.

The strength of a memory may also be linked to specific contexts; the brain may be set up to facilitate strong, long-lasting memory of important items. Specific aversions, for example, are easily learned and difficult to forget. Humans and animals probably are primed in similar ways to place important information about social interactions, such as information used to recognize close relatives, into long-lasting memory quickly and with little or no reinforcement. Another example of strong memories comes in the context of imprinted information, discussed later; both the narrow time period in which imprinting occurs and the subsequent accurate recall point to the involvement of neural priming in the easy creation of strong, long-lasting memories.

> ### BRINGING ANIMAL BEHAVIOR HOME: WHY ARE CATS PICKY EATERS?
>
> Domestic cats are known for being finicky, a likely result of a mix of aversions and preferences for foods experienced when young; this behavior is notoriously challenging for the human caretakers of cats.[26] No one is willing to test whether his or her cat would rather starve to death than eat a particular brand of food. Exposure of cats to a wide range of foods during and immediately after weaning makes them more accepting of foods later in life. Could advertising this fact to the general public result in a change of feeding practices for cats?

Of course, memories do not always get stronger; sometimes they weaken. When items are lost from memory, they seem to disappear at an exponential rate.[27] *Forgetting* is largely a function of failure to reinforce a memory through repeated experiences; as indicated previously, reinforcement is critical in the transition of memory from short term to long term and from long term to long lasting. Forgetting is measured by observing whether the animal expresses the learned behavior over time using periodic assessment. Although forgetting is not exactly a mirror image of learning, in some animals, the rate of forgetting may be more easily studied than the rate of learning. For instance, it is difficult to study the rate at which caching animals learn their cache locations, but the rate at which they forget them can be more easily documented. *Extinction*, however, occurs when the expression of a learned response is suppressed. Skinner box experiments with rats and mice sometimes employ an electrical shock to suppress a previously learned behavior. Dog trainers may attempt to use extinction to eliminate unwanted behaviors by training an animal to perform the negative behavior in association with a stimulus and then to remove the stimulus.

Memory Capacity

Memory imposes costs on an animal; both the brain size and the metabolic energy used to support memory can be expensive.[12,28] Discarding unneeded memory reduces this cost. Although even relatively small-brained animals, such as the honeybee, can have impressive capacity for memory,[18] storage selectivity leads to efficient use of neural capacity. Honeybees are very good at learning landmarks, which are essential for finding the way home, but rather less adept in learning experiments irrelevant to their natural setting.

Animals foraging on flowers learn how to "handle" the flower to get at the nectar or pollen. Many plants present complicated flowers that serve as puzzles for the forager to solve. A study of cabbage butterflies showed that these butterflies could learn a limited number of flower types; when presented with novel flowers, they "forgot" previously learned flowers. Subsequent studies have shown some ability of butterflies to generalize handling learned on one flower type to another flower.[29] They may even use what they have learned about places to lay eggs in their choice of flowers. In other cases, the inability of many foragers on flowers to learn how to handle multiple flower types helps to ensure fidelity to one or a few plant species; this, in turn, improves pollination, as pollen is carried to flowers of the appropriate species.

Less is known about memory and learning capacities in birds and mammals. Although some investigators have postulated that learning capacity may be roughly linked to brain size, this assertion has been controversial. Across primate species, innovative behavior, social learning, and tool use are correlated with the size of the part of the brain devoted to "executive function" (neocortex and striatum). It is probably true that across a wide range of species, brain size is a meaningful measure of brain capacity. However, within species or when making comparisons between brains with relatively small differences in size, factors other than size have a greater influence on the animal's comparative abilities to learn and remember.

Of all the questions that Tinbergen asked, causation is perhaps the most difficult to answer when considering the physiology of learning and memory. Causation can also be addressed with models for learning—the ways in which animals learn—and these process-oriented approaches hold more promise. Because models for learning involve external events and experiences that are, in some cases, time- or stage-sensitive, they can also address Tinbergen's question about ontogeny.

5.3 BASIC MODELS FOR LEARNING

This section addresses the major ways that animals learn, that is, alter behavior in response to experience. Although that sounds simple enough, learning takes many forms. It occurs across animal taxa and includes behaviors ranging from *Paramecium* avoiding shock to dogs sitting for treats, from goslings recognizing mothers to squirrels locating their winter hoard. Learning happens in many ways.

Imprinting

In 1873, Douglas Spalding first described *imprinting*, which was identified as an important form of learning by Konrad Lorenz (see Chapter 1). Lorenz is best known among biologists for his pioneering work on imprinting in young animals. During a *critical period* early in their lives, many young animals learn the identities of their mother and father. Once learned, this information is firmly fixed and may be used later in life in identifying mates, in forming flocks, and in other social interactions. Lorenz found that by substituting himself for the mother during this critical period, he could induce young geese to imprint on him. Famous photographs of Lorenz show him being followed by geese imprinted in this way. Indeed, the word *imprint* refers to the seal of an important person placed on molten wax. The wax hardened and the image of the seal remained—permanent evidence of the authoritative origin of the document. Examples of imprinting include the mother's identity in geese and ducks, species-specific characteristics of birdsong, and chemical characteristics of water in natal streams for salmon.

> **KEY TERM** Imprinting is the learning of a critical feature in the environment, such as the identity of a parent, at a young age and the retention of this knowledge for later use.

> **KEY TERM** A critical period is a short, defined period of time during which imprinting may take place.

Habituation and Sensitization

Much of the sensory input an animal receives is ignored—filtered out because it is unimportant. *Habituation* is one way this filtration can happen (Table 5.1 and Figure 5.6). It is an extremely

TABLE 5.1 Animals Are Most Likely to Habituate in Response to
Weak stimuli
Less important (relevant) stimuli
Repeated stimuli
Frequent stimuli

153

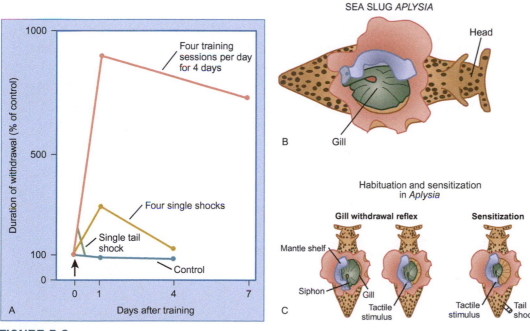

FIGURE 5.6

(A) Habituation and sensitization in the sea slug, *Aplysia californica*. (B) The gill and siphon are delicate areas and tactile stimulation results in a quick withdrawal of these structures. If stimulated 10–15 times in rapid succession, the gastropod soon habituates and the withdrawal response diminishes. The withdrawal response can be restored and the animal sensitized by one strong stimulus (e.g., shock) to the region. Sensitization can also occur independently of habituation, but the sea slug offers a clear juxtaposition of both (C). *Adapted from Nobel Prize Lecture, Eric Kandel, 2000.*

simple form of learning, in which an animal stops responding to a stimulus after repeated exposure. Habituation is a key mechanism in filtering important from irrelevant information. It can function at the level of the receptor or within the central nervous system.[30–32]

How does habituation occur at different levels in the nervous system? After a while, sensory systems may stop sending signals to the brain in response to a continuous or repetitive stimulus.[33] Lack of continued response to strong odors is a common example of sensory adaptation or fatigue, which is not habituation, but which has a similar effect, in that it causes an animal to ignore a stimulus. Habituation occurs at the level of the brain and may involve more complex stimuli; the stimulus is still perceived, but the animal has simply "decided" to no longer pay attention. Sensory adaptation and fatigue are temporary phenomena that disappear after recovery. True habituation can be reversed through new experiences that change the interpretation of the stimulus or through forgetting over time (http://www.ncbi.nlm.nih.gov/pmc/articles/PMC2754195/).

Habituation is important in filtering the large amounts of environmental information bombarding an animal. By habituating to less important signals, an animal can focus its attention on the most important features of its environment. A good example of this is species that rely on alarm calls to convey information about predators. Alarm calls cease when animals become familiar with other species in their environment that turn out not to be predators. Habituation is an important component of "not crying wolf" when nonthreatening animals come close.

FIGURE 5.7
In prairie dog towns near busy trails, the rodents may habituate to the presence of people and their pets. They may even learn the individual identities of frequent human visitors. *Photo: Michael Breed.*

Prairie dogs give alarm calls when mammals, large birds, or snakes approach. Individual prairie dogs are particularly susceptible to becoming food for a coyote, hawk, or rattlesnake, but collectively they are quite well defended because their alarm calls facilitate escape in burrows (Figure 5.7). When prairie dog towns are located near trails used by humans, giving alarm calls every time a person walks by is a waste of time and energy for the group. Habituation to humans is beneficial in this context. By habituating and not responding to animals that have proven not to be a threat, prairie dogs save time for foraging and display defensive behavior only when needed.

In animal behavior studies in the field, investigators often rely on the study animals becoming habituated to the presence of the investigator. Jane Goodall's famous studies of chimpanzees, for example, depended on the chimpanzees learning to tolerate her presence. This complex level of habituation differs greatly from learning to ignore an odor, for instance, but has a similar role in helping the animal ignore irrelevant stimuli.[34,35]

Sensitization looks like the opposite of habituation: as a result of experience, a sensitized animal becomes more responsive to a stimulus. In the case of sensitization, the stimulus is usually relevant and has potential consequences for the animal. Although sensitization and habituation seem like opposites, the neural pathways differ.

Conditioning (Associative Learning)

Much of animal learning is captured by the *conditioning* paradigm. Conditioning is the building of a learned association between two events. *Associative learning*—that is, associating certain conditions or actions with certain outcomes—allows birds to efficiently find bugs under rocks and bees to find nectar in specific flowers. This simple type of learning increases opportunities for animals to behave efficiently, seeking resources where they have been found before, or collecting them in ways that have worked previously. It is reasonable to hypothesize that animals may also associate certain locations or conditions with unpleasant events and thus hone their predator avoidance abilities. This seems to be true for honeybees, which will avoid feeding at stations if they drink a nonlethal (but possibly unpleasant) pesticide there.[36]

Not all animals rely on associative learning to the same extent, and life history traits may influence the importance of learning in a given species. Associative learning takes time, and "fast-lived" animals (animals that have short life spans during which they produce many young) have much less time for learning than "slow-lived" animals do. Natterer's bats (*Myotis nattereri*) have extremely "slow" life histories, living 30 years compared to the single year or so of common shrews (*Sorex araneus*). These animals differ in many respects, including some that are thought to influence intelligence, such as their solitary and social behavior, respectively, so they may seem like odd choices for comparison, and in some ways, they are. However, they offer a rare opportunity to explore learning in animals that share traits such as body size, brain size, and diet, and that are nonetheless vastly different in terms of life history. When placed in a maze that contained a food reward (mealworms), the bats soon learned to associate the odor from the mealworm box with the presence of mealworms and they headed directly for the food; the shrews made no such association but continued to employ a methodical pattern of searching the maze for food. The bats not only exhibited associative learning but remembered what they had learned a month later. Similar experiments with other species may shed light on the role that life history plays in the ability to learn.[37]

> **KEY TERM** Conditioning is a form of learning in which an association is made between two stimuli or between an action and a consequence of that action.

Much insight into animal learning and memory has been gained from the study of conditioning. Ethologists and behavioral ecologists often dismiss laboratory tests of conditioning as being so far removed from the animal's "natural" biology as to be irrelevant. Observations of animals in the field, though, suggest the trial-and-error learning (see Section 5.3) that often is used to gain experience with the environment is identical to associative learning. There are two major types of associative learning: classical conditioning and operant conditioning.

CLASSICAL CONDITIONING

Classical conditioning occurs when an animal associates a relevant stimulus, such as food, with a stimulus that would be irrelevant, were it not for the food. The animal is learning that two things are often associated; the animal has no control over either of them.

The most famous example of classical conditioning is the case of Pavlov's dog (Figure 5.8). Nobel Laureate Ivan Pavlov was a gastrointestinal researcher studying insulin production in dogs. When a helper entered the kennel with food for the dog, a bell was sounded. After a while, the dog began to salivate at the sound of the bell, whether or not food was available. What happened to cause the dog to salivate at something as irrelevant as the sound of a bell? To begin with, consider a normal stimulus, such as the odor of food; this is likely to elicit salivation in dogs. Food (or its odor) is a highly relevant stimulus; a bell, of course, is not and means nothing to a dog unless something happens that causes the animal to associate the bell with food. In this situation, as classical conditioning begins, the odor is termed the *unconditioned stimulus* (UCS) and the response (salivation) is called the *unconditioned response*. No learning is required for the dog to salivate at the odor of food. When Pavlov rang the bell (a *neutral stimulus*—NS) at the same time he presented food (a UCS), and did so repeatedly and consistently, the dog formed an association between the bell and the food

155

FIGURE 5.8
Although bells are normally irrelevant to a feeding dog, the dog in Pavlov's laboratory associated the bell with the food. When this type of association is made, the bell alone, with no food present, can stimulate salivation. The formation of the association—a result of conditioning—transforms the neutral stimulus (the bell) into the conditioned stimulus and salivation becomes the conditioned response.

Saliva

and began to salivate when it heard the bell. At that point, the bell became what is called a *conditioned stimulus* and salivation became the *conditioned response*. Anyone who has kept pet guinea pigs is familiar with the shrill and noisy excitement that can accompany the sound of an opening refrigerator door; guinea pigs routinely associate refrigerator doors and the goodies within.

OPERANT CONDITIONING

In classical conditioning, the animal has no control over its environment; it is simply making associations between things that happen at the same time in that environment. In operant conditioning, the animal learns to operate an environmental feature to produce a consequence, often a reward. For example, people often think of rats and pigeons that press levers in boxes (the "Skinner box") when they think of operant conditioning; a hand-shaking dog has probably been operantly conditioned as well.

In laboratory studies of operant conditioning, an animal is typically allowed to explore its environment and to perform an action, such as pressing a lever. This can be viewed as a trial-and-error phase of learning. If the animal chooses the "correct" operation, it is rewarded with a food item. Thus, its action has a consequence. Repetition of the operation and the resulting reward cause the animal to associate the two. More complex series of actions can be chained together by requiring the animal to add additional actions to what it has already learned. For example, two lever presses might be required before the reward is delivered.[38]

> **KEY TERM** An operant is a behavior that results in a consequence that can be learned. The word "operant" was first used in this context by B.F. Skinner himself in 1937 in the *Journal of General Psychology*.

Much as a rat that has learned to press a lever for food, many animals can associate a reward with a seemingly irrelevant action if the reward is received immediately after the action. Dogs can be taught to "shake" hands based on their expectation of receiving a food reward after shaking. Once the dog associates the hand signal and the extension of its paw, the food reward is no longer necessary. Such behavior is built on positive reinforcement—association of the desired action with the receipt of a reward.

Not all animals can learn something as irrelevant to their biology as shaking hands. Ease of conditioned learning is often dictated by the importance or relevance of the events to the animal species' evolutionary history and ecological conditions. Animals may be *prepared* to learn a task that is relevant to their evolutionary history (e.g., food choices in rats), they may be *unprepared* to learn a task that has no relevance (e.g., lever-pressing), and they may be *contraprepared* to learn a task if their evolutionary history inclines them in the opposite direction.[39–41] By the same token, although food rewards are frequently invoked in this discussion for convenience and because many animals will work for food, in fact, there is much individual variation in what animals value. For some domestic animals, a favorite toy is more important than food and thus may yield better results in training.

Classical conditioning and operant conditioning are used together in a training method known as clicker training. In such training, an additional stimulus is delivered at the same time as a food reward. This is often employed with dogs, horses, and other domestic animals, as well as with marine animals (Figure 5.9). Initially, the sound of the clicker occurs at the same time as a food reward. After a while, the food can be omitted from the sequence, and the animal accepts the "click" as the reward in place of the food. Clicking can then be used to modify the animal's behavior in the same way as the initial reward might have been used. In clicker training, as with most associative learning, timing is everything: if the click is not delivered at the right instant, that is, nearly simultaneously

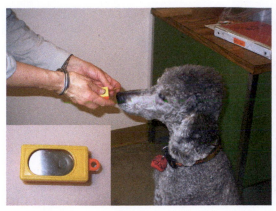 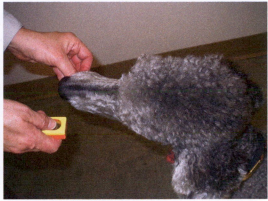

FIGURE 5.9

Clicker training of dogs (and dolphins and chickens and just about anything that can hear!) combines both classical and operant conditioning. First, the dog is taught to associate something very good (e.g., a treat reward) with the sound of a clicker (inset); this depends on classical conditioning. Eventually, the association between "reward" and the click is strong enough that the click can be used instead of the reward in operant conditioning, reinforcing desired behaviors by "rewarding" them with clicking. This is thought to be more effective than verbal praise, for instance, because it is more consistently delivered (the click always sounds the same) and delivered more quickly. Timing and consistency are essential in effective training. *Photos: Janice Moore.*

with the food, the animal will not associate the click with the food reward. Try *shaping* the behavior of a friend or roommate by using only a clicker. It can be a simple behavior, such as touching a light switch. The difficulty and frustration that can be encountered in this game teaches the importance of appropriate timing, an essential aspect of animal training. Clickers also implement another important aspect of conditioning: consistency. Unlike a human voice, the sound that a clicker makes does not vary. Consistency and timing are crucial to successful conditioning of animals. Clickers are especially useful in conditioning large animals and marine animals.

> **KEY TERM** A clicker is a device used in animal training. The trainer builds an association between the reward and the sound of the clicker so that the click itself becomes rewarding. The clicker can then be used to help the trainer get the animal to understand what behavior is desired.

> **KEY TERM** Shaping is the step-by-step training of an animal to perform a specified set of complex movements. Training a dog to roll over or do a headstand involves shaping.

157

OF SPECIAL INTEREST

Motorboats are a threat to Florida manatees (*Trichechus manatus latirostris*); the boats crash into them, slashing them with propellers, breaking ribs, and puncturing lungs. Because sound carries so well in water (see Chapter 7), one might wonder if the manatees can hear approaching motorboats at all. Scientists used operant conditioning to answer this question, offering rewards to manatees that approached a station in response to different sound frequencies. In this way, they discovered that manatees can hear a wide range of frequencies, leaving scientists to wonder even more about manatee–motorboat collisions.[42]

Training by Positive Reinforcement and Negative Punishment

Much learning results in a change in the probability of a behavior occurring. Behaviorists use a specific vocabulary for the discussion of this sort of learning, as summarized in Table 5.2.

TABLE 5.2	The Terminology of Punishment and Reinforcement
Punishment	• Always decreases the likelihood of a behavior
	• Is positive if it causes a decrease in the probability of the target behavior because of the addition of an aversive stimulus immediately following the target behavior
	• Is negative if it results in a decrease in the probability of the target behavior by removing a desired stimulus immediately following the target behavior
Reinforcement	• Always increases the likelihood of a behavior
	• Is positive if it causes an increase in the probability of the target behavior when a stimulus is added immediately following the target behavior
	• Is negative if it causes an increase in the probability of the target behavior when an aversive stimulus is removed immediately following the target behavior

Thus, giving a treat to a dog when it sits on command is *positive reinforcement*: it increases the probability of the "sit" response (i.e., it reinforces the behavior or makes it stronger) and does so by adding a treat or something else that is pleasant. The *addition* of the treat (or anything else) is deemed positive. Turning away from a dog that is obnoxiously demanding attention is *negative punishment*; it is punishment because by involving a consequence that is unpleasant, it reduces the likelihood of the behavior in the future, and it is negative punishment because something (attention) has been removed, or *subtracted*.

These distinctions do not trip off the tongue and can degenerate into semantic squabbles. As a result, many behaviorists are now often content to use *reinforcement* and *punishment* without adding qualifying terms such as *positive* or *negative*.

Positive punishment might seem to be as effective as positive reinforcement. Indeed, many animals can be taught boundaries or limits on behavior based on painful experiences, such as electric fences or shock collars. Most training using positive punishment, though, is less effective than positive reinforcement. The reason is that when positive reinforcement is used, an animal is encouraged to associate an act with the reward; in addition, the trainer is planning in advance to elicit the desired response. Thus, the trainer is prepared and the reinforcement occurs in a timely fashion—almost immediately. Returning to the example of a dog shaking hands, the training is likely to begin by showing food to the dog. This gets the dog's interest and helps it to associate the desired behavior and subsequent reward.

Positive punishment, however, generally occurs after an act that the trainer cannot predict and therefore is unprepared to condition. There is often a significant lapse between the behavior and the consequence, making association difficult. For instance, if a dog urinates on the floor and then later gets hit, will it associate the punishment with the bad behavior? Probably not, because unlike the attention the dog might devote to a treat, there is no anticipated punishment, nor is the trainer prepared to deal with the dog in a timely fashion. Even if the association is made, how does the dog know what alternative behavior will receive a positive reward? Most dogs that intend to urinate become restless and sniff around their environments. Moving such a dog to an appropriate location, combined with positive reinforcement when urination occurs in the desired place, is much more likely to allow the dog to associate urination with an approved location and a job well done. Understanding these critical differences between reinforcement and punishment (both positive and negative) is an important step in learning how to train animals.

Positive punishment has been used more successfully in scientific investigations of learning, particularly in the realm of discrimination tasks that involve shock avoidance. In such cases, the experimental apparatus is set up so that sloppy timing, so often a problem when using positive punishment in training, is avoided, that is, the shock is automatically administered when the animal makes the "wrong" choice. In such a situation, the shore crab *Carcinus maenas* demonstrated that it could discriminate between two shelters, one of which it associated with shock. The crabs did not pay much attention to visual cues when they learned this; instead, the evidence points to their simply learning to walk to the left or right to find the safe shelter. The evidence also shows that these crustaceans can perceive pain and seek to avoid it.[43]

BRINGING ANIMAL BEHAVIOR HOME: POSITIVE PUNISHMENT AND COMPANION ANIMALS—DOES IT WORK?

The plastic sheet shown in Figure 5.10 is an example of a positive punishment device used in training dogs to stay off furniture. It has embedded wires that deliver a slight electric shock when the plastic is touched. Although not harmful to the animal, the shocks are annoying and cause the animal to avoid the plastic; it is not a comfortable place to nap. As a result (and because, again, the shock is delivered in a timely fashion), the dog learns to avoid the plastic but not the furniture. This pattern of learning is very typical of punishment training; the punishment is learned, but the animal fails to form an association between the punishment and the act that the trainer is trying to discourage and has no opportunity to learn a behavior that the trainer might want to encourage. The association is between the plastic and the shock, not between the furniture and the shock. Consequently, when the plastic is not present, the animal uses the furniture as a bed. Long-term experience with the shocking device does not lead the dog to generalize from the punishment to the principle that the trainer is trying to achieve. In the same manner, dogs that are trained with shock collars often must wear a "dummy" shock collar when not in training because they quickly associate the collar with the shock. The absence of the collar may not result in the desired behavior.

When a dog is presented with an alternative bedding location, in combination with experiencing positive punishment associated with bedding on furniture, it learns to use the alternative location (as shown in Figure 5.10). The positive reinforcement—a warm, comfortable location, presented as an alternative to the punishment–enhances the animal's ability to learn appropriate behavior. A cautionary note here: it is important to note that although some dog trainers persist in techniques based on the idea that dogs want to be "dominant," and therefore humans must "dominate" a dog to train it, behavioral scientists reject this approach, which is likely to make a dog fearful and perhaps even aggressive.[44]

In general, much training of domestic animals has moved toward a "pure positive" stance, based on the reality that animals work harder and faster to gain rewards than they do to avoid unpleasantness. The agility competitions that you may have seen on television follow this method. The dog must do the course correctly but must also do it at great speed. To encourage that speed, agility trainers never use punishment.

FIGURE 5.10
An electrical punishment mat can be used to prevent animals from making themselves comfortable on furniture. The soft bedding on the right serves as a positive reinforcement. *Photos: Michael Breed.*

Trial-and-Error Learning

Trial-and-error learning allows learning of optimal responses for conditions by attempting a solution to a problem; if that solution is incorrect, then another is tried until the correct solution is discovered (Figure 5.11). The operant conditioning model fits well for this type of learning; the animal learns the behavior that is reinforced by the solution to the problem. Note that trial-and-error learning can be costly, depending on the nature of the "error." In Chapter 6, we will explore the possibility that some animals can do trial-and-error learning mentally, without taking risks. This is what humans do when they think through consequences for actions before performing the actions.

> **KEY TERM** Trial-and-error learning occurs when an animal attempts a series of solutions of a problem, eliminating possible solutions that do not work.

FIGURE 5.11
A Steller's Jay (*Cyanocitta stelleri*), a common bird in mountain habitats of the western United States. The Steller's Jay can foil ingenious devices meant to keep it out of camp food, and it is an aggressive predator on the eggs of other birds. Its ability to obtain food in the face of strong defenses has gained this species a reputation for cleverness and an ability to use trial-and-error learning in solving problems. *Photo: Michael Breed.*

160

FIGURE 5.12
Unlike bats that eat fruit or insects, vampire bats cannot learn taste aversion. Their food—blood from living animals—is unlikely to be toxic. *Photo: J. Scott Altenbach.*

Taste Aversion Learning

By far, the most important context in which an animal needs to learn quickly occurs when a behavioral choice has put that animal's life in danger. Eating a poisonous or tainted meal can lead to the formation of a specific aversion to that type of food based on a single experience. *Emesis*, the process of vomiting a poisonous food item, costs an animal not only that specific meal but other food items and often the ability to eat for a period of time. For animals that are on tight energy budgets, this loss can be costly; thus, selection has favored a route for quickly learning to avoid foods that result in illness and for retaining this information as strong, long-lasting memory.

Taste aversion learning is widespread among animals. It is a learned pattern of aversion to a specific food. The ability to learn food aversion has been favored by natural selection and helps animals avoid poisonous foods. Although mice vary in their ability to taste and respond to aversive stimuli, some species demonstrate an ability to form strong specific aversions.[45] Poisonous marine larvae can elicit taste aversion in their predators.[46] Interestingly, vampire bats, which feed on blood and would not normally encounter toxic food, do not demonstrate an ability for taste-aversive learning to avoid unpalatable food, although other bats have that ability[47,48] (Figure 5.12). For humans, eating a food and experiencing strong nausea and vomiting shortly thereafter result in an aversion to that food that often lasts for decades. Rats also have excellent abilities to form long-standing food aversions, a facility that allows them to avoid human attempts at poisoning them. Birds that eat insects quickly learn to avoid the poisonous ones, and that information is retained over the life of the bird.

Taste aversion learning facilitates the evolution of chemical defense by plants and animals. A plant or animal that can produce or obtain a toxin that causes emesis has an excellent chance to avoid being eaten because its potential consumers will develop specific aversions to the food type (Figure 5.13). Perhaps the best-studied examples of this are those remarkable omnivores, the rats, as well as birds that encounter toxic butterflies.[49–54]

Some prey animals do not simply make their would-be predators sick—they are toxic enough to kill a potential predator. The invasive cane toad

(*Bufo marinus*) is one such toxic animal, and its advance in Australia has been associated with spreading local extinction of a native marsupial predator, the quoll (*Dasyurus hallucatus*). In a desperate attempt to save this cat-sized predator, conservation biologists in Australia have experimented with taste aversion learning. They infiltrated dead cane toads with a nausea-inducing substance (thiabendazole) and gave them to 31 radiocollared quolls; another 31 radiocollared quolls remained naïve and had no exposure to toads. When released, the "toad-smart" quolls had higher survival rates than the naïve quolls. Clearly, such a strategy is not meant to treat all conservation ills, and implementing it on a large scale could present challenges. This field test does offer hope that an experience as straightforward as taste aversion learning can be another tool in conservation biology's toolbox.[55]

Taste aversion learning may appear identical to associative learning, but it is remarkably different. First, the aversive substance needs to be experienced only once for the aversion to be formed; in most associative learning tasks, repeated exposure is needed. Second, the taste-aversive substance can cause illness up to 24 h after ingestion and still form the aversion; in classical conditioning, a gap of only seconds between the NS and the UCS will slow the learning process. In fact, taste aversion learning can even occur under anesthesia. To summarize, taste aversion learning and associative learning are vastly different, and there is a high likelihood that they use different neural mechanisms. This may be why vampire bats, given their restricted diet, have lost the unnecessary ability to do taste aversion learning while retaining other forms of learning; they may well have lost those specialized neural pathways.

Cache Retrieval

Many animals *cache* (store) food items, leaving them for later retrieval. Some animals, such as mountain lions and ravens, leave spoilable food (e.g., deer carcasses) for relatively short periods (hours or days) before retrieval and consumption. Others, particularly rodents (squirrels and chipmunks; Figure 5.14) and birds (chickadees, jays), store seeds for weeks or even months before using them later. Seeds are particularly appropriate for long-term caching, and many seed-eating ants, birds, and mammals cache seeds as they drop from plants.[56–61]

Caching animals often *scatter hoard*—hide food items throughout their home range. Dispersing food items in this manner has the advantage of not creating a single large food resource in one location, which might be quite attractive to other animals, but has the disadvantage of requiring retrieval of food from dozens, or even hundreds, of locations. The alternative to scatter hoarding is to assemble the food items in a central place, such as a hollow tree or a cavity in the ground. This is called a *larder hoard*. Such a cache does not present great challenges in finding the food at a later time but may require defensive efforts.

Obviously, then, cached food is valuable only if it can be retrieved later. The most commonly hypothesized mechanism for efficient cache retrieval invokes learning cache locations and returning later to those spots. Some animals actually do this,

FIGURE 5.13
A corvid exhibiting taste aversion learning upon beginning to consume a monarch butterfly.

FIGURE 5.14
Tree squirrels cache food for use during the winter.
Photo: Jeff Mitton.

161

but two other hypotheses must also be considered: reforaging and searching by rule. Each of these hypotheses can reasonably explain cache retrieval.

LEARNED CACHE RETRIEVAL

Psychologists refer to the type of memory used in learned cache retrieval as *episodic memory* (see Chapter 6). This means that the memory is attached, or associated, with a specific experience in an animal's life. In this model, each caching event is an episode that is stored in the caching animal's memory. Birds in the families Paridae (chickadees and their relatives) and Corvidae (jays, crows, and their relatives) are champions at retrieving cached food from remembered locations (involving spatial memory).[62] Moreover, these abilities differ within species. In the case of mountain chickadees (*Poecile gambeli*), for instance, cache retrieval, hippocampus size and the number of hippocampal neurons were greater in higher elevation birds than in lower elevation ones; higher elevations and their harsh climates make greater demands on chickadee foraging and caching.[63] Rodents, such as squirrels, may rely partially on memory but also have other strategies that will be discussed later.[64]

> **KEY TERM** Scatter-hoarding animals distribute their cached food throughout their home range, so no one cache contains a large proportion of their food store.

> **KEY TERM** A larder hoard is a large cache of food, often all or a large fraction of an animal's stored food placed in one location. The honey stored by a honeybee colony is a larder hoard.

The easiest experimental approach to determine if learning is involved in cache retrieval is to allow an animal to cache food in an observation area. (This could be a natural home range, if the behavior can be seen throughout the area, or an enclosure.) That animal's rate of cache recovery is then observed and compared with the ability of a second animal that is allowed to search the area but that has not cached food there. To prevent observational learning of cache locations by the second animal (see Section 5.4), the animals are allowed to forage for cached items separately. In this type of experiment, which has been conducted with both birds and mammals, the caching animal has a significant advantage over the other animal in finding food if it learned cache locations; such a result is interpreted as support for the learned retrieval hypothesis. The learned retrieval hypothesis is not the only possible explanation for this behavior, however. Whenever an experiment tests the hypothesis of learned cache retrieval, two alternative hypotheses, *reforaging* and *searching by rule*, must always be considered.

> **KEY TERM** Reforaging is finding cached food using cues produced by the cached items. Decomposing animal flesh may produce odors that are used in reforaging for cached items.

> **KEY TERM** Searching by rule uses a simple rule to guide a search for cached items. For example, if food is always cached under rocks of a certain size, then the animal looking for cached food may be able to find the food by turning up rocks of that size. In this case, no learning of cache locations is necessary.

REFORAGING

Simply put, reforaging is a strategy in which an animal places food items in its home range, where they may be easily found, but it does not actually learn the locations of the food items. Instead, it searches them out. Although not as efficient as learned retrieval, reforaging has the advantage of not requiring elaborate mental capabilities. In rodents, reforaging may supersede learned retrieval under moist conditions, when seeds can be retrieved by their scent.

SEARCH BY RULE

Another strategy for finding hidden food stores is to have rules by which they are cached and then to use those rules as clues in a retrieval search. For example, the rule may be "always cache food under rocks." The caching animal then knows to look under rocks for food, even if it does not remember the locations of the food. Probably, all cache retrieval involves search by rule at some level; "reasonable" (in the animal's view) cache locations are considered when looking for a cached item, whereas "unreasonable" locations need not be included in the search. The role of search by rule in discovery of cached items is difficult to quantify, however, and it is simply a factor that should be considered in design of cache retrieval experiments. Human filing systems—at least ones that work—are based on search by rule. "Strong" computer passwords are ones that are not easily searched by rule.

PILFERAGE

Pilfering of caches by other animals—both of the same species and other species—is an important problem for caching animals. Even for scatter hoarders, *pilferage* may substantially reduce the cache. Pilfering animals have two strategies. One is to search for food in another animal's home range. Interestingly, weather conditions greatly affect the effectiveness of pilferage. When it is dry, yellow pine chipmunks and deer mice use spatial memory to find their stored food, and a caching animal has a substantial advantage over a pilferer in finding food. Moist conditions bring out the scents from cached seeds, enabling pilferers to sniff them out. Under such conditions, the cacher has little or no advantage over the pilferer in discovering seeds.[65]

> **KEY TERM** Pilferage is stealing food from another animal's cache.

The second strategy employed by a pilfering animal involves observational learning. Scrub Jays, when observed caching by another bird, are likely to recache the food at a later time. Similarly, Pinyon Jays and Mexican Jays may watch other birds and learn their cache sites so that they can pilfer the sites later. Ravens display elaborate avoidance behavior to keep other ravens from observing their cache sites. These observations invite consideration of the next topic, observational learning.[66,67]

5.4 SOCIAL LEARNING: TRADITIONS AND "CULTURAL" TRANSMISSION OF INFORMATION IN ANIMALS

Observational learning occurs when one animal watches the actions of another and learns from those actions. This may be as simple as learning the location of a food source or as complicated as learning a sequence of actions that leads to a reward. Many animals can do this; good experimental evidence is available for a wide range of animals, including the common octopus, quail, rats, and a variety of primates. Sometimes, social learning can help an animal to avoid a risky location.[68] This chapter focuses on some examples of observational learning in which a task or preference is learned. In later chapters dealing with cognition (Chapter 6) and social behavior (Chapter 13), observational learning is placed into the contexts of social groups and animal culture.[69-71]

Observational Learning in Octopi

The common octopus provides a classic example in the study of observational learning (Figure 5.15). When one octopus is trained to perform a task with a reward, another observing octopus can duplicate the task. This example demands attention on several levels. First, it shows that a passive observer can note the details of another animal's behavior, see that the behavior is rewarded, and then perform the behavior in an attempt to receive the same reward. Second, the 1992 discovery of this ability in a mollusk stretched the limits of how scientists imagined the operation of nonhuman animal minds. This study was on the leading edge of a remarkable scientific expansion in considering learning and memory and how they might function in a broad range of animals.[72] It helped open the door to an increased appreciation for complexity of learning and memory across animal taxa.

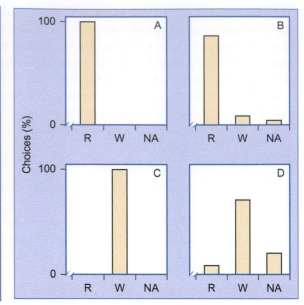

FIGURE 5.15
Observational learning in octopi. (Left) In this experimental setup, two octopi have shelters in enclosures that are separated by a glass divider. One octopus watches, whereas the other chooses between retrieving a red or white ball and is rewarded for retrieving a ball of a specified color. (Right) (A) When a demonstrating animal has been trained to retrieve the red ball, it always does so. (B) The observing octopus almost always follows the behavior of the demonstrator and retrieves the red ball. (C) The demonstrator retrieves the white ball. (D) The observer almost always does likewise. R, red ball; W, white ball. *Adapted from Fiorito, G., Scotto, P., 1992. Observational learning in* Octopus vulgaris. Science 256 (5056), *545–547.*[72]

Bats and Tethered Mealworms

Not surprisingly, big brown bats must be taught to "catch" mealworms that are dangling from a thread in an observation chamber. Most (64%) of the naïve bats that were then allowed to forage with these trained bats learned to emit a feeding buzz and then capture the tethered mealworms; naïve bats that flew with untrained bats emitted relatively few feeding buzzes and did not capture any mealworms. Once the naïve bats began to learn the task (i.e., they began to buzz), they flew closer to the trained bats and seemed to interact more. Both adults and juveniles were able to learn this foraging task. Now consider the fact that most of the 1100+ bat species are gregarious. Opportunities for social learning in these inhabitants of the evening skies are almost without limits.[73]

Birds and Milk Bottles

FIGURE 5.16
A Blue Tit robbing cream from a milk bottle. The short lives of these small birds combined with the increasing dependence on supermarket purchase of milk—and low-fat milk, at that—mean that the behavior is rarely observed in modern Britain.

The most famous putative example of observational learning was the spread of the ability to open milk bottles among Blue Bits and European Robins in Great Britain (Figure 5.16). These birds learned to rob cream from the tops of milk bottles during the early part of the twentieth century, and the Blue Tits later adapted to the use of aluminum foil seals on the bottles, learning to tear them to access the cream. There are actually two hypotheses that could explain the increase in these behaviors: First, birds might observe other birds feeding in this manner and adopt the behavior. Second, each bird might, independently, discover this feeding option. David Sherry and Jeff Galef, leading scientists in the study of observational learning and cognition, argued for just this (second) interpretation of this classic example.[74]

Although Sherry and Galef's reinterpretation is controversial, this example illustrates the need for carefully controlled experiments and a mind open to considering alternative explanations. Indeed, this same alternate hypothesis is the nemesis of many who would conclude that observational learning occurs in a species or that cultural transmission of a behavior has taken place. If one animal has performed the novel behavior, then it is obviously possible for such a novelty to appear in the population. At that point, how does a scientist determine—especially under field conditions—that it has not appeared multiple times? The Blue Tits offer a mystery, and one that now defies investigation, because this intriguing behavior has disappeared from birds in the United Kingdom, even as milk left on doorsteps is becoming a rarity. People now rely more on buying milk in stores than on delivery, and scientists are left with no way to explore the mystery of the avian milk thieves.

Cache Raiding

Ravens are very clever birds and are highly capable of observational learning. One context in which they demonstrate this is in pilferage of each other's caches; they keenly observe each other's cache locations and pilfer if given the opportunity (Figure 5.17).[75] Observational learning can play a role in exploiting other species' food caches as well.

Migration

Migratory behavior is complex, involving genetics, individual learning, and social learning. Whooping cranes present a special opportunity to investigate the role of learning in migration, because each bird is individually tracked and its history is well known. The eastern migratory population, which summers at the Necedah National Wildlife Refuge in Wisconsin and winters near the Chassahowitzka National Refuge in Florida, is captive-bred, so the genetic relationships among the birds are known. In addition, all individuals perform their first migration with the leadership of human-piloted ultralight aircraft; they all fly the same route the first time. After that first experience, 1-year-old birds that migrated with older birds did not deviate as far from the straight-line migratory path compared to 1-year-old birds that flew with their own age group. The latter managed to add half again as many kilometers to their trip; they flew 34 additional kilometers. Each year of experience mattered. The age of the oldest individual in any given group was clearly related to efficiency, with each year of age reducing deviation from the straight path by a little over 4 km (~5.5%).

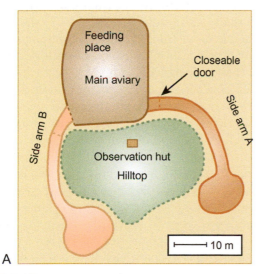

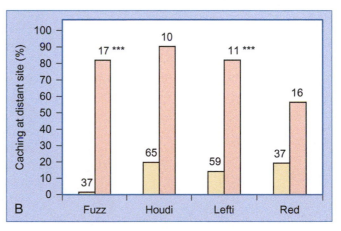

FIGURE 5.17

When ravens kept in a group in an aviary cache food, they respond to the presence of other, potentially raiding, ravens. Each of four ravens (names across the bottom of B) flew much further to cache food when other ravens were present (pink bars) than when other ravens were absent. This demonstrates the potential of observational learning by ravens stealing other birds' caches and suggests that the presence of potential pilferers causes ravens to be more devious in their selection of cache locations. (A) Map of experimental area. (B) Frequency of caching at distant sites in the presence and absence of other ravens. *** = P < 0.005. *Adapted from Heinrich, B., Pepper, J.W., 1998. Influence of competitors on caching behaviour in the common raven, Corvus corax. Anim. Behav. 56, 1083–1090.*[67]

As the older birds gain experience with each passing season, the migration is expected to become increasingly efficient. Relatedness did not seem to play a role in this behavior, but the importance of genetics in crane migration should not be ruled out. Other bird species do show evidence of inherited navigation ability (see Chapter 8), and juvenile cranes migrating on their own for the first time are able to initiate northward travel in the absence of older cranes.[76]

Teaching and Learning

In Chapter 12, we will touch upon teaching as one aspect of parenting behavior, but teaching is not limited to parents and offspring. Teaching goes beyond other forms of observational learning, in that naïve "pupils" learn as a result of unique teacher behavior that occurs at some cost to the teacher, or at least carries no immediate benefit to the teacher.[77] In addition to tandem-running ants (see Case Study), African meerkats (*Suricata suricatta*) have been shown to teach young meerkats how to safely forage on dangerous prey (scorpions), and pied babblers (*Turdoides bicolor*) use a form of operant conditioning (yes!) to teach nestling babblers a specific, energetically costly call. The nestlings initially associate this call with food, but later, as fledglings, will follow it, and thus move away from danger.[78,79]

5.5 PLAY, LEARNING, AND DEVELOPMENT

Play is not only about learning, and because of that, the topic of play could easily be placed in at least three other chapters. It is the engagement of an animal in seemingly purposeless activity that has no immediately apparent survival value. It may help to develop reasoning skills, so a discussion of play could be considered in Chapter 6 on cognition. Some kinds of play may relate to developing skills for hunting and might go into Chapter 9 on foraging. Still other types of play seem essential for developing social skills and could easily find a home in Chapters 13 and 14, which deal with social behavior. Play is discussed in this chapter on learning but note the links between play and other topics and refer to this chapter as needed.[80]

Play is much more common in mammals than in other kinds of animals, but arguments have been made for the existence of play in birds and in some invertebrates. Perhaps, play is more easily recognized in mammals because there are striking similarities among mammals in play behavior; rat or monkey play exhibits the same basic patterns. These play patterns in mammals are mirrored in human play. Playing animals appear to be deeply engaged in an activity without immediate survival value. They may use an object, such as a stick or a fragment of a dead animal, in their play. Oftentimes, play involves social signals of aggression, courtship, copulation, or parenting, but the signals are incomplete, muted, or delivered by animals that are physically incapable of performing the complete behavior because they are too young or not fully developed. At other times, play resembles hunting, but prey are absent. Signals such as play bows (Figure 5.18A), invitations to chase, and solicitations for tug of war games are seen in many species.

> **KEY TERM** Autoplay is a behavior in which an animal plays alone, sometimes using an object to play with. A kitten with a ball of string is engaging in autoplay.

> **KEY TERM** Social play is a behavior in which animals play in pairs or groups; interaction with other animals is a component of the play behavior.

Play serves critical roles in the behavioral development in many animals. It is often divided into two distinct categories: *autoplay*, or playing alone, and *social play*. Autoplay and social play can both serve as platforms on which an animal's brain development is stimulated. An animal that plays gains motor skills, fosters muscular development and coordination, learns important details about its environment, and can experiment while being protected

FIGURE 5.18
(A) A dog's playbow. Dogs preface (and punctuate!) their play with this unmistakable signal. (B) Dogs at play.
(C) Dogs seem to play "fair," at least where puppies are concerned. In this tug of war, the adult dog could have
easily won, but did not. *Photos: (B) Michael Breed; (C) Janice Moore.*

167

by parents or other group members. Social play can have the added functions of helping
animals to learn social skills, to form bonds with other group members, and to establish
social relationships that may persist into adulthood.[81]

In chimpanzees, play appears to allow an individual to assess the potential for cooperative
behavior in itself and other members of its social group.[82] Self-assessment may be an
important motivation for play, although it is difficult observationally to separate self-
assessment from self-improvement (getting better at a motor skill by practicing it).

Play can be risky: animals may be injured or exposed to predation while playing. The cost
of play implies significant benefits, even if they are difficult to demonstrate with any rigor.
Rough physical play is common in carnivores and primates. Social play is often repressed in
habitats in which risk is high.[83,84] Interestingly enough, animals whose play is restricted in
natural habitats appear to develop "normally," but highly restrictive captive environments
result in adults with abnormal social behavior.[85] Recently, play has been proposed as a
measure of animal welfare; animals maintained in good conditions are thought to play more
than animals under stressful conditions.[86]

Do animals play fair? The root of fair play lies in the hypothesis that a "social contract" keeps
play within bounds so that injury is rare. In fair play, boundaries with respect to potential
injury, turn-taking, and resource sharing are respected. Observations of play suggest that,
indeed, young animals do tend to play fair. Why might this be? A simplistic explanation
is that animals have a "theory of mind," which allows them to imagine the emotions

or feelings of other animals and act in a sympathetic manner. This type of hypothesis is discussed in more detail in Chapter 6. Alternatively, does play generally occur with close relatives? Kin selection could put bounds on rough play. There is strong evidence in primates that animals "self-handicap" in play, evening the playing field among young of different sizes, ages, and abilities.[87]

DISCUSSION POINT: PLEASURE IN ANIMALS

The concept of "pleasure" is subjective and difficult to define for a human, much less for an animal. Do you think your dog or cat derives pleasure from play? Is it a reasonable hypothesis that a sensation of pleasure serves as an internal reinforcer for behavior that is important for survival but that is not otherwise directly reinforced? In other words, because play does not result in immediately receiving food, mating, shelter, or the like, has "pleasure" evolved as a means to ensure that play occurs? How comfortable are you with attributing the subjective state of "pleasure" to a dog or cat?

Play bouts involving more than one species of animal are intriguing. Sometimes, interspecific play is pretty one-sided; it amounts to harassment. Magpies appear to be playing when they pluck hairs from the tail of a tethered dog, and certainly cats "playing" with a mouse before finishing it off is not play for the mouse. Other cases seem more nearly mutual. Squirrels descend to lower branches on trees and chatter at dogs, which respond with jumping and barking. Neither animal seems to take these bouts seriously; the squirrels are safely out of the reach of the dog, and after some experience, the dog may learn that the squirrel is uncatchable. Yet both animals persist in their contest, engaging in occasional bouts for weeks or months when a dog is fenced near trees. It is not clear why they do this, but perhaps the close but safe encounter gives each a pleasurable adrenaline rush, or maybe they both benefit from the experience gained in these encounters when a more risky situation arises.

OF SPECIAL INTEREST: THE RAVEN AND THE COYOTE

Ravens can use observational learning to exploit the caches of other animal species, such as coyotes (Figure 5.19). We have just such a story for you. The actors in this scenario are a coyote and a raven that have been eating a deer. Birds in the family Corvidae—crows, ravens, magpies, jays, and their relatives—have a reputation for high intelligence. Humans have this impression because

FIGURE 5.19
A coyote and a raven, frequent competitors for food. *Photo: Lynne K. Stone.*

these birds are adept problem solvers, good at caching and finding food, and they even seem to have a sense of humor. There is much to admire about corvid behavior. Consider this natural history description:

> On the morning of January 15, 1938, I saw a coyote trotting along the base of Mount Everts on the margin of a wide flat. Across the flat, a raven was standing on a snowdrift. When the coyote had trotted to a point opposite the raven, about 200 yards away, it turned its course directly toward the bird. By that time, a second raven had alighted beside the first one to feed on a tiny food morsel it had been carrying. When the coyote was somewhat less than 10 yards from the feeding raven, it made a quick dash. The raven easily escaped and lit again a few yards to one side. The coyote sniffed the spot where the raven had been feeding and then made another dash for it. These tactics were repeated six or seven times before the bird flew off about 250 yards. After peering at the departing bird, and seeming to hesitate whether or not to follow, the coyote trotted after it. When the coyote had covered half the distance, the raven circled back, wheeling 15 or 20 feet over the coyote, which looked up. The raven lit on the snow again to feed on its morsel, and the coyote trotted along as if to pass it, but suddenly turned to make another quick charge. The rushes, as before, were repeated five or six times. Once the coyote leaped high in the air toward the raven and rolled over twice when it hit the snow. The raven finally flew away along the river and the coyote disappeared in a draw. It appeared that both animals were enjoying the fun, for the raven could easily have flown away to escape if it were annoyed, and it would seem that the coyote, which was probably well fed by the abundant carrion, would not have been so persistent unless he were enjoying the play.[88]

The behavior is so human like, on the part of both the raven and the coyote, that it is hard not to conclude that these are thinking, intelligent animals. As scientists, we admit that this account lacks rigorous evidence. As humans, we believe that few days could have a better beginning than this one day in January did.

169

Survival Value of Learning

Tinbergen asks about the survival value of behavior, and the aspects of learning clearly address that question. An animal that learns can avoid dangerous situations and can use a variety of cues, ranging from environmental correlates to the behavior of other animals, as a way to find food and shelter.

Tinbergen's question about evolution of behavior is a bit trickier in the world of learning, but it can nonetheless be addressed. Some workers have examined fossil skulls in an effort to determine the brain size or shape. Others look for phylogenetic patterns in the distribution of learning abilities. There may also be life history correlations: learning takes time, so longer lived animals are more likely to make use of learning. Social animals also have a propensity for learning; they have many social rules and relationships to understand and many group members to remember.

SUMMARY

Learning provides animals means for coping with elements in their environments. In the process of learning, animals can acquire information about their environment and update that information as conditions change. This flexibility is both the great advantage of learning and its downfall. Learning allows instantaneous adaptation, but an inexperienced animal may be at real risk because it has not had the opportunity to learn the key facts about its environment. In this chapter, our approach has been to distill a complex and difficult academic subject into basic principles that apply to the study of animal behavior in field settings.

Is the study of learning and memory in the domain of psychology or biology? Like the nature–nurture issue discussed in Chapter 3, the question of which academic discipline "owns" these topics is an interesting part of the history of the scientific study of animal behavior. This book adopts a more biological approach, and this is reflected in somewhat limited coverage of the topic of conditioning and the use of laboratory rats, mice, and pigeons in testing learning paradigms. For the student who is interested in highly structured laboratory studies or the physiological and molecular bases for learning, psychology departments offer entire courses on these topics. Human learning and the adaptive behavior of humans find their way into the curricula of psychology, sociology, economics, and education departments.

The biology or psychology student whose primary interest is animal behavior, though, should not underestimate the value of laboratory studies of learning. Such studies provide the basic framework for understanding learning and memory, and they are included in this chapter because they are clearly relevant to understanding the behavior of animals in the field. Other behaviors, such as learned cache retrieval and observational learning of food preferences, illustrate the importance of learning as a topic in behavioral biology. Learning is one of the areas of animal behavior where laboratory and field are wonderfully complementary.

This chapter began with the cellular changes, including protein synthesis and remodeling of synapses, that underlie memory formation. This then led into a discussion of habituation and sensitization, which are simple mechanisms for sorting important from unimportant information. From there, the discussion moved into conditioning, a set of learning phenomena that have been developed in laboratory studies but have broad application in the interpretation of learning in field studies. Social learning, in which animals learn by observing one another, is an important and multifaceted topic. Finally, play is considered in the context of learning. Play is a component of learning and development, and understanding play sets the stage for discussions of a variety of topics, including cognition, foraging, and social behavior.

STUDY QUESTIONS

1. Members of a bird species learn their songs (which vary among individuals) at an early age and do not change their songs later in life. How would you test to determine whether this species has a critical period for song learning?
2. Killer whales "play" with their sea lion food, much in the same manner that domestic cats play with small rodents or birds. Propose one or two reasonable hypotheses for the existence of this behavior, and briefly describe how you would test your hypothesis or hypotheses.
3. What is the difference between classical and operant conditioning? How might both play roles in the life of an animal in a natural environment?
4. What is the role of forming specific aversions in shaping the diet of animals? Why is this type of learning an important mechanism?
5. What is a "critical period" for learning? If a male duck is raised by a female of another species, what happens to his courtship behavior when he is an adult and why?
6. A squirrel buries nuts in a scattered pattern throughout its home range. It could find the nuts either by reforaging (randomly digging or smelling them) or by learning and remembering the locations of the cached nuts. How would you design an experiment to test between the hypotheses of reforaging and learning?
7. The cream-thieving birds of Britain set up an intriguing problem: have they learned from each other, or have they independently invented this foraging tactic? If they were still involved in cream thievery, how would you determine the relative amount of support for these two hypotheses?
8. Survey your friends and discover the methods they use for remembering various passwords. (We are not advising that you poach their passwords!) How do these methods compare to cache retrieval methods? Do they use rules? Do they use obvious passwords (reforaging)?
9. Why is "fair play" and suppression of the use of weaponry important in social play?

Further Reading

Meck, W.H., Church, R.M., Matell, M.S., 2013. Hippocampus, time, and memory—a retrospective analysis. Behav. Neurosci. 5, 642–654.

Pearce, J.M., 2008. Animal Learning and Cognition, third ed. Psychology Press, Florence, KY, pp. 432.

Rudy, J.W., 2008. The Neurobiology of Learning and Memory. Sinauer, Sunderland, MA, pp. 380.

Thornton, A., Raihani, N.J., 2008. The evolution of teaching. Anim. Behav. 75, 1823–1836.

Notes

1. Franks, N., Richardson, T., 2006. Teaching in tandem-running ants. Nature 439, 153.
2. Minter, N.J., Sendova-Franks, A.B., Franks, N.R., 2013. Nest-seeking rock ants (*Temnothorax albipennis*) trade off sediment packing density and structural integrity for ease of cavity excavation. Behav. Ecol. Sociobiol. 67, 1745–1756.
3. Franklin, E.L., Franks, N.R., 2012. Individual and social learning in tandem-running recruitment by ants. Anim. Behav. 84, 361–368.
4. Franklin, E.L., 2014. The journey of tandem running: the twists, turns and what we have learned. Insect. Soc. 61, 1–8.
5. Capaldi, E.J., Neath, I., 1995. Remembering and forgetting as context discrimination. Learn. Mem. 2 (3–4), 107–132.
6. White, K.G., 2001. Forgetting functions. Anim. Learn. Behav. 29 (3), 193–207.
7. Sangha, S., McComb, C., Lukowiak, K., 2003. Forgetting and the extension of memory in *Lymnaea*. J. Exp. Biol. 206 (1), 71–77.
8. Sargisson, R.J., White, K.G., 2003. The effect of reinforcer delays on the form of the forgetting function. J. Exp. Anal. Behav. 80 (1), 77–94.
9. Veit, L., Hartmann, K., Nieder, A., 2014. Neuronal correlates of visual working memory in the corvid endbrain. J. Neurosci. 34, 7778–7786.
10. Pflueger, L.S., Valuch, C., Gutleb, D.R., Ansorge, U., Wallner, B., 2014. Colour and contrast of female faces: attraction of attention and its dependence on male hormone status in *Macaca fuscata*. Anim. Behav. 94, 61–71.
11. Yorzinski, J.L., Patricelli, G.L., Babcock, J.S., Pearson, J.M., Platt, M.L., 2013. Through their eyes: selective attention in peahens during courtship. J. Exp. Biol. 216, 3035–3046.
12. Dukas, R., 1999. Costs of memory: ideas and predictions. J. Theor. Biol. 197 (1), 41–50.
13. Zhang, W., Linden, D.J., 2003. The other side of the engram: experience-driven changes in neuronal intrinsic excitability. Nat. Rev. Neurosci. 4 (11), 885–900.
14. Holland, P.C., Bouton, M.E., 1999. Hippocampus and context in classical condition. Curr. Opin. Neurobiol. 9, 195–202.
15. Brodin, A., Lundborg, K., 2003. Is hippocampal volume affected by specialization for food hoarding in birds? Proc. R. Soc. Lond. B 270 (1524), 1555–1563.
16. Bliss, T.V.P., 2003. A journey from neocortex to hippocampus. Philos. T. Roy. Soc. B 358, 621–623.
17. Griffiths, D.P., Clayton, N.S., 2001. Testing episodic memory in animals: a new approach. Physiol. Behav. 73 (5), 755–762. (special issue).
18. Menzel, R., Mueller, U., 1996. Learning and memory in honeybees: from behavior to neural substrates. Annu. Rev. Neurosci. 19, 379–404.
19. Rosenkranz, J.A., Grace, A.A., 2002. Dopamine-mediated modulation of odour-evoked amygdala potentials during pavlovian conditioning. Nature 417 (6886), 282–287.
20. Pare, D., 2002. Mechanisms of pavlovian fear conditioning: has the engram been located? Trends Neurosci. 25 (9), 436–437.
21. Daly, K.C., Christensen, T.A., Lei, H., Smith, B.H., Hildebrand, J.G., 2004. Learning modulates the ensemble representations for odors in primary olfactory networks. Proc. Natl. Acad. Sci. USA 101, 10476–10481.
22. Anderson, P., Sadek, M.M., Larsson, M., Hansson, B.S., Thoming, G., 2013. Larval host plant experience modulates both mate finding and oviposition choice in a moth. Anim. Behav. 85, 1169–1175.
23. Blackiston, D.J., Silva Casey, E., Weiss, M.R., 2008. Retention of memory through metamorphosis: can a moth remember what it learned as a caterpillar? PLoS One. http://dx.doi.org/10.1371/journal.pone.0001736.
24. McGaugh, J.L., 2000. Neuroscience–memory—a century of consolidation. Science 287 (5451), 248–251.
25. Barrett, M.C., Sherry, D.F., 2012. Consolidation and reconsolidation of memory in black-capped chickadees (*Poecile atricapillus*). Behav. Neurosci. 126, 809–818.
26. Bradshaw, J.W.S., Goodwin, D., LegrandDefretin, V., Nott, H.M., 1996. Food selection by the domestic cat, an obligate carnivore. Comp. Biochem. Physiol. A. Physiol. 114, 205–209.
27. White, D.J., Galef, B.G., 1998. Social influence on avoidance of dangerous stimuli by rats. Anim. Learn. Behav. 26 (4), 433–438.
28. Weiss, M.R., Papaj, D.R., 2003. Colour learning in two behavioural contexts: how much can a butterfly keep in mind? Anim. Behav. 65, 425–434.
29. Lewis, A.C., 1986. Memory constraints and flower choice in *Pieris rapae*. Science 232, 863–865.
30. Rose, J.K., Rankin, C.H., 2001. Analyses of habituation in *Caenorhabditis elegans*. Learn. Mem. 8 (2), 63–69.

31. Deshmukh, S.S., Bhalla, U.S., 2003. Representation of odor habituation and timing in the hippocampus. J. Neurosci. 23 (5), 1903–1915.
32. Burrell, B.D., Sahley, C.L., 2001. Learning in simple systems. Curr. Opin. Neurobiol. 11 (6), 757–764.
33. Cohen, T.E., Kaplan, S.W., Kandel, E.R., Hawkins, R.D., 1997. A simplified preparation for relating cellular events to behavior: mechanisms contributing to habituation, dishabituation, and sensitization of the *Aplysia* gill-withdrawal reflex. J. Neurosci. 17 (8), 2886–2899.
34. Werdenich, D., Dupain, J., Arnheim, E., Julve, C., Deblauwe, I., van Elsacker, L., 2003. Reactions of chimpanzees and gorillas to human observers in a non-protected area in South-Eastern Cameroon. Folia Primatol. 74 (2), 97–100.
35. Van Krunkelsven, E., Dupain, J., Van Elsacker, L., Verheyen, R., 1999. Habituation of bonobos (*Pan paniscus*): first reactions to the presence of observers and the evolution of response over time. Folia Primatol. 70 (6), 365–368.
36. Abramson, C.I., Singleton, J.B., Wilson, M.K., Wanderley, P.A., Ramalho, F.S., Michaluk, L.M., 2006. The effect of an organic pesticide on mortality and learning in Africanized honey bees (*Apis mellifera* L.) in Brasil. Am. J. Environ. Sci. 2, 33–40.
37. Page, R.A., von Merten, S., Siemers, B.M., 2012. Associative memory or algorithmic search: a comparative study on learning strategies of bats and shrews. Anim. Cogn. 15, 495–504.
38. Turnbough, P.D., Lloyd, K.E., 1973. Operant responding in Siamese fighting fish (*Betta splendens*) as a function of schedule of reinforcement and visual reinforcers. J. Exp. Anim. Behav. 20, 355–362.
39. Domjan, M., Cusato, B., Krause, M., 2004. Learning with arbitrary versus ecological conditioned stimuli: evidence from sexual conditioning. Psychon. Bull. Rev. 11, 232–246.
40. Johanson, I.B., Hall, W.G., 1982. Appetitive conditioning in neonatal rats: conditioned orientation to a novel odor. Dev. Psychol. 15, 379–397.
41. Cleaveland, J.M., Jäger, R., Rößner, P., Delius, J.D., 2003. Ontogeny has a phylogeny: background to adjunctive behaviors in pigeons and budgerigars. Behav. Processes 61, 143–158.
42. Gaspard III, J.C., Bauer, G.B., Reep, R.L., Dziuk, K., Cardwell, A., Reed, L., et al., 2012. Audiogram and auditory critical ratios of two Florida manatees (*Trichechus manatus latirostris*). J. Exp. Biol. 215, 1442–1447.
43. Magee, B., Elwood, R.W., 2013. Shock avoidance by discrimination learning in the shore crab (*Carcinus maenas*) is consistent with a key criterion for pain. J. Exp. Biol. 216, 353–358.
44. Bradshaw, J.W.S., Blackwell, E.J., Casey, R.A., 2009. Dominance in domestic dogs—useful construct or bad habit? J. Vet. Behav. 4, 135–144.
45. Glendinning, J.I., 1992. Effectiveness of cardenolides as feeding deterrents to *Peromyscus* mice. J. Chem. Ecol. 18 (9), 1559–1575.
46. Lindquist, N., Hay, M.E., 1995. Can small rare prey be chemically defended—the case for marine larvae. Ecology 76 (4), 1347–1358.
47. Paradis, S., Cabanac, M., 2004. Flavor aversion learning induced by lithium chloride in reptiles but not in amphibians. Behav. Processes 67, 11–18.
48. Ratcliffe, J.M., Fenton, M.B., Galef, B.G., 2003. An exception to the rule: common vampire bats do not learn taste aversions. Anim. Behav. 65, 385–389.
49. Mikulka, P., Klein, S., 1980. Resistance to extinction of a taste aversion: effect of level of training and procedures used in acquisition and extinction. Am. J. Psychol. 93, 631–641.
50. Rusiniak, K.W., Palmerino, C.C., Rice, A.G., Forthman, D.L., Garcia, J., 1982. Flavor-illness aversion: potentiation of odor by taste with toxin but not shock in rats. J. Comp. Physiol. Psychol. 96, 527–539.
51. Brower, L.P., Fink, L.S., 1985. A natural toxic defense system—cardenolides in butterflies versus birds. Ann. N Y Acad. Sci. 443, 171–188.
52. Malcolm, S.B., Brower, L.P., 1989. Evolutionary and ecological implications of cardenolide sequestration in the monarch butterfly. Experientia 45 (3), 284–295.
53. Ritland, D.B., 1995. Comparative unpalatability of mimetic viceroy butterflies (*Limenitis archippus*) from 4 southeastern United States populations. Oecologia 103 (3), 327–336.
54. Galef, B.G., 1989. Socially mediated attenuation of taste-aversion learning in Norway rats—preventing development of food phobias. Anim. Learn. Behav. 17 (4), 468–474.
55. O'Donnell, S., Webb, J.K., Shine, R., 2010. Conditioned taste aversion enhances the survival of an endangered predator imperiled by a toxic invader. J. Appl. Ecol. 47, 558–565.
56. HadjChikh, L.Z., Steele, M.A., Smallwood, P.D., 1996. Caching decisions by grey squirrels: a test of the handling time and perishability hypotheses. Anim. Behav. 52, 941–948.
57. Moller, A., Pavlick, B., Hile, A.G., Balda, R.P., 2001. Clark's nutcrackers *Nucifraga columbiana* remember the size of their cached seeds. Ethology 107 (5), 451–461.
58. Tomback, D.F., 1980. How nutcrackers find their seed stores. Condor 82 (1), 10–19.
59. Vander Wall, S.B., 2000. The influence of environmental conditions on cache recovery and cache pilferage by yellow pine chipmunks (*Tamias amoenus*) and deer mice (*Peromyscus maniculatus*). Behav. Ecol. 11 (5), 544–549.
60. Cristol, D., 2005. Walnut-caching behavior of American Crow. J. Field Ornithol. 76, 27–32.
61. Devenport, J.A., Luna, L.D., Devenport, L.D., 2000. Placement, retrieval and memory of caches by thirteen-lined ground squirrels. Ethology 106, 171–183.
62. Gibson, B., Kamil, A., 2009. The synthetic approach to the study of spatial memory: have we properly addressed Tinbergen's "four questions"? Behav. Processes 80, 278–287.

63. Freas, C.A., Bingman, K., LaDage, L.D., Pravosudov, V.V., 2013. Untangling elevation-related differences in the hippocampus in food-caching Mountain Chickadees: the effect of a uniform captive environment. Brain Behav. Evol. 82, 199–209.

64. Balda, R.P., Kamil, A.C., 1992. Long-term spatial memory in Clark's nutcracker, *Nucifraga columbiana*. Anim. Behav. 44, 761–769.

65. Leaver, L.A., Daly, M., 2001. Food caching and differential cache pilferage: a field study of coexistence of sympatric kangaroo rats and pocket mice. Oecologia 128 (4), 577–584.

66. Emery, N.J., Clayton, N.S., 2001. Effects of experience and social context on prospective caching strategies by scrub jays. Nature 414 (6862), 443–446.

67. Heinrich, B., Pepper, J.W., 1998. Influence of competitors on caching behaviour in the common raven, *Corvus corax*. Anim. Behav. 56, 1083–1090.

68. Griffin, A.S., Boyce, H.M., MacFarlane, G.R., 2010. Social learning about places: observers may need to detect both social alarm and its cause to learn. Anim. Behav. 79, 459–465.

69. Altshuler, D.L., Nunn, A.M., 2001. Observational learning in hummingbirds. AUK 118 (3), 795–799.

70. Bednekoff, P.A., Balda, R.P., 1996. Observational spatial memory in Clark's nutcrackers and Mexican jays. Anim. Behav. 52, 833–839.
    Lindberg, A.C., Kelland, A., Nicol, C.J., 1999. Effects of observational learning on acquisition of an operant response in horses. Appl. Anim. Behav. Sci. 61 (3), 187–199.

71. Sasvari, L., Hegyi, Z., 1998. How mixed-species foraging flocks develop in response to benefits from observational learning. Anim. Behav. 55, 1461–1469.

72. Fiorito, G., Scotto, P., 1992. Observational learning in *Octopus vulgaris*. Science 256 (5056), 545–547.

73. Wright, G.S., Wilkinson, G.S., Moss, C.F., 2011. Social learning of a novel foraging task by big brown bats, *Eptesicus fuscus*. Anim. Behav. 82, 1075–1083.

74. Sherry, D.F., Galef, B.G., 1984. Cultural transmission without imitation—milk bottle opening by birds. Anim. Behav. 32, 937–938.

75. Bugnyar, T., Kotrschal, K., 2002. Observational learning and the raiding of food caches in ravens, *Corvus corax*: is it "tactical" deception? Anim. Behav. 64, 185–195.

76. Mueller, T., O'Hara, R.B., Converse, S.J., Urbanek, R.P., Fagan, W.F., 2013. Social learning of migratory performance. Science 341, 999–1002.

77. Caro, T.M., Hauser, M.D., 1992. Is there teaching in nonhuman animals? Q. Rev. Biol. 67, 151–174.

78. Thornton, A., McAuliffe, K., 2006. Teaching in wild meerkats. Science 313, 227–229.

79. Raihani, N., Ridley, A., 2008. Experimental evidence for teaching in pied babblers. Anim. Behav. 75, 3–11.

80. de Oliveira, C.R., Ruiz-Miranda, C.R., Kleiman, D.G., Beck, B.B., 2003. Play behavior in juvenile golden lion tamarins (Callitrichidae: Primates): organization in relation to costs. Ethology 109, 593–612.

81. Spinka, M., Newberry, R.C., Bekoff, M., 2001. Mammalian play: training for the unexpected. Q. Rev. Biol. 76, 141–168.

82. Palagi, E., Cordoni, G., Tarli, S.M.B., 2004. Immediate and delayed benefits of play behaviour: new evidence from chimpanzees (*Pan troglodytes*). Ethology 110, 949–962.

83. Harcourt, R., 1991. Survivorship costs of play in the South American fur seal. Anim. Behav. 42, 509–511.

84. Caro, T.M., 1995. Short-term costs and correlates of play in cheetahs. Anim. Behav. 49, 333–345.

85. Boissy, A., Manteuffel, G., Jensen, M.B., Moe, R.O., Spruijt, B., Keeling, L.J., et al., 2007. Assessment of positive emotions in animals to improve their welfare. Physiol. Behav. 92, 375–397.

86. Marquez-Arias, A., Santillan-Doherty, A.M., Arenas-Rosas, R.V., Gasca-Matias, M.P., Munoz-Delgado, J., 2010. Environmental enrichment for captive stumptail macaques (*Macaca arctoides*). J. Med. Primatol. 39, 32–40.

87. Petru, M., Spinka, M., Charvatova, V., Lhota, S., 2009. Revisiting play elements and self-handicapping in play: a comparative ethogram of five old world monkey species. J. Comp. Psychol. 123, 250–263.

88. Murie, A., 1940. Ecology of coyotes in the Yellowstone. National Park Service Fauna Series No. 4. US Government Printing Office, Washington, DC, pp. 33–34.

Cognition

CHAPTER 6

175

LEARNING OBJECTIVES

Studying this chapter should provide you with the knowledge to:

- Apply the concept that "brain" and "mind" are different. While the brain is the physical structure that coordinates behavior, the mind is where thought occurs.

- Interpret cognition as the ability to visualize the self in the context of the larger world, to forecast the results of actions and emotional states, to solve complex problems, and to interpret the emotional states of other animals.

- Explain how cognition reduces impulsive behavior, and allows reflection, planning, and foresight—in other words, nonimpulsive behavior.

- Consider the historical unwillingness of many biologists to credit nonhumans with cognition and how modern insights have broken down this cognitive/noncognitive divide.

- Link how cognition intertwines with perceptions of intelligence, problem solving, and insight.

Animal Behavior. DOI: http://dx.doi.org/10.1016/B978-0-12-801532-2.00006-4

- Understand personality as a noncognitive expression of an animal's strategy for survival. When animals interpret each other's personalities, this becomes a part of the cognitive picture. Behavioral syndromes are tools for understanding and measuring variation in animal personalities.
- Appreciate cognition as a mechanism by which an animal assesses its own emotional state and the feelings of surrounding animals; it is not necessary for emotional life or personality.

6.1 INTRODUCTION: WHAT IS COGNITION

Of all the topics covered in animal behavior, *cognition* is the trickiest. Scientists struggle with defining cognition, much less measuring it. Cognition involves the concept of self as a separate identity from the surrounding environment, and at its highest levels involves the imaginative ability to bring seemingly unrelated facts or ideas together to create a novel solution to a problem or a unique plan for future action. A consistent thread through this chapter is the question of how scientists can test animals to determine if they use cognitive processes like those that are so obvious from human perspectives? What serves as evidence of cognition in an animal that cannot tell us what, if anything, the animal is thinking?

> **KEY TERM** Cognition is the ability of an animal to separate itself from the moment in which it is living and to contemplate the past, predict the future, and act accordingly.

Cognition is particularly tricky because of the history of how cognition has been approached in behavioral studies of animals. The study of cognition is a vibrant, edgy science that asks what animals think and how they think it. The history of cognitive science itself is rich with ideas and examples that extend far beyond a creaky collection of factoids and dates. Victorians had ideas of clever foxes and diligent insects, and displays of early twentieth century horses that do arithmetic (e.g., Clever Hans, see Chapter 1). Early psychologists had a fondness for rats in boxes equipped with levers (and not much else) and this approach still has a strong effect on scientists today, who are only now emerging from those shadows. The reaction of the first modern behavioral scientists such as Morgan (of Morgan's Canon; see Chapter 1) to the *anthropomorphism* of earlier generations was in many ways appropriate: anthropomorphic imaginings have no place in science.

> **KEY TERM** Anthropomorphism is the uncritical attribution of human characteristics to nonhuman animals.

As is true with many human endeavors, however, the rejection of anthropomorphism led to overreaction. In their zeal to avoid further anthropomorphism, many behavioral biologists rejected any and every suggestion of cognition in the animal world. A review of Chapter 1 will reveal that only courageous resistance to this overreaction, led by scientists such as Donald Griffin, has caused a recent increase in acceptance of cognition as an important and fruitful part of animal behavior.[1]

We introduce this topic with a discussion of some simple ways to assess cognition, including an examination of the importance of language in such assessments. This is followed by the concepts of self and *self-awareness*, which are common to most, if not all, attempts to define cognition. From there, the subject expands to include foresight and planning, including developing the idea of *mental time travel*. Next is a discussion of intelligence. Because cognition is often interpreted as intelligence, difficulties of cross-species comparisons of intellectual

> **KEY TERM** Self-awareness is the ability of an animal to assess its own condition and to contrast that condition with the states of other animals in its population.

capacity are important. The final sections deal with the cognitive aspects of variation among animals; personality, impulse control, and emotion all play important roles in cognition and behavioral variation within animal groups (Figure 6.1).

Intelligence, insight, personality, and *emotion* are among the most intriguing of animal traits that we study, because of their central role in our own lives. These attributes are difficult to separate from cognition because human perceptions of cognition are intertwined with how smart, engaging, and emotional the public thinks an animal is. In addition, there are reasonable arguments that the emotional lives of animals reflect the fact that they have cognitive lives. Cognition is closely associated with some of the material devoted to learning and memory (covered in Chapter 5) because using information about past events in planning for the future is key to many aspects of cognition. Finally, while reading about animal cognition, consider whether cognitive abilities elevate humans above other animals and how explorations of animal cognition should inform the ongoing discussion of the relationships between humans and the rest of the natural world. Recall the evolutionary linkages

KEY TERM Mental time travel is the ability to use past experiences to forecast the future.

KEY TERM Intelligence is a measure of the ability to learn, to remember, and to solve problems.

KEY TERM Insight is the ability to consider a set of information and to derive from that information a novel solution to a problem.

KEY TERM Personality is the sum of behavioral tendencies of an individual that make it unique among animals of its species.

KEY TERM Emotions are mental states, or "feelings," that humans and perhaps other animals experience. Examples include joy, sadness, anger, and love.

177

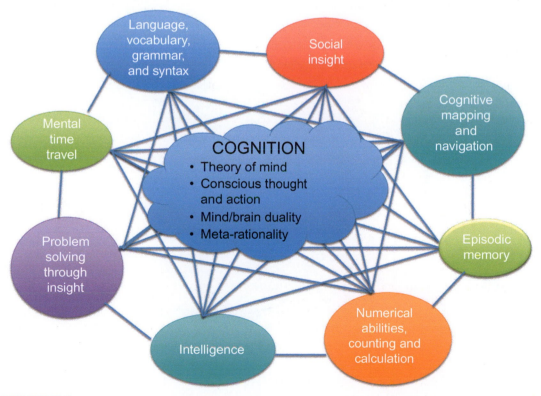

FIGURE 6.1
Cognition is difficult to define as it encompasses many different possibilities in animal behavior. This figure attempts to capture the ideas, theories, and hypotheses that come under the general intellectual umbrella of cognition.

that unify all living things, and ask yourself the following question: if humans are the only animals with cognitive abilities, how did that unique trait evolve?

Cognition is a reasonable hypothesis that explains how at least some animal species mentally operate. That reasonable hypothesis yields a difficult question: If animals are cognitive, how should this affect human behavior toward animals? Humans have, at times, awarded themselves a privileged position in the natural world based on the argument that only humans have cognitive and emotional lives. After reading this chapter, students should question those assumptions, recalling that no one knows for sure what another person is thinking, much less what a walrus might imagine.

CASE STUDY: COGNITIVE MOLLUSCS?

In a chapter about cognition, you'll encounter fascinating examples drawn from a range of animals, from crows to elephants. Those animals all have something in common—they have backbones, like humans do. The veined octopus (*Amphioctopus marginatus*) challenges us to broaden this perspective. The octopus and its cephalopod relatives are all molluscs—members of the same phylum that contains clams, snails, and a variety of other organisms that might glide along on mucous trails—if they move at all. Unlike the majority of other molluscs, however, the cephalopods are active predators: they move quickly, they have big brains, and we are beginning to realize that they may even plan ahead.

The intelligence of cephalopods, especially octopi, is legendary; we have long realized that they can learn to distinguish colors and patterns, and when it comes to leaving laboratory aquaria, they are formidable escape artists.[2] It was nonetheless something of a surprise to Julian Finn and other scientists[3] when they discovered that the veined octopus not only uses tools, but carries the tools around, prepared for future use (Figure 6.1).

The tools in question are halves of coconut shells, and the octopus uses them as a defense against predators and as a container for clusters of eggs.[4] The octopus manipulates two shells so that the concave surfaces face each other, creating a protected spherical space that can contain the animal. The octopus seems to prize these shells; when it moves to a new location, it stacks the shells concave surfaces up and then—arms to the outside of the shells—does what Finn calls "stilt-walking," using its rigid arms to walk (or tiptoe) away. This form of octopus locomotion has not been seen in any other context and appears to be both inefficient and perhaps risky, as a stilt-walking octopus is exposed to predation. Octopi have been seen to stilt-walk up to 20 m. The result of all this effort is that the octopus is ready to use the coconut shells when it arrives at a new location (Figure 6.2).

FIGURE 6.2
The veined octopus not only manipulates coconut shells and uses them as shelter, but carries them around for use in the future. Such mental time travel is a component of cognition.

CASE STUDY (CONTINUED)

FIGURE 6.2
(Continued)

Of course, coconut halves are recent additions to the ocean, the result of human coconut use. Finn and co-workers hypothesize that the octopus used large clam shells in a similar manner before humans began contributing coconut shells to the marine environment. Octopi have long been known to use rocks to barricade the entrances to their refuges,[5] but the movement of coconut shells for future use is an even more sophisticated environmental manipulation.

The study of cognition continues to create tempting opportunities to use words laden with anthropomorphic sentiment.[6] Scientists often have a difficult time coming up with concise nonanthropomorphic descriptors of cognitive processes. Table 6.1 provides some adjectives describing cognitive processes; can these be expressed as testable hypotheses, or are these terms too closely allied with a description of the human mind to be useful for a scientist?

TABLE 6.1 Adjectives Describing Cognitive Processes

Self-awareness
Forecasting
Problem solving
Empathy
Thinking
Intuition
Insight

Tests for Cognition

Just as there is no simple definition of cognition, there is no single easy test for cognitive abilities in animals. A perfectly motionless animal may still be performing wonderful cognitive feats. The problem for the scientist studying cognition is how to "get inside the animal's mind" and measure what, if any, cognition is occurring. This chapter covers a wide range of experimental approaches, from measuring brain waves to having animals look in mirrors. Scientists have used all of these approaches and more to try to capture the workings of an animal's mind. Not one of them is a perfect test of cognition, but each approach gives unique insight into the mental lives of animals.

OF SPECIAL INTEREST: MORE READING ON ANIMAL COGNITION

Animal cognition is one of the great puzzles of modern science and may have stimulated more book writing than the rest of the field of animal behavior combined. There are the numerous books on cognition; *Animal Thinking* by Donald Griffin is considered by many to be one of the field's classics. The following are among those that are useful reading:

- *Animal Cognition in Nature*, edited by Russell Balda, Irene Pepperberg, and Alan Kamil
- *Minding Animals: Awareness, Emotions and Heart* by Marc Bekoff
- *Through Our Eyes Only: The Search for Animal Consciousness* by Marian Stamp Dawkins
- *Social Intelligence: From Brain to Culture* by Nathan Emery, Nicola Clayton, and Christopher Frith
- *Wild Minds: What Animals Really Think* by Marc Hauser
- *Comparative Cognition*, edited by Edward Wasserman and Thomas Zentall
- *Born That Way: Genes, Behavior, Personality* by William Wright
- *Animal Cognition: The Mental Lives of Animals* by Clive Wynne
- *Beyond the Brain: How Body and Environment Shape Animal and Human Minds* by Louise Barrett
- *Comparative Cognition* by Mary C. Olmstead and Valerie A. Kuhlmeier
- *Dog Behaviour, Evolution, and Cognition* by Adam Miklosi

180

External Representations of Internal States: Does Behavior Represent Thought?

In most behavioral studies, the scientist's clues to an animal's thoughts or feelings come from observations of that animal's responses to a situation. Most mammals, and certainly primates, carnivores, and ungulates, behave in ways that seem to reflect cognitive attributes. Even Charles Darwin was intrigued enough to write a book on the topic, *The Expression of the Emotions in Man and Animals*.

BRINGING ANIMAL BEHAVIOR HOME: DARWIN, DOGS AND CATS, AND EMOTIONS

Darwin was fascinated by the commonality of emotional expression between humans and animals (Figure 6.3). Like most of us, he had experiences with dogs and cats; his empathy with these animals allowed him to hypothesize that similarities of emotional expression reflected shared evolutionary roots. The following short passage, excerpted from the last chapter of his book about emotion, expresses exactly this reasoning. He also used these observations to support the idea that emotional expression is innate rather than learned, and that it is the same in young and old animals. Behavior as an innate characteristic, of course, supports his central theme of evolution by natural selection. His Victorian prose seems almost languid compared to that used in our instant communication society, but there is nothing languid about his thinking, and the images he takes the time to create only reinforce his argument. Darwin's observations fit perfectly into current scientific debates about animal expression, emotion, and cognition. In Darwin's own words:

> "Take, for instance, the oblique eyebrows of a man suffering from grief or anxiety…That the chief expressive actions, exhibited by man and by the lower animals, are now innate or inherited, — that is, have not been learnt by the individual, — is admitted by every one… We are so familiar with the fact of young and old animals displaying their feelings in the same manner, that we hardly perceive how remarkable it is that a young puppy should wag its tail when pleased, depress its ears and uncover its canine teeth when pretending to be savage, just like an old dog; or that a kitten should arch its little back and erect its hair when frightened and angry, like an old cat."[7]

FIGURE 6.3
These drawings from Darwin's book on animal and human emotion show a cat in three unmistakably different emotional states. *Adapted from Darwin, C., 1872. The Expression of the Emotions in Man and Animals. John Murray, London, pp. 350.*[7]

Modern approaches try to "see" what is going on inside a brain by using external measures, such as functional magnetic resonance imaging (fMRI), to measure brain activity. This technique uses large magnets outside an animal's head to measure changes in the magnetic state of molecules in the brain. With this tool, scientists can assess how the inside of the brain changes during activity. fMRI scans the entire head and gives scientists an image of the intact brain and its activity without requiring surgery. Animals can even be asked to perform tasks during the procedure, which then constructs images that show what parts of the brain are used for these tasks. Thus, scientists can discover if nonhuman animal brains process a given type of information in the same part of the brain as humans do. If the same part of the brain is used in both organisms as they perform similar tasks, this supports the hypothesis that the two brains are working in similar ways.

For example, numerical assessments and judgments of relative numerical values are thought to take place in the intraparietal sulci of the brain. Among nonhuman animals some kind of numerical reasoning has been demonstrated in corvid birds (jays and crows), rats, and primates, and may involve the same region of the brain.

fMRI has also been used in studies of gaze following.[8] Gaze following is the ability of an observer to note that another animal is looking or staring in a particular direction or at a specific location and to follow that animal's gaze so that the observer locates the same object of interest. fMRI studies in humans show that gaze following is associated with the superior temporal sulcus. The *homologous* area in the brain of a rhesus macaque is activated when it follows another's gaze. This suggests that the human ability to follow gazes is shared with other primates as a result of descent from a common ancestor. Gaze following is discussed in more detail later in this chapter.

fMRI has also yielded intriguing results in the study of voice perception. Dogs and people last shared a common ancestor over 90 million years ago; on the other hand, they have lived together for at least 18,000 years. Scientists in Hungary trained 11 dogs to lie still in an fMRI scanner and studied their reactions to both human and dog vocalizations, as well as to nonvocal sounds. The vocal sounds had emotional valence that ranged from highly negative to highly positive. These fMRIs were compared to those from 22 humans subjected to the same sounds. Both species devote similar parts of their brains to vocal processing (a new finding for non-primates), and the area around the auditory cortex of both dogs and people reacted more strongly to sounds with positive emotional content than negative ones. What is not clear is whether these similarities are the result of shared ancestry or the domestication process.[9] This question might find resolution if more species could be studied in this way, but the researchers were dubious about training other species to relax in the fMRI scanner.

> **KEY TERM** Homologous is a comparative term used to indicate that two structures are alike in developmental (and evolutionary) origin. A bat's wing is homologous (same evolutionary root) to the forelimb of a mouse, and is analogous to an insect's wing (same function, but different evolutionary root).

Language and Cognition

Language is a special case in considering the evidence for cognition. It is a special case in part because of its long, rocky history in debates about cognition and in part because of its special role in human cognition. Cognition does not require *language*; it can be expressed in actions. The existence of language in humans, however, means that cognition is much easier to assess in humans than in other animals. Humans are the one species known to have highly symbolic language (at least language that human investigators can recognize and understand). This means that a person can be asked directly what he or she is thinking or feeling, but humans can only pose this question to a nonhuman animal indirectly. Note that one result of this difference is that humans can attribute a wide range of thinking and feeling (or absence of those traits) to an animal without much objection or affirmation from the animal itself. This is the central problem of cognitive studies.

> **KEY TERM** Language is an abstract representation of objects, actions, and emotions. Language uses symbols, novel combinations, and syntax to allow communication among animals.

Humans use an abstract, symbolic language in both internal thought processes and external verbal communication with others. When people stray from language-based thought, this human process is characterized as "intuitive." If human behavior serves as the basis for defining language, three attributes of language can be identified:

1. The ability to assign meaning to sounds, gestures, or symbols that have no direct relationship to the object or action being represented.
2. The ability to draw on the symbols to create novel combinations that are appropriate to a particular need to communicate.
3. The use of grammatical rules that govern the structure of expression. This is called *syntax*.

The second attribute—novel combinations—distinguishes language from ritualized displays (see Chapter 7), such as mating signals in birds. Ritualized communication, such as that which often accompanies mating or aggression, for instance, consists of repetition of the same signals. It is the accuracy of this *invariable* repetition, not new and potentially confusing combinations, that determines in part the success of that communication. In contrast, if an animal can use language, it has a set of symbols that can be combined in various ways. "The snake is in the tree" has different meaning than "The snake is in the grass" or "the tree is in the snake." Some combinations, such as "The snake is in the sky," make no sense, and the cognitive animal can separate sense from nonsense.

The dance "language" of bees provides an interesting point for discussion. Bees have a relatively small set of symbols available to them, including the angle of the dance relative to gravity, the speed of the dance, the number of times the dance is performed, vibrations made when dancing, and perhaps a few other features. The information being communicated is moderately complex; it is predominantly distance and direction to the food, and perhaps the quality of the food. To accommodate all the permutations of direction and distance, a large number of dances are possible (Figure 6.4). But is this really language? The lack of novelty of information communicated—bees never stray beyond food sources and nesting sites in their use of this "language"—suggests it is not. Then again, what else are they likely to talk about, and are there other channels for that sort of communication, thus far unrecognized by humans?

In contrast, male Campbell's monkeys (*Cercopithicus campbelli*) use six call types in a variety of sequences to communicate a wide variety of events ranging from falling trees and nonpredatory animals to specific information about predators in general, leopards, and crowned eagles. They do this by using the sequences in different combinations. Thus, one sequence might be a call to group cohesion, another sequence might be a general predator alert, and used together, those sequences might refer to a falling tree. Yet another sequence, when prefacing a call, may indicate the absence of predator threat. Although there is no evidence of any symbolic associations in this communication, there seems to be a tendency toward what the authors of the study call "proto-syntax."[10]

A variety of animals have been taught to use basic elements of human-developed symbolic language. Probably the most famous examples are chimpanzees that use sign language to communicate with their handlers and African grey parrots that can communicate using both vocalizations and symbolic objects provided by their handlers. These examples have been questioned for two reasons: first, "Clever Hans" (see Chapter 1) effects are often not fully controlled: the people doing the experiments argue that the "special" relationship between the handler and animal creates an emotional environment in which communication can take place, but this eliminates the possibility of controls for subtle (and perhaps unintentional) signals from handler to animal that may stimulate behavior that resembles language use but is not. Second, even though animals may use something that looks like language in communicating with their handlers, this communication does not appear to translate to the use of the same symbols in communication among animals in their own social contexts. Critics of these objections (remember that this is a lively and contentious field of study) have pointed out that the

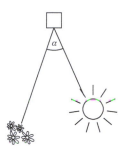

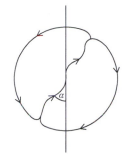

FIGURE 6.4
The upper drawing shows a bee hive (the square at the top), the location of the sun, and the location of flowers, which provide food for the bees. The angle formed by the flowers, the hive and the sun in the upper drawing is then communicated among bees by dances that are performed within the hive. This angle is important information in helping bees find food sources. Bees use a variety of dance attributes, including the angle of the dance relative to gravity (shown here by the vertical line), to inform the nestmates about the location of food.

183

symbols taught to animals in order to investigate their use of those symbols are not natural to the animal. In many instances, perhaps, the animal may well be "unprepared" (see p. 156) to learn this sort of communication. Thus, using the symbols with a conspecific might be reasonably unlikely.

DISCUSSION POINT: NIM CHIMPSKY

For a fascinating read on the subject of teaching animals language and the ethical complexities of involving long-lived animals in human research, take a look at *Nim Chimpsky: The Chimp Who Would Be Human*, by Elizabeth Hess. Nim Chimpsky was taught sign language and assimilated into human culture, but eventually was returned to a research colony of chimpanzees, then sold for tuberculosis research, only to be ultimately rescued. (His name is a play on the name of famous linguist Noam Chomsky, discussed later in this section.) For a class discussion, perform a web search on Nim Chimpsky; a wealth of information will appear about his abilities with sign language and the ethical dilemmas that emerged throughout his life.

Evidence of symbolic communication among animals in natural social groups, combined with novel combinations of symbols, would provide a more powerful argument for language use by animals. As might be imagined, this information is much harder to come by, and while arguments for this type of communication have been made for wolves, dolphins, and chimpanzees, the actual evidence is thin. This is a reminder of the judgment that every scientist must make, weighing the contrasting costs and benefits of laboratory and field (Chapter 1). Perhaps the most intriguing example of symbolic communication in animals comes from vervet monkeys in social groups, which may come to use different vocalizations to specify what type of predator (snake, hawk, leopard) is in the area (Figure 6.5).[11,12] Recently, African elephants, which can distinguish subsets of humans using color or smell (see Chapter 10), have been reported to use different alarm calls when confronted by humans or by bees, and they responded differently to those threats.[13]

FIGURE 6.5
A vervet monkey. The alarm calls of vervet monkeys are predator-specific, as shown by the reactions of monkeys that hear the calls. Depending on the predator, monkeys may look up (hawk) or quickly climb a tree (leopard). *Photo: Allyson Woodard.*

184

Language has served as the platform for arguments favoring human uniqueness; people who want to establish that somehow humans are elevated above the rest of the animal world often use language as a last resort for support of their ideas. This is a "last resort" because other perceived distinctions have evaporated as scientific knowledge of animal behavior has expanded. Thus, humans are no longer the only animals that use tools, they are no longer the only animals that care for nonrelatives…. Indeed, the more that is learned about animals, the more some standard-bearers for human superiority focus on redefining the defining traits of humans!

In addition to asking if other species have language, there are equally puzzling questions about cognition that reside squarely within the confines of nonhuman animal behavior. For instance, if a nonhuman animal is thinking, in a cognitive way, what system of internal representation does it use? It is certainly possible that an animal would have a fully developed internal symbolic language that is never expressed externally, but how could this be discovered? Nonhuman primates exhibit human processes in mirror tests and gaze following, topics that are discussed in following sections; it is

KEY TERM Universal grammar is the hypothetical grammatical structure of human language that is based on the nature of the human brain. Universal grammar, if it exists, would be identical among all human languages.

OF SPECIAL INTEREST: TOOL-MAKING BIRDS

Rooks are corvids, a group of birds known for intelligence and creativity that also includes jays, crows, and magpies (see "problem solving" in Section 6.3). In contrast to New Caledonian crows, which can make tools in the wild and in the laboratory, rooks do not seem to make tools in the wild. Therefore, it surprised scientists to see them do it in the laboratory (Figure 6.6).[14] The rooks were presented with a delicious waxworm (a caterpillar) in a difficult-to-access location in the laboratory. When given a variety of choices among potential tools, they selected the correct tool, they used the wrong tool to acquire the correct tool, and they made tools for retrieving the caterpillars. Why rooks show this ability in captivity, but apparently not in the wild, is a mystery. The investigators argue that their findings demonstrate that the animals show insight—the ability to put together the pieces of a mental puzzle.[15–20]

FIGURE 6.6

A rook using a piece of metal, curved into a hook, to retrieve food. *Photo: Nathan Emery.*

very likely that similarities to human language also are found in nonhuman primates. The hypothesis that language is used entirely internally—as a facilitator of thought, rather than as a communicatory mechanism—cannot be refuted.

Following this idea about human language, consider for a moment its nature. The linguist Noam Chomsky developed the concept of *universal grammar*, which hypothesizes that underlying neural features of the human brain drive similarities in structure among all human languages.[21,22] He reasoned that human language must function within the structure of the human brain (Figure 6.7). As human brain structures are derived from the brain structures of primate ancestors, this suggests that the basis for language may be shared with at least some other primates. In the words of Chater et al., "language has evolved to fit the human brain."[23]

Returning to the larger issue of language and cognition, the definition of cognition presented in this chapter involves planning and a sense of self. Neither of these requires language, and it is up to the creative mind of humans to imagine whether nonhuman animals have internal language-based reasoning or if their thinking is entirely nonlinguistic. Discovering the extent of language use by animals is one of the great challenges for the next generation of animal behaviorists.

DISCUSSION POINT: DO HUMANS STAND ABOVE ANIMALS?

This discussion of the relative status of humans and other animals raises a bothersome question: Why would a person want to elevate humans above other animals? If we are simply looking for distinctive traits, then every species, by definition, is distinct from every other. Moreover, as superior as we would like to feel, it is difficult to imagine a human being that matches the filter-feeding talents of, say, a clam or the persistent architectural genius of an orb-weaving spider. Why might humans be compelled to feel superior? Do humans have a special place in the universe? Are some species better than others? Are humans co-equals in a web of life that defies imagination? What do you think? Why?

FIGURE 6.7
Can Chomsky's idea of "universal grammar" for humans be extended to other species?

> **KEY TERM** Theory of mind is the ability to form hypotheses about the thoughts of surrounding animals.

6.2 THE CONCEPT OF SELF
Minds and Bodies

It is widely accepted that cognitive animals use their minds in ways that may be at least partially separate from events that affect their brains and bodies. The *theory of mind* features prominently in studies of animal cognition and intelligence. An animal with a theory of mind can form hypotheses about the thoughts of surrounding animals. This means it can discern another individual's intentions, reflect on the other's desires, and understand its beliefs. People hope that others in their group have theories about their minds and that they act on these theories in beneficial ways. But when a dog looks at a person, is it forming a theory about that person's mind? Crows can observe people and remember which person is a threat and should be avoided;[24] this is an example of making predictions about behavior based on a "theory" of another animal's mind.

Consciousness of self is a key characteristic of humans as a cognitive species. A person separates him- or herself, mentally, from other humans, from the larger living world, and from the physical environment by self-recognition—that is, by the recognition of him- or herself as a unique entity. People examine themselves, mentally, and reflect on their individual experiences. This mental life is distinct, in many ways, from human physical life, and includes conjectures, dreams, and fantasies that merge the past, present, and future. Indeed, humanity has at times been defined by the fact that humans are thinking animals, with a mental reality that separates mind from brain and body.

It is possible to imagine that a broad range of animals live cognitive lives much as humans do. Because human communication by gesture, expression, and voice is deeply rooted in the evolution of vertebrates, much of the human range of communication can be observed among other animals, particularly mammals. This does not, however, reveal what is going on in the animal's brain while it is communicating. Does it experience the same mind/body dichotomy that humans do? Philosophical thought on the difference between the mind and the brain extends back at least to Plato, and remains a point of fascination for us. As might be imagined, considering this dichotomy in other species becomes more difficult as phylogenetic distance from humans increases. In perhaps a related vein, that distance also reduces the risk of anthropomorphism.

Self-Awareness and Mirror Tests

Some general tests for cognition can be found in the introduction to this chapter. Other tests address self-awareness in particular, and they are covered here. For instance, a simple form of cognition is the ability to differentiate "self" from others. Some of the earliest explorations in cognition used mirror tests to determine if animals showed evidence of self-awareness, that is, an ability to separate their concepts of their own bodies (self) from the bodies of others (Figure 6.8). When a dog or cat walks in front of a mirror, it may show some interest in the image in the mirror and in some cases become aggressive or territorial toward it, as if the image were an interloper in the home. Ultimately, the dog or cat habituates to the image (see Chapter 5), perhaps because it does not respond in interesting ways, perhaps because it has no smell. There is nothing about the dog's or cat's behavior that says "that's me in the mirror." How might a scientist determine if an animal has some perception of self when it looks at its image in a mirror? One way to do this is to put a mark on the animal or otherwise change its physical appearance. Then if the animal responds to the difference, this could be interpreted as evidence for self-awareness. The use of mirror tests as a measure of self-awareness is controversial and should be approached as a topic for critical thinking about the concept and its application.[25]

FIGURE 6.8
Do elephants have a theory of mind? They do if mirror tests are indicators of such a state. Although previous mirror tests had not revealed such behavior on the part of elephants, when the setting was made more elephant-friendly, elephants made it clear that they understood that THEY were the elephants in the mirror! *Photo: Richard Byrne.*

187

OF SPECIAL INTEREST: MIRROR NEURONS

Mirror neurons activate both when an animal expresses a behavior and when the animal observes that same behavior performed by another animal. Found in the brains of vertebrates and best studied in humans and monkeys, mirror neurons provide a mechanism that allows comparison of an action with its targeted result. Mirror neurons located in the premotor cortex and the parietal lobe have important roles in object manipulation and grasping.

Some scientists have hypothesized much deeper functions for mirror neurons, including roles in imitation and cognition. The concept behind these arguments is that mirror neurons could be key in cognitive aspects of language and empathy. The large imitative component of mirror neuron function suggests that information from the mirror neurons could be used in "projecting" another animal's emotional state. Studies of mirror neuron function in people with autism spectrum conditions show that sometimes (but not always!) these individuals have a mirror neuron deficit. In a monkey, the rhesus macaque, mirror neurons are involved in shaping and imitating grasping movements and have a larger role in the development of the animal's cognitive understanding of the principle of grasping. In other words mirror neurons help in forming generalizations.

The possible role in emotional cognition for mirror neurons is controversial. Advocates of this interpretation point to emotional and cognitive disorders associated with damage to the part of the brain, the parietal lobe, where some of the mirror neurons are found. Critics point out that imitation of object manipulation does not establish empathy. If a monkey's mirror neuron fires when it touches another monkey, and that same neuron fires when the monkey watches another monkey touching, is that empathy? Vilayanur Ramachandran, a neuroscientist who has publicized his ideas in a TED talk and other media, has argued just this. Further research is needed to resolve the question of involvement of mirror neurons in emotion and empathy.[26,27]

The basic procedure for a mirror test is to provide a mirror in a focal animal's habitat, and to allow it to observe and explore the image in the mirror. Given some time, the animal either figures out that it is looking at itself, or it continues to perceive the image as something other than itself. After a defined interval, the observer changes the appearance of the *focal animal*, perhaps by dyeing a patch of fur on its head a contrasting color that the animal can see. After this manipulation, if the focal animal sees the image as its own reflection, it may respond by touching the dyed patch, or otherwise acknowledging that its own appearance has been changed. This acknowledgment supports the hypothesis that the animal has a sense of self and therefore is cognitive.

To date, dolphins, magpies, elephants, and nonhuman apes such as chimpanzees have all shown positive responses in mirror tests.[28] The ability of the great apes to perform well in mirror tests is not surprising, given their evolutionary proximity to humans; the expectation is that minds of animals that are closely related to humans work similarly to human minds. While very young human infants do not respond positively in mirror tests, the response develops by the time a child is 18–30 months old. The major drawback to these tests is that a negative result cannot be interpreted as a lack of cognition. An animal may perfectly well recognize, internally, that it is seeing itself in the mirror and that its appearance has changed, but it may not be primed to display any overt response. This is a good example of why scientists are cautious when they get negative results from an experiment; in the long run, more refined tests of the same hypothesis may give a positive result.

> **KEY TERM** The focal animal is the animal under observation. Focal animal sampling involves extended observation of one animal, in contrast to scan sampling, which is intermittent and repeated observation of many animals.

The importance of design and its effect on results can be seen especially clearly in the case of elephant mirror self-recognition. Initial mirror tests with elephants showed no self-recognition; the mirrors were relatively small and were kept out of range of the animals' trunks. When these two design features were changed, and when large (elephant-proof) mirrors were placed close enough for elephants to touch and explore the mirrors, results changed, and elephants displayed self-recognition. The three female elephants began by trying to look behind the mirror and continued with a variety of behaviors, including ones that they did not perform in any other location, such as exploring their oral cavities in front of the mirror and moving food to a location immediately in front of the mirror, where they consumed it. One of the three showed a dramatic increase in head-touching (where the mark was located) in the mark test.

Does this concept of self as revealed in mirror tests really demonstrate that an animal has a theory of mind—for example, does it suggest a cognitive ability to predict how the animal's own actions might then translate into subsequent events within a social group? This issue is certainly open to debate; an alternative hypothesis is that mirror tests measure problem-solving ability and insight learning, rather than cognition driven by a theory of mind. This uncertainty plagues research on animal cognition because it is always difficult to separate the result of associative thinking from actual cognition. For now, mirror test results are intriguing and suggestive but do not conclusively support arguments for cognition. Other tests of theory of mind hypotheses have yielded equal uncertainty about how the results should be interpreted.[29]

> **KEY TERM** Insight learning is the spontaneous ability to solve problems or make novel connections without a process of trial and error. The experiences that we call "'aha!' moments" are examples of insight learning.

Gaze Following

As described previously, gaze following is another basic measure of an animal's ability to separate self from others. Humans are very sensitive to gaze direction; a person can easily observe the direction that another person is looking and identify the object of that person's attention. Human infants develop the ability to follow gazes at 6–8 months old, much younger than their development of responses in mirror tests. For many years, investigators thought that only humans could follow gazes; newer studies demonstrate gaze following in a variety of primates, suggesting that, like numerical assessments, gaze following is deeply rooted in primate evolution. Rhesus monkeys can follow gazes by looking at the orientation of another monkey's head, rather than focusing on the eyes. Most dog owners believe

FIGURE 6.9

Gaze following experiments may be used as evidence—albeit controversial—of theory of mind. These animals are looking intently at something, and humans can follow their gaze, but can they follow each other's gazes? This is an interesting, and difficult to test, question in animal cognition. Likely the chimpanzee (right) can follow other chimpanzee's gazes, while the hypothesis is untested in ring-tailed lemurs (left). *Photos: Michael Breed (left) and John Mitani (right).*

that their dogs can follow their gazes, but this ability has only recently been established scientifically for dogs;[30] studies of wolves suggest that gaze following may be a unique result of the domestication of dogs, along with artificial selection that favors dogs who can follow human instructions. The fact that dogs are more likely to follow a human gaze if the human first looks at or speaks to the dog supports the artificial selection hypothesis (Figure 6.9).[31,32]

Like mirror tests, positive results of gaze following studies can be interpreted to support a concept of self for an animal that follows a gaze; animals that do not follow a gaze may not have a theory of mind—the ability to project the thoughts of other animals. Returning to the definition of cognition presented earlier in this chapter, humans use gaze following to assess intent and forecast the future. But do other animals, even chimpanzees, use gaze in the same way? Perhaps gaze following is merely a functional adaptation that allows an organism to compete for food, spot potential predators, and perform similar simple and not necessarily cognitive operations. As with the mirror tests, the interpretation of gaze following as evidence of cognition remains controversial.

Self and Self-Consciousness

A level of self-consciousness is implicit in a discussion of self and self-awareness. Self-consciousness involves an ability to judge one's own actions in the context of values or traditions within a social community. Self-conscious judgments involve knowledge of normative or standard behavior and a realization of how others would view an individual's behavior.

How would a scientist know if an animal is self-conscious? The central argument for self-consciousness in animals derives from behavior that humans interpret as representing embarrassment or remorse. When a person leaves the room and his or her dog filches food from a table or visits the kitchen garbage, the person may later see behavior—tail down, eyes drooping, ears low, slinking away—that could be interpreted as embarrassment or remorse. This interpretation carries a high risk of anthropomorphism (see Chapter 1) and is not viewed as strong evidence for cognition.

BRINGING ANIMAL BEHAVIOR HOME: "GUILTY" DOGS

Many dog owners think their dog looks "guilty" if the animal violates a household rule. "Guilty" looks in dogs include submissive facial expressions, lowered tails, and cowering. Common violations include taking food from a counter or rummaging in the garbage when no one is in the room. Alexandra Horowitz tested this perception using a simple but clever experimental design.[33] The experiment focused on dogs in their familiar household environment. The dog's owner left a room, and then for half the dogs, the experimenter gave an illicit treat, while half the dogs received nothing. The owners were then either told that their dog had violated the family rule or that the dog had complied. Half of the owners received the truth, while half were told a lie.

In two types of dog-owner pairs, the owner knew the truth:

1. the dog received a treat and the owner knew it, and
2. the dog did not receive a treat and the owner knew that.

In the other two types, the owner had false information:

3. the dog received a treat but the owner thought it had complied with the rule, or
4. the dog did not receive a treat and the owner thought it had broken the rule.

In conditions 1 and 4, the owners perceived that their dog was a violator, even though only half of these dogs had actually violated the rule. In conditions 2 and 3, the owner thought the dog had followed the rules, when in fact in half of the cases it had not.

Owners judged their dogs as looking equally guilty in conditions 1 and 4, whether or not the dog had actually followed the rule. Owners judged their dogs as looking not-guilty in conditions 2 and 3, even though half of these dogs were guilty. The driving factor was the owners' punitive demeanor when told that their dog had violated the rules. This illustrates how sensitive dogs are to human moods, and argues against dogs feeling a moral sense of regret when they take an opportunity to violate a rule.

OF SPECIAL INTEREST: EMPATHY

Theory of mind is seen as a prerequisite for *empathy*. Are animals empathetic? How can this be tested? Some workers have suggested that self-recognition (see the earlier discussion of mirror tests) is related to empathetic behavior; they point out that species that can recognize themselves in mirrors have also displayed behaviors that can be interpreted as empathetic. In more nearly direct tests, other scientists have examined facial mimicry, a behavior that is closely related to "emotional contagion," the human tendency to share the emotions of others. Facial mimicry (smiling when smiled at, yawning in the presence of yawning people) can be voluntary or involuntary; the latter can occur within a second or so of the initial event. In orangutans, the open mouth face is an affiliative expression and is mimicked within one second (Figure 6.10). Scientists concluded that this building block of empathy is exhibited by a great ape that last shared an ancestor with humans at least 12 million years ago.[34]

FIGURE 6.10

The open-mouth face of orangutans (shown here) is mimicked within one second by another orangutan; facial mimicry is one facet of empathetic behavior.

Although rats are typically given a bad rap by most humans, recent work has pointed to empathetic behavior by these rodents. When confronted with a cagemate that has been trapped in a restrainer, free rats circled the restrainer. They contacted their cagemate through holes in the restrainer, bit the restrainer, and performed a variety of activities that eventually resulted in their opening the door and liberating their cagemate. When faced with the choice of liberating a cagemate or "liberating" chocolate (a preferred treat) in a second restrainer, the rats did both and often shared the chocolate with the rat that they liberated.[35] The study of empathy in animals is gaining traction within the scientific community, and it is satisfying to us that even animals as maligned as rats may be helping to light the way.

> **KEY TERM** Empathy is the ability to project or "feel" the emotions of another animal.

BRINGING ANIMAL BEHAVIOR HOME: COMPANION ANIMALS

Facial Mimicry in Dogs?
It is easy enough to recognize facial mimicry in the great apes, which have faces that are similar to ours, but what about our "best friends?" Several primate species exhibit "contagious yawning," the tendency to yawn when witnessing conspecific yawning. Dogs, however, may cross species lines in this ability. They not only yawn when humans yawn, but they yawn at the sound of human yawns, especially the sound of yawns from humans that they know. This field of research has its share of controversy; not everyone is convinced that dogs are sensitive to human yawns. Contagious yawning has been investigated with a variety of methods, including audio- and video-only, and some of the variation in results may be due to the ways in which yawning is presented to dogs. Because of that, this corner of animal behavior is a microcosm of many of the issues that arise when we investigate cognition: What is the role of experimental design? To what extent is yawning related to empathy?[36,37]

191

6.3 THOUGHT, FORESIGHT, AND PROBLEM SOLVING

Another way of considering cognition and a theory of the mind is by using the concept of mental time travel, that is, thinking about the past and using that information to form plans for the future. Humans reflect on the past and hope that knowledge gained from those experiences helps in coping with the future. Mental time travel is more complex than simply benefiting from trial-and-error learning (see Chapter 5 on learning); it involves memory, and a good deal more, in that it brings together seemingly unrelated pieces of information, from experiences that perhaps occurred at different times, to anticipate the solution to a problem. *Chronesthesia* is a term related to mental time travel. A chronesthetic animal is aware of past and future, and its life in the present is shaped by that awareness. As with the theory of mind concept, testing animals for chronesthesia is a challenge; experiments must ask animals to respond to complex tasks in a way that could be possible only if the animal has chronesthetic awareness.

> **KEY TERM** Chronesthesia is the awareness of the past and the future, and the use of that information in the present.

Memory and Cognition

The differences among semantic, procedural, and episodic memories are important to understanding mental time travel. *Semantic memory* is the abstract mental representation of concrete identification or concept. The ability to learn and remember language is semantic memory. *Procedural memory* is the ability to remember and apply a series of steps to a task. *Episodic memory* places learning into specific contexts of what, when, and where, allowing cognitive activities such as reflection and forecasting.[38]

192

SEMANTIC MEMORY

The average dog can learn and react correctly to 100 verbal/commands, such as "walk," "sit," "stay," "doggie," "squirrel," and so on, but can also respond to the specific names of the people with whom it interacts. Rico, a border collie, learned over 200 items and could associate each item with a spoken word.[39] This suggests a degree of semantic memory that extends beyond the range variation in canine vocalizations. Alex, an African grey parrot (this species is shown in Figure 6.11) who worked with Irene Pepperberg, knew 150 words and could identify 50 objects. While 150 or 200 words is impressive, it pales in comparison with the 60,000-word vocabulary of a human high school graduate. But perhaps animals such as dogs and parrots have much larger internally held "vocabularies" that are not represented in the tests that humans have devised. Semantic memory is not considered cognitive, because even though the animal may draw an abstraction (such as "doggie" as a general term for dogs), no mental time travel or creative use of combined facts is necessary for semantic memory.

Experiments that feature a single animal, such as Rico or Alex, are subject to considerable suspicion. Are the experimenters' claims really supported, or is there a design flaw? Of particular concern is the Clever Hans effect. This effect, detailed in Chapter 1, means that it is possible for animals to produce the "right" answer by responding to unintentional cues from their human trainers. In the experiments with Rico, the dog was given the command to retrieve an item, and then went into an adjacent room to select the object from an array of objects; no one was in the room with him, so a Clever Hans effect was impossible. With Alex, his ability to interact and respond to a variety of humans argues against a Clever Hans effect in his use of vocabulary and object identification. The assessments of the semantic abilities of Rico and Alex are well verified.

PROCEDURAL MEMORY

Procedural memory is memory of how to perform a task. The steps of a task, such as a bee working a complicated flower to obtain nectar or a monkey opening a difficult fruit, are remembered in detail and in order. Procedural memories are formed by trial and error and can be modified and improved over time. Despite the impressive accomplishments that can result from procedural memory, it is not cognition because it does not require reflection and integration of information.

EPISODIC MEMORY

Episodic memory captures individual experiences of particular times and places. Episodic memory associates what, when, and where, and is the basis for cognition, because the animal that remembers episodes can use them to recall previous events, or to project forward, predicting future events. These are key elements of mental time travel. For example, if an animal has previously forded a rapidly running stream, it might reflect on that experience and form a plan for fording a newly encountered stream. Sometimes reference to an occurrence at a specific time is useful, as in time–place learning (see the following section on time–place learning). While humans seem to refer to specific times during the circadian cycle (the time of day that an event happened), rats measure the time elapsed since the event happened.[40] Either approach is useful in cognitive processes. Experiments designed to test for cognition in animals focus on episodic memory.

Time–Place Learning

Many ecological events occur at predictable times and places. Often, predators appear, flowers produce nectar, and potential mates display at predictable times and places. *Time–place learning* is the ability to remember both the location of a resource and the time at which the resource is present. The honeybee was the first well-studied example of time–place learning; foragers can learn to return to feeding stations at the same time on a daily schedule. Studies of time–place learning usually involve training the animal to a feeding location where food is provided for a limited interval, such as 15 min, once a day. After a few days—usually two or three are sufficient—the food is withheld. If time–place learning has occurred, the animals will show up at the appointed time and place, even though the food is not there to attract them. Control measurements, taken through the day, will show few, if any, visits to the feeding location at nonfeeding times, with the exception of times just prior to feeding. The appearance of trained animals prior to the exact feeding time, termed *anticipation*, is typical of time–place learning, as anyone who feeds a pet on a schedule can verify (Figure 6.12).

> **KEY TERM** Time–place learning is the ability to associate a reward or behavioral consequence with a specific location and time, and to use that information to return to the location at the specified time in the future.

Time–place learning brings together two remarkably different pieces of neural machinery: the animal's internal clock and the animal's spatial memory of its landscape. To the best of human knowledge, all organisms have a circadian clock—even fungi have one!—and many animals can learn key aspects of their environments and navigate repeatedly to specific locations. This is particularly true of central place foragers; these are animals such as social insects, birds, and mammals that leave their nesting locations to forage and then return to the nest. (See Chapter 9 for more information about central place foraging.) Time–place learning can be an important component of feeding for central place foragers.

193

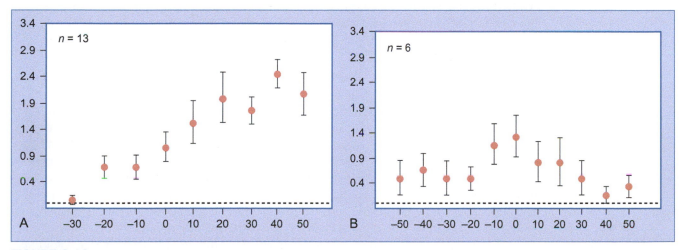

FIGURE 6.12
Time–place learning in a tropical stingless bee (*Trigona fulviventris*). Both of these graphs show the number of bees responding over time (horizontal axis). (A) For several days prior to the day that these data were collected, sugar water was provided at time "0." As a result, some bees are showing up as early as 20 min before feeding time. This is called "anticipation." (B) In this case, no food was offered to bees that had previously been trained to arrive at a given time. Bees showed up early, but gave up quickly after no sugar water appeared at time zero. *n* = number of bees. *Adapted from Murphy, C.M., Breed, M.D., 2008. Time-place learning in the stingless bee,* Trigona fulviventris. *J. Kans. Entomol. Soc. 81, 73–76.*[41]

Although fish tend not to be central place foragers, some fish (e.g., rainbow trout) can also time–place learn.[42] Mutations of circadian clock genes in mice disrupt time–place learning; this is not surprising, but confirms the importance of the circadian clock in time–place learning.[43]

Time–place learning can be remarkably complex; honeybees can learn up to nine different time–place associations and can hold this memory for about a week. For a bee, the ability to retain the memory for several days is important because rain or cold may cause bees to skip a day or two of foraging.

Is time–place learning episodic (mental time travel) memory? The ability to associate time and place is complex only in that it unites two separate neural processes involving time and space; the actual act of associating the time, place, and resource is simple. But if episodic memory is defined as the ability to remember what, when, and where, and to use that information to shape future behavior, time–place learning merges with episodic memory. Time–place learning nonetheless falls a bit short of the definition of cognition because it lacks the necessity for mental reflection. This illustrates how tricky it is to separate examples of what–when–where learning from truly cognitive behavior.

Caching and Thievery

Collecting more food than can be immediately consumed leads to an interesting set of problems. The food can be actively defended from competitors, or it can be hidden to prevent discovery by competitors. The general term for hiding food is *caching*, a topic introduced in Chapter 5 and dealt with in more detail in Chapter 9 on foraging. Hiding (caching) presents two problems of its own; the first is finding the food at a later time, and the second is reacting appropriately to the presence of potential thieves.

The Western Scrub Jay provides a classic example of cognitive processes.[44,45] Focusing on the problem of cache robbers and using the logic of "it takes a thief to know a thief," the experiment compared nonplunderers with experienced cache plunderers and asked if experienced birds behaved in ways that better protected their own cache. Indeed, birds with experience as thieves displayed greater alertness when a potential robber observed their caching behavior, and they responded by later recaching the food (Figure 6.13). This is consistent with the idea that recaching happens more frequently in birds that rob caches because those birds foresee that other birds may behave like thieves, and use this mental forecast to shape their own behavior.[46]

Cognitive Maps

Some investigators have argued that the use of *cognitive maps* is evidence for cognition in honeybees.[47] A cognitive map is a mental representation of landscape that an animal can use to calculate optimal routes between locations. Because calculating a route from an internally held map involves reflection, problem solving, and foresight, an animal that plans its routes in this way could be considered cognitive. (See Chapter 8 on navigation for more detailed information about the experiments that test for cognitive maps in bees and other animals.) This is all the more impressive when we consider that Egyptian fruit bats (*Rousettus aegyptiacus*) are able to form such maps in three dimensions as they fly![48] While the use of cognitive maps would reflect sophisticated thinking and a certain type of mental time travel, it does not require a theory of mind or involve personality, emotion, or empathy, which are other features often used to characterize cognitive animals. As a result, skeptics feel that even if bees use cognitive maps, this does not establish them as cognitive animals in the same sense as has been argued for birds and mammals.[49]

KEY TERM A cognitive map is a mental representation of an animal's landscape used for calculating optimal routes.

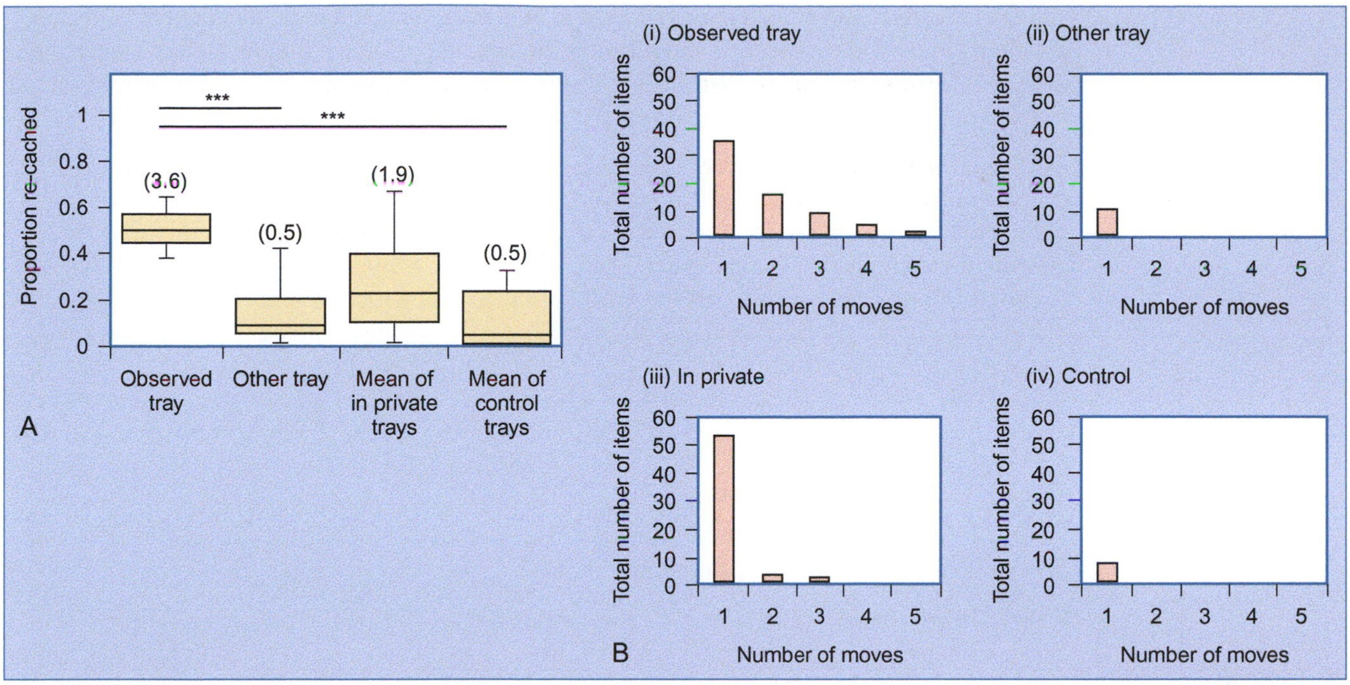

FIGURE 6.13
Cache-hiding in Scrub Jays (*Aphelocoma californica*). In (A), focal birds have foraged on food while being observed by other birds in an enclosure. These birds frequently move their caches, suggesting an awareness of their competitors and the use of a caching strategy to avoid loss of food. (B) is a condition in which the focal birds foraged in private; in this circumstance they rarely moved their cache. *** indicates $P < 0.001$. *Adapted from Daily, J.M., Emery, N.J., Clayton, N.S., 2006. Food-caching Western Scrub-Jays keep track of who was watching when. Science 312, 1662–1665.*[44]

DISCUSSION POINT: COGNITIVE MAPS

One of us lives with a yellow Labrador retriever named Clyde in a house with front-facing windows on the first and second floors. When the doorbell rings, Clyde looks out the ground-level windows to see who is at the door. If the shutters are closed, he runs upstairs and looks out the second floor window. There are two possible explanations for his "knowledge" that if he runs upstairs, he will see the same view from a different perspective. He could be using a complex cognitive map of his world to forecast that both perspectives will allow him to see what he wants. Or, he could be using simple associative learning so that he identifies the two views as essentially the same, without referring to anything resembling a cognitive map. This simple example illustrates the conundrum of cognition: How can an experiment be designed to separate simple association from more complex cognitive processes?

Experimental Approaches to Mental Time Travel and Foresight

A study of bonobo chimpanzees and orangutans investigated tool use, intelligence, and ability to plan for the future.[50,51] The animals first learned to use an object as a tool to obtain a reward. They then were allowed to choose among suitable and unsuitable tools for this task and to take one into a waiting room. An hour later, they were allowed back into the test room. In almost half of the trials, subjects selected a suitable tool, transported it out to the waiting room, and then brought it back into the test room to get the reward. Subsequent experiments showed the same ability to plan ahead when the time interval between tool choice and reward was much longer. The experiment also demonstrated that animals vary in cognitive ability. The star performer was the orangutan Dokana, who in the first study succeeded 15 out of 16 times—more than twice as often as the second best performing animal.

Like many of the questions raised in this text, intraspecific differences among animals pose problems that reach far beyond cognition itself. Because selective pressures vary in space and time, many traits should exhibit this kind of intraspecific variation. As Dokana shows us, care must be taken when speaking of a "typical" orangutan…or a "typical" anything else!

Does such planning happen spontaneously, or is it restricted to carefully designed experiments? If he could talk, a 30-year-old male chimpanzee in the Furuvik Zoo (north of Stockholm) would emphatically agree that planning does happen, and punctuate his statement by flinging a stone projectile in our direction. When visitors are absent, this dominant male calmly collects stones and even pieces of concrete that he modifies into discs. He caches these objects along the shoreline of his exhibit, facing the area where visitors appear. He then throws the stones at visitors during highly agitated dominance displays. The evidence for planning is clear in this case: the stones are only cached in areas where they can best be hurled at visitors, they are collected hours earlier, in the absence of visitors, and when the stones are collected, the chimpanzee is always in a calm (not agitated) state. The caches have numbered in the hundreds, and the gathering behavior itself has been recorded over 50 distinct times. The chimpanzee has even been observed to knock on concrete rocks and break off pieces that were then cached.[52]

Keep in mind that the only reason caretakers noticed this particular planning behavior was because visitors were getting rocks hurled at them. In other words, it was a highly conspicuous behavior. This suggests that more subtle forms of planning may abound in nature, and that our experimental demonstrations of animal planning are only one more small glimpse of the vast unknown that is animal behavior. What kinds of animal planning *don't* we notice!?

Another approach to the detection of mental time travel involves asking an animal an "unexpected" question that requires it to reflect on its experience and, if it can, to generate a novel answer. Zentall et al.[53] took this approach with pigeons. Facing an unexpected question calls on episodic memory, the type of memory used to make time associations and engage in mental time travel. Pigeons were trained to peck at locations to receive rewards and then had to refer to that previous experience when confronting a new task. The behavior of the pigeons suggested that they do, indeed, use episodic memory to inform answers to unexpected questions.

Problem Solving

There are numerous examples of clever problem solving by animals;[54] Corvids (the jays, crows, magpies, and ravens) are notoriously intelligent, and have frustrated as many people who have tried to outwit them as they have amused others. A clear tube 42 mm wide containing some water, along with a delicious caterpillar floating in the water, provided the setting in which four rooks (an Old World corvid) could demonstrate their problem-solving abilities. There was not enough water in the tube for the caterpillar to float within reach of the rooks, but there were some stones provided, and by the second trial, all of the rooks used stones to raise the water level, bring the caterpillar within reach. One rook later became ill after eating one of the caterpillars, refused to approach the tube again, and was removed from the study (Are you surprised? See Chapter 5). The birds not only used stones to raise the water level, they displayed an impressive accuracy in their estimates of the number of stones needed.[55]

An impressive array of animals have been shown to be good problem-solvers, ranging from a variety of corvids to chimpanzees and capuchin monkeys.[17,20,51] Some of these events have been interpreted as a form of insight learning.[15] In the case of rats, such learning is accompanied by a noticeable shift in neural activity. Scientists first gave rats two levers, each with a light; the rats learned that if one of the two lights came on and they pressed the lever associated with that light, they would be rewarded with food. Then the tricky scientists changed the rules: No matter what the lights did, food appeared only when one of the levers—left for some rats, right for others—was pressed. This revealed that neural activity in the prefrontal cortex of a rat shifts abruptly as the rat suddenly realizes what the new task requires and modifies its behavior accordingly. In other words, problem solving, especially that involving insight learning, is accompanied by definite change in prefrontal cortex neuronal activity.[56]

DISCUSSION POINT: TOOL USE

"Everyone knows" what tool use is, correct? It's one of those behaviors that doesn't require much definition. Or does it? In a comprehensive review of tool use among animals, Crain et al. note that while definitions abound—and the number of examples grows—there is no universally accepted definition of the term. This means, among other things, that some potential examples of tool use remain unrecognized and other alleged examples may simply be cluttering the landscape. Review some examples of tool use reported in this book and elsewhere and attempt to craft and defend a definition of the term. Would you include the use of living organisms as tools? Is the type of task important? What about the way in which the tool is used? Is a rat using a tool when it presses a lever?[57] How do we interpret major differences in tool use between captive and wild individuals?[58]

Counting

Animals often need to determine how many items are present. Counting and the ability to make calculations using numbers are strongly related to the issue of problem solving. While the ability to count is not necessarily evidence of cognition, the ability to make comparisons, such as "more than" or "less than," and the ability to predict future rewards based on current numerical values do suggest cognitive processes.

Determining whether one quantity is more or less than another quantity can also provide critical information. Counting the number of eggs or offspring being tended can provide much needed information to a parent. Assessing the number of food items, and comparing relative numbers of food items between patches of food, is elemental to efficient foraging. A reasonable hypothesis is that counting improves survival in many ways, and that an ability to count, when it occurs, should be favored by natural selection. Is counting a cognitive process? Counting often allows animals to gather information that shapes their future decisions. In the sense that counting is part of forecasting the future, counting is a cognitive process.[59,60]

Testing the hypothesis that an animal can count is nonetheless difficult, and adequate tests have been performed on only a handful of animals. Tests for counting are similar to tests for color vision, in that they usually involve a two-step process: first giving the test animal the opportunity to learn (count) objects and then giving the test animal a choice between sets, one of which has the same number of objects as the learned set.

Using honeybees (*Apis mellifera*), Chittka and Geiger trained workers to feed at a station along a 300 m route from their hive.[61] Small tents, which the bees could use as landmarks to guide their flight, were set up along the route. Bees that learned the route using a fixed number of tent landmarks changed their foraging distance—the location where they searched for the feeding station—if the investigators changed the number of landmarks. Inserting landmarks between the hive and the feeding station caused bees to fly a shorter distance before searching for food, while removing landmarks caused them to fly farther. This suggests that flying bees could count the number of landmarks and that they extended their flight in search of additional landmarks when they did not find as many as they expected.

Coots face the challenge of preventing other birds from laying eggs in their nest. Female American coots (*Filuca americana*) often attempt to lay eggs in the nests of unsuspecting conspecifics. If they succeed, their young will be reared by another bird, benefitting the cheating female. Coot eggs vary in the amount of speckling on their surface, a feature that sometimes allows coot females to tell their own eggs from eggs that have been imposed upon them by other females. Coot females who reject the eggs of cheaters must do so based on maintaining a count of their own eggs in the nest.[62]

197

Tests of monkeys (rhesus macaques, *Macaca mulatta*) and chimpanzees (*Pan troglodytes*) suggest not only an ability to count, but also abilities to add and subtract. In Hauser's (a well-known scientist in the field of cognition at Harvard University) tests of counting and numerical representations in macaques, the monkey was shown a small number of objects that were then hidden with a screen.[63] While the monkey could not see the objects, the investigator could leave the same number (this is the control) or change the number by adding or subtracting objects (this is the experimental treatment). The screen was then removed, and the amount of time the monkey spent looking at the objects was recorded. If the number did not change (control), the monkey spent only a second or two looking at the objects, but if the number did change, the monkey stared at the objects for 3 or 4 s. This and other tests suggest that macaques and chimpanzees can count at least small numbers of objects and perform simple addition and subtraction (Figure 6.14).

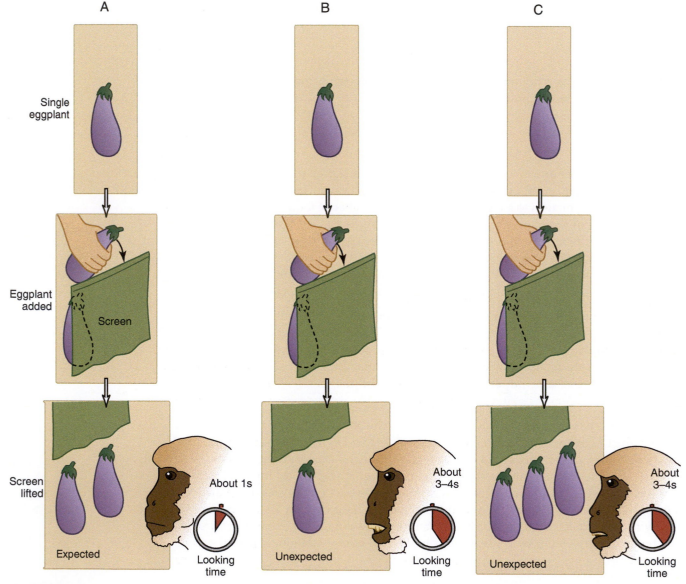

FIGURE 6.14
Some nonhuman primates may be able to count. In tests where a monkey observed two eggplants being placed behind a screen, the monkey stared for a longer period of time if an unexpected number of eggplants (one or three) was revealed. *Adapted from Hauser, M.D., 2000. What do animals think about numbers? Am. Sci. 88, 144–151. Hauser, M.D., MacNeilage, P., Ware, M., 1996. Numerical representations in primates. Proc. Natl. Acad. Sci. USA 93, 1514–1517.*[63]

198

Rats and pigeons, when tested using conditioned responses, showed similar counting abilities as those of rhesus macaques.[64] They could also generalize; rats trained to respond to a certain number of light flashes responded to the same number of bell rings. Clever experiments are almost certain to reveal numerical abilities in a wide variety of animals.

Researchers at Kyoto University's Primate Research Institute showed sequences of numerals (1–9) to three mother–offspring pairs of chimpanzees.[65] The chimpanzees learned the routine sequence (1-2-3, etc.) and transferred this knowledge to sequences that were missing numbers (e.g., 2-3-5-8-9). The researchers then showed the chimpanzees touchscreens with numbers appearing in a variety of locations. For each screen, when the first number was touched, the remaining numbers were masked. The chimps had to recall the location of each number, and touch each masked location in numerical order. The results of these experiments show that chimpanzees can learn—at a glance—the location of nine numerals on a screen. The best young chimpanzee and the best mother chimpanzee were then compared to human subjects (nine university students). The university students did slightly better than the adult chimpanzee, with performance in both declining as duration of exposure to the screen decreased. The young chimpanzee, on the other hand, did better than either university students or the adult chimpanzee, in terms of both speed and accuracy, and its performance did not decline with duration of exposure to the screen. This may suggest that young chimpanzees have stronger attentiveness than the other test groups.

6.4 INTELLIGENCE AND SOCIAL COGNITION

The definition of intelligence is fraught, as it influences the content of intelligence tests and therefore influences attempts at measuring a valued trait among humans. Humans usually think of intelligence as a combination of quickness to learn, ability to retain information, and facility in solving problems. An animal can have all of these and not be particularly cognitive, but intelligence certainly helps an animal to be cognitive. Intelligence tests for humans are notoriously unreliable because they are subject to cultural and socioeconomic biases, and often favor test-takers who have strong learned skills in multiple-choice questions.

Human intelligence tests are certainly difficult to design, but it is more challenging, by far, to create intelligence tests for nonhuman animals. As in humans, if an investigator presents an animal with an intelligence test, the test should (1) give the same result if it is taken again by the same animal; (2) in a demonstrable way, measure a combination of quickness, retention, and problem solving; and (3) have good value for predicting future performance in tasks. In addition, given the principles outlined in the initial discussion of *umwelt* in Chapter 1, problems or challenges in animal intelligence must also include measures that are relevant to that species' ecological context and evolutionary history.

What kinds of animals are generally regarded as "smart?" The list includes primates, mammalian carnivores (members of the Order Carnivora, such as canids, felids, hyenas, raccoons, seals, and their relatives), corvids (members of the bird family Corvidae, such as jays, magpies, crows, and ravens[66]), dolphins, and cephalopod mollusks (particularly octopi). What do these animals have in common that might predict intelligence? They are only distantly related evolutionarily, so the trait called intelligence cannot be interpreted as being monophyletic, that is, as having evolved only once among the animals on a single branch of the evolutionary tree. Intelligence must have evolved more than once (Figure 6.15).

What conditions favor this evolution? Two commonalities stand out: predatory life styles and living in social groups. Some species in this list are both predatory and social; others are one or the other. British psychologist Nicholas Humphrey receives credit for the social intelligence hypothesis—the idea that animal societies are highly shaped by previous social

FIGURE 6.15

Animals that are credited with "intelligence" because of their problem-solving abilities and the possibility that they may show social cognition: (A) Steller's Jay, (B) magpie, and (C) orangutan. The assertion of intelligence in a nonhuman animal can stimulate interesting discussion about how intelligence is defined and measured. *Photos: Jeff Mitton (A), Michael Breed (B and C).*

interactions, and that animals have much to gain if they can predict the outcomes of future interactions and recall previous ones, that is, if they can engage in mental time travel.[67] The social intelligence hypothesis has had strong backing, although some investigators, such as Kay Holekamp of Michigan State University,[68] have questioned the validity of sociality as an explanatory factor in intelligence. Successful carnivory, independent of sociality, can also require quickness, retention, and problem solving; many carnivorous animals are social and gain advantages over larger prey when they hunt in groups. This makes it difficult to disentangle the roles of sociality and carnivory in the evolution of intelligence.

OF SPECIAL INTEREST: SMART BEARS

The cognitive abilities of bears have rarely been studied. They are much more given to omnivory than carnivory and are certainly not social, so the two reigning hypotheses for why animals are intelligent do not pertain to them. They do, however, have very large brains relative to body size and have been reported to use tools.[69] Three bears in the Mobile Zoo in Wilmer, Alabama, have gone where no bear has gone before, at least in the area of cognitive studies, and in so doing, have offered us yet more ideas about the adaptive framework surrounding how animals think.

Using paws or noses to work with a touch screen, the bears demonstrated that they could distinguish smaller and larger numbers of dots (see "Counting," above), and could discriminate areas covered by the dots as well. The bears were also successful at categorization tasks, distinguishing between types of animals, and between animal and nonanimal. In their ability to handle these tasks, the bears are comparable to great apes and other species that we consider intelligent. Although bears are not social, their generalized diet, requiring that they make a multitude of choices, may make intelligence as advantageous for them as it is for social animals.[70,71]

Interestingly, the social insects—the termites, ants, bees, and wasps—have been omitted from the list of smart animals. These insects can be impressively clever, but their problem-solving abilities are limited to highly specific contexts, such as learning how to get nectar from a complicated flower or remembering how to travel from the nest to food and back, and any social politics are worked out over evolutionary time, rather than within the moment-to-moment culture of the colony. (See Chapter 13 on social behavior for a more detailed discussion of the evolution of cooperation.)

The junction of social cognition and apparent intelligence provides some interesting examples.[72] Because of the unique and widespread nature of human interactions with dogs, people probably think more about dog intelligence than that of any other nonhuman species. Dogs can seem hugely clever and act in ways that appear to express social cognition, both with humans and with other dogs. Nonetheless, attempts to design actual intelligence tests for dogs are confounded by scientific inability to distinguish activity and inquisitiveness from problem solving, by human tendencies to project feelings and abilities onto canine companions, and by a desire to see dogs solve problems that are outside the range of conditions their species would have experienced over evolutionary time.

Lemurs present an intriguing test case of social cognition, because within the lemur clade, social group sizes vary considerably. Representatives of six species of lemurs with similar-sized brains and dissimilar social group sizes were given a test of social cognition: Would they try to steal food while the human competitor was looking at them, or would they be more likely to do it when the human's back was turned? Group size was strongly correlated with a seeming talent for theft; species with larger social groups (e.g., ring-tailed lemurs) were much more likely to wait till the human's back was turned before stealing food than were species with small social groups. In contrast, cleverness that did not involve social cues was not correlated with group size. When shown food in a plastic cylinder with open ends, lemurs are understandably tempted to reach directly toward food; they must learn to reach around the cylinder and into the open ends in order to acquire the food. This is a nonsocial cognitive task, and the lemurs' ability to perform it had nothing to do with their social group size.[73]

Punishment in a social context is sometimes called *vengeance* and should not to be confused with punishment in the context of learning (see Chapter 5). It provides an excellent final example of how social cognition, requiring some level of intelligence, plays out. To exact punishment, an animal must understand that another animal did something to it, it must project that there are ways of punishing that animal, and it probably needs to consider the possible response if the punishment is meted out (counter-punishment). Punishment is the social opposite of reciprocal altruism, discussed in Chapter 13. (*Spite*, by the way, is behavior that damages both the recipient and the conveyor of the spiteful behavior; spite is unlikely to evolve because there is no evolutionary benefit to the spiteful animal.)

> **KEY TERM** Punishment in a social context is retaliation by one animal as a result of the action of another animal. Punishment benefits the animal delivering the punishment, to the detriment of the recipient.

> **KEY TERM** Spite is an action which is detrimental to both the animal that acts and the animal that receives the act.

In a test of punishment in chimpanzees, one chimp was given the opportunity to take food from another one that was feeding.[74,75] The chimpanzee whose food was taken could then punish the other chimpanzee by causing the food to drop out of reach. Chimpanzees did not make the food vanish if it was removed by an unseen manipulator or by a human, but they clearly punished other chimpanzees for removing the food. This experiment contained an interesting special circumstance: there was no cost to the punisher for making the food vanish. This design probably maximized the chances that theft would be followed by punishment; nonetheless, the punished animal did not escalate the conflict. Conflict and reconciliation are normal parts of wild chimpanzee behavior.[76]

6.5 THE FRONTAL LOBE AND IMPULSE CONTROL

The ability to control impulses is a key aspect of personality in social groups, particularly when uninhibited behavioral expression might damage either the animal's future in the group or the survival of the group. *Social restraint* and related behaviors such as submission and appeasement are most commonly exhibited by the lower-ranking animals in a social group.[77] One hypothesis is that the enlarged frontal lobes of humans evolved partly to enhance social restraint (impulse control) in human societies. The frontal lobes have a large number of other critical functions, such as problem solving, memory, and language, so judgment and impulse control are not the only selective factors. The primate frontal lobe is larger than that of other mammals and is larger in apes than in monkeys. The volume of the human frontal lobe is about three times larger than what would be expected from looking at the brains of other apes.[78,79]

Social restraint, submission, and appeasement have costs for animals in lower social ranks. In vertebrates, stressful conditions result in lowered levels of the neurotransmitter serotonin and other neurotransmitters in the brain. (See Chapter 2 for a more detailed discussion of neurotransmitters and mood.) Low serotonin stimulates the hypothalamus of the brain to release corticotrophin releasing factor and vasopressin into the circulation. These hormones act on the adrenal glands (small glands associated with the kidneys) so that stress hormones, called *corticosteroids* (also called *glucocorticosteroids*), are released. The steroid hormone cortisol is particularly important in stress responses. Under normal circumstances, corticosteroids prepare the animal physiologically to cope with stressful conditions by increasing blood flow and mobilizing the body's energy resources. This can be part of the "fight-or-flight" syndrome that is key to survival under dangerous conditions.

But what happens to personality if an animal is chronically stressed, such as when a low-ranking animal must continuously restrain its behavior? In a wide range of animals, cortisol levels are much higher in low-ranking animals.[80] Conveniently, cortisol levels can be measured from fecal samples, meaning that the social stress can be measured without the added stress of capture and blood sampling. Hypothetically, an animal's cognition of its low rank may be mediated by feelings that are analogous to depression and anxiety, although this can be inferred only from external attributes, such as self-directed behaviors (see Chapter 4). The effect of stress on personality leads into the next section, which focuses on emotions in animals.

6.6 ANIMAL EMOTIONS

Can a dog love a person? A discussion of emotion logically follows a consideration of personality. Emotion is at the core of human cognitive life. Human self-perception of emotions is one of the roots of the human separation of mind from body. Human theory of mind depends on the ability to perceive the emotions of other humans, the ability to project the effect of those emotions, and the ability to understand how the consequences of human actions depend on the emotional state of the recipient. This leads many humans, scientists and nonscientists alike, to conclude that emotion is a fundamental aspect of cognition. Many humans long to believe that nonhuman animals have emotional lives parallel to human emotional lives. Is this true? Or is it simply a projection of human emotional lives onto animals that in reality have no emotional lives of their own? (Figure 6.16) (For that matter, many people need to believe that animals have no emotional lives.)

In reality, the core question is whether emotions such as joy, sorrow, love, and sadness can be ascribed to a dog (or any animal) as feelings that underlie behavioral actions. Beyond the person's own emotional response—"Yes, of course, my dog loves me"—can the hypothesis that animals feel emotion be tested? A direct test is probably impossible; exactly what a dog thinks may never be known, although brain scans like those discussed at the beginning of this chapter may offer a window into other species' thought processes. However, there is a line of evidence that can be considered as support for the hypothesis that animals have emotions of the human sort.

FIGURE 6.16
Hundreds (and thousands) of years of natural and artificial selection have resulted in dogs that people love and that seem to love people. Who can know the mind of a dog? *Photo: Ross Madden.*

First, the neural architecture that regulates behaviors and their underlying emotions in humans did not arise *de novo* when humans evolved. Humans share brain structure, neurotransmitters, and hormones with the entire vertebrate lineage and beyond. Vertebrate homologies, that is, attributes that are shared as the result of descent from a common ancestor, extend beyond having four limbs and similar bone structures. For instance, human emotions are greatly affected by neurotransmitters such as serotonin; likewise, treatment of animals such as dogs with drugs such as Prozac (fluoxetine) has behavioral effects on dogs consistent with the behavioral effects that Prozac has on humans. Even mice exhibit behavior that reflects "anxiety."[81] This supports the argument that emotions could have deep evolutionary roots within the vertebrates, if not deeper.

In addition, consider the logic of how emotions compel human behavior. In "normal" behavior, human emotions serve as motivators to express evolutionarily appropriate behavioral responses. (See the discussion of motivation in Chapter 4 for more on this issue.) An excellent example of this is the role of emotion in stimulating parental care of infants. If emotion is the great motivator of human behavior, would it be surprising that animals with similar brains are motivated in the same way? If emotion does not drive the behavior of animals with physiology that is similar to humans, what does?

Ultimately, emotionality in animals may be an untestable hypothesis. Descriptors of emotion, such as fear, sadness, loneliness, and affection, are subjective representations of a human state. They are difficult enough to quantify in humans and to compare among humans, much less to document in animals. While joy and sorrow are qualitatively different, there is no way of knowing whether everyone's experience of an emotion is the same. Indeed, some rather incendiary miscommunication can occur if one assumes that human emotional lives are the same. If emotionality is difficult to document in animals that are similar to us, how much more difficult is it to document in, say, an octopus?

In some ways, this chapter has come full circle, in a return to the challenge of anthropomorphism. For the past century, biologists have been trained to avoid anthropomorphism, probably rightly so, because ascribing human emotions and motivations to animals interferes with objective hypothesis testing. It is certainly unscientific to thoughtlessly apply subjective descriptors to the behavior under observation. It is equally unscientific to ignore reasonable hypotheses because of a well-established taboo among scientists. Enough is known now about the neurochemical bases of behavior that the brain is no longer a black box; an extensive fund of knowledge concerning the roles of hormones and neurotransmitters in behavioral modulation has replaced pure imagination.

What should students and researchers of animal behavior do? When appropriate, emotion should be considered as a hypothetical shaper of behavior. In manipulative experiments and clinical treatment of animal behavior, matching subjective descriptions of human emotion with an animal's state gives excellent clues about how pharmacological treatment might affect the animal's behavior. This approach, though, needs to be tempered with the recognition that such subjective representations can lead to gross misinterpretations or oversimplifications of animal behavior and social systems. As with any hypothesis, emotion should not be thoughtlessly denied or affirmed, but should be seen as a potentially testable statement, rather than as an assumption.

DISCUSSION POINT: DO ANIMALS HAVE EMOTIONAL LIVES?

Let us consider the behavior of a female baboon whose offspring has just died. The baboon may stop eating, she may carry the dead baby around for a while, and even when she no longer carries the dead infant, she may appear to be looking for it in familiar places. Eventually, she will resume her previous social interactions, she will have more babies, and behaviorally, she will be indistinguishable from other female baboons.

Now let us consider the behavior of a human mother whose child has died. Unbearably sad, she cannot eat, she does not want to go out in public, and for quite a while, it seems that she hears the child's voice at odd intervals; she is distracted, and may unconsciously look for the child before she remembers how futile that is. Eventually, she will resume her previous social interactions, and she may even have more children. She will behave much like other women.

Both of these scenarios are accurate descriptions of what happens to many individuals—baboon and human—when an offspring dies. What do you make of them? Can you conclude that the baboon's grief is comparable to that of the human's? Is there a way to test that? Given the similarity in behavior, why might a person think that the experience of grief differs between the species? What are the ethical implications of the study of animal emotion? What are the potential pitfalls?[82]

6.7 ARE COGNITIVE ABILITIES UNDER- OR OVER-ATTRIBUTED TO ANIMALS?

Human history is littered with attempts to draw a distinguishing boundary between animals and humans. This is an impossible task because, of course, humans are animals, and whatever makes humans feel special, relative to the rest of the animal world, is simply a set of adaptations for the peculiarities of human life. In a sense, these adaptations create "human-ness," but any other species under observation has its own unique set of adaptations; other species are different, but are not lesser forms of life. In a recent book, Clive Wynne argued: "the psychological abilities that make human culture possible—enthusiasm to imitate others, language, and the ability to place oneself imaginatively into another's perspective on events—are almost entirely lacking in any other species."[83] This is a common point of view, and the contrast between this conclusion and the arguments of some scientists in favor of more widespread cognition in animals is part of what makes this such an interesting field of study. In grade schools in the 1960s, students were taught that specific characteristics—language, tool use, upright posture, and government—separate humans from nonhuman animals. Scientific experience over the intervening 50 years seems to have dissolved all of these barriers. Humans may be better at performing some tricks than nonhuman species, but other species have capabilities humans can only dream about.

Human-centric thinkers tend to give animal species less credit for cognition than they deserve. In contrast, other humans claim to have special empathy for animals, thereby feeling that they are elevated on a "moral" platform above most of their fellow humans. Over evolutionary time, human interactions with animals have passed from coexistence and relationships in which sometimes humans were prey and sometimes were predators, to the domestication and employment of animals as companions, food, and beasts of burden. In the process, humans have become the greatest threat to the continued existence of many animal species. As these relationships have changed over time, humans, as thinking, contemplative animals, have been challenged to reshape ethical and moral views of their interactions with animals.

One of the major points of contention raised by scientific findings about animal cognition is whether humans should then use the discovery of the cognitive lives of animals to inform choices about treatment of those species. Bluntly put, does knowing an animal is cognizant of how it is being treated make humans, as a society, want to treat it differently? If so, how should it be treated? This question is as difficult to answer as others in this chapter because there is good evidence that (not surprisingly) different species are not identical in the ways that they experience "comfort," and even if they were, the conditions that create comfort are often not "natural," nor do they necessarily result in good health.[84] There are no good, definitive, answers to those questions, but they certainly will be the subject of lively debate for the foreseeable future.

Humans can be quite calculating when it comes to assessing the ability of an animal to perform a task, but that facility at predicting animal behavior is not the same as a window into the animal's actual thought processes. In other words, even though humans can project their interpretations onto animal behavior, that does not mean that animals think the same way that humans do.

OF SPECIAL INTEREST: ANECDOTE AS EVIDENCE IN STUDIES OF COGNITION

One of the biggest scientific barriers to acceptance of animal cognition, particularly of personality and emotion as components of the cognitive life of nonhuman animals, is that much of the evidence comes from anecdotes. Typical of this type of information are stories about dolphins rescuing drowning humans by supporting them in the water. If true, this could be argued as evidence for dolphins living an empathic life. But how strong is the evidence? Were the observers reliable? Was the story changed or amplified through multiple retellings? Could the circumstances be replicated? What about all the times when no dolphin appeared? Reliability and replicability are hallmarks of scientific veracity.

Donald Griffin (see p. 256), argued that accumulated anecdotal evidence for cognition could develop into compelling scientific evidence. The current scientific community is divided on this issue, with probably the majority of animal behaviorists having some healthy skepticism about anecdotal evidence for cognition. Accepted scientific standards require that results be repeatable and adequately controlled, a set of criteria not often met by anecdotal accounts supporting cognitive events in animals' lives.

What if the argument that cognitive abilities are widespread among other animal species is accepted? Do scientists also accept the corollaries that personality is a reflection of cognition and that emotions are part of the lives of animals? The question of whether this informs/changes human interactions with animals is still open. The debate over animal cognition reaches into human religion, spirituality, and legal systems. An interesting approach taken in Boulder, Colorado, and other cities has been to legislate that humans are "guardians" rather than "owners" of animals. *Companion animal* has become the preferred term over *pet*. Some people argue that this change in terminology can modify, in a meaningful way, the relationship between humans and nonhuman animals, whereas others view this as semantic silliness. (This calls up the previous discussion of language and mental symbols!)

Following this line of reasoning, critics of the maintenance of animals in zoos sometimes argue that part of the cruelty of animals' captivity stems from animal cognition of surroundings and the conditions of confinement. Chapter 4 presents methods of measuring the stress level of animals and strategies for changing the conditions of captivity to at least eliminate the behavioral manifestations of stress. But the argument against captivity based on hypothesized cognition reaches more deeply, suggesting that it is immoral to maintain an animal whose cognitive abilities allow it to understand that it is captive. For some people, this argument outweighs the positive roles of zoos in education and conservation. Others argue that hypotheses about cognition are nowhere near as well supported as the risk of extinction. Is it better to allow extinction than to use captive breeding? Can animals acclimate to captivity? What about captive-born animals?

SUMMARY

What is cognition? This proves to be a difficult concept to define, and it is fair to say that cognition means different things to different scientists. The separation of the brain as the physical structure that coordinates behavior and the mind as a philosophical concept is central because cognition is a property of the mind. This then leads to questions about tests for cognition. One key element of many definitions of cognition is the ability to visualize oneself in the context of the larger world. Following this line of reasoning, simple behaviors such as gaze following and mirror tests offer some guidance about the cognitive abilities of an animal. Harder to test but also important is how animals forecast social relationships and solve complex problems. Time–place learning, caching, and thievery may also require foresight and planning, making them cognitive processes.

Intelligence, personality, and emotion also play central roles in discussions of cognition. Intelligence is as hard to define as cognition, but it is clear that phenomena that can be interpreted as reflecting intelligence often are cognitive processes. When animals interpret each other's personalities, personality becomes a part of the cognitive picture. The concept of the behavioral syndrome has yielded a somewhat broader tool than personality to understand variation in the behavioral tendencies of animals. Emotionality is, in itself, not a cognitive process, but empathy for the emotions of others is such a process, and the discussion of emotions in this chapter helps in understanding impulsivity, impulse control, and empathy. Cognition separates an animal from being a purely impulsive creature, in that cognition allows reflection, planning, and foresight—in other words, nonimpulsive behavior.

Do humans under- or over-attribute cognitive abilities to animals? People seem to be pulled in both directions. It is tempting to project thought and emotional processes onto animals without critically testing hypotheses of cognition. On the other hand, as more data have been collected, particularly over the past decade, scientific arguments credit animals with much greater levels of cognition than previously suspected. Modern insights have broken down the cognitive/noncognitive divide that many humans have constructed between animals and humans.

What is the endpoint for this evolving topic? Developing an understanding of animal cognition is a blossoming field in animal behavior, and of all the topics covered in this book, it is the one for which scientific understanding is mostly likely to dramatically change over the next one or two decades. For students of animal behavior, the topic of cognition should prove a rich area for debate and discussion. Evolutionary theory compels us to consider whether cognition, like any other suite of traits, is unique to humans or is perhaps shared in some form with other organisms. Scientific knowledge in this area is far from definitive, and the potential ethical ramifications seem to have no end.

STUDY QUESTIONS

1. What is cognition?
2. How might the roles of sociality and carnivory in the evolution of intelligence be disentangled?
3. If animals are cognitive, how should this affect human behavior toward animals?
4. How could a test be devised for the hypothesis that a nonhuman animal has a theory of mind?
5. In the laboratory manual that accompanies this text, students are challenged to develop reliable intelligence tests for dogs (Chapter 15 in the laboratory manual[85]). This is harder than it might seem at first glance!
6. Dog owners often boast of the intelligence of their dog and sometimes show a startling tendency to view their dog's emotional state as a reflection of their own emotions. Dogs can appear quite empathic, but are they truly so? Can empathy be formulated as a testable hypothesis and be subjected to the same sorts of rigorous tests that are required in other areas of animal behavior? How can humans separate their projection of human behavior from what the animal is actually doing?
7. Recall Tinbergen's questions. Although cognition is a new and unsettled field of animal behavior, some topics nonetheless might claim to answer all of his questions. Review this chapter and find a topic or two that you think might be fruitful in that regard.

Further Reading

Balda, R.P., Pepperberg, I.M., Kamil, A.C. (Eds.), 1998. Animal Cognition in Nature, Academic Press, New York, NY.
Bekoff, M., 2002. Minding Animals: Awareness, Emotions and Heart. Oxford University Press, Oxford, UK, pp. 256.
Bekoff, M., Allen, C., Burghardt, G., 2002. The Cognitive Animal. MIT Press, Boston, MA, pp. 504.

Emery, N., Clayton, N., Frith, C., 2008. Social Intelligence: From Brain to Culture. Oxford University Press, Oxford, UK, pp. 432.

Gregg, J., 2013. Are Dophins Really Smart? The Mammal Behind the Myth. Oxford University Press, Oxford, UK, pp. 320.

Griffin, D., 1985. Animal Thinking. Harvard University Press, Boston, MA, pp. 265.

Hauser, M., 2001. Wild Minds: What Animals Really Think. Holt. pp. 336.

Hess, E., 2008. Nim Chimpsky: The Chimp Who Would Be Human. Bantam. pp. 384.

Marler, P., Ristau, C.A., 2013. Cognitive Ethology: Essays in Honor of Donald R. Griffin. Psychology Press, Taylor and Francis, London.

Stamp Dawkins, M., 1998. Through Our Eyes Only?: The Search for Animal Consciousness. Oxford University Press, Oxford, UK, pp. 206.

Wasserman, E., Zentall, T., 2009. Comparative Cognition. Oxford University Press, Oxford, UK, pp. 720.

Wright, W., 1999. Born That Way: Genes, Behavior Personality. Routledge. pp. 304.

Wynne, C., 2006. Do Animals Think?. Princeton University Press, Princeton, NJ, pp. 288.

Wynne, C., 2002. Animal Cognition: The Mental Lives of Animals. Palgrave Macmillan. pp. 231.

Notes

1. Willemet, R., 2013. Reconsidering the evolution of brain, cognition, and behavior in birds and mammals. Front. Psychol. 4, 396.
2. Wells, M.J., 1962. Brain and Behavior in Cephalopods. Stanford University Press, Stanford, CA, pp. 171.
3. Finn, J.K., Tregenza, T., Norman, M.D., 2009. Defensive tool use in a coconut-carrying octopus. Curr. Biol. 19, R1069–R1070.
4. Sreeja, V., Bijukumar, A., 2013. Ethological studies of the veined octopus *Amphioctopus marginatus* (Taki) (Cephalopoda: Octopodidae) in captivity, Kerala, India. J. Threat. Taxa 5, 4492–4497.
5. Mather, J.A., 1994. "Home" choice and modification by juvenile *Octopus vulgaris* (Mollusca: Cephalopoda): Specialized intelligence and tool use? J. Zool. 233, 359–368.
6. Bahlig-Pieren, Z., Turner, D.C., 1999. Anthropomorphic interpretations and ethological descriptions of dog and cat behavior by lay people. Anthrozoos 12 (4), 205–210.
7. Darwin, C., 1872. The Expression of the Emotions in Man and Animals. John Murray, London, pp. 350.
8. Kamphuis, S., Dicke, P.W., Thier, P., 2009. Neuronal substrates of gaze following in monkeys. Eur. J. Neurosci. 29, 1732–1738.
9. Andics, A., Gacsi, M., Farago, T., Kis, A., Miklosi, A., 2014. Voice-sensitive regions in the dog and human brain are revealed by comparative fMRI. Curr. Biol. 24, 574–578.
10. Ouattara, K., Lemasson, A., Zuberbuhler, K., 2009. Campbell's monkeys concatenate vocalizations into context-specific call sequences. Proc. Natl. Acad. Sci. USA 106, 22026–22031.
11. Struhsaker, T.T., 1977. Infanticide and social organization in the redtail monkey (*Cercopithecua ascaniua schmidti*) in the Kibale Forest Uganda. Z. Tierpsychol. 45, 75–84.
12. Owren, M.J., 1990. Acoustic classification of alarm calls by vervet monkeys (*Cercopithecus aethiops*) and humans (*Homo sapiens*): II. Synthetic calls. J. Comp. Psychol. 104, 29–40.
13. Soltis, J., King, L.E., Douglas-Hamilton, I., Vollrathand, F., Savage, A., 2014. African elephant alarm calls distinguish between threats from humans and bees. PLoS One 9, e89403. http://dx.doi.org/10.1371/journal.pone.0089403.
14. Emery, N.J., Seed, A.M., von Byern, A.M.P., Clayton, N.S., 2007. Cognitive adaptions of social bonding in birds. Phil. Trans. R. Soc. B. 362, 489–505.
15. Bird, C.D., Emery, N.J., 2009. Insightful problem solving and creative tool modification by captive nontool-using rooks. Proc. Natl. Acad. Sci. USA 106, 10370–10375.
16. Visalberghi, E., Addessi, E., Truppa, V., Spanoletti, N., Ottoni, E., Izar, P., et al., 2009. Selection of effective stone tools by wild bearded capuchin monkeys. Curr. Biol. 19, 213–217.
17. Sanz, C.M., Shoning, C., Morgan, D.B., 2009. Chimpanzees prey on army ants with specialized tool set. Am. J. Primatol. 72, 17–24.
18. Pruetz, J.D., Bertolani, P., 2007. Savanna Chimpanzees, *Pan troglodytes verus*, hunt with tools. Curr. Biol. 17, 1–6.
19. Mercader, J., Barton, H., Gillespie, J., Harris, J., Kuhn, S., Tyler, R., et al., 2007. 4,300-year-old chimpanzee sites and the origin of percussive stone technology. Proc. Natl. Acad. Sci. USA 104, 3043–3048.
20. Kenward, B., Weir, A.A.S., Rutz, C., Kacelnik, A., 2005. Tool manufacture by naïve juvenile crows. Nature 433, 121.
21. Nowak, M.A., Komarova, N.L., Niyogi, P., 2001. Evolution of universal grammar. Science 291, 114–118.
22. Henry, C., 1995. Universal grammar. In: Rocha, L. (Ed.), Communication and Cognition—Artificial Intelligence. 12 (1–2), 45–61. Special Issue. Self-Reference in Biological and Cognitive Systems.
23. Chater, N., Reali, F., Christiansen, M.H., 2009. Restrictions on biological adaptation in language evolution. Proc. Natl. Acad. Sci. USA 106, 1015–1020.
24. Marzluff, J.M., Walls, J., Cornell, H.N., Withey, J.C., Craig, D.P., 2010. Lasting recognition of threatening people by wild American crows. Anim. Behav. 79, 699–707.
25. Suddendorf, T., Butler, D.L., 2013. The nature of visual self-recognition. Trends Cogn. Sci. 17, 121–127.

26. Oztop, E., Kawato, M., Arbib, M.A., 2013. Mirror neurons: functions, mechanisms and models. Neurosci. Lett. 540, 43–55. http://dx.doi.org/10.1016/j.neulet.2012.10.005.

27. Casile, A., 2013. Mirror neurons (and beyond) in the macaque brain: an overview of 20 years of research. Neurosci. Lett. 540, 3–14.

28. Plotnik, J.M., de Waal, F.B.M., Reiss, D., 2006. Self-recognition in an Asian elephant. Proc. Natl. Acad. Sci. USA 103, 17053–17057.

29. Penn, D.C., Povinelli, D.J., 2007. On the lack of evidence that non-human animals possess anything remotely resembling a "theory of mind." Philos. Trans. R. Soc. B. 362, 731–744.

30. Byrne, R.W., 2003. Animal communication: what makes a dog able to understand its master? Curr. Biol. 13, R347–R348.

31. Wynne, C.D.L., Dorey, N.R., Udell, M., 2008. Wolves outperform dogs in following human social cues. Anim. Behav. 76, 1767–1773.

32. Teglas, E., Gergely, A., Kupan, K., Miklosi, A., Topal, J., 2012. Dogs' gaze following is tuned to human communicative signals. Curr. Biol. 22, 209–212.

33. Horowitz, A., 2009. Disambiguating the "guilty look": salient prompts to a familiar dog behavior. Behav. Processes 81, 447–452.

34. Marina, D., Menzier, S., Zimmerman, E., 2008. Rapid facial mimicry in orangutan play. Biol. Lett. 4, 27–30.

35. Bartal, I.B., Decety, J., Mason, P., 2011. Empathy and pro-social behavior in rats. Science 334, 1427–1430.

36. Yoon, J.M.D., Tennie, C., 2010. Contagious yawning: a reflection of empathy, mimicry or contagion? Anim. Behav. 79, e1–e3.

37. Silva, K., Bessa, J., de Sousa, L., 2012. Auditory contagious yawning in domestic dogs (*Canis familiaris*): first evidence for social modulation. Anim. Cogn. 15, 721–724.

38. Brown, M.F., Farley, R.F., Lorek, E.J., 2007. Remembrance of places you passed: social spatial working memory in rats. J. Exp. Psychol. Anim. Behav. Process 33, 213–224.

39. Kaminski, J., Call, J., Fischer, J., 2004. Word learning in a domestic dog: evidence for "fast mapping." Science 304, 1682–1683.

40. Roberts, W.A., Feeney, M.C., MacPherson, K., Petter, M., McMillan, N., Musolino, E., 2008. Episodic-like memory in rats: is it based on when or how long ago? Science 320, 113–115.

41. Murphy, C.M., Breed, M.D., 2008. Time-place learning in the stingless bee, *Trigona fulviventris*. J. Kans. Entomol. Soc. 81, 73–76.

42. Heydarnejad, M.S., Purser, J., 2008. Specific individuals of rainbow trout (*Oncorhynchus mykiss*) are able to show time–place learning. Turk. J. Biol. 32, 209–229.

43. Van der Zee, E.A., Havekes, R., Barf, R.P., Hut, R.A., Nijholt, I.M., Jacobs, E.H., et al., 2008. Circadian time–place learning in mice depends on *Cry* genes. Curr. Biol. 18, 844–848.

44. Daily, J.M., Emery, N.J., Clayton, N.S., 2006. Food-caching Western Scrub-Jays keep track of who was watching when. Science 312, 1662–1665.

45. Raby, C.R., Alexis, D.M., Dickinson, A., Clayton, N.S., 2007. Planning for the future by Western Scrub-Jays. Nature 445, 919–921.

46. Clayton, N.S., Dally, J.M., Emery, N.J., 2007. Social cognition by food-caching corvids. The Western Scrub-Jay as a natural psychologist. Philos. Trans. R. Soc. B. Biol. Sci. 362, 507–522.

47. Gould, J.L., 1986. The locale map of honey-bees—do insects have cognitive maps. Science 232, 861–863.

48. Yartsev, M.M., Ulanovsky, N., 2013. Representation of three-dimensional space in the hippocampus of flying bats. Science 340, 367–372.

49. Collett, M., Collett, T.S., 2006. Insect navigation: no map at the end of the trail? Curr. Biol. 16, R48–R51.

50. Mulcahy, N., Call, J., 2006. Apes save tools for future use. Science 312, 1038–1040.

51. Fragaszy, D.M., Pickering, T., Liu, Q., Izar, P., Ottoni, E., Visalberghi, E., 2010. Bearded capuchin monkeys' and a human's efficiency at cracking palm nuts with stone tools: field experiments. Anim. Behav. 79, 321–332.

52. Osvath, M., 2009. Spontaneous planning for future stone throwing by a male chimpanzee. Curr. Biol. 19, 190–191.

53. Zentall, T.R., Singer, R.A., Stagner, J.P., 2008. Episodic-like memory: pigeons can report location pecked when unexpectedly asked. Behav. Processes 79, 93–98.

54. Gruber, T., Muller, M.N., Strimling, P., Wrangham, R., Zuberbuhler, K., 2009. Wild chimpanzees rely on knowledge to solve experimental honey acquisition task. Curr. Biol. 19, 1806–1810.

55. Bird, C.D., Emery, N.J., 2009. Rooks use stones to raise the water level to reach a floating worm. Curr. Biol. 19 1410–1414.

56. Durstewitz, D., Vittoz, N.M., Floresco, S.B., Seamans, J.K., 2010. Abrupt transitions between prefrontal neural ensemble states accompany behavioral transitions during rule learning. Neuron 68, 438–448.

57. Crain, B.J., Giray, T., Abramson, C.I., 2013. A tool for every job: assessing the need for a universal deintion of tool use. Int. J. Comp. Psychol. 26, 281–303.

58. Haslam, J., 2013. "Captivity bias" in animal tool use and its implications for the evolution of hominin technology. Philos. Trans. R. Soc. Lond., B, Biol. Sci. 368, 20120421. http://dx.doi.org/10.1098/rstb.2012.0421.

59. Hyde, D.C., Spelke, E.S., 2009. All numbers are not equal: an electrophysiological investigation of small and large number representations. J. Cogn. Neurosci. 21, 1039–1053.

60. Gallistel, C.R., Gelman, R., 2005. Non-verbal numerical cognition: from reals to integers. Trends Cogn. Sci. 4, 59–65.

61. Chittka, L., Geiger, K., 1995. Can honeybees count landmarks? Anim. Behav. 49, 159–164.

62. Andersson, M., 2003. Behavioural ecology: coots count. Nature 422, 483–485. Lyon, B.E., 2003. Egg recognition and counting reduce costs of avian conspecific brood parasitism. Nature 422, 495–499.

63. Hauser, M.D., 2000. What do animals think about numbers? Am. Sci. 88, 144–151. Hauser, M.D., MacNeilage, P., Ware, M., 1996. Numerical representations in primates. Proc. Natl. Acad. Sci. USA 93, 1514–1517.

64. Scarf, D., Hayne, H., Colombo, M., 2011. Pigeons on par with primates in numerical competence. Science 334, 1664–1666.

65. Inoue, S., Matsuzawa, T., 2007. Working memory of numerals in chimpanzes. Curr. Biol. 17, R1004–R1005.

66. Seed, A., Emery, N., Clayton, N., 2009. Intelligence in corvids and apes: a case of convergent evolution? Ethology 115, 401–420.

67. Humphrey, N.K., 1976. The social function of intellect. In: Byrne, R.W., Whiten, A. (Eds.), Machiavellian Intelligence: Social Expertise and the Evolution of Intellect in Monkeys, Apes and Humans. Clarendon Press, Oxford, UK, pp. 285–305.

68. Holekamp, K.E., 2007. Questioning the social intelligence hypothesis. Trends Cogn. Sci. 11, 65–69.

69. Bentley-Condit, V.K., Smith, E.O., 2010. Animal tool use: current definitions and an updated comprehensive catalog. Behaviour 147, 185–221.

70. Vonk, J., Jett, S.E., Mosteller, K.W., 2012. Concept formation in American black bears, *Ursus americanus*. Anim. Behav. 84, 1–12.

71. Vonk, J., Beran, M.J., 2012. Bears "count" too: quantity estimation and comparison in black bears, *Ursus americanus*. Anim. Behav. 84, 231–238.

72. Hare, B., Brown, M., Williamson, C., Tomasello, M., 2002. The domestication of social cognition in dogs. Science 298, 1634–1636.

73. MacLean, E.L., Sandel, A.A., Bray, J., Oldenkamp, R.E., Reddy, R.B., 2013. Group size predicts social but not nonsocial cognition in lemurs. PLoS One 8 (6), e66359. http://dx.doi.org/10.1371/journal.pone.0066359.

74. Silk, J., 2007. Chimps don't just get mad, they get even. Proc. Natl. Acad. Sci. USA 104, 13537–13538.

75. Jensen, K., Call, J., Tomasello, M., 2007. Chimpanzees are vengeful but not spiteful. Proc. Natl. Acad. Sci. USA 104, 13046–13050.

76. Subiaul, F., Vonk, J., Okamoto-Barth, S., Barth, J., 2008. Do chimpanzees learn reputation by observation? Evidence from direct and indirect experience with generous and selfish strangers. Anim. Cogn. 11, 611–623.

77. Liang, Z.S., Nguyen, T., Mattila, H.R., Rodriguez-Zas, S.L., Seeley, T.D., Robinson, G.E., 2012. Molecular determinants of scouting behavior in honey bees. Science 335, 1225–1228.

78. Semendeferi, K., Lu, A., Schenker, N., Damasio, H., 2002. Humans and great apes share a large frontal cortex. Nat. Neurosci. 5, 272–276.

79. Passingham, R.E., 2002. The frontal cortex: does size matter? Nat. Neurosci. 5, 190–192.

80. Ostner, J., Heistermann, M., Schulke, O., 2008. Dominance, aggression and physiological stress in wild male Assamese macaques (*Macaca assamensis*). Horm. Behav. 54, 613–619.

81. Nyberg, J., Vekovischeva, M.O., Sandnabba, N.K., 2003. Anxiety profiles of mice selectively bred for intermale aggression. Behav. Genet. 33, 503–511.

82. Moore, J., 2000. Expression of emotions. BioScience 50, 843.

83. Wynne, C.D.L., 2006. Do Animals Think?, 7. Princeton University Press, Princeton, NJ, pp. 288.

84. Fraser, D., 2008. Understanding Animal Welfare: The Science in its Cultural Context. UFAW; John Wiley & Sons, Chichester, UK.

85. Tillberg, C.V., Breed, M.D., Hinners, S.J., 2007. Field and Laboratory Exercises in Animal Behavior. Academic Press, London, pp. 232.

209

Communication

CHAPTER 7

211

LEARNING OBJECTIVES

Studying this chapter should provide you with the knowledge to:

- Use the concept that communication involves an evolved signal produced by a sending animal and interpreted by a receiving animal.

- Explain the co-evolution between the sender (or signaler) and receiver as a necessary part of the communication process. Evolution refines both the sender's ability to send a meaningful signal and the recipient's ability to perceive and understand the signal.

- Employ the basic principles of co-option, ritualization, stereotypy, and redundancy to explain the evolution signals used in communication.

- Identify the circumstances favoring honest and deceitful signaling in animal populations.

- Integrate how public information from communication that is available both to the intended recipient and to other animals can work to the benefit or the detriment of the sender.

Animal Behavior. DOI: http://dx.doi.org/10.1016/B978-0-12-801532-2.00007-6

- Apply the effects of noise to the evolution of communication.
- Consider the concept that, in interspecific signaling, signals evolve to influence the behavior of other species.

7.1 INTRODUCTION: COMMUNICATION THEORY

Communication can be as simple as the stare of a challenging wolf or as complicated as honeybee dances conveying information about food locations.[1,2] *Communication* always involves the transmission of information from one animal to another. It differs from other forms of information because it involves *signals* sent from one animal to another.

> **KEY TERM** Communication is the transmission of signals from a sender to a receiver.

> **KEY TERM** A signal is produced by one animal and carries a specific message to another animal. Evolution has shaped the signal for its specific function in communication.

Humans are highly communicative organisms, with lives so embedded in communication that we rarely think about it as a phenomenon; nonetheless, communication is also foundational for much of the behavior that occurs among animals in the natural world. It is difficult to imagine the existence of social behavior, for instance, without communication (Figure 7.1).

212

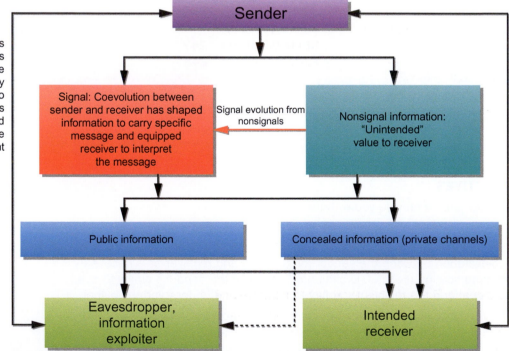

FIGURE 7.1
The flow of information among animals is complex as it includes both signals that have evolved for communication purposes and non-signal information. Both types of information can have considerable value to a receiver. In some instances a sender attempts to conceal the information it is broadcasting; this is termed private information. Other information can be public. Both intended and unintended receivers can take advantage of information opening the door for eavesdropping and exploitation.

CASE STUDY: HEADBANGING

If you Google the word "headbanger," you do not turn up anything at all about termites on the first page, or even on the second, and more's the pity, for when termites headbang, other termites "listen." This is because headbanging is one way that termites communicate danger. Termites of the genus *Macrotermes* build large, durable mounds (6 m high or more);[3,4] these mounds may have the appearance of earth, but they have the consistency of well-dried cement. (One of us has an acquaintance who, as he rode past such a mound on a motorcycle, decided it would be fun to kick the mound. The mound survived intact; he wound up in the hospital.) The mounds are at the center of much larger subterranean foraging galleries that can cover an area of 2000 m². They can persist over decades and benefit the colony in many ways, including colony defense and thermoregulation. When the mound is damaged (and some specialized predators with powerful legs and claws can successfully attack a mound), air currents in the nest change and termites sense this. What they do next is more impressive than anything we found in our Googling of "headbanger."

The soldier termite caste is in charge of defending the colony, and when soldiers of the South African termite *M. natalensis* sense changes in air currents, they hit their heads on the substrate with great impact, creating a series of vibrational pulses. The soldiers are more likely (67%) to respond to air with CO_2 in it than to pure air (44%); this may indicate that while air currents themselves are potentially dangerous, signaling damage to the mound, predator breath is worse. In the first (fastest) strike, the soldier raises its head almost 12 mm above the substrate, then smashes into the substrate at speeds up to 1.5 m/s. Foraging worker termites leave the damaged site, but soldiers in the vicinity begin drumming, thus spreading the signal over the vast expanse of these termite colonies; colony response over tens of meters can occur within a few seconds as a result of this social amplification of the drumming signal (Figure 7.2).

The drumming may have more than one function; an alternate hypothesis is that the drumming could warn away predators, much as bumblebee hissing repels mice. This hypothesis is as yet unconfirmed, but it is clear that *M. natalensis* colonies initiate and amplify this vibrational alarm over distances that could not be achieved with pheromones. By amplifying this signal, these colonies extend the reach of vibrational alarm signaling well beyond that described in individual treehoppers, ants, and other insects.[5]

FIGURE 7.2
(A) A *Macrotermes natalensis* soldier. These insects use their large heads to drum a warning to the rest of the colony when the nest is damaged. (B) A *M. natalensis* nest. *Photos: Wolfgang Kirchner.*

What Is the Adaptive Significance of Communication?

This is a question straight out of Niko Tinbergen's playbook: Why has animal communication evolved? Thinking broadly, perhaps finding a mate is the most elemental and universal reason that communication exists. Finding a mate means being able to identify another animal's location, its species, and sometimes its quality. Such broad need for information favors the evolution of communication.

As if the advantages that communication confers upon mating behavior were not beneficial enough, parents may communicate key survival information to offspring. Within social groups, animals sometimes communicate status, needs for care or desire to play, alarm when predators or competitors are present, and information about the location and quality of food or shelter. For individuals or social groups, communication plays key roles in establishing and maintaining territories.

Many of the studies in the current animal behavior literature ask if signals are honest or dishonest.[6] Honesty and deception in communication revolve around whether the animals share interests in the outcome of their interaction, or if their interests differ. Signals are expected to be honest in the case of shared interests; if interests differ, signals might evolve toward dishonesty.

What Is a Signal?

Simply put, a signal is an animal product that has evolved to carry a specific meaning to another animal. Signals can be external attributes of the animal, such as color; actions, such as waving an arm or producing a sound; a chemical compound, such as an alarm pheromone; or an energetic output, such as an electrical discharge or a magnetic field. The scientific literature offers somewhat more complex definitions. Maynard Smith and Harper define a signal as "any act or structure which alters the behaviour of other organisms, which evolved because of that effect, and which is effective because the receiver's response has also evolved."[7] If we step back from the apparent complexity of these definitions, the key feature is that signals have evolved specifically for communication.

This feature, then—that signals are the result of natural selection that favors information content—is what differentiates a signal from other types of information. Virtually everything an animal does creates information in the environment. Even as signals have been honed by natural selection to carry information, so it is that animals able to receive the signals have also been favored by natural selection. What adaptive value would there be in producing a signal that cannot be received or being primed to receive a signal that never arrives? The definitions are complicated only because there are so many types of signals. Some structural signals are produced almost continuously, whereas other behavioral signals are produced at certain times, when they are likely to be received. Thus, a male Canyon Wren (Figure 7.3A) is brown and speckled all of the time, even though that signal is not always needed. In contrast, it sings when it "wants" to be heard (see Figure 7.3B). Both the color and the song are signals, but one is a persistent part of the physical nature of the animal and the other is a product that is periodically broadcast into the world around the animal.

DISCUSSION POINT: WHAT IS A SIGNAL?

It is sometimes difficult to test the hypothesis that a given attribute is a signal, that is, that a structure or behavior has evolved for communication. This creates some uncertainty about whether or not some information is a signal. To put this into an easily understood framework, consider that you can probably recognize your friends from a great distance by how they walk; in fact, recognizing someone's gait may be your first clue as to his or her identity. Clearly, there is information in the gait, but is it a signal? It would only be a signal if gait has evolved for the specific communicative function of identifying a person. Can you design an experiment to test whether gait is a signal?

Senders and Receivers

Most of the animals we study in this book have, over evolutionary time, modified receptors that originated early in the history of life (see Chapter 2). Regardless of size and complexity, the sender has a structure that constitutes the signal (e.g., the peacock's tail) or that produces the signal (such as vocal cords, scent glands, and electric organs). The receiver has receptors, which are usually modified nerve cells. These cells, such as rods and

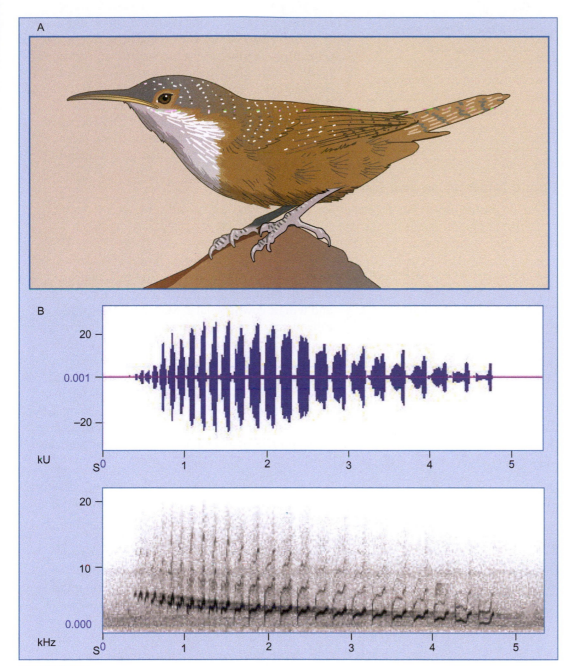

FIGURE 7.3
(A) A male Canyon Wren and (B) a sonogram of its song. Wrens are known for the complexity of their songs.

cones in the eyes, hair cells in ears, and the olfactory epithelium, *transduce* the signal into nerve impulses that are processed by the nervous system. (See Chapter 2.) Decoding sensory information is a critical task of the nervous system; noise must be separated from meaningful signals. In some cases, an animal communicates with itself; bats do this when they echolocate. Communication with oneself is called *autocommunication*. In the study of communication, *noise* refers to a disturbance or variation in a stimulus that interferes with a signal. Noise is not limited to auditory signals but refers to stimuli of any sort.

> **KEY TERM** Autocommunication, or communication with oneself, most commonly occurs in two contexts. The first is echolocation. The second is the use of territorial or trail markers for orientation; an ant might leave a pheromone trail, which it then uses later to return to a feeding location. If no other ants use the trail, then this is autocommunication.

215

> **KEY TERM** Noise is unwanted information that interferes with signals. Acoustic noise, for example, from other animals or from human sources, such as traffic, may interfere with sounds produced as signals.

In sum, these are the building blocks of communication: The sender creates a signal by converting energy into something—a molecule, a movement, an electric field, for example—that can be transmitted into the environment. The receiver uses receptors to transduce the signal into a form that can be used inside the receiver. The information content of a signal can be measured by its effect on the behavior of the receiver. (See Chapter 2 for more information about transduction.)

Public Information and Eavesdropping

When an animal broadcasts a signal, there are "intended" recipients, or targets. A female emitting sex pheromone is luring males of her own species. A bull elk bugling in the fall is signaling his dominance to elk males and his attractiveness to females. When the information is broadcast, however, it becomes *public information*. For instance, eavesdropping female cowbirds can listen to the interactions between male and female cowbird pairs. Male cowbirds produce calls to attract mates, and when a female chooses a male (presumably based on his quality as a mate), she responds to his vocalizations with "chattering" calls. Eavesdropping females learn to prefer male calls that have already elicited chattering by other females.

> **KEY TERM** Public information is signals or attributes of an animal that are generally available to other animals in its environment. Public information can be exploited by other animals, sometimes to the detriment of the animal that has produced the information.

Thus, the eavesdropper could potentially take advantage of public information to find a high-quality mate without having to go through the process of assessing a large number of males (Figure 7.4).[8]

Eavesdropping can be of great value in the presence of predators. Consider the fact that many lizards have ears, but do not seem to vocalize. What, then, is the use of ears? For the Madagascan spiny-tailed iguana, having ears means that it can hear the alarm calls of the Madagascar paradise flycatcher, with which it shares some predators, and then increase its vigilance. When recorded alarm calls were played to spiny-tailed iguanas, the lizards became very still except for increased head movements; song playback did not cause these behavioral shifts. The lizards did not run away or perform other defensive behaviors that are best executed when the location of the threat is known. Eavesdropping on birds' information about predators expands the scope of the lizards' awareness. Ears can be advantageous even in the absence of intraspecific audible communication.[9] (See Chapter 10 for much more information about self-defense behavior.)

FIGURE 7.4
Bettas, *Betta splendens*, are known to eavesdrop on dominance interactions among other bettas and to use this information in their own interactions. *Photo: Michael Breed.*

7.2 EVOLUTION OF COMMUNICATION
Evolution of Signals

How does communication evolve? That is a reasonable question; after all, if communication requires a sender and a receiver, how likely is it that two such organisms would accidentally be able to share information? At first glance, this seems to be an unlikely coincidence, but it makes good sense to return to basic principles: All living organisms require energy sources, and the ability to locate those energy sources has always been strongly favored by natural selection. Thus, photosynthetic organisms must locate light sources, chemosynthetic organisms must locate sources of appropriate chemicals, and heterotrophic animals must locate other organisms for their nutrition. Because all living things emit metabolic waste and byproducts, an organism that could respond to such chemicals in the environment would be at an evolutionary advantage—it could find more energy and do so more quickly than organisms lacking such ability. The ability to sense and respond to energy sources—light, chemicals, prey—means that corresponding receptors exist for those sources—photoreceptors, chemoreceptors, and the like.

However, this correspondence is not yet communication. A rabbit in hiding that has to sneeze is not communicating with the predator; she is just having a bad day. What is important here is that efficient receptors have been favored by natural selection since the early days of living things—as evidenced by the chemical receptors of protists, for instance. These efficient receptors, combined with the ability of a receiver to glean information about a living information source, are the necessary ingredients for the evolution of communication. Indeed, receptors on one-celled organisms also paved the way for intercellular communication in multicellular organisms; as a result, the evolution of signaling systems between organisms and the evolution of signaling systems within organisms are intertwined. Given the right variation, animals can evolve receptors for any form of energy in the environment, and once these receptors have evolved, they become fair evolutionary game for use in communication systems.

Now consider an animal that is ready to reproduce. Mate-finding can be hazardous and time consuming. If a reproductively receptive organism has an attribute that can be perceived by a receiver that is ready to mate and that then responds to the attribute, both animals benefit from the increased reproductive efficiency. To generalize, if selection favors changes or variation in the sender's attributes, and favors the receiver's ability to ascribe meaning to this variation, then communication is taking place. At this point, the sender's attribute can be called a *signal*.

217

Co-option, Ritualization, and Stereotypy

The evolutionary course of *co-opting* an attribute to form the basis of a signal is called *ritualization*. Ritualization is adaptive because in communication, both parties stand to benefit from increased efficiency and accuracy. Therefore, in many cases, selection favors reduction in variation in the signal; this is the evolutionary process of *stereotypy*. Note that this is different from stereotypical, repetitive behaviors covered in Section 4.2, which are often caused by poor welfare; the similar terminology may be initially confusing. Reduced variation helps minimize uncertainty about the meaning of the signal by reducing possible overlap with other signals; it allows the evolution of both finely tuned receptors and specific neural mechanisms for recognizing the signal. For example, birds that use different calls for mating and territoriality gain an advantage in clarity of communication, which comes from having each type of call highly stereotyped and distinctive.[10]

> **KEY TERM** Co-option is the evolutionary adoption of something the animal already does, or has, for use in a different form of communication.

> **KEY TERM** Ritualization is the association, through evolution, of a meaning with a signal.

> **KEY TERM** Stereotypy is the evolutionary reduction of the variation of a signal so that its meaning is more easily understood.

> **KEY TERM** Redundancy is the use of multiple signals that have the same meaning; this reinforces a message and reduces possible confusion.

FIGURE 7.5
This flicker reveals redundant information to potential mates through two communication channels: its drumming ability (auditory) and its colorful patterning (visual). *Photo: Michael Breed.*

218

Redundancy

The final key element in the evolution of communication is *redundancy*. This is the use of multiple signals, often in different modes, to communicate the same message. Dogs mark their territories with odors from urine but also use visual displays such as leaving scratch marks on the ground near their scent mark. Redundant signals such as these help to ensure that the recipients will get the message (Figure 7.5).

7.3 MODES OF COMMUNICATION
Choosing the Type of Signal

The evolution of communication allowed for a wide variety of *communication modes*, ranging from ones familiar to humans (e.g., vision and sound) to those that are far outside the human experience (e.g., electrical discharges, seismic waves). The goal in this section is to develop an understanding of the circumstances that cause evolution to favor a particular mode of communication over the other possible modes and to show how each mode presents special opportunities for communication, while at the same time presenting limitations.

> **KEY TERM** Communication mode is linked to sensory mode—the type of sensory receptor used to receive the signal. The primary communication modes are visual, auditory (sound), touch, and olfactory (taste and smell). Other modes, such as electricity, are also possible routes for communication.

These opportunities and limitations arise because different environments and different messages are best communicated with different types of signals. Disparate modes are favored by natural selection, driven by divergences in how directional and specific a signal might be, how far it might travel, and how fast it might travel. In some cases, signals that travel rapidly over long distances are favored, but distance and speed are not always important or desirable attributes of a signal; some signals are best restricted to close-by individuals and may actually attract predators or parasites if they travel too far. The function and context of the signal have a strong influence on the mode that is used.

Scientists have sometimes been fairly anthropocentric when assessing how animals use signals; they have assumed that the animal world is much like the human world when it comes to communication. It does not take deep thinking, though, to realize that dogs rely more on olfaction than humans do, and that fish swimming in muddy water face very different communication challenges than either dogs or humans. This fits with the overall theme of *umwelt*, and the unique perceptual world of each animal species (see Chapter 2).

The general lesson in sensory perception and communication is that any form of energy in the environment has the potential to be perceived and that anything that is perceived has the potential for use as a signal. The challenge to human scientists is to consider channels for which we, as a species, are receptor-poor. To have even a small chance at filling such a tall order, we must again address first principles: What abiotic and biotic environments favor what modes of communication? The following sections address major communication modes, including signal production in that mode, limitations on use and perception of those signals, and examples of signals using that particular mode. Sensory perception is discussed in detail in Chapter 2; refer to that chapter for a basic review of how each signal mode is perceived.

Chemical Signals: Messages by Smell and Taste

Although humans tend to think of visual and audible signals first when considering communication, chemical signals may have been the first signals to evolve, if signaling began with increasing sensitivity to chemicals in the environment that indicated resources. Chemical signals are certainly ubiquitous, used by many animals. Chemical signals work well in dark environments and are not impeded by obstacles. Depending on their composition, they may linger in the environment or travel large distances. They are usually inexpensive to produce.

If the production of a molecule has been shaped by evolution so that it has a signaling function within a species, then it is a *pheromone*. Animals also produce defensive molecules; this use is discussed in Chapter 10. Humans tend to differentiate between smell (molecules carried in the air and perceived on the olfactory epithelium of the nose) and taste (molecules dissolved in liquid and perceived on the tongue), but both of these senses fall into the broad category of chemoreception.

> **KEY TERM** A pheromone is a chemical signal used in transmitting information within a species.

Virtually all pheromones are molecules with carbon chain backbones. The major differences among pheromone molecules are size and polarity; these two features are matched with the function of the pheromone. Small molecules and less polar molecules are relatively volatile; an airborne pheromone, such as a moth sex attractant, is likely to be a smaller molecule. A pheromone that does well to persist in the environment, such as scent that marks a territory, is more likely to be higher molecular weight and relatively more polar. The reason is that higher molecular weight molecules require more energy (heat) to evaporate, and polarity makes molecules "stick" (due to hydrogen bonding) to each other. Thus, the main alarm pheromone of the honeybee, isopentyl acetate, has only seven carbons and evaporates almost immediately after release. Rapid evaporation enhances quick defensive responses by bee colonies. The main component in honeybee queen pheromone is a larger, more polar molecule, (E)-9-oxodec-2-enoic acid, which is passed from the queen to workers, and among workers, via contact (Figure 7.6). This lower volatility causes it to persist within the nest and retains the signal among the bees.

Pheromones provide valuable channels for communication when audible or visual signals might fail; thus, a nocturnal animal might use chemicals to mark its territory. In addition to functioning well in the dark, pheromones can be deposited on surfaces where they remain after the animal has left; the signal retains its information value over time and in the absence of the sender. Potential disadvantages of chemical signals are their slow and uncertain dispersal relative to visual or auditory signals (no faster or more predictable than the air currents that carry them) and the difficulty of conveying complex information with a molecule. That difficulty is especially apparent when considering directional information. In a situation in which consistent wind or water currents carry the signaling molecules, the direction the signal is coming from may be obvious to the receiver. In other cases, gusts or

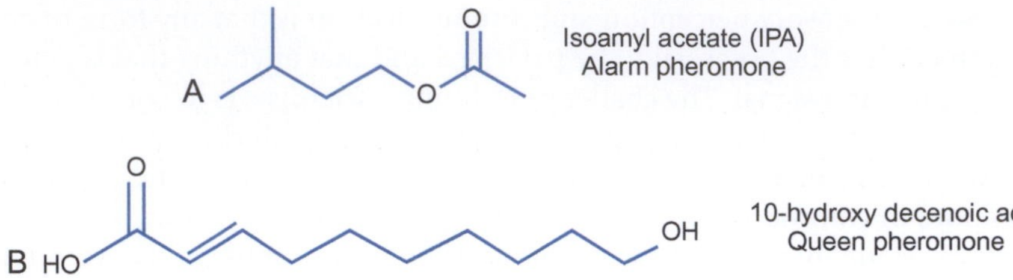

FIGURE 7.6
Structures of honeybee alarm pheromone and honeybee queen pheromone. These chemical structures of honeybee pheromones are instructive not only in their basic chemistry but also what they tell us about the pheromones' functions. The alarm pheromone is a small molecule, meant to disperse rapidly and warn of danger. The queen pheromone is larger and used for information within the nest.

sidedrafts can break up the odor stream, forcing the receiver to use more complex search strategies to discover the directional information (see Chapter 8). Pheromone trails on surfaces, such as those laid by ants from the nest to food, do not convey information about direction; an ant encountering the trail must gain directional information from direct contact with other ants on the trail or by using landmarks that establish the direction of the nest.

In contrast to visual or vibrational communication, noise is not a major issue in olfactory communication because natural selection favors unique pheromones (and receptors) not otherwise found in the environment. The most likely sources of interference with olfactory communication are similar species in the same environment; in such instances, sex pheromones are commonly blends of several compounds that together convey the full message. Differences in these blends distinguish sex pheromone messages of moths such as the tobacco budworm (*Heliothis virescens*) and its close relative *Heliothis subflexa*.[12]

There is much that remains to be learned about odors and what they communicate. For instance, the administration of hormonal contraceptives to female ring-tailed lemurs resulted in changes in a variety of odors related to a wide range of information about individual recognition, fertility, and kinship.[13] Recently, a research group at McGill University found that gender-related attributes of observer odor (e.g., t-shirts worn overnight by males or females) caused stress-induced analgesia and related behavior in rodents.[14] Much remains to be investigated about this intriguing result, but it could have profound consequences for behavioral research methods.

BRINGING ANIMAL BEHAVIOR HOME: SCENT MARKING IN DOGS AND CATS

The wild ancestors of dogs and cats used territorial markings to announce their presence in an area, and this behavior has carried over into domestic dogs and cats. The territorial markings of dogs are a composite signal, using both olfactory and visual cues. Dogs distribute olfactory cues by urinating small amounts on objects in their home range. This behavior is most marked in male dogs, which lift one hind leg—a behavior that develops as the male matures—and spray urine on an object. Males urinate frequently and do so more often in areas that are less familiar to them. Females urinate less frequently, but also direct their urine at objects and increase the frequency of urination when in less familiar territory; females may also lift their legs when placing urine on objects. Some objects—the bases of certain trees, lampposts, or fire hydrants—become olfactory signposts that attract much sniffing and marking. In habitats with winter snows, the urination sites are recognizable as "yellow snow." For students of animal behavior, these sites provide an opportunity for experimentation with canine territoriality because patches of yellow snow can easily be moved using a shovel.[15]

The development of raised leg urination is slowed and may never develop fully in castrated males. The younger the age of castration, the less likely the behavior will develop. Dogs, particularly males, sometimes also scratch the ground with their hind paws after urinating; this happens more often when other dogs are in view and may serve as a visual signal of presence of the scent mark. Ground scratching may also leave a lasting visual signal on the ground.[16,17] In the case of giant pandas, males even do handstands in order to leave urine in high places, a marking strategy that seems to intimidate subadult males. Is this deceit? Could only a REALLY large panda mark that high? Or does it communicate competitive determination?[17]

Cats have a similar system of territorial marking, in which males, and sometimes females, spray urine on vertical surfaces. Cats differ in a critical way from dogs, though; cats concentrate their urine spraying at the core of their territory. For domestic cats, this means spraying on surfaces, such as walls, doorways, and furniture, within their households. About 10% of castrated male cats spray and about 5% of female cats spray. Compounding the problem, cat urine has a much stronger odor to humans than does dog urine. The most effective prevention of spraying is castration of males at a young age. Sometimes spraying results from behavioral problems related to stress or anxiety, and veterinarians occasionally suggest behavioral therapy, perhaps in combination with antianxiety medication, for cats that spray excessively. Cats may experience more anxiety than owners realize. Up to 67% of cats examined for urinary tract problems are diagnosed with idiopathic cystitis—that is, cystitis with no known cause—consistent with the hypothesis of anxiety-induced urinary problems. For such cats, a reprieve from having to interact with cats that make them uncomfortable, even if they are other cats in the same home, can relieve the problem.[18]

Tactile Signals

Tactile signals (touching) can be remarkably efficient when animals are close together and do not require ambient light or air currents. Their major limitation is the requirement of proximity; they are ineffective for distant communication and, for that matter, do not work well in the presence of obstacles.

Tactile senses are an important source of information for animals. Touch provides information when light is unavailable, when noise interference obscures sound, or when vocalizations might attract predators. In mammals, touch is a significant aspect of affiliative behaviors in social groups. Touch also provides important information about proximity of food, predators, and other environmental features.

Touch functions in both aggressive and affiliative interactions. In any aggressive encounter, direct physical force is the most escalated and threatening form of communication; bites, slashes, and kicks carry powerful messages. Touch can also facilitate mutualistic interactions. Consider tandem running, explored in the beginning of Chapter 5. Ants are famous for leaving chemical trails that other ants follow to foraging sites, but some ants also lead their nestmates to forage for food. In tandem running, the leading ant moves toward the goal while the following ant maintains antennal contact with the leader. In this manner, chains of ants might be following a leader. This phenomenon illustrates two principles of animal behavior: first, the signal is honest—the two ants have the same goal, to provide for the nest; and second, the signal is redundant, as it involves pheromones as well as tactile sensation. Grooming is another powerful affiliative behavior for many animals. Allogrooming (mutual grooming) is widespread in mammals, birds, and social insects. Social groups are a fertile breeding ground for external parasites, and one driving force, evolutionarily, for all grooming is the removal of parasites. Might unusually frequent grooming be an (inadvertent) signal of high parasite loads?

> **KEY TERM** Honest signals accurately reflect the meaning and intent of the signaler. Honest signals reflect shared interests of the sender and receiver.

Audible Signals

Audible signals work well in the dark and are not impeded by obstacles such as rocks and trees. They do not linger in the environment; the distance that they travel depends in part on the medium and the type of sound (wavelength) involved. The production of audible signals tends to be energy intensive.[19]

WHAT IS SOUND?

Sound is vibration. Some sound is audible to a given species and some is not, depending on the frequency of the vibration. Humans usually sense some vibrations in air or water as sound, but other species sense vibrations that people do not perceive, including vibrations in soil and rock. The physical characteristics of the medium through which the sound travels have a major effect on how sound can be used. The density of the medium is particularly important and affects both the energy required to cause the vibration and the speed at which the vibration travels. For instance, it requires less energy to make air vibrate than water, and making the ground vibrate takes substantial energy. Thus, the use of vibrations in communication depends on the ability of an animal to make a material vibrate, so not surprisingly, only very large animals, such as elephants, can use the soil as a transmitting medium for vibrational signals. The density of the medium also affects speed. In air, sound travels at about 334 m/s (at 20°C). This seems fast, but the rate of travel in surface ocean water is nearly five times faster, 1520 m/s, and in rock and soil it can even be faster, up to 13,000 m/s. Again, all of these speeds depend on the density of the medium, which changes with temperature, salinity of water, and type of rock or soil. In some ocean depths, cold temperatures and moderate pressures yield ideal conditions for transmission, resulting in a SOund Fixing And Ranging (SOFAR) channel. Sounds can travel thousands of kilometers in this SOFAR channel, and marine mammals can communicate across oceans in this way. The frighteningly rapid travel of tidal waves (tsunamis) and even faster rate of travel of earthquake tremors through the earth's crust reflect the speed at which a vibration will travel in these dense media.

OF SPECIAL INTEREST: PSYCHOPHYSICS AND NONLINEAR SENSORY RESPONSES

Our discussion of auditory communication merits a small detour into psychophysics. *Psychophysics* is the study of how the physical properties of a stimulus affect the interpretation of the stimulus by the nervous system. Psychophysics adds several important elements to our understanding of how perception works in communication. For instance, psychophysics tells us about the threshold of perception for a stimulus, the animal's ability to judge differences in intensity between stimuli, and the energy needed to produce a stimulus that will result in a given type of receptor response. A basic principle is that receptor response is not linearly related to stimulus strength.

Sound is a great example of this *nonlinearity*. To us, 0 dB is the threshold for hearing. A 10 dB sound is barely perceptible, the tiniest of rustles, whereas a 130 dB sound is immense, deafening. If you ask people to rate intensities of sound, most would tell you the sound roughly doubles with every 10 dB increase. (The relationship is actually more complex and is captured by the sone scale; Figure 7.7.)

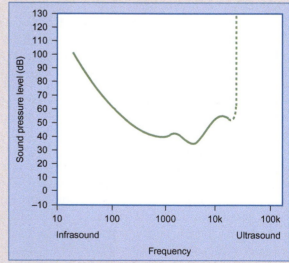

FIGURE 7.7

The line represents perception of loudness across a range of sound frequencies (x-axis). This curve, developed from human data, is representative of how bird and mammal ears function. Sounds at the low (infrasound) and high (ultrasound) end of the hearing range require more energy (dB) to be heard.

Yet the decibel scale is logarithmic, so each 10 dB interval corresponds to a 10-fold increase in the intensity (energy) of the sound. From a 10 dB rustle to a 20 dB whisper is a 10-fold increase in sound energy; from 10 to 30 dB is 100-fold; from 10 to 40 dB is 1000-fold. Normal conversation, 60 dB, is 100,000 times more intense than the 10 dB rustle. In communication, this means that at the low end of the loudness range, relatively small differences in sound energy can be perceived, but as loudness increases, only major differences in intensity can be perceived (see Figure 7.7). In other words, the perception of loudness may double with every 10 dB increase, but the additional energy required for this increase is far greater than a mere doubling.

> **KEY TERM** Nonlinearity occurs in the relationships between two variables when one variable increases at a faster rate than the other. This results in something other than a straight line when the two variables are graphed. Nonlinear relationships are often described as exponential, logarithmic, or power curves.

As we move through our discussion of communication, keep in mind that the nonlinear nature of receptor systems means that the cost of producing more intense signals may not pay off, in terms of perceived increases in the intensity of the signal. For a male frog, high-*amplitude* (loud) calls may be better, but at

> **KEY TERM** Amplitude is the intensity, or loudness, of a sound.

the upper end of his ability to make high-amplitude sounds, the female's perception of loudness may not allow her to tell the difference.

There are a few other relevant points about sound as a mode of communication. *Dissipation* is the loss of intensity as sound travels. One important contributor to this loss is the effect of spreading; as the sound spreads out over a larger area, less energy is present at any one point. Mathematically, the inverse-square law describes dissipation: sound energy decreases exponentially over distance. Another cause of loss of sound is absorption and scattering by the medium in which the sound is traveling. Low-*pitched* signals have considerable energy and carry well, making them suitable for territorial, group-cohesion, and other long-distance calls; "famous" low-pitched signals include the vocalizations of whales[20-23] and the seismic-wave communication of elephants.[24-26] Higher-pitched signals are employed in echolocation systems.

Sound is also subject to *reflection* and *refraction*. When sound hits a denser medium, some of the energy is reflected; it echoes, off the surface of the denser medium. Echoes can be useful, as in the echolocation system of bats, or can result in confusing noise. When sound moves from a medium of one density to a medium of a different density, it refracts, or changes wavelength as well. This results in distortions of the sound.

> **KEY TERM** Dissipation is the loss of intensity as sound travels.

> **KEY TERM** Pitch is the tone of a sound. Bass notes are low pitched, treble notes are high pitched. Pitch is also termed *frequency* because it reflects the number of sound oscillations during a given period of time. Human hearing responds to pitches from about 50 cycles per second to about 20,000 cycles per second.

> **KEY TERM** Reflection is the bouncing of sound or light waves off dense objects. Echoes result from reflection and may confuse communication using sound.

> **KEY TERM** Refraction is the change in frequency of waves when they move into a medium of different density. Refraction of sound changes its pitch, which sometimes can affect its signal value.

SOUND PRODUCTION

There are four common sound production mechanisms. Essentially, any movement or action that causes molecules to vibrate or makes pressure waves produces sound. Most animals use special structures to make sound, but a few simply take advantage of a resonant substrate by slapping, tapping, or drumming on it. Specific sound production mechanisms include:

- Vibrating a drum-like membrane. Vibrating membranes are used by cicadas to produce their mate attraction calls. The membrane, or tymbal, is held in place by rigid exoskeletal structures and moved by a muscle attached directly to the membrane.
- Stridulating with a file and scraper. In crickets, grasshoppers, some ants, and a variety of other insects, one body part is equipped with a scraper (imagine something like a guitar pick slightly protruding from the animal), and an adjacent body part has a rough or file-like structure. The animal makes a chirp or whir by moving the two body parts against each other, dragging the scraper across the file. The entire structure (file and scraper) is called a *stridulatory organ*. The frequency (pitch) changes depending on how fast the two structures are moved. Because insects are ectotherms, their stridulations may vary in pitch depending on the temperature.
- Vibrating a membrane in an air flow. Like the reed in a saxophone or clarinet, the larynx of frogs, toads, and mammals and the syrinx of birds feature a membrane that vibrates as air from the lungs is pushed past it. The rate of vibration, and hence the frequency (pitch) of the sound, can be regulated by muscles that tighten or loosen the membrane.
- Hitting a substrate. Beavers slap their tails on the surface of the water when alarmed. Because sound travels much faster in water than air, this is an effective warning signal. Some termites, such as the Formosan subterranean termite (*Coptotermes formosanus*), bash their heads against substrate when alarmed, producing sound and substrate vibrations. Literally millions of termites in a colony may do this synchronously, producing a sound that is audible for several meters around the nest. Woodpeckers, such as the flicker (*Caleptes cafer*), knock on hollow trees (or buildings that resonate like hollow trees) as a way of attracting mates.

OF SPECIAL INTEREST: FRT SIGNALS IN HERRING

Animal communication has given us the Ig Nobel Prize in Biology (2004) (this is a prize given for research that is both humorous and thought-provoking) with a study of burst pulse sounds in Pacific and Atlantic herring (*Clupea pallasii* and *C. harengus*), respectively. Although these fish are commercially important, their sound production has been poorly studied. Wild-caught Pacific herring produced a train of pulses (1.7–22 kHz) that lasted up to 7.6 s. This occurred mostly at night, and video analysis of Atlantic herring showed that it was associated with air release from the anal duct region. These are termed *Fast Repetitive Tick* (FRT) sounds. Because the *per capita* rates of FRT production increased with fish density, Wilson et al.[27] hypothesized that these sounds are involved in social behavior in these fish.

REGULATING PITCH AND AMPLITUDE

Each of the first three mechanisms (tymbal, stridulatory organ, larynx/syrinx) offer the animal a way to modulate pitch, making it easier to produce calls that are distinctive for a species, sex, or individual. In these cases, amplitude (loudness) can be varied as well, by changing the amount of energy with which the sound is made. Because of the relationship between energy and amplitude, body size limits sound amplitude. In an absolute sense, strength is a function of body size, so the loudest sounds are reserved for the largest animals.

Special adaptations help some organisms surmount the size–strength–sound limitation. One such adaptation involves using a resonant structure to amplify the sound. This can be external to the animal, such as a cricket calling from inside a hollow stem, or internal to the animal, such as the hyoid cartilage of the howler monkey. With vocal sacs, animals like howler monkeys and many frogs can force larger volumes of air through the larynx. While these alone do not amplify the sound, they allow the production of longer-duration sounds than would be produced by air from the lungs alone.

BRINGING ANIMAL BEHAVIOR HOME: GROWLING DOGS AND PURRING CATS

Although dogs vocalizing at 2 a.m. may well be a nuisance, and most people do not consider barking to be terribly informative, behavioral biologists are learning that those people are wrong. A bark is not always simply a bark, nor is a growl one-size-fits-all. When listening to playbacks of barks that were recorded in six situations (the approach of a stranger, Schutzhund training, a walk, alone, a game of ball, and owner–dog play), both nondog-owners and dog owners could differentiate among the categories and could correctly assign emotional context, especially at the extremes (e.g., aggressive/ approach of a stranger, playful/owner–dog play). The playful sounds tended to have higher frequencies associated with them, while the aggressive sounds had lower ones, and this seems to be true across many species. Not surprisingly, dogs can also discriminate among dogs and among situations. Finally, dogs can even distinguish among growls (aggressive vs. play growl) and respond appropriately. Dogs seem to be more generous with their barks than their nearest relative, the wolf, and the fact that even humans without experience with dogs can glean correct information from barks may reflect events throughout domestication, during which dogs came to depend on humans more and more, and humans depended on dogs that offered reliable and distinctive information.[28–31]

The ontogeny of purring in the domestic cat begins in kittenhood, during suckling. The adult cat purrs in many circumstances, and this is often taken to be a sign of contentment. Some cats, however, also purr when they are hoping (insisting?) to be fed. Recordings and analyses of the spectra of these purrs have shown that the latter purr—called the "solicitation" purr—has a different component, one that is higher frequency and that is comparable to the cry of a human infant. While no one would confuse a cat purr with a baby's cry, the inclusion of that particular frequency may mean that the solicitation purr is harder for humans to ignore. Indeed, both cat-owners and people with little experience with cats found that recordings of solicitation purrs including that high-frequency peak seemed more urgent than the same purrs in which that peak had been synthetically removed.[32]

INFRASOUND AND COMMUNICATION

Infrasound and ultrasound are not audible to humans. Respectively, they are below and above the normal range of human hearing. The scientific exploration of infrasound communication is just beginning, and the realization that some animals use this channel has alerted us to potential (and previously unsuspected) social complexity in those animals. Elephants and killer whales are two such species.

Infrasound has the special characteristic of traveling well in ground or water; in fact, the waves of an earthquake can be thought of as a form of infrasound. Because sound travels much faster in ground than in air (or water), ground-borne vibrations, if perceived, can serve as an early-warning system, arriving well before airborne sound from the same source arrives. Infrasound dissipates less rapidly in air, making it ideal for long-distance communication. Perception of infrasound, however, presents some specific problems. An object smaller than the distance between waves is a poor receiver for infrasound waves. Thus, infrasonic receivers need to be large and tend to be found on the large animals able to generate infrasound. This is probably the reason that infrasonic communication is used by only a few animals, and the best-understood infrasonic communication system is that of the African elephant (Figure 7.8). The elephant's large pinnae (external portion of the ears) may play an important role in perception of low-frequency sounds, which are significant in elephant communication.

FIGURE 7.8
African elephants have large pinnae (external ears) that assist in perception of infrasound, which is important in elephant communication. *Photo: Pen-Yuan Hsing.*

FIGURE 7.9
Killer whales use infrasound for communication and group cohesion. Some calls are stereotyped while others are highly variable. *Photo: Jennifer Marsh.*

Infrasonic calls, because they travel great distances, provide an important mechanism for communication and affiliation in these wide-ranging animals. By remembering the identity of different callers, elephants can maintain contact with specific individuals. Female African elephants use "contact calls" to communicate with other elephants in their bands (usually a family group). These infrasonic calls, with a frequency of about 21 Hz and a normal duration of 4–5 s, carry for several kilometers, and help elephants determine the location of other individuals. Calls vary among individual elephants, so responses to familiar and unfamiliar calls differ; perhaps elephants can recognize the identity of the caller.

The social use of individual identification is not known. Nonetheless, as little as we understand about elephant communication, we suspect that it functions at least in part in elephant social behavior. African elephants have a social structure best described as fluid; animals move freely over wide areas, sometimes affiliating with other animals. Female members of a family tend to stay together, and of course their juveniles travel with them. These female-centered groups may merge with other such groups periodically. Adult males are less likely to join groups.

Killer whales (*Orcinus orca*) also live in matrilineal (mother plus her offspring) social groups (Figure 7.9). In the summer, these groups sometimes come together with up to seven other matrilineal groups to form a *pod*. Unlike many other mammals, male killer whale offspring tend not to disperse; they remain with their mothers. Female offspring ultimately have offspring of their own and eventually split from their maternal group to form their own group. Killer whales fitting this social pattern are termed *residents*. Transient killer whales show fewer group affiliation tendencies; their social behavior is not as well understood.

Killer whales make calls and echolocation clicks. Calls are repeated (pulsed) sounds in the 1–10 kHz range; between seven and 17 calls can be recognized by scientists, depending on the matrilineal group under study. Scientists recognize calls by listening to them, or by examining a graphical representation of the physical characteristics of the call, known as a *sonograph*. The number of different calls that the whales (as opposed to scientists) actually perceive is more difficult to determine, and this remains an area of active scientific investigation.

Calls may be stereotyped so that a call is quite similar each time it is given and varies little among individuals who give it. Stereotyped calls can be interpreted based on genetically coded information in the receiver, and are thus consistent with part of the model for innate releasing mechanisms of classical ethology (see Chapter 1). Call stereotypy is particularly valuable when the communicatory context is critical and brooks no misinterpretation. Alarm calls are especially suitable for call stereotypy. Male humpback whales have a stereotyped song that they use as part of sexual display. As songs change (and they do so rapidly, on the order of 2–3 months), they spread to other populations, with other groups picking up those modifications. For whales in the South Pacific Ocean basin, these changes almost always move from west (eastern Australia) to east (French Polynesia). Of course, this raises many questions (e.g., why does it almost always begin in the western part of the basin?), most of which cannot be answered, given the limited knowledge we have about whales. What is clear is that these are culturally transmitted changes; they happen too rapidly to be explained by genetic changes. Thus, in the case of the humpback whale, the song is both stereotyped and subject to modification over time.[33]

In contrast, other calls may vary among individuals or social groups. Call variation can encode information about age, sex, social group membership, or other aspects of social status. To take advantage of variable calls, animals must be able to perceive call differences, associate a behavioral context with the call, learn that association, and remember it for future use. Variable calls are particularly useful in more complex aspects of social interactions. Killer whale calls were more similar within matrilineal groups than between groups and, during one study, groups actually became more distinctive in their calls.[21] Thus, killer whale calls contain information that can be used to identify social groups and may play a function in maintaining kin groups over time.

ULTRASOUND AND COMMUNICATION

Ultrasound presents two challenges for the recipient animal. First, recall that high frequency translates to short wavelengths; the hearing organ must be miniaturized to accommodate the wavelength. Second, high-frequency sounds tend to be supported by little energy. They not only dissipate rapidly as the sound travels, making them relatively faint even close to the source, but also may be absorbed by the hearing organ without being transduced into a neuronal signal. To accommodate the lower energy of ultrasound, the hearing membrane, or tympanum, is typically thinner in animals that rely on ultrasound for communication or navigation. The outer ears (pinnae) of mammals that can perceive high-frequency sound may be quite complex; bat ears are characterized by grooves and channels that help carry sounds to the tympanum, as well as maintain small differences in frequency (pitch) and amplitude (volume) which can be used to localize sound sources.

Ultrasonic signals are produced in two major contexts. First, several species use ultrasound in social communication. Bats, murid rodents (rats and mice), and various sorts of moths use ultrasounds to communicate with mates, and rodent pups use it to call their mothers if they become isolated. The second context—echolocation—is more famously associated with bats producing high-pitched sounds that are reflected off objects in the bat's flight path (Figure 7.10).[34–39] Because bats prey on insects, many insect species are attuned to bat echolocation calls and take evasive measures if they hear a bat call. Males produce a calling song to attract females in greater wax moths (*Galleria mellonia*) and lesser wax moths (*Achroia grisella*) stop calling if a calling bat approaches; presumably the bat can orient to the mating call of the moth.

High-pitched sounds (ultrasound) have several advantages in echolocation: (1) they take relatively little energy to produce; (2) because of their short wavelength, they are more likely to bounce back to the bat, which is essential for echolocation; and (3) they dissipate rapidly, reducing confusion from "old" sounds that otherwise could still be bouncing around an area.

227

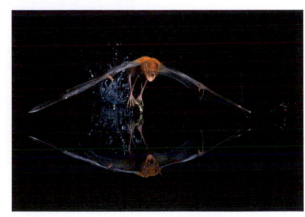

FIGURE 7.10
Bats are famous for their ability to catch aerial prey such as insects with echolocation. Echolocation can be used to locate other types of prey as well. *Noctilio*, a fishing bat, uses echolocation to sense ripples in the water that indicate a fish prey is near the surface. *Photo: J. Scott Altenbach.*

OF SPECIAL INTEREST: HOW SENSITIVE ARE ANIMALS TO DIFFERENCES IN SIGNALS?

Psychophysics is the study of how the physical characteristics of signals affect their use in communication. One principle of psychophysics, Weber's law, tells us how different stimuli need to be in order for an animal to differentiate between the stimuli. The first step is to determine the minimum detectable, or just noticeable, difference between stimuli for the animal. How different do two playbacks of a noise need to be, in terms of volume (amplitude), for the animal to tell them apart? A shorthand way of referring to this *just noticeable difference* is JND. When humans assess the weights of objects, the JND is constant at about 10% of the weight of the smaller object. Thus the JND for a 5 g object is 0.5 g (i.e., you can tell the difference between a 5 and 5.5 g object but not between a 5 and a 5.3 g object) and the JND for a 10 kg object is 1 kg.

Weber's law is important in animal communication because it tells us that as signals become more intense the differences between signals needs to be larger to be perceived. Weber's law is a loose approximation for light and sound, and does not to apply for extremely weak or extremely strong stimuli, but it gives a basis for understanding why a bigger difference is required to distinguish between strong as opposed to weak signals.[40] In algebraic terms, Weber's law is:

$$\Delta I / I = k$$

where I is the initial signal strength, ΔI is the JND for a signal being compared to the initial signal, and k is a constant that expresses the proportional difference needed to discriminate between the two signals. Weber's law tells us that it is easier to tell weak signals apart, meaning a whisper can be more effective than a shout in conveying nuance.

CHORUSES

What do many frogs, insects, and birds have in common? They use acoustical signals—sound—to attract mates. Generally, the male sings mightily, sometimes all day or all night, in hopes of attracting an interested female. Males may sing to demonstrate their strength compared to other males, or to attract females to a territory. If intrasexual competition pushes males to perform to their maximum capabilities, then these songs become an honest signal of male condition (see Chapter 11 on mating).

Insect and frog "choruses" are common features of summer nights in many locations; all of the animals seem to sing synchronously (Figure 7.11).[41] Why might they do this? Greenfield and his research group have led the way in answering this question.[42] Perhaps the simplest model comes from species in which females have a preference for the first male in a group to sing. This "leading" male receives the female's attention. Other males are then pressured to sing ahead of the "leader." Ultimately, the males sing in chorus because each is trying to be first. The amount of time between songs is governed by the males' recovery time and by the female's response to repeated songs. If the lag between a pair of songs is too short, then the female regards the second song as a less attractive "following" song. Selection favors males which wait long enough that their call is distinctive from the previous calls. Once enough time has elapsed, the female is receptive to a "leading" song. These factors combine to push the males into synchronous singing.

Although we frequently think of anthropogenic light pollution as a source of confusion in visual signals, its occurrence in the form of street lights can affect chorusing in four species of songbirds (blue tits, great tits, blackbirds, and robins). Males that nest near streetlights begin to sing earlier than other males; depending on the species, this advance can range from a few minutes (blue tit) to almost an hour (robin), with earlier-singing species being more strongly affected. In the case of blue tits, which were studied in detail, this can have remarkable consequences. Females prefer to mate with older males, and older males establish territories in the central forests, leaving the edges (and in some cases, light pollution) to yearling males. Females tend to wander (and participate in extra-pair copulations (EPCs); see Chapter 11) early in the morning, and they are attracted to the earliest-singing males. As a result, yearlings that establish near streetlights increase their fitness; other yearlings that also hold edge territories but that are not near light do not sing early and do not have more EPCs. This is powerful evidence for the competitive pressure that produces the dawn chorus. (By the way, females were not immune from the effects of artificial light; females on edge territories near lights began to lay eggs 1.5 days earlier than other females.) This may seem like a small thing, but the shift in mate preference that results from early singing may mean that traits that females have used over evolutionary time as indicators of mate quality are no longer as reliable in a modern, well-lit environment. Could such a shift be maladaptive? How would you find out?[43]

FIGURE 7.11
During calling, the frog's vocal sac inflates and then forces air through the larynx to produce sound. *Photos: Matthew Bolek.*

BRINGING ANIMAL BEHAVIOR HOME: VOCAL MIMICRY IN BIRDS

Most birds must sing the songs they inherited, or songs they learned from their parents at a young age. Others—15–20% of songbird species—are remarkable vocal imitators, able to listen to sounds, learn them, and reproduce complex vocal patterns.[44] This ability stays with the bird through its life. Parrots, parakeets, and mynahs are impressive mimics and are often kept in human households. (Mynahs are more commonly kept in households in South Asia.) Good imitators not usually kept in households are starlings and catbirds. The evolutionary and functional reasons for vocal mimicry in birds are not well understood. These birds respond well to repetition and reward when imitating human voices; parrots are most adept at this. One of the more disappointing possibilities for a child is to bring home a female parakeet and to try to teach it to talk. Only male parakeets are vocal mimics!

Chorusing (synchronous singing) has its disadvantages. If dozens, hundreds, or even thousands of males participate in a chorus, auditory confusion may result for females. Consequently, in at least some species, females have evolved mechanisms for selectively listening to a nearby subsample of the males.[45] This ability allows a female to evaluate the signals of individual males and use the information in choosing and locating her mate. Other animals that chorus include katydids,[46] humpback whales,[20] and chimpanzees.[47]

Male crickets such as *Acheta domesticus*, the common household cricket, also chorus to attract females. Producing the monotonous chirps for long periods of time must be taxing to the male crickets, making persistent calling an honest signal of male energetic capabilities.

Females can discriminate male condition based on the calls and are preferentially attracted to certain males.[48] Other males listen in and are attracted as well, preferring the same, generally larger, males.[49] Why might males be attracted to the songs of other males, particularly those males preferred by females? Occasional chirps—advertisement calls—are cheaper for the males than walking around searching for mates, but trilling—continuous calling used in courtship—is twice as expensive as chirping.[50] Meanwhile, by increasing proximity to a trilling male, a chirping male may intercept a female. Many other insects, such as periodical cicadas (*Magicicada*), also engage in acoustical contests to attract females.[51,52]

AUDITORY NOISE

Most environments have lots of auditory noise. The wind whistling, rain dripping, trees rustling, and water splashing can all obscure airborne animal-produced sounds. Rain, waves, and currents produce equally omnipresent underwater noise. Add all the birds and whales singing, frogs croaking, lions roaring, and so on, and the auditory world is noisy, indeed. Hearing is difficult when the signal-to-noise ratio is low, as when ocean surf overwhelms shorebird calls. Animals cope with this noise using three basic strategies. One strategy is to produce signals in a limited range of pitches and to hear them selectively with ears that respond only in the acoustic range of the signal. Another strategy is the neural ability to sort meaningful sound from background noise. Finally, acoustic signaling can be partitioned by time of day so that species evolve to use certain times and not others for signaling. In this way, they avoid each other's signals. Evolution is a powerful tool for overcoming problems like noise, and each of these solutions help animals to communicate within life's cacophony.

The "cocktail party effect" is a good example of how animals can evolve to deal with noise using the second strategy. This phenomenon gained its name from the ability of humans to isolate a single voice in a crowded, noisy room, and to focus on the signal content of that one voice. Birds[53] and frogs[54] are known to have the same abilities to separate relevant sounds from background chatter (Figure 7.12).

Of greater concern is anthropogenic noise—sounds produced by humans. In urban environments and in areas around roadways, noise pollution can overwhelm animal sounds. Rustling leaves measure about 20 dB, whereas cars, motorcycles, and trucks can produce 80–100 dB within 25 m of a roadway. Remember that the decibel scale is logarithmic, so motorized vehicles are producing approximately 10^6 more energetic sounds than soft natural noises. Recent studies have shown that noise pollution does actually affect how birds and frogs communicate.[55–58] Such noise pollution can have consequences for fitness; in Virginia, the production of Eastern Bluebird fledglings was negatively correlated with anthropogenic noise.[59] In the vicinity of gas compressor stations, which produce noise around the clock, bats with low-frequency echolocation calls reduced their activity and modified their search calls at louder sites.[60] Similarly, noise from ships, SONAR, and other types of location systems have the potential to disrupt auditory communication in the oceans. In addition, recently developed underwater communication devices have the potential for disrupting animal communication in the SOFAR channel. Worryingly, most studies of anthropogenic noise and its effects have focused on vertebrates; impacts on invertebrates (over 98% of all animals on earth) are likely to be equally severe, and the cascade of these effects through food webs and ecosystem services could have widespread negative consequences.[61] In sum, the effects of noise on animal communication present a major conservation concern.

Vibrational Communication

Many animals lack ears—that is, they do not have a membrane that vibrates in sympathy with the surrounding air—but they are

230

FIGURE 7.12
A colony of pelicans can be cacophonous. Birds in such environments are able to screen sounds and pay attention to relevant information, much as humans can at loud cocktail parties. *Photo: David Hall/US Fish and Wildlife Service.*

still able to gather vibrational information.[62] Infrasound is a common type of vibrational information carried in rock and soil instead of air. (Signals carried in the ground are sometimes called *seismic signals*.) Predators can detect movements of large herding animals by using vibrational information carried in the soil, much as a person might detect a passing car radio that has the bass turned up. Most examples of direct communication via vibration come from the invertebrate world. Some termites communicate alarm by whacking their heads on the substrate in unison; the unified effort of many termites causes the substrate to vibrate. Wolf spiders, which do not build webs, drum the ground during courtship; signals reach their potential mate through a combination of vibrations in the ground and in the air. Orb-weaving spiders use vibrations in their web to alert them to the presence and location of prey, and in some species, male spiders intentionally vibrate the web to gain the attention of their potential mate on the web.[63,64] Vibrational information also plays an important role in communication in the honeybee dance. Vertebrates do use vibrational communication. Treefrogs vibrate their perches in aggressive displays that elicit responses from other male treefrogs, and snakes, which have lost the external ear and tympanum (ear drum), can sense airborne vibrations transmitted directly to their skulls.[65,66]

Visual Signals

VISUAL SIGNAL PRODUCTION

Visual signals work well in the presence of light and in the absence of objects (such as vegetation) that block transmission. Visual signals reach their destination quickly—at the speed of light. Visual signals can be fairly inexpensive to produce, but they do not linger in the environment in the absence of the action or structure that produces them.

Light moves through water, as well as air, but visual signals in water face three related challenges. First, water absorbs light, so with increasing depth less light is available for animals to see signals. Some waters are "crystal clear," and good illumination is present to depths of 20 m or more, but other waters have suspended particles of clay, dirt, or sand (called *turbidity*) that greatly reduce visibility even near the surface. Second, water absorbs orange, red, and yellow wavelengths, so as depth increases, these colors become much less useful. Colorful fish and other aquatic animals are usually found in clear waters at shallow depths; fish from deeper waters are often more drab. Finally, light entering water is refracted, so interpretation of cues and movements at the interface between water and air can require experience; this is probably more important for predators than it is in visual signaling (Figure 7.13). (See Chapter 9 on foraging.)

Visual signals can be assigned to one of three categories. First, visual signals can be patterns on the surface of the animal, or patterns combined with color.[67,68] Coral reef fish present many outstanding examples of combinations of color and pattern to produce signals (Figure 7.14). In fact, there is evidence that the Ambon damselfish (*Pomacentrus amboinensis*) can make species discriminations based on detailed facial patterns that reflect ultraviolet light.[69] Second, visual signals can involve movement. One example of this would be a fiddler crab waving its large claw; some fiddler crab species choose display areas with backgrounds that highlight that claw movement. Not surprisingly, many visual signals combine the two elements of pattern/color and movement, exemplified when a peacock spreads its tail (movement) to reveal the magnificent pattern and color (see Section 7.5). Signals in these first two categories require sunlight or moonlight to be functional. The third category does not require an external light source because it involves animals that produce their own light for visual signals and do not need sunlight to illuminate a visual display. A few animals, such as fireflies and deep-sea organisms, fall into this category.

FIGURE 7.13
Lizards use the dewlap, a flap of skin that can be extended and retracted, as a visual signal in territorial and sexual displays. *Photo: Jonathan Losos.*

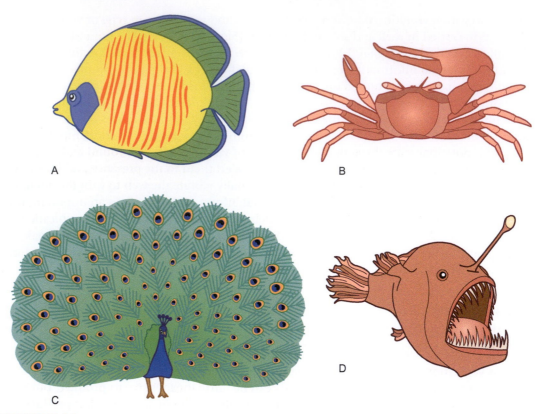

FIGURE 7.14

Visual signals in the animal world range across all major taxa that have image-forming eyes. (A) Colorful reef fish advertise sexual and dominance status. (B) Fiddler crabs (so named for their large, signaling claw) position themselves against backdrops that increase their visibility and wave their major claw in the air. (C) Peacocks are legendary for using their colorful, large tail to signal to peahens. (D) Some fish have bioluminescent extensions that lure other fish near their mouths.

OF SPECIAL INTEREST: GLOWING IN THE DARK

Light production by an animal can be a powerful signal. Animal species that make their own light include fireflies (lightning bugs, beetles in the Family Lampyridae), larvae of fireflies and some other insects, some species of click beetle (Family Elateridae), and a large number of deep-sea animals. In bioluminescent beetles (e.g., fireflies), light production clearly has a signal function in attracting mates. Deep-sea organisms may use luminescence as a signal, as a lure for prey, or as illumination so they can see their underwater world.

Male fireflies in the genus *Photinus* fly above the ground and flash signals of their mating availability; the signals vary so that the intensity of the flash and delays between flashes identify the males' species. Females on the ground respond with a single flash; the male and female then signal back and forth, a prelude to the female choosing a male. Better-quality males produce longer flashes, so male signaling advertises more about the male than just his species.[70] The male gives the female a "nuptial gift" of a nutritious glandular secretion (see Chapter 11 on mating) that nourishes the female, so finding better males affects a female's reproductive ability. Fireflies in the genus *Photuris* prey on male *Photinus* fireflies by mimicking *Photinus* signals; they draw in unsuspecting males using this form of aggressive mimicry and then eat them (Figure 7.15).

Bioluminescent marine animals often use their light as a defense, startling or confusing predators, or as an offensive weapon, illuminating or attracting prey. In fact, a vast preponderance of deep-sea animals use bioluminescence in some offensive or defensive aspect of their lives. In some cases, though, bioluminescence in the ocean serves as a signal to attract mates. A good example of this is a small crustacean, *Photeros annecohenae* (Class Ostracoda). (An ostracod looks like a tiny clam with a shrimp inside.) *P. annecohenae* males use bioluminescence as a signal to attract females for mating. Males of this species swim together, with many males flashing at the same time, producing a spectacular miniature lighting display. Some males "sneak," swimming near flashing males while not using their own signal. Bioluminescence works well as a communication mode in dark environments, opening the door for these interesting interactions.

FIGURE 7.15
This female *Photuris* firefly has mimicked the flashing signal of a female *Photinus* firefly, luring a male *Photinus* firefly to his death.

Conceptually, visual signals are not difficult to understand. As covered in Chapter 2, receptive cells that transduce light (generally considered to range from long-wave infrared radiation to shortwave ultraviolet radiation) are widespread in animals. In many animals, light-receptive cells are concentrated in a light-receptive surface, the retina, and light is focused on the retinal surface by a lens or lenses. The processing of light information is complex and involves filtering and enhancement of the image in several stages, beginning at the level of individual receptor cells and proceeding to interactions among adjacent receptors. The central nervous system integrates the information about image formation and pattern recognition. (See Chapter 2 for more details.)

Recall that in communication, the key element is that the signal and the receptor have co-evolved, so the intended target can perceive the signal. This may seem obvious, but evolution can yield elegant variations on this theme. Many deep-sea organisms are bioluminescent, for example, but most produce blue-green light. Loosejaw fish (Family Malacosteidae; Figure 7.16) make red light, instead. This means they can signal to each other without interference by predators, but to accomplish this, both the light production system and the eyes had to evolve to use red wavelengths.

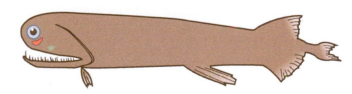

FIGURE 7.16
Loosejaw "stoplight" fish produce and perceive red bioluminescence, an unusual signaling adaptation. The red light is produced by a bioluminescent organ beneath the eye.

As indicated previously, pattern, color, and bioluminescence are three categories of visual signals. Color will be discussed in more depth here. There are two major ways to produce color that do not involve bioluminescence; color can be produced with pigments and with physical structure. Of these, pigments are more commonly used. Most colors result from pigments stored on or underneath the animal's surface. A pigment that absorbs all visible (to that species) wavelengths is black; a pigment that reflects all wavelengths is white. Pigments give animals color by reflecting specific wavelengths. Generally, pigments are chemicals with double-bonding structures among carbons or between carbon and nitrogen; the longer the series of such double bonds within a molecule, the darker the pigment. Most animal pigments are pterins (white, yellow, or red), quinones (yellow, red, or orange), melanins (generally yellowish-brown, brown, or black) and carotenoids (yellow, orange, or red). Animals usually produce pterins, quinones, and melanins themselves, but carotenoids often

233

FIGURE 7.17
Blue occurs less commonly among animals, and is most likely to be found in birds. *Photos: (A and C) Jeff Mitton; (B and D) Michael Breed.*

come from their diet, so the quality of food available to an animal, or its ability to compete for food, can affect its coloration. The bright red of male house finches in mating season depends on having carotenoid pigments in their diets. Structural color can also be highly important, however (see "Of Special Interest" below), and none other than Isaac Newton was among the first to investigate this source of color.[71]

Unlike the blues and greens commonly produced by bioluminescence, blue and green have been scarce in this discussion of pigments. A few blues, such as bird eggshell blues, are pigmentary, but blues and greens are often nonpigmentary colors, such as Tyndall blues, which are discussed later (Figure 7.17).

In the case of fluorescence, light is absorbed by a pigment and then retransmitted at a different wavelength; this is known to provide color in some coral reef fish. Fluorescent pigments are interesting because they allow an animal to produce a color that may not be present in the ambient light; for instance, by the time light has traveled a few meters through water, the red wavelengths have been absorbed, resulting in a blue-green underwater world. Fluorescence allows fish living at greater depths to produce red coloration by absorbing the predominantly blue and green wavelengths and emitting the energy as red wavelength light.[72]

OF SPECIAL INTEREST: IS RED REALLY A WARNING COLOR?

Red is used as an exclamation point signal; the color red stands out against the green, gray, brown, and black backgrounds that are common in nature (Figure 7.18). Sometimes red is produced as a signal within a species, such as the red chevrons of the male red-winged blackbird. From a human perspective, however, we often think of red as a generic warning color, a thought that is culturally

FIGURE 7.18
The color red is ubiquitous among major groups of living things, from berries to ladybeetles, birds to grasshoppers. As a signal, it communicates many things. *Photos: (A) Pine grosbeak, Jeff Mitton; (B) chili peppers, Michael Breed; (C) lubber grasshopper, Jeff Mitton; (D) poison dart frog, Michael Breed.*

reinforced by red stop signs, poison labels, and the like. Is red generically used as an interspecific warning signal?[73] This could clearly fall into our definition of communication, because a red signal might evolve specifically to warn away potential predators, and selection could easily favor receivers who use this information. The existence of red-striped coral snakes, red poison frogs, and red-marked black widow spiders all seems to support this thought. But apples, cherries, and strawberries are all also red and are hardly poisonous; in these instances, the red seems to have evolved as an attractive signal. In some cases, such as birds feeding on insects, the bird has the opportunity to sample and build an association among color, pattern, and palatability. One could then hypothesize that red is a color that is easily learned, perhaps. In contrast, a bird bitten by a coral snake often does not get a second chance; this suggests that sometimes learning may not be important in recognizing warning colors. There is at least some evidence that in birds color-based aversions are inherited, then reinforced by experience. In chicks, red-colored fruit shapes are attractive, but red-colored insect shapes are repellent;[74,75] thus, context (shape) interacts in an important way with color.[76] To answer our original question, there is at least some evidence for innate avoidance of red by some animals, but there is more evidence that aversion to red is learned and in some cases is cultural. The general topic of warning colors is treated in more depth in Chapter 10 on self-defense.

Colors can result from the structure of an animal's surface rather than pigments. Blues and greens in animals are almost always structural colors. Iridescent colors of butterfly wings are the product of thin layers within the scales on the wings that selectively reflect wavelengths in the blue or green color range (Figure 7.19). The blue of the iris of some human and animal eyes is a Tyndall blue, which is produced by the scattering of blue wavelengths of light by

235

FIGURE 7.19
The iridescence of (A) butterfly wings, (B) a peacock's tail, and (C) a damselfly results from the structure of the scales on the wings. *Photos: (A and C) Jeff Mitton; (B) Michael Breed.*

FIGURE 7.20
These cabbage butterflies (family Pieridae) occur in the same habitat but have divergent wing coloration. *Photos: Michael Breed.*

Note
In 1859, John Tyndall realized that small particles in suspension scattered short wavelengths more than longer ones, thus explaining why the sky is blue.

small particles in liquid in the iris. The blue of the sky is also a Tyndall blue, produced by the scattering of light from molecules.

VISUAL NOISE AND THE DISRUPTION OF VISUAL SIGNALS

Visual noise includes irrelevant movements, patterns, and colors in an animal's environment. In natural habitats, animals can separate visual noise from signal, often through natural selection acting to produce novel combinations of pattern and color; birds, coral reef fish, and butterflies all present examples of signal divergence to overcome problems of visual noise (Figure 7.20). Camouflage (see Chapter 10 on self-defense) can use visual noise to hide an animal against its background.

Human activities complicate the visual environment and can result in various types of visual noise that disrupt animal behavior. Animals may mistake colors on signs, automobiles, and the like for signals. Tinbergen's famous observation of three-spined stickleback fish responding to red British mail vans in the distance is a classic example of this. Reflective surfaces, such as windows or mirrors, may cause an animal to attempt to court or fight its image, which it mistakes for the visual signals of another animal. Streetlights disrupt the commuting routes of bats between perches and feeding areas.[77] Porchlights and streetlights at night may be mistaken for the moon or stars in navigation (see Chapter 8); while navigation suffers the direct effect of light pollution, these lights may also disrupt visual or pheromonal communication among potential mates.

Electrical Signals

Electrical signals are largely limited to fish swimming in muddy waters, where visual signals are not useful. Sharks may also use sensitivity to electrical impulses to detect prey. Like chemical cues inadvertently released as a result of metabolic activity (discussed earlier in this chapter), all multicellular animals emit some form of electrical signal. The reason is that nerves and muscles use changes in membrane polarity to function. A few organisms (e.g., electric fish) have enhanced this ability with the evolution of electric organs that can produce well-defined electric signals.[78,79]

Receiving electrical signals can be valuable to an animal in either of two ways:

1. Receipt of communication from other animals.
2. Location of other animals, particularly potential prey, which unintentionally produce electrical fields.

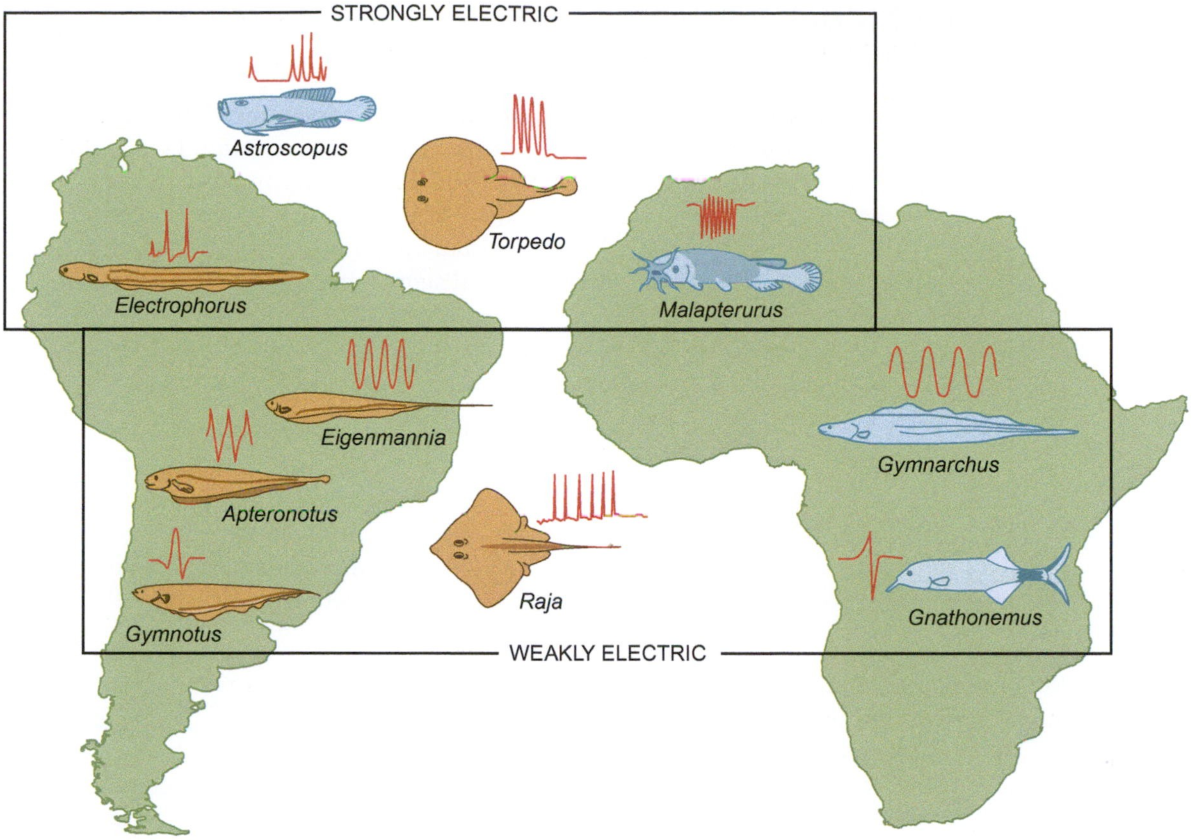

FIGURE 7.21
The geographic distributions of strongly and weakly electric fish and the broad use that such electric signals may serve. Strongly electric fish use their electric emanations for predation; weakly electric fish use them for communication. *Adapted from Moller, P., 1995. Electric Fishes: History and Behavior. Chapman & Hall., 583 pp.*[80]

Electroreception is common in aquatic organisms such as fish and amphibia, but unknown in terrestrial vertebrates and invertebrates. Because air is a poor conductor of electricity, such electrical discharges are limited to animals that live in water; they would be of little use to a terrestrial organism. Usually, there are enough ions in water to make the medium a good electrical conductor, and electrical currents and fields are excellent sources of information. Additionally, the interaction of water currents and the earth's magnetic field produces electrical fields that can be used in navigation.

Electrical fish come in strongly electric forms, which use their electricity primarily for predation and defense, and weakly electric forms, which use electrical signals for communication and electrolocation, similar to the use of sound in echolocation (Figure 7.21). Electroreceptors of fish include ampullary receptors, canals opening from the surface of the fish into cavities lined with nerve cells, and tuberous receptors, found on weakly electric fish (mormyriforms and gymnotoforms). The tuberous receptors are structurally similar to the ampullary receptors but are designed to reduce the loss of the incoming signal.[81,82] Electrosensors have recently been discovered in dolphins, which probably use them for prey detection. Dolphins are the first placental mammals known to have electroreceptors, which are located on the snout, and were once thought to be vestigial whiskers. A dolphin with a covered snout cannot perceive electrical signals; otherwise, dolphins sense signals weak enough to be imperceptible to humans.[83]

7.4 MULTIMODAL SIGNALING AND ENCODING COMPLEX MESSAGES

A *multimodal signal* is exactly what the term implies: a signal that uses more than one mode to convey its message. Sometimes the multiple modes establish signal redundancy, which involves repeating the same message in more than one mode. Primates, for example, recognize each other using both visual and olfactory signals. In other cases of multimodal signaling, each mode contributes a unique element to the overall meaning. In a video clip, the soundtrack and the video itself each contribute uniquely to the information gained.

> **KEY TERM** A multimodal signal is composed using two or more signaling modalities. When a male grasshopper jumps to attract females, it makes a clicking sound and flashes its colorful wings; this is a multimodal signal.

> **KEY TERM** Encoding is the translation of information into a signal that can be decoded by the recipient.

The more complex the message, the more likely an animal is to use multimodal signaling. The dance language of the bees is a good example of multimodal signaling. In their dance language, bees combine olfaction, vibration, touch, and sometimes visual signals to *encode* detailed navigational information.

Von Frisch observed that once one honeybee finds a feeding station, many others soon appear at the same station. This suggests that the first bee recruits other bees to the food. How might honeybees recruit help in collecting food? Von Frisch's discovery of the dance language of the honeybee required careful determination of the correlations between movements of bees inside the hive and the locations of feeding stations. He found two types of dance. The round dance (Figure 7.22A) causes bees to look for food a short distance (up to about 50 m) from the hive. The waggle dance (see Figure 7.22B and C) tells bees the direction and distance to fly to find more distant food sources. Scout bees use these dances

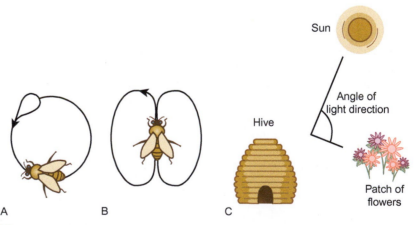

FIGURE 7.22

(A) Diagram of the round dance. This alerts bees to food near the hive but does not convey directional information. (B) Diagram of the waggle dance. The tempo of the dance tells recruits how far to fly (the slower the dance, the greater the distance) and the angle of the straight part of the dance tells them the direction to fly. (C) A hive forms the base of an angle between the sun and the food source. Because the inside of the hive is dark and the comb is vertical, bees convert the angle of the dance on the vertical comb to the angle formed by the feeding station, the hive, and the sun. A dance straight up the comb, as illustrated in (B), tells recruits to fly toward the sun. A straight-down dance tells them to fly directly away from the sun. Dances at angles to the vertical indicate intermediate flight directions. The bees' circadian clocks allow them to correct their dances for the movement of the sun in the sky. *Adapted from von Frisch, K. 1967. The Dance Language and Orientation of Bees. Harvard University Press.*[1]

to recruit assistance in collecting food.[84] Similar dances are used to help find a new home when bees swarm. In this case, scouts dance to direct bees in the swarm to hollow trees, caves, or other likely nesting sites. After a number of bees have visited each nesting site, a "voting" process takes place, until one site (generally the best available location) wins out by having more bees dance for it (Figure 7.23).

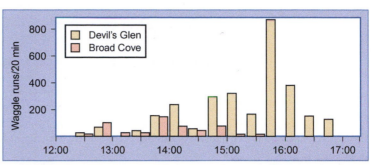

FIGURE 7.23
Honeybees in a swarm "vote," choosing between two potential nest sites. At the beginning of the process, scouts dance equally for two sites, Devil's Glen and Broad Cove, but later in the day one of the sites, Devil's Glen, wins out and the swarm departs for that location. *Adapted from Seeley, T.D., Visscher, P.K., 2003. Choosing a home: how the scouts in a honey bee swarm perceive the completion of their group decision making. Behav. Ecol. Sociobiol. 54, 511–520.[85]*

239

OF SPECIAL INTEREST: THE DANCE LANGUAGE CONTROVERSY

Some controversy has surrounded the dance language hypothesis. Wenner et al.[86] argued that the correlations between bee dances and information conveyed about floral locations is spurious and that the bees are really cueing on odors associated with landscape and food. Wenner felt that von Frisch had not adequately controlled for the possible influence of information other than that from the dance. As a result of Wenner's critique, two types of experiments were conducted to once again test the dance language hypothesis. The first type was misdirection experiments, that is, experiments that asked if the bee dance or landscape odors mattered more to bees experiencing the dance.[87] One component of the dance language hypothesis states that in their dance, returning foragers use the angle of the sun to inform nestmates about the location of food. By shifting the nestmates' circadian clocks, scientists demonstrated that those bees used the dance to identify the location of food. The workers went to the location predicted by the dances, rather than landscape odors. The second type of experiment used mechanical robot bees to simulate dances; again, the workers attending the dances responded as predicted by the information in the dances.[88] The overwhelming scientific evidence supports the dance language hypothesis (Figure 7.22). Wenner and colleagues' hypothesis was a reasonable alternative and warranted testing because olfaction is important in the behavioral biology of honeybees.

OF SPECIAL INTEREST: INFORMATION THEORY

How can we know the extent to which communication actually occurs between a pair of interacting animals? This is a tricky business, because the only scientific approach to this question is to measure the effects of one of the animal's signals on the subsequent behavior of the second animal. In more formal terms, a *bit* is the amount of information needed to answer a single yes/no or on/off question. If an animal is making a complex decision, then it may need more than one bit of information; the minimum number of bits to make a decision (H) among M possibilities is

$$H = \log_2 M$$

Another way of looking at this is to ask how much information is contained in an array of sender behaviors by looking at the correlation between sender behaviors and receiver responses. This one aspect of communication—sender effect on receiver behavior—can reveal the extent of communication that is occurring. For instance, if the sender's action results in a high diversity of receiver responses with no evidence of pattern, then little or no communication has taken place. On the other hand, if the sender's action restricts the receiver's response to one or two behaviors from a large list of possibilities, then there is evidence for influential communication.

Information theory provides an equation for measuring the diversity of responses to sender signals:

$$H(X) = -\sum_{i=1}^{n} p(X_i) \log_b p(X_i)$$

In this formula, there are i possible behaviors. The probability (p) of each behavior is calculated and multiplied by the log of that probability. (In ecology, this same equation is often used as a measure of species diversity in communities.) The next step is to look at sender–receiver interactions as Markov chains, in which the dependence of receiver response on each signal is measured.

Information theory measures of communication are particularly useful when the communication value of a potential signal is unknown. If a bird, for example, does something that seems odd, does the odd behavior have a communicatory function? This hypothesis can be tested by creating a list of receiver behaviors (an ethogram; see Chapter 4). One then compares the diversity of behavior in the receiver in response to a variety of sender behaviors, including the "odd" behavior in question. Highly diverse responses suggest no communication; restricted responses suggest a strong signal value. Clearly, this approach is very helpful in analyzing behavioral interactions. Information theory analyses can be coupled with the construction of a sociogram, that is, a diagram of social interactions.

You will recognize the human game "Twenty Questions" in this discussion. How much information is needed to make a decision? For humans, any arbitrarily selected object can be identified with yes or no answers to 20 well-chosen questions. This approach is also similar to the concept of "Six Degrees of Separation," in which one human can find any another human based on six social linkages. The reason there are fewer than 20° of separation is that each question can narrow more than half the field of people. If you are looking for social linkages to someone in Moscow, you might ask, "Do you know anyone in Moscow?" If the response is positive, you have narrowed the earth's population considerably with a single social link.

7.5 RUNAWAY SEXUAL SELECTION AND SIGNALING

Most signals are optimized for their purpose. They are produced to balance cost with benefit, so they are strong enough to convey the message over the needed distance without wasting effort by broadcasting information beyond the distance at which the signal is useful. The value of communication is also balanced against the risk of attracting a predator. Evolution, as a process, is outstanding at finding cost-efficient ways of conveying information.

In the light of such efficiency, mate choice has a peculiar effect on signaling. If one sex (generally the female) prefers mates to have a strongly expressed signal, selection will favor making that signal bigger, longer, or stronger, with the upper limits well beyond what would be minimally necessary to convey the message. Hence, animals have bright colors, huge horns and antlers, and impressive tails. This is sometimes called *runaway sexual selection*, and occurs when traits in one sex are magnified beyond any reasonable scale as a result of preferences of the other sex for very large, loud, or bright signals (Figure 7.24). Such

FIGURE 7.24
Signals that may have resulted from runaway sexual selection. Contrast the large horns of the male bighorn sheep (on the left) with the smaller horns of the female. The male mallard (above) is brightly adorned while the female (below) is drab. *Photos: bighorn sheep, Jeff Mitton; mallards, Michael Breed.*

FIGURE 7.25
As Tinbergen observed, a gull will react more strongly to an oversized egg than to one of its own, even though the oversized egg cannot possibly be its own egg. A "larger than life" stimulus that elicits such a response is called a *supernormal stimulus*.

preferences may have their roots in animals' heightened responses to supernormal stimuli; these are stimuli that are greater in magnitude than a normal stimulus and also more attractive (Figure 7.25).

OF SPECIAL INTEREST: WHAT YOU LOOKIN' AT? THE PEACOCK'S TAIL

The peacock's tail—huge and extravagantly colored—is a symbol of runaway sexual selection. The tail is important in mate choice, especially the eyespots on that tail. For over two decades, behavioral biologists have compiled evidence supporting the hypothesis that male peacocks with more eyespots attracted more mates.[89] But what is it about the eyespots (a.k.a. "ocelli") that females find so compelling? The eyespot is not a solid color; instead, blue-green and bronze-gold areas surround a purplish black center, and all three are iridescent. Because of this, the angle of observation might be important in how the eyespot is perceived and evaluated. Indeed, males assume an angle of about 45° to the right of the arc of the sun when displaying to a female in front of them, so investigators asked how the eyespots would look at angles of 30°, 45°, and 60°. In an effort to measure what the peahen might perceive, the investigators used a model that was based on the sensitivities of the cones in the peafowl eye. They covered different elements of the eyespot with patches and in the end, determined that the iridescent blue-green portion accounted for more than half of the attractive power of the eyespot; the other two colors did little to interest females. The functions of the other colors may be lost to us—perhaps they were attractive at one time, but no more. Or perhaps they simply intensify the appearance of the blue-green areas?[90]

Other investigators have used miniaturized gaze-trackers to learn exactly what the peahens are paying attention to when they look at a peacock's tail. The peacock display, so lovely to us as a unified vision, has several components—the lower part of the train, the upper part of the train, and the rattling of feathers and shaking of wings. The females seem to notice the upper portion of the train when they are at a distance, and pay more attention to the lower areas when they are nearer the male; this could often be the case in nature, when the lower parts might be obscured by vegetation if the female is some distance away. Wing-shaking and feather-rattling increased attentiveness of females.[91] When taken together with the eyespot study, these results remind us that a single overall communication attempt—in this case, the attempt to attract a female—may have use of more than one mode and have multiple components (see Section 7.4).

7.6 DECEIT VERSUS HONEST SIGNALING ✕

Natural selection should favor any signal that enhances an animal's likelihood of survival and reproduction. Honest signals in communication are given when both the sender and receiver have an interest in the same result.[92] *Deceit* occurs when one animal can exploit another in order to improve its fitness.[93] The most obvious arena for deceit is mate choice. Natural selection strongly favors any device that increases an animal's reproduction; if a dishonest or deceitful signal does that, then genes for the signal will spread rapidly in a population. Another context in which deceit is often found is in the area of parental care, in which young may use signals that exaggerate their actual needs so that they can extract more than necessary care from a parent.

DISCUSSION POINT: DECEIT

While deceit is most common in mate choice, complex social systems, such as those found in apes, provide opportunities for deceitful signaling within the social group. Certainly deceit is a noticeable part of human social behavior. Can you find examples of deceit in our great ape relatives in the literature or on the Web? How credible, scientifically, are these reports? How is deceit signaled, or how is concealment of signals used in deceit?

Given the apparent advantages of deceptive behavior, studies that demonstrate intraspecific deceit are surprisingly rare, with the exception of EPCs (see Chapter 11). This rarity may be due to the fact that the receiver has the upper hand in choosing whether or not to respond to a signal, and consequently can enforce honesty. Scientists have only recently recognized the high frequency of EPCs in seemingly monogamous species of birds and mammals. These EPCs often result in extrapair paternity of offspring. In species that have apparently monogamous mating systems, so that a male and female are pair-bonded, the individuals may nonetheless seek copulations with other partners, doing so at times or locations that make detection by their pair-bonded mate unlikely. This phenomenon is best studied in apparently monogamous birds; averaged over a large number of species, 13% of the offspring were sired by nonpair males. Behavioral deception (secretively copulating) plays an important role in facilitating this behavior.

Costly Signals as a Way of Ensuring Honesty

When a mate attraction signal is costly to produce, as are cricket trills or frog calls, males that are unable to produce attractive, yet costly, signals sometimes adopt a satellite strategy. Satellite males lurk near calling males, intercept females approaching the attractive male, and attempt, sometimes successfully, to mate. This type of deceit takes advantage of a calling males' inability to guard females as they approach and was mentioned briefly in the discussion of auditory signals.[94,95]

Why produce a costly mating attraction signal? As previously indicated, deceit is uncommon in part because the receiver can impose stringent requirements upon the sender; natural selection favors a receiver that is not easily deceived. Thus, communication can be an evolutionary conflict between the sender and the receiver, in which one is trying to take advantage of the other. If the signal is costly to the sender and if the receiver responds only to costly signals, then the communication evolves to reflect the sender's ability to pay the cost; that is to say, senders that can produce costly signals will be favored by natural selection. This can lead to examples of the *handicap principle*,[96] which is the second major explanation for expensive signals and is especially relevant to mate choice and parental care communication. The ability of a receiver to impose a cost on the signals of the sender by preferentially responding to expensive signals is a key factor in the evolution of signaling; the response of the receiver (e.g., mating receptivity in this case) must be worth the cost to the sender (Figure 7.26).

Does the fact that communication often involves individuals who share a common goal, such as mating, feeding, or caring for their young, mean that communication evolves as simple sharing of information? In some cases, this is correct. For instance, this sharing of information would certainly enhance the pair's ability to reach their commonly held goal. In other cases, however, there are underlying conflicts of interest and asymmetries. Knowledge of animal behavior is not sophisticated enough to allow recognition of all the costs and benefits involved in communication; in some circumstances, deceit may be more common than honest signaling. Interacting animals generally act in their own best interest and convey the information, honest or not, needed to enhance their fitness. From the sender's point of view signals should be advantageous (whether they are honest or not) and from the receiver's point of view the challenge is to get as much information as possible, while parsing honesty and dishonesty in the signal.

From the sender's point of view, communication evolves:

- to minimize the time and effort spent on communication,
- to present a message (signal) advantageous to the sender.

From the receiver's point of view, communication evolves:

- to enhance acuity so that the maximum information can be obtained from the signal,
- to separate honest from deceitful signals.

If a signal is used intraspecifically, then it is vulnerable to exploitation by another sender. The sender can be a different species, provided that sender can produce the signal. Because of this, deceit is common in signals that function between species. Carnivorous fireflies mimic the mate attraction signals (light flashes) of other firefly species, attracting males looking for would-be mates; the encounter ends in death instead (Figure 7.15).[97] Bolas spiders use moth sex pheromones in a similar manner.[98] "Guests" in ant nests acquire the chemical signature used by the ants to discriminate nest members from nonmembers.[99] Signals such as these are aggressive mimicry; they take advantage of the victim's need to communicate.

Studies of Deceit

An anecdotal account reports that a female marsh harrier courted a male to be able to obtain access to food he had stored.[100] She then took this food and fed it to chicks that had been fathered by another male. More extensive studies focus on possibly deceitful behavior in the pied flycatcher (*Ficedula hypoleuca*), a species in which males may possess more than one territory simultaneously. In this species, females gain from mating with a male that has no other mates; males may attempt to deceive females about their mating status (mated or unmated). Females assess whether a male has already mated; if he is alone on a territory during repeated visits by the female, then he is probably unmated. Mated males will be absent from the territory (presumably because they are at another territory with their mate). By repeated sampling of male behavior, females are usually able to avoid mating with previously mated males.[101–103]

Topi antelopes provide another example, one in which the deceit is a bit more circuitous; these males "lie" about the risks of leaving the area, not about their own suitability as a mate. A male antelope gains an advantage if he retains a female on his territory for the one day during which she is sexually receptive; the longer she stays around him, the greater the likelihood that he will successfully mate with her. Topi antelopes can often outrun their predators if they are not surprised at close range, so it is to the antelopes' advantage to let a stalking predator know that they've discovered it and an attack will bring no benefit. The antelopes do this with an alarm snort. They also give these snorts when no predators are in the area but when a female seems likely to move on. This causes the female to remain in the area (ignoring an alarm snort could be lethal), while costing the male very little. Although false alarms are not particularly

FIGURE 7.26
Male moose may be examples of the handicap principle. The heavy antlers may be an honest representation of the male's ability to carry weight and maneuver. *Photo: Jeff Mitton.*

243

rare among animals (the "broken wing" display of some parent birds comes to mind), and although many animals might attempt deception in signaling their quality as mates (hence the selection for honest, costly signals for quality), this is an unusual case of deceptively using an alarm signal in order to enhance mating prospects.[104]

7.7 GAME THEORY AND COMMUNICATION

Game theory uses mathematical or logical models to predict how animals with mutual interests should behave.[105] Game theory is particularly appropriate for modeling and analyzing interactions involving communication, because the exchange of information between the two animals can be manipulated in the model and provides a very useful framework for analyzing communication. Remember that when an animal communicates, it has a goal, which may be in conflict with the goals of other animals.

To be successful within a game, an animal needs an overall strategy. The strategy defines the goal and a general plan for achieving the goal. In a Prisoner's Dilemma game, the strategy may be to induce the other animal to cooperate while benefiting more than the other animal from the cooperation.[106] Within the overall strategy, tactics are employed to achieve the goal. In Prisoner's Dilemma,[107] there are various tactics, such as "always defect," "alternate defect and cooperate," and "tit for tat." This makes it sound as though animals that engage in games must be very cognitive—able to remember the responses of the other animals in the game and to calculate and adjust their responses as the game proceeds. This is not necessarily true, because evolution can take the place of cognition in producing this or most other types of behavior. Natural selection will shape strategy and tactics in communicatory games if the reproductive success of the animal is influenced by the outcome of the game. Clearly, strategy and tactics play important roles in animal communication.

An important consideration in these games in nature is the potential for death or serious injury when communication escalates into outright fights. The misconception that animals do not fight to death is widely held. This is simply not true; when the stakes are high for both individuals, such as two males competing for breeding rights with a harem of females, death or serious injury are common outcomes. We may also underestimate the cost of injuries; those that result only in veterinary care for domestic animals may be much more costly in wild animals.

DISCUSSION POINT: THE PRISONER'S DILEMMA

The Prisoner's Dilemma is a two-player game in which the object is to gain more points than your opponent by inducing your opponent to cooperate with you. The players take turns; in each turn a player chooses to either cooperate or defect and the other player responds. The payoff depends on how the response matches the initial play. Table 7.1 shows the payoffs for the two players, given the choice of cooperating or defecting for each player. If both cooperate, each gets 3 points. If both defect, each gets

TABLE 7.1 Prisoner's Dilemma Payoff Matrix

Prisoner's Dilemma

	Cooperate	Defect
Cooperate	3 Reward	0 Sucker
Defect	5 Cheat	1 Punishment

1 point. The highest payoff, 5, comes if your opponent cooperates and you defect. In this case, your opponent gets the "sucker's payoff" of 0. The number of turns in the game is not known to the players, although you do know that at some point the game will be stopped.

Think about playing the game. Can you devise a strategy (a combination of cooperation and defection) that will induce your opponent to cooperate? After you have thought about your strategy for a while, check these strategies.

What is your best strategy if you know the next turn is the last turn of the game?

Try the game a few times, and see which of your strategies gives you the best result. To see how various strategies fare in computer simulations, search the Web for "Prisoner's Dilemma"; you will find several fun sites, such as http://www.iterated-prisoners-dilemma.net/.

Game theory can be taken one step further. Given the variety of tactics available to an animal in a game and the variety of tactics that can be used in response by the other animal, some tactical choices will be poor under any conditions, whereas in the right conditions, others will perform optimally. These optimal choices are called *evolutionarily stable strategies* or ESSs. An ESS is a strategy that cannot be displaced in a population by a new or mutant strategy. If Prisoner's Dilemma is played repeatedly between the same two players with no definite endpoint (this is called *reiterated Prisoner's Dilemma*), then tit for tat is an ESS, and no new strategy in the population will succeed over tit for tat.

To apply game theory to animal behavior, a behavioral biologist goes through the following steps: (1) Identify possible strategies. They may be based on observation (what do the animals actually do?) or logic (what alternatives can the scientist think of?). (2) Determine the outcomes when different tactics are employed by each animal. For instance, in the Prisoner's Dilemma, there were payoffs for cooperating and defecting, depending on the behavior of the opponent. (3) Through repetitions of the game, determine whether one strategy is always superior (an ESS). (4) Observe animals in the field to see if their communicatory strategies fit game theory predictions.

In addition to the Prisoner's Dilemma, Hawk versus Dove is another game that is often used in the analysis of signaling in animal behavior.[108] This game is used primarily to model whether animals should fight or not in disputes such as those over territory. In a simple Hawk versus Dove game, two tactics are available to the participants. A hawk always attacks (aggressive signaling), and a dove always retreats or yields (submissive or passive signaling).[109]

Try this with a classmate: Put out a number of food items. (Ten candy bars is a good starting point.) In the first round of the game, put all the food in one pile. Assign one person to play a hawk strategy and the other to play a dove strategy, but both are to try to capture the food. In this instance, the hawk should quickly come into possession of the food and will continue to possess it because the hawk can efficiently signal aggression and drive the dove away. Now spread the 10 pieces of food so that they are a meter or two apart. When the hawk tries to signal defense at all of these food locations, it will end up spending all of its time running after the dove, which should have an easy time collecting the food. This is an excellent example of the application of game theory in the analysis of communication because it reveals that a strategy that works well under one set of environmental conditions may not fare well in a different environment. Thus, animal species that have evolved in variable environments are expected to have alternate tactics that they can employ, depending on the circumstances. Again, using these tactics does not require cognition or deep thinking on the part of the animal; alternative courses of action can be hard-wired, and the appropriate action can be triggered by environmental conditions.

Game theory can also be used to explore the possibility of behaviors that may be difficult to observe in nature. Martin Nowak has summarized his work with game theory in an engaging and popular book about how cooperation might be more common than we think. In that book, called *SuperCooperators*, he reviews some of the better known aspects of game theory and discusses how it can be applied to social behavior (see Chapter 13).[110]

7.8 INTERSPECIFIC SIGNALING

Most signals evolve for intraspecific perception. Some signals, though, have clear interspecific functions. All of the principles of signaler/receiver coevolution that apply in intraspecific signals are relevant to interspecific signals as well.

Signals between species can be divided into three basic groups. First are attractive devices, used by one species to lure members of another species. Examples of attractive devices include floral attractants (odor, color, and shape) for pollinators and baits or lures used by predators to attract prey, such as angler fish. In some cases, attraction can be related to shared interests, such as interspecific mutualisms, but clearly luring prey is a selfish, one-sided piece of communication.

Cleaner fish represent a very neat example of signaling in an interspecific mutualism.[111] Cleaning has evolved several times; the most commonly mentioned cleaners are wrasses and gobies (Figure 7.27). Cleaner species eat ectoparasites and dead skin from the surface of

FIGURE 7.27
Cleaner wrasses and gobies, with clients. *Photos: Alexandra Grutter.*

other, larger species of fish. Both species benefit; cleaners obtain food and clients suffer from few parasites. For this relationship to work, the cleaner must signal in a way that the client recognizes the cleaner as a mutualist, rather than a prey item. The client must also signal its willingness to be cleaned so that the cleaner is assured it will not be the victim of predation or a defensive action. There is plenty of room for cheating in this relationship—cleaners may take a bite out of the client, or the client, after being cleaned, may eat the cleaner—and signaling helps to keep the participants' intentions clear. Cleaner species may engage in dancing displays and tend to have distinctive colors and stripes, which make them stand out; this is exaggerated by the tendency of cleaners to group at "cleaning stations." The stripes are often blue and yellow;[112] through the convergence of many cleaner species on the same type of signal, the effectiveness of the signal is increased.

The second group of interspecific signals is repellent or defensive signals, such as distinctive sounds (rattlesnakes, for example), colors (red, in some cases; see "Of Special Interest: Is Red Really a Warning Color?"), and patterns (the striped abdomen of bees and wasps, the striped body of coral snakes) that serve to cue potential predators about the dangerous prospects. In an odd way, warning signals used as repellants serve a mutual goal between the two species: the sender and receiver each avoid damage. Finally, many interspecific cues are given off inadvertently, or are eavesdropped by animals wishing to exploit the sender. While these are not signals in the sense of having evolved for the resulting interspecific communication, they merit mention here because eavesdropping can place strong selective pressure on how signals are used.[113]

SUMMARY

Communication is the foundation for mating behavior (Chapter 11) and social behavior (Chapters 13 and 14)—in fact, any behavior routinely involving more than one animal. Communication involves a signal that is produced by a sending animal and received by another animal that interprets the signal. Senders (or signalers) and receivers co-evolve in the communication process. Evolution refines both the sender's ability to send a meaningful signal and the recipient's ability to perceive and understand the signal. Signals used in communication evolve following the basic principles of co-option, ritualization, stereotypy, and redundancy.

Modes, or channels, such as sight, sound, or touch are used in animal communication. Each mode has advantages and disadvantages, depending on the environmental conditions and the signaling and receiving capacities of the species. Signals are often honest, but sometimes deceitful signaling is advantageous, which is especially important in mate choice. Once broadcast, signals become public information, available both to the intended recipient and to other animals, which might take advantage of the signal.

Noise interferes with signaling and influences the evolution of communication by causing selection to favor sensory mechanisms that filter noise or behavior that avoids noise. The potential for human-caused noise to disrupt animal communication is a major conservation concern. Interspecific signaling occurs when a signal has evolved to influence the behavior of other species; examples include attractive signals between mutualists and warning signals produced to avoid predation. Communication is the keystone in the arch of behavior from the individual to its participation in behavioral interactions with other animals.

STUDY QUESTIONS

1. What are ritualization and stereotypy? Why are these concepts important in understanding the evolution of communication?
2. Why might animals signal honestly? When and why might dishonest signals be expected?

3. Under what circumstances should animals present honest signals about their own condition (gender, strength, ovulatory status, and so on)? Are there circumstances under which it might be to an animal's advantage to deceive other animals? Why might deception be advantageous? Design an experiment to test the hypothesis that males attempt to deceive females when females are choosing mates.

4. Watch a honeybee dancing. It is moving very slowly and the straight part of the dance is horizontal, from left to right. What information about the food location can be gleaned from the dance?

Further Reading

Davis, M.D., 1997. Game Theory: A Nontechnical Introduction. Dover Publications, Mineola, NY, 252 pp.

Dugatkin, L.A., Reeve, H.K., 1998. Game Theory & Animal Behavior, xiv. Oxford University Press, New York, NY, 320 pp.

Lucas, J., Freeberg, T.M., 2010. Communication: an overview. In: Breed, M.D., Moore, J. (Eds.), Encyclopedia of Animal Behavior, vol 2, Academic Press, Oxford, pp. 337–339. http://www.sciencedirect.com/science/referenceworks/9780080453378.

Morley, E.L., Jones, G., Radford, A.N., 2014. The importance of invertebrates when considering the impacts of anthropogenic noise. Proc. R. Soc. B 281, 20132683. http://dx.doi.org/10.1098/rspb.2013.2683.

Ord, T.J., 2010. Evolution and phylogeny of communication. In: Breed, M.D., Moore, J. (Eds.), Encyclopedia of Animal Behavior, vol. 2, Academic Press, Oxford, pp. 652–660. http://www.sciencedirect.com/science/referenceworks/9780080453378.

Ryan, M.J., 1998. Sexual selection, receiver biases, and the evolution of sex differences. Science 281, 1999–2003.

von Frisch, K., 1956. Bees: Their Vision, Chemical Senses, and Language. Cornell University Press, Ithaca, NY, 157 pp.

von Frisch, K., 1993. The Dance Language and Orientation of Bees. Harvard University Press, Boston, MA, 592 pp.

Wenner, A.M., Wells, P.H., 1990. Anatomy of a Controversy: The Question of a "Language" Among Bees, xiv. Columbia, University Press, New York, NY, 399 pp.

Notes

1. von Frisch, K., 1993. The Dance Language and Orientation of Bees. Harvard University Press; Esch, H.E., Zhang, S., Srinivasan, M.V., Tautz, J., 2001. Honeybee dances communicate distances measured by optic flow. Nature 411, 581–583.

2. Nieh, J.C., 2004. Recruitment communication in stingless bees (Hymenoptera, Apidae, Meliponini). Apidologie 35, 159–182.

3. Meyer, V.W., Crewe, R.M., Branck, L.E.O., Groeneveld, H.T., van der Linde, M.J., 2000. Intracolonial demography of the mound-building termite *Macrotermes natalensis* (Haviland) (Isoptera, Termitidae) in the northern Kruger National Park, South Africa. Insect. Soc. 47, 390–397.

4. Leuthold, R.H., Triet, H., Schildger, B., 2004. Husbandry and breeding of African giant termites (*Macrotermes jeanneli*) at Berne Animal Park. Der Zool. Garten 74, 26–37.

5. Hager, F.A., Kirchner, W.H., 2013. Vibrational long-distance communication in the termites *Macrotermes natalensis* and *Odontotermes* sp. J. Exp. Biol. 216, 3249–3256.

6. Searcy, W.A., Nowicki, S., 2005. The Evolution of Animal Communication: Reliability and Deception in Signaling Systems, cxii. Princeton University Press, Princeton, NJ, 270 pp.

7. Maynard Smith, J., Harper, D., 2003. Animal signals Oxford Series on Ecology and Evolution. Oxford University Press, Oxford, 3 pp.

8. Freed-Brown, G., White, D.J., 2009. Acoustic mate copying: female cowbirds attend to other females' vocalizations to modify their song preferences. Proc. R. Soc. B Biol. Sci. 276, 3319–3325.

9. Ito, R., Mori, A., 2009. Vigilance against predators induced by eavesdropping on heterospecific alarm calls in a non-vocal lizard *Oplurus curieri curieri* (Reptilia: Iguania). Proc. R Soc. B Biol. Sci. 277, 1275–1280.

10. Stoddard, P.K., Beecher, M.D., Horning, C.L., Campbell, S.E., 1991. Recognition of individual neighbors by song in the song sparrow, a species with song repertoires. Behav. Ecol. Sociobiol. 29, 211–215.

11. Oxford English Dictionary Second Edition on CD-ROM (v. 4.0.0.3), 2009. Oxford University Press, Oxford, UK.

12. Groot, A.T., Ward, C., Wang, J., Pokrzywa, A., O'Brien, J., Bennett, J., et al., 2004. Introgressing pheromone QTL between species: towards an evolutionary understanding of differentiation in sexual communication. J. Chem. Ecol. 30, 2495–2514.

13. Crawford, J.C., Boulet, M., Drea, C.M., 2011. Smelling wrong: hormonal contraception in lemurs alters critical female odour cues. Proc. R. Soc. B. Biol. Sci. 278, 122–130.

14. Sorge, R.E., Martin, L.J., Isbester, K.A., Sotocinal, S.G., Rosen, S., Tuttle, A.H., et al., 2014. Olfactory exposure to males, including men, causes stress and related analgesia in rodents. Nat. Methods. http://dx.doi.org/10.1038/nmeth.2935.

15. Bekoff, M., 2001. Observations of scent-marking and discriminating self from others by a domestic dog (Canis familiaris): tales of displaced yellow snow. Behav. Processes 55, 75–79.

16. Bekoff, M., 1979. Ground scratching by male domestic dogs: a composite signal? J. Mammal. 60, 847–848.

17. White, A.M., Swaisgood, R.R., Zhang, H., 2002. The highs and lows of chemical communication in giant pandas (Ailuropoda melanoleuca): effect of scent deposition height on signal discrimination. Behav. Ecol. Sociobiol. 51, 519–529.

18. Bradshaw, J., 2013. Cat sense. Basic Books, New York, NY, 307 pp.

19. Gammon, D.E., Baker, M.C., 2004. Song repertoire evolution and acoustic divergency in a population of black-capped chickadees, Poecile atricapillus. Anim. Behav. 68, 903–913.

20. Au, W.W.L., Mobley, J., Burgess, W.C., Lammers, M.O., Nachtigall, P.E., 2000. Seasonal and diurnal trends of chorusing humpback whales wintering in waters off western Maui. Mar. Mammal. Sci. 16, 530–544.

21. Miller, P.J.O., Bain, D.E., 2000. Within-pod variation in the sound production of a pod of killer whales, Orcinus orca. Anim. Behav. 60, 617–628.

22. Ford, J.K.B., 1989. Acoustic behavior of resident killer whales (Orcinus orca) in coastal waters of British Columbia. Can. J. Zool. 67, 727–745.

23. Ford, J.K.B., 1991. Vocal traditions among resident killer whales (Orcinus orca) in coastal waters of British Columbia. Can. J. Zool. 69, 1454–1483.

24. Moss, C.J., Poole, J.H., 1983. Relationships and social structure of African elephants. In: Hinde, R.A. (Ed.), Primate Social Relationships: An Integrated Approach Blackwell Scientific, Oxford.

25. Poole, K., Payne, K., Langauer Jr., W.R., Moss, C.J., 1988. The social contexts of some very low frequency calls of African elephants. Behav. Ecol. Sociobiol. 22, 385–392.

26. McComb, K., Moss, C., Sayialel, S., Baker, L., 2000. Unusually extensive networks of vocal recognition in African elephants. Anim. Behav. 59, 1103–1109.

27. Wilson, B., Batty, R.S., Dill, L.M., 2004. Pacific and Atlantic herring produce burst pulse sounds. Proc R. Soc. Lond. B 271 (Suppl.), S95–S97.

28. Pongracz, P., Molnar, C., Mikosi, A., Csanyi, V., 2005. Human listeners are able to classify dog (Canis familiaris) barks recorded in different situations. J. Comp. Psych. 119, 136–144.

29. Maros, K.P., Pongracz, G., Bardos, C., Molnar, T., Farago, Miklosi, A., 2008. Dogs can discriminate barks from different situations. Appl. Anim. Behav. Sci. 114, 159–167.

30. Molnar, C., Pongracz, P., Farago, T., Doka, A., Miklosi, A., 2009. Dogs discriminate between barks: the effect of context and identity of the caller. Behav. Processes 82, 198–201.

31. Farago, T., Pongracz, P., Range, F., Viranyi, Z., Miklosi, A., 2010. "The bone is mine": affective and referential aspects of dog growls. Anim. Behav. 79, 917–925.

32. McComb, K., Taylor, A.M., Wilson, C., Charlton, B.D., 2009. The cry embedded within the purr. Curr. Biol. 19, R507–508.

33. Garland, E.C., Goldizen, A.W., Rekdahl, M.L., Constantine, R., Garrigue, C., Hauser, N.D., et al., 2011. Dynamic horizontal cultural transmission of humpback whale song at the ocean basin scale. Curr. Biol. 21, 687–691.

34. Blake, B.H., 2002. Ultrasonic calling in isolated infant prairie voles (Microtus ochrogaster) and montane voles (M. montanus). J. Mammal. 83 (2), 536–545.

35. Greenfield, M.D., Baker, M., 2003. Bat avoidance in non-aerial insects: the silence response of signaling males in an acoustic moth. Ethology 109 (5), 427–442.
Greenfield, M.D., Tourtellot, M.K., Tillberg, C., Bell, W.J., Prins, N., 2002. Acoustic orientation via sequential comparison in an ultrasonic moth. Naturwissenschaften 89 (8), 376–380.

36. Jones, G., Barabas, A., Elliott, W., Parsons, S., 2002. Female greater wax moths reduce sexual display behavior in relation to the potential risk of predation by echolocating bats. Behav. Ecol. 13 (3), 375–380.

37. Moles, A., D'Amato, F.R., 2000. Ultrasonic vocalization by female mice in the presence of a conspecific carrying food cues. Anim. Behav. 60, 689–694.

38. Surlykke, A., Yack, J.E., Spence, A.J., Hasenfuss, I., 2003. Hearing in hooktip moths (Drepanidae: Lepidoptera). J. Exp. Biol. 206 (15), 2653–2663.

39. Thornton, L.M., Hahn, M.E., Schanz, N., 2003. Genetic and developmental influences on infant mouse ultrasonic calling: patterns of inheritance of call characteristics. Behav. Genet. 33, 721–772.

40. Ryan, M.J., Cummings, M.E., 2013. Perceptual biases and mate choice. Annu. Rev. Ecol. Evol. Syst. 44, 437–459.

41. Gerhardt, H.C., Klump, G.M., 1988. Masking of acoustic signals by the chorus background noise in the green tree frog: a limitation on mate choice. Anim. Behav. 36, 1247–1249.

42. Greenfield, M.D., 1994. Synchronous and alternating choruses in insects and anurans—common mechanisms and diverse functions. Am. Zool. 34, 605–615.
Snedden, W.A., Greenfield, M.D., 1998. Females prefer leading males: relative call timing and sexual selection in katydid choruses. Anim. Behav. 56, 1091–1098.
Minckley, R.L., Greenfield, M.D., Tourtellot, M.K., 1995. Chorus structure in tarbush grasshoppers—inhibition, selective phonoresponse and signal competition. Anim. Behav. 50, 579–594.
Snedden, W.A., Greenfield, M.D., Jang, Y.W., 1998. Mechanisms of selective attention in grasshopper choruses: who listens to whom? Behav. Ecol. Sociobiol. 43, 59–66.

43. Kempenaers, B., Borgstrom, P., Loes, P., Schlicht, E., Valcu, M., 2010. Artificial night lighting affects dawn song, extra-pair siring success, and lay date in songbirds. Curr. Biol. 20, 1735–1739.

44. Hindmarsh, A.M., 1984. Vocal mimicry in starlings. Behaviour 90, 302–324.
45. Snedden, W.A., Greenfield, M.D., 1998. Females prefer leading males: relative call timing and sexual selection in katydid choruses. Anim. Behav. 56, 1091–1098.
46. Tauber, E., 2001. Bidirectional communication system in katydids: the effect on chorus structure. Behav. Ecol. 12, 308–312.
47. Mitani, J.C., Gros-Louis, J., 1998. Chorusing and call convergence in chimpanzees: tests of three hypotheses. Behaviour 135, 1041–1064.
48. Gray, D.A., 1999. Intrinsic factors affecting female choice in house crickets: time cost, female age, nutritional condition, body size, and size-relative reproductive investment. J. Insect Behav. 12 (5), 691–700.
Gray, D.A., 1997. Female house crickets, *Acheta domesticus*, prefer the chirps of large males. Anim. Behav. 54, 1553–1562.
49. Kiflawi, M., Gray, D.A., 2000. Size-dependent response to conspecific mating calls by male crickets. Proc. R. Soc. Lond. B 267, 2157–2161.
50. Hack, M.A., 1998. The energetics of male mating strategies in field crickets (Orthoptera: Gryllinae: Gryllidae). J. Insect Behav. 11, 853–867.
51. Cooley, J.R., Marshall, D.C., 2001. Sexual signaling in periodical cicadas, *Magicicada spp.* (Hemiptera: Cicadidae). Behaviour 138, 827–855.
52. Wells, M.M., Henry, C.S., 1992. The role of courtship songs in reproductive isolation among populations of green lacewings of the genus *Chrysoperla* (Neuroptera: Chrysopidae). Evolution 46, 31–42.
53. Aubin, T., Jouventin, P., 1998. Cocktail-party effect in king penguin colonies. Proc. R. Soc. Biol. Sci. 265, 1665–1673.
54. Bee, M.A., Michey, C., 2008. The "cocktail party problem": what is it? How can it be solved? And why should animal behaviorists study it? J. Comp. Psychol. 122, 235–251.
55. Bee, M.A., Swanson, E.M., 2007. Auditory masking of anuran advertisement calls by road traffic noise. Anim. Behav. 74, 1765–1776.
56. Pohl, N.U., Slabbekoorn, H., Klump, G.M., Langemann, U., 2009. Effects of signal features and environmental noise on signal detection in the great tit, *Parus major*. Anim. Behav. 78, 1293–1300.
57. Swaddle, J.P., Page, L.C., 2007. High levels of environmental noise erode pair preferences in zebra finches: implications for noise pollution. Anim. Behav. 74, 363–368.
58. Slabbekoorn, H., Peet, M., 2003. Birds sing at a higher pitch in urban noise. Nature 424, 267.
59. Kight, C.R., Saha, M.S., Swaddle, J.P., 2012. Anthropogenic noise is associated with reductions in the productivity of breeding Eastern Bluebirds (*Sialia sialis*). Ecol. Appl. 22, 1989–1996.
60. Buckley, J.P., McClure, C.J.W., Kleist, N.J., Francis, C.D., Barber, J.R., 2015. Anthropogenic noise alters bat activity levels and echolocation calls. Glob. Ecol. Conserv. 3, 62–71.
61. Morley, E.L., Jones, G., Radford, A.N., 2014. The importance of invertebrates when considering the impacts of anthropogenic noise. Proc. R. Soc. B. 281, 20132683. http://dx.doi.org/10.1098/rspb.2013.2683.
62. Hill, P.S.M., 2009. How do animals use substrate-borne vibrations as an information source? Naturwissenschaften 96, 1355–1371.
63. Maklakov, A.A., Bilde, T., Lubin, Y., 2003. Vibratory courtship in a web-building spider: signalling quality or stimulating the female? Anim. Behav. 66, 623–630.
64. Hebets, E.A., 2008. Seismic signal dominance in the multimodal courtship display of the wolf spider *Schizocosa stridulans* Stratton 1991. Behav. Ecol. 19, 1250–1257.
65. Christensen, C.B., Christensen-Dalsgaard, J., Brandt, C., Madsen, P.T., 2012. Hearing with an atympanic ear: good vibration and poor sound-pressure detection in the royal python, *Python regius*. J. Exp. Biol. 215, 331–342.
66. Caldwell, M.S., Johnston, G.R., McDaniel, G., Warkentin, K.M., 2010. Vibrational signaling in the agonistic interactions of red-eyed treefrogs. Curr. Biol. 20, 1012–1017.
67. Cuthill, I.C., Partridge, J.C., Bennett, A.T.D., 2000. Avian UV vision and sexual selection. In: Espmark, Y., Amundson, T., Rosenquist, G. (Eds.), Animal Signals Tapir Academic Press, Trondheim, Norway, pp. 61–82.
68. Senviratne, S., Jones, I., 2010. Origin and maintenance of mechanosensory feather ornaments. Anim. Behav. http://dx.doi.org/10.1016.
69. Siebeck, U.E., Parker, A.N., Sprenger, D., Mathger, L.M., Wallis, G., 2010. A species of reef fish that uses ultraviolet patterns for covert face recognition. Curr. Biol. 20, 1–4.
70. Cratsley, C.K., Lewis, S.M., 2003. Female preference for male courtship flashes in *Photinus ignitus* fireflies. Behav. Ecol. 14 (1), 135–140.
71. Newton, I. 1704. Opticks or a Treatise of the Reflexions, Refractions, Inflexions and Colours of Light. Also Two Treatises of the Species and Magnitude of Curvilinear Figures. London: Smith and Walford [cited May 11, 2014] see <http://archive.org/details/opticksortreatisnewt>.
72. Michiels, N.K., Anthes, N., Hart, N.S., Herler, J., Meixner, A.J., Schleifenbaum, F., et al., 2008. Red fluorescence in reef fish: a novel signalling mechanism? BMC Ecol. 8, 16.
73. Pryke, S.R., 2009. Is red an innate or learned signal of aggression and intimidation? Anim. Behav. 78, 393–398.
74. Gamberale-Stille, G., Tullberg, B.S., 2001. Fruit or aposematic insect? Context-dependent colour preferences in domestic chicks. Proc. R. Soc. Lond. B Biol. Sci. 268, 2525.
75. Aronsson, M., Gamberale-Stille, G., 2009. Importance of internal pattern contrast and contrast against the background in aposematic signals. Behav. Ecol. 20, 1356–1362.
76. Skelhorn, J., Griksaitis, D., Rowe, C., 2008. Colour biases are more than a question of taste. Anim. Behav. 75, 827–835.

77. Stone, E.L., Jones, G., Harris, S., 2009. Street lighting disturbs commuting bats. Curr. Biol. 19, 1123–1127.

78. Hopkins, C.D., 1999. Design features for electric communication. J. Exp. Biol. 202, 1217–1228.

79. Dunlap, K.D., Zakon, H.H., 1998. Behavioral actions of androgens and androgen receptor expression in the electrocommunication system of an electric fish, *Eigenmannia irescens*. Horm. Behav. 34 (1), 30–38.

80. Moller, P., 1995. Electric Fishes: History and Behavior. Chapman & Hall., 583 pp.

81. Hagedorn, M., Heiligenberg, W., 1985. Court and spark: electric signals in the courtship and mating of gynotid fish. Anim. Behav. 33, 254–265.
    Hagedorn, M., Zelick, R., 1989. Relative dominance among males is expressed in the electric organ discharge characteristics of a weakly electric fish. Anim. Behav. 38, 520–525.

82. Heiligenberg, W., 1990. Electrosensory systems in fish. Synapse 6, 196–206.

83. Czech-Damal, N.U., Liebschner, A., Miersch, L., Klauer, G., Hanke, F.D., Marshall, C., et al., 2012. Electrorecption in the Guiana dolphin (*Sotalia guianensis*). Proc. R. Soc. B Biol. Sci. 279, 663–668.

84. Matilla, H., Burke, K., Seeley, T., 2010. Genetic diversity within honeybee colonies increases signal production by waggle dancing foragers. Proc. Natl. Acad. Sci. USA 275, 809–816.

85. Seeley, T.D., Visscher, P.K., 2003. Choosing a home: how the scouts in a honey bee swarm perceive the completion of their group decision making. Behav. Ecol. Sociobiol. 54, 511–520.

86. Wenner, A.M., 1967. Honey bees: do they use the distance information contained in their dance maneuver? Science 155, 847–849.

87. Gould, J.L., 1975. Honey bee recruitment dance language. Science 189, 685–693.

88. Michelsen, A., Andersen, B.B., Kirchner, W.H., Lindauer, M., 1989. Honeybees can be recruited by a mechanical model of a dancing bee. Naturwissenschaften 76, 277–280.

89. Petrie, M., Haoliday, T., Sanders, C., 1991. Peahens prefer peacocks with elaborate trains. Anim. Behav. 41, 323–331.

90. Dakin, R., Mongomerie, R., 2013. Eye for an eyespot: how iridescent plumage occelli influence peacock mating success. Behav. Ecol. 24, 1048–1057.

91. Yorzinski, J.L., Patricelli, G.L., Babcock, J.S., Pearson, J.M., Platt, M.L., 2013. Through their eyes: selective attention in peahens during courtship. J. Exp. Biol. 216, 3035–3046.

92. Smith, J.M., 1991. Honest signaling—the Philip Sidney game. Anim. Behav. 42, 1034–1035.

93. Dawkins, M.S., Guilford, T., 1991. The corruption of honest signaling. Anim. Behav. 41, 865–873.

94. Tibbetts, E.A., 2014. The evolution of honest communication: integrating social and physiological costs of ornamentation. Integr. Comp. Biol. 54, 578–590.

95. Biernaskie, J.M., Grafen, A., Perry, J.C., 2014. The evolution of index signals to avoid the cost of dishonesty. Proc. R. Soc. B Biol. Sci. 281, 20140876.

96. Zahavi, A., 1975. Mate selection—selection for a handicap. J. Theor. Biol. 53 (1), 205–214.
    Zahavi, A., 1977. Cost of honesty (further remarks on handicap principle). J. Theor. Biol. 67 (3), 603–605.

97. Lloyd, J.E., 1975. Aggressive mimicry in *Photuris* fireflies: signal repertoires by femmes fatales. Science 197, 452–453.

98. Eberhard, W.G., 1977. Aggressive chemical mimicry by a bolas spider. Science 198 (4322), 1173–1175.

99. Vandermeer, R.K., Wojcik, D.P., 1982. Chemical mimicry in the myrmecophilous beetle *Myrmecaphodius excavaticollis*. Science 218 (4574), 806–808.

100. Simmons, R., 1992. Brood adoption and deceit among African marsh harriers *Circus ranivorus*. Ibis 134, 32–34.

101. Searcy, W.A., Eriksson, D., Lundberg, A., 1991. Deceptive behavior in pied flycatchers. Behav. Ecol. Sociobiol. 29 (3), 167–175.

102. Getty, T., 1996. Mate selection by repeated inspection: more on pied flycatchers. Anim. Behav. 51, 739–745.

103. Davis, M.D., 1997. Game Theory: A Nontechnical Introduction. Dover Publications.
     Dugatkin, L.A., Reeve, H.K., 1998. Game Theory & Animal Behavior, xiv. Oxford University Press, New York, NY, 320 pp.

104. Bro-Jorgensen, J., Pangle, W.M., 2010. Male topi antelopes alarm snort deceptively to retain females for mating. Am. Nat. 176, E33–E39.

105. Lomborg, B., 1996. The Structure of Solutions in the Iterated Prisoner's Dilemma. (Center for International Relations Series). University of California, Los Angeles, CA.

106. Axelrod, R.M., 1984. The Evolution of Cooperation, x. Basic Books, New York, NY, 241 p.

107. Dale, S., Slagsvold, T., 1994. Polygyny and deception in the pied flycatcher—can females determine male mating status. Anim. Behav. 48 (5), 1207–1217.

108. Crowley, P.H., 2001. Dangerous games and the emergence of social structure: evolving memory-based strategies for the generalized hawk-dove game. Behav. Ecol. 12 (6), 753–760.

109. Enquist, M., 1985. Communication during aggressive interactions with particular reference to variation in choice of behavior. Anim. Behav. 33, 1152–1161.

110. Nowak, M., Highfield, R., 2011. SuperCooperators: Altruism, Evolution, and Why We Need Each Other to Succeed. Free Press, New York, NY.

111. Lettieri, L., Cheney, K.L., Mazel, C.H., Boothe, D., Marshall, N.J., Streelman, J.T., 2009. Cleaner gobies evolve advertising stripes of higher contrast. J. Exp. Biol. 212, 2194–2203.

112. Cheney, K.L., Grutter, A.S., Blomberg, S.P., Marshall, N.J., 2009. Blue and yellow signal cleaning behavior in coral reef fishes. Curr. Biol. 19, 1283–1287.

113. Stowe, M.K., Turlings, T.C.J., Loughrin, J.H., Lewis, W.J., Tumlinson, J.H., 1995. The chemistry of eavesdropping, alarm, and deceit. Proc. Natl. Acad. Sci. USA 92, 23–28.

251

CHAPTER 8

Movement: Search, Navigation, Migration, and Dispersal

253

LEARNING OBJECTIVES

Studying this chapter should provide you with the knowledge to:

- Apply the techniques animals use to integrate navigational information from genetic and environmental sources.

- Analyze how basic navigational mechanisms, including kineses (undirected movements) and taxes (directed movements) allow movement to resources.

- Examine how directional information comes from sequential or simultaneous comparisons of environmental stimuli.

Animal Behavior. DOI: http://dx.doi.org/10.1016/B978-0-12-801532-2.00008-8

- Construct models for search behavior that starts with undirected or looping movements, and shifts to goal-directed movements as cues are gained from the environment.
- Organize the complex mechanisms that animals can use, including path integration and compasses, to set their course.
- Apply how homing behavior allows animals to return to a central place, such as a nest or den.
- Understand that migration gives animals the opportunity to exploit more than one habitat during their life cycles; knowledge of migratory routes may be innate, learned, or a mixture of the two.
- Relate dispersal behavior to avoidance of competition and to colonization of new habitats.

8.1 INTRODUCTION

Animals move to find better conditions. Movement is a prelude to life-sustaining behaviors such as feeding, avoiding predators,[1] and mating. A variety of motivations prompt animals to move. The ability to change locations at almost any time during their lives distinguishes most animals from fungi and plants, which, once established, are more or less stationary in their environments.

Movement is one of the most fascinating topics in animal behavior because it requires an exquisite integration of sensory inputs, neural calculation, and execution of behavior. This chapter begins with a look at the tools animals use to shape their movements. It then follows the progression that an animal might use in its own daily and seasonal routines. This progression starts with *searching* movements[2] and then continues to goal-directed *navigation*, which includes *homing* and migration.[3] The chapter concludes with a discussion of dispersal—movements that take an animal from its birthplace to a different population or that lead to colonization of new habitats—and how dispersal can prevent inbreeding and reduce competition within populations.

Scientific explorations of animal movement have generated clever and intriguing experimental designs that provide windows into the use of unexpected sensory inputs, such as geomagnetic fields,[4,5] and that reveal navigational abilities of seemingly simple animals that far exceed human capabilities. Complex movements, such as homing and migration, are the culmination of interactions between sensory inputs and innate abilities to calculate how to reach a goal (Figure 8.1).

KEY TERM Search is movement to find a goal. Search can be entirely random, can be patterned to cover space efficiently, or it can be shaped by new information as it comes in.

KEY TERM Navigation is guided movement from one location to another, typically using a compass or landmarks.

KEY TERM Homing is movement with the goal of reaching a known resource, such as a shelter, that is used as the central place within an animal's home range.

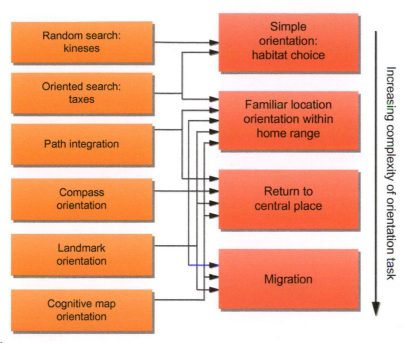

FIGURE 8.1

Animals combine orientation strategies with information in order to reach a spatial goal. In the simplest case a random walk (called a kinesis) can lead an animal to a desired habitat. Directional information from the environment allows oriented movements to the goal. Complexity is added when an animal has a home range and is familiar with landmarks in the area, has a nest, or migrates in order to find seasonally appropriate habitat. Path integration, compasses, landmarks, and cognitive maps all aid animals in achieving more complex orientational goals.

255

CASE STUDY

The movement of bats has long fascinated people (Figure 8.2). The story of how we came to understand the unerring way they avoid obstacles as they move through darkness is stranger than you might imagine. And yes, the unexpected twists and turns

FIGURE 8.2

"It would perhaps be best not to delve too deeply into Cuvier's triumph of logic over experimentation."—Robert Galambos, 1942. The development of tools that allowed scientists to enter the auditory world of bats made bats' use of ultrasound undeniable. *Corynorhinus townsendii* in flight. *Photo: J. Scott Altenbach.*

that the story contains are there because of that central concept in animal behavior—*umwelt*.

For our purposes, this story begins with an eighteenth century Italian scientist-priest named Lazzarro Spallanzani who is perhaps best known for his experiments refuting the idea of spontaneous generation. Toward the end of his career, he investigated the flight of bats. He began by blinding them, and noting that blind bats were just as skilled at avoiding obstacles as sighted bats were. (It is apparently not clear from his extant writings why he chose sight, and not some other sense, as a place to begin his investigations.) Other investigators of the time confirmed his observations, and all wondered what senses were used by bats. The hypothesis that bats used their sense of touch quickly became the reigning explanation, and was almost as quickly dispatched by Spallanzani, who set up experiments such as those showing that bats could follow the twists and turns in a tunnel without ever coming close to its walls. Perhaps his most definitive experiment refuting the touch hypothesis was one in which he coated blind bats with a kind of varnish—and indeed followed that with an experiment that involved covering a blind bat with flour paste! The bats' ability to avoid obstacles was unimpaired.

CASE STUDY (CONTINUED)

After eliminating the hypotheses of taste and smell, Spallanzani and his colleagues turned to covering the heads of the bats with hoods; hooded bats, blind and sighted, crashed into things, even when the hoods were nearly transparent. The investigators then attempted to interfere with bats' hearing, and this yielded mixed results. The sum of all of these experiments seemed to eliminate touch, hearing, sight, taste and smell, and some of the best scientists of the age were cautiously leaning toward a hypothesis that involved some heretofore-unknown sixth sense. Spallanzani himself suggested this, but also urged his colleagues to repeat and refine his own experiments. In Geneva, Louis Jurine did just that, plugging bats' ears more tightly and producing totally disoriented bats. Upon reviewing Jurine's results, Spallanzani performed more experiments, taking more care with the bats' ear plugs; he found that these bats collided with obstacles in both light and dark. Spallanzani immediately embraced the "ear hypothesis." Galambos (1942, p. 135) quotes Spallanzani: "…I say only that deaf bats fly badly and hurtle against obstacles in the dark and in the light, and that blinded bats avoid obstacles in either light or dark."

This should have been the end of this discussion, but the "sixth sense" hypothesis was out of the bag. In part, this was because the famed anatomist Georges Curvier seized upon the sixth sense hypothesis, primarily in order to refute it; he turned not to the ear hypothesis as an alternative, however, but simply claimed that the bat's success in avoiding obstacles must be attributed to an enhanced sense of touch, an explanation he continued to promote for decades despite remarkable lack of experimental evidence. Indeed, throughout the nineteenth century, the touch hypothesis was received wisdom, the explanation for bat navigation.

Well over a century had passed since Spallanzani's death in 1799 when George Pierce, a physicist, and Harvard undergraduate

Donald Griffin "heard" the ultrasonic sounds of bats via Pierce's parabolic ultrasonic detector, the first apparatus that could do so. Because of the limits of the detector, however, Pierce and Griffin did not detect these sounds when the bats were flying. This remained for Griffin, then a graduate student, and another graduate student, Robert Galambos, to investigate. Using Pierce's equipment and Galambos's skill with cochlear microphonics, they showed that bats respond to the echoes of their own ultrasonic cries as they navigate a room full of wires. Later, scientists would realize that bats also use ultrasound to catch their food.

During the long interval between Spallanzani and a room containing bats and a parabolic ultrasonic detector, additional evidence did implicate sound as an influence in bat navigation. For 140 years, it had been clear that, blind or not, deafened bats crashed into obstacles and hearing bats did not. Why had the ear hypothesis languished? Beyond the unfortunate weight of Cuvier's unsupported sarcasm, there was the simple fact that the ear hypothesis rested on the existence of sound that humans could not hear, and experimental evidence was no match for human handicap. Indeed, even after Griffin and Galambos's discovery, bat echolocation was initially hard for scientists to believe; according to Galambos's obituary in the New York Times (July 15, 2010), after one early presentation of the bat research, a distinguished scientist approached Galambos, took him by the shoulders, and shook him, exclaiming, "You can't really mean that!" Like so many discoveries in science, what now seems obvious awaited the right tools, in this case, tools that allowed humans to enter the auditory world of bats.[6–9]

Bats are wonderful organisms in their own right, no doubt about that. But they bring a more powerful lesson to those of us who would study the behavior of animals: As much as we love what we see, touch and hear, *umwelt* remains our prison, and we should not underestimate it.

Another way of looking at animal movement is to think of a progression of types of movements. This progression starts with actions that have no set direction. Sometimes *random movement* accomplishes an animal's goals of finding food or shelter. However, in most instances, animals use *directional movements* to travel from one place to another. A directional movement is oriented with respect to input from the animal's external world, such as traveling up or down by assessing gravitational input.

Navigation itself is movement with a specific destination. Movement from one point to another is essential to survival in most animal species. The roots of the word navigation can illuminate this concept. For ancient people, travel over long distances often involved the sea; the Latin roots of navigation refer to ships (navis) and guidance (agere). For animals, navigation can occur at vastly different scales, ranging from a movement of a few body lengths that takes the animal to food or shelter to the thousands of kilometers (40,000, in fact) that an Arctic Tern spans on its nearly pole-to-pole migration (Figure 8.3).

FIGURE 8.3
The Arctic Tern accomplishes the longest migration of any organism as it flies pole to pole. They often return to the breeding ground where they hatched in order to reproduce. *Photos: (A) Ben Pless; (B) Tim Bowman/US Fish and Wildlife Service.*

Navigation seems like an obvious enough process—a simple task—but what does organizing movement from one place to another entail? How do animals gather the information they need for navigation, and what do they do with it? These are the central questions for this chapter, which covers the full range of possible travels, from short trips around nests or dens to long journeys across oceans and continents.

To be able to navigate, an organism needs to know at least two things: its current location and the location of its goal. Thus, navigation is not the same as simply moving in a randomly chosen direction, even if that movement is in a straight line. In this regard, it is useful to note the distinction between *navigation* and *orientation*. *Orientation* is a word that indicates movement in a given direction—classically, movement in a compass direction, although a variety of stimuli that have nothing to do with compass direction may also elicit orientation. (*Orient* itself is from a French word meaning "east.") However, moving in a direction is not the same as moving toward a goal or destination. If an *orienting* animal is moving north and it is displaced to the east, it will start from its new location and continue to move due north. If a *navigating* animal is moving north and is displaced to the east, it will adjust its direction of movement toward the northwest and correct its path, moving to the northwest; in so doing, it will reach the same destination that it had prior to being displaced (Figure 8.4).

Navigational movements often involve an environmental cue that allows setting the direction of movement. This can be a cue such as landmark, the sun, or the earth's geomagnetic field. In some cases navigation involves only internal calculation of the distance and direction to be traveled; this interesting process is called *path integration*. Studies of navigation often focus on discovering how animals obtain the information needed to direct their movements (Figure 8.5); such investigations have led to spectacular findings of homing pigeons using odors to navigate home from distances of hundreds of kilometers and of sea turtles using magnetic fields in the ocean to orient their long-distance migrations.

As noted previously, navigation is central to survival in many species. Probably because of this, there are reinforcing linkages among many of the concepts discussed in this chapter, such as search and navigation to and from a central place, and elements of Chapter 9 on foraging. The discussion of cognitive maps in Chapter 6 comes into play as well; refer to that discussion to assist with topics in this chapter, such as homing and migration. The maintenance of

KEY TERM Random movements cannot be predicted by either the animal's previous movement or by assessing the surrounding environment.

KEY TERM Directional movements are oriented with respect to the surrounding environment.

KEY TERM Path integration is the ability to track an outward path and to internally calculate the shortest possible path back to the starting location.

257

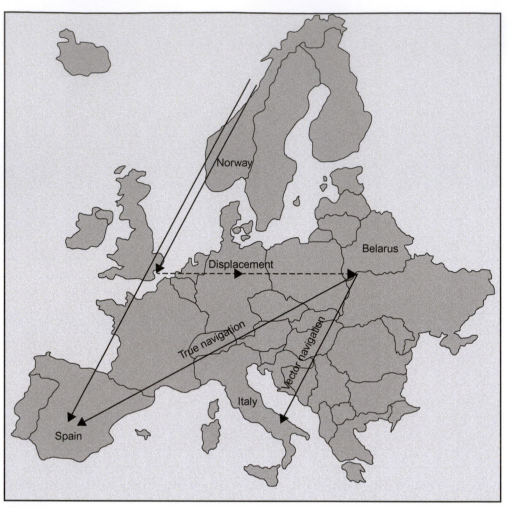

FIGURE 8.4

In this figure, an organism begins its travel off the western coast of Norway. If it is *orienting* toward, say, southwest, it will move in a southwesterly direction, eventually arriving in Spain; if displaced hundreds of kilometers to the east (Belarus; see dotted line), it will continue to move southwest, hundreds of kilometers east of the path it would have taken prior to displacement—in this case, arriving in Italy with what is called vector navigation. If an animal is truly *navigating* toward a goal in the southwest (e.g., Spain) and is displaced hundreds of kilometers to the east, it will compensate for the displacement, shift its path west, and continue to move toward the goal, arriving in Spain from Belarus in this example. *Adapted from Berthold, P., 2001. Bird Migration: A General Survey, second ed. Oxford University Press. pp. 272; Berthold, P., Gwinner, E., Sonnenschein, E. (Eds.), 2003. Avian Migration Springer, Berlin, Germany.*[10]

homeostasis described in Chapter 4 and the nature–nurture collaboration explored in Chapter 3 on behavioral genetics are never far away. Continue to find and explore linkages to other topics while reading this chapter; this is how understanding of animal behavior grows.

Early ethologists took a highly mechanistic approach to animal movement, emphasizing the classification of movements and how sensory inputs and movements are integrated. This led to intriguing revelations about the relationships between sensory systems and navigation; animals without highly complex nervous systems, such as earthworms, for example, can make orientation decisions that bring them to appropriate habitat sites. More contemporary approaches to animal movement led scientists to ask questions such as the following: Are there hidden (to humans) sensory processes involved in collecting navigational information? How are multiple, sometimes contradictory, information inputs integrated to make navigational decisions? Can neural network models be used to analyze these decisions? What role do cognitive processes play in navigational decision making?

FIGURE 8.5
A bird's search for appropriate habitat or food shifts among levels, including a very local level (upper left), an intermediate landscape level (perched on a tree, center), and a larger landscape level (lower right).

8.2 SOURCES OF NAVIGATIONAL INFORMATION

It seems obvious that most navigation is shaped by several sources of information. One form of information is a product of evolution—if the best shelter, such as hiding under rocks (dark, humid places for cockroach shelter, for example), remains in the same location generation after generation or if the same migratory route works well over evolutionary time, then if the appropriate genetic variation exists, natural selection may favor coding this information into a species' DNA. Another form of information is learned information about the animal's landscape. Squirrels, for example, may be genetically compelled to build nests, but a good nest location will be specific to an individual squirrel and likely changes with time, generation to generation. Thus, while the motivation to build a nest may have a large genetic component, the location of the nest itself depends strongly on learned information. The need to orient and navigate continues through an animal's life. For instance, once its nest is built, a squirrel needs to learn how to leave its nest, forage, and return. Learning provides a bridge between elements of an environment that may be unpredictable or changeable during an animal's lifetime and information that has been shaped over the species' evolutionary history. When making a decision about movement in "real time," an animal needs to integrate its genetically encoded information, what it has learned during its lifetime, and sensory inputs about what is happening at that moment. Those sensory inputs of the moment are determined by its condition and where it is located:

- Is it hungry?
- Does it need shelter?
- Is it fearful?
- Is it looking for a mate?
- Has it reached sexual maturity so that it is time to disperse from its parents' nest?
- Has a change of season triggered the initiation of migration?

As discussed in Chapter 4, changes in neurotransmitter and/or hormonal levels trigger these behavioral decisions that culminate in the impulse to move; refer to that chapter for a more detailed discussion of these changes. To repeat, information about movement can be either genetic, as in sea turtle migration, or learned, as in homing in pigeons.

DISCUSSION POINT: CLIMATE CHANGE, HOMING, DISPERSAL, AND MIGRATION

Based on what you have learned so far about animal behavior and information that you can glean from online sources, how might global climate change affect movements such as homing, dispersal, and migration? Form your own hypotheses at this point, as this will give you the basis for critically approaching the rest of this chapter.

FIGURE 8.6
The geographic range of pea aphids is broad, covering continents, and their host plant preferences differ across populations. This translates into differing search strategies among populations. *Photo: Bernie Roitberg.*

Genetically determined habitat preferences play central roles in an animal's search for optimal habitat in its landscape. Studies of the genetics of habitat preferences are easier to conduct in insects than in birds or mammals because of insects' short generation times and the ease with which some species can be bred and tested for habitat preferences. The examples here are insects, but the same general principles apply to all animals.

Habitat preferences can involve simple information, such as the cockroach preference for dark places mentioned previously. More complex examples of stored information come from herbivorous insects that must find their host plant for themselves or their offspring. (In the case of host plants, "habitat" and "food" are near-synonymous.) Genetics provides information about host plant odors for some insects, so inexperienced individuals can use those odors to find the right plant. Pea aphids, for example, are widely distributed in North America and England (Figure 8.6) and suck the juices from a number of plant species; populations of pea aphids are genetically differentiated in their preference for host plants such as clover and alfalfa.[11] Likewise, picture-winged flies (*Rhagoletis pomonella*, apple maggots) can live in a variety of fruits, including apples, hawthorns, and dogwoods. The adult flies find the appropriate fruit by its odor and lay their eggs on the fruit.

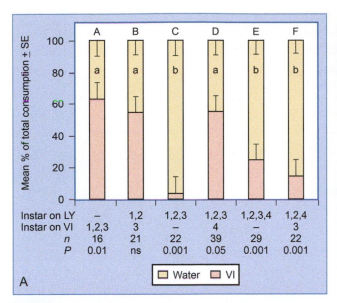

FIGURE 8.7

(A) Changes in tobacco hornworm diet preferences as a result of experience. LY, *Lycopersicon esculentum* (tomato, the natural host plant), VI, *Vigna unguiculata* (cowpea, an acceptable nonhost plant). The letters atop the bars indicate the instar exhibiting the preference; A, B, C = fourth instar and D, E, F = fifth (ultimate) instar. The two rows of numbers just below each bar indicate, for that preference test, the food that earlier instars were reared on. Thus, the fourth instars represented by bar A were fed tomato leaves during their first three instars. They exhibited a preference for tomato plants. This was decidedly different for the fourth instars represented by bar C, which were fed cowpea during their first three instars; those animals preferred moist filter paper to tomato. The fifth instars represented by bar D show that even if the first three instars are raised on cowpea, they will prefer tomato if they are given that plant during the fourth instar. (B) Tobacco hornworm. *Adapted from de Boer, G., 2004. Temporal and developmental aspects of dietary-induced larval food preferences in the tobacco hornworm. Entomologia Experimentalis et Applicata 113, 197–204.*[13] *Photo: Shelly Adamo.*

Preferences among fruit species are genetically determined, and this fly species is divided into host plant races based on these preferences. Matings between host plant races of the fly result in offspring that are poor at locating fruit for oviposition.[12]

In other cases, host plant information comes from environmental sources. For instance, some insects learn host plant preferences by feeding on those plants as larvae. The common tobacco hornworm has a range of acceptable food plants, including members of the nightshade family like tobacco and tomato, but accept a variety of other plants as food. Once a caterpillar has eaten a certain plant, it develops a preference for that plant type over other potentially acceptable plants (Figure 8.7). This shows how learned information can shape orientation decisions that drive simple habitat preferences.

Host plants are not the only source of environmental information. Many animals can use cues, learning critical *landmark* features or using *compass* information (from the sun, stars, patterns of polarization in the sky, or the earth's magnetic field) as referents so they can revisit locations;[14] this is particularly important for homing and migrating animals.[15] In addition, animals can remember information about previous movements to set a route in the absence of environmental cues. This is the strategy used in path integration, a phenomenon in which an animal returns to a beginning point (perhaps its nest) by extrapolating the direct return path from the twists and turns of its outbound journey.[16]

> **KEY TERM** A landmark is a significant environmental feature that is stable in location (typically these are visual features) that can be used in orientation.

> **KEY TERM** A compass is a physiological or mechanical device that allows assessment of angular direction in the 360° landscape. The most common compasses rely on the earth's magnetic field, the sun, the moon, or the stars to establish this frame of reference.

OF SPECIAL INTEREST: THE USE OF THERMAL DETECTION BY PIT VIPERS TO LOCATE PREY

Precise location of prey is extremely important for predators; predator strikes require precision navigation on a microcosmic level. Pit vipers, venomous snakes that include the copperhead, rattlesnakes, the bushmaster, and the fer-de-lance, have a pair of indentations, or pits, just below and slightly in front of their eyes (Figure 8.8). The pits contain hundreds of infrared (temperature)-sensitive cells, which form a thermal image of the area in front of them. This image is fairly low resolution, and the pit viper relies on choosing a hunting site with a cool background so warm-blooded prey will stand out. Because the pits are paired, there may be a stereo effect, helping localize the prey and increasing strike accuracy. For this reason, field biologists (and anyone else!) should avoid wearing open-toed shoes or sandals in areas where pit vipers are common.

FIGURE 8.8

The pits (for which pit vipers like this one are named) are small indentations just in front of the eyes. They are lined with infrared heat sensors that enable the animal to locate warm prey with amazing precision. *Photo: Jeff Mitton.*

8.3 SENSING THE ENVIRONMENT IN TIME AND SPACE

Unless an animal uses only *idiothetic* information to shape its navigational decisions, it needs directional information to get to the correct location in the environment. Some of this information may be easily interpreted; sensing the goal itself and moving toward it usually is not complicated. Using only the cues that emanate from the goal and then using extrapolation to find the source of those cues is more difficult. Imagine, for instance, trying to find a distant object in total darkness; we are reduced to using senses such as hearing and smell for which our cue-localization is not well developed. This human analogy captures the central issue for animals in gaining sensory information from the environment; it is easy to perceive the presence and intensity of a cue such as a light, an odor, or a sound. The task of extrapolating the source of the cue is much more difficult, but if the animal is to organize its movements to reach the stimulus (or to use the stimulus as a landmark or compass), it must perform that extrapolation and find the source.

> **KEY TERM** Idiothetic information is internal to the animal; path integration and homing are two examples of movement that uses idiothetic information. If an animal remembers the direction it traveled from its nest to food, and then uses the reverse route to return home, it has used idiothetic information.

> **KEY TERM** Allothetic information is obtained from the environment. Stimuli such as a light or sound source are allothetic.

How do animals locate the source of a stimulus? Generally, animals triangulate; that is, they measure the intensity of the stimulus from two different locations in some combination of time and/or space. To triangulate on something, an animal must perceive it from two different locations. This perception can be done simultaneously, as in stereoptic vision used by many animals, or sequentially, so one leg of the triangle is perceived and then the animal moves (or at least moves its head) to establish the other leg of the triangle. For instance, animals triangulate and locate a stimulus by simultaneously using two sensory organs (two eyes, two ears) to perceive the stimulus, and then use the difference in perception between the two organs to calculate distance/direction.[17–19]

Visual triangulation confers the ability to tell how far away an object is. In the context of vision, triangulation is called *depth perception, stereopsis,* or *binocular stereopsis.* Stereopsis is the use of two simultaneous sensory inputs to triangulate on the distance of a stimulus. Binocular vision gives us depth perception because it allows stereopsis. Stereopsis is very important for predators that strike at their prey, because to strike accurately, they need to know the distance to the prey. Such animals tend to have forward-directed eyes that overlap considerably in the visual field; the overlap facilitates the comparisons needed for stereopsis (Figure 8.9). Animals also use additional strategies to refine this distance information, such as relating the perceived size of objects to their distance from the animal.

Many animals (grasshoppers and lizards are great examples) combine simultaneous with sequential sensory strategies to maximize their accuracy in stereopsis (Figure 8.10). By moving their heads back and forth while triangulating on a prey item or a spot to jump toward, they increase the angle with which they triangulate, making their measurement more accurate, much as might happen if their eyes were further apart.[20]

In theory, animals with spatially separated odor detectors, such as the antennae of insects, could use that spatial separation to triangulate in a way that is analogous to binocular stereopsis. However, the common reaction of insects attempting to find an odor source, particularly a sex pheromone source, is to walk or fly upwind when they encounter an odor plume (Figure 8.11).[21,22] Experimentally, it is difficult to know if this response is based on spatial odor comparisons between the antennae or sequential comparisons, but the generally

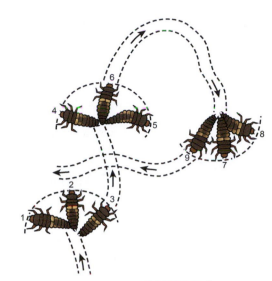

FIGURE 8.9
A ladybird beetle larva moves its head from side to side, gathering sensory information sequentially. This is a form of triangulation. *Adapted from Bell, W.J., 1991. Search Behaviour. Chapman & Hall, London.*[2]

263

FIGURE 8.10
Lizards use triangulation in prey (or perch) location. This is the result of simultaneous input from both eyes (binocular vision) and is called simultaneous stereopsis. Lizards also use sequential stereopsis; they move their heads from side to side, thus increasing the angle of triangulation, and the accuracy of location. *Photo: Michael Breed.*

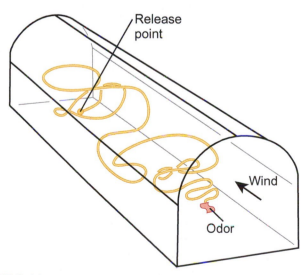

FIGURE 8.11
Insects are thought to move upwind when in an odor plume and then crosswind when they lose the plume. This allows them to follow an odor, such as a sex pheromone, that may shift with the wind.

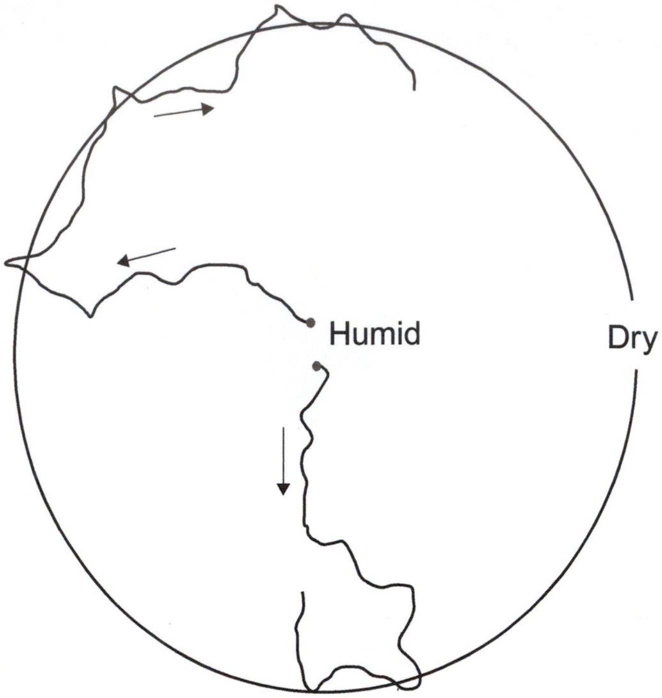

FIGURE 8.12
How a taxis keeps an animal (or bacterium) in favorable habitat conditions. As long as the animal is in favorable conditions, it moves in a straight line or randomly, but when it passes a threshold (low humidity as indicated in the figure, or for a bacterium, low concentration of a nutrient), it turns more sharply, keeping it in the favorable habitat. *Adapted from http://www.rowland.harvard.edu/labs/bacteria/index_movies.html; http://jb.asm. org/cgi/content/full/182/24/6865.*

held view is that walking or flying insects make sequential comparisons of odor concentrations and change their movements if they begin to find lower concentrations of the attractive odor.

In contrast, bacteria (not exactly animals, but interesting enough to include in the discussion of navigation) are thought to sense and use microscale differences and gradients in concentrations of molecules in their surroundings. Many bacteria have a flagellum—a thread-like organ that can propel the bacterium in a more or less straight line. (A bacterial flagellum should not be confused with a eukaryotic flagellum.) When a change of direction is needed, the flagellar motion changes to produce a tumbling movement that reverts to straight line swimming when the new direction is established (Figure 8.12).[23,24]

The most important point to remember when we are considering how animals sense the environment is that *locating the source of an environmental cue requires a mechanism, either sequential or simultaneous, for comparing the strength of stimuli.* Comparisons allow triangulation on the cue source. Whether organisms use sequential or simultaneous comparisons depends in part on the structure of their nervous and sensory systems.

OF SPECIAL INTEREST: PERCEIVING MAGNETIC FIELDS

The study of the response of animals to magnetic fields has not been easy, in part because the receptors for these fields are not all that obvious. Because magnetic fields pass through the tissues that form animals, the receptors are not restricted to the outside of the animal. Magnetic fields are not gathered by special structures such as eyes or eardrums—in fact, the receptors for such stimuli might be very small and scattered throughout an animal. In other words, when we look for a magnetic field receptor, we could be looking for next to nothing, almost anyplace. In the case of magnetite, involved in one hypothesized mechanism for magnetoreception, we do know the location of the mineral in some animals, but it does not tend to be in a specialized structure, and has been found in a variety of places across animal taxa.

Not surprisingly, understanding the mechanisms involved in the perception of magnetic fields is equally challenging. Some hypotheses include systems that involve special chemical reactions that respond to magnetic fields or that involve magnetite crystals. Some marine animals that possess electroreceptors may use those in electromagnetic induction.

The elusive nature of magnetic field receptors means that the evidence for their existence comes from the behavior of the animals that perceive and react to magnetic fields. Much of this evidence stems from laboratory experiments that use strong magnets to disorient homing pigeons, for instance, or that show changes in neurological activity in animals exposed to magnetic fields. In some ways, this kind of evidence is akin to watching the changes in a moth's behavior in order to assess the presence or absence of a moth pheromone. The evidence is at once powerful and humbling for the observer-scientist, who depends on an animal's reaction for information about something that, to a human, is soundless, tasteless, invisible.

A few animals, such as sea turtles and spiny lobsters (see Section 8.4), can perceive several geomagnetic parameters and can learn how these parameters vary in space; this gives them true navigational tools; these animals are said to possess a magnetic map. A greater number of animals may not have such a map at their disposal, but can use magnetic information to determine compass direction, and thus have a magnetic compass,[25] perhaps calibrated by another cue such as the sun.[26] While the actual receptors for magnetic fields remain unidentified (see Chapter 2 for more discussion of hypothesized receptors), four areas of the pigeon brain have shown activity when artificial magnetic fields change; these are the dorsal thalamus, the hippocampus, visual areas of the cortex, and the vestibular region of the brainstem. In addition, specific neurons in the vestibular region have been identified that are activated in response to changes in an artificial magnetic field.[27] These are not receptors, but it is reasonable to hope that the discovery of such brain activity means that the discovery of the receptors themselves is not far behind.

> **KEY TERM** A magnetic compass allows an animal to use geomagnetic fields as directional information.

> **KEY TERM** A magnetic map allows an animal to determine its geographic location and navigate to a goal with that information.

8.4 HOW TO RESPOND TO SENSORY INFORMATION: A TOOLBOX FOR FINDING THE WAY

After the previous review of the ways in which animals gather and process information about movement, the next step is to learn how they use it and to learn about the tools they have to convert sensory information into locomotor decisions. These tools range from deceptively "simple" (e.g., "move" and "don't move") decisions to amazingly complex ones, involving mental maps and geomagnetic fields.

Kinesis

Early ethologists wanted to be able to describe animal behavior by classifying instinctive patterns of behavior. This approach was particularly common in ethological analyses of animal movements. The simplest locomotor responses to the environment, *kineses* (plural), involve changing velocity or altering turning rates depending on environmental stimuli. Indeed, the word *kinesis* (singular) comes from a Greek word that means "motion." Kinesis does not include orientation to environmental stimuli and, consequently, requires only very simple sensory and neural systems. If an animal has a specific need, such as food, a more humid environment, or shelter from the sun, but lacks the sensory ability to gather information about the location of the needed resource, it may engage in an undirected search, or kinesis; that is, it will start moving (Table 8.1). Such a first-order response works very well for initial analyses of the behavior of many animals, including most invertebrates and cold-blooded vertebrates, such as amphibia and snakes.

Taxis

The integration of orientation (directional response to a stimulus) with kinesis (locomotor response to a stimulus) yields a *taxis* (pronounced *tax-sis*, not to be confused with cars for hire). The plural of *taxis* is *taxes* (*tax-eez*). In biology, a taxis is a movement that is directed with respect to a stimulus, such as an object, a light, or an odor source.

TABLE 8.1 Definitions of Simple Movements and Orientation Patterns

Kineses	Changes in speed of movement without orientation to a stimulus source.
Orthokinesis	Motion is in a straight line but is undirected with respect to the source of the stimulus.
Klinokinesis	Turning movements that change in rate as stimulus strength changes. Examples: looping behavior of beetles during search, tumbling behavior of bacteria.
Taxes	Changes in direction of movements that are oriented to a stimulus source.
Klinotaxis	Turning that is oriented to a stimulus. A classic example is the negatively phototactic (moving away from light) response of blowfly larvae.
Tropotaxis	Moving to a point between two stimuli as a result of averaging the input. For instance, in the case of two lights—one perceived by each eye—the animal averages the input and moves to a point between the two light sources. Many insects display this response.
Telotaxis	Moving directly to one of two stimuli when two stimuli are presented. A positively phototelotactic animal would choose the stronger of two lights.

Kinesis and *taxis* are often paired with prefixes to indicate the nature of the stimulus that motivates the response. Attractive stimuli are termed positive, repulsive stimuli are negative. The nature of the stimulus is used as a prefix for the type of movement (*anemo*=wind, *geo*=gravity, *photo*=light, etc.). Hence, positive phototaxis occurs as a result of attraction to light. This can be figured out by looking at the parts of the phrase. "Positive" refers to attraction, so that indicates movement toward. "Photo" refers to light, and "taxis" means oriented movement: so "positive phototaxis" means "movement toward light." In the same vein, negative anemotaxis means moving downwind, and so on. A moth that flies toward a light is positively phototactic. A cockroach moving away from light is negatively phototactic. Any environmental stimulus with a directional source can be used to orient a taxis.

Simultaneous or sequential comparisons of stimulus strength can be used to establish the direction of movement. While many organisms display simple positive and negative taxes, such as an insect moving directly to a light, more complex navigation like that which occurs in homing and migration can require the ability to keep the stimulus at a fixed angle somewhere between 0° (toward the stimulus) and 180° (away from the stimulus). Moving at an angle relative to the stimulus, rather than directly toward or away from the stimulus, is called *menotaxis*, which is useful because it can be applied to homing. An animal that leaves its nest and moves at an angle with respect to a landmark needs only to reverse that angle to return to its nest.

OF SPECIAL INTEREST: NATURAL (?) SELECTION AND THE ANATOMY OF MOVEMENT

The shape of a bird's wing dictates, in large part, whether it will be a "sprinter," good at short, explosive flight, or a long-distance flier. To see this, one has only to contrast the short, stubby wings of a quail or pheasant, as it blasts out of its cover, with the almost impossibly long wings of a soaring albatross. These physical constraints have taken a toll on cliff swallows in southwestern Nebraska. Charles Brown and Mary Bomberger Brown included road kill in a long-term (30 years) study of cliff swallows in that area, and found that despite an overall increase in the cliff swallow population, road kill swallows declined over time. In addition, wing length changed during the time of the study, with surviving birds' wings becoming shorter and road kill birds' wings becoming longer. Brown and Bomberger Brown suggest that these morphological changes reflect the fact that birds with longer wings are less able to escape oncoming traffic. Given the estimated 80 million birds that are killed by cars every year in the United States alone, it is intriguing to consider the role of the automobile as an agent of selection.[28]

Counterturning

Straight movements are often thwarted by obstacles. If the animal is an ant, these obstacles could be small pebbles; for larger animals, trees and boulders are effective barriers. How can an animal continue to move in a relatively fixed direction when its path is frequently deflected? Counterturning is one simple mechanism used by many animals to maintain their course of movement.[29] In counterturning, each change of direction is balanced by a movement in the opposite (counter) direction when it becomes possible to do so. The classic test for counterturning forces a test animal to establish a movement direction, running or walking down the arm of a maze to a point where it must turn (Figure 8.13). After following the second branch of the maze, it reaches a point at which it can choose to counterturn, thus re-establishing its original course, or choose to make a reverse turn. In an experimental context, a number of animals can be asked to move through the maze, and the frequency of counterturns then can be compared with random turning, which is characterized by a 1:1 ratio of counterturns and reverse turns. The distance of the second arm of the maze and the time taken to traverse it affect the likelihood of counterturning; as distance and time increase,

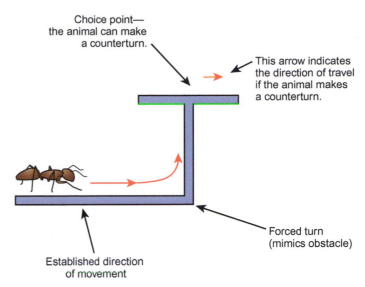

FIGURE 8.13

In this classic test for counterturning, a test animal is forced to establish a movement direction, running or walking down the arm of a maze. It then reaches a point where it is forced to turn; this mimics obstacles the animal might encounter in the natural world. After traveling down the second branch of the maze, perpendicular to its original direction, it reaches a decision point where it can make a counter turn, which reestablishes the original direction of travel, or make a reverse turn.

the animal "forgets" the needed course correction and is more likely to make a random choice. Early work on counterturning in milkweed bugs set the stage for consideration of counterturning as a basic behavioral mechanism in animal navigation. Most of the recent literature that highlights counterturning as an important mechanism deals with how moths find odors,[30] but counterturning is probably displayed by most animals, if they are given the opportunity.

Landmarks

Animals can use prominent physical features in the landscape (a mountain, large trees, rocks, etc.) as landmarks in orientation. The classic demonstration of landmark orientation is an experiment by Niko Tinbergen on digger wasps. When a digger wasp leaves its nest, it flies several loops around the nest; this allows it to learn the landmarks it needs to know to return to the correct location.[31] To demonstrate this, Tinbergen first arranged a circle of pinecones around the wasp's nest entrance (a hole in the ground); by moving or altogether removing the cones, he discovered that he could change the return of the wasp. If the circle of cones was moved a few centimeters from the nest, the wasp searched, in vain, for the nest inside the relocated circle.[32]

Many mammals, such as pika, deer, and members of the cat and dog families, create their own landmarks by scent marking. By placing glandular secretions on key locations, such as nest sites or territorial boundaries, the animal creates a guidepost for itself and a signal to other animals that the area is occupied. Scent marking was briefly discussed in Chapter 7, on communication, and is covered in more detail in Chapter 11, in the context of territoriality.

OF SPECIAL INTEREST: WHY DO MOTHS FLY TO A FLAME?

At night, a candle or light bulb can attract hundreds or thousands of moths, along with a few beetles, planthoppers, and occasionally lesscommon insects. Homeowners don't want a flock of moths hovering around their patio, landing in drinks, and flying into guests' mouths and noses. Bats and spiders also move in to take advantage of feasting on the insects, adding to the homeowners' distress. A partial cure comes from employing a bug-zapper, usually a device that combines an ultraviolet light (which is more attractive to moths than a regular incandescent light) and an electric grid that shocks the insect. The bug-zapper is no respecter of insects, frying beneficial bugs along with nuisance and harmful ones, so the environmental effects of a bug-zapper must be balanced against the urge to enjoy an insect-free patio.

All this leaves open the question of why the insects are flying toward the light. The likely answer is that under natural conditions the insects use a distant light, such as the moon or a star, as a cue for their flight direction. Moving in a line at a constant angle relative to a distant landmark (menotaxis) results in more or less straight flight—the farther away the landmark, the straighter the movement (Figure 8.14A). A closer landmark, for example, a porch light, has a different effect. Flight that maintains the same angle relative to the landmark results in an inward spiral, leading the moth to its death dance with the flame or hot light bulb (Figure 8.14B). Because moths fly at night largely to find mates, the conservation effects of this type of light pollution—never mind bug-zappers—are likely immense, but few studies have addressed this question.

While watching moths come to a light, one can enjoy the company of other animals, such as a cat-faced spider (*Araneus gemmoides*) who may set up shop (and web) across the top of the patio door, near a light. The moths provide good "hunting" for the spider, and good spider-watching for a biologist!

268

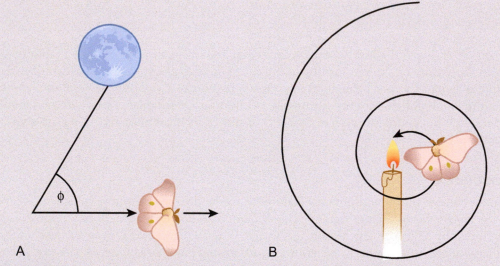

FIGURE 8.14
(A) In the natural, preelectrified world, insects use light (always from distant sources) as a guide for straight line movement. By flying at a constant angle relative to a very distant light source, they effectively move in a straight line. (B) The advent of electricity made nearby light sources abundant. (Candles, being expensive, were never so abundant.) When insects fly at a constant angle relative to a light bulb or candle flame, their flight path describes a spiral, ever closer to the light source, with no good effect for the insect.

Learning the Landscape

As an animal develops, its need to know how to orient in its environment is likely to increase. Many young animals make exploratory trips from their home base, walking or flying short distances and returning to the nest or den. In mammals, young making such trips are often accompanied by adult members of their social group that provide protection and facilitate learning. Significant landscape features, such as the location of trees, large rocks, and more distant objects such as mountains can all become critical pieces of information in orientation. These orientation trips are well studied in ants and bees, which progressively add to their store of information about the surrounding landscape in a series of excursions from their nests.[33–35] Learning the landscape is essential for snapshot orientation, which is discussed next.

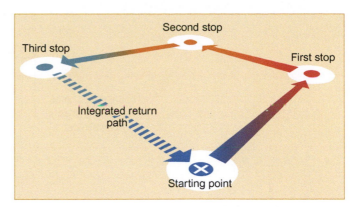

FIGURE 8.15
Path integration over three stops. The animal changes direction at each stop, and distances among the stops vary. The distance and direction information is integrated to allow the animal to return directly to the starting place from the third stop.

Snapshot Orientation

Snapshot orientation also uses landmarks but is more sophisticated than directional movement toward a stimulus. In snapshot orientation the animal remembers a visual image of the landscape, or at least of a composite of significant landscape features. It can then compare the landscape it sees at any given moment with a "library" of memorized landscape snapshots. If the current landscape matches the snapshot, then the animal is at the correct location. If the landmarks appear smaller than expected, this indicates to the animal that it needs to move closer to the goal, and if they appear too large, then the animal needs to move further away. If one landmark is at the wrong angle with respect to another landmark, then movements across the landscape can reduce this angular error, bringing the animal to the proper orientation. This is an interesting and simple solution to the problem of adjusting orientation that is based on the animal's perspective of the landscape.

The concept of snapshot orientation was first developed in studies of honeybees (see p. 275) and has been applied in bees and ants. Humans seem to be able to perform this sort of orientation, although differentiating between use of snapshot orientation and the use of a cognitive map (see pp. 194 and 272) is difficult.

Path Integration

The ability to start at a fixed point, visit several locations, and return directly to the original point is fundamental to animal orientation. To accomplish path integration, the animal simply computes its return distance and direction vector from the vectors joining the locations on its route (Figure 8.15).[36,37] This saves retracing steps. Honeybees, ants, dogs,[38] rats,[39] and humans are all known to have excellent path integration abilities. While a variety of animals can path integrate reasonably accurately in the absence of landmarks or other cues, long-distance orientation using path integration usually incorporates corrections from landmarks or celestial cues (Figure 8.16).[41]

DISCUSSION POINT: CAN YOU PATH INTEGRATE?

Have a little fun with your friends and roommates by testing their abilities to path integrate. Blindfold your "victims" and take them over a route with two or three turns (walking 10–20 steps between turns is good). When you have them at the end of this route, turn them around in a circle two or three times, and then let them try to path integrate to find their starting point.

Compass Orientation

The position of the sun or the stars in the sky can be used as a compass. Because of the earth's rotation, the sun appears to move through the sky from the eastern to the western horizon. At the equator, this path travels through the vertical azimuth; away from the

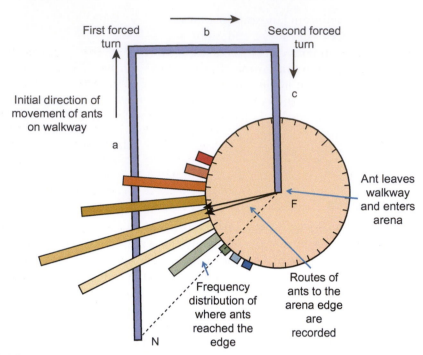

First forced turn   b   **Second forced turn**

Initial direction of movement of ants on walkway

a

c

F

Ant leaves walkway and enters arena

Frequency distribution of where ants reached the edge

Routes of ants to the arena edge are recorded

N

FIGURE 8.16

This diagram represents path integration in the desert ant *Cataglyphis fortis*. The ant begins at the nest (N), proceeds along path a, and is forced to turn twice, moving until it arrives at point F, a feeding station. The results are shown in this figure, where the length of the bars radiating away from the feeding station indicates the number of ants moving in that direction. The ants do not integrate precisely, but their approximation of the nest location is sufficient to return them to the nest, along with the use of landmarks and other environmental cues. *Adapted from Müller, M., and Wehner, R., Proc. Natl. Acad. Sci. USA 85 (1414), 5287–5290.*[40]

equator, the sun's path is not vertical, but the sun nonetheless remains a strong directional cue in most locations except those very near the poles.

To the earth-bound observer, the stars appear to be fixed on the inner surface of a rotating sphere, much like the view in a planetarium. It is similar to standing in the center of a rotating ball; the stars are points of light on the inside of the ball. The ends of the celestial axis of rotation (the southern and northern points in the ball at which there is no apparent movement of the stars) correspond to the North Star (Polaris) and the Southern Cross. Sailors, of course, used these points to help determine their position prior to the invention of modern navigational methods, such as Global Positioning Systems. Other animals are equally capable of using the sun or the stars as navigational reference points.

How do animals compensate for the apparent motion of the stars or the sun? The spin of the earth creates a problem; reference points may appear to move through the daily cycle of rotation. This apparent motion of the sun and stars is predictable on a daily basis, and animals can compensate for it by integrating a clock mechanism with their navigational system; the result is easily demonstrated by time-shifting animals in artificial light environments (habituating them to a day–night cycle that is a few hours displaced from the natural cycle) and then observing their orientation behavior when they are moved to a field situation.[14,42,43] In addition, long-distance navigators, such as migrating birds, use geomagnetic cues to calibrate their compass. Through the use of multiple, interacting cues, a more accurate course can be set than could be obtained by using any one cue.

OF SPECIAL INTEREST: DUNG BEETLES LOOK TO THE STARS

Never underestimate any animal, even one that specializes in finding, rolling, eating, and/or squabbling over feces. Along with many animals, dung beetles use the sun and the moon for orientation (see Section 8.7). Orientation is important to dung beetles because competition for feces is intense, and once a ball-rolling dung beetle forms its ball, it is better off if it decamps with its ball, preferably directly away from the dung pile (straight line orientation). The African dung beetle (*Scarabaeus satyrus*) does this reasonably well even on moonless (but starry) nights, which led investigators to wonder if the beetles were using stars. This is unlikely, given the limitations of the beetles' compound eyes and the dimness of the stars, making the beetles' feat that much more intriguing. The obvious experiment is to test the beetles when they could not see the night sky at all; under those conditions, the beetles took almost three times as long to roll their balls to the edge of their test arena. How could the beetles be using a moonless night sky?

Further experiments in a planetarium tested the beetles under five conditions: ranging from the full complement of stars and Milky Way to total darkness, and including settings that allowed beetles to see only the Milky Way, only dim stars, and only bright stars. The presence of the Milky Way, with or without other stars, resulted in the fastest exit from the arena, comparable to beetle performance under a starry night sky. In the planetarium, if the Milky Way is removed but other stars are left in place, the area occupied by the Milky Way still has more stars and while the beetles take a bit longer to exit (1.5 times longer), this is not significantly different from their best performance. On the other hand, if the beetles can only see bright stars or none at all, their exit is significantly slowed. Unlike the fickle moon, the Milky Way is a dependable cue, probably perceived by the beetles as a relatively bright, linear blur in the sky. With this set of simple, but creative, experiments, dung beetles became the first animals shown to orient using the Milky Way. They probably won't be the last.[44]

Odometers and Measuring the Distance Traveled

For an animal to find its way to a location, whether it is a bee returning to its hive or a salmon finding its stream after life at sea, it needs a measure of its desired travel distance. In a car, the *odometer* gauges distance traveled; ultimately, the car odometer is linked to the number of times the car wheels turn. Animals display a wide variety of strategies for gathering the same sort of information. Odometer mechanisms have the greatest importance when landmarks are unavailable (as in a fairly homogeneous desert environment) or when an animal is using genetic information to complete a migration. The alternatives available to animals to assess distance traveled are step-counting,[45] visual flow (the rate at which the landscape image moves across the animal's eyes),[46] and effort expended in travel.

> **KEY TERM** An odometer is any physiological or mechanical device that allows measurement of the distance traveled.

Desert ants (*Cataglyphis cursor*) use something analogous to the car odometer.[47] They use step counting to assess distance traveled, much like a person counts paces to guess the dimensions of a field. In a clever experiment, scientists put stilt-like extensions on the legs of these ants before the ants' return to their nests. These ants had used path integration during an outward trip from the colony to "know" the distance to be traveled. Stilted ants moved the number of steps that would have been required to find their nest; because their legs were unexpectedly long, they ended up moving farther than their goal. In a reverse of the experiment, ants moving outward with stilts did not travel far enough to reach the nest after the stilts were moved. In other words, ants use a pedometer to gauge their progress.

Animals that fly across the landscape require different odometer mechanisms. Because of the effect of wind, effort alone is not a good measure of distance traveled. For instance, a headwind can result in much expenditure of effort but little progress, whereas a tailwind can greatly ease the difficulty of a trip. Eastbound passenger airplanes crossing North America fly with the jet stream as a tailwind, but the same air currents are headwinds for westbound planes; a cross-continent westbound trip takes about an hour longer than an eastbound trip because of these wind effects.

Migratory birds, because they encounter predictable upper air currents akin to the jet stream, may be able to use effort as an indicator of distance, with the expectation of quite different efforts for northbound and southbound trips. Migrating fish and sea turtles may encounter these difficulties, but their mechanisms for measuring distance traveled are less well understood. Honeybees gauge the rate of movement of landscape across their visual field; a bee flying into a headwind experiences a slow rate of change of visual images. This compensates for wind effects and results in an accurate odometer, a key component for accurately following dance information. Thus, animals that know the local landscape can also use landmarks to judge distances.

Cognitive Maps

A cognitive map is an internal neural representation of the landscape in which an animal travels. Animals that use cognitive maps can "visualize" the landscape and solve orientation problems by referring to these maps. While it is generally accepted that birds and mammals can form cognitive maps, and that the hippocampus is the most important part of the brain in their formation, considerable controversy has centered around whether other animals, such as honeybees, can form similar maps.[48] An animal with a cognitive map should be able to assess landmarks and compass information and then calculate its travel path to any location within its mapped area. In practice, though, it is difficult to design experiments that differentiate among landmark orientation, path integration, and use of a cognitive map. The strongest evidence for cognitive processes in orientation decisions comes from experiments that require animals to solve multiple problems, particularly to navigate to more than one point in the area around their home.[49] One particularly creative approach to this problem was applied to pigeons that had learned the locations of both their home and a feeding area. When taken to a release site that they had never visited, pigeons that had not been fed flew directly to the feeding location, whereas pigeons that had been fed flew home, demonstrating that the pigeons knew where they were, where they wanted to go, and how to get there—the elements of a cognitive map.[50] Bats are also excellent navigators and seem to learn maps quickly; when faced with a mirror image of an obstacle course they had learned, they learned the new course more rapidly; the intrigued investigators wondered if they had somehow used the original spatial information to enhance learning the new map.[51]

Cognitive maps are discussed in more detail in Chapter 6; remember that the use of landmarks and compasses alone does not establish the use of a cognitive map. If you are feeling slightly unsteady with the notion of maps, you are not alone. As James Gould[52] says, "The map sense remains animal behaviour's mystery of mysteries. No other set of questions takes us so far from human experience and analogy."

All of these data point to the fact that cattle and deer orient on a north–south axis in line with the magnetic poles of the earth. Because of the broad range of data collection, Begall and colleagues could eliminate many other hypothesized environmental causes of orientation, such as avoiding wind, sun-basking, and the like. Although the Google Earth images were too unresolved to allow anterior/posterior distinctions in the case of cattle, the field observations of deer indicated that they tended to face north, not south. Moreover, cattle and deer in the vicinity of power lines did not conform to this pattern,

> **KEY TERM** Scan sampling is observing and recording the behavior of multiple animals repeatedly, at specified intervals for specified amounts of time. Thus, it is like a series of snapshots of a given animal, taken at known intervals.

and indeed, displayed different alignment patterns, probably because power lines generate extremely low-frequency magnetic fields that can disrupt the animals' ability to sense the magnetic field of the earth.[53]

> ### OF SPECIAL INTEREST: PERCEIVING MAGNETIC FIELDS—INSIGHTS FROM GOOGLE EARTH™
>
> On any given day, a scientist—or a farmer—might be able to observe a herd of cows and determine which direction they tended to face. This one observation would not allow us to determine the cause of their alignment—it could be a response to sun, wind, or a number of factors, depending on conditions—and would not give us numerous observations under a wide range of geographic and meteorological conditions. In other words, we could not generalize from such an observation, and the number of observations required for such generality would be daunting.
>
> Begall et al.[54] solved this problem by using three sources of data that addressed the orientation of cattle and deer. First, they used satellite images from Google Earth that showed the body alignment of 8510 cattle in 308 pastures on six continents; in this manner, they could *scan-sample* numerous herds, reducing the likelihood that the same weather events influenced them all at the time of the sample. Second, they observed alignment in over 2000 red and roe deer under a variety of conditions and at various times and places in the Czech Republic. Third, they used data from the body impressions of red and roe deer in snow.
>
> > **KEY TERM** Scan sampling is observing and recording the behavior of multiple animals repeatedly, at specified intervals for specified amounts of time. Thus, it is like a series of snapshots of a given animal, taken at known intervals.

FIGURE 8.17
Examples of search patterns: (A) undirected search; (B) straight line search, covering a large area; (C) spiral or looping search; (D) another type of straight line (large area) search.

273

8.5 SEARCH

Movement often starts with a search, which may look random or only slightly systematic to the casual observer.[2] Searching behavior is critical to resource location and is an example of a behavior that uses some of the simple responses to environmental information that are discussed in Section 8.4. In a search for food, shelter, or a mate, an animal may initially have little or no information about the location of its goal. The first step for the animal is to employ a strategy that will bring it that information, such as an odor, a visual sighting, a sound, or another cue coming from its goal. Search may be undirected, but even undirected search is usually not random; undirected search should be structured to maximize the amount of environment sampled (Figure 8.17, strategy A, Figure 8.18). A carnivore in search of prey typically walks in a fairly straight route, allowing it to sample as much of the environment as possible. (Turning or looping movements would likely cause it to cover less ground.) If the search is for an odor, visual, or sonic cue that can travel a considerable distance, then movement in a straight line allows sampling of a large habitat area (Figure 8.17, strategies B and D).

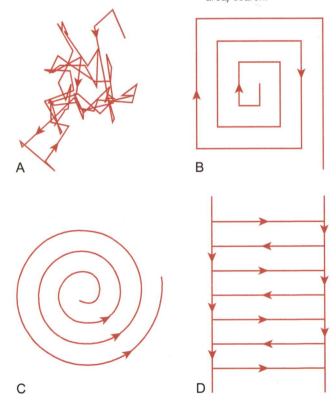

FIGURE 8.18
Homing pigeons exhibit large-scale central place homing, returning to home nests from hundreds of miles away.

Looping or spiral movements (Figure 8.17, strategy C) concentrate search and are effective when the target may be found only if it lies directly in the searcher's pathway. Long-distance fliers such as the albatross may concentrate on relatively small areas, turning repeatedly, and for that matter, concentrate on even smaller patches nested within those small areas. These searches are punctuated by long flights without turning. Overall, this pattern of movement is called a Levy flight (or Levy walk, if one is walking), named after the French mathematician, Paul Levy. Albatrosses consume approximately four times their daily energy budget when using these search methods, demonstrating that Levy flight is a very efficient way to hunt for the kinds of resources that albatrosses seek—sparsely distributed prey in an open sea with few clues.[55] The Hadza hunter-gatherers of Tanzania also use Levy walks in almost half of their searches for food.[56]

Once the animal finds a cue, kinetic or tactic orientation mechanisms can kick in (Figure 8.18). In many cases the first cue is not from the target itself, but rather is information about the environment that helps in directing the search. Animals searching for odor cues, such as a moth searching for sex pheromone, typically fly at right angles to the direction of the wind so that they sample the greatest number of odor plumes possible. If a cue is detected, but its location is uncertain, the animal may then change its direction of travel, turning more frequently (klinokinesis), so that it makes random loops in a fairly well-defined area; this behavior is likely to bring it back into contact with the cue. Some animals use a layered, sequential search, employing a hierarchy of cues. A parasite seeking a specific snail host will follow salinity gradients to get to the salinity preferred by that species of snail; it will then use chemical cues from snail mucus to identify the correct species. Specialized types of searching behavior include sit-and-wait predation and the use of search images (both discussed in Chapter 9) (Figure 8.18).

8.6 HOMING
The Concept of a Central Place

Many animals center their activities around a nest or shelter. As these animals range out from their central place, finding the way back becomes a critical problem. Animals "home" when they return to a central place, such as their nest or their territory. Homing often occurs after foraging bouts or other relatively local movements. Homing is therefore a repetitive act, and this distinguishes it from dispersal and migration, which are long-distance movements that usually occur seasonally in an animal's life. (These movements are discussed later in this chapter.) Understanding how animals glean the directional information needed for movement is critical to understanding homing behavior, and as will become apparent, homing may involve most of the toolbox covered so far.

Animals that have a central place are, in a sense, bound to that location; their movements are centered around a fixed spot, and their searches for food and mates are limited by the distance that they can travel from that central place before needing to return. Solving this problem is complex because animals usually search for food by using random circular movements or by traveling between a succession of spots where food is likely to be found (refer to Section 8.5 on search). A bird searching for insect larvae might fly from tree to tree, moving among a large number of trees before finding enough food to take back to its nest. Ultimately, its path probably has involved many loops and turns, and retracing its steps would be difficult indeed for the bird; path integration solves this problem. In contrast, most honeybee foraging trips involve traveling in a straight line (a beeline!) from the nest to the food source; in this case orientation using a compass or landmarks is the typical solution. In yet other examples, some animals leave a pheromone trail that guides the return to the nest. As with searching, animals frequently use such information hierarchically, so that one type of environmental information is important when they are distant from their home, and others become more important as they approach their goal.

Homing in Bee Foraging

Honeybees (*Apis mellifera*) are central-place foragers and may forage several kilometers from their hive, making them a good model species for studying navigation in central-place foragers. Honeybees primarily use path integration in making their way to and from foraging sites. Dance information (see Chapter 7) provides outgoing bees with a distance and direction to be traveled. Flight direction is set by sun compass orientation, and the distance of flight by an internal "odometer" that measures the rate at which visual images flow past the eyes.[57] Other inputs, such as odors, provide supplementary information. Once a route is learned, bees incorporate visual and olfactory landmarks when repeating visits to a foraging site. The return trip is governed by path integration as well, but also may be informed by landmarks.

How do bees incorporate landmarks into their orientation? Two basic models, snapshot memory and cognitive maps, have been proposed. The simplest, and probably correct, model is the snapshot model, which calls for the bee to remember a series of visual images ("snapshots") of the landscape as it passes. The bee also remembers images of particularly prominent landscape features. These images can then be compared with the actual landscape surrounding the bee at any given moment. Ultimately, the bee may be able to use the landmark snapshot information in an "allocentric" manner, projecting its position when it is displaced to an unfamiliar location but still in view of an array of familiar landmarks (i.e., viewing them from a novel angle).

The cognitive map model is more complex and requires the bee to construct a relatively complete neural representation of the landscape based on its experiences while flying. Tests of the cognitive map model demand that displaced bees calculate a novel route home, based on their memory of the landscape map (as humans might). Experimental support of the cognitive map explanation of bee navigation has been put forward, but critics have noted simpler explanations for these results. The allocentric model for snapshot use begins to converge with the cognitive map model but remains a simpler explanation for orientation.

Homing in Pigeons

Homing pigeons (*Columba livia*) carry almost the entire animal movement toolbox with them, and they may have tools yet to be discovered (Figure 8.19). As will become apparent, the mechanisms with which they navigate are highly redundant. Redundancy is common in many essential biological processes, but pigeons are nonetheless impressive. Ever since ancient times, people have been fascinated by these birds, which find their way home from unfamiliar sites up to thousands of kilometers from their roost. Pigeon races may feature releases of birds from France, for example, which then successfully fly home to sites in England or the Netherlands. The extraordinary reliability of homing pigeons, under artificial selection since antiquity, makes them excellent subjects for studies of navigation. This is central-place homing, albeit on a large scale, as it involves learning a location and returning to it. Genetic information is unlikely to be useful because the location of the home may change from generation to generation.

275

FIGURE 8.19
This simple diagram illustrates the basics of an undirected search. The animal, traveling from left to right in the diagram, moves in a more or less straight line through unsuitable habitat. *Adapted from Bell, W.J., 1991. Search Behaviour. Chapman & Hall, London*[2]; *Figure 6.1.*

How do pigeons find their way home when deposited in an unfamiliar location? To do this, they must have "map sense," that is, knowledge about their geographic location, and they must have "compass sense," that is, knowledge about the direction they need to fly from their new location in order to reach their home. In other words, they need to know where they are and what direction to fly. If either information source is disrupted, then homing fails or is delayed.

In familiar surroundings—locations from which pigeons have previously homed or landscapes through which they have flown—landmarks play a predominant role in homing. Pigeons learn visual features of the landscape and use these visual features to determine their current position (map location) relative to their roost.

Pigeons clearly use visual landmarks, based on the fact that pigeons orient better in familiar landscapes even when other sensory inputs, such as olfaction, are eliminated. Direct tests of landmark usage are difficult, however, because experiments manipulating visual landmarks are generally not feasible. One can hardly bulldoze mountains or cut forests as part of an experimental design, and interference with the eyes, such as using contact lenses, may be so much of a general disruption to the pigeon that it confounds tests of landmark perception in orientation.

Landmarks cannot be used in completely new locations—they are not landmarks under such novel conditions—and it turns out that in such situations, pigeons use olfaction to produce a map. In their roost, they associate odors with wind directions. When released, they assess the odor of their new location and extrapolate the map location from their roost-gained knowledge of winds and odors. Familiarity may enhance olfaction-based homing; pigeons may home better if they have some time to use olfaction to experience their new surroundings prior to release.

Given this ability, it is not surprising that pigeons in visually unfamiliar territory whose sense of smell has been disrupted (by cutting olfactory nerves or treatment of the nasal passages with zinc sulfate solution) have a great deal of difficulty homing. Similarly, homing fails if the roost is blocked from winds and provided with filtered air.

Path integration is one possible orientation mechanism that seems to have been experimentally eliminated as a potential tool used in pigeon homing. There is no evidence that pigeons track their movement when being transported in cages and then use path integration to determine the distance and direction needed to reach home. Tests of this possibility involve eliminating sensory inputs during transport. Because this conclusion relies on a negative result (continued orientation ability in the absence of a sensory input), there is a small possibility that pigeons use an unknown (and consequently experimentally uncontrolled) sensory input for path integration during transport.

The primary compass information of pigeons comes from the position of the sun in the sky. By integrating their internal clock with the sun's position, they compensate for the apparent movement of the sun across the sky. Pigeons whose circadian rhythm is shifted by keeping them under an artificial light:dark cycle display incorrect orientations when released. For example, if artificial "sunrise" comes for the pigeons 6 h prior to actual sunrise, then their orientation is shifted counterclockwise. If their "sunrise" is later than the actual sunrise, then their orientation shifts clockwise. Like the sun compass of migrating birds, the pigeon's sun compass interacts with a magnetic compass. Under some conditions, experimental modification of the magnetic field around pigeons causes problems in homing.

Experiments with clock shifts and magnetic disruption often do not interfere as much as expected with homing. The reason is the redundancy in pigeon navigation systems; olfactory and landscape information used in establishing their map sense can also be used to correct for compass misinformation. Another possible source of landscape information is the pattern of reflecting sound from mountains. Ocean waves generate infrasound, which then reverberates through the atmosphere. However, the only evidence for use of infrasound by pigeons is the hypothesis that sonic booms from jets may have disrupted pigeon races by impairing the pigeons' hearing, causing them to be unable to use ultrasound in orientation. This is a far cry from the successful use of infrasound in homing, but then again, the pigeon's navigation toolbox probably holds many more surprises.

HOMING IN SHEARWATERS

Breeding on the remote Portuguese Azones, Cory's shearwaters fly many miles over the Atlantic to forage before returning to their colony and their incubating mates. Indeed, they return to that island every year to breed. This is no mean trick, given the fact that the Atlantic Ocean offers nothing in the way of landmarks. How do these birds find their way home?

The clever use of magnets, GPS trackers and zinc sulfate offers a strong indication. Over a period of 2 years, a total of 24 birds were studied; they all had recently returned from foraging and were highly motivated to find their nests. Sixteen of them either were outfitted with GPS trackers and magnets or they were given satellite signalers and rendered temporarily unable to smell (anosmic) by a zinc sulfate wash of their nostrils. One third of the birds were not manipulated, but were controls. The only birds unable to navigate directly back to the nest were the birds without a sense of smell. Disrupting perception of the earth's magnetic field did not affect the shearwaters' ability to navigate.[58] Clearly, the navigation toolbox is not only diverse, but different animals, even within the same clade, make diverse use of that toolbox.

8.7 MIGRATION

Most migrations are movements of animal populations between seasonally appropriate habitats, but migration can occur between any two habitats and can occur more frequently than the typical annual cycle.[59-61] The fact that it is between habitats hints at one important difference between dispersal and migration: dispersal is a one-way ticket, whereas migratory behavior involves a return, on the part of descendants, if not the migrants themselves. Migratory behavior is found in many kinds of animals, ranging from monarch butterflies to birds and mammals, and allows animals to take advantage of habitats that are favorable in different seasons, or that are favorable for different stages in an animal's life cycle.[62] Another major difference between dispersal and migration, then, is that the migrating animal must "know" the route.[63] Some animals rely entirely on inherited, genetic information; in migrants such as monarch butterflies and sea turtles, each young animal must find its own way to the next habitat, without guidance from experienced adults. In other species, young animals migrate alongside experienced migrators and can learn route details—distance, direction, stopover points—by following the choices of animals that have already completed the route one or more times. Investigations of migration typically begin with determining the extent to which learned and innate information is used to shape routes.

Traveling from point A to point B requires two basic pieces of knowledge: distance and direction. These topics were covered earlier in this chapter; review those sections for more detail. Direction is normally set by a compass; the sun, stars, or the earth's magnetic field are the typical referents for the compass.[64-66] To assess distance, one needs an "odometer"; typically, this is measured as physiological effort. Recall that bees measure the rate at which the landscape moves past their eyes; this suggests there may be other types of odometers in migratory animals as well. Again, direction and distance knowledge can be a function of genetics, experience, or both. For longer-distance migrants, stopover points are often essential; they must have features such as food, water, and shelter.[67] Social learning of stopover points in group migrants is probably a key element for migratory success for many birds and mammals.

Not surprisingly, most migratory animals seek favorable conditions for travel—open, unobstructed routes for terrestrial animals, air currents for flying animals, and water currents for swimming animals. Routes that require less effort, even if they are longer, are typically preferred. Migrating birds must balance water loss with flight speed in their choice of flight altitude;

FIGURE 8.20
Migratory birds vary widely in the details of how and where they migrate, but the underlying physiology that makes migration possible is similar across avian species. These are sandhill cranes, which migrate from northern Canada to the Gulf of Mexico. *Photos: upper, Jeff Mitton, middle and lower, Ben Pless.*

higher, stronger wind currents provide a speed advantage but may expose the birds to higher rates of water loss. Generally, birds choose rapid transit over slower routes, even at the expense of water conservation; this means that migrating birds often fly quite high above the ground (Figure 8.20).

Migratory animals present many of the world's most critical conservation problems. Migrants can be vulnerable in any or all of their three key habitats: their nesting or spawning areas, their transit routes, and their overwintering (or unfavorable season) habitats. Change in any one of the habitats can reduce migratory populations and affect the number of animals observed in all of the habitats, and conservation efforts in one habitat may be at least partially thwarted by failure to manage the species in another. Dams reduce populations of migrating salmon by affecting their transit routes, rainforest harvest decreases wintering habitat for songbirds, and beachfront development eliminates nesting habitat for sea turtles. Understanding the behavioral mechanisms behind migration helps in the development of solutions for the conservation of migratory animals.

Little is understood about the evolution of migration. How did Arctic Terns evolve to migrate from pole to pole, further than any other animal in the world?[68] Or how did sea turtles come to migrate from beach to open ocean and back? Perhaps exploratory behavior takes some animals to a new habitat, where they are successful, but then seasonal changes or physiological needs drive them back to their old habitat. If animals that carry out these movements have better reproductive success, then whatever genes predispose them to the movements would increase in frequency in succeeding generations. Some scientists have proposed hypotheses involving continental drift or species range expansion/contraction. These verbal models are a beginning, but the evolutionary roots of intricate long-distance migrations are still a mystery.

Bird Migration

We know more about bird migration than other animal migrations.[10,69–72] Some bird migrations are particularly spectacular because of the immensely long distances traveled. The migratory feats of very small birds, such as hummingbirds, capture the imagination as extraordinary behavioral and physiological feats.

Because bird species vary in the details of their migratory mechanisms, an in-depth analysis of migratory mechanisms in birds would be quite complicated. Fortunately, the basic mechanisms are the same in all birds, so a generalized picture of bird migration is easy to construct.

Seasonal migration in birds, which occurs twice yearly, is closely linked to their biological rhythm (see Chapter 4). There seem to be two distinct stages; there is a preparatory stage followed by the actual initiation of migration. Photoperiod is the most important external cue for the preparatory stage, while the initiation of migration that follows this preparation frequently depends on more variable cues, such as weather conditions. For temperate birds, photoperiod is a dependable cue for a major behavioral shift such as migration. Not only is it linked to more predictable long-term conditions than, say, daily temperature fluctuations, but it also allows the organism to "anticipate" annual change. The physiological shifts that accompany migration or breeding occur over several weeks, before the (average) weather permits such behavior. In the tropics, photoperiod stays much more constant: at the equator there are 12 h of light and 12 h of dark year-around, so birds that have migrated to the tropics must rely on cues other than photoperiod to time their annual return to their temperate breeding grounds. Much less is known about migration triggers that involve equatorial habitats.

Two crucial physiological events occur during the preparatory stage: fat deposition and *migratory restlessness*, or *zugunruhe*.[73] Some species fly nonstop (when possible) to their destinations, and individuals of these species may lose up to 40% of their body weight during this arduous process; clearly, fat deposition is critical to successful completion of the journey. For species that stop along the way (see following text), fat is perhaps less important. For any bird to be able to deposit so much fat, hormonal changes must occur. These are mediated by the pituitary gland (see Chapter 2), which controls metabolism and sex hormone production.

As preparation for migration continues, birds increase their activity levels. This has been monitored with electronic perches, and Figure 8.21 shows the dramatic difference after

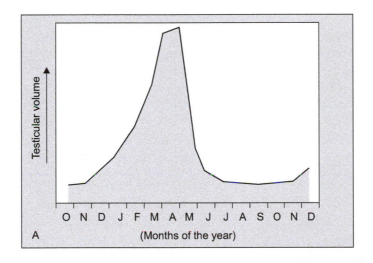

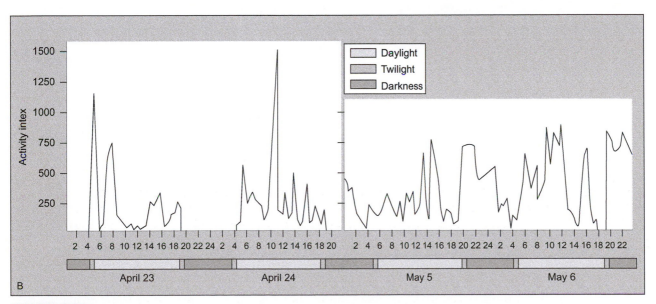

FIGURE 8.21

(A) In male birds, an annual cycle of testicular development and regression is associated with the migratory cycle. (B) Immediately prior to migration, birds display migratory restlessness, or *zugunruhe*. Before the onset of *zugunruhe*, birds exhibit a typical circadian activity cycle, with much diurnal activity and no nocturnal activity. After the onset of *zugunruhe*, birds remain active throughout the night. *Adapted from Farner, D.S., 1955. The animal stimulus for migration: experimental and physiologic aspects. Recent Studies in Avian Biology. University of Illinois Press.*[76]

the onset of migratory restlessness. Note that restless birds not only hop more on their perches, but do this all day and all night, in seeming disregard for their usual circadian rhythm.

> **KEY TERM** *Zugunruhe* is migratory restlessness, the behavioral manifestation of the physiological changes leading up to departure on a migration.

With the accumulation of sufficient fat reserves, good weather is the proximate cue for initiating migration. Indeed, migratory restlessness does not increase smoothly over time, but reflects weather conditions; restless birds are more restless in good (migration-friendly) weather.

While the image of bird migration is shaped by seeing waterfowl—ducks, geese, and cranes, in particular—migrate during the daytime, most avian migration is nocturnal. This fact is especially intriguing because when not migrating, most birds are diurnal. Why shift activity time during migration? Birds face two critical physiological problems in their migration: energy usage and loss of water. Flying at night reduces their thermal load (less heating from the sun, thus less desiccation) and may provide some protection from predators at takeoff and landing areas.

Although many of these night-flying birds stop on a daily basis during the daylight hours, there is another category of pausing during migration called a *stopover* that differs from these daily stops. For the many migrant birds that lack enough fat to fuel their entire migration, stopover locations become a necessary component of successful migration. Moreover, habitats that provide food, water, and shelter—essential for surviving migration—may be infrequent on the migratory route. Time spent at a stopover can vary greatly among birds within a species; those with lower fat reserves spend a longer time feeding, sometimes

FIGURE 8.22
Sandhill cranes stopping over in the Platte River Valley. *Photo: Ben Pless.*

280

as long as several weeks, before moving on. In addition, depending on the migratory route, some species routinely employ stopovers; waterfowl migrating on the central flyway in North America pass over vast expanses of dry upland and must take advantage of the relatively few low areas with water on their route, using areas such as the Platte River valley in western Nebraska for stopovers (Figures 8.22, 8.23). In such gathering places, many birds may be landing in the water at any given time; no matter what their initial approach, on a windless day, they are likely to land in a surprisingly orderly fashion, aligned with geomagnetic north.[75] Stopovers also frequently occur prior to crossing important barriers, such as large bodies of water or mountain ranges.

OF SPECIAL INTEREST: HAS GLOBAL CLIMATE CHANGE AFFECTED ARRIVAL AND DEPARTURE DATES FOR MIGRATING BIRDS?

For migration to work, the arrival of birds in their seasonal habitat needs to be synchronized with good weather and the availability of food. You can imagine that a bird which arrives too early or too late will be at a serious, perhaps fatal, disadvantage. For birds leaving a tropical habitat and flying to the temperate zone, the timing of their arrival is partly controlled by the timing of their departure and partly by their ability to wait along the route if they encounter unseasonal conditions as they make their way into the temperate zone. Ultimately, though, a predictable set of conditions needs to be present in their nesting habitat. What if global climate change causes flowers to emerge at the "wrong" time, or causes insect populations to peak too early? What if the birds themselves adapt to changing conditions and arrive early, only to find that the flora did not respond in the same way to the altered climate? There is much to be concerned about here, because migratory patterns may be disrupted and populations diminished. Although not all bird species are affected, clear patterns of early arrival have been demonstrated recently in some species (Figure 8.23). How these changes in timing affect the survival and reproduction of bird species remains to be discovered.[74,77]

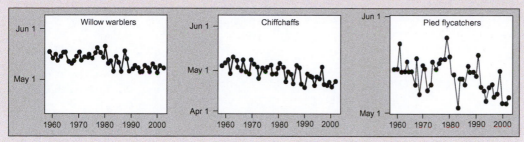

FIGURE 8.23

Changes over a 40-year time span in the arrival date of three species of bird at the Rybachy biological station on the Baltic Sea. (Left) Willow warblers; (center) chiffchaffs; (right) pied flycatchers. All three species show substantially earlier arrival dates as global climate conditions have changed. *Adapted from Sparks, T.H., Bairlein, F., Bojarinova, J.G., Huppop, O., Lehikoinen, E.A., Rainio, K., et al., 2005. Examining the total arrival distribution of migratory birds. Glob. Change Biol. 11, 22–30.*[74]

Typically, information about migratory direction is innate (genetically coded). This is known from displacement experiments, in which naive birds (young birds that have never migrated) are transported a substantial distance, usually hundreds of kilometers, from their birthplace. When seasonal cues trigger migration, these birds follow the same flight direction as they would have had they not been displaced; the displacement does not change their flight direction. Cross-fostering of birds between populations with different migration routes yields the same result; the cross-fostered birds fly the direction inherited from their biological parents, not in the direction flown by their foster parents. The innate directional information is coordinated with a compass: the sun for diurnal migrants and a celestial compass for nocturnal ones. The moon is used as a reference point less commonly. The solar or celestial compass can be complemented by information from a geomagnetic compass.

A series of classic experiments used a planetarium to show that birds, indeed, can orient to the stars in the sky when they fly at night; they are genetically predisposed to learn the night sky that they see as nestlings and use that learned map to determine the direction they should fly at migration time. For instance, in the northern hemisphere, the North Star is fixed; if the planetarium constellations rotated around a different star, the young birds used that star as an indication of north when migration time arrived. Nestlings that were not allowed to see the night sky did not learn the stellar map and were unable to identify north (Figure 8.23).[78]

DISCUSSION POINT: WHAT ABOUT BATS?

Many bats survive the winter by hibernating, but others may migrate long distances to warmer areas. While much is known about bird migration, flying mammals have not been studied as much. What similarities and differences would you predict, based on what you know (or can find out) about bats and birds? How does evolutionary history limit the ways that different animals solve the same problem?

Migration in Salmon

Salmon undergo a remarkable odyssey during their lives (Figure 8.24).[79–82] Young salmon leave the freshwater streams in which they hatched, move thousands of kilometers into the open ocean, and then return as adults to spawn in their natal stream. The question of how

281

FIGURE 8.24
(A) A salmon ladder on the Columbia river. Salmon, migrating upstream, navigate these ladders to return to upriver spawning grounds. (B) Spawning sockeye salmon in Alaska. *Photos: Michael Breed (A) and USFWS (B).*

they find their way back after many months and across an enormous distance has intrigued scientists for decades.

Starting with an adult fish initiating its journey to freshwater, the migration can be broken down into two phases. The fish must first locate the coastline or general area of the freshwater source it is seeking. The second task is the specific identification of the stream from which it migrated earlier in life. The first task—finding the coastline—does not require the fish to experience the outward swim, but that experience is essential for the second task—locating the natal stream. Thus, it appears the fish learns some of the information about its course on its outward migration.

There are many possible tools that salmon might use to find the coastline. For instance, they contain magnetite and may be able to use geomagnetic information in orientation; such information might allow them to find a coastline in the absence of other information about the location of the coast. As in honeybees and pigeons flying home, salmon need a direction in which to swim, and perhaps an expected travel distance, to achieve the coastline. Salmon have several possible sources of compass information for orienting their travel (Figure 8.24). These sources include polarized light, the sun, and geomagnetic information. Ocean temperatures and currents may also be informative. The salmon may use path integration, or the direction may simply be genetically encoded. For Chinook salmon (*Oncorhynchus tshawytscha*), magnetic fields seem to play a role. Juvenile salmon were exposed to magnetic fields that were typical of those at the boundaries of their range in the ocean. These salmon were less than 1-year old and had not commenced their journey to the ocean. The naïve salmon tested in the northern magnetic field oriented toward the south-southwest, and those tested in the southern magnetic field oriented north-northeast. This is consistent with the hypothesis that Chinook salmon find their way in the ocean using an inherited magnetic map, much as loggerhead sea turtles do.[83,84]

Finding the natal stream requires that the salmon experience that natal stream early in life. When the salmon are close to their natal stream, they identify the appropriate freshwater source by olfaction. Young salmon, perhaps primed by the hormone thyroxine, imprint on a stream-specific chemical signature. This imprinting may take place during the transition from the parr to the smelt stage in development. (*Parr* and *smelt* are terms for immature salmon.) Understanding the imprinting period is critical to developing strategies for restocking salmon streams from hatchery stock; this is an area of current research interest.

Migration in Monarch Butterflies

Monarch butterflies are an excellent example of a long-distance migrator that uses genetically derived information to determine its migratory route. The butterflies in the southward-moving generation have had no contact with their parents; a different generation migrates each direction. Clearly, in this case there is no opportunity for learning the migratory route. Instead, ideothetic information, combined with orientation based on the sun compass and prevailing winds,[85] must be sufficient to guide the butterflies, each weighing less than a gram, thousands of kilometers to their overwintering sites in southern California and Mexico (Figure 8.25).[86] The butterflies shift to a northern orientation in response to colder temperatures; in the laboratory, only 24 days of 4° "nights" and 11° "days" were needed for shift in orientation. Without cold temperatures, the monarchs continued south.[87]

Experiments with monarch butterflies provide a real-life illustration of the difference between navigation and orientation introduced early in this chapter. Monarchs in eastern North America fly in a southwest direction and thus reach their overwintering site in Mexico.

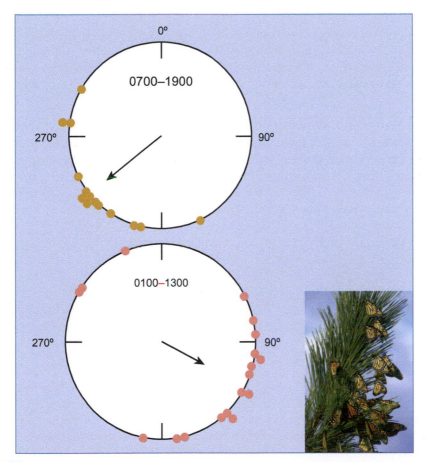

FIGURE 8.25
Insects like the monarch butterfly, which rely on a sun compass to orient their migration, can be fooled by shifting their circadian clocks. The circles in this diagram represent the horizon. The center of the circle is the point at which butterflies were released, and each dot represents the point at which a butterfly disappeared from view as it flew toward the horizon. This type of diagram is the standard way of representing flight direction of animals from a release point. The arrows show the mean flight direction for each treatment. When maintained in photophase from 7:00 a.m. to 7:00 p.m. (upper circle), the butterflies flew in a southwesterly direction. Shifting the light part of their cycle from 1:00 a.m. to 1:00 p.m. resulted in flight in a southeasterly direction. This effect of clock shifting suggests that the clock shifted animals misinterpret the position of the sun.[91] *Adapted from Froy, O., Gotter, A.L., Casselman, A.L., Reppert, S.M., 2003. Illuminating the circadian clock in monarch butterfly migration. Science 300, 1303–1305.*[91] *Inset photo: US Fish and Wildlife Service.*

FIGURE 8.26
Like the great migration of mammals in Africa, some North American mammals also have impressive migrations. For example, Caribou migrate over 1000 km from summer to winter foraging ranges. *Photo: Aaron Collins, US Fish and Wildlife Service.*

284

When displaced 2500 km to the west, they persist in flying in a southwest direction; they do not adjust their route, given their new location. Orientation seems to be their primary means of reaching their overwintering grounds,[88] although some researchers disagree[89] with that assessment.

Another invertebrate, the spiny lobster (*Panulirus argus*), is a true navigator. When displaced up to 37 km, the lobsters orient toward the capture site. They do this when deprived of visual, chemical, and magnetic cues during displacement. The lobsters were transported in opaque containers lined with magnets through a series of sharp turns and circles, and tested the next day with eyestalks covered; none of this deterred the lobsters from moving in the direction of the site where they had been captured. Finally, they were tested in two different magnetic fields, one replicating a site approximately 400 km to the north, and one replicating a site approximately 400 km to the south. (Spiny lobsters move over 400 km in their travels.) Once again, the lobsters began walking in what would have been the direction of their capture site had they been displaced to the real sites that were replicated in the test. These experiments support the hypothesis that spiny lobsters not only navigate, but do so using a magnetic map.[90]

Migration in African Ungulates

In contrast, some of the most picturesque images of animal behavior come from the great migrations of hoofed mammals in Africa. Vast seas of wildebeest, zebra, and giraffes move in concert with the seasonal availability of water and forage (Figure 8.26). Ungulate migrations tend to follow "traditional routes" along which there is much opportunity for learning. The group dynamic shapes the migratory route and from one year to the next, experienced animals remember routes of travel, locations of watering holes, and destinations. Young animals have the opportunity to learn routes and migratory cues from older animals, and movement in herds provides protection from the numerous predators that live along the migratory route or follow the migrating animals. North American ungulate populations, such as bison, elk, and pronghorn, engaged in similar migrations before they were decimated. Disruption of the African migration—"the last great migration"—by anthropogenic (human caused) changes in the landscape is a looming conservation issue; changes in vegetation patterns in East Africa likely mean that the current generations of these species will be the last to engage in one of the most spectacular wildlife sights on the globe.

Turtle Migration

Sea turtles have a migration pattern somewhat similar to that of salmon, with a life cycle that takes them from hatching on beaches to feeding waters in the ocean and back to the beaches to lay the next generation of eggs.[92,93] Adult females migrate to their natal beach and lay eggs in the sand (Figure 8.27). Hatchlings undertake a short journey to the shoreline (they orient to the horizon), fraught with danger from predators.[94] They then swim to the open ocean, initially orienting into the waves, which near the shore provide an indication of the direction of the open ocean. Their migration then continues for hundreds or thousands of kilometers into the ocean, where they probably use the earth's magnetic field and the direction of the currents to find their way into favorable feeding grounds.

Even less is known about how sea turtles accomplish this feat than is known for salmon. Turtles may actually use the geomagnetic field to tell them their location. (Of course, the

open ocean has no landmarks that humans know about.) If this is the case, then their magnetoperception system seems to work on a more refined level than that of birds, which is thought to provide directional but not locational information. Some scientists think that sea turtles probably use the geomagnetic field to orient their swim in the deep ocean, but they would also have celestial cues available to them when close to the surface. One classic test for the use of magnetic fields in navigation and orientation consists of fitting magnets to an animal's head; the magnet will interfere with geomagnetic cues. However, just such an experiment with green turtles migrating between the open Atlantic and Brazil found that the turtles with magnets performed as well as their counterpart controls. Another (untested) hypothesis is that the directional vector of sea turtles is determined genetically. Also intriguing but unexplained is the ability of the turtles to return to their natal beach over their vast travel distance. Do they imprint, as salmon do, on olfactory features in the water? Or is the location pinpointed using geomagnetic information?

Migratory routes and distances vary among species of sea turtles. Green sea turtles migrate primarily along the coasts from nesting to feeding grounds. However, turtles from some populations will travel approximately 2100 km (1300 miles) across the Atlantic Ocean from the Ascension Island nesting grounds to the Brazilian coast feeding grounds. Loggerheads leave foraging areas and travel on breeding migrations that range in distance from a few to thousands of kilometers. Leatherbacks have the longest migration of all sea turtles. They have been found more than 4800 km (3000 miles) from their nesting beaches.

8.8 DISPERSAL

Dispersal is a topic in both behavior and ecology; many of the ecological principles concerning dispersal apply to both plants and animals, but of course here animal dispersal is of primary interest.[95-100] An ecologist is most likely concerned with the numbers of animals that disperse and how movement into and out of populations affects the growth of those populations. Animal behaviorists look at dispersal on a different scale, asking different questions: What causes an animal to decide to disperse? Is dispersal behavior linked to age, sex, or behavioral history? Within a population are there behavioral syndromes that make some animals more likely to disperse than others?

Dispersal allows animals to avoid competition, avoid inbreeding,[101] and to colonize new habitats. Animals disperse by leaving their natal area and finding new territories or home ranges. The dispersing animal, like the migrating one, is attempting to improve its lot in life by finding a suitable habitat. (As will become apparent, sometimes it is attempting to improve its lot in life by leaving increasingly hostile adult relatives!) Unlike migration, dispersal is usually a one-way trip to an unknown destination. Perhaps the greatest difficulty for a dispersing animal is finding unoccupied suitable habitat. In most species, all the best locations will

FIGURE 8.28
Animals move at many scales, from monarch butterflies, whose migration distributes them over large geographic areas, to elk, who may range hundreds of kilometers, to honeybees, which may fly from a few meters to a few hundred meters.

already be occupied, and a disperser is faced with an uncertain future, including the possibility of starvation in an unsuitable habitat. For the vast majority of animal species, dispersal is the simple act of finding suitable habitat after hatching from an egg. Indeed, the choice of a location for oviposition is one of the most important decisions most mothers make.

Who Disperses?

PARENTS VERSUS OFFSPRING

Dispersers in birds and mammals are likely to be adolescents or young adults. Lacking the experience or strength to wrest a favored location from their parents, they leave, or are driven, from parental territories. Female dispersers are unlikely to remain without suitable habitat for long, as males with territories or harems will typically accept multiple females into their area. A striking counterexample to the view that younger animals always suffer the risks associated with dispersal is the honeybee, in which the reigning queen in a colony leaves with a swarm; this allows one of her daughters to inherit the nest, food, and the portion of the worker force that the queen leaves behind (Figure 8.28).

MALES VERSUS FEMALES

In most mammals, males are the dispersing sex,[102] whereas in most birds, females are more likely to disperse.[103] The general argument is that mammals tend to be polygynous—live in social groups in which one male is associated with many females. Wild horses are a familiar example of this type of system. (See Chapter 11 on mating systems for more details on polygyny.) Birds, on the other hand, tend to be monogamous and to have territorial males, leading to a situation in which females disperse to search out male territory holders. When exceptions occur, such as brown jays, in which males are more likely to disperse,[104] or hamadryas baboons, in which females are more likely to disperse,[105] complex social dynamics seem involved in the evolution of these exceptions to the general rules about gender-specific dispersal.

The example of prairie dogs is consistent with the conventional wisdom about mammalian dispersal. Prairie dogs live in dense colonies, which are subdivided into burrow systems occupied by families (coteries; Figure 8.29). Two species of prairie dog, Gunnison's and the black-tailed prairie dog, have been well studied; the biology of the species differs, but the overall pattern of dispersal behavior and colony structure is similar between the species. Young males are the primary dispersers and can move a few kilometers to join a new colony; young females remain in their natal colony. This model of male dispersal and female philopatry (a tendency for an animal to remain close to, or return to, the animal's birthplace) facilitates inbreeding avoidance.[106] Prairie dogs tend to decimate the vegetation in their colony; because of this, sites on the edges of the colony may be favored because the forage is better. Perhaps for this reason, some dispersal of older animals also takes place, with older males and females equally likely to move.

Young dispersing male mammals sometimes form coalitions with other males in similar circumstances, forming "bachelor herds." Males in bachelor herds benefit from shared vigilance and shared information about feeding and watering locations, and also serve as foils in practice combat. Bachelor herds are common in ungulates, such as horses, mountain sheep, and deer, but are also found in marine mammals and in lions.

Mammalian females may leave social groups to find better locations for the birth and feeding of their young. This type of dispersal may protect the young from potential infanticide by adult males, and is well known in a variety of primates. Maternal dispersal is usually reversible, and once the juvenile is old enough, the mother and infant return to the social group. In some cases, though, dispersing pregnant females are responding to competitive conditions and, while pregnant, search for new suitable habitat.

Why Disperse?

COMPETITION DISPERSAL

Competition dispersal is probably a major cause of dispersal in plants; light, root space, and nutrients are depleted or dominated by the parental plant. In animals, it also plays a key role in many species and recurs as a theme in the discussion of dispersal. In competition dispersal, movement is typically an uneven affair, with gender and age biases. Typically, it is the young males that bear the brunt of the risk of dispersal. As a juvenile male approaches reproductive age, it may threaten the status of the dominant male in the group, or it may simply be an extra mouth to feed.

KIN COMPETITION AVOIDANCE

Kin competition avoidance is dispersal or other behavior that keeps sibs or parents and their offspring from competing with each other. Similarly, natural selection may favor animals that disperse and thus avoid competition with kin. This point is important to consider even though it is difficult to discriminate between avoidance of competition with kin and inbreeding avoidance (see later section). In both cases, the result is essentially the same: close kin end up not living near one another.

INFANTICIDE VERSUS DISPERSAL

Infanticide, the killing of a young animal by an older animal, is performed by females and males that are unrelated to or are unfamiliar with the pups (see the section on kin recognition in Chapter 13). In crowded populations of related mice and voles, infanticide is much more likely than it is in less-crowded circumstances. (As an aside, infanticide is a common response to habitat depletion in rodents and can be thought of as the result of resource competition.)[107,108] Infanticide is an alternative to dispersal, albeit an unpleasant one, which maintains resource conditions for the remaining animals in the population.

INBREEDING AVOIDANCE

Another result of dispersal is inbreeding avoidance. Inbreeding avoidance is any behavioral or physiological mechanism that prevents closely related animals from mating. If young of one sex remain near their birthplace and the other sex disperses fairly widely, the chances of inbred matings become low. For inbreeding avoidance, it does not matter which sex does the dispersing. Dispersal can help animals avoid kin associations; meadow voles are more likely to disperse from plots occupied by siblings than from plots occupied by nonsiblings.

DISPERSAL FOR COLONIZATION

Colonizing species succeed by establishing populations in new habitats. Inseminated females are the most powerful dispersers for colonization; by mating before dispersing, they eliminate the need to find a mate in the new habitat (or to disperse in pairs).

FIGURE 8.29
Young male prairie dogs disperse from their natal group; females remain, so prairie dog colonies are matrilineal. *Photo: Jeff Mitton.*

287

Invasive insects are particularly effective at exploiting this strategy to gain footholds in new locations; the red imported fire ant, which has successfully invaded ecosystems in the southeastern United States, Arizona, and California, is an excellent example of a species that combines strong dispersal and colonization abilities. Among mammals, Norway rats are highly effective dispersers and colonizers of new habitat. In the western United States, changing land-use patterns and perhaps climate change have opened habitats for house finches, which have proven to be good dispersers into previously unoccupied areas.

Recent experience with reintroductions of mammalian carnivores illustrates the complexity of thinking about dispersal. Wolves, which have been reintroduced into the greater Yellowstone ecosystem in Wyoming and Montana, have shown the ability to move hundreds of miles relatively rapidly. Thus, the occasional wolf ends up far from the introduction site and well away from the intended population area. Ultimately, this is promising for re-establishment of wolves in the western United States, but in the short term this behavior fuels the anxiety of some ranchers and residents of rural areas.

DISCUSSION POINT: BEHAVIORAL SYNDROMES AND DISPERSAL

In Chapter 4 we talked about behavioral syndromes—sets of behavior that are relatively constant within an animal but vary among animals in a population. Many documented behavioral syndromes range along a continuum from shy to bold. Can you postulate adaptive values for both shyness and boldness in dispersal?

SUMMARY

The study of animal movements in the environment is a journey in itself, starting with the integration of simple sensory inputs and basic movements. For many animals, the ability to discover and orient to food, shelter, and mates is all that is needed for evolutionary success. Kineses, or undirected movements, and taxes, or directed movements, provide the basic mechanisms by which many animals find their appropriate habitats. The discovery of the direction or location of resources is often key to navigating to those resources. Search behavior typically begins with undirected or looping movements and, as cues are gained from the environment, shifts to goal-directed movements. Sensory systems, such as stereoptic vision, are needed to allow goal-oriented movements.

For some animals, navigational needs extend far beyond these abilities. Navigation sometimes relies primarily on genetically sourced information, the product of generations of evolutionary trial and error. In other cases, learning plays a larger part in navigation, giving the animal the flexibility to accommodate its current location and to exploit information obtained from other population members. Path integration and compasses are often used in homing. Finding home requires using sophisticated mechanisms such as path integration, landmarks, and snapshot memory.

Migration, moving seasonally between widely separated habitats, gives animals the opportunity to exploit more than one habitat during their life cycle. Migration means that animals must use highly intricate compass and odometer mechanisms. Migratory routes may be innate or learned; again, many animals rely on a mixture of these two types of information to direct their movements. The diversity of organisms that migrate and the vulnerability of the migratory lifestyle make migrating animals a particularly important conservation concern. Finally, dispersal is what allows animals to establish new populations and avoid the negative outcomes of inbreeding and competition. By dispersing, animals avoid competition and may colonize new habitats.

STUDY QUESTIONS

1. As part of an experiment, pigeons are kept in a room where the lights are controlled by a timer. For a week, the timer is adjusted so that the lights go on 3 h before the natural sunrise at this location and go off 3 h before the natural sunset. In this particular experiment, when released some distance from their roost, these "clock-shifted" pigeons fly in the same direction as control pigeons. Are they using a sun-compass? Why or why not?

2. A repeating theme in this book has been the interaction of genetic and environmental information in shaping behavior. As we discussed migration in this chapter, you saw examples of genetic shaping of migratory movements as well as learned migration information. Under what circumstances would you expect a species to rely on genetic information for migration and under what circumstances would learning be dominant?

3. What are the problems that animals face in orientation at night, and how have they overcome those problems? Subdivide your answer by types of orientation, such as search for food, homing, and migration. Why do most bird species migrate at night?

4. How would you test the hypothesis that humans use a magnetic compass in their orientation?

5. What circumstances would favor males as the dispersing sex? Females? Can you model a circumstance under which both sexes would be equally likely to disperse?

Further Reading

Gauthreaux Jr., S.A., 2010. Bird migration. In: Breed, M.D., Moore, J. (Eds.), Encyclopedia of Animal Behavior, vol. 2, Academic Press, Oxford, pp. 211–219.

Gould, J.L., Gould, C.G., 2012. Nature's Compass: The Mystery of Animal Navigation. Princeton University Press. 320 pp.

Lohmann, C.M.F., Lohmann, K.J., 2010. Sea turtles: navigation and orientation. In: Breed, M.D., Moore, J. (Eds.), Encyclopedia of Animal Behavior, vol. 2, Academic Press, Oxford, pp. 101–107.

McGuire, L.P., Fenton, M.B., Guglielmo, C.G., 2013. Phenotypic flexibility in migrating bats: seasonal variation in body composition, organ sizes, and fatty acid profiles. J. Exp. Biol. 216, 800–808.

Reppert, S.M., 2014. Monarch butterfly migration: from behavior to neurons to genes. Annu. Rev. Entomol. 60 (1)

Tape, K.D., Gustine, D.D., 2014. Capturing migration phenology of terrestrial wildlife using camera traps. BioScience 64, 117–124.

Notes

1. Lindeyer, C.M., Reader, S.M., 2010. Social learning of escape routes in zebrafish and the stability of behavioural traditions. Anim. Behav. 79, 827–834.
2. Bell, W.J., 1991. Search Behaviour. Chapman & Hall, London.
3. Bingman, V.P., Cheng, K., 2005. Mechanisms of animal global navigation: comparative perspectives and enduring challenges. Ethol. Ecol. Evol. 17 (4), 295–318.
4. Phillips, J.B., Borland, S.C., 1992. Behavioural evidence for use of light-dependent magnetoreception mechanism by a vertebrate. Nature 359, 142–144.
5. Sandberg, R., Backman, J., Moore, F.R., Lohmus, M., 2000. Magnetic information calibrates celestial cues during migration. Anim. Behav. 60, 453–462.
6. Galambos, R., 1942. The avoidance of obstacles by flying bats: Spallanzani's ideas (1794) and later theories. Isis 34, 132–140.
7. Griffin, D.R., 2001. Return to the magic well: echolocation behavior of bats and responses of insect prey. BioScience 51, 555–556.
8. Gross, C.G. 2005. Donald R. Griffin. Biographical Memoirs of the National Academy of Sciences, National Academies Press, Washington, DC. 86, 1–20.
9. Griffin, D.R., 1998. In: Squire, L. (Ed.), The History of Neuroscience in Autobiography, vol. 2 Academic Press, San Diego, CA, pp. 68–93.
10. Berthold, P., 2001. Bird Migration: A General Survey, second ed. Oxford University Press. pp. 272; Berthold, P., Gwinner, E., Sonnenschein, E. (Eds.), 2003. Avian Migration Springer, Berlin, Germany.
11. Ferrari, J., Godfray, H.C.J., Faulconbridge, A.S., Prior, K., Via, S., 2006. Population differentiation and genetic variation in host choice among pea aphids from eight host plant genera. Evolution 60 (8), 1574–1584.

12. Dambrowski, H.R., Linn, C., Berlocher, S.H., Forbes, A.A., Roelofs, W., Feder, J.L., 2005. The genetic basis for fruit odor discrimination in *Rhagoletis* flies and its significance for sympatric host shifts. Evolution 59, 1953–1964.

13. de Boer, G., 2004. Temporal and developmental aspects of dietary-induced larval food preferences in the tobacco hornworm. Entomologia Experimentalis et Applicata 113, 197–204.

14. Wiltschko, R., Wiltschko, W., 1999. The orientation system of birds—I. Compass mechanisms. J. Ornithol. 140 (1), 1–40.

15. Whishaw, I.Q., Brooks, B.L., 1999. Calibrating space: exploration is important for allothetic and idiothetic navigation. Hippocampus 9, 659–667.

16. Biegler, R., 2000. Possible uses of path integration in animal navigation. Anim. Learn. Behav. 28 (3), 257–277.

17. Rossel, S., 1983. Binocular stereopsis in an insect. Nature 302 (5911), 821–822.

18. Kral, K., Vernik, M., Devetak, D., 2000. The visually controlled prey-capture behaviour of the European mantispid *Mantispa styriaca*. J. Exp. Biol. 203 (14), 2117–2123.

19. Steck, K., Markus, K., Hansson, B.S., 2010. Do desert ants smell the scenery in stereo? Anim. Behav. 79, 939–945.

20. Kral, K., 2003. Behavioural-analytical studies of the role of head movements in depth perception in insects, birds and mammals. Behav. Processes 64 (1), 1–12.

21. Willis, M.A., Avondet, J.L., 2005. Odor-modulated orientation in walking male cockroaches *Periplaneta americana*, and the effects of odor plumes of different structure. J. Exp. Biol. 208 (4), 721–735.

22. Grasso, F., Basil, J., 2002. How lobsters, crayfishes , and crabs locate sources of odor: current perspectives and future directions. Curr. Opin. Neurobiol. 12, 721–727.

23. Thar, R., Kuhl, M., 2003. Bacteria are not too small for spatial sensing of chemical gradients: an experimental evidence. Proc. Natl. Acad. Sci. USA 100, 5748–5753.

24. Mitchell, J.G., Kogure, K., 2006. Bacterial motility: links to the environment and a driving force for microbial physics. FEMS Microbiol. Ecol. 55 (1), 3–16.

25. Lohmannm, K.J., 2010. Magnetic-field perception. Nature 464, 1140–1142.

26. Holland, R., Borissov, I., Siemers, B., 2010. A nocturnal mammal, the greater mouse-eared bat, calibrates a magnetic compass by the sun. Proc. Natl. Acad. Sci. USA 107, 6146–6151.

27. Wu, L., Dickman, J.D., 2012. Neural correlates of a magnetic sense. Science 336, 1054–1057.

28. Brown, C.R., Bomberger Brown, M., 2013. Where has all the road kill gone? Curr. Biol. 23, R233–R234.

29. Dingle, H., 1962. The occurrence of correcting behavior in various insects. Ecology 43, 727–728.

30. Vickers, N.J., 2000. Mechanisms of animal navigation in odor plumes. Biol. Bull. 198, 203–212.

31. Capaldi, E.A., Smith, A.D., Osborne, J.L., Fahrbach, S.E., Farris, S.M., Reynolds, D.R., et al., 2000. Ontogeny of orientation flight in the honeybee revealed by harmonic radar. Nature 403, 537–540.

32. Collett, M., Chittka, L., Collett, T.S., 2013. Spatial memory in insect navigation. Curr. Biol. 23, R789–R800.

33. Norgaard, T., Gagnon, Y.L., Warrant, E.J., 2012. Nocturnal homing: learning walks in a wandering spider? PLoS One 7, e49263.

34. Menzel, R., 2014. The cognitive structure of visual navigation in honeybees. In: Werner, J.S., Chalupa, L.M. (Eds.), New Visual Neurosciences. MIT Press, Cambridge, MA, pp. 1179–1189.

35. Capaldi, E.A., Dyer, F.C., 1999. The role of orientation flights on homing performance in honeybees. J. Exp. Biol. 202, 1655–1666.

36. Menzel, R., Geiger, K., Joerges, J., Muller, U., Chittka, L., 1998. Bees travel novel homeward routes by integrating separately acquired vector memories. Anim. Behav. 55, 139–152.

37. Etienne, A.S., Maurer, R., Berlie, J., Reverdin, B., Rowe, T., Georgakopoulos, J., et al., 1998. Navigation through vector addition. Nature 396 (6707), 161–164.

38. Seguinot, V., Cattet, J., Benhamou, S., 1998. Path integration in dogs. Anim. Behav. 55, 787–797.

39. Benhamou, S., 1997. Path integration by swimming rats. Anim. Behav. 54, 321–327.

40. Müller, M., and Wehner, R., Proc. Natl. Acad. Sci. USA 85 (1414), 5287–5290.

41. Lebhardt, F., Ronacher, B., 2014. Interactions of the polarization and the sun compass in path integration of desert ants. J. Comp. Physiol. A Neuroethol. Sens. Neural. Behav. Physiol. 200, 711–720.

42. Dickerson, J., Dyer, F., 1996. How insects learn about the sun's course: alternative modelling approaches In: Maes, P., Mafaric, J.J., Meyer, J., Pollock, J., Wilson, S. (Eds.), Animals to Animats, vol. 4 MIT Press, Cambridge, MA.

43. Dyer, F.C., 1991. Bees acquire route-based memories but not cognitive maps in a familiar landscape. Anim. Behav. 41, 239–246; Dyer, F.C., Berry, N.A., Richard, A.S., 1993. Honeybee spatial memory—use of route-based memories after displacement. Anim. Behav. 45 (5), 1028–1030.

44. Dacke, M., Baird, E., Byrne, M., Scholtz, C.H., Warrant, E.J., 2013. Dung beetles use the Milky Way for orientation. Curr. Biol. 23, 298–300.

45. Walls, M.L., Layne, J.E., 2009. Direct evidence for distance measurement via flexible stride integration in the fiddler crab. Curr. Biol. 19, 25–29.

46. Ortega-Escobar, J., Ruiz, M.A., 2014. Visual odometry in the wolf spider *Lycosa tarantula* (Araneae: Lycosidae). J. Exp. Biol. 217, 395–401.

47. Collett, M., Collett, T.S., Wehner, R., 1999. Calibration of vector navigation in desert ants. Curr. Biol. 9 (18), 1031–1034.

48. Gould, J., 1986. The locale map of honey bees: do insects have cognitive maps? Science 232, 861–863.

49. Menzel, R., De Marco, R.J., Greggers, U., 2006. Spatial memory, navigation and dance behaviour in *Apis mellifera*. J. Comp. Physiol. A Neuroethol. Sens. Neural. Behav. Physiol. 192 (9), 889–903.

50. Blaser, N., Dell'Omo, G., Dell'Ariccla, G., Wolfer, D.P., Lipp, H.-P., 2013. Testing cognitive navigation in unknown territories: homing pigeons choose different targets. J. Exp. Biol. 216, 3123–3131.
51. Barchi, J.R., Knowles, J.M., Simmons, J.A., 2013. Spatial memory and stereotypy of flight paths by big brown bats in cluttered surroundings. J. Exp. Biol. 216, 1053–1063.
52. Gould, J.L., 2014. Animal navigation: a map for all seasons. Curr. Biol. 24, R153–R155. http://dx.doi.org/10.1016/j.cub.2014.01.030.
53. Burda, H., Begall, S., Cerveny, J., Neef, J., Nemec, P., 2009. Extremely low-frequency electromagnetic fields disrupt magnetic alignment of ruminants. Proc. Natl. Acad. Sci. USA 106, 5708–5713.
54. Begall, S., Cerveny, J., Neef, J., Vojtech, O., Burda, H., 2008. Magnetic alignment in grazing and resting cattle and deer. Proc. Natl. Acad. Sci. USA 105, 13451–13455.
55. Humphries, N.E., Weimerskirch, H., Queiroz, N., Southall, E.J., Sims, D.W., 2012. Foraging success of biological Levy flights recorded *in situ*. Proc. Natl. Acad. Sci. USA 109, 7169–7174.
56. Raichlen, D.A., Wood, B.M., Gordon, A.D., Mabulla, A.Z.P., Marlowe, F.W., Pontzer, H., 2014. Evidence of Levy walk foraging patterns in human hunter-gatherers. Proc. Natl. Acad. Sci. USA 111, 728–733.
57. Si, A., Srinivasan, M.V., Zhang, S., 2003. Honeybee navigation: properties of the visually driven "odometer." J. Exp. Biol. 206, 1265–1273.
58. Gagliardo, A., Bried, J., Lambardi, P., Luschi, P., Wikelski, M., Bonadonna, F., 2013. Oceanic navigation in Cory's shearwaters: evidence for a crucial role of olfactory cues for homing after displacement. J. Exp. Biol. 216, 2798–2805.
59. Dingle, H., 1996. Migration: The Biology of Life on the Move. Oxford University Press, New York, NY.
60. Dingle, H., Alistair Drake, V., 2007. What is migration? Bioscience 57, 113–121.
61. Fitzsimons, J.M., Parham, J.E., Nishimoto, R.T., 2004. Similarities in behavioral ecology among amphidromous and catadromous fishes on the oceanic islands of Hawai'I and Guam. Environ. Biol. Fishes 65, 123–129.
62. Baker, R.R., 1978. The Evolutionary Ecology of Animal Migration. Hodder and Stoughton, London.
63. Åkesson, S., Hedenström, A., 2007. How migrants get there: migratory performance and orientation. Bioscience 57, 123–133.
64. Wiltschko, W., Weindler, P., Wiltschko, R., 1998. Interaction of magnetic and celestial cues in the migratory orientation of passerines. J. Avian Biol. 29 (4), 606–617.
65. Weindler, P., Liepa, V., Wiltschko, W., 1998. The direction of celestial rotation affects the development of migratory orientation in Pied Flycatchers, *Ficedula hypoleuca*. Ethology 104, 905–915; Weindler, P., Bohme, F., Liepa, V., Wiltschko, W., 1998. The role of daytime cues in the development of magnetic orientation in a night-migrating bird. Behav. Ecol. Sociobiol. 42, 289–294.
66. Mouritsen, H., 1998. Redstarts, *Phoenicurus phoenicurus*, can orient in a true zero magnetic field. Anim. Behav. 55, 1311–1324.
67. Aborn, D., 2004. Activity budgets of summer tanagers during spring migratory stopover. Wilson Bull. 116, 64–68.
68. Egevang, C., Stenhouse, I.J., Phillips, R.A., Petersen, A., Fox, J.W., Silk, J.R.D., 2010. Tracking of Arctic terns *Sterna paradisaea* reveals longest animal migration. Proc. Natl. Acad. Sci. USA 107, 2078–2091.
69. Moore, F.R., 1987. Sunset and the orientation behaviour of migrating birds. Biol. Rev. 62, 65–86.
70. Holberton, R.L., 1993. An endogenous basis for differential migration in the dark-eyed junco. Condor 95, 580–587.
71. Greenberg, R., Marra, P.P. (Eds.), 2005. Birds of Two Worlds: The Ecology and Evolution of Migration. Johns Hopkins University Press, Baltimore, MD.
72. Pulido, F., 2007. The genetics and evolution of avian migration. Bioscience 57, 165–174.
73. Ramenofsky, M., Wingfield, J.C., 2007. Regulation of migration. Bioscience 57, 135–143.
74. Sparks, T.H., Bairlein, F., Bojarinova, J.G., Huppop, O., Lehikoinen, E.A., Rainio, K., et al., 2005. Examining the total arrival distribution of migratory birds. Glob. Change Biol. 11, 22–30.
75. Hart, V., Malkemper, E.P., Kusta, T., Begall, S., Novakova, P., Hanzal, V., et al., 2013. Directional compass preference for landing in water birds. Front. Zool. 10, 38.
76 Farner, D.S., 1955. The animal stimulus for migration: experimental and physiologic aspects Recent Studies in Avian Biology. University of Illinois Press.
77. Lehikoinen, E., Sparks, T.H., Zalakevicius, M., 2004. Birds and climate change. Adv. Ecol. Res. 35, 1–31.
78. Emlen, S.T., 1970. Celestial rotation: its importance in the development of migratory orientation. Science 170 (3963), 1198–1201.
79. Berman, C.H., Quinn, T.P., 1991. Behavioural thermoregulation and homing by spring Chinook salmon, *Oncorhynchus tshawytscha* (Walbaum), in the Yakima River. J. Fish. Biol. 39, 301–312.
80. Thorrold, S.R., Latkoczy, C., Swart, P.K., Jones, C.M., 2001. Natal homing in a marine fish metapopulation. Science 291, 297–299.
81. Scholz, A.T., Horrall, R.M., Cooper, J.C., Hasler, A.D., 1976. Imprinting to chemical cues—basis for home stream selection in salmon. Science 192 (4245), 1247–1249.
82. Dittman, A.H., Quinn, T.P., Nevitt, G.A., 1996. Timing of imprinting to natural and artificial odors by coho salmon (*Oncorhynchus kisutch*). Can. J. Fish. Aquat. Sci. 53 (2), 434–442.
83. Putman, N.F., Scanlan, M.M., Billman, E.J., O'neil, J.P., Couture, R.B., Quinn, T.P., et al., 2014. An inherited magnetic map guides ocean navigation in juvenile Pacific salmon. Curr. Biol. 24, 1–5.
84. Gould, J.L., 2011. Animal navigation: longitude at last. Curr. Biol. 21, R225–R227.
85. Heinze, S., Reppert, S.M., 2011. Sun compass integration of skylight cues in migratory monarch butterflies. Neuron 69, 345–358.

291

86. Roff, D.A., Fairbairn, D.J., 2007. The evolution and genetics of migration in insects. Bioscience 57, 155–164.

87. Guerra, P.A., Reppert, S.M., 2013. Coldness triggers northward flight in remigrant monarch butterflies. Curr. Biol. 23, 419–423.

88. Mouritsen, H., Derbyshire, R., Stalleicken, J., Mouritsen, Ø.O., Frost, B.J., Norris, D.R., 2013. An experimental displacement and over 50 years of tag-recoveries show that monarch butterflies are not true navigators. Proc. Natl. Acad. Sci. USA 110, 7348–7353.

89. Oberhauser, K.S., Taylor, O.R., Reppert, S.M., Dingle, H., Nail, K.R., Pyne, R.M., et al., 2013. Are monarch butterflies true navigators? The jury is still out. Proc. Natl. Acad. Sci. USA 110, E3681.

90. Boles, L.C., Lohmann, K.J., 2003. True navigation and magnetic maps in spiny lobsters. Nature 421, 60–63.

91. Froy, O., Gotter, A.L., Casselman, A.L., Reppert, S.M., 2003. Illuminating the circadian clock in monarch butterfly migration. Science 300, 1303–1305.

92. Lohmann, K.J., Hester, J.T., Lohmann, C.M.F., 1999. Long-distance navigation in sea turtles. Ethol. Ecol. Evol. 11 (1), 1–23; Lohmann, K.J., Lohmann, C.M.F., 1998. Migratory guidance mechanisms in marine turtles. J. Avian Biol. 29 (4), 585–596.

93. Papi, F., Luschi, P., Akesson, S., Capogrossi, S., Hays, G.C., 2000. Open-sea migration of magnetically disturbed sea turtles. J. Exp. Biol. 203 (22), 3435–3443.

94. Goff, M., Salmon, M., Lohmann, K.J., 1998. Hatchling sea turtles use surface waves to establish a magnetic compass direction. Anim. Behav. 55, 69–77.

95. Waser, P.M., 1985. Does competition drive dispersal? Ecology 66, 1170–1175.

96. Lena, J.-P., Clobert, J., de Fraipont, M., Lecomte, J., Guyot, G., 1998. The relative influence of density and kinship on dispersal in the common lizard. Behav. Ecol. 9, 500–507.

97. Dobson, F.S., 1982. Competition for mates and predominantly juvenile male dispersal in mammals. Anim. Behav. 30, 1183–1192; Dobson, F.S., 1985. Multiple causes of dispersal. Am. Nat. 126, 855–858.

98. Bowen, B., Koford, R.R., Vehrencamp, S.L., 1989. Dispersal in the communally breeding groove-billed Ani (Crotophaga sulcirostris). Condor 91, 52–64.

99. Clobert, J., Danchin, E., Dhondt, A.A., Nichols, J.D. (Eds.), 2001. Dispersal Oxford University Press, Oxford.

100. Holekamp, K., 1984. Natal dispersal in Belding's ground squirrels (Spermophilus beldingi). Behav. Ecol. Sociobiol. 16, 21–30.

101. Bollinger, E.K., Harper, S.J., Barrett, G.W., 1993. Inbreeding avoidance increases dispersal movements of the meadow vole. Ecology 74, 1153–1156.

102. Greenwood, P.J., 1980. Mating systems, philopatry and dispersal in birds and mammals. Anim. Behav. 28, 1140–1162.

103. Clarke, A.L., Saether, B.E., Roskaft, E., 1997. Sex biases in avian dispersal: a reappraisal. Oikos 79 (3), 429–438.

104. Williams, D.A., Rabenold, K.N., 2005. Male-biased dispersal, female philopatry, and routes to fitness in a social corvid. J. Anim. Ecol. 74, 150–159.

105. Hammond, R.L., Handley, L.J.L., Winney, B.J., Bruford, M.W., Perrin, N., 2006. Genetic evidence for female-biased dispersal and gene flow in a polygynous primate. Proc. R. Soc. Lond., B, Biol. Sci. 273 (1585), 479–484.

106. Lucas, J.R., Waser, P.M., Creel, S.R., 1994. Death and disappearance: estimating mortality risks associated with philopatry and dispersal. Behav. Ecol. 5, 135–141.

107. Morelli, T.L., King, S.J., Pochron, S.T., Wright, P.C., 2009. The rules of disengagement: takeovers, infanticide, and dispersal in a rainforest lemur, Propithecus edwardsi. Behaviour 146, 499–523.

108. Sherman, P.W., 1981. Reproductive competition and infanticide in Belding's ground squirrels and other animals. In: Alexander, R.D., Tinkle, D. (Eds.), Natural Selection and Social Behavior Chiron Press, New York, NY.

109. Zeil, J., 1998. Homing in fiddler crabs (Uca lactea annulipes and Uca vomeris: Ocypodidae). J. Comp. Physiol. A 183 (3), 367–377.

Foraging

293

LEARNING OBJECTIVES

Studying this chapter should provide you with the knowledge to:

- See how dietary requirements, natural selection for dietary specialization, and the evolution of foraging behavior interact.
- Understand the diversity of dietary strategies employed by animals.
- Be able to use optimality theory to test and refine hypotheses about important resources—the currencies of animal lives. In the study of foraging, this is frequently energy.
- Understand the limits of optimality theory; animals are not expected to optimize every behavior. These constraints are the same as those that make adaptation and natural selection works in progress rather than finished portraits.
- Apply the marginal value theorem as an economic concept that makes predictions about where to forage.
- Realize that risk sensitivity, along with whether an animal is risk-averse or risk-prone, can influence foraging behavior.

Animal Behavior. DOI: http://dx.doi.org/10.1016/B978-0-12-801532-2.00009-X

9.1 INTRODUCTION

It is a truism that all animals have to eat. After all, the act of eating is part of what distinguishes animals from other living things. Animals do not photosynthesize; many other types of organisms capture energy from the sun and use solar energy to turn basic elements such as carbon, oxygen, hydrogen, and nitrogen into organic molecules. Animals then exploit these molecules for their own nutritional needs. Foraging is collecting food, and the fact that most potential food items resist being eaten makes this a dynamic and complex topic (Figure 9.1).

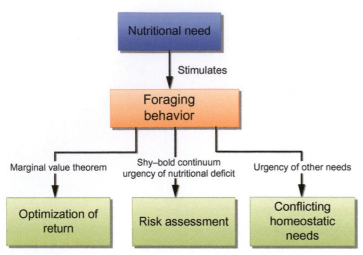

FIGURE 9.1
Foraging behavior is driven by nutritional need. When making decisions about foraging strategy, a foraging animal must balance the need to optimize its return from foraging effort with the risks involved in foraging and conflicting homeostatic needs.

Foraging involves dietary choice, and this is the first topic of discussion in this chapter. Diet depends on what can be consumed, on the nutritional needs of the forager, and ultimately, on specialized adaptations for collecting and digesting specific types of food. Obtaining energy is usually the primary goal of foraging, but essential nutrients must also be consumed, and animals sometimes make surprising dietary selections to gain needed nutrients. Selection to avoid competition may drive animals to choose foods that are relatively undigestible, and this may result in behavioral adaptations related to digestion, such as passing digestive symbionts among animals or gardening fungi.

Foraging behavior defines an animal's trophic niche, such as carnivory, within its ecological community. Foraging specializations involve behavioral adaptations that coordinate with morphology and physiology. Big cats, such as lions and tigers, have stealthy predatory behavior and strong pursuit abilities, adaptations that are accompanied by cryptic coloration, sharp claws, slicing teeth, short digestive tracts, and a digestive physiology evolved to handle a diet high in protein and fats. Herbivores are no less well equipped with behavior and physiology appropriate to their diets.

Once dietary choice is established, food must be located. The first step is search, and the second step is handling the food. For herbivores, food handling is often as simple as consumption, but plant material can be difficult to digest and sometimes extra behavioral steps prepare the food for digestion. For carnivores, handling requires first capturing and then subduing and killing the prey. Complex games of "cat and mouse" have evolved as selection favors prey that are hard to locate, catch, and subdue. At the same time, selection favors countermeasures in predators. These evolutionary arms races are central to understanding foraging; over evolutionary time, adaptations that make foragers more efficient are met with counteradaptations of food species.

Food is not usually evenly distributed in the environment; it occurs in patches, and animals must make decisions about when to leave patches as they become depleted. A patch may be a small meadow in which a certain type of plant is found, the space under a rock in a stream, or a flock of birds. Optimal foraging theory (OFT), mathematical models that describe how an animal gains the maximum possible yield from its foraging efforts, yields important tools for understanding discovery and exploitation of food. For a sit-and-wait predator such as a snake or an orb-weaving spider, a patch is where prey are likely to walk or fly. Successful foragers must be able to find patches, exploit food in that patch, and then assess when the patch is depleted and move on to more profitable locations. From digestive physiology to changing dietary needs over a lifetime, and from the work of natural

CASE STUDY: SLOTH, MOTH, AND ALGAE

Not surprisingly, three-toed tree sloths (Figure 9.2) live in trees, and eat leaves. This is not as easy a life as it sounds, because life in the canopy requires low body weight, and life as a foliovore (leaf-eater) has some nutritional challenges, as leaves are not easy to digest. These animals are restricted even compared to their relatives, the two-toed sloths, which have broader diets and larger home ranges. Indeed, no mammal digests its food more slowly than a three-toed sloth, which has—again, not surprisingly—a remarkably low metabolic rate. Nonetheless, a three-toed sloth will predictably descend to the ground about once per week; once there, it creates a depression, defecates, covers the depression, and then returns to the canopy. This is risky business; it is the most dangerous time in a three-toed sloth's life, given the impact of predation, and in addition, it requires a lot of energy. In comparison, two-toed sloths simply defecate from the canopy.

Jonathan Pauli and his colleagues wondered why a three-toed sloth would do such a thing. Several hypotheses had been set forth, including communication and tree-tending. Given the sloths' restricted diets, Pauli and co-workers wondered if nutritional issues might be influencing this behavior.

In order to fully appreciate this hypothesis, one has to recognize that a sloth, no matter how many toes, is a [slowly] walking garden. Sloth fur is its own ecosystem, containing such organisms as algae, fungi, and arthropods. The algae are important supplements to the sloth's nutrient-poor diet, having much more fat than leaves do; when sloths groom their questionable-looking fur, they are also eating algae. The algae thrive in sloth fur, because the fur is structured in such a way that it catches rainwater, which benefits the algae. So the algae and the sloth have a mutually beneficial relationship. This is a *mutualism*, an interspecific association in which both participants benefit (see Chapter 13). This, however, does not explain why the sloth climbs down the tree.

For that, we must turn to the moth component of the ecosystem—moths that, when they die in the sloth's fur, are

FIGURE 9.2

The three-toed sloth carries a small ecosystem in its fur. Algae thrive there because the fur catches and holds moisture; the algae, in turn, supplement the sloth's nutrient-poor, folivorous diet. The moths in the sloth's fur nourish the algae when they die and decompose. In order to have good supply of moths, the sloth cannot defecate from high in the tree, as would be convenient and safe, but must descend to the ground, where its feces provide a nursery for moth larvae. Upon becoming adult moths, those larvae will fly up into the tree, find a sloth to live on, and eventually nourish its algae. *Photo: Dominique Keller.*

decomposed by the fungi there. The nitrogen they contain is then used by the algae. These pyralid moths do not complete their life cycle in the sloth; rather, adults live in the sloth fur, mate, and female moths oviposit in fresh sloth feces when the sloth defecates on the ground. The larvae feed and pupate in the feces, and then the adults fly up into the canopy and find a sloth. In this way, the sloth engages in moth husbandry by defecating on the ground—producing moths and moth offspring that will eventually fertilize the algae, which in turn, provide the sloth with much needed nutrients.[1]

295

selection in predator–prey interactions to the information we glean from fossil teeth, foraging is an area in which all of Tinbergen's questions can be clearly identified, if not always answered.

9.2 DIET CHOICE AND FOOD SELECTION

This section provides a brief overview of animal nutrition. Foraging behavior is the bridge between an animal's physiological needs and the potential foods in the environment that will meet those needs. A basic conclusion from this section is that a diverse diet is more likely to meet an animal's full nutritional needs. Given the advantages of a diverse diet, why do animals specialize? In some cases, natural selection favors dietary specialization because focused adaptations to collect specific food types increase foraging efficiency.

What Is an Adequate Diet?

Foraging behavior must fulfill two goals. The first is to gain enough energy to support growth, development, and reproduction. The second is to gain the right nutrients. The obvious energy-rich foods are carbohydrates and fats. Protein can also be used as an energy source and has the added benefit of being composed of nitrogen-containing amino acids. Thus, the bulk of most animal diets consists of carbohydrates, fat, and protein, and most of this chapter is about behavior that leads an animal to the energy it needs. In addition, for growth and development, an animal's diet must be much broader than just energy-rich sources; essential macro- and micronutrients are also targeted in foraging. For animals with carbohydrate-rich or fat-rich diets—including animals that eat foliage, fruits, seeds, and nuts—special strategies may be needed to gain enough dietary nitrogen. Sources of minerals such as sodium, potassium, and phosphorus are also important. For animals with protein-rich diets—carnivores and animals that eat decaying flesh—nutrients such as vitamin A may be in short supply. Behavioral strategies play important roles for animals as they seek to fulfill their dietary needs.

Essential Nutrients

All animals have specific organic molecules or minerals that must be in their diets for survival. Some such nutrients, particularly minerals such as iron, sodium, magnesium, and copper, are essential for all animal life. Nearly all animals require specific amino acids in their diets, but the list of essential amino acids differs among types of animals. Vitamins, such as β-carotene (vitamin A), are also essential to many animals. Essential dietary components can be surprising; insects need cholesterol in their diet because they lack the ability to synthesize this compound. From a behavioral point of view, essential dietary elements are important because sometimes animals go to great lengths, and perhaps considerable risk, to gain needed nutrients. Elemental nutrient deficiency may not be all bad; for instance, phosphorus-limited *Daphnia* were also poorer hosts for a pathogenic bacterium.[2]

The sodium-limited world of herbivores has been selected for some odd behaviors. Sodium is particularly valuable for herbivores because plant tissues have relatively little sodium, and an entirely plant-based diet is usually poor in sodium. Because of this, human perspiration attracts sweat bees. These small black or green bees lick the sweat, thereby gaining sodium and other minerals. Insects such as butterflies often gather at the edges of puddles where evaporation has concentrated the minerals.[3] This behavior is called "puddling." Salt licks, exposed patches of crystallized salt, attract mammalian herbivores such as cattle and deer. Nutritional constraints are revisited in Section 9.10, where salt again takes a starring role.

Digestion of Cellulose

Cellulose may be the most abundant organic macromolecule on the planet. It is also one of the least digestible. Like starch, cellulose is a string of glucose molecules, but unlike starch, the bonds between the cellulose glucoses are strong; they require considerable energy to create and equal energy to break down. (These are termed *ß bonds*.) Few animals have the enzyme needed to break these strong bonds, so gaining access to the energy tied up in cellulose is challenging. Animals that eat cellulose commonly solve this problem by having microbes in the gut that do have the enzyme. For this strategy to work, young animals must be "infected" with the symbiotic microbes; the obvious route is consumption of feces from infected animals. This is taken to an extreme in termites, in which anus-to-mouth feeding is a key element of their social structure (Figure 9.3). Another solution is employed by leafcutting ants, which collect leaves and feed them to a fungus garden. The

FIGURE 9.3
Termites practice anus-to-mouth feeding as a way to share essential intestinal symbionts.
Photo: USDA/ARS Formosan termite image gallery.

296

fungus digests the cellulose, and the ants then consume the fungus. (See Chapter 14 for a more complete discussion of fungus-gardening ants.)

9.3 HOW ANIMALS GET FOOD

When animals are feeding, the first step is to find food through search behavior, covered in Chapter 8. This section will expand on those fundamental ideas about search and introduce the basic types of feeding strategies. Search strategies vary among species, but within a species, a given strategy may be invariable. These species-specific search behaviors may be genetically determined or influenced, but individuals can frequently become better foragers by incorporating experience into their species-specific strategy. Also, animals often take advantage of whatever foraging opportunities present themselves, so sometimes an animal takes unusual (for its species) prey or eats an unusual plant. The world of eating is so broad that feeding strategies defy easy categorization. Indeed, Manuel Molles suggested a broader term in place of typical categories of food acquisition (e.g., herbivory, predation);[4] that term—*exploitation*—refers to the habit of an organism living at another organism's expense.

> **KEY TERM** Exploitation refers to an animal living at another organism's expense. Almost all animals are exploiters in this sense.

Trophic Strategies and Styles of Foraging

An examination of diet first leads to a discussion of the ways in which animals capture energy from other organisms. The diversity of feeding styles among animals is almost intimidating. Foraging animals can be classified by their trophic roles:

- *Herbivores* (animals that eat plants)
- *Carnivores* (animals that eat other animals)
- *Saprophages* (animals that scavenge or eat nonliving organic matter)

These categories do not describe the full range because each of these can be broken down further if desired: *frugivores* eat fruit, *insectivores* eat insects, and so on. Another way of looking at this diversity places the focus on the effect that foraging has on the prey: there are predators—animals that kill and eat their prey—and there are parasites—animals that eat tissues but usually do not kill their hosts. Sometimes animals that eat seeds are called *seed predators*, expanding the usual meaning of the word. Often, though, seeds are not destroyed when consumed; they are dispersed when the animal defecates.[5] *Hematophagous animals*, such as mosquitoes, are an interesting variant; these species drink the blood of other animals. (Hematophagous animals are discussed in more detail later in this chapter.) If the unwilling (or unwitting) blood donor is large enough, these organisms are simply parasitic, but small animals may be exsanguinated and thus killed.

> **KEY TERM** A hematophagous animal consumes blood.

Mobility is a key issue in trophic behavior. Thus, many filter-feeders sit in one place to feed. They strain organic particles and small prey out of the water. They are very good at competing for space, and almost un-animal-like in their disinclination to move. Predators that chase prey are quite the opposite. They frequently attack very large prey, especially if they hunt cooperatively, and they run like the deer, which they must do, if they want to catch one. Predators that ambush prey are usually very good at hiding, and so on (Figure 9.4).

Styles of Herbivory

Browsers glean leaves, bark, and green stems from plants, whereas grazers clip vegetation at or near ground level. Deer, such as the white-tailed deer in the Rocky Mountains, are browsers, a distinct advantage when grasses and other ground-level vegetation are covered

---

FIGURE 9.4
The ways that animals forage are diverse beyond imagining, as these examples illustrate. (A) Nectar-feeding, (B) tissue parasitism, (C) active hunting, (D) carrion feeding, (E and F) sit-and-wait prey capture. *Photo credits: (A) Michael Breed, (B) Jeff Mitton, (C) Ken Keefover-Ring, (D, E) Ben Pless, (F) Thomas Barnes, US Fish and Wildlife Service.*

by deep, wet snow. The disadvantage of browsing is that height may make vegetation inaccessible, and oftentimes browsers must glean parts of the vegetation that are low in nutrients, chemically defended, or both.

Grazers, such as sheep and cattle, can feed on the more nutrient-rich meristematic regions of the grasses when those grasses are available. The differentiation between browsers and

FIGURE 9.5

These species represent a contrast in primate feeding styles. Howler monkeys (A) are thought to eat primarily leaves, a relatively poor diet, while white-faced monkeys (B) eat fruits, a richer but more difficult to discover food source. *Photo: Paul Bolstad, University of Minnesota, Bugwood.org.*

grazers is one way in which ungulates (hoofed mammals) can partition food resources.[6] Resource partitioning, in response to competition for food, ultimately allows species to exploit different diets and to co-exist without competing.

This type of differentiation also occurs in the behavioral and digestive adaptations of monkeys in Central America. The howler monkey mostly eats leaves (folivory) but will take fruits if available. Leaves are nutritionally difficult because most of the sugars are tied up in cellulose, which has already been discussed as a problematic nutrient. Folivores tend to be slower and less active; they must also have a longer intestinal tract to maximize extraction of energy from their food. White-faced monkeys live in many of the same habitats but prefer fruits (frugivory), which are rich in more easily digested carbohydrates. Frugivores are more active and have shorter intestinal tracts (Figure 9.5).

299

FIGURE 9.6

The spines on this locust tree are examples of the defenses that plants employ to resist the attentions of herbivores. *Photo: Michael Breed.*

Overcoming Plant Defenses

Plants have an impressive array of defenses to deter herbivores. Plant defenses include tough, thick, indigestible leaves, along with spines, thorns, stinging hairs (Figure 9.6), and poisons (secondary compounds) produced in the plant tissues. Not surprisingly, herbivores evolve countermeasures to these defenses. Tough leaves may be countered by stronger jaws, muscular tongues, and large grinding molars. Spines and thorns can be overcome by thick skin. Chemical defenses may lead to specialized mechanisms for detoxification or to sequestration (storage) of the toxins in specialized body compartments. Sequestered plant toxins can actually become defenses for the animals that consume them, such as the cardiac glycosides produced by milkweeds and sequestered by monarch butterflies. Different types of foragers (e.g., carnivores, grazers) may have different tolerances for bitter (potentially toxic) compounds.[7]

Behaviorally, one of the primary methods of overcoming plant defenses is dietary selectivity. This involves finding the species of plant that an animal is adapted to eat; again, the example of milkweeds and monarchs comes to mind. Adult monarchs seek out milkweed plants when they lay their eggs; this means the developing caterpillars will feed on plants that fit their metabolic adaptations. Selectivity may also involve choosing susceptible plants from the population of possible targets; some individual plants may be less well defended than others of the same species. Pine bark beetles weaken host plants by attacking *en masse*; a large number of beetles burrowing into a single tree overwhelms the tree's ability to defend itself.

Some herbivores manipulate plants in a way that interferes with the delivery of defensive chemicals. Thus, the way a squash beetle "trenches" the leaves it eats does not trigger the host plant's defenses. This behavior allows leaves to remain palatable.[8,9]

Styles of Carnivory

Because carnivory involves consuming another animal that is highly motivated to avoid being consumed, and will therefore hide, run, fight and even play dead (see Chapter 10), the task of being a carnivore can be a challenge, and carnivores as a group respond to this challenge in many ways. These strategies involve behaviors ranging from active pursuit to patient waiting and pouncing. Some carnivores are highly specialized in their hunting techniques, and these adaptations can influence other areas of their lives. Herons, for instance, are stealthy, sit-and-wait predators, standing motionless in the water for long periods, waiting for prey to swim by. This is probably why they are much less defensive against biting flies than other birds with more active habits.[10]

OF SPECIAL INTEREST: A CHEETAH'S PREY CHOICE

How do animals choose what to eat? For most animals, and for predators like cheetahs in particular, foraging is not just one decision, but a series of discrete decisions; in this case study, those decisions involve whether to hunt and what to hunt. Discrete choice models, borrowed from economists, give us a way to ask how animals make these decisions. Cooper et al. did this with cheetah foraging in the Serengeti.[11] This technique calculates the probability of the choice to hunt, and given that choice, the probability of choosing a particular prey species in the context of prey on offer, environmental conditions, and predator conditions. Cheetahs hunt diurnally (during the day) and in the open, so observational data can be collected with some accuracy. Cooper and coworkers tested the following hypotheses about conditions that should positively or negatively influence the decision to hunt:

1. The presence of preferred prey (Thomson's gazelle in particular)—a positive influence
2. The presence of competitors (e.g., lions and hyenas)—a negative influence
3. Energy needs as reflected in the presence of cubs or hunger level (indicated by belly size)—a positive influence
4. Cover availability—a positive influence
5. Time of day (preferred times are 700–800 h and 1600–1800 h)

Additional predictions emerged from the facts that preferred prey are seasonal and male cheetahs hunt larger species than females do.

Cooper and colleagues found broad agreement with their hypotheses, but there were some interesting nuances. For instance, cheetahs were reluctant to hunt in the presence of lions, but hyenas had no effect. Thomson's gazelles motivated hunting, but other prey such as wildebeest and Grant's gazelles did not increase the probability of hunting. The presence of cubs increased the probability of hunting, but hunger level did not. Moreover, although hyenas did not affect the decision to hunt, they did affect decisions about what prey to take, with Grant's gazelles increasing in preference. These and other results show us both the complexity of decision making by predators and the extent to which models can help us appreciate those decisions.

BRINGING ANIMAL BEHAVIOR HOME: PLAYING WITH *UMWELT* AND FORAGING

Most of us with canine companions enjoy doing "fun" things with the furry set…but sometimes those activities are geared more to human enjoyment than anything a dog can naturally revel in. This brings us to K9 Nose Work®, a new trend in human–dog interactions that is built around the canine *umwelt*. In K9 Nose Work®, the dog first learns to sniff out hidden treats. Then, through associative learning (see Chapter 5), the dog begins to associate a specific odor with the treats. Along the way, the hiding places become less predictable, and another odor is added to that on offer. The dog works in an unrestrained fashion (off-leash in a safe environment), and the searches are structured so that the dog is challenged, but not frustrated. In other words, everything is geared to the dog's strengths and its success, and not to the sometimes quirky tasks that we humans set for our dogs. At the least, almost all dogs (even very old or injured ones) are engaged in a task that is built for their own sensory systems; in addition, it is quite possible that such activities may enhance canine confidence and the human–animal bond. Whatever the benefit, the foundation of this activity is recognition of the dog's sensory-perceptual world (Figure 9.7).

FIGURE 9.7
In K9 Nose Work®, dogs use associative learning and their powerful sense of smell to first find treats, and then sources of specific odors. (A) This Shetland sheepdog is locating a source of odor within a box, ignoring other boxes without odor. (B) Aldo the pit bull progressed to searching for odor outdoors, in a variety of locations. *Photo: (A) Lisa Kretner, Tail Wagging Photos, (B) Karen Connell, Karen Connell Photography.*

FIGURE 9.8
Adult tiger beetles (Family: Cicindelidae) visually track their insect prey, making short runs which are punctuated by stops that allow them to recalculate prey position and accommodate for prey movement. They quickly zero in on prey location using this process. Tiger beetle larvae (called *doodlebugs*) are also predators, but they use a sit-and-wait strategy, living within burrows in the ground. The larva blocks the burrow with the top of its head and the back of its thorax. When a prey animal comes close, it flips its head backward, exposing powerful jaws that grab and subdue the prey. As with the adult, visual tracking of the prey is an important component of being a successful predator. *Photo: Jeff Mitton.*

301

Active Hunters

Catching prey is not a simple matter, as potential prey species are under constant pressure to evolve mechanisms to avoid becoming prey items (Figures 9.8–9.10). In order to survive, predators must evolve in response to adaptations of their prey. One consequence of this co-evolutionary race is that predators are often endowed with excellent sensory capabilities.

Some carnivores eat, if they can, other carnivores in their home range. When a carnivore consumes another carnivore, it both obtains food and eliminates a competitor. One important implication of these observations is that links in food webs may often include carnivore–carnivore consumption, as well as consumption of herbivores by carnivores. For instance, carnivorous house pets, domestic dogs and cats, often become victims of wild carnivores, such as coyotes and mountain lions, in suburban areas; these pets are particularly vulnerable because they have lost much of their self-defense ability in the domestication process.

FIGURE 9.9

Tigers (*Panthera tigris*) are an interesting example of active predators because they quite willingly include humans in their menu. They seem to prefer medium to large prey; a study in India showed that half of the diet of tigers in a forest reserve was composed of chital (a spotted Indian deer species), sambar (another common species of deer in India), and wild pigs. Excellent camouflage and stealthy movements combine with powerful jaws and claws to make this a highly feared predator. As humans encroach on tiger habitats, the threat of tigers to human life creates a particularly difficult conservation challenge. Conservation plans that include multiple forest uses—usually low-level logging or harvesting of other forest products by humans—may be thwarted by tiger predation on human users of the forest. This creates pressure on conservation officials to remove the tigers from the forest. Modeling tiger/human movement patterns and probabilities of encounters may provide a tool for minimizing these problems. This tiger is maintained in an animal rescue facility. *Photo: Michael Breed.*

302

Cheetahs in Africa are the target of predation by lions and hyenas; this contributes to the precarious status of cheetah populations. Finally, wolves were absent from the Yellowstone ecosystem for over half a century. During their absence, coyotes expanded their dietary range to include larger animals, such as deer, which formerly were preyed upon by wolves. The reintroduction of wolves into Yellowstone in the 1990s brought these two carnivore species back together. While the two species co-exist in many habitats, wolves prey upon coyotes, given the opportunity. After the wolf reintroduction, coyote diet preferences shifted back to smaller prey, probably as a result of competitive pressure from the wolves.

FIGURE 9.10

Not all spiders build webs for catching their prey. Hunting spiders use a variety of strategies, including active search and ambush, to obtain their food. This spider, a common resident of neotropical wet forests, hunts at night, seeking prey in the rich environment of the forest leaf litter. *Photo: Jeff Mitton.*

OF SPECIAL INTEREST: SPEED AND MANEUVERABILITY

Speed is another adaptation for prey capture, and cheetahs excel at this, hitting speeds of 25.9 m/s. Studies of hundreds of chases performed on five cheetahs with collars that recorded location (GPS) as well as motion (e.g., acceleration) showed that the mean sprint speed of a cheetah is considerably less than the maximum, averaging almost 15 m/s. During the final approach to prey, a cheetah will dodge and weave, maneuvers that do not work well at top speed. Cheetahs can brake very quickly immediately prior to such movements, reducing turning radius from over 19 m to a little more than 1 m and in addition, their rough footpads and strong claws give them considerable traction.[12]

BRINGING ANIMAL BEHAVIOR HOME: LIVING WITH PREDATORS

Many human companion animals are predators; this obviously includes dogs and cats, but extends to other commonly kept animals such as ferrets, frogs, snakes, and some species of fish. Some companion animals that are largely herbivorous, such as iguanas, eat meat when presented with the opportunity to do so. In isolated instances, people maintain larger mammalian predators such as wolves and lions in their households. Animal shows, zoos, and aquaria manage large predators like orca (killer) whales,[13] tigers, and lions and incorporate interactions between trainers and the predators into their shows. What are the risks to humans from co-existence with these carnivores?

One unfortunate possibility is for a person to be bitten by someone's companion animal. Bites may result from territoriality, defensive behavior, anxiety, or predatory instinct. For the human victim of a bite, the animal's motivation matters less than the injury, but understanding the circumstances that lead to bites can prevent injuries. In a sense, *umwelt* is as meaningful in this context as in the rest of nature; our perception of the events surrounding our attempt to pet a dog, for instance, may not match up with those of the dog.

Thousands of years of artificial selection have moderated carnivorous instincts in dogs and cats to present acceptable behavioral traits: dogs and cats are domesticated. Anxious, fearful, and defensive dogs often can be recognized by their posture—teeth bared, hackles raised, tail lowered, ears back. But sometimes a dog's reactions are so quick that they cannot be predicted, and all dogs do not express fear or defensiveness in the same ways. For this reason, never greet a dog, particularly a dog that is not familiar with you, with your face. Too many people rapidly move their face toward a dog's muzzle; if the dog is fearful or interprets this movement as a threat, the resulting bite to the human can be tragic, leading to plastic surgery and permanent facial scars. The dog often suffers a tragedy as well, either being euthanized as a "vicious animal" or being condemned to spend the rest of its life in a cage. It is best to let an unfamiliar dog take the lead in greeting and to let the dog sniff you where it wants.

Cats are less worrisome in this regard, perhaps because of their smaller size. That said, it is still much better to allow a frightened cat to run away and hide than to attempt to hold it. When biting and clawing happen, it is usually because an anxious cat is being restrained against its will. As a group, felids are committed carnivores, and many of their physical and behavioral traits, ranging from their remarkable tactile sense to their legendary spatial ability, are the products of natural selection acting on a stealthy predator.[14]

Ferrets, frogs, and small non-poisonous snakes are too outsized by humans to pose threats beyond bites that potentially could become infected, and small carnivorous fish are also benign except for the occasional piranha bite. Iguanas have strong enough jaws to seriously injure a person, including completely severing a finger. Common sense in handling can help to avoid bites from these animals. With small- to medium-sized snakes and lizards, making sure one's hands do not carry a meat scent helps. In other words, do not eat a ham sandwich and then put your hand into a snake's or lizard's living quarters.

Large constrictors, such as boas, pythons, and anacondas are dangerous predators, and large captive snakes have caused human deaths from crushing and suffocation. Many cities forbid keeping snakes above a certain size because of the hazards. Crocodiles and alligators also outgrow their interest as pets and become dangerous housemates. Unauthorized releases of these animals when they outgrow their homes extends the danger to the general public. It is hard to imagine the circumstances under which bringing a young, wild predator into the household would be a good idea; the outcome is usually bad for animal and human alike.

Sit-and-Wait Predation

Omnivory

Omnivory, the consumption of both animal and plant tissues, is an excellent strategy for obtaining the correct mix of nutrients, but it is a difficult lifestyle in terms of morphological and behavioral adaptations. An omnivore is often a "jack of all trades, but master of none." Some primates, including humans, successfully occupy omnivorous niches, largely because they can use their cooperative social skills in predation (see following sections and Chapter 13; Figure 9.12). Some ants employ another successful omnivorous strategy, in which they prey on insects and other small-bodied animals, but also collect nectar or *honeydew*, which are carbohydrate-rich foods. When compared with trophic styles that allow more specialization, omnivory is relatively rare among animals.

Mimicry and Luring

Mimicry will be discussed in more detail in Chapter 10, which focuses on antipredator behavior, because many animals use a variety of mimetic devices to help them avoid discovery by

FIGURE 9.12
Chimpanzees are omnivorous; these are sharing a meal of meat. They cooperate to hunt larger prey such as bush pigs and monkeys. Jane Goodall was one of the first primatologists to observe such a hunt, which ended in the death of a colobus monkey. *Photo: John Mitani.*

predators. In a fascinating symmetry, predators, especially predators that engage in ambush techniques, often employ mimicry as a way to disguise the threat they present. While many ambush predators simply hide and then pounce, some actually look like completely different objects, or use lures to decrease the distance between themselves and their prey.

Male burrowing owls line their burrows with shredded mammalian feces, which they also distribute around the opening of the burrow (Figure 9.13). This habit has inspired numerous hypotheses and investigations about why they might do this. Smith and Conway[20] tested four of these hypotheses:

1. Were males trying to attract mates?
2. Were males trying to signal that the burrow was occupied?
3. Were males trying to camouflage the burrow, protecting it from predators?
4. Were the males trying to attract prey arthropods?

FIGURE 9.13
Burrowing owls attract arthropods to their burrows by lining the burrows with feces.
Photo: Ben Pless.

The first three hypotheses were not supported. The manure scattering behavior often began after pair formation and almost a month after the burrow was occupied. Predation did not differ between nests that were festooned with manure and those that were not. However, nests with manure attracted 69% more arthropods than those without manure, and included many arthropods reported from burrowing owl diets.

Meanwhile, 1600–2300 m below the surface of the sea, a siphonophore (a colonial hydrozoan) may use bioluminescence to attract prey.[21] The behavior and sensory physiology of deep sea organisms is difficult to investigate, but the fact that this siphonophore eats fish (most siphonophores eat crustaceans) is probably relevant. The bioluminescence is emitted from side branches of tentacles.

As noted earlier, predators that sit and wait for prey are usually pretty good at hiding. In fact, in the case of orb-weaving spiny spiders (*Gasteracantha fornicate*), bright colors actually attract prey to the vicinity of the web! Hauber investigated the role of color when he dyed some of the spiders black, covering their conspicuous yellow and black dorsum.[22] Spiders that were dyed black caught fewer prey.

Cooperative Hunting

When two or more animals forage together, they are foraging socially. Social behavior is covered in more detail in Chapters 13 and 14, but it has special relevance to foraging; some behavioral biologists see foraging as a major selective force favoring social groups, whether they are loosely knit colonial nesters that share (or parasitize) information about food sources, predators that can take prey far larger than themselves by cooperative effort, or tightly regimented eusocial insects.[23]

Animals may cooperate to trap elusive prey, or to take down larger prey. While social bonds derived from kinship or mating may enhance cooperation, unrelated animals can also benefit from cooperation. An extreme example of unrelated animals cooperating to overwhelm a food item is the case of pine bark beetles, which alone cannot overcome a tree's defenses but can do so in a mass attack. More familiar examples of cooperative hunting involve felids and canids (wild species of cats and dogs, respectively). Once the prey item is secured, competition for food among the hunters can be severe and may be regulated by dominance relationships in the group.

One of the more unusual instances of cooperative hunting occurs off the coast of Brasil, in Laguna. A group of bottlenose dolphins cooperates with local fisherman who cast nets from

305

the beach. The dolphins herd schools of mullet toward the beach where, using stereotyped head- or tail-slaps, they show the fishermen where to cast the nets. This results in more food for both dolphins and humans. Oddly, only one closely-knit group participates in this behavior; two other groups of dolphins in the area do not. It may be that the close social ties of the cooperative group facilitate learning about this complex form of foraging.[24]

Uninvited Guests During Foraging

Uninvited guests can be members of the same or different species. When one animal observes another eating, it is attracted and attempts to share the food. This behavior can lead to outright competition for food and to the loss of the food item by its original discoverer. For the discoverer, the problem is that adequate defense of the food takes time away from eating, even to the point of not eating at all. The animal must choose a balance between defense and consumption. For the guest, the advantages are numerous; it may get to eat without searching and may avoid the risks involved in capturing and killing prey.

9.4 WILLING FOOD
Predator Saturation

Sometimes animals are victims of their parents' reproductive strategies. From the point of view of the parent, it may be a better strategy to produce many defenseless offspring, or to fail to defend them, than to produce a few better-defended offspring. Selection may lead a species to produce so many seeds or offspring that consumers cannot possibly eat them all. This is reinforced by making the vulnerable stage available for short periods of time, separated by long periods of unavailability. This strategy is called *predator saturation*, because during the bursts of reproduction, predators cannot possibly eat all the offspring. The strategy might equally be termed *predator starvation*, because specialized predators may go long periods of time with nothing to eat.

> **KEY TERM** Predator saturation (also called *predator swamping*) is a reproductive strategy in which synchronized reproduction produces so many offspring that predators cannot consume all of them.

306

Examples of predator saturation include both plant and animal prey. For instance, fruit- and seed-producing trees such as bamboos in southeast Asia "mast," producing large crops of fruits and/or seeds at several-year intervals. The adult stage of many mayfly species is present for only a few days in the spring; a very large number of individuals emerge. The mayflies are tasty to a variety of birds, but the birds cannot eat them all, leaving enough to reproduce and lay eggs (Figure 9.14). Likewise, cicada species are hidden underground through most of their life cycle. Species that emerge as adults once every few years (a 7-year cycle is common) saturate their potential predators.

Saprophagy

Animals that feed on dead or decaying material do not have to overcome their prey's resistance, but likely encounter substantial competition for the food. Saprophages feed on carrion and on more decayed plant and animal material. Competition among animals such as crows, vultures, eagles, coyotes, and wolves (in North America), or vultures, hyenas, wild dogs, and jackals (in Africa) is obvious and is usually resolved with aggressive interactions around the carcass. Carrion flies, burying beetles, and ants enter the competition; they are less obvious but equally important. Finally, microbes play vital but often underappreciated roles in the competitive landscape, secreting chemicals that make dead material poisonous, noxious, or unpalatable. Microbes wage chemical warfare among themselves and with animals that seek to consume dead materials.

FIGURE 9.14
Mass emergence of cicadas. Many insects emerge simultaneously from their larval or pupal states, thus swamping predators. *Photo: John Cooley.*

The behavioral result for saprophagous animals is *scramble competition*; often the first animal to find carrion "wins" or at least gains a disproportionate share of the carrion before being displaced by larger or more aggressive competitors. This puts a premium on the discovery of carrion, hence the classic image of vultures circling a dying animal.

> **KEY TERM** In scramble competition, the animal that reaches the resource is the successful competitor; it is an ecological version of "first come, first served."

9.5 MANIPULATION OF PREY

After an animal's discovery of food, there are often behavioral steps before food can be actually consumed. These actions are termed *handling*. This discussion focuses on carnivores handling prey, but the same principles apply to bees handling flowers or leafcutter ants handling the plants they collect. The important aspects of handling are rapid physical control of the prey, avoidance of injury, and efficient use of time.

Two examples illustrate how prey can be rapidly controlled (Figure 9.15). Mountain lions (also known as puma) are common in western North America and, while they can take a variety of prey, they specialize on white-tailed deer. The lions are well armed, with large canine teeth, powerful claws, and strong body musculature. The deer have surprisingly effective countermeasures; given the opportunity, their hooves are sharp and their antlers can be intimidating defenses. Mountain lions quickly subdue deer by severing the cervical spinal cord with their canines; the teeth have sensitive nerves that allow the lions to find the lethal spot.

> **KEY TERM** Handling refers to all of the activities involved in converting a resource into usable nutrition; these activities can be everything from subduing prey to digesting food.

307

FIGURE 9.15
These examples hint at the wide range of food-handling challenges that confront animals, whether they eat plants or other animals. Upper left, gull cracking a crab; lower left, bird preying on an egg; right, squirrel harvesting seeds. *Photos: Ben Pless (gull) and Jeff Mitton.*

Pit vipers, such as rattlesnakes, copperheads, and cottonmouth water moccasins, inject venom deep into their victim's tissues with long fangs through which the venom flows. The interval between the snakebite and the immobilization of the prey can range from a few seconds to a few minutes; during this time the snake could be bitten or stomped by the prey. To avoid this, the snake recoils and waits. The challenge for the snake, then, is to find the wounded prey, which may have traveled some distance. Heat and odors coming from the prey assist in this search.[25] In this case the snake sacrifices some time and even risks losing the prey to be able to protect itself from injury. This illustrates the often-underestimated threat of prey defensive behavior (see Chapter 10). Boa constrictors, on the other hand, are sit-and-wait predators that have no venom. Instead, they wrap themselves around prey and apply increasing, suffocating pressure. If they release the prey too soon, they risk being injured by the recovering animal. They avoid this by responding to the prey's heartbeat and releasing the compression only when the heartbeat ceases.[26]

9.6 PARASITIC LIFE CYCLES

Much like herbivores and carnivores, parasites exploit other organisms (hosts) to gain nutrients. Many parasites have complex life cycles—that is, they involve at least two hosts: a final host, in which the parasite reaches adulthood and reproduces, and an intermediate host, in which the immature parasite develops. In such life cycles, the intermediate host must often be eaten by the final host for the parasite to complete the life cycle. Consider intestinal worms with complex life cycles: their eggs pass out with final host feces. When the egg is consumed by an intermediate host, it hatches and the immature parasite invades that host and develops to an infective stage. When the intermediate host, along with the parasite that it contains, is eaten by the final host, the parasite takes up residence in the final host intestine and the life cycle continues.[27]

Animals, including animals that might become intermediate hosts, have been under strong selection to avoid being eaten, and this caution on the part of the intermediate host would seem to interfere with a parasite's goal of getting into the intestine of the final host. A large number of parasites have solved this problem by altering the behavior of their intermediate hosts. For instance, in the case of acanthocephalans ("thorny-headed worms" that use arthropod intermediate hosts; Figure 9.16), every acanthocephalan investigated has been shown to modify the behavior of its arthropod host. These parasites modify the host's responses to environmental stimuli, making the host more vulnerable to predation, and increasing the likelihood that the parasite will get transmitted to the next host in the life cycle. Other parasites (e.g., nematodes, tapeworms, trematodes, and a variety of protists) have similar effects. In some cases, there are obvious physical reasons for the modification; it makes sense that sheep with cestodes in their brains might stagger around and fall prey to wild canids (the final host), and it makes sense that cestodes in the lungs of moose might make the moose winded during a chase. But many parasites do not produce such obvious morphological damage, yet they profoundly alter their hosts' responses to environmental cues. In so doing, they change habitat choice in ways that can be critical to predator–prey encounters. For instance, terrestrial isopods that are host to an acanthocephalan of songbirds spend more time on light surfaces and in areas of lower humidity than do uninfected isopods. They ignore overhanging shelter and tend to be more active. All of this may contribute to the fact that they suffer disproportionally high levels of predation (compared to uninfected isopods) in the field and in laboratory tests. Amphipods infected with an acanthocephalan of dabbling ducks have an equally hard

FIGURE 9.16
This is a typical acanthocephalan life cycle, with a vertebrate final host and an arthropod intermediate host. In the life cycle illustrated here, the juvenile acanthocephalan lives in a terrestrial isopod and alters its behavior in ways that make it more likely to encounter the avian final host.

308

time: they move into the lighted zone of the water column. When the water is disturbed, they swim to the surface (uninfected ones dive), and skim around in circles, attaching to floating debris and even to ducks (Figure 9.17). Alterations such as these can have wide-ranging effects that go well beyond parasite transmission itself. For instance, altered habitat preference can affect mating possibilities, and altered dietary preference can influence food webs.

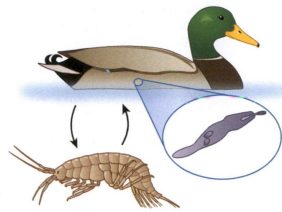

Of course, not all parasites have complex life cycles involving more than one host. Some complete their entire life cycle using only one host (a simple life cycle), but these, too, can affect host behavior and, as a result, affect their own transmission. Fungal infections of cicadas are lethal, but initially, the fungus restricts its growth to the abdominal area of the host; thorax muscles remain strong and functional, and the cicada is able to fly around, spreading fungal spores. Many insects infected with fungi or with viruses crawl up to high places, where they may serve the source of a rain of infectious particles. Paradoxically, high places also tend to be warm places, so this could be a form of anti-parasitic therapy (see Chapter 10). For an even more dramatic example, turn to the arthropod hosts of juvenile horsehair worms (Phylum Nematomorpha). When the horsehair worms mature, they must leave the host in order to mate in the water or in damp soil. This could seem like a challenge for those that live in terrestrial arthropods such as crickets, a challenge that is neatly overcome when the worm causes the cricket to jump into nearby water, be it a swimming pool, a dog water dish, or a toilet. The worm then emerges.[28,29] Horsehair worms resemble horsehairs to some extent and, not surprisingly, can sometimes be found in horse troughs.

FIGURE 9.17
An acanthocephalan of ducks, showing the spine-covered proboscis from which the worm gets its name (*acantho* = thorny, *cephala* = head). The life cycle shown here also involves behavioral alteration of the invertebrate host. To date, all acanthocephalans seem to be able to induce behavioral change in their arthropod hosts.

309

OF SPECIAL INTEREST: BLOOD SUCKERS

Hematophagous (blood-sucking) insects are prime candidates to transmit blood-borne parasites such as malaria and sleeping sickness (Figure 9.18). They are living vectors of disease. (A vector, in biological terms, is something that transmits disease.) Biting flies (mosquitoes, horseflies, blackflies, and the like) are particularly well known for this, and they are also good candidates for manipulation by their protistan parasites. In general, they acquire the parasites when they ingest them while imbibing blood from a vertebrate host. The parasites alter their insect host's behavior in a variety of ways, but the end result of most of their machinations is a

FIGURE 9.18
A blood-sucking arthropod. *Photo: Jeff Mitton.*

reduction in the host's ability to feed. This may seem counterintuitive; after all, wouldn't the parasite "want" the host to feed and thus be transmitted to the next host? That is correct, but when it interferes with feeding, the parasite does not inhibit it altogether. Instead, it simply causes feeding to be difficult, resulting in increased probing by the fly, and in the best case (for the parasite), probing multiple hosts in an attempt to get an adequate blood meal.

How can this happen? Most blood-feeding animals have a concoction of salivary chemicals that would make a pharmacist jealous. Blood-feeding is a dangerous way to make a living; for instance, host defense is thought to be the primary source of mortality in adult female mosquitoes. (Think of all the mosquitoes you have tried to slap and squash.) To the extent that the blood-feeder can probe, feed, and leave—preferably without discovery—it can live to reproduce and feed another day. Thus, the chemicals of the salivary glands contain anesthetics, blood thinners, and all manner of compounds that facilitate the feeding process. As a result, there are abundant opportunities for parasites to alter blood-feeding, and there are many examples of vector-borne parasites doing so in a variety of ways. For example, plague can interfere with flea feeding by blocking the anterior intestine some protists transmitted by tsetse flies interfere with the receptors that tell the fly about how much blood it has consumed.

One of the most sophisticated forms of interference is seen in *Plasmodium gallinaceum*, a malarial parasite. Most mosquitoes have an enzyme, an apyrase, that inhibits platelet aggregation and thus inhibits blood clotting. As you might imagine, this makes for much easier blood acquisition for the mosquito; if blood clots, the general outcome must be very much like trying to suck an old-fashioned malted milkshake, made with real ice cream, through a straw—a lot of work, as the delicious ice cream "clots" slow the progress of milkshake up the straw.
P. gallinaceum interferes with the production of this enzyme by damaging the region of the salivary glands that produces it. The rest of the salivary gland is left untouched, all the better to transmit the infective malarial propagules. But the enzyme that would make feeding fast and efficient—the apyrase—is not present in the infected mosquito's saliva. As a result, blood does not pool under the skin where the mosquito is biting, and feeding is difficult and time consuming.[30]

Given the staggering importance of hematophagous animals in the spread of some of the most devastating diseases we know (malaria, dengue fever, West Nile fever, several versions of encephalitis, sleeping sickness, and more), and given the fact that parasites in general are known to alter animal behavior, it is surprising that we know so little about how parasites of blood-feeding animals alter behaviors other than blood-feeding itself. There is some record of changed activity levels, but beyond that, our knowledge is thin. Recently, *Plasmodium falciparum*, the most virulent malaria in humans, was shown to alter its mosquito host's response to nectar sources, an important source of carbohydrates for the mosquito. Mosquitoes harboring developing, non-infective stages of this parasite were more attracted to nectar sources than uninfected mosquitoes were. The hypothesized fitness benefit for both parasite and host is that this removes the mosquito from the potentially lethal realm of feeding on the blood of a defensive host until the parasite is ready for transmission; at the same time, it allows the mosquito to consume needed nutrients.[31]

Are all of these behavioral alterations (and there are more than can possibly be described here) simply a result of "pathology"? That is unlikely; behaviors that involve hyperactivity or that cause an animal to alter its responses to environmental stimuli are not usually termed *pathology*; the host is not too sick to move or run away. Instead, in some cases, parasites may be actively manipulating the behavior of their hosts by secreting or interfering with neurologically active compounds. The parasite *Toxoplasma gondii* is a good case in point: rats infected with this parasite are attracted to the odor of the felid final host; they do not lose aversion to odors from mink or dog, which are other potential predators of rats, nor are many other behaviors altered. *T. gondii* seems very targeted in its approach to behavioral change. Finally, the distinction between pathology and active manipulation may be less important than it might seem. Pathology itself is subject to natural selection, and the evolution of virulence of a pathogen depends in large part on what that virulence does to the pathogen's fitness.[32] Thus, any pathology that is induced by a parasite may be favored by natural selection on the parasite if it improves parasite transmission. It is not necessarily a side effect of infection, somehow invisible to selection.

This discussion gets into fascinating but dimly lit territory. The evolutionary origins of parasite-induced behavioral alterations are not well understood; neither are the mechanisms that cause them. Some may benefit the parasite; others, the host (see Chapter 10). In the case of parasites with intermediate hosts, some behaviors that benefit the host (e.g., crawling into exposed areas to bask and induce fever) may be difficult to discern from behaviors that benefit the parasite (e.g., crawling into exposed areas and encountering more final hosts). The one clear thing is that an animal with a parasite is unlikely to behave like an animal without that parasite, and that makes parasites a consideration in every aspect of animal behavior. Beyond that, much remains to be learned.[33,34]

9.7 FORAGING AND OPTIMALITY THEORY

How does food distribution affect food discovery? In 1966, two independent papers published in *The American Naturalist* set the stage for optimal foraging theory (OFT). Written by R. H. MacArthur and E. R. Pianka and by J. M. Emlen, they asked how animals made decisions about food preferences and places to seek food.[35,36] In general, the expectation of OFT is that by making adaptive foraging decisions, an animal will seek to maximize its inclusive fitness. (*Optimal* comes from a Latin word meaning "best.") Before continuing this consideration of feeding, however, a longer visit with optimality theory is in order. This theory can be applied to any one of a number of behaviors, but it is covered here because historically it has been most closely associated with foraging.[37]

> **KEY TERM** In OFT, the currencies of time or energy usually are the parameters being optimized.

The application of optimality theory to foraging or any other behavior does not mean that all scientists expect animals to optimize fitness outcomes. There are many *constraints* that can prevent an animal from achieving an optimal state. These are typically the constraints that prevent natural selection from producing perfectly adapted organisms, but they bear special consideration in a discussion of optimality because they limit what can be expected as optimal behavior. For instance, because of heterogeneity in time and space, what is optimal in one location or at one time may not be optimal under different circumstances. The environment may change too rapidly, evolution may proceed too slowly, or genetic variation may be insufficient. There may be conflicting selective forces that exclude optimal adaptations. The animal itself may simply be unable to gather the information that is required to make optimal decisions. In other words, the use of optimality theory to predict what animals should do if natural selection optimized their behavior is not the same as testing for natural selection, nor is it the same as expecting a perfect outcome.[38]

> **KEY TERM** A constraint is a limitation imposed by existing behavioral, morphological, or physiological adaptations. For example, an animal that is unable to learn how to efficiently open a certain kind of seed cannot be a completely optimal forager on those seeds.

311

Instead, optimality theory provides a way to test hypotheses about what is important in an animal's behavioral decisions. This involves identifying the *currency* used to investigate behavior. For instance, if animals are assumed to maximize energy intake (and this assumption is a common approach when investigating foraging), then energy acquisition is also assumed to be crucial to the animal's fitness and its foraging decisions should reflect that.

What if these predictions are not met by the data? If the experiment is well designed and the resulting data are nonetheless inconsistent with predictions about energy acquisition based on optimality theory, then a return to the assumptions on which the predictions were based is in order. How can they be changed or refined to better reflect reality? Of course, this general approach is true of most hypothesis testing, but the added advantage of bringing optimality theory to one's investigation is that it provides a fairly clear road map for generating predictions. There are usually a limited number of ways in which an animal can reach an optimal outcome and an infinite number of ways to achieve suboptimal ones.

Optimality theory—and modeling in general—has benefits beyond those of generating predictions and testing assumptions. It demands clear thinking about the ideas involved in investigating a behavior. For instance, an animal may be expected to feed its young high-energy prey that are easy to digest, but as those prey are depleted, at what point should that animal switch to a different prey? How long should the forager keep looking in one area for a diminishing number of food items? Under what conditions should it begin to look elsewhere? (Please note that for these purposes, prey can be any food item, not only a prey animal.) These and other refinements are core concepts in the optimal foraging literature and introduce more precision into thinking and investigation than a vague notion about yummy caterpillars.

OFT concerns itself primarily with where an animal eats (patch choice) and what it eats when it gets there (prey choice). These appear to be separate considerations, and they frequently are. However, keep in mind that in some cases (e.g., a relatively large prey that takes a long time to eat), *prey item* and *patch* can become the same thing. For practical purposes, prey and patch will be considered separately here. Within these two seemingly simple decisions lies a wealth of discovery about how animals sustain themselves.

OF SPECIAL INTEREST: HOLLING'S DISC EQUATION

Studies of both prey choice and patch choice emerge from a surprisingly simple experiment conducted by C. S. Holling.[39] In this experiment, he asked how predators respond to changes in prey density. After all, such changes are inevitable if the predator is successful at finding and consuming prey.

To investigate this question, Holling tacked 4 cm (diameter) sandpaper discs to a 3 × 3 ft table. A blindfolded assistant (the "predator") was given 1 min to locate the discs by tapping her finger across the table; as each disc was encountered, it was removed and the search continued. The initial density of discs varied from 4 to 256, and the experiment was replicated eight times for each density.

Holling found that as the food supply increases, an animal increases the rate at which it feeds, but this eventually levels off (Figure 9.19). The relationship between the amount of food and the rate at which it is consumed is called a *functional response*. The most common type of functional response—and one that was demonstrated by Holling's "predator"—results in a reduction in the rate of feeding as the discs are collected. Ecologists have investigated perhaps thousands of exploiter–prey relationships, and most fall into this category.

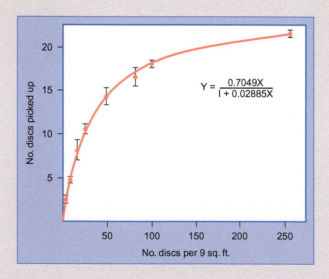

$$Y = \frac{0.7049X}{1 + 0.02885X}$$

FIGURE 9.19
The *functional response* is the relationship between the amount of food and the consumption rate of that food. This curve shows the rate at which Dr. Holling's "predatory" (and blindfolded) assistant located and removed sandpaper discs spread across a table. When more discs are available, the rate at which they are located increases; eventually, this levels off. *Adapted from Holling, C.S., 1959. Some characteristics of simple types of predation and parasitism. Can. Entomol. 91, 385–398,*[39] *p. 386.*

9.8 OPTIMAL PATCH CHOICE

Different types of predators have different challenges facing them as they find prey, but for predators that seek habitat-specific prey in a heterogeneous ("patchy") environment, their first decision must be where to look.[40,41] This decision requires that they know quite a bit about their environment so that they can decide which patch might be most profitable. As a result of this, scientists interested in optimal foraging have investigated how animals sample their environment when given a variety of patches. They have done this in the laboratory under fairly artificial conditions. For example, Smith and Sweatman[42] used trained birds foraging in grids with different densities, but even under those conditions, birds are able to comprehend which patches are better for foraging and when to switch patches. *How* animals discover and process this information in the field is not well understood, but learning about the environment is a big part of the business of being an animal. Animals that can do so have many advantages. They can avoid predators and avoid abiotic extremes by knowing about hiding places, and they can certainly find food more efficiently by knowing where and how much of it there is.

Given the fact that animals have some information about their environments and about the relative foraging value of different patches, an animal might be expected to choose a particularly rich patch and stay there, eating to its heart's content. In many instances, however, this will not work out for very long; as the animal consumes prey, the value of the patch—that is, the number of prey it contains—will diminish. At some point, the patch that was so rich and attractive will become less so. Assuming that an animal knows about the richness and location of surrounding patches, what influences an animal to leave the patch it has been exploiting and enter another one? Eric Charnov[43] proposed an answer to that question in the marginal value theorem, a solution that bears closer examination.

The Marginal Value Theorem

Imagine a berry-picking expedition, complete with a bush that is laden with delicious berries. After a spree of rapid collection (and eating), things begin to slow down—the remaining berries become harder and harder to find. All such berry collectors and eaters have faced this quandary: when to leave and find another bush and when to persist at the current one.

Making this decision requires the same information that other animals probably employ when they forage in patches. Foraging in patches is not that unusual. Many food items, from termites in nests to fruit on trees, are patchily distributed. To decide whether or not to move on to the next patch, an animal needs to know (at minimum) how far away the next patch is. The further away the next patch is, the longer the forager will stay in the current patch. (It also helps to know how rich the next patch is, but for now, assume all patches are equivalent. In Charnov's model, the comparison is the average amount of food in the other patches.)

In essence, an animal should move to the next patch when the net gain from remaining in the current patch decreases so that it equals the net average gain available from the environment. This rule is modified somewhat when the cost of traveling to and exploiting the next patch are considered. The berry harvester should stay at one bush until it can do better elsewhere. To do this, animals must be able to learn the average value of patches in the environment or to be equipped with this information as a result of natural selection. This requirement is usually assumed in foraging studies but has been tested in only a few species, such as bumblebees (Figure 9.20).[44]

This decision point is illustrated in Figure 9.21. The decision about the berries may seem obvious, but the beauty of Charnov's approach is that his graphical model allows for both precision and a large variety of conditions. For instance, if travel time is

FIGURE 9.20
Bumblebee foraging in a patch of flowers. Bumblebees use experience gained over a few hours of foraging to inform decisions about patch choice and time spent in patches. *Photo: Michael Breed.*

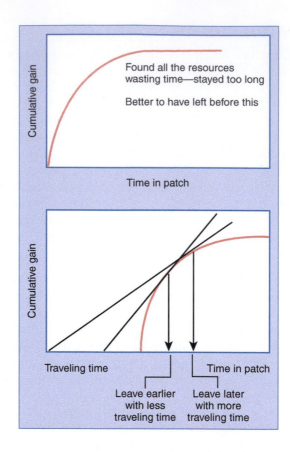

FIGURE 9.21

When to leave a patch? The solid curved line represents the cumulative energy gain. In the top figure, the flat cumulative energy gain means that it is past time to leave the patch. In the bottom figure, the greater the distance to the next patch, the greater the transit time. Drawing a line from the origin so that it is tangential to the gain curve reveals the point in the gain curve at which the animal should shift to the next patch. Note that the greater the transit time, the more time the animal is expected to remain in the initial patch.

FIGURE 9.22

The large-scale foraging movements of bison are influenced by such things as abiotic factors and travel time. *Photo: Phyllis Cooper, US Fish and Wildlife Service.*

decreased, the slope of the tangent is increased and optimal time in the patch is reduced. In the berry-picking example, if the next bush is a few steps away and equally rich, the current bush will not be attractive for very long. If travel time is increased, then a slower energy gain will be tolerated before moving on; the tangent will intersect the energy gain curve after it begins to flatten. Animals tend to linger a bit longer in a patch than theory would predict, perhaps because of the likely variability in travel time to discover the next patch.

A surprising variety of animals have been tested and found to conform to the expectations of the marginal value theorem. Of course, there is no reason to expect animals to have perfect knowledge of their environments. There may be patches that they have not sampled and, as the next chapter reveals, prey are under strong selection not to be found. Nonetheless, the marginal value theorem provides a framework for exploring how animals make decisions about where to forage.

Some environments will present greater challenges in obtaining needed levels of energy. Forage quality and availability may also change with season. American bison search for high-energy patches of sedge in the summer,[45] but dietary preferences change seasonally as food availability and snow cover impact movement and foraging. Across the broader landscape, choice of meadow by bison herds is less influenced by the presence of particular plant species and more influenced by abiotic ecological factors, travel distance among meadows, and probably the presence of other bison herds (Figure 9.22).

OFT can also be used to estimate resource abundance at a landscape level. Assuming the marginal value theorem— that is, that animals leave patches when the yield declines below the average yield of the environment—the route to this calculation is simple. The food abundance in a patch, as measured when animals depart the patch, should represent a slight underestimate of the food availability in the landscape. (The underestimate is caused by the effects of travel time and conservative decisions by foraging animals.) This gives ecologists a tool for evaluating landscape-level resource levels that otherwise might be difficult to measure.

Another approach to optimal foraging is to assess how an animal decides to give up on a patch and leave in search of the next patch. In some cases this is time dependent, so the appropriate measure is giving-up time. In other cases, density, and not time, is the relevant measure; the animal appraises the density of food items in the patch, so giving-up density is measured. In a few cases, animals assess changes in the food composition of the patch, so there is a giving-up composition (Figure 9.23).[47]

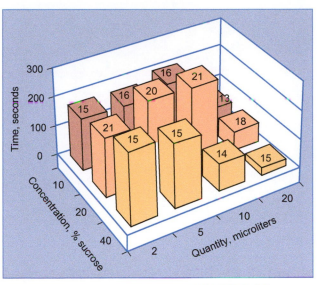

FIGURE 9.23
Giving-up times for ants foraging on sugar solutions. Ants linger the longest when presented with high concentration rewards of intermediate amounts; these factors predict that the patch is worth exploiting. At the highest quantity (20 µL) the ants have collected all they can carry and quickly return to their colony with the food.[46]
Adapted from Breed, M.D., Bowden, R.M., Garry, M.F., Weicker, A., 1996. Search time variation in response to differences in nectar quantity and quality in the giant tropical ant, Paraponera clavata. J. Insect Behav. 9, 659–672.[46]

Central Place Foraging

Central place foraging addresses the fact that many organisms, for all or part of their life cycles, leave a central place to forage and return to that place with provisions. Central places can be as diverse as bee hives, wolf dens, squirrel caches, and bird nests. This topic is an extension of the marginal value theorem, in that the animal must know not only when to leave a patch, but how much to carry back to the central place. Travel distance is the significant variable. The further the animal has to carry the load, the longer it will remain in the patch and the more it will carry.

DISCUSSION POINT: CENTRAL PLACE FORAGING

You can think of central place foraging in terms of your own shopping. It costs more energy and time to shop at distant stores than to shop nearby. If you have a corner grocery, you may go there almost every day for something; keeping an organized grocery list is unnecessary. But if only certain stores carry the dog food that your best friend clearly prefers, and if those stores are further away, you will go to them infrequently and stock up on everything you think you might need for a while. When you do that, you are conforming to the predictions of central place foraging. Can you think of other instances (not related to food) in which the principles of central place foraging might apply? How do you make decisions in those instances?

Risk Sensitivity

What if patches are variable, and what if animals can comprehend that? There is good evidence that both are true. Charnov's marginal value theorem dealt with the average amount of food in patches, but average values can obscure some information. Imagine a choice of two games at an arcade. Either game is included in admission to the arcade, but only one can be played. In one, the prize is guaranteed to be a $10 gift card. (Admittedly, this is sort of a boring game.) In the other game, there is an equal chance of winning a $20 gift card, or absolutely nothing. If each game is played many times, the average prize in either place would be the same: $10. But because only one game is allowed, the player is faced with a decision that causes acute awareness of the variability in these two patches. What is better—to take the $10 and run, or to hope for the best with the highly variable version of the game?

FIGURE 9.24

Dark-eyed juncos are risk-sensitive. Whether they are risk-prone or risk-averse depends on their nutritional state. *Photo: Lee Karney, US Fish and Wildlife Service.*

Researchers have asked animals to make similar decisions. For instance, a tray that always has five seeds in it has an average of five seeds. So does a tray that alternates seed number so that it has 0 seeds for one trial and 10 the next. If a little bird like a junco (Figure 9.24) is exposed to both sets of trays and given an opportunity to learn that the two trays differ, it can distinguish between the two types of trays. An animal that can make this distinction is called *risk-sensitive*. The junco can then be asked to choose between the trays. A junco (or any other animal) that chooses the tray that always has the same amount of food in it is called *risk-averse*. If the junco prefers the variable tray, it is considered *risk-prone*. Note that in this context, risk is related to uncertainty, not necessarily danger.

Caraco et al.[48] showed that juncos can tell the difference between the trays, but they also discovered that the junco's response varied depending on its nutritional state. Hungry, cold juncos were more risk-prone than well-fed juncos. This response would be expected as a survival strategy. If the stable tray offered enough food to get by on, then the juncos chose that tray. When it did not, under adverse conditions, they were willing to gamble. The interaction of risk perception and foraging has been investigated in a wide range of organisms, from humans to bees.[49,50]

The consideration of risk sensitivity leads to a discussion of *utility functions*. This is another term that sounds more complicated than it is. A utility function is simply a graph of a resource's value (utility) to an animal plotted against its abundance (Figure 9.25). A linear utility function, say, a steadily sloping line, indicates that the value of the resource in question (liters of water, number of seeds, units of protein, etc.) does not change with its amount. This is not the most common case. It is more common to see curved utility functions. When the curve is "concave down," the utility of a resource for an animal diminishes as more is

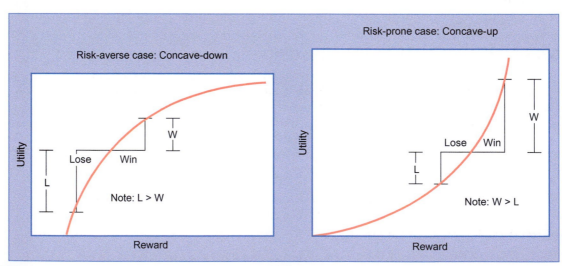

FIGURE 9.25

These two curves plot the value of a resource to an animal (its utility) against units of resource. In the case of concave-down functions (left), the more units of resource the organism consumes, the less valuable those units are. This is a common occurrence when foraging on a batch of morsels. After a certain point, eating another one is not all that beneficial. In the process of such foraging, risking the current morsel (and losing) costs more than gaining the next one. In that case, the lost utility (L) exceeds the later gains (W). Animals faced with this situation should be risk-averse. The curve on the right shows the opposite situation: units of a resource increase in value (again, utility) as they are acquired, a situation that often occurs when some threshold must be attained before the resource can be useful. In this case, an animal is likely to risk losing a unit of resource in order to gain more, because the potential gain (W) outweighs the possible loss (L). Such animals are risk-prone. *Adapted from Stephens, D., Krebs, J., 1986. Foraging Theory. Princeton University Press, Princeton, NJ,[51] p. 131.*

consumed. Ingested food is a good example of this; under most conditions, it is less useful as more is consumed. On the other hand, a "concave-up" utility function indicates that each unit of resource is more valuable than the last. This can happen when an animal needs a certain number of resource units to reach a threshold; as that threshold is approached, each unit is more valuable than the previous one. In the case of the juncos, the well-fed birds that chose the stable tray had a concave-down utility function; each seed was less valuable than the previous seed. On the other hand, the cold, hungry juncos had a concave-up utility function because each additional seed was more valuable.

> **KEY TERM** A utility function is a graph that plots the value of a resource against its abundance.

Discussions of risk and utility functions reveal that foraging decisions can be nuanced. They may change over a lifetime or as a result of shifting environmental conditions. Much like human decision makers, animals may balance long-term and short-term gain.[51]

9.9 OPTIMAL PREY CHOICE

Should animals be less selective about food choice and save time and energy in their search for food? In most cases, that expectation would be wrong. Animals do not eat food in proportion to its availability in the environment. Some foods may be abundant but less nutritious or require more effort to eat and digest, and there is clear evidence that animals have definite preferences in the food that they eat. How can OFT be used to address prey choice?

To begin with, whether an animal forages for seeds or antelope, there are some basic elements in a foraging event. Some of these elements are the amount of time it takes to find the food item, how frequently the item might be encountered, the energy in the food item, and how long it takes to eat the item. The list would be very long, of course, but the goal of using models based on theory is to identify the most influential elements.

To understand such models, begin with the currency—again, using energy (E). Of course, prey do not give up their energy willingly; they require some chasing and subduing—that is called handling time (H). In addition, there are search times (S) associated with prey. Beginning with the first two variables, the profit (P) that a predator experiences upon eating the prey can be described as:

$$P = E/H \qquad (9.1)$$

Now assume that there are several types of prey. Each prey will have its own energy value, its own handling time, and its own search time, so to make these equations informative in all cases, we must identify these different types of prey for comparison. Subscripts are handy for that. Thus, prey type i has a profitability of P_i, an energy value of E_i, a handling time of H_i, and a search time of S_i. When we compare two prey types (designated 1 and 2), a notation that $P_1 > P_2$ simply means that prey type 1 is more profitable than prey type 2.

Assume that the predator knows these values; again, given the knowledge that animals have about their world, that is not an unreasonable assumption, even if how they comprehend such information is not precisely understood.

Suppose that a predator has indeed encountered prey type 1. What should it do? Given the inequality of the two prey items, the answer is easy: it should eat the more profitable item. So the *first prediction* of the model is not terribly difficult to embrace: the most profitable prey, when encountered, should always be eaten.

What happens, however, if the predator encounters prey type 2? Answering this question is more difficult. Prey type 2 is within the predator's grasp and requires no searching, but there still may be handling time. In contrast, prey type 1 will require a search, so

$$P_1 = E_1/(S_1 + H_1) \qquad (9.2)$$

This yields a *second prediction*, that prey type 2 should be eaten when its profitability exceeds that of prey type 1, something that can occur if H_1 is large enough. In that case,

$$E_2/H_2 > E_1/(S_1 + H_1) \tag{9.3}$$

Clearly, the search time for prey type 1 is highly influential in this decision. If prey type 1 is easy to find, then S_1 will be very small, and profitability of prey type 1 will remain high. If that profitability is high enough, then it should remain the preferred prey, and prey type 2 should be ignored, no matter how often it is encountered. The reason is that given the small value of S_1, the profitability of prey type 1 still exceeds the profitability of prey type 2, no matter which one was encountered. This yields a *third prediction*: The foraging choice is solely based on the encounter rate with the most profitable prey; the abundance of the less-profitable prey is irrelevant.

The concept of optimal foraging is not without its critics, and it is not clear if all examples of optimal foraging in the literature are good representations of what might happen in a natural setting. (In fact, it is pretty clear they are not.) Lest these predictions seem too simple or unrealistic, remember the benefits of using models. If organisms ignore the predictions generated by theory, then there is an opportunity to revisit and refine the assumptions of the model being tested, not to mention the experimental design. Of course, one goal is that with increased understanding, models approximate reality, but even if they do not, they offer useful insights into problem-solving techniques used with models.

Of course, many things can impinge on an optimal diet; a few of those constraints were discussed in the introduction to optimality. Three deserve special mention: nutritional constraints are discussed in the following section, whereas predator risk and self-medication are explored in Chapter 10. Search image formation can also affect the discovery of prey items.

OF SPECIAL INTEREST: POLAR BEARS AND GOOSE DINNER

Polar bears famously hunt and consume ringed seal pups. As seal hunting diminishes with the Arctic ice pack, scientists are left to wonder how a polar bear supplements its diet; at some point, fat reserves will be insufficient. To complicate things, a polar bear on the hunt is not easily observed, but polar bear fecal analysis does suggest that polar bears consume water birds. While other terrestrial foods that do not flee (e.g., berries, eggs, vegetation) may contain little energy but cost little to harvest, water bird consumption presents a question for OFT. Rare observations of polar bears chasing snow geese during their flightless period (they molt 2 weeks after goslings hatch) indicate that these chases average twice as long as predicted by OFT; in other words, they expend more energy than they gain from the geese. Some hypotheses aimed at explaining this behavior include the benefits of learning how to get a snow goose (subadult bears had the longest chases), and the possibility of simultaneous cooling and hunting. Most of the chases occurred in shallow water and in some cases, the bear seemed to chase the goose into the water.

Are polar bears actually incurring an energy deficit when they hunt geese? Are unique micronutrients or temperature advantages mitigating this deficit? Is the potential for learning efficient ways to hunt geese worth the temporary disadvantage of a longer chase? This is an example of the power of OFT: it asks us to look deeper and work harder to understand polar bears, and a great deal more.[52]

9.10 NUTRITIONAL CONSTRAINTS

The classic study of nutritional constraints in foraging was done not with a highly specialized insect or rodent devoted to a limited range of dietary items, but with a moose! Vertebrates have important dietary requirements for sodium, and moose are no exception. In addition, moose face a daunting trade-off: Aquatic plants contain the most sodium, but have little energy, whereas terrestrial plants offer the greatest energy, but are sodium-poor. In addition, aquatic plants are not available in the winter when the lakes are frozen. Therefore, a moose must acquire all of its winter sodium during summer foraging bouts. All of this adds up to a critical foraging decision for a moose.

Gary Belovsky used a linear programming model to investigate this decision.[53] He included variables such as the moose's daily energy requirement, the energy in the plants, the sodium needed, and time for digestion. His model predicted that 18% of a moose's summer foraging schedule should be devoted to aquatic plants, despite the fact that they are a poor source of energy—and this was borne out by observations of moose. Clearly, energy is not the sole concern of a foraging moose and a foraging model that did not take sodium into account would have been off the mark.

OF SPECIAL INTEREST: FOSSIL FORAGING

Foraging models can be used to explore the diets of long-extinct organisms. Sauropod dinosaurs hold the record for being the largest herbivores (up to 70 tons), yet they lived on plant taxa such as gymnosperms and ferns whose modern relatives are thought to have little nutritional value. By using analytical techniques that have been developed to investigate modern herbivore nutrition, Hummel and colleagues discovered that (assuming fermentation) at least some of the plants descended from those that coexisted with sauropods could have yielded enough energy to support the giant animals.[54] While this study certainly didn't address optimal foraging in dinosaurs, it set forth interesting questions: *Araucaria* is a large tree that had global distribution during the Jurassic Period. Hummel and collaborators discovered that modern relatives ferment slowly, but when they do, they yield significant energy. Other plants such as *Ginkgo* offer more protein. In contrast, yet other plants available to the sauropods (e.g., cycads, podocarp conifers, tree ferns) would not have sustained the large animals. Given the nutritional information that we can gather from the descendants of plants that were available to sauropods, how do you think that those dinosaurs made foraging decisions? What might the currency have been?

The late Pleistoscene extinction claimed all but two of the large cats inhabiting North America. One of those, the cougar (*Puma concolor*), may have persisted while the sabertooth cats (*Smilodon fatalis*) and the American cave lion (*Panthera atrox*) perished because of the cougar's more generalized dietary habits. Teeth of all three species from the La Brea tar pits reveal that the cougar had a fairly generalized diet compared to the other two species, and this could have contributed to its survival.[55] The sabertooth cat *S. fatalis* was specialized in the way it killed prey as well. Its impressive teeth were oval in cross-section; they were good at slicing, but not as strong as those of modern cats. In contrast, sabertooth cats had unusually robust forelimbs compared to any other cat, and used those to subdue prey so that they could execute a killing bite with minimal risk to their teeth.[56] As you can tell from these and other reports, paleoethologists put a wide range of tools to creative use as they tease behavioral secrets from ancient bones.

319

SUMMARY

Foraging is a fundamental behavior that defines animals. Animals do not photosynthesize; we capture energy from other organisms. There are many ways to forage, but many of them can be placed into the framework of optimality theory. This theory is not limited to foraging behavior—almost any decision can be optimized—but it has perhaps had the biggest impact there. When we consider optimality and animal behavior, or optimality and any product of evolution, it is important to understand that many constraints can prevent optimal outcomes in nature. Nonetheless, models are useful for what they teach about the assumptions on which they are based and for how they demand clarity of thinking.

In the study of foraging, optimality models have classically been applied to two large areas: decisions about where to forage (patch choice) and decisions about what to eat (prey choice). Both of these areas are based on several assumptions, including the fact that animals are likely to have sampled their environment and be familiar with the choices on offer.

The marginal value theorem predicts that moving from patch to patch should be a function of both the average richness of the patches in the area and the distance to the nearest patch that equals the current one in richness. A related concept, central place foraging, is concerned with the fact that many animals must make foraging decisions about how to provision core areas that can be some distance from foraging locations. Perception and response to risk can influence an animal's choice of where to forage. Theory about prey choice yields some simple, testable predictions that may be influenced by constraints such as nutritional demands.

Parasites can manipulate hosts in ways that take advantage of host foraging behavior. One common manipulation involves changing intermediate host responses to environmental stimuli in ways that make them more conspicuous to predators (final hosts). Vectors of blood-borne parasites also exhibit altered patterns of blood-feeding that may well favor parasite transmission.

The ways that animals get food are so diverse as to defy description or categorization. From fungal gardens to central place foraging (and hoarding), from antelope grazing to cheetah decisions, this diversity is both a wonderful showcase for how evolution encourages animals to convert nutrients into offspring and also a schoolhouse, teaching the importance of clear assumptions and careful experiments.

STUDY QUESTIONS

1. Using the marginal value theorem, we have examined the influence of travel time on the decision to leave a patch and go to another one. What is the influence of patch quality? How would you investigate this aspect of patch choice and time spent in patch? Can you apply the marginal value theorem to your own foraging behavior?
2. We have discussed some of the more widely acknowledged constraints on optimizing prey choice. Choose an animal in which you are particularly interested. Read about that animal and see if you can come up with a list of constraints that are likely to influence that animal's foraging. Your list may be very different from our general one!
3. What is your preferred method of categorizing the ways in which animals get food? Why do you like this method? How does it help you think about foraging? Are some categorization methods better at stimulating thoughts and hypotheses than others?
4. Foraging can, at times, trump phylogeny. By that, we mean that certain foraging tactics can promote convergent evolution, with animals using specific resources converging on traits that help them exploit those resources. Consider the following groups: blood-drinkers, wood-eaters, intestinal parasites, filter-feeding invertebrates. Do some research on any one of them and discover what foraging traits they have in common!

Further Reading

Abramson, C.I., Singleton, J.B., Wildon, M.K., Wanderley, P.A., Ramalho, F.S., Michaluk, L.M., 2006. The effect of an organic pesticide on mortality and learning in Africanized honey bees (*Apis mellifera* L.) in Brasil. Am. J. Environ. Sci. 2, 33–40.

Ahearn, S.C., Smith, J.L.D., Joshi, A.R., Ding, J., 2001. TIGMOD: an individual-based spatially explicit model for simulating tiger/human interaction in multiple use forests. Ecol. Modell. 140 (1–2), 81–97.

Biswas, S., Sankar, K., 2002. Prey abundance and food habit of tigers (*Panthera tigris tigris*) in Pench National Park, Madhya Pradesh, India. J. Zool. 256, 411–420.

Borrell, B.J., 2007. Scaling of nectar foraging in orchid bees. Am. Nat. 169, 569–580.

Catania, K.C., 2006. Olfaction: underwater 'sniffing' by semi-aquatic mammals. Nature 444, 1024–1025.

Clark, R.J., Harland, D.P., Jackson, R.R., 2000. Speculative hunting by an araneophagic salticid spider. Behaviour 137, 1601–1612.

Darmaillacq, A.-S., Chichery, R., Dickel, L., 2006. Food imprinting, new evidence from cuttlefish *Sepia officinalis*. Biol. Lett. 2, 345–347.

Dees, N.D., Hofmann, M., Bahar, S., 2010. Physical constraints and the evolution of different foraging strategies in aquatic space. Anim. Behav. 79, 603–611.

Filippi, L., Hironaka, M., Nomakuchi, S, 2005. Kleptoparasitism and the effect of nest location in a subsocial shield bug *Parastrachia japonensis* (Hemiptera: Parastrachiidae). Ann. Entomol. Soc. Am. 98, 134–142.

Gilbert, C., 1997. Visual control of cursorial prey pursuit by tiger beetles (Cicindelidae). J. Comp. Physiol. A 181 (3), 217–230.

Gillette, R., Huang, R.-C., Hatcher, N., Moroz, L.L., 2000. Cost-benefit analysis potential in feeding behavior of a predatory snail by integration of hunger, taste and pain. Proc. Natl. Acad. Sci. USA 97, 3585–3590.

Giraldeau, L.-A., Caraco, T., 2000. Social Foraging Theory. Princeton University Press, Princeton, NJ.

Jacobs, L.F., Spencer, W.D., 1994. Natural space-use patterns and hippocampal size in Kangaroo rats. Brain. Behav. Evol. 44, 125–132.

Kaushik, M., Knowles, S.C.L., Webster, J.P., 2014. What makes a feline fatal in *Toxoplasma gondii's* fatal feline attraction? Infected rats choose wild cats. Integr. Comp. Biol. 54, 118–128.

Kevan, P.G., Greco, C.F., 2001. Contrasting patch choice behaviour by immature ambush predators, a spider (*Misumena vatia*) and an insect (*Phymata americana*). Ecol. Entomol. 26 (2), 148–153.

Molles, M., 2008. Ecology: Concepts and applications, fourth ed. WCB/McGraw-Hill, Boston, MA.

Mull, J.F., MacMahon, J.A., 1997. Spatial variation in rates of seed removal by harvester ants (etc). Am. Midl. Nat. 138, 1–13.

Nelson, X.J., Garnett, D.T., Evans, C.S., 2010. Receiver psychology and the design of the deceptive caudal luring signal of the death adder. Anim. Behav. 79, 555–561.

O'Brien, T.G., Kinnaird, M.F., Wibisono, H.T., 2003. Crouching tigers, hidden prey: Sumatran tiger and prey populations in a tropical forest landscape. Anim. Conserv. 6, 131–139.

Okamura, J.Y., Toh, Y., 2001. Responses of medulla neurons to illumination and movement stimuli in the tiger beetle larva. J. Comp. Physiol. A 187 (9), 713–725.

Pasquet, A., Ridwan, A., Leborgne, R., 1994. Presence of potential prey affects web-building in an orb-weaving spider *Zygiella x notata*. Anim. Behav. 47 (2), 477–480.

Pyke, G.H., 2010. Optimal foraging theory: introduction. In: Breed, M.D., Moore, J. (Eds.), Encyclopedia of Animal Behavior, vol. 3, Academic Press, Oxford, pp. 601–603.

Raubenheimer, D., 2010. Foraging modes. In: Breed, M.D., Moore, J. (Eds.), Encyclopedia of Animal Behavior, vol. 3, Academic Press, Oxford, pp. 749–758.

Samelius, G., Alisauskas, R.T., Hobson, K.A., Larivière, S., 2007. Prolonging the arctic pulse: long-term exploitation of cached eggs by arctic foxes when lemmings are scarce. J. Anim. Ecol. 76, 873–880.

Schmalhofer, V.R., 2001. Tritrophic interactions in a pollination system: impacts of species composition and size of flower patches on the hunting success of a flower-dwelling spider. Oecologia 129 (2), 292–303.

Tanaka, K., 1989. Energetic cost of web construction and its effect on web relocation in the web-building spider *Agelena limbata*. Oecologia 81 (4), 459–464.

Watwood, S.L., Miller, P.J.O., Johnson, M., Madsen, P.T., Tyack, P.L., 2006. Deep-diving foraging behaviour of sperm whales (*Physeter macrocephalus*). J. Anim. Ecol. 75, 814–825.

Wirsing, A.J., Cameron, K.E., Heithaus, M.R., 2010. Spatial responses to predators vary with prey escape mode. Anim. Behav. 79, 531–537.

Notes

1. Pauli, J.N., Mendoza, J.E., Steffan, S.A., Carey, C.C., Weimer, P.J., Peery, M.Z., 2014. A syndrome of mutualism reinforces the lifestyle of a sloth. Proc. R. Soc. B 281, 20133006.
2. Frost, P.C., Ebert, D., Smith, V.H., 2008. Responses of a bacterial pathogen to phosphorus limitation of its aquatic invertebrate host. Ecology 89, 313–318.
3. Adler, P., Pearson, D.L., 1982. Why do male butterflies visit mud puddles? Can. J. Zool. 60, 322–325.
4. Molles, M., 2008. Ecology: Concepts and Applications, fourth ed. WCB/McGraw-Hill, Boston, MA.
5. Sork, V.L., 1983. Mammalian seed dispersal of pignut hickory during three fruiting seasons. Ecology 64, 1049–1056.
6. Post, D.M., Armbrust, T.S., Horne, E.A., Goheen, J.R., 2001. Sexual segregation results in differences in content and quality of bison (*Bos bison*) diets. J. Mammal. 82, 407–413.
7. Glendinning, J.I., 1994. Is the bitter rejection response always adaptive? Physiol. Behav. 56, 1217–1227.
8. Tallamy, D.W., 1985. Squash beetle feeding behavior: an adaptation against induced cucurbit defenses. Ecology 6, 1574–1579.
9. Dussourd, D.E., 2009. Do canal-cutting behaviours facilitate host-range expansion by insect herbivores? Biol. J. Linn. Soc. 96, 715–731.

10. Edman, J.D., Day, J.F., Walker, E.D., 1984. Field confirmation of laboratory observations on the differential antimosquito behavior of herons. Condor 86, 91–92.
11. Cooper, A.B., Pettorelli, N., Durant, S.M., 2007. Large carnivore menus: factors affecting hunting decisions by cheetahs in the Serengeti. Anim. Behav. 73, 651–659.
12. Wilson, A.M., Lowe, J.C., Roskilly, K., Hudson, P.E., Golabek, K.A., McNutt, J.W., 2013. Locomotion dynamics of hunting in wild cheetahs. Nature 498, 185–189.
13. Goley, P.D., Straley, J.M., 1994. Attack on gray whales (*Eschrichtius robustus*) in Monterey Bay, California, by killer whales (*Orcinus orca*) previously identified in Glacier Bay, Alaska. Can. J. Zool. 72, 1528–1530.
14. Bradshaw, J., 2013. Cat Sense. Basic Books, New York, NY, pp. 307.
15. Venner, S., Pasquet, A., Leborgne, R., 2000. Web-building behaviour in the orb weaving spider *Zygiella x-notata*: influence of experience. Anim. Behav. 59, 603–611.
16. Eisner, T., Nowicki, S., 1983. Spider web protection through visual advertisement—role of the stabilimentum. Science 219 (4581), 185–187.
17. Blackledge, T.A., Wenzel, J.W., 1999. Do stabilimenta in orb webs attract prey or defend spiders? Behav. Ecol. 10 (4), 372–376.
18. Schoener, T.W., Spiller, D.A., 1992. Stabilimenta characteristics of the spider *Argiope-argentata* on small islands—support of the predator-defense hypothesis. Behav. Ecol. Sociobiol. 31 (5), 309–318.
19. Eberhard, W.G., 1990. Function and phylogeny of spider webs. Annu. Rev. Ecol. Syst. 21, 341–372.
20. Smith, M., Conway, C.J., 2007. Use of mammal manure by nesting burrowing owls: a test of four functional hypotheses. Anim. Behav. 78, 65–73.
21. Haddock, S.H.D., Dunn, C.W., Pugh, P.R., Schnitzler, C.E., 2005. Bioluminescent and red-fluorescent lures in a deep-sea siphophore. Science 309, 263.
22. Hauber, M.E., 2002. Conspicuous colouration attracts prey to a stationary predator. Ecol. Entomol. 27, 686–691.
23. Kurland, J.A., Beckerman, S.J., 1985. Optimal foraging and hominid evolution: labor and reciprocity. Am. Anthropol. 87, 73–93.
24. Daura-Jorge, F.G., Cantor, M., Ingram, S.N., Lusseau, D., Simoes-Lopes, P.C., 2012. The structure of a bottlenose dolphin society is coupled to a unique foraging cooperation with artisanal fishermen. Biol. Lett. 8, 702–705.
25. Dole, J.W., Rose, B.B., Tachiki, K.H., 1981. Western toads (*Bufo boreas*) learn odor of prey insects. Herpetologica 37, 63–68.
26. Boback, S.M., Hall, A.E., McCann, K.J., Hayes, A.W., Forrester, J.S., Zwemer, C.F., 2012. Snake modulates constriction in response to prey's heartbeat. Biol. Lett. 8, 473–476.
27. Papaj, D.R., Vet, L.E.M., 1990. Odor learning and foraging success in the parasitoid, *Leptopilina heterotoma*. J. Chem. Ecol. 16, 3137–3150.
28. Thomas, F., Schmidt-Rhaesa, A., Martin, G., Manu, C., Durand, P., Renaud, F., 2002. Do hairworms (Nematomorpha) manipulate the water-seeking behaviour of their terrestrial hosts? J. Evol. Biol. 15, 356–361.
29. Ponton, F., Otalora-Luna, F., Lefevre, T., Guerin, P.M., Lebarbenchon, C., Duneau, D., et al., 2011. Water-seeking behavior in worm-infected crickets and reversibility of parasitic manipulation. Behav. Ecol. 22, 392–400.
30. Rossignol, P.A., Ribeiro, J.M.C., Spielman, A., 1984. Increased intradermal probing time in sporozoite-infected mosquitoes. Am. J. Trop. Med. Hyg. 33, 17–20.
    Rossignol, P.A., Ribeiro, J.M.C., Spielman, A., 1986. Increased biting rate and reduced fertility in sporozoite-infected mosquitoes. Am. J. Trop. Med. Hyg. 35, 277–279.
31. Nyasembe, V.O., Teal, P.E., Sawa, P., Tumlinson, J.H., Borgemeister, C., Torto, B., 2014. *Plasmodium falciparum* infection increases *Anopheles gambiae* attraction to nectar sources and sugar uptake. Curr. Biol. 24, 217–221.
32. Anderson, R.M., May, R.M., 1992. Infectious Diseases of Humans: Dynamics and Control. Oxford University Press, Oxford.
33. Moore, J., 2002. Parasites and the Behavior of Animals. Oxford Series in Ecology and Evolutionary Biology. Oxford University Press, New York, NY.
34. Camp, J.W., Huizinga, H.W., 1979. Altered color, behavior and predation susceptibility of the isopod *Asellus intermedius* infected with *Acanthocephalus dirus*. J. Parasitol. 65, 667–669.
35. MacArthur, R.H., Pianka, E.R., 1966. On the optimal use of a patchy environment. Am. Nat. 100, 603–609.
36. Emlen, J.M., 1966. The role of time and energy in food preference. Am. Nat. 100, 611–617.
37. Schmidt, K.A., 1999. Foraging theory as a conceptual framework for studying nest predation. Oikos 85, 151–160.
38. Cooper Jr., W.E., 1995. Foragine mode, prey chemical discrimination and phylogeny in lizards. Anim. Behav. 50, 973–985.
39. Holling, C.S., 1959. Some characteristics of simple types of predation and parasitism. Can. Entomol. 91, 385–398.
40. Waddington, K., 2004. Flight patterns of foraging bees relative to the density of artificial flowers and distribution of nectar. Oecologia 44, 199–204.
41. Raguso, R., Wills, M., 2002. Synergy between visual and olfactory cues in nectar feeding by naïve hawkmoths, *Manduca sexta*. Anim. Behav. 64, 684–695.
42. Smith, J.N.M., Sweatman, H.P., 1974. Food searching behaviour of titmice in patchy environments. Ecology 55, 1216–1232.
43. Charnov, E.L., 1976. Optimal foraging: the marginal value theorem. Theor. Popul. Biol. 9, 129–136.
44. Biernaskie, J.M., Walker, S.C., Gegear, R.J., 2009. Bumblebees learn to forage like Bayesians. Am. Nat. 174, 413–423.

45. Fortin, D., Fryxell, J.M., O'Brodovich, L., Frandsen, D., 2003. Foraging ecology of bison at the landscape and plant community levels: the applicability of energy maximization principles. Oecologia 134, 219–227.
46. Breed, M.D., Bowden, R.M., Garry, M.F., Weicker, A., 1996. Search time variation in response to differences in nectar quantity and quality in the giant tropical ant, *Paraponera clavata*. J. Insect Behav. 9, 659–672.
47. Barrette, M., Boivin, G., Brodeur, J., 2010. Travel time affects optimal diets in depleting patches. Behav. Ecol. Sociobiol. 64, 593–598.
48. Caraco, T., Martindale, S., Whittam, T.S., 1980. An empirical demonstration of risk-sensitive foraging preferences. Anim. Behav. 28, 820–830.
49. Ludvico, L.R., Bennett, I.M., Beckerman, S., 1991. Risk-sensitive foraging behavior among the Bari. Hum. Ecol. 19, 509–516.
50. Perez, S.M., Waddington, K.D., 1996. Carpenter bee (*Xylocopa micans*) risk indifference and a review of nectarivore risk-sensitivity studies. Am. Zool. 36, 435–446.
51. Stephens, D., Krebs, J., 1986. Foraging Theory. Princeton University Press, Princeton, NJ.
   Stephens, D.W., Brown, J.S., Ydenberg, R.C. (Eds.), 2007. Foraging: Behavior and Ecology University of Chicago Press, Chicago, IL.
52. Iles, D.T., Peterson, S.L., Gormezano, L.J., Koons, D.N., Rockwell, R.F., 2013. Terrestrial predation by polar bears: not just a wild goose chase. Polar Biol. 36, 1373–1379.
53. Belovsky, G.E., 1978. Diet optimization in a generalist herbivore: the moose. Theor. Popul. Biol. 14, 105–134.
54. Hummel, J., Gee, C.T., Sudekum, K.-H., Sander, P.M., Nogge, G., Clauss, M., 2008. *In vitro* digestibility of fern and gymnosperm foliage: implications for sauropod feeding ecology and diet selection. Proc. R. Soc. 275, 1015–1021.
55. DeSantis, L.R.G., Haupt, R.J., 2014. Cougars' key to survival through the Late Pleistocene extinction: insights from dental microwear texture analysis. Biol. Lett. http://dx.doi.org/10.1098/rsbl.2014.0203.
56. Meachen-Samuels, J.A., Van Valkenburgh, B., 2010. Radiographs reveal exceptional forelimb strength in the sabertooth cat, *Smilodon fatalis*. PLoS One 5, e11412. http://dx.doi.org/10.1371/journal.pone.0011412.

Self-Defense

CHAPTER OUTLINE

325

LEARNING OBJECTIVES

Studying this chapter should provide you with the knowledge to:

- Remember the three elements of predator avoidance: avoiding detection, evading capture, and fighting back
- Understand how animals hide from predators, including background matching, countershading, brokenness, and disruptive coloration
- Evaluate a prey animal's options if discovered by a predator: fleeing confusing predators, engaging in flash behavior, taking evasive action, or advertising prowess at fleeing
- Understand how prey fight back, and the models of self-defense that incorporate the costs incurred by predators at these times
- Remember other ways to escape predator notice, including diversion and types of mimicry
- Understand the importance of vigilance behavior and the role of group formation in vigilance
- Understand the parallels between the avoidance and deterrence of predation and the ways that animals cope with parasites and pathogens.

Animal Behavior. DOI: http://dx.doi.org/10.1016/B978-0-12-801532-2.00010-6

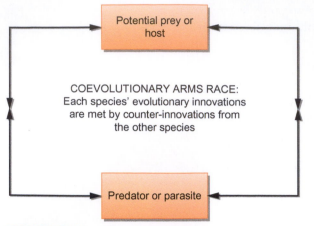

FIGURE 10.1
Consumer/food evolution revolves around a continuous arms race between species, in which innovations that make better predators and better-defended prey are countered by the other species. Even though sometimes predators do not look completely efficient or prey are apparently not fully defended, likely evolution has taken each species to the optimum it can achieve, given the limits of morphology and physiology.

10.1 INTRODUCTION

Self-defense is about not being eaten; put another way, it is about not becoming lunch. There are three key elements to self-defense. The first is avoiding detection, which typically involves either hiding or camouflage. The second is evading capture; if an animal is detected, then it must take evasive actions to avoid capture. The third is fighting back if an animal is in the grasp of a predator (Figure 10.1).

Self-defense is a deadly serious game, similar to a lethal form of "Hide and Seek" between predators and prey, that plays itself out over evolutionary time. The current strategies that potential prey use to avoid being eaten reflect countless generations of evolution. The evolutionary game results in adaptation, counteradaptation, counter-counteradaptation, and so on, leading to ever more sophisticated defenses and counterdefenses. Coevolution between the hunter and the hunted is the powerful force underlying self-defense.

Much of self-defense is about an animal's structure (morphology) and its coloration,[1] and a major portion of this chapter is necessarily devoted to defenses that are largely structural or visual (e.g., color), with behavior playing a smaller, but nonetheless essential, role. Often behavior makes a morphological adaptation effective. Looking like a dead leaf works for a moth only if it locates itself among dead leaves and is motionless, and looking like a stick is protective for a walking stick only if it blends in among actual sticks (Figure 10.2).

The section at the end of this chapter addresses how animals avoid and respond behaviorally to diseases and parasites. There are some similarities between antipredator and antiparasite behavior that connect these two topics. After all, parasites and predators are on a continuum defined by exploiting another animal. Predators do so directly and with lethal results; parasites, more often, do so by consuming host resources or host tissues. Once parasites are included in the crowd of animals that exploit other organisms, a consideration of self-defense must include the ways that animals behaviorally defend themselves against parasites and pathogens. Responses to parasitism and disease complete the picture of how animals defend themselves.

Each of these elements of self-defense will be examined in greater detail in the pages that follow. All of the strategies are subject to a number of constraints that influence all of behavior. Those constraints have appeared in this text before—the realities of trade-offs and

FIGURE 10.2
Dead leaf moths and walking sticks are almost impossible to see if they choose the right backgrounds. *Photos: Left and right, Jeff Mitton; center, Michael Breed.*

the limits of evolution. Large animals may resist some predators, but they have a hard time being cryptic, and spines and toxins are not equally available, evolutionarily speaking, to every kind of animal.

Finally, it is useful to recall that self-defense—especially predator avoidance—can have profound effects on the distribution of prey species. Most animals base habitat selection and activity patterns at least in part on the presence, absence, and habits of their predators.[2] Humans have probably never appreciated the fine points of predator avoidance as thoroughly as do field mice, many caterpillars, or a variety of zooplankton (to name a few dietary staples of the natural world). After all, we are relatively large animals, and size limits the kinds of predators that can successfully attack us. Moreover, stretching back far into prehistory, humans and their ancestors have lived in social groups, which are effective predator detection/deterrent units, as we will learn in Chapter 13. Finally, our ancestors minimized the threat of predation many generations ago, modifying the surrounding environment to discourage such activity. In short, once again we are bumping into the concept of *umwelt*, which should never be far from our consciousness as behavioral biologists. When it comes to the threat of being pounced upon, shredded, eaten, devoured, and/or consumed, most humans do not have a clue, relatively speaking. This makes what we will learn about how other animals cope with predation that much more amazing.

CASE STUDY: BEHAVIORAL FEVER

When most people think about self-defense in the natural world, they might imagine a motionless, well-camouflaged rabbit or the last-ditch race of an impala and a hungry cheetah across the African plains. In addition to those dramatic images, animals defend themselves against things that are far smaller than cheetahs; mammals do this every time they become feverish in response to some pathogen (see Section 10.7), because metabolic fever is an effective mechanism with which to combat disease. This is beneficial for the few thousand species that are endotherms. What about the vast majority of animals in the world that cannot generate metabolic fever because they are ectothermic?

Behavioral fever was first experimentally induced in the 1970s in the desert iguana,[3] by injecting killed bacteria. Injected lizards, when given a choice of temperatures, moved to warmer environments and raised their body temperatures by 2°C. Hissing cockroaches were the first insects to exhibit a similar response to an injection of endotoxin,[4] although other insects had been suspected of feverish responses to fungi years before.[5] We now know that many ectotherms move to hot places when faced with an infection, and that this is associated with physiological changes that can mitigate the effects of infection.[6] They may bask, crawl up high on vegetation, or become hyperactive. In one particularly dramatic example, a fungus (*Entomophthora muscae*) that kills houseflies (*Musca domestica*) 5 days after infection can itself be killed if the fly can get to a hot enough environment during the first 3 days of infection, and as expected, infected flies do bask and move toward sources of heat.[7] The way that a behaviorally feverish animal raises its temperature differs from that seen in endotherms, but the end

FIGURE 10.3
This anthomylid fly has been killed by the fungal parasite, *E. muscae*. It is not known if the fly attempted to generate fever when it was first infected, but in Colorado, where this photograph was taken, it may be difficult or even impossible to generate fever on a cool day. Climate can be a significant constraint for behavioral fever, and may even limit the efficacy of some forms of biological control. *Photo: Whitney Cranshaw, Bugwood.org.*

result is the same and, in the case of the fly and the fungus, is a matter of life and death[7] (Figure 10.3).

Although fever is a near-ubiquitous occurrence among animals, it is not universal, and some host–parasite combinations and environmental conditions do not seem to produce a change in thermal preference.[8] In addition, some fungi of locusts can carry the host's behavioral response and inhibit it.[9] Behavioral chills have

(Continued)

327

also been observed; in some host–parasite associations, the host may move to cool places and delay development of the parasite.[10]

Altered thermal choice in the face of challenges from pathogens has intriguing implications for behavioral biologists and managers alike. For instance, adaptation is not always what it may seem in the war between parasite and host: Is a basking ectotherm trying to generate a fever that can kill a parasite, or (if it is an intermediate host) is it the victim of parasite manipulation[11] (see Chapter 9)? What is the fitness cost of such conspicuousness?[12] And if animals can use high temperatures to combat infection, should weather forecasts be studied before distributing fever-sensitive biocontrol agents?[13] Finally, given the energetic cost of endothermy, is the evolution of endothermy itself a consequence of the arms race between pathogen and host?[14]

10.2 CRYPTIC BEHAVIOR: CAMOUFLAGE
Camouflage: An Introduction

The first step in avoiding the predicament of becoming someone else's lunch or dinner is fairly simple: do not be discovered by a predator. *Crypsis*—being hard to see—is a successful strategy for many animals. Being cryptic includes both hiding and *camouflage* strategies; whatever makes an animal hard to find contributes to crypsis. Predators take advantage of camouflage as well; sit-and-wait predators such as mountain lions and rattlesnakes blend into their backgrounds so that unsuspecting prey can wander dangerously close without detecting the predator.

KEY TERM Crypsis is behavior, color, or shape that makes an animal difficult for predators to detect. It comes from a Greek word meaning "hidden," and shares its root with "encrypt."

KEY TERM Camouflage is color or shape that helps to hide an animal from visual predators when it is on its normal substrate.

KEY TERM Mimicry occurs when one species evolves to look, sound, smell, or act like another species.

Many small animals and some large ones (herd animals being one notable exception) have nests, caves, and crevices in which they hide. Although there are certainly predators that enter burrows in search of prey, the use of hiding places is on average a good predator-avoidance strategy; it does impose a cost, however, in that most animals cannot forage or seek mates while hiding. Darkness provides shelter from predators that rely on vision, and for nocturnal animals, foraging and other activities often peak during dark nights and wane on moonlit nights, all the better to avoid detection, especially by visual predators.

Many animals "hide" in plain sight by using camouflage to become more or less invisible or by taking on the appearance of other organisms or objects that are less interesting to the predator. (This is *mimicry*, from the word *mime*.)

The word *camouflage* is an unexceptional one—common in everyday use. It is a military term and refers to disguising something so that it is hidden from an enemy. Its military origins are hardly surprising; they are consistent with its appearance in a chapter about predator–prey arms races. Surprisingly, the first record of the word in the English language is in 1917, when the editors of the *Daily Mail* felt the need to define the term for readers. Digging into its French origins reveals that during World War I (1915), the French military actually hired artists to help camouflage equipment and people.[15]

Is it curious that armies had to wait until World War I to discover camouflage? Was this some tragic oversight? Paintings of warfare before the twentieth century are awash with colorful uniforms and brilliant battle standards; there are no muted greens and browns. This great divide in military attire may have more biological significance than is initially apparent.

328

FIGURE 10.4
A male and female Red-Rumped Tanager in Costa Rica. Note the strikingly different visibility of the camouflaged female and the male, dramatically apparent against the green background. Sexual dimorphism in camouflage is quite common in birds.

World War I was an unusual war in that it was largely stationary; countless people fought and died in trenches. (Fatality estimates are in the vicinity of 9 million, but no one knows the precise number.) Prior to that time, the flashy appearance of uniforms points to another great biological function of color: communication (see Chapter 7). The brilliant battle standards and the brightly colored jackets identified the troops to friend and foe at a time when there were no telephones of any sort, much less wireless communication. In the trenches, on the other hand, invisibility was more important than identification. The analogy between human uses of camouflage and crypsis in animals is clear and direct. Humans have improved their own camouflage strategies by copying what animals do.

One irony to consider in this discussion of camouflage is that natural selection must balance ability to hide from predators with the ability to attract mates. This may happen at an individual level,[16] but more often results in species-level changes, such as sexual dimorphism in camouflage; one sex in a species (usually the females) is cryptic, whereas the other sex (usually the males) is showy (Figure 10.4). The evolutionary reasons for this await further illumination in Chapter 12, but for now, consider that dimorphism means that males and females differ dramatically not only in appearance but also in risk from predation.

The following sections analyze the different ways in which natural selection has favored the use of physical appearance to limit detection. These adaptations are largely a matter of color and pattern—and again, although color and pattern may seem to be primarily morphological properties, camouflage works only when facilitated by behavior. Throughout these sections, think about the way behavior interacts with morphology to provide efficient protection from predators.

Types of Camouflage

Camouflage can act in several ways, including *countershading, background matching*, and *disruptive coloration*. Countershading is the gradation from dark color on the top of an animal to a light-colored underside. Animals use background matching when their color resembles that of their normal substrate. Disruptive coloration (or disruptive patterning) is a pattern that obscures an animal's shape, making it more difficult to see; a zebra's stripes are the classic example of disruptive coloration (Figure 10.5).

FIGURE 10.5
These stripes on a zebra (top), and an okapi (bottom)—so noticeable at close range—make them difficult to see at a distance. *Photos: Pen-Yuan Hsing (top) and Michael Breed (bottom).*

CASE STUDY: ZIG-ZAGS AND VENOMOUS SNAKES

Vipers often have zig-zag patterns on their backs. When the snake is coiled, the zig-zags merge to make it cryptic against the background. The patterning is effective alone, but when combined with color that matches the background, the snake may become almost impossible to see[17,18] (Figure 10.6). An alternative hypothesis suggests that banding and zig-zags on the backs of venomous snakes are aposematic warnings of the poisonous defenses of the snakes (Figure 10.7). Scientific studies support cryptic rather than aposematic functions for these markings.

FIGURE 10.6
Crypsis and warning coloration in deadly snakes. (A) The fer-de-lance is a deadly poisonous snake that is common in Central and South America. The diagonal slashes across this snake's body break up the edges and, particularly when the snake is coiled, make it cryptic against a background of dead leaves. (B and C) These eyelash vipers occur in two strikingly different color morphs (types) in Central America. In (B), the snake's green and brown tones, combined with brokenness that disrupts the animal's edges, are effective in making this snake cryptic. The morph in (C) is highly visible, probably to warn away birds that might attempt to prey on it. Notice how the edges of this morph clearly stand out against the background. *Photos: Michael Breed.*

330

  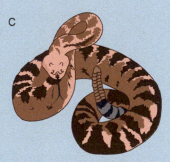

FIGURE 10.7
Venomous North American snakes showing zig-zag patterns on their backs. As with the fer-de-lance (Figure 10.6) these patterns may serve as disruptive coloration, providing camouflage so they blend in with their background. Alternatively, the zig-zags could be aposematic (warning) coloration. If this is the case then these snakes may be Mullerian mimics (Section 10.4).

Countershading

In the late nineteenth century, the artist Abbott Thayer noticed that many animals show a gradation in color, top to bottom, from dark to light.[19] This sort of color differential is called *countershading*. Because light comes from above, shadow is cast on the ventral surface of an animal. A light ventral surface mitigates this and makes the overall outline of the animal more difficult to see. This coloration works

KEY TERM Countershading is a gradation of color from the dorsal (top side) to ventral (underside) surface of an animal.

FIGURE 10.8
Countershading can be seen in these antelope and penguins. It is a form of camouflage that works well on land or in the sea. *Photos: left, Jeff Mitton; right, Michael Breed.*

not only for animals such as pronghorn antelope but also for aquatic animals and flying animals, which may be observed from below. In these cases, countershading effects may overlap with those of background matching (Figure 10.8). Again, because color can serve several biological functions, it is possible that countershaded animals are also using their color patterns for thermoregulation or signaling.

Countershading is a simple form of camouflage that allows an animal to blend into the darker backgrounds (soil, rock, dried leaves) when viewed from above and into the lighter backgrounds presented by sky or vegetation when viewed from below. Fish that swim upside down, such as some catfish, have reverse countershading. Flatfish, such as flounder and halibut, which have both eyes on one side of the body, have dark shading on their "up" side and light on their "bottom," thus following the general rule for countershading (Figure 10.9).

The flounder is an extreme example of countershading. These flatfish lie on muddy or sandy bottoms and wait for prey to come along. The fish lies on its side; in the course of evolution the eyes have shifted to be on the side of the fish that faces up. The upper side is cryptic with the background habitat and the lower side is light colored so if the flounder is swimming and a potential predator looks up at it the flounder blends into the brightness from the water's surface. Various species in the fish order Pleuronectiformes show *convergent evolution* on this type of camouflage (Figure 10.9).

> **KEY TERM** Background matching occurs when an animal has evolved to have the same color as its normal substrate.

> **KEY TERM** Disruptive coloration is patterning of an animal that disrupts its larger outlines.

331

Background Matching

If an animal's habitat is generally green, as is true for animals that occupy living leaves, or if the habitat is generally brown, as is the case for animals that crawl on bare soil, rock, or sand (Figure 10.10), then the behavioral challenge for the animal that exploits background matching is to end up on the appropriate background. A bright green caterpillar may disappear against a healthy leaf but be as conspicuous as bright green can be if it crawls onto a branch (Figure 10.11).

An animal can simply look like its background; this is called *general background matching*. Sometimes animals that match backgrounds change color with the seasons; polar and high-altitude animals seem particularly adept at this, perhaps because winter in these locations is predictably white. Of course, background matching means that the animal must choose a

FIGURE 10.9
These colorful flounders are disruptively colored on top so that they are camouflaged against their background. On the underside they are pale and lack markings. *Credit: Renard's fanciful fish, Dover Pictura.*

FIGURE 10.10
For this animal, choosing the appropriate background color is critical to its successful crypsis. *Photo: Jeff Mitton.*

background that matches itself, for in most cases, that choice is going to be faster and less complicated than changing appearance to match a variety of backgrounds. (Some intriguing exceptions to this will be discussed later in this section.) Indeed, many moths and other insects, including the famous peppered moth (*Biston betularia*) of freshman biology fame, not only choose their background, but choose specific locations that enhance their ability to blend into their surroundings (Figure 10.11). Although crypsis and camouflage in general are elements of appearance, the success of such appearance depends on appropriate behavior. A wriggly, twitchy caterpillar that matched its background would probably fare worse than a less rambunctious one.

Disruptive Coloration

Vision—this applies to nearly all visual systems—emphasizes edges. Long straight or slightly curved edges, such as the outlines of an animal, are particularly noticeable to the eye and the brain. One of the keys to being hidden is to break up the edges; this makes it more difficult for the visual system to discover the outline of an animal (Figure 10.12).

Disruptive coloration hides an animal's outline and therefore is another form of camouflage (Figure 10.12). Unlike an animal that looks like its background (background matching), a disruptively colored animal exhibits patterns that serve to hide it in a different way, by making its outline difficult to perceive.[20,21] Disruptive coloration differs from a random assortment of contrasting patterns because it works best when the disruptive patches are at the margins of the animal (as it might be viewed by a predator) and not in the middle of the body.

DISCUSSION POINT: CAMOUFLAGE

Find images of military camouflage from various time periods (WWI, WWII, etc.) on the Web. How has the concept of effective camouflage changed over time? Can you critique contemporary camouflage strategies that are digitally generated? How might these be better than older patterns, and how might they be improved? What might an animal behaviorist bring to a discussion about effective camouflage?

Changing Colors and Patterns

Recall that cryptic animals often choose backgrounds that match their own appearances. However, it is possible to change appearance to match a changing background and achieve the same undetectable result. As mentioned earlier, many residents of high mountains and the far north, such as snowshoe hares (Figure 10.13), do this by gradual seasonal color change, turning white in the winter. However, changing appearance can be rapid as well. For this, consider cephalopods, those mollusks that can learn to unscrew jars for food and to escape from their aquaria. They can also match the pattern, color, intensity, and even texture of their backgrounds and, in so doing, become functionally invisible.

As if that were not enough, octopus can do all of this quickly, within milliseconds![22] How? *Octopus vulgaris* is a typical cephalopod in that it has tiny chromatophores. A chromatophore is a cell in an animal's surface that contains pigment and that has contractile fibers that can expand the cell, thus increasing that pigment across the surface. These contractile fibers are innervated by neurons that proceed directly from the octopus brain to the chromatophore. There may be hundreds of chromatophores in a square millimeter of octopus dermis; any given chromatophore contains one of several pigments. Thus, if a dark patch occurs that is the result of the contraction of fibers around the margins of chromatophores that contain dark pigments. The pigment cell expands, and the dark color spreads. In this way, the octopus can match the background precisely and quickly ... and if that were not enough, an octopus can do all of this in three dimensions! The octopus skin papillae can change shape or disappear entirely, all the better to match background texture as well as color.[23]

Keep in mind that while the cephalopod is matching its background, it is aware of potential predators and aware of potential prey. It may be moving slowly, foraging, as it shifts its appearance slightly but continuously. How does it know how to match its background? What neurophysiology surrounds this ability to perceive the environment in a way that allows an animal to almost instantly disappear, for all practical purposes? Hanlon and coworkers discovered that all the great variety of cephalopod camouflage patterns can be sorted into three categories: uniform patterns (very little variation), mottle patterns (light–dark contrast on a small scale), and disruptive patterns (larger scale contrast).[22] They raised an intriguing question: If the Masters of Camouflage use only three basic camouflage patterns, are predators that easy to fool? (Would Professor Tinbergen be pleased with the number of answers we can bring to his questions?)

Finally, cephalopods can be used to introduce another strategy for hiding: mimicry (Section 10.4). The octopus *Macrotritopus defilippi* not only uses camouflage (in the form of background matching) to its advantage but also employs mimicry to escape notice by predators. Recall that background matching usually works best when the animal does not move. In the open habitat frequented by *M. defilippi*, swimming would destroy the benefits of camouflage. When *M. defilippi* swims, it takes on the shape and locomotor behavior of a flounder. Flounders are not toxic, but their skeletal components may make them less desirable prey than a soft-bodied octopus, at least to small predators.[24]

FIGURE 10.11
Tobacco hornworms are almost invisible against their host plants, as anyone can attest who has tried to find the thing that is taking big chunks out of the tomato plants! (Hint: The caterpillar is remarkably conspicuous from *below*, an angle that most of its predators do not employ.) *Photo: R.J. Reynolds Tobacco Company Slide Set, R.J. Reynolds Tobacco Company, Bugwood.org.*

FIGURE 10.12
Zebra stripes are effective protection against biting flies, and that seems to be their primary benefit (see Section 10.7), but they can also serve as disruptive coloration. In the distance, individuals are difficult to see and to tell apart. *Photo: Mike Huffman.*

FIGURE 10.13

Seasonal color change is common in high-elevation and high-latitude animals such as the snowshoe hare and Arctic hare. In summer, the brown, mottled fur is effective camouflage in the subalpine woods. In the winter, the white fur blends well with the snow. *Photos: left, Michael Breed; right, US Fish and Wildlife Service.*

334

FIGURE 10.14

Bull snakes can be mistaken for rattlesnakes. Their behavior encourages such mistakes; they coil and behave as if they will strike…but unlike rattlesnakes, have no venom.

Mimicry differs from background matching; although in a stretch, an animal engaged in background matching might be said to mimic the background, mimicry usually involves looking like some other organism. In background matching, the animal often escapes notice. A mimic, on the other hand, may be noticed, but it is perceived to be something it is not. Whether engaged in background matching or mimicry, dephalopods are remarkably good at both!

Tests for the Effectiveness of Camouflage

Given the problems just described, scientists have turned to artificial prey in an attempt to ask questions about the function of camouflage. One such test was conducted by Schaefer and Stobbe, who used artificial moths to determine that in a heterogeneous environment, disruptive coloration was a more effective method of concealment than background matching.[20] Indeed, because disruptive coloration obscures an animal's outline, it is effective on many backgrounds, whereas background matching requires a specific background to work well. In the case of disruptive coloration, they found that the same color patterns, when distributed on the interior of the "moth" wings rather than the margins of the wings, did not conceal the animal as well.

There are surprisingly few tests of the efficacy of camouflage. Remember that what seems almost invisible to a person might be obvious to a predator with a different visual system, so what a scientist imagines as camouflage may be less effective than it seems. Another problem with testing hypotheses about camouflage is the fact that color itself can have several functions, ranging from concealment to thermoregulation, intraspecific communication to interspecific warning. It is difficult to devise experiments that sort out the relative importance of these influences. Another drawback is the difficulty of experimental comparison; if an animal is conveniently polymorphic, the morphs can be compared under different conditions, but if such variety is absent from natural populations, then scientists are faced with comparing the naturally camouflaged animal to a conspecific that they have somehow manipulated. In addition, not all color is equal; that is, a given hue may have a wide range of properties (e.g., reflectance) that affect its visibility. Again, humans may not be aware of all of this variation.

Distraction and Dazzle

Before leaving camouflage, let us consider how an animal can discourage predators by making itself more, rather than less, obvious. This is called *startle behavior*. Good examples are nonvenomous snakes, such as bull snakes (Figure 10.14), which rattle their tails in dead leaves when disturbed. Bull snakes also do a good job of coiling and gaping their mouths; these behaviors attract attention, but because the bull snakes use behavior to disguise themselves as venomous snakes, they gain protection against predators (see Figure 10.14). Another way to startle a predator is to have a feature that is hidden but surprising when revealed. The eyespots on the wings of many kinds of moths may fit into this category (Figure 10.15). A momentary pause on the part of the startled predator may provide just enough time for the prey to escape. Squeaks, screams, and buzzes produced by captured animals may also have startle value.

> **KEY TERM** Startle behavior is the sudden production of movement, sound, or another stimulus that distracts a predator.

FIGURE 10.15
The eyespots of this *polyphemus* moth are hidden until the wings are spread; this is thought to startle predators. As one of us can attest after an encounter with a *polyphemus* moth in the laboratory, it certainly startles humans. *Photo: Lacy L. Hyche, Auburn University, Bugwood.org.*

Aposematism

Chemical deterrents can be used before or after a predator's attack. Poisons, along with repellants, are effective. They can favor the evolution of aposematic coloration, in which a brightly colored, easy-to-remember appearance, when combined with the disastrous experience of tangling with such a noxious prey, causes predators to avoid an animal with that appearance in the future. *Aposematism* (Greek, *apo* = away, *sematic* = sign) is the use of warning coloration to inform potential predators that an animal is poisonous, venomous, or otherwise dangerous. Oftentimes orange or red patterns may be warnings (as in coral snakes), but do not assume that red is always a warning (see Chapter 7). Examples of warning colors and patterns are shown in Figure 10.16.

The blue-ringed octopus combines startle behavior and aposematism to warn away predators. When disturbed, this venomous octopus can flash iridescent blue rings at the rate of three flashes per second. The blue rings are patches of special cells called iridophores; they contain layers that reflect light in the form of iridescent colors. The rapid flashes are the result of muscular contraction and relaxation of pouches of skin that surround the iridophores; when the pouch opens, there is a flash. And that is the true story of how the blue-ringed octopus got its flashing blue rings.[25]

Although many foul-tasting prey are aposematic, some animals use chemical defense as a last resort and do not advertise it. This seems to be true in the well-camouflaged silkmoth

335

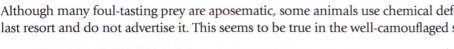

FIGURE 10.16
Three examples of potentially aposematic coloration. Right: The poison frog is mildly toxic and certainly stands out against the monotonous greens and browns of the rainforest. Center: The seed-feeding bug (family *Coreidae*) has red and yellow patterning that may warn that it contains toxins. Left: A toxic tetrio sphinx caterpillar. This caterpillar may also mimic the color pattern of a coral snake! *Photos: left and center, Michael Breed; right, Jeff Mitton.*

FIGURE 10.17
Prairie dogs chatter, whistle, and sound the alarm when predators are in the neighborhood. Prairie dog coteries are matrilineal, and animals that give alarm calls tend to be animals that are related to others in the area. *Photo: Jeff Mitton.*

caterpillars. When discovered and attacked, they emit a series of clicks using their mandibles, accompanied by regurgitation of deterrent liquid. This phenomenon was first discovered by neuroethologist Jayne Yack,[26] who brought caterpillars home with her when no one was on campus to care for them. In truth, it was first discovered by Yack's cat, who was, in turn, discovered (by Yack) gagging in the presence of a caterpillar and its regurgitant. This led to a series of experiments that confirmed the hypothesis that both acoustic aposematic signals and chemical deterrents are used by these caterpillars when camouflage fails. (The cat, which gagged for years thereafter any time it saw a similar caterpillar, also seemed to confirm the hypothesis.)

10.3 VIGILANCE AND ALARM

When a predator is nearby, it helps tremendously to know where the predator is and, if possible, what its intentions are. Behavioral biologists call this *vigilance*.[27–29] Much of what is known about vigilance is related to social behavior; for instance, vigilance is usually more effective in groups than for solitary animals. Groups have a hard time hiding, but they are good at detecting predators. Nonetheless, seeing and observing a predator is only the beginning of how animals cope with predation. Clearly, it is advantageous for the (vigilant) prey if the predator is not equally aware of the prey. A vigilant animal is attentive to predators or parasites. Being vigilant takes effort away from other important tasks, such as feeding, but it is a critical part of the time budget of many animals (Figure 10.17). Because vigilance plays a key role in self-protection, it is a central topic that crops up in several chapters. Here, vigilance is placed into the array of self-defense behaviors that an animal might employ. Chapter 13, on social behavior, explores how vigilance is shared among members of social groups. Chapter 15, on conservation, discusses how measuring vigilance helps to assess the effects of human disturbances on animal populations.

> **KEY TERM** Vigilance is the awareness that an animal has of its surroundings and the potential presence of predators. "Vigilant" comes from a Latin word that means "to keep awake."

To further consider vigilance, imagine camping out—alone—in a wilderness area, one of the few remaining parcels that is home to wolves, bears, and other carnivorous megafauna. It is dark; the fire (if it is even allowed) is not going to last the night; and the tent is a wonderfully lightweight, high-tech marvel that is great for rain but no match for teeth and claws. And by the way, the wilderness area abuts a maximum security facility, and there is an escaped convict on the loose. Now imagine camping with friends in a circle of 10 such tents. Does it feel safer to be camped alone or as part of the circle?

Some trainers recommend clicker training as a way to deal with barking behavior. They encourage training the dog to bark on command, that is, putting the bark on cue. If the command to bark is then rarely or never given, the dog's impulse to bark may be extinguished. "No bark" may also be taught once the bark can be cued; the dog must understand what the behavior of interest is before it can learn commands aimed at ceasing the behavior.

BRINGING ANIMAL BEHAVIOR HOME: VIGILANCE AND HOUSEHOLD ANIMALS

What are the results of vigilance in household animals? Dogs and cats are both vigilant about intrusions into their space, but the results differ dramatically between the species. Both dogs and cats can distinguish familiar animals and humans from unfamiliar intruders. Watchful cats often freeze or flee when a stranger enters a room. Within the long list of animals kept in people's houses, only one, the dog, shows a strong desire to communicate alerts about intrusions by strangers to other members of the household. This is hypothesized to have been one of the benefits for early humans as they began to form associations with dogs, and continues to this day.

Despite the security that a vigilant dog can offer, barking at intruders often becomes an issue for dog owners. While vigilance and barking are a normal part of dog behavior, dogs that bark can annoy neighbors. Stranger intrusion is only one possible stimulus for vigilance barking, but this can be a particularly difficult issue in managing dog behavior. Barking can be discouraged by distraction, followed by praise and rewards when the dog does not bark; owners can "stage" situations in which friends walk their dogs by the owners' house and the dog can be trained to associate calmer reactions with reward. Punishments are not very effective in this context, and yelling at the dog may simply raise the excitement level.

The urge to be among more tents reflects one of the benefits of group living—enhanced vigilance, also known as the "many eyes" effect. Assuming that the escaped convict is not among the 10 friends inhabiting that group of tents, being with those friends is a much safer situation than being alone. Humans are not the only organisms to have figured this out. A 1970 classic study with wood pigeons and a trained goshawk showed that birds in larger flocks respond more quickly to the appearance of the predator than ones in smaller flocks; that is, they take off when the goshawk is farther away.[30] Indeed, there is good evidence that in many group-foraging birds (e.g., geese), the amount of time that each organism is vigilant (head up and watching versus head down and foraging) is inversely proportional to the size of the flock. Animals in groups can share vigilance and can profit if another animal becomes alarmed, or sends out an alarm signal. Antipredator systems that rely on vigilance and alarm are greatly enhanced if animals are near to each other.

OF SPECIAL INTEREST: ELEPHANT VIGILANCE

Vigilance allows an animal to perceive a threat, but assessing the likelihood or magnitude of the threat is a more refined task. Elephants live in groups of female relatives, and it is well known that the group follows the lead of the oldest female, called the matriarch, when foraging or locating water. The matriarch also assesses threats; while all of the groups reacted more strongly to playbacks of three lions than to playbacks of one, indicating their ability to quantify threat, older matriarchs (>59 years) also discriminated between the sounds of males and female lions, and reacted more strongly to male lion roars. This is appropriate, because male lions pose a greater threat, and even one male lion can overcome a young elephant—something that elephant matriarchs apparently learn with age.[31] Elephants also discriminate between groups of humans, reacting fearfully to the smell and color of Maasai men's clothing; Maasai men are more likely to spear elephants than Kamba men, and clothing from the latter group did not elicit a negative reaction.[32] In addition, the elephants were more defensive when they heard adult male Maasai voices compared to voices of females or boys, or to the voices of Kamba men.[33] These types of distinctions are examples of how some of the attributes covered in the chapters on learning and cognition (see Chapters 5 and 6) can have important survival value.

Obviously, alarm is an important result of vigilance. When animals in groups perceive danger, they frequently behave in ways that alert other animals to the danger.[34] Here it is important to understand the difference between cues picked up from other animals and actual alarm signals, which evolved in response to selection that favored conveying alarm. For instance, sometimes the alert is simply a matter of being tuned in to neighboring animals; in a feeding flock of geese, one or more birds will be vigilant and the rest feeding. If a goose becomes alarmed, its head pops up, then the goose next to it looks up, and so on, so the heads-up alarm passes as a wave through the flock.[35,36] Indeed, the trade-off between feeding and vigilance is one of the major constraints of antipredatory behavior.[37–39] In other cases, alarm is signaled—using ways of communicating alarm that have evolved specifically for that purpose (see Chapter 7). Alarm signals can be chemical, visual, or auditory.

Signaling alarm can be costly; when an animal makes itself more visible, makes a sound, or emits an odor, it may attract the predator to itself. Because alarm signaling is often costly, alarm signals are expected to have evolved only when there is an evolutionary benefit to the signaler. Generally this benefit comes from alerting closely related animals—siblings or offspring (see Chapter 13). In other words, overt alarm signals should occur in family groups. Often animals can "eavesdrop" on alarm signals, gaining valuable information that may not have evolved for their benefit. Eavesdropping can happen between family groups or even across species lines.

DISCUSSION POINT: HUMAN VIGILANCE

Even humans display vigilance behavior in ways that might be surprising. Of course, we are alert in dark alleys and strange, forbidding places, but there is evidence that we are routinely vigilant in ways that perhaps elude consciousness. For instance, when eating in a group setting (e.g., cafeteria), individual humans looked up (vigilance scanning) less often and for shorter time periods as the size of their group(s) increased. People did not alternate scans or show any other evidence of coordinated scanning, which is consistent with our understanding of scanning behavior in animals that have fluid societies.[40] How might scanning be compared between groups of friends and groups of strangers? Should friendship make a difference in shared scanning?

10.4 MIMICRY AND DIVERSION

Unpleasant experiences and even noxious tastes have given rise to one of the most fascinating areas of mimicry because animals experience lower predation risk when they look like bad tasting, dangerous, or poisonous animals. This mimicry seems to work whether or not the mimic itself tastes bad. Henry Walter Bates, a contemporary of Darwin's, first described the mimicry that bears his name—*Batesian mimicry*. In this type of mimicry, a palatable prey (the mimic) is favored by natural selection if it resembles an unpalatable species (the model); it will benefit if predators learn that an animal that looks like the model is not worth eating. For this to occur, the model needs to be relatively abundant. At the other extreme, *Muellerian mimicry*, named after Fritz Mueller, occurs when two unpalatable species converge on an appearance; both may benefit by increasing the number of "teachers" that are informing predator-"students" about these bad-tasting prey. In sum, mimicry covers a broad range of evolutionary possibilities (Figure 10.18).

Batesian mimicry, in which a benign food item looks like or behaves like a distasteful or poisonous species, and Muellerian mimicry, in which noxious animals converge on the same appearance or behavior, are important self-defenses; examples range throughout the animal world. One hypothesis for the brightly colored but nontoxic king snake is that it may mimic the highly poisonous coral snake. Both have red, yellow, and black bands in differing arrangements: "Red and yellow, kill a fellow; red and black, venom

KEY TERM Batesian mimicry is a behavior in which a benign food item (prey) looks like or behaves like a distasteful or poisonous species.

KEY TERM Muellerian mimicry is a behavior in which noxious animals converge on the same appearance or behavior.

FIGURE 10.18
Stinging insects and their mimics. (A) This is a yellowjacket, a wasp with a very painful sting. (B) A honeybee.
(C and D) Flies that mimic stinging insects. *Photos: (A) Whitney Cranshaw, Colorado State University, Bugwood.org;*
(B) Deng Xiaobao; (C) Susan Ellis, Bugwood.org; (D) Johnny N. Dell, Bugwood.org.

lack." Many butterflies are Batesian mimics of other butterflies, whereas yellow or orange
stripes on the abdomen denote Muellerian mimicry among stinging insects.

Probably every student has experienced a wasp or bee sting and learned, through that
experience, to associate yellow/orange and black banding in a flying insect with pain. Birds
and nonhuman mammals make the same association; this results in protection for stinging
species with similar appearances (Muellerian mimicry) and for nonstinging species that look
like a bee or wasp (Batesian mimicry). Color and shape tell part of the story, but for a species
to be a really convincing mimic of a stinging insect, behavior comes into play. Flies that look
like bees add to their deceptive story by being around flowers. Flies give themselves away
to knowledgeable humans (particularly entomologists) by having a different pitch to their
buzz and a propensity to hover rather than to dart when they fly. The extent to which a good
morphological mimic is also a good behavioral mimic often determines its success.

DISCUSSION POINT: BATESIAN OR MUELLERIAN MIMICS?

Experimentally, it can be difficult to discover whether a mimic is Batesian or Muellerian.[41] For many years,
scientists thought that Viceroy butterflies were Batesian mimics of Monarch butterflies (Figure 10.19).
More recently, studies have shown that Viceroys can be quite unpalatable and that some populations
of Monarchs can be tasty to birds. There are actually many instances all along this continuum between
palatable/unpalatable and unpalatable/unpalatable pairings in which mimics may be unpalatable, but
not as unpalatable as models. In fact, palatability can vary even within a population, so the story in at
least some cases is not as neat as that presented here. Nevertheless, mimicry does happen, and in most
cases, the thing that is mimicked is the model's aposematic advertisement of unpalatability. Do your own
literature search to explore this issue further, and discuss with your classmates the interaction between
Batesian and Muellerian mimicry.

FIGURE 10.19
A monarch (left) and a Viceroy butterfly (right), long thought to be a tasty mimic of the noxious Monarch butterfly. Recent research has blurred this distinction. *Photos: left, Jeff Mitton; right, Thomas G. Barnes/US Fish and Wildlife Service.*

340

FIGURE 10.20
In addition to *Hemeroplanes*, other caterpillars such as this spicebush swallowtail (*Papilio* sp.) have benefitted from looking like snakes. *Photo: Ronald F. Billings, Texas Forest Service, Bugwood.org.*

Sometimes animals defend themselves by imitating dangerous things. A number of caterpillars have developed an alarming resemblance to snake heads. Perhaps the most realistic example of this is the *Hemeroplanes* caterpillar (also known as a "viper worm"). When startled, it inflates its head and appears remarkably snake-like (Figure 10.20). The mimic octopus (see earlier description) comes to mind again.[42] When disturbed or in the presence of predators, this octopus is known to mimic other animals, such as sea snakes; tellingly, the animals it mimics are often dangerous ones. Indeed, even top predators such as sea snakes might benefit from imitating dangerous animals—in this case, imitating themselves. Despite their extremely toxic venom, they nonetheless are at risk from sharks and other large predators. When they forage and explore crevices, they must relax vigilance. During these times, they use both behavior and appearance to ward off danger. In both color and pattern, the tail of the sea snake *Laticauda colubrine* looks remarkably similar to the head, especially from the side. When the snake is probing crevices, it slowly twists its tail, thus offering the appearance of the head to all comers. In this way, the sea snake mitigates the reduced vigilance that frequently accompanies foraging.[43]

A rather dramatic octopus was recently discovered in a decidedly undramatic habitat. Indeed, the ordinary and even uniform appearance of the background (silt and sandy littoral areas near the mouths of rivers in Indonesia) may account for both the reason that the octopus remained undiscovered until 2001 and also the adaptive influences that cause the octopus's mimicry strategy to be so unusual. This mimic octopus (*Thaumoctopus mimicus*) mimics a variety of poisonous vertebrates, ranging from poisonous fish to sea snakes, possibly because camouflage against such a uniform background might not be highly successful. (Be ready to consider the phenomenon of mimicking dangerous animals later, in the context of deterrence.) In addition, however, when fleeing, this octopus can mimic a flounder, but then match its background (camouflage) when it stops, much like *M. defilippi* (see Section 10.2). Clearly, although predator avoidance strategies can be categorized for ease of organized discussion by humans, to the animal doing the predator avoidance, they blend together with remarkable effectiveness.

Mimicry extends beyond adopting another animal's appearance. Some caterpillars look deceptively like bird droppings, others like twigs, leafhoppers may look more like leaves than like insects, and a variety of treehoppers mimic thorns. Perhaps even more deceptive, some animals disguise parts of their bodies so that they resemble other parts.[44] This behavior is called *diversion*. For instance, some Lepidoptera (e.g., *Thecla togarna*) have wing markings that look like heads (complete with long "antennae") on the posterior parts of their wings. There is some debate about how redirecting attack benefits the prey, but the consensus is that the predator's attack is indeed redirected by such mimicry.

In lizards, tail-flicking may serve a similar diversionary purpose. In this case, the tail does not mimic any other part of the body, but because it moves more than the rest of the animal, it draws the predator's attention away from more vulnerable areas. If the tail is seized by the predator, the lizard can autotomize (break off) the tail and escape.[45,46] In at least one species of lizard, *Acanthodactylus beershebensis*, this diversion is correlated with ontogenetic changes in foraging behavior.[47] Young (<3 weeks old) lizards are active predators, moving about more than older lizards and frequenting more open microhabitats. Because of this, young lizards probably expose themselves to predators more than older lizards do, which tend toward sit-and-wait foraging behavior. The young lizards also have blue tails, which they wiggle and wave more often than adults do; the blue color of the tails fades with age, along with the wiggling and waving behavior. Thus, conspicuous tail color, conspicuous tail movement, and conspicuous foraging habits are all common in young lizards, and not common in older lizards.

Even scent mimicry is possible. It is thought that ground squirrels engage in defensive scent mimicry when they chew rattlesnake skins that have been shed and then lick their fur.[48]

OF SPECIAL INTEREST: MASS OVERWINTERING BY SNAKES

Garter snakes (*Thamnophis sirtalis parietalis*) overwinter in dens. When they first emerge, they are cold and sluggish. In addition, emerging females are almost immediately set upon by hordes (>100) of males in mating balls. Emerging males may mimic females, producing a pheromone that is typical of a female. It was first thought that this behavior was a reproductive tactic that allowed the female-mimicking males (called *she-males*) increased access to females. Further research with larger sample sizes and temperature-sensing devices revealed that this was a male strategy that could have two benefits: protection from predators in the mating ball and faster posthibernation warm-up. The reproductive males seeking females have a body temperature of over 25°C; the ground is more than 15°C cooler. Females that are the subject of male attention increase their body temperatures from 4°C to 20°C in 30 min. Males in a similar situation receive similar advantages. The she-male behavior is transient and disappears as a she-male warms up.[49]

Animals may even mimic a wounded and vulnerable version of themselves. Killdeers are well known for their "broken wing" displays. In this display, a parent bird, upon spotting a predator, will move away from its nest, dragging an "injured" wing. It will do this, continuing to move just out of the predator's reach, until it is a safe distance from the nest; then it will fly away. The predator, in the meantime, has been lured away from the bird's offspring (Figure 10.21).

Finally, some animals go so far in self-mimicking as to mimic their dead selves. This behavior is called *thanatosis*, from a Greek word meaning "a putting to death," and is common not only among insects but also vertebrates. Virginia opossums (*Didelphis marsupialis*) are perhaps most famous for this behavior and have given rise to the expression "playing possum," in reference to not only death-feigning, but almost any kind of duplicity, especially that involving health.[50] The adaptive benefits of the behavior have been rigorously tested in flour beetles. Researchers selected two strains of *Tribolium* beetles for long and short death-feigning episodes, and showed that after 10 generations of selection, which resulted in a clear thanatosis difference between the two lines, those selected for thanatosis were far more likely to survive the attentions of a predatory spider.[51]

OF SPECIAL INTEREST: THANATOSIS IN SHAKESPEARE—THE BETTER PART OF VALOR

Humans are among the vertebrates that occasionally feign death, and in the world of theater, there are few death-feigning scenes more famous than that of Falstaff in Henry IV Part I. *After escaping almost certain death by playing dead, Falstaff says that playing dead in order to live is "no counterfeit." He then goes on to advise caution, speaking one of those many phrases from Shakespeare's work that has remained in use in the English language for hundreds of years:*

> *To die is to be a counterfeit, for he is but the counterfeit of a man who hath not the life of a man; but to counterfeit dying when a man thereby liveth, is to be no counterfeit, but the true and perfect image of life indeed. The better part of valor is discretion…. (Scene 4: 114–118)*

10.5 EVASION

Not all prey are cryptic, and not all crypsis succeeds. In this event, when a prey animal attracts the unwelcome attention of a predator, two major strategies remain: fleeing and fighting. These strategies are not mutually exclusive; fleeing offers fewer choices than fighting, which can assume many forms, ranging from outright combat to chemical defenses (e.g., venom and repellants), morphological defenses (spines, armor), and simply very bad taste.

In the case of fleeing, the choices are usually fairly straightforward, although mimicry can once again play a role, usually by confusing a predator. Cephalopods (again!) are experts at this confusion strategy, using not only chromatophores but also ink sacs. These ink sacs are a modification of part of the lower intestine; they empty near the siphon, so when the mollusk jet-propels itself to safety, the ink (if released) squirts out with the mantle cavity water. The large dark cloud created by the ink may have several effects on the predator. First, recall that the cephalopod can change colors rapidly; if the cephalopod turns a very light color as it flees (after the ink release), the predator may be more attracted to the dark image where the cephalopod was (and what the cephalopod looked like a moment ago) than to the light-colored apparition receding in the distance. Even if there is no color change, the dark mass of ink may be confused with the cephalopod, by now rapidly decamping. Second, the ink has been hypothesized to contain chemicals that can hamper the olfactory abilities of the predator or be otherwise unpleasant. Some workers have suggested that the ink also acts as an alarm pheromone, but this is difficult to distinguish from other antipredatory properties of the ink; if it is chemically unpleasant, other cephalopods, as well as predators, may react negatively to it. The ink of another mollusk, the sea hare, is viscous and sticks to predator sensory appendages such as lobster antennae. In so doing, it may physically prevent the antennae from sensing the presence of prey. It can also interfere with neuronal responses to prey chemicals.[52]

"Flash" behavior also mixes predator confusion with fleeing. In flash behavior, an animal flees and then freezes. One common example of this is the flight of grasshoppers. When fleeing a disturbance, a grasshopper shows different colors (e.g., colored second wings), which then disappear when the insect alights and remains motionless.

Animals that run away without freezing often tend to do so in an erratic, zig-zag fashion, introducing an element of unpredictability into their escape. This makes it difficult for the predator to know where the prey is and also makes the prey difficult to chase with great speed, even if it can be followed. Noctuid moths are some of the champions of this behavior. They have membranes on each side of the thorax that act as "ears" that are connected to the A1 and A2 neurons. The A1 neuron alone is stimulated when the predator, in this case a bat, is far away and the sound is not loud; under these conditions, the moth simply flies away, a feat that is made possible by a comparison of input from right and left neurons. When the right and left A1 neurons are responding equally, then the moth is moving directly away from the bat. If the sound from the bat is loud, however, both A1 and A2 neurons fire; in this case, the moth wingbeats are desynchronized, and the moth flies erratically and falls through the air in an unpredictable pattern, perhaps the ultimate in evasive action.[53] True to the "arms race" framework for predator–prey interactions, the contest between moths and bats does not end with an erratic crash landing on the part of the moth. At least one bat species has evolved to emit lower amplitude echolocating calls; in so doing, it sacrifices relatively long distance prey location for the advantage of a stealthy approach to moths that might otherwise hear its cries.[54] Yet again, some moths can produce ultrasonic clicks that interfere with bat echolocation and reduce the bats' foraging success.[55] The arms race does not end.

Absent more sophisticated responses, many long-legged animals simply run very fast. Deer and their relatives are excellent examples of this strategy. Although unadorned running away is an obvious escape strategy, its importance should not be underestimated. The intense selection on predators for speed and stealth (e.g., cheetah, leopard) is testimony to the value of a fast getaway.

Stotting is an intriguing behavior in which the prey is thought to actually advertise its readiness to take on a race with a predator.[56] Ungulates such as antelope and deer perform this behavior by jumping, stiff-legged, with all four feet leaving the ground simultaneously. Of course, like many behaviors, it might also function as a warning display or a diversion tactic if young are in the area, but observations of the effect of Thomson's gazelle stotting on cheetahs showed that cheetahs were less likely to pursue a gazelle that had stotted, and if they did pursue it, they were more likely to fail in the attempt to catch it. Stotting therefore combines potential fleeing behavior with a deterrent; most predators do not voluntarily waste time or energy in high-speed chases. They may also be deterred if the prey seems dangerous.

10.6 PREDATOR DETERRENCE AND FIGHTING BACK
Physical Deterrence

Spines, horns, teeth, large size, and a variety of other weaponry can keep predators at bay. Many animals use hair, feathers, or posture to appear larger than they actually are; they do this in intraspecific conflict (see Chapter 11) and when confronted by other threats, including those from predators. Although large animals such as porcupines and elk are usually associated with this defensive equipment, even small crustaceans can be armored. One of the functions of spines on cladocerans may be to thwart predators; in some experimental situations, cladocerans grow longer spines in the presence of predators.[57,58] Likewise, spiny rotifers are also less likely to be eaten (by other rotifers) than less well-protected conspecifics.

Predators do well to be cautious of armored or aggressive prey. If predators are injured in the process of acquiring a meal, such wounds may, at best, hamper future foraging and, at

343

worst, cause death if infected. Since the discovery of penicillin, humans have lost much of their awareness of the dangers of infection, yet it was always a lurking danger prior to the discovery of antibiotics, and may be again if we continue to ignore the impact of widespread antibiotic use on the evolution of drug resistance in pathogens. Death from infection, or from the debilitation associated with infection, is not an uncommon occurrence in a world without antibiotics, and that is the predator's world.

OF SPECIAL INTEREST: SPINY EXOSKELETONS

As we have emphasized, few benefits are without their costs, and this is certainly true of the arthropod exoskeleton. Although its tough construction is thought to have played a major role in making arthropods the most successful phylum on the planet, the fact that it is hard and largely inflexible requires that it be molted periodically if the animal is to grow. The molt is a hazardous time for arthropods; emergence from the old exoskeleton may be fraught, and even if that proceeds smoothly, their new body armor is soft and easily punctured until it hardens. For this reason, many arthropods behave secretively and hide during the time surrounding their molt.

Spiny lobsters (*Panulirus argus*) can use their exoskeletons for defense even when they are newly molted (Figure 10.22). When alarmed by a predator, they make a loud rasp that is thought to startle the predator. Unlike many arthropods that make sounds by rubbing hard ridges of the exoskeleton together (see Chapter 7),

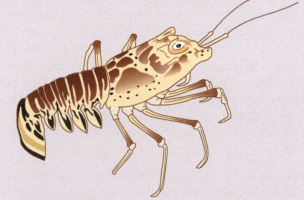

FIGURE 10.22
Spiny lobsters defend themselves with their hardened exoskeleton, but when it is soft and newly molted they make a noise that repels predators, rubbing an extension of the antenna across a "file"-like area near their eyes.

spiny lobsters use structures that do not need to be hard in order to produce sound. They rub an extension of the antenna across microscopic shingles on a "file" located below the eye, producing sound in much the way a violin does, although perhaps with less pleasant results for the predator.[59]

Individual animals are known to react aggressively toward predators; a variety of birds and small mammals are noticeably unfriendly to predatory snakes. A special case of physical deterrence involves group defense; this is often called *mobbing*. Smaller birds, for example, may mob a crow or hawk attempting to steal eggs or hatchlings from a nest.[60] In turn, crows sometimes use mobbing against larger birds, such as eagles (Figure 10.23). Each of the individual prey may be relatively defenseless on their own, but as a group they can put up an effective defense, dive-bombing and harassing the predator. Some behavioral biologists see mobbing as a type of aggression (see Chapter 12).

Chemical Deterrence

There are two fundamental kinds of chemical deterrence: those that repel a predator before it attacks and those that create unpleasant experiences for the predator after an attack. Many organisms secrete, emit, squirt, or exude what collectively amounts to a pharmacopoeia of noxious repellants. The skunk is perhaps the most dramatic of the group. Although biologists have dutifully described four genera and several species of skunk, the names of the family (Mephitidae) and type genus (*Mephitis*) translate from Latin as "noxious vapor"

344

FIGURE 10.23
Mobbing is a common occurrence among birds. Smaller birds form groups and harass larger birds, including birds of prey, making it difficult for the larger bird to hunt or remain unnoticed.

and leave no doubt about the skunk's chemical deterrence strategy. As none other than Darwin observed, "Conscious of its power, it roams by day about the open plain, and fears neither dog nor man. …Whatever is once polluted by it, is forever useless."[61] When disturbed, a skunk stomps its feet, hisses, growls, and displays its vivid striped pelage, all prior to an accurate 2–5 m squirt of an obnoxious mixture of sulfides and mercaptans. Recent observations of responses of carnivores to models of skunks and gray foxes showed that mammalian carnivores approached skunk-like models more cautiously than they approached gray fox models (both pattern and shape were important to the carnivores), and that the degree of caution is positively related to the abundance of skunks in the area. Thus, skunk appearance qualifies as aposematism, that is, a conspicuous warning from a distance that this animal might be best left alone.[62] Aposematism is due more consideration shortly, in a discussion of negative postattack experiences.

345

FIGURE 10.24
A bombardier beetle spraying noxious, heated, defensive secretions. This behavior effectively deters ants and other predators. *Photo: Copyright (1999) National Academy of Sciences, USA.*

Animals other than skunks use chemicals to fend off attackers. Beetles, in general, need good defenses because, given their elytra, they are frequently slower to take flight than some other insects. For example, carabid beetles have paired abdominal glands containing chemicals that are mixed upon release to impressive effect. The bombardier beetle (e.g., *Stenaptinus insignus*) stores hydroquinones and hydrogen peroxide in one chamber and enzymes including peroxidases in the other. When these come together, a rather intense chemical reaction takes place, with oxygen released from the hydrogen peroxide and reacting with the hydroquinones to produce both great heat (100°C) and noxious chemicals (benzoquinones) that can be aimed from the abdominal tip with surprising accuracy (Figure 10.24). Attacking a bombardier beetle cannot be a good experience, and eating one (should a predator somehow persist through this deterrence) must be downright miserable. Bombardier beetles are high on the list of Animals We Do Not Wish to Consume.[63]

OF SPECIAL INTEREST: CHEMICAL DEFENSES

A wide assortment of poisonous or distasteful animals gets noxious chemicals from diets. For instance, when feeding on milkweed, monarch butterfly caterpillars sequester the cardiac glycosides and other poisons that the host plant contains, thus discouraging many (but not all!) predators. Lovely poisonous frogs get their toxic compounds from their arthropod diets. A variety of nudibranchs have special evaginations (external pouches called *cerata*) of the dorsal and lateral body wall that contain the undischarged stinging cells from the corals on which they graze. These cerata—along with other noxious substances—discourage and distract predators and provide a good deal of protection for these mollusks that have (over evolutionary time) lost their shells.

The use of another animal's chemical defense can be related to other aspects of antipredator behavior. The Asian snake *Rhabdophis tigrinus* has special nuchal glands that contain bufadienolides (cardiotonic compounds) in the skin of the dorsal neck area. (*Nuchal* indicates the nape of the neck.) When confronted by a predator, the snake actually directs this area of the neck toward the attacker. Oddly, the nuchal glands do not possess secretory cells, and few of their cells contain organelles associated with secretory activity (e.g., Golgi apparatus and endoplasmic reticulum); they are, however, heavily vascularized. This histological evidence led researchers to hypothesize that the glands store chemicals acquired from an external source.

By comparing snakes from an island that had abundant toads with snakes from a toad-free island, Deborah Hutchinson and colleagues showed that the nuchal glands of the former group were rich in bufadienolides, whereas the latter group had none. However, the snakes from the toadless island had not lost their ability to sequester these poisonous compounds. They, too, could accumulate bufadienolides when fed toads. In addition, female Asian snakes with a good supply of such compounds can pass them on to their offspring. Finally, while snakes from the toad-rich island used their nuchal glands as part of a defensive posture, snakes from the toad-free island were more likely to flee from a predator.[64]

Venom, Stings, and Defensive Bites

Predators often become prey. As discussed in Chapter 9 on foraging, predators that are higher in the food web benefit from consuming lower predators because this eliminates some of the potential competition. Ecologists refer to middle-sized predators as *mesopredators*; coyotes, for example, consume domestic dogs, foxes, and smaller predators but are themselves vulnerable to predation by wolves. Top predators, such as mountain lions in North America, are typically not potential prey because of their size and armament, although all animals, no matter what their size, may be exploited by parasites (see Section 10.7). Much of the weaponry used by predators can be turned to defensive purposes. A classic example is the venom of pit vipers, which evolved to allow the snakes to immobilize prey, but which is also a potent deterrent against potential predators (Figure 10.25). Sometimes defensive bites by vipers are "dry," meaning that no venom is transferred; the dry bite has the effect of inducing fear in the potential predator while at the same time preserving venom for later use in predation.

> **KEY TERM** A mesopredator is a middle-sized predator that occupies an intermediate position within a food web. Mesopredators are the potential prey of top predators as well as being predators themselves. The prefix "meso" comes from a Greek word for "middle."

10.7 PATHOGEN AVOIDANCE/DETERRENCE AND SICKNESS BEHAVIOR

Although many behavioral biologists like to think about lions and tigers more than tapeworms and ticks, parasites, by common definition, have negative effects on fitness, and the ways in which animals cope with the threat of disease are rich beyond imagining. They are only now coming to light, because until recently, most behavioral biologists did not recognize that coping with parasites and pathogens played a major role in animal behavior.[65] Chapter 9 briefly summarized some of the ways that natural selection has favored parasites that can use behavior to get transmitted and, once in a host, to survive. This chapter addresses how natural selection favors hosts that can use behavior to fight back.[66]

Pathogen Avoidance

Humans tend to think of host defense as something that happens after an animal acquires a parasite, when the immune system is activated. That certainly occurs, and it has behavioral components that will be addressed shortly. However, even as hiding successfully from a predator may be less risky than fighting against one, so, too, the use of behavior to avoid infection altogether can be at least as valuable as fighting off the infection. After all, as humans, our first line of defense is not the pharmacy or the doctor, but the simple act of washing our hands, a form of behavioral immunity.

Ectoparasites provide some of the best examples of parasite avoidance. Although humans are inclined to view ectoparasites as more of an annoyance than a serious threat, this (again) is not representative of the vast majority of animals that have no shelter from, say, biting flies, and no medicines with which to combat the diseases that they transmit. For smaller animals, blood loss itself can be a significant health hazard; biting flies can have a negative impact on milk production and weight gain in animals as large as cattle. To most animals, ectoparasites can be a substantial threat and are worth avoiding.

Indeed, fly swatting by howler monkeys (*Alouatta palliata*) proves this point nicely: these monkeys engaged in over 1500 slaps or other avoidance maneuvers during a 12-h period, using 24% of their metabolic budget in the process (less basal metabolism).[67] Wood rats can use bay leaves—nibbling them so as to release the volatile chemicals they contain—to "fumigate" their nests, probably reducing the number of flea larvae in the nest.[68]

> **KEY TERM** An ectoparasite lives on the outside of its host; a flea is an example of an ectoparasite. Endoparasites, such as tapeworms, live within their host.

Parasite avoidance could have some serious implications for behavioral ecology. It is clear, for instance, that organisms may change habitat preference in an attempt to avoid parasites. Like many animals, juvenile stickleback fish (*Gasterosteus* spp.) frequented areas near the bottom and near vegetation in a tank that was devoid of the branchiuran parasite *Argulus canadensis*. In the presence of the parasite, however, the fish spent more time near the surface and in open areas, where the parasite was less abundant.[69] Clearly, the presence or absence of the parasite would influence a study of stickleback habitat preference.

Social behavior is affected by parasites, much as it is affected by predators. Both the dilution effect and the selfish herd principle (see Chapter 13) can be seen in animal responses to biting flies. During fly season, many animals "bunch," that is, form larger and tighter herds. In experimental tests of bunching, repellant-sprayed Holstein heifers did not form groups as often as control (water-sprayed) ones did.[70] Group formation may lower the likelihood of any one individual being bitten.

Even coloration and camouflage, so prominent in predator avoidance, may function in parasite avoidance as well. In 1981, Jeffrey Waage hypothesized that the stripes of zebras may contribute to protection from tsetse flies, which have distinct preferences for some patterns over others and, in general, prefer large, dark objects to lighter ones. (The boldest zebra patterns co-occur with tsetse.[71]) A recent study has explored Waage's hypothesis in more detail and found that the geographic occurrence of stripe patterns and anatomical locations of stripes on zebra subspecies correspond to the distribution and biting habits of major biting flies (tabanids and tsetses) in the zebras' geographic range. In contrast, there was little congruence between stripe occurrence and measures that reflected other hypotheses about zebra stripes, such as camouflage, thermoregulation, and predator confusion.[72,73] Of course, many adaptations serve multiple functions, and zebras may be another reminder that science is often not about finding a single explanation.

Avoidance of microscopic parasite propagules is more difficult to study; if nothing else, time and extent of exposure are frequently unknown. Nonetheless, parasite avoidance has been hypothesized to account for xenophobia; when strangers are kept away from the group, so are their pathogens. Hygienic behavior is also widespread, and many animals are fastidious about where they deposit urine or feces. This may be in part related to the function of each in identification and territory marking, but in addition, some species will not forage where they have defecated. Social insects have developed special alarm behaviors to alert nestmates to contagion.[74] Dying social insects often leave the colony whether or not they are diseased[75] and in the absence of such behavior, some social insects are adept at recognizing dead nestmates and removing them.[76] Some termites actually use their own feces as part of nest construction; the fecal material supports a bacterium that has antifungal properties, reducing the likelihood of fungal infection in the colony.[77]

Sexual selection (see Chapter 11) can be influenced by parasite avoidance. There are several reasons why choice of a parasite-free mate is beneficial: first, the absence of parasites may indicate "good genes," that is, a genetically based resistance to parasites. Second, a parasite-free mate will not transmit parasites to mates or offspring. Parasite-free mates may also be better at parental care, if such exists.

OF SPECIAL INTEREST: PARASITES AND ANTI-PREDATOR BEHAVIOR

As we have noted (see Chapter 9), parasites are adept at manipulating host behavior, and they often do so in ways that enhance parasite transmission. Perhaps the most surprising manipulation is that of reduced antipredator behavior. This has been seen in a wide range of parasite and host taxa, ranging from isopods parasitized by acanthocephalans (they are more likely to be around a fish predator) to fish with cestodes (they recover more quickly from fright response). However, in no system has this phenomenon been more thoroughly investigated than in rodents with *Toxoplasma gondii*. This protistan parasite can infect a variety of hosts but undergoes a sexual phase in felids. We have known since 1980 that mice with *Toxoplasma* were less fearful, but recent studies on rats have revealed an even more amazing scenario: rats infected with *Toxoplasma* are not only unafraid of cat urine, but are actually attracted to it. Indeed, they are attracted to the urine of wild felids (cheetahs and pumas) more than to the urine of domestic cats![78] In addition, this is not because of a general reduction in fear; the infected rats continued to display more general fear responses (e.g., edge preference, shock avoidance) and spatial learning was unaffected. Although the parasites were encysted in a variety of areas of the brain, some studies have found them to be most concentrated in the amygdala (see Chapter 2).[79]

Parasites with complex life cycles are known to manipulate predator avoidance in at least one more way: some can maximize exposure to appropriate hosts and minimize risk from predators that are not suitable hosts. This occurs in snails infected with *Microphallus* trematode larvae that mature in waterfowl. The infected snails forage on the tops of rocks when waterfowl are feeding, but move to more protected places when fish predation is heaviest.[80]

Sickness and Fever

What happens if, despite an animal's best efforts, avoidance fails, and it becomes parasitized? Behavior is also important in this event, and being "sick" is often just a part of getting well. Humans who are sick are frequently sleepy, depressed, inactive. Food may not be very attractive. Pets behave in a similar manner; they groom themselves less, sleep a lot, and are not interested in the food bowl. These behaviors, hereafter called *sickness behavior*, are frequently seen as undesirable, even "pathological," and when viewed as symptoms of disease, they are subject to "treatment," with the goal of "feeling better." Veterinary medicine professor Ben Hart realized that, far from being signals of physiological systems gone awry, these behaviors are adaptive.[81] They can be viewed as a syndrome, much of which functions to conserve energy so that the fever response can be supported. Fever helps the body defend against pathogens by both inhibiting their reproduction and stimulating the immune system.

> **KEY TERM** Sickness behavior is a set of behavioral adaptations that cause an animal to conserve energy and protect itself while it is fighting infection or parasitism.

Elements of the immune system, such as fixed tissue macrophages and lymphocytes, release cytokines when they are exposed to foreign substances such as bacteria. These cytokines raise the thermoregulatory set point of the organism, causing the generation of fever. The cost of raising body temperature varies with the animal and its environment; in humans, a temperature increase of $1\,°C$ can cost a 13% increase in metabolism. Remaining inactive can furnish the extra energy needed for such an increase. Moreover, by not foraging, the animal can sequester its existing iron and keep new iron sources from pathogens that are inhibited by iron deficits. By not grooming, the animal can reduce the substantial water loss, as well as evaporative cooling, that is associated with grooming. Of course, in the absence of grooming, ectoparasites will inevitably increase, but they can be managed by resumed grooming after the threat of illness has passed.

Not every type of animal can generate fever metabolically like an endotherm can. For ectotherms, there is *behavioral fever*, a particularly fascinating phenomenon because the beneficiary of behaviors that lead to fever in ectotherms can be ambiguous. Of course, it can be powerfully adaptive for the infected animal, as reviewed in the case study for this chapter. On the other hand, the feverish animal is exposed to the sun, and thus exposed to predators, a potentially adaptive outcome for any parasites with complex life cycles that the host might harbor. In addition, in the case of social insects, behavioral fever can remove infected individuals from the colony as they seek hot places, thus reducing exposure of relatives to pathogens. Finally, if the infected animal crawls high, it can disperse infective propagules over a wide area (see Chapter 9). Indeed, the phenomenon of insects infected with some pathogens crawling to high places is so widespread and well known that it garnered its own special designation over 100 years ago: *Wipfelkrankheit*, or "tree-top disease."[82] Much like the zebra's stripes, this behavior offers a wealth of hypotheses in pursuit of the answers to Tinbergen's questions.

Self-Medication

Animals have a rich arsenal of pathogen deterrents, but one of the more intriguing ones is self-medication. Some of the best evidence for self-medication comes from two far-flung organisms: chimpanzees and wooly bear caterpillars. Chimpanzees are known to eat a variety of plants that have medicinal properties, but perhaps the most interesting example is whole-leaf swallowing. Whole-leaf swallowing is unlike other instances of chimpanzee self-medication, such as bitter-pith chewing, during which the juice of medicinal plants is consumed. In whole-leaf swallowing, the entire leaf of certain plants that have "hairy" (trichomes) leaves is folded and swallowed whole. Michael Huffman

FIGURE 10.26
This tropical ant is infected by a *Cordyceps* fungus. The fungus has commandeered the ant, stimulating it to walk up a twig and then to fasten its jaws firmly on the twig. The fungal fruiting bodies have then grown out of the ant—the behavior of the ant, under the influence of the fungus, has perfectly placed the fruiting bodies for dispersal of spores in the wind (see Chapter 29). *Photo: Erich G. Vallery, USDA Forest Service-SRS-4552, Bugwood.org.*

350

examined the dung produced by a whole-leaf swallowing chimpanzee and discovered that living pathogenic roundworms (*Oesophagostomum stephanostomum*) had passed out with the leaves, often entangled in their rough surfaces. It may be that animals use a variety of mechanical "medications" to clean out intestinal passages.[83]

In the case of wooly bear caterpillars, *Platyprepia virginalis*, unparasitized caterpillars thrived on the host plant lupine (*Lupinus arboreus*). In the field, however, these caterpillars forage on hemlock (*Conium maculatum*) if they are parasitized by the tachinid fly *Thelairas americana*.[84] Hosts that were allowed to exert this preference and eat hemlock could survive the tachinid infection; moreover, the fly itself did better than it did in those caterpillars forced to eat lupine. The lethal effects of tachinids are legendary, but part of the reason for that reputation may be that when field-collected caterpillars are reared in the lab, they are not given a choice of food, but are given their assumed host plant (lupine). That means that if they are parasitized, they are not allowed to display the food choice behavior that could inform scientists about self-medication and about the fact that the parasitized caterpillars need different nutrients.[85]

The issue of medicinal plants presents something of a puzzle to conservation biologists. If such plants are important to animal species that are the targets of conservation efforts, then clearly, these plants need to be present in conserved habitats…if only they could be identified. Scientists have just begun to investigate what may be the rich natural pharmacy that surrounds animals challenged by pathogens.

Host Behavior Controlled by Parasites

In some fascinating cases, an animal behaves at the command of the infectious organism or parasite that has attacked it (Figure 10.26). This was discussed in greater detail in Chapter 9 on foraging. In the case of a directly transmitted parasite, its host may move into closer contact with other animals of the same species (encouraging the passage of infection) or move to a place from which the disease or parasite is best dispersed into the environment. When hosts crawl up, and disperse pathogen propagules, such behavior can be seen to aid transmission. However, one also wonders if it is an unsuccessful attempt at defense, a fever response perhaps too late, too little, or a response captured by the parasite?

In diseases that are transmitted via sexual contact, the animal in some cases may be driven to have unusually frequent sexual contacts.[86,87] Sexually transmitted diseases (STDs) often sterilize the host, but there is little evidence that potential mates are able to avoid copulation with carriers of STDs. One reasonable hypothesis is that STDs are under strong selection NOT to interfere with traits that attract mates.

These are only a few samples of a rich literature that addresses how animals cope with parasites and pathogens. The altered behaviors range from reduction in dung processing by dung beetles to changed habitat and food preferences by ungulates and snails. Each host–parasite association has its own circumstances, its own dynamic arms race. What is clear is that parasites and the threat of parasites are ubiquitous, and behavior is a fundamental part of the host–parasite interaction. Embedded in this broad spread of behavioral change is the fact that hosts use behavior to avoid and deter parasites just as they use it to avoid and deter predators.[88]

SUMMARY

This journey through self-defense shows that evolution has used many different strategies and tactics to provide animals with tools for avoiding predation. One important theme has been that behavior interacts with nonbehavioral defenses. For example, hiding is behavior, but being well hidden requires the adaptations of color and shape involved in camouflage. Being a convincing mimic may start with color and shape, but for mimicry to work, the mimic needs to behave like its model.

Defense against predators has three elements: avoiding detection, evading capture, and fighting back. Crypsis (camouflage) can employ background matching, countershading, and disruptive coloration. These different patterns of color and form typically have the same result: they make the animal difficult to see.

If a potential prey fails to escape notice, it may run away from predators. In so doing, it may confuse predators, engage in flash behavior (sudden stops) or evasive action, or produce displays such as stotting that advertise prowess at fleeing.

Mimicry and diversion are other ways to escape notice. Mimicry differs from camouflage in that the mimic usually appears to be another living organism. Diversion may direct the predator toward a different (and more expendable) prey body area, or it may direct predators away from nest locations and young. Batesian mimics are palatable or defenseless prey that mimic unpalatable or dangerous animals. Muellerian mimics are noxious, unpalatable prey that mimic other noxious, unpalatable prey, thus increasing the negative experience a predator is likely to have with these organisms.

Vigilance behavior is best studied in groups. An alarm can be raised on any one of a number of communication channels. The existence of a group can minimize the risk of a given individual through the dilution effect or via the structure of the selfish herd.

Many of the behaviors that are expressed in the avoidance and deterrence of predation have parallels in animals' relationships with parasites and pathogens. If parasitized or infected with a pathogen, behavioral responses of the host can play an important role in survival.

351

STUDY QUESTIONS

1. Camouflage: If we come to these phenomena thinking about predators, we can begin to see considerable adaptive significance in the appearance of many animals, but once again, we must weigh alternate hypotheses before concluding that an animal's appearance is a result of natural selection favoring invisibility to predators. In many cases, we do not know the adaptive significance of an animal's color. For instance, in the case of diurnal animals and for animals that live in open areas, color may be important in thermoregulation. Color also serves important roles in communication. Tim Caro has written a thought-provoking review that proposes and, when possible, tests, several hypotheses for mammalian coloration.[1] Can you develop a group of hypotheses for color in other groups?

2. Tests of camouflage: Kettlewell, in his famous experiments on the peppered moth, assumed that birds would see what we see when they foraged for moths.[89] They seem to do just that. Can you think of a different predator that might not fit such anthropocentric assumptions? How would you devise a predation test of camouflage for a nonvisual predator, or for a predator that sees wavelengths for which we are blind? Think of forms of camouflage that involve sensory information other than vision.

3. In an entry on mimicry in the *Encyclopedia of Evolution*, John Turner says, "Mimicry is the parasitic or mutualistic exploitation of a communication channel."[90] What do you think he means by that?

4. Review the material about *A. beershebensis*, the lizard with blue-tailed young, in Section 10.4. Two other lizards in the same family do not exhibit such changes, nor do they shift foraging behavior with age. How does this help us address Tinbergen's question about the evolution when we consider this trait? Why is the comparison with other lizards in the same family (rather than with lizards in general) particularly informative?

5. Should sickness behavior invariably accompany an infection? Can you think of social, seasonal, or reproductive circumstances that might mitigate the behavior? Develop one such hypothesis and describe how you might test it. Would some species be more informative than others in this regard? Can you think of human diseases in which it would be helpful to mitigate the sickness behavior syndrome?

Further Reading

Forbes, P., 2009. Dazzled and Deceived: Mimicry and Camouflage. Yale University Press, New Haven, CT. 300 pp.

Hart, B.L., 2010. Beyond fever: comparative perspectives on sickness behavior. In: Breed, M.D., Moore, J. (Eds.), Encyclopedia of Animal Behavior, vol. 2, Academic Press, Oxford, pp. 205–210.

Hart, B.L., 2011. Behavioural defences in animals against pathogens and parasites: parallels with the pillars of medicine in humans. Philos. Trans. R. Soc. B 366, 3406–3417.

Rowe, C., 2010. Defense against predation. In: Breed, M.D., Moore, J. (Eds.), Encyclopedia of Animal Behavior, vol. 3, Academic Press, Oxford, pp. 106–111.

Stevens, M., Merilaita, S., 2009. Defining disruptive coloration and distinguishing its functions. Philos. Trans. R. Soc. B 364, 481–488.

Stoner, C.J., Caro, T.M., Graham, C.M., 2003. Ecological and behavioral correlates of coloration in artiodactyls: systematic analyses of conventional hypotheses. Behav. Ecol. 14, 823–840.

Wickler, W., 1968. Mimicry in Plants and Animals. McGraw-Hill, New York, NY.

Notes

1. Caro, T., 2005. The adaptive significance of coloration in mammals. BioScience 55, 125–136.
2. Lima, S.L., 1998. Nonlethal effects in the ecology of predator–prey interactions. BioScience 48, 25–34.
3. Vaughn, L.K., Benheim, H.A., Kluger, M.J., 1974. Fever in the lizard *Dipsosaurus dorsalis*. Nature 252, 473–474.
4. Bronstein, S.W., Conner, W.E., 1984. Endotoxin-induced behavioural fever in the Madagascar cockroach, *Gromphadorhina porentosa*. J. Insect Physiol. 30, 327–330.
5. Marikovsky, P.L., 1962. On some features of behavior of the ants *Formica rufa* L. infected with fungous disease. Insect. Soc. 9, 173–179.
6. Boltana, S., Rey, S., Roher, N., Vargas, R., Huerta, M., Huntingford, F.A., et al., 2013. Behavioural fever is a synergic signal amplifying the innate immune response. Proc. R. Soc. B 280. http://dx.doi.org/10.1098/rspb.2013.1381.
7. Watson, D.W., Mullens, B.A., Petersen, J.J., 1993. Behavioral fever response of Musca domestica (Diptera: Muscidae) to infection by Entomophthora muscae (Zygomycetes: Entomophthorales). J. Invertebr. Pathol. 61, 10–16.
8. Stahlschmidt, Z.R., Adamo, S.A., 2013. Context dependency and generality of fever in insects. Naturwissenschaften 100, 691–696.
9. Hunt, V.L., Charnley, A.K., 2011. The inhibitory effect of the fungal toxin, destruxin A, on behavioural fever in the desert locust. J. Insect Physiol. 57, 1341–1346.
10. Moore, J., Freehling, M., 2002. Cockroach hosts in thermal gradients suppress parasite development. Oecologia 133, 261–266.
11. Moore, J., 2002. Parasites and the Behavior of Animals. Oxford Series in Ecology and Evolution. Oxford University Press, New York, NY.
12. Otti, O., Gantenbein-Ritter, I., Jacot, A., Brinkhof, M.W.G., 2012. Immune response increases predation risk. Evolution 66, 732–739.
13. Ignoffo, C.M., 1981. The fungus *Nomuraea rileyi* as a microbial insecticide. In: Burges, H.D. (Ed.), Microbial Control of Pests and Plant Diseases 1970–1980 Academic Press, New York, NY, pp. 513–538.
14. Casadevall, A., 2012. Fungi and the rise of mammals. PLoS Pathog. 8 (8), e1002808. http://dx.doi.org/10.1371/journal.ppat.1002808.
15. Rankin, N., 2009. A Genius for Deception: How Cunning Helped the British Win Two World Wars. Oxford University Press, Oxford, p. 24.
16. Steinberg, D.S., Losos, J.B., Schoener, T.W., Spiller, D.A., Kolbe, J.J., Leal, M., 2014. Predation-associated modulation of movement-based signals by a Bahamian lizard. Proc. Natl. Acad. Sci. 111, 9187–9192.

17. Santos, X., Vidal-Garcia, M., Brito, J.C., Fahd, S., Llorente, G.A., Martinez-Freiria, F., et al., 2014. Phylogeographic and environmental correlates support the cryptic function of the zigzag pattern in a European viper. Evol. Ecol. 28, 611–626.

18. Niskanen, M., Mappes, J., 2005. Significance of the dorsal zigzag pattern of *Vipera latastei gaditana* against avian predators. J. Anim. Ecol. 74, 1091–1101.

19. Thayer, G.H., 1909. Concealing Coloration in the Animal Kingdom. Macmillan, New York, NY.

20. Schaefer, H.M., Stobbe, N., 2006. Disruptive coloration provides camouflage independent of background matching. Proc. R. Soc. B 273, 2427–2432.

21. Dimitrova, M., Stobbe, N., Schaefer, H.M., Merilaita, S., 2009. Concealed by conspicuousness: distractive prey markings and backgrounds. Proc. R. Soc. Lond., Biol. Sci. 276, 1905–1910.

22. Hanlon, R., 2007. Cephalopod dynamic camouflage. Curr. Biol. 17, R400–R404.

23. Finn, J.K., Tregenza, T., Norman, M.D., 2009. Defensive tool use in a coconut-carrying octopus. Curr. Biol. 19, 1069–1070.

24. Hanlon, R.T., Watson, A.C., Barbosa, A., 2010. A "mimic octopus" in the Atlantic: Flatfish mimicry and camouflage by *Macrotritopus defilippi*. Biol. Bull. 218, 15–24.

25. Mathger, L.M., Bell, G., Kuzirian, A.M., Allen, J.J., Hanlon, R.T., 2012. How does the blue-ringed octopus (*Hapalochlaena lunulata*) flash its blue rings? J. Exp. Biol. 215, 3752–3757.

26. Brown, S.G., Boettner, G.H., Yack, J.E., 2007. Clicking caterpillars: acoustic aposematism in *Antheraea polyphemus* and other Bombycoidea. J. Exp. Biol. 210.

27. Haff, T.M., Magrath, R.D., 2010. Vulnerable but not helpless: nestlings are fine-tuned to cues of approaching danger. Anim. Behav. 79, 487–496.

28. Makowska, J.J., Kramer, D.L., 2007. Vigilance during food handling in grey squirrels, *Sciurus carolinensis*. Anim. Behav. 74, 153–158.

29. Périquet, S., Valeix, M., Loveridge, A.J., Madzikanda, H., Macdonald, D.W., Fritz, H., 2010. Individual vigilance of African herbivores while drinking: the role of immediate predation risk and context. Anim. Behav. 79, 665–671.

30. Kenward, R.E., 1978. Hawks and doves: factors affecting success and selection in goshawk attacks on wood pigeons. J. Anim. Ecol. 47, 449–460.

31. McComb, K., Shannon, G., Durant, S.M., Sayialel, K., Slotow, R., Poole, J., et al., 2011. Leadership in elephants: the adaptive value of age. Proc. R. Soc. B 278, 3270–3276.

32. Bates, L.A., Sayialel, K.N., Njiraini, N.W., Moss, C.J., Poole, J.H., Byrne, R.W., 2007. Elephants classify human ethnic groups by odor and garment color. Curr. Biol. 17, 1938–1942.

33. McComb, K., Shannon, G., Sayialel, K.N., Moss, C., 2014. Elephants can determine ethnicity, gender, and age from acoustic cues in human voices. Proc. Natl. Acad. Sci. 111, 5433–5438.

34. Childress, M.J., Lung, M.A., 2003. Predation risk, gender and group size effect: does elk vigilance depend upon the behaviour of conspecifics? Anim. Behav. 66, 389–398.

35. Blumstein, D.T., Anthony, L.L., Harcourt, R., Ross, G., 2003. Testing a key assumption of wildlife buffer zones: is flight initiation distance a species-specific trait? Biol. Conserv. 110, 97–100.

36. Abbott, K.R., 2006. Bumblebees avoid flowers containing evidence of past predation events. Can. J. Zool. 84, 1240–1247.

37. Clark, J., 1983. Moonlight's influence on predator/prey interactions between short-eared owls (*Asio flammeus*) and deermice (*Peromyscus maniculatus*). Behav. Ecol. Sociobiol. 13, 205–209.

38. Maerz, J.C., Panebianco, N.L., Madison, D.M., 2001. Effects of predator chemical cues and behavioral biorhythms on foraging activity of terrestrial salamanders. J. Chem. Ecol. 27, 1333–1344.

39. Arenz, C., Leger, D., 2000. Antipredator vigilance of juvenile and adult thirteen-lined ground squirrels and the role of nutritional need. Anim. Behav. 59, 535–541.

40. Wawra, M., 1988. Vigilance patterns in humans. Behaviour 107, 61–71. Stable URL: http://www.jstor.org/stable/4534719.

41. Ritland, D.B., Brower, L.P., 1991. The viceroy butterfly is not a Batesian mimic. Nature 350, 497–498; Ritland, D.B., 1991. Revising a classic butterfly mimicry scenario: demonstration of Mullerian mimicry between Florida viceroys (*Limenitis archippus floridensis*) and queens (*Danaus gilippus berenice*). Evolution 45, 918–934.

42. Norman, M.D., Finn, J., Tregenza, T., 2001. Dynamic mimicry in an Indo-Malayan octopus. Proc. R. Soc. B 268 (1478), 1755–1758.

43. Rasmussen, A.R., Elmberg, J., 2009. "Head for my tail": a new hypothesis to explain how venomous sea snakes avoid becoming prey. Mar. Ecol. 30, 385–390.

44. Rassmussen, A.R., Elmberg, J., 2009. Head for my tail—venomous sea snakes avoid becoming prey. Mar. Ecol. 30, 385–390.

45. Punzo, F., 1997. Autotomy and avoidance behavior in response to a predator in the wolf spider, *Schizocosa avida* (Araneae, Lycosidae). J. Arachnol. 25, 202–205.

46. Formancowicz, D.R., 1990. The antipredator efficacy of spider leg autotomy. Anim. Behav. 40, 400–401.

47. Hawlena, D., Boochnik, R., Abramsky, Z., Bouskila, A., 2006. Blue tail and striped body: why do lizards change their infant costume when growing up? Behav. Ecol. 17, 889–896.

48. Clucas, B., Rowe, M.P., Owings, D.H., Arrowood, P.A., 2008. Snake scent application in ground squirrels, *Spermophilus* spp.: a novel form of antipredator behaviour? Anim. Behav. 75, 299–307.

49. Shine, R., Phillips, B., Waye, H., LeMater, M., Mason, R.T., 2001. Benefits of female mimicry in snakes. Nature 414, 267.

353

50. Gerald, G.W., 2008. Feign versus flight: influences of temperature, body size and locomotor abilities on death feigning in neonate snakes. Anim. Behav. 75, 647–654.

51. Miyatake, T., Katayama, K., Takeda, Y., Nakashima, A., Sugita, A., Mizumoto, M., 2004. Is death-feigning adaptive? Heritable variation in fitness difference of death-feigning behaviour. Proc. R. Soc. B 271, 2203–2296.

52. Love-Chezem, T., Aggio, J.F., Derby, C.D., 2013. Defense through sensory inactivation: sea hare ink reduces sensory and motor responses of spiny lobsters to food odors. J. Exp. Biol. 216, 1364–1372.

53. Fullard, J.H., Dawson, J.W., Jacobs, D.S., 2003. Auditory encoding during the last moment of a moth's life. J. Exp. Biol. 206, 281–294.

54. Goerlitz, H.R., ter Hofstede, H.M., Zeale, M.R.K., Jones, G., Holderied, M.W., 2010. An aerial-hawking bat uses stealth echolocation to counter moth hearing. Curr. Biol. 20, 1568–1572.

55. Corcoran, A.J., Barber, J.R., Conner, W.E., 2009. Tiger moth jams bat sonar. Science 325, 325–327.

56. Caro, T., 1986. The functions of stotting in Thomson's gazelles: some tests of the predictions. Anim. Behav. 34, 649–662.

57. Relyea, R.A., 2001. Morphological and behavioral plasticity of larval anurans in response to different predators. Ecology 82, 523–540.

58. Dewitt, T.J., Sih, A., Hucko, J.A., 1999. Trait compensation and cospecialization in a freshwater snail: size, shape and antipredator behaviour. Anim. Behav. 58, 397–407.

59. Patek, S.N., 2001. Spiny lobsters stick and slip to make sound. Nature 411, 152–153.

60. Shedd, D.H., 1982. Seasonal variation and function of mobbing and related antipredator behaviors of the American Robin (*Turdus migratorius*). Auk 99, 342–346.

61. Darwin, C., 1839. The Voyage of the Beagle. John Murray, London, Chapter 4.

62. Hunter, J., 2009. Familiarity breeds contempt: effects of striped skunk color, shape and abundance on wild carnivore behavior. Behav. Ecol. 20, 1315–1322.

63. Eisner, T., Aneshansley, D.J., 1999. Spray aiming in the bombardier beetle: photographic evidence. Proc. Natl. Acad. Sci. 96, 9705–9709.

64. Hutchinson, D.A., Mori, A., Savitsky, A.H., Burghardt, G.M., Wu, X., Meinwald, J., et al., 2007. Dietary sequestration of defensive steroids in nuchal glands of the Asian snake *Rhabdophis tigrinus*. Proc. Natl. Acad. Sci. USA 104, 2265–2270.

65. Moore, J., 2012. History of parasites and hosts, science and fashion. In: Hughes, D.P., Brodeur, J., Thomas, F. (Eds.), Host Manipulation by Parasites Oxford University Press, Oxford, pp. 1–13. Ch. 1.

66. Lopes, P.C., 2014. When is it socially acceptable to feel sick? Proc. R. Soc. B Biol. Sci. 281, 20140218.

67. Dudley, R., Milton, K., 1990. Parasite deterrence and the energetic costs of slapping in howler monkeys, *Alouatta palliata*. J. Mammal. 71, 463–465.

68. Hemmes, R.B., Alvarado, A., Hart, B.L., 2002. Use of California bay foliage by wood rats for possible fumigation of nest-borne ectoparasites. Behav. Ecol. 13, 381–385.

69. Poulin, R., Fitzgerald, G.J., 1989. Risk and parasitism and microhabitat selection in juvenile sticklebacks. Can. J. Zool. 67, 14–18.

70. Schmidtmann, E.T., Valla, M.E., 1982. Face-fly pest intensity, fly-avoidance behavior (bunching) and grazing time in Holstein heifers. Appl. Anim. Ethol. 8, 429–438.

71. Waage, J.K., 1981. How the zebra got its stripes—biting flies as selective agents in the evolution of zebra coloration. J. Entomol. Soc. South Afr. 44, 351–358.

72. Egri, A., Blaho, M., Kriska, G., Farkas, R., Gyurkovszky, M., Akesson, S., et al., 2012. Polarotactic tabanids find striped patterns with brightness and/or polarization modulation least attractive: an advantage of zebra stripes. J. Exp. Biol. 215, 738–745.

73. Caro, T., Izzo, A., Reiner Jr., R.C., Walker, H., Stankowich, T., 2014. The function of zebra stripes. Nat. Commun. 5 Article number: 3535. http://dx.doi.org/10.1038/ncomms4535.

74. Rosengaus, R.B., Jordan, C., Lefebvre, M.L., Traniello, J.F.A., 2000. Pathogen alarm behavior in a termite: a new form of communication in social insects. Naturwissenschaften 86, 544–548.

75. Heinze, J., Walter, B., 2010. Moribund ants leave their nests to die in social isolation. Curr. Biol. 20, 249–252.

76. Choe, D.-H., Miller, J.G., Rust, M.K., 2009. Chemical signals associated with life inhibit necrophoresis in Argentine ants. Proc. Natl. Acad. Sci. 106, 8251–8255.

77. Chouvenc, T., Efstathion, C.A., Elliott, M.L., Su, N.-Y., 2013. Extended disease resistance emerging from the faecal nest of a subterranean termite. Proc. R. Soc. B 280.

78. Kaushik, M., Knowles, S.C.L., Webster, J.P., 2014. What makes a feline fatal in *Toxoplasma gondii*'s fatal feline attraction? Infected rats choose wild cats. Integr. Comp. Biol. 54, 118–128.

79. Vyas, A., Kim, S.-K., Giacomini, N., Boothroyd, J.C., Sapolsky, R.M., 2007. Behavioral changes induced by *Toxoplasma* infection of rodents are highly specific to aversion of cat odors. Proc. Natl. Acad. Sci. USA 104, 6442–6447.

80. Levri, E.P., 1998. The influence of non-host predators on parasites-induced behavioral changes in a freshwater snail. Oikos 81, 531–537.

81. Hart, B.L., 1988. Biological basis of the behavior of sick animals. Neurosci. Biobehav. Rev. 12, 123–137.

82. Goulson, D., 1997. *Wipfelkrankheit*: modification of host behaviour during baculovirus infection. Oecologia 109, 219–228.

83. Huffman, M.A., 1997. Current evidence for self-medication in primates: a multidisciplinary approach. Yearb. Phys. Anthropol. 40, 171–200.

84. Karban, R., English-Loeb, G., 1997. Tachinid parasitoids affect host plant choice by caterpillars to increase caterpillar survival. Ecology 78, 603–611.

85. De Moraes, C.M., Mescher, M.C., 2004. Biochemical crypsis in the avoidance of enemies in an insect herbivore. Proc. Natl. Acad. Sci. USA 101, 8993–8997.

86. Knell, R.J., Webberley, K.M., 2004. Sexually transmitted diseases of insects: distribution, evolution, ecology and host behaviour. Biol. Rev. 79, 557–581.

87. Abbot, P., Dill, L.M., 2001. Sexually transmitted parasites and sexual selection in the milkweed leaf beetle, *Labidomera clivicollis*. Oikos 92, 91–100.

88. Moore, J., 2002. Parasites and the Behavior of Animals. Oxford University Press, Oxford, 338 pp.

89. Cook, L.M., 2003. The rise and fall of the Carbonaria form of the peppered moth. Q. Rev. Biol. 78, 399–417.

90. Turner, J.R.G., 2002. Mimicry In: Pagel, M. (Ed.), The Encyclopedia of Evolution, vol. 2, Oxford University Press, Oxford.

Mating Systems

357

LEARNING OBJECTIVES

Studying this chapter should provide you with the knowledge to:

- Analyze the importance of gametic investment in predicting behavior.
- Predict how and why animals choose mates.
- Classify mating systems by observing social interactions.
- Discover how gametic investment can help in predicting the choosy sex.
- Appreciate the effects of intrasexual competition for mates.
- Assess why animals that do not succeed in obtaining mates may opt for alternate strategies to gain reproductive success.

Animal Behavior. DOI: http://dx.doi.org/10.1016/B978-0-12-801532-2.00011-8

11.1 INTRODUCTION

A mating system is a pattern of male–female pairings. The term *mating system* captures the broad sweep of reproductive biology, including the nature of the *gametes*, location of mates, strategies for ensuring the correct choice of a mate, patterns of male–female pairings, and behavioral strategies for individuals to attain reproductive success and mate even if they are not chosen. In short, mating systems exhibit all the dimensions of sexual reproduction and offer a rich field of inquiry. Mating systems continue to be a major focus of study in animal behavior.

> **KEY TERM** A mating system describes how male and female interactions are built around choosing mates.

> **KEY TERM** A gamete is the haploid reproductive cell produced in meiosis. Two gametes must merge to produce a zygote, which develops into a new animal.

Understanding mating systems starts with the underlying asymmetry between females, which are defined as the sex that invests heavily in each of a relatively small number of gametes, and males, which are defined as the sex that invests little in each of a large number of gametes. This fundamental difference between the sexes means that males and females approach mating with differing and sometimes conflicting interests. This chapter begins with a hypothesis about how this disparate condition came to be, and continues with descriptions and examples of the different types of mating systems, such as *monogamy* (a male and a female), *polygyny* (one male, many females), and *polyandry* (one female, many males). It then delves deeply into the intriguing issue of how animals choose their mates (Figures 11.1 and 11.2).

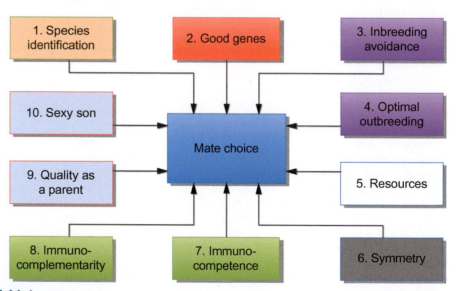

FIGURE 11.1

Mate choice is often a key element of mating systems in animals. Mate choice is also one of the most complicated and fascinating areas of study in biology. One or more of the factors illustrated here can influence why and how animals choose their mates. An obvious factor (number 1 in the figure) is mating with the correct species, but determining this may be more complicated for an animal than it first seems. Good genes (factor 2) seem like an obvious qualification in a mate but there are many ways in which genes can be "good." Inbreeding can be deleterious and should be avoided (factor 3), but if parents are too mismatched genetically then the mating may disrupt gene combinations that are well suited for the environment (factor 4). Resources held by a potential mate, or just the ability of the mate to gain the resources necessary for survival may also be a sign of good genes (factor 5). Other types of good genes may be signified by the symmetry of a potential mate's body (factor 6). For example, symmetry is often correlated with immune function of the mate. The immune system of the offspring is also shaped by the degree of match between the genes for immunity carried by each parent (factors 7 and 8). If the potential parenting of the mate (factor 9) can be assessed offspring survival might be maximized. Finally (factor 10) attractive mates can yield attractive offspring, meaning that selection can favor attractiveness even if it is not correlated with good genes or good parenting.

FIGURE 11.2
This chapter is all about very complicated questions that come from a simple act—mating. Look at this pair of beetles. It seems as though it would be an easy picture to interpret; they are mating, and the male is on top of the female. Are they paired for life, or will each have other partners over time? Did the male choose the female? Did the female choose the male? Will the male behave in ways that limit whether the female can mate with other males? This chapter focuses on how and why biologists ask Tinbergen's questions about mating in animals. *Photo: Jeff Mitton.*

DISCUSSION POINT: POTENTIALLY CONFUSING TERMINOLOGY

In this book the word "sex" refers to three distinct processes: meiosis, physiological and anatomical maleness/femaleness, and the act of mating. "Gender" is a word that, in humans, captures the sociocultural construct of sexual identity and we have avoided using gender as a synonym for sex.

CASE STUDY: MATE CHOICE, MONOGAMY, AND HUMAN SEXUAL BEHAVIOR

Human behavior has not been a primary focus in this text, but everyone is curious about how humans fit into the mating behavior picture. Perhaps most obvious from a physical perspective is that sexual selection has affected both male and female human phenotypes. The presence of secondary sexual features in both males and females suggests that both sexes actively choose their mates. Scientific studies and personal experience support the idea that each person has a distinct view of what makes an "optimal" mate.[1]

Individual perspectives on what constitutes "beauty" and "sexiness" also vary and are strongly influenced by genes and by cultural background.[2] Some scientists have argued that people judge more symmetrical faces ("even" or "balanced" features) to be attractive. Others note that symmetry seems to have little to do with actual mate choice. Symmetry may convey information about genetic variation and a person's health; the reasons for this are presented in this chapter. A counterargument is that asymmetries bring character and interest to a person's appearance. Mate choice revolves not only around physical appearance, but involves perceptions of future fecundity (ability to have children) and abilities as a parent, including possession of resources that will aid in rearing children.

The question of whether or not humans are naturally monogamous is fascinating and definitely serves as a flashpoint in moral, legal, and religious debates. The two ape species most closely related to humans, chimpanzees and bonobos, are strikingly promiscuous in their sexual behavior, with sex playing roles in bonding, social affiliations, and perhaps appeasement in both species.[3,4] Copulation serves a variety of nonreproductive social functions in these apes. The validity of looking to chimpanzees and bonobos to understand human sexual behavior is an open question because the ecology of humans differs substantially from that of chimpanzees and bonobos, and ecology plays a major role in mating systems.

Across a broad range of human cultures, the most common mating systems are monogamy and polygyny; the latter occurs when a male has multiple female mates (sometimes this is called a *harem mating system*). However, from a lifetime perspective, both males and females are likely to have multiple mates, and serial monogamy—pair bonds that are nonpermanent—is a common mating system in humans. Extrapair copulations (EPCs)—sexual encounters outside a pair bond—are certainly frequent in humans. EPCs may allow a person to produce genetically diverse offspring while avoiding the costs of parental care.

Human mating patterns resist analysis because biological influences are difficult to separate from social and cultural influences. Morals, laws, and religion often seek to modify or restrict what might be natural impulses. From evolutionary and sociocultural perspectives, control over the reproduction of others is an extreme power. In this chapter, the same biological dynamics that are so much a part of human life play themselves out across the animal world. In many ways animal mating systems are much easier to understand than human mating systems because insofar as we know, biology alone is the driving force.

11.2 EVOLUTION OF SEX: WHY SOME ANIMALS ARE CALLED MALE AND OTHERS FEMALE

Biologists define *maleness* and *femaleness* by the relative size of the gametes that individuals produce. Species in which all gametes are the same size are termed *isogamous* and lack identifiable males and females. Isogamy is common in algae and protists, but virtually all animal species are *anisogamous*, producing small motile gametes, or sperm, and large gametes, or eggs. By convention, organisms producing large, nutrient-rich gametes are termed *female*, and organisms producing small, motile gametes are termed *male*. The isogamous nature of many unicellular organisms leads to the inference that isogamy may be the ancestral state for gametes. One look at an ostrich egg or a brooding octopus reveals that much has changed since that time. What caused this change? What caused the evolution of male and female?

> **KEY TERM** By definition, a male produces small, motile gametes called *sperm*. Males produce large numbers of these gametes, with small investment in each.

> **KEY TERM** By definition, a female produces small, motile gametes called *eggs*. Females produce small numbers of these gametes, with large investment in each.

> **KEY TERM** Isogamous species produce gametes that are all the same size. There is no differentiation of male and female. Anisogamous species produce gametes of two distinct sizes; these species have males and females.

As is the case in all evolution, variation played a significant role in the evolution of gametes. Gametes that were slightly smaller than average were more nimble and faster than other gametes; they could travel quickly to other gametes and fuse to produce diploid zygotes. In the world of motile gametes, if everything else were equal, small would often win the race to fertilization. True, the small gametes had reduced nutrient content for the zygote, but that could be balanced by fertilizing larger, nutrient-rich gametes. Meanwhile, large, nutrient-rich gametes also could influence the development of the zygote and could exert control over the fate of the offspring. Intermediate-sized gametes were neither nutrient-rich, nor were they particularly speedy. Thus, natural selection on gametes favored the extremes—small and large. This sort of selection, called *disruptive selection*, is thought to be the selective force that resulted in the sexes (Figure 11.3).

Some animals are capable of producing both eggs and sperm; these are *hermaphrodites*. Hermaphroditism can be further subdivided into animals such as some sea slugs that are simultaneously male and female (simultaneous hermaphrodites) and others that are first one sex and then the other (sequential hermaphrodites). For example, clownfish are males early in adult development and become females as they grow larger; this is called *protandrous*, or male-first, *hermaphroditism*. In contrast, wrasses are females when smaller and become males as they grow larger; this is called *protogynous*, or female-first, *hermaphroditism*.

> **KEY TERM** A hermaphrodite is an animal that has both male and female reproductive organs and produces both eggs and sperm.

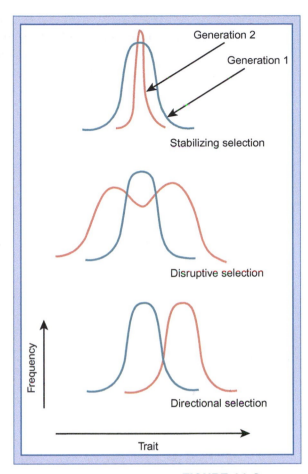

Stabilizing selection

Disruptive selection

Directional selection

Frequency

Trait

Generation 2

Generation 1

361

> **KEY TERM** Protandrous hermaphrodites are male early in their lives and females later. Protogynous hermaphrodites are female early in their lives and males later.

Why does sex exist? Superficially, it may seem obvious that organisms should reproduce sexually, but sex carries a high evolutionary cost. Recall freshman biology lessons about *diploid* reproductive cells becoming *haploid* through a special cell division process called *meiosis*; in other words, each time meiosis occurs, half the genome is "lost." This is called the *cost of meiosis*. As a result, an *asexually reproducing* animal is able to pass along all of its genes to each of its offspring, but animals participating in *sexual reproduction* pass only half of their genes to each offspring. In addition, there is a cost associated with producing male and female offspring (both are necessary in sexually reproducing species), instead of putting all reproductive effort into asexually reproducing female offspring. Given this reasoning, why does evolution so often fail to favor organisms that transmit 100% of their genes to each offspring—that is, organisms that asexually reproduce? What forms of counterbalancing selection could favor sexual reproduction?

> **KEY TERM** The cost of meiosis is the loss of half an animal's genetic material when it produces haploid gametes.

> **KEY TERM** Asexual reproduction is the production of offspring that are genetically identical to a single parent.

> **KEY TERM** Sexual reproduction is the production of offspring by the combination of gametes from two parents.

In other words, why be sexual? The major argument for evolution and maintenance of sex is that the selective advantages from *genetic recombination* and *genetic diversity* among the offspring outweigh the costly nature of sex. In addition, deleterious genes do not accumulate in a sexual lineage. Indeed, this genetic diversification is so beneficial that it occurs in many organisms that do not have "male" and "female" forms, such as mating strains of protists. In a strict sense, then, sex refers to the mixing of genetic material, either within an organism (e.g., recombination) or between organisms. It is this mixing that proves to be highly advantageous. Gorelick and Heng[5] argue that the main functions of sexual reproduction are essentially genetic editing. Meiosis eliminates harmful changes and mutations and allows the repair of damage to DNA. Meiosis also fosters resetting accumulated epigenetic signals that limit gene expression. In Gorelick and Heng's analysis, sexual reproduction through meiosis may actually reduce genetic variation by eliminating major chromosomal changes while maintaining small mutations. In either case, the recombinatorial phase of meiosis reshuffles genetic combinations, with the potential of producing offspring that are either more poorly or better suited for the environment.

Sexual reproduction is a fact of life for birds and mammals, which, with one or two rare exceptions, reproduce sexually. Hypothetically sexual reproduction is the ancestral state in

FIGURE 11.3
This diagram shows the effects of disruptive, stabilizing, and directional selection. The *x*-axis represents a measure of the trait—in this case, gamete size. The blue curve shows the distribution of the trait before selection; the red curve shows the distribution of the trait after selection has had an effect. Under disruptive selection, animals producing either large or small gametes are favored by selection, while animals producing middle-sized gametes are selected against. Stabilizing selection reduces the variation in a trait by favoring a narrow range of phenotypes. Directional selection shifts the phenotype of the population in one direction.

vertebrates. There are nonetheless numerous examples of asexual reproduction in fish and a few examples in amphibia and reptiles.

Asexual reproduction is likely to evolve in environments that change little from generation to generation, because genetic recombination is not an advantage in predictably constant environments. Sometimes asexual reproduction crops up in invasive or pioneering species, in which single individuals may migrate to a suitable habitat; if no potential mates are present, then asexual reproduction is obviously the only way to produce offspring, and selection will favor it. (Indeed, unless it arrived with fertilized gametes, a solitary invader that was incapable of asexual reproduction would leave few traces and no descendants.) It is possible that due to physiological and/or neurobiological constraints that tightly link reproductive physiology to survival, birds and mammals may have lost the ability to evolve to asexual reproduction.

> **KEY TERM** Genetic diversity results from different offspring having different genetic combinations. Each egg from a female is genetically unique, and each sperm from a male is genetically unique. Thus, when a male and female mate, if they have multiple offspring, their offspring will differ genetically; in other words, their offspring are genetically diverse.

> **KEY TERM** Basal taxa are types of organisms that arose earlier in evolution. We use this term to avoid using the misleading words *primitive* or *lower* when describing animals.

Among invertebrates asexual reproduction is more common. In some invertebrate taxa there appears to be considerably more flexibility in terms of evolutionary switching between sexual and asexual patterns of reproduction. Given the ubiquity of sexual reproduction in animals and the fact that the more *basal animal taxa* reproduce sexually, the hypothesis that sexual reproduction is basal for animals as a whole is reasonable. Sea anemones produce dispersing offspring sexually, but when one of those offspring arrives on a rock and develops into an anemone, it can also reproduce by *budding*, producing clones of itself. This alternation of sexual and asexual reproduction allows a species to take advantage of maintaining genetic combinations in stable environments through asexual reproduction, while using sexual reproduction for recombination and offspring diversity when colonizing new habitats (Figure 11.4). Similarly, in parts of their life cycles, many aphid species reproduce asexually (Figure 11.5). In these organisms, sex seems

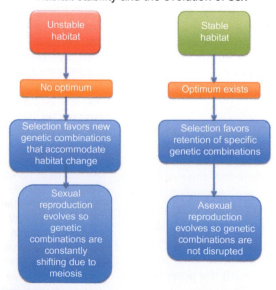

Habitat stability and the evolution of sex

Unstable habitat → No optimum → Selection favors new genetic combinations that accommodate habitat change → Sexual reproduction evolves so genetic combinations are constantly shifting due to meiosis

Stable habitat → Optimum exists → Selection favors retention of specific genetic combinations → Asexual reproduction evolves so genetic combinations are not disrupted

FIGURE 11.4
The effects of habitat stability on the evolution of reproduction. Stable habitats may favor asexual reproduction, while unstable (or unpredictable) habitats may favor sexual reproduction.

FIGURE 11.5
Aphids such as these may reproduce asexually, resulting in clones of the insect. *Photo: Michael Breed.*

to be *facultative*; that is, it may or may not occur, depending on environmental stability. Stable environments favor asexuality.

Having established that sexual reproduction is common, and probably basal, in animals, it should be clear that the evolution of mating systems probably began before the evolutionary origin of animals, among their *protistan* predecessors in which proto-sperm swam in the ocean searching for proto-eggs to fertilize. This means that the use of chemical cues to find mates and competition among sperm are ancient behavioral features.

> **KEY TERM** An egg is a gamete that is large and relatively immobile.

Much of the theory that underlies the understanding of mating systems derives from the fact an *egg* is more costly to produce than a *sperm*. This means that females produce a few individually expensive gametes, while males can produce a huge number of relatively cheap gametes. Theoretically, then, eggs are valuable and worth protecting, while sperm are

> **KEY TERM** A sperm is a small and relatively mobile gamete.

much less valuable. Another way of looking at this is to contemplate the numerical disparity between eggs and sperm: because fewer eggs are made than sperm, many, perhaps most, eggs should be fertilized, while many sperm will never fertilize an egg. The result of this observation is the evolutionary argument that, in general, females should protect their investment in eggs by caring for them, while males should be less likely to care for either their sperm or for their offspring. This is the evolutionary argument for the root of differences between the sexes. Laying aside sociocultural arguments about sex and gender roles in humans (recall the discussion of nature and nurture in Chapter 3), this characterization is a good starting point in exploring the evolution of mating systems across animal taxa.

These concepts lead to the often repeated prediction that females should be very choosy about who fertilizes their gametes, whereas males should be relatively indiscriminate in their mating. But is this really always true? Males that invest substantially in offspring care may also exert careful choice of mates. Either sex may have secondary sexual characteristics such as ornamentation or weaponry that exists for combat or territoriality. These characteristics may distract from parental care because of the hormonal state that underlies combat and territoriality does not prime the animal for parental care.

This is an important point that will form the core for a large part of the discussion in this chapter. Indeed, when these predictions are not met, there are frequently other forces at work, such as circumstances that limit male options for mating; with limited mating options, the number of potential matings for the males is more or less equal to the number of potential matings for females, and male choosiness about mates should equal that of females. Males that invest more than females do in offspring are often more choosy than females (Figure 11.6).

363

FIGURE 11.6
Female mormon crickets eat spermatophores such as this one. Spermatophores are sperm containers that males of many arthropod species produce; in the case of the mormon cricket, the spermatophore is nutrient-rich and left with the female after mating. Because the mormon cricket spermatophore is costly to make, a male carefully chooses his mate. *Photos: Darryl Gwynne (left) and Michael Breed (right).*

OF SPECIAL INTEREST: THE RED QUEEN AND THE EVOLUTION OF SEX

How can sex as a reproductive mechanism be explained in the face of the cost of meiosis? The key may lie in the evolutionary arms race between species and their diseases and parasites. Sex provides new genetic combinations to fight disease, but does a species ever truly get ahead of its diseases and parasites?

In Lewis Carroll's *Through the Looking Glass*, Alice finds herself in the company of the Red Queen (Figure 11.7). The charming story of Alice climbing a hill with the Red Queen is accompanied by the Red Queen's confusing explanation of how hills can become valleys can become hills. In the same conversation Alice learns that you have to run faster and faster just to keep up with the competition:

> Alice didn't dare to argue the point, but went on: "And I thought I'd try and find my way to the top of that hill — "
>
> "When you say 'hill,'" the Queen interrupted, "I could show you hills, in comparison with which you'd call that a valley."
>
> "Well, in our country," said Alice, still panting a little, "you'd generally get to somewhere else—if you ran very fast for a long time, as we've been doing."
>
> "A slow sort of country!" said the Queen. "Now here, you see, it takes all the running you can do, to keep in the same place. If you want to get somewhere else, you must run at least twice as fast as that!"

Leigh van Valen introduced the Red Queen to evolutionary biology, pointing out that a race which becomes faster and faster without a sign of a winner is a perfect analogy for coevolutionary arms races (Figure 11.8).[6] In such coevolutionary races, for example, a prey species evolves a defense, the predator evolves to defeat the defense, the predator evolves another defense, and on and on.... Moving fast seems to never do more than get you back to the starting point, but stopping during an evolutionary arms race leads to extinction. The same principle can apply to the evolution of sex.

Many evolutionary biologists see the Red Queen's tale as relevant to the evolution of sex, explaining that the value of sexual reproduction fluctuates over time; genetic recombination gained from sexual reproduction produces new defenses against diseases and parasites. When disease and parasite pressure are low, the cost of meiosis and the benefit of proven genetic combinations favor asexual reproduction, but when diseases and parasites surge, sexual reproduction is favored. Thus, the race goes up and down hill, and the fast-moving species are again condemned to run in the same place.

Evolutionary Arms Race

Step 1 — Host defenses improve due to new genetic combinations generated through sexual reproduction

Step 2 — Parasite or disease species evolves countermeasures

Step 3 — Host overcomes the parasite or disease countermeasures

And so on....

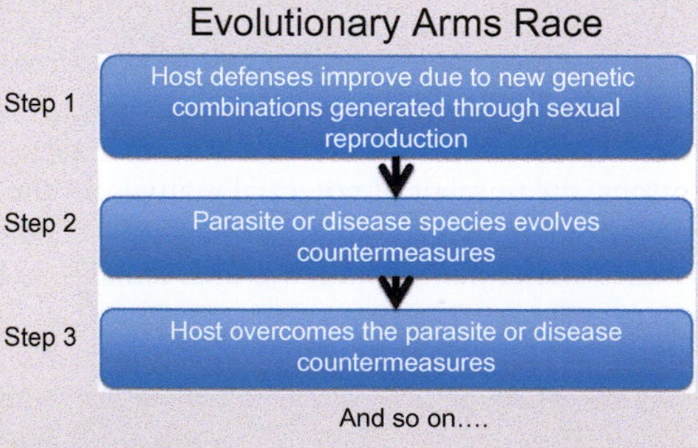

FIGURE 11.8

The steps involved in an evolutionary arms race. Each time a species evolves an innovation (step 1) the other species evolves a counter-innovation. The first species evolves a counter-counter-innovation. This continues over evolutionary time repeating itself so that neither species ever truly wins the race. In this chapter the red queen hypothesis applies to the evolution of sex, but it can also apply to predator–prey coevolution.

FIGURE 11.7

The Red Queen lectures Alice about the difficulties of running as fast as you can, only to realize that you are not really moving forward relative to your surroundings.

Why are sex ratios often 1:1 in populations? Evolutionary theory developed by R. A. Fisher[7] and expanded by R. L. Trivers[8] states that within a population the investment in male and female gametes should be equal. This theory concerns sex ratios in populations and applies to both parental investment in caring for offspring in species that do so and to parental investment in sperm and eggs alone for species that provide no additional care. At the population level, equal time and energy should be invested in male and female offspring, but the number of male and female offspring may differ. The amount of investment in males and females at the population level is what matters. Because this equality is at the population level, rather than the individual level, individuals may produce male- or female-biased broods, as long as across the population, the ratio is equal. As Fisher explained, if the ratio of males to females deviates very much from 1:1, the rarer sex becomes more valuable, and selection favors individual parents increasing production of the rarer sex. This causes an evolutionary balancing effect which maintains sex ratio to 1:1.

Equal investment in males and females is based on the assumption of equal genetic relationship of the parent to its male and female offspring. Nature does not always honor this assumption. There are at least three major exceptions to this condition of equal genetic relationship:

1. In ants, bees, and wasps, females are more related to their sisters than to their daughters (Chapters 13 and 14). This disparity in genetic relationship leads to possible asymmetries in investment in males and females.
2. Local mate competition happens when brothers end up competing with each other to mate. In this case theory predicts that the population sex ratio should be strongly female biased, so that only enough males are produced to ensure that every female mates.
3. Local resource competition occurs when females remain near their birth site and males disperse. The females end up competing with sisters for resources, while males probably do not compete with their brothers. This results in male-biased sex ratios.

11.3 SEXUAL SELECTION

Darwin argued for two kinds of selection: natural and sexual. Sexual *selection* is divided into *intersexual selection*—the effects that mating preferences of one sex have on characteristics of the other sex—and *intrasexual selection*—the effects of competition within a sex for mates. (Darwin did not use these terms, but he did allude to the existence of these phenomena in The Descent of Man and Selection in Relation to Sex.) Intersexual selection can result in the emphasis of features that attract mates, such as vocalizations or the peacock's tail. Intrasexual selection can yield features such as horns or antlers that are used in intrasexual combat. However, sometimes intersexual and intrasexual selection can both act on a trait, so the antlers of red deer serve in both combat (intrasexual) and female choice among males (intersexual; Figure 11.9). For that matter, some traits used in combat can also be favored by natural selection if they enhance predator defense, for instance. Either intersexual and intrasexual selection can result in sexual dimorphism—different "morphs"—or appearances—for males and females.[9,10]

Runaway sexual selection refers to circumstances in which intersexual preference is so consistent that selection on the trait produces an extreme form (Figure 11.10; Section 7.5). Natural selection can also limit the power of sexual selection,[11] especially in the case of runaway sexual selection. If females persist in choosing males with large tail feathers, for instance, then the males

> **KEY TERM** Sexual selection results from differential reproduction of males or females within a species. Male deer, for example, may have large antlers as a result of sexual selection.

> **KEY TERM** Intersexual selection results when males choose female mates (or vice versa) based on specific traits.

> **KEY TERM** Intrasexual selection results from mating-related competition within a sex. Horns, antlers, and other weaponry may result from intrasexual selection.

FIGURE 11.9
Right, male and female Canada Geese, which differ little in appearance. Left, male and female ducks, showing stronger sexual dimorphism, which may be the result of either intrasexual selection (competition among males) or intersexual selection (female mate choice). *Photos: Michael Breed.*

FIGURE 11.10
Examples of runaway sexual selection. Left, a male earwig, with large forceps used in mating; right, a male bighorn sheep with large horns used in combat. This evolution is a result of intrasexual selection. *Photos: Jeff Mitton. Tomkins, J.L., Simmons, L.W., 1998. Female choice and manipulations of forceps size and symmetry in the earwig* Forficula auricularia *L. Anim. Behav. 56:347–356.*[107]

> **KEY TERM** Runaway sexual selection occurs when one sex chooses mates based on a characteristic of the other sex, resulting in an extreme expression of that characteristic. Over generations this causes that trait to become amplified, as in the peacock's tail.

with the largest tail feathers will leave the most offspring. These offspring will then have very large tail feathers (if male) or prefer males with very large tail feathers (if female). As can easily be seen, this could eventually result in gargantuan tail feathers that, past a point, would be detrimental were there any predators in the vicinity. Natural selection curbs the excesses of runaway selection, although not before some spectacular traits appear.[12–14] The costs of producing sexually selected signals can be high, making these signals an honest measure of the condition of the animal. The value and dependability of honest signals resulting from runaway sexual selection reinforces their use in mate choice.

DISCUSSION POINT: SEXUAL SELECTION

How do you think intersexual and intrasexual selection have affected phenotypes in a range of animals with which you are familiar? If you compare humans to common domestic animals, such as dogs and cats, are humans more or less sexually dimorphic than those species? (Hint: Can you determine the sex of a dog without looking at its genitalia?) Can a dog determine the sex of another dog without sniffing and looking? Are any human, dog, or cat traits the result of sexual selection? What might prevent runaway sexual selection from taking over a trait used in mate choice?

The Handicap Principle

If the sexually selected trait is actually a handicap for the animal carrying it, then it exacts a price in energy or vulnerability to predation. In this case, the sexually selected trait, such as huge antlers, becomes an honest advertisement of the male's ability to carry the handicap. Zahavi argued in the mid-1970s that sexually selected traits were handicaps for the animal carrying them and that, as such, sexually selected traits reflect good genes even if the trait itself has been selected to the point of having no residual additive genetic variance on which selection might work.[15]

If one sex consistently chooses mates with extravagant decoration (like a peacock's tail), the costs of producing the decoration can become very high. The ornament may be metabolically costly to produce and difficult to carry. Ornaments may also put an animal at risk for predation by making it more conspicuous or making it move awkwardly. The ability of an animal to pay these costs may then be an honest measure of that animal's metabolic and physical capabilities. It is "honest" because the animal cannot pretend to pay the costs; they are real. This is the "handicap principle" developed by Zahavi and explored later in the chapter. For the purposes here, the important point to understand is that runaway sexual selection can ultimately be limited by natural selection acting on the costs of signal production and that natural selection defines the cost of a handicap.

OF SPECIAL INTEREST: MATE CHOICE IN PEACOCKS

In a classic series of experiments, Petrie explored the function of the peacock's tail in mate choice.[16] Peafowl are highly sexually dimorphic; the peahen is relatively drab, whereas the peacock carries a large, beautifully colored tail that is decorated with an eyespot pattern (Figure 11.11). Petrie and coworkers discovered that removing eyespots from a male's tail made him less attractive to females, and adding eyespots made him more attractive. Recently, Petrie's results have been criticized by Takahashi and his colleagues as not fully explaining female choice in peafowl.[17] Petrie and her colleagues responded by noting that Takahashi had not actually provided measures of the density of eyespots in the males' tails in his study.[18] On balance, the evidence seems to favor Petrie's interpretation. Nonetheless, Takahashi's point is interesting: might females have abandoned (over evolutionary time) their response to the males' tails because all of the males have them? A trait that all of the animals in a population express is not informative for choosing a mate or, indeed, for making any sort of decision. In addition to using the male's tail in choosing mates, peahens also consider the male's vocal and physical displays, as well as his location. Mate choice is complex, and these studies show that oversimplifying the analysis by focusing on single features or behaviors of animals during mate choice may be misleading (see page 241).[19] That said, the peacock tail is emblematic of sexual selection; it would be the odd biologist who would not respond to it with ardent curiosity.[20]

FIGURE 11.11

A peacock. The long tail with prominent eyespots arose as a result of intersexual runaway sexual selection. *Photo: Marion Petrie.*

The Effects of Sexual Selection on the Heritability of Traits

Strong *directional selection* usually exhausts *additive genetic variance* for a trait in three to five generations. (In this context, this means traits governed by *polygenic inheritance*, or *quantitative trait loci*; see Chapter 3 on genetics.) This means that the proportion of variation in the phenotype due to genetic variation, or heritability, approaches zero. After that, there can be no further response to selection because the remaining phenotypic variation is

> **KEY TERM** Directional selection causes one form of a trait in a population, over generations, to be favored (see Figure 11.3).

from either environmental or nonadditive genetic variation. In theory, sexual selection on a trait such as antler size should rapidly eliminate the additive genetic variance for the trait. In other words, the trait will be genetically fixed. In practice, many traits that seem to be under strong sexual selection still have considerable heritability.[21]

There are a number of possible explanations for why selection does not eliminate all of the additive genetic variance for traits involved in mate choice. They include the following:

1. Sexual selection is strong only under extreme environmental conditions in which survivorship is low. Variance is maintained during periods of relaxed selection.
2. Interactions with other traits (e.g., linkage effects, viability effects) limit sexual selection before the additive variation is exhausted.
3. Mate choice relies on many factors, rather than one trait. When selection acts on multiple traits, they limit each other's evolution so that variation remains for each of the traits.
4. Counterbalancing selection for factors like protection from predators maintains additive genetic variance by limiting the elaborateness of a signal.[22,23] It is hard to overemphasize the complexity of mate choice and the need to consider multiple factors involved in any mate choice decision.

Courtship: The First Step in Sexual Behavior

In most animals courtship precedes mating. Courtship is an exchange of information between a male and a female, and sexual selection often acts on signals used in courtship. Typically, courtship includes signals that allow the male and female to ensure that they are mating with an animal of the same species.[24] It may also include information that would prevent mating with a closely related animal (inbreeding). Finally, courtship gives animals the opportunity to present information about their quality as a potential mate.[25] The issue of mate quality is complex, and there is no simple answer to the question: Which potential mate is best?[26] Through the rest of this chapter, the assessment of mate quality comes up repeatedly, offering opportunities to form a synthesis of how animals assess the quality of potential mates.

Honesty and Dishonesty in Mating Systems

The mating arena promotes considerable *dishonesty*. In many species, males run a high risk of never having the opportunity to mate, and females suffer from the risk of choosing a low-quality mate (this is a restatement of Bateman's principle, which is discussed above). This situation leads to strong selection for two types of dishonesty. The first is misrepresentation of quality in mate choice. This is the animal version of human male misrepresentation of monetary and social status. In species in which males participate in parental care, males may also misrepresent whether they already have a mate and are spending time caring for young from that mating. Mated males would be expected to conceal this fact, as their attractiveness to additional mates depends in part on their potential to contribute to offspring care.

> **KEY TERM** Dishonesty is a false representation of an animal's genotype or of resources it holds.

> **KEY TERM** Infanticide occurs when an adult animal kills a young animal.

Females show the second type of dishonesty. Generally speaking, females should represent themselves as mating only with one male. Having chosen a mate, however, a female might be somewhat uncertain that her choice was correct, or unpredictable conditions might make it impossible for her to know which male is appropriately matched to the environment their offspring will encounter. Such uncertainties about her choice may favor surreptitious matings with other males. Females hide these matings to protect their offspring from the loss of male interest in caring for young in the nest and from *infanticide*; males of some species often kill young if they are uncertain about paternity. (See Chapter 12 on nesting and parental care for more information on this topic.) These two forms of dishonesty combine to make the study of mating systems in animals full of surprises.[27]

In some cases, of course, male and female roles are more alike, or the stereotype of philandering males and choosy females may be reversed. The more similar male and females roles are, the more similar male and female deceptive practices will be (Figure 11.12). If male and females roles are reversed, then patterns of deception and cheating may also be reversed.

FIGURE 11.12
In fiddler crabs, the enlarged claw can be an honest signal of male fighting ability. However, if a male loses a claw, it may regrow a large, but weak claw that is a dishonest signal of fighting ability. *Photo: Allen Bridgman, South Carolina Department of Natural Resources, Bugwood.org.*

369

BRINGING ANIMAL BEHAVIOR HOME: MATING SYSTEMS IN DOGS AND CATS

Dog and cat mating systems contrast in interesting ways. When it comes to mating behavior, domestic cats are not much different from their closest relatives, the European wildcat. Tom cats are not involved in parental care; they associate with a female only during her estrous period, when they compete ferociously for her. Sometimes cats that have been released on farms or in urban areas (these are called *feral cats*) congregate around food, and then adult females may show more mutual tolerance. Copulation stimulates ovulation in cats, favoring tom cats that kill kittens in order to father a new litter. Females become receptive again shortly after copulation, and litters of mixed paternity are relatively frequent in cats.

Life for dogs is more pack-oriented, and wolves, the closest wild species to the domestic dog, have a mating system much like dogs. A pair-bonded male and female typically head a wolf pack, and their offspring form the core of the pack. Male wolves participate in parental care, at least to the extent that the pack shares food and that the male wolf is active in defending the pack. Perhaps because humans often have only one dog as a member of their household, the mating system of domestic dogs is not so strongly based on pair bonds. Of course, domestic dogs are often neutered, so they do not participate in mating. Estrous females attract a number of males, and, as in cats, a female may mate with more than one male while she is in season and mixed litters are common. The dog's penis expands during copulation, tying the male and female together, so that an attempt to forcibly separate copulating dogs can injure the animals. This may be a form of mate guarding, helping to ensure that some of the female's eggs are fertilized from that copulation.

11.4 VARIANCE IN MATING SUCCESS

Classic experiments on fruit flies by Angus Bateman extended our knowledge well beyond what might be applied to these useful little insects. Bateman learned that variation among females in mating success is likely to be low, while male variance is high. The reason is that one mating is likely to be sufficient to fertilize the eggs of a female, while male reproductive success increases each time the male mates (Figure 11.13). As a result of this difference, competition for mates should be higher among males than females, and in general, sexually selected traits, such as antlers in deer or plumage in birds, should be more elaborate in males than females. The relationship between the number of matings and offspring produced (fecundity) is sometimes called a Bateman gradient.

FIGURE 11.13
These figures illustrate Bateman's rule, the idea of sexual differences in mating success as a function of number of mates (horizontal axis). Many biologists think that while a male's reproductive success increases with the number of females he mates with (A), a female's success would not increase as females have a limited supply of gametes (B). One outcome of this reasoning is that mating success may be highly variable among males, with a few males in a population accounting for most of the mating, while variance in female mating success would be lower. If male mating success is highly variable then sexual selection would be expected to act very strongly on males. Subsequent research has shown that there are many exceptions to this rule.

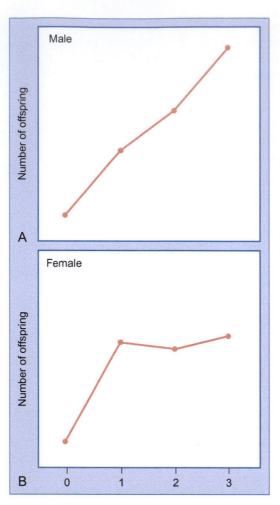

Are a female's best interests really served by mating only once? Is mating actually so cheap for males that they can mate repeatedly? As will become apparent in this chapter, there are ample exceptions to these "rules." For instance, natural selection can favor females mating with a number of males, and mating can be costly enough to limit variance in male reproductive success. Females may mate with more than one male to ensure fertility (the first male may be infertile), to increase genetic variation among their offspring, or to seek a better genetic match. The effort of locating and courting a mate can be high for males; the energy expenditure involved in achieving a mating may be so high that efforts to find subsequent mates are limited.

Recently, criticisms by Gowaty and her colleagues have brought Bateman's principle into question. Gowaty was unable to replicate Bateman's original study on fruit flies[28] and has also proposed theoretical arguments showing that variance in male and female mating success should be similar. Other new studies, such as one by Collet and colleagues on jungle fowl[29] find evidence in support of Bateman's principle, although they propose statistical refinements in how data on mating variance are analyzed.

11.5 MATE CHOICE

Most of the rest of this chapter deals with how mates are chosen. This section introduces the basic principles of mate choice. To give a concise preview, animals can choose mates based on:

1. What the potential mate possesses, such as a territory.
2. The genetic quality of the potential mate; this is used to forecast the genetic qualities of the pair's offspring. The genetic quality of a mate can be viewed fairly broadly to include factors such as the ability of the potential mate to avoid parasites.
3. A tangible nutritional offering by the potential mate; these are called nuptial gifts.[30]

Some scientists argue that all of these factors basically reduce to genes: for instance, an animal cannot hold a territory if it does not have the genes to make it strong enough to do so.

Female Mate Choice

For most species, females exert more choice concerning their mates than do males. The theoretical basis of female choosiness was explained at the beginning of this chapter and is simple: females invest more in their gametes (eggs), make fewer gametes, and are more likely to be present when the young emerge than are males. The difference between female and male gametic investment and the relative rarity of female gametes, combined with the likelihood that in many species the female provides parental care, favor females that carefully select their mates.

What effect does female mate choice have on males? Any trait that is preferred by females will be strongly favored. This type of selection, called *sexual selection*, explains the elaborate ornamentation of males of many species, such as the peacock's tail (see Section 11.3). With some exceptions,[31] females should make their choices among males based on the likelihood that mating with the chosen male will enhance the survival of the female's offspring. This could be because the male possesses resources, such as food within a territory, or because the male has "good genes," which will then be passed on to their offspring. "Good genes" models are complicated because any trait chosen by females will be under strong selection, and the usefulness of the trait for assessment will be lost if selection results in all males being the same. Also, separating honest signals representing good genes from deceitful signals may be difficult for females. Finally, some female animals are known to copy other female choices;[32,33] males do this as well.[34]

Another prediction about mate choice states that females should prefer to mate with males that will father sons that will be attractive to females in the next generation. In other words, nothing matters except male attractiveness—not gene quality, not potential as a parent, and not resources held by the male. This is known as the *bad boy* phenomenon or the *sexy son* principle.[35] It does not take too much imagination to apply this line of reasoning to a broad range of animals.

A meta-analysis of data on mate choice suggests that sexy sons are more of a driving factor in female mate choice and sexual selection than are good genes.[36] This makes intuitive sense because if male signaling evolves to take advantage of female perceptual biases then male signals are likely not to be linked with the quality of the male in terms of fitness attributes (other than sexiness!) that it will pass on to its offspring. Selection favors females that can attract social mates (i.e., mates that stay within the pair bond) that are good fathers but this can translate into the male in the social bond being cuckolded due to EPCs by the female, so that he rears at least some young that are not genetically his. Remarkably natural selection has not provided males of most species methods of assessing their relatedness to the young that they are rearing. Males of some species modify their parental behavior based on assessments of the fidelity of the female with which they are socially bonded.[37] As an extreme case, blue-footed booby males push eggs out of the nest if the female in the pair has been away from the nest long enough to suggest she may have copulated with another

> **KEY TERM** A social mate is affiliated with their mate for a substantial amount of time, such as the duration of a breeding season. In monogamous species this is termed pair bonding. The same level of social fidelity can exist in polygynous or polyandrous species. Social mates are confirmed by behavioral observations of associations between animals.

> **KEY TERM** A genetic mate is the actual parent of a particular offspring. This may be the same as the social mate or may be another animal that has mated outside of the social bond. Genetic mates are typically confirmed by DNA analyses.

FIGURE 11.14
A pair of blue-footed boobies. These birds construct rudimentary nests along shorelines in the eastern Pacific. Both partners care for the eggs, but males may kill an egg if the female has had the opportunity to mate with other males. *Photo: Michael Breed.*

KEY TERM A meta-analysis takes the results of several published studies and aligns those results so the accumulated data can be studied statistically. The power of a meta-analysis lies in the ability to exploit larger sample sizes (because data are added together from the original studies); it may then separate more probable hypotheses from hypotheses that may have been spuriously supported in one of the original studies.

male[38] (Figure 11.14). Of course the other side of the coin is that males which spend time looking for mates outside their social bond may have less time and energy to devote to parenting; this puts evolutionary pressure on females that participate in biparental care to choose faithful males for their mates.[39] The importance of the male as a resource-provider differs depending on the level of parental care.

KEY TERM The sexy son model argues that females should prefer to mate with males whose sons are likely to mate. In other words, sexy fathers produce sexy sons. This could lead the female to ignore traits that suggest the father would be a good parent.

Male Mate Choice

Male mate choice is less well documented than female mate choice, but plays an important role in some mating systems. Male mate choice is likely to appear when the male has substantial involvement in parental care or when females vary greatly in their qualities as mates (see the account of Mormon crickets, Figure 11.6). If males of a species are choosy about their mates, females should develop secondary sexual characteristics and ornamentation, much as males do in other species (Figure 11.15). Close observation of hyenas revealed that high-ranking males associated most closely with high-ranking females and exhibited mate choice, favoring fecund females.[40]

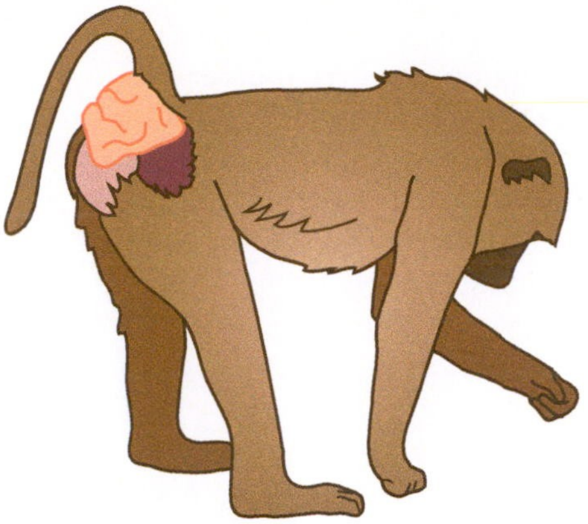

FIGURE 11.15 Female baboons are famous for their ornamentation; male primates are known to choose females on the basis of qualities that indicate high fertility.

11.6 MATING SYSTEMS: HOW MANY MALES, HOW MANY FEMALES?

Early animal behaviorists observed that some species lived in male–female pairs, whereas others lived in groups with a single, dominant male and several females. Orians and Emlen and Oring systematized ways of looking at how animals come together to mate.[41,42] This section elaborates on these mating systems.

For several decades, animal behaviorists and ethologists thought the key to understanding mating systems lay in documenting the number of males and females in a group of mating animals. For example, wild horses live in herds with one stallion and many mares; informally, this might be called a *harem*; more formally, this is a called a *polygynous* (many females) *mating system*. Mating systems tend to be a species-level characteristic, although alternative mating strategies may be expressed by individuals within a species.[43]

KEY TERM A social mating system is measured by the time spent by animals in pairs or groups.

Contemporary biology has shown that there are dramatic differences between observations of *social mating systems* and *genetic mating systems*. Until the 1980s, animal behaviorists assumed that social bonding translated directly into mating and a genetic relationship with juveniles associated with the bond. As is often the case in science, the advent of sophisticated genetic techniques and the widespread use of these new tools, particularly genetic techniques for testing paternity, reveal that this assumption is often far from the truth.

In many species the social father may not be the biological father of the young, and surreptitious matings, usually called EPCs, are unexpectedly frequent.[44-47] Because EPCs are often difficult to observe directly, investigations of mating systems prior to DNA testing generally credited animals with a higher level of fidelity to their mate than they deserved.

> **KEY TERM** A genetic mating system is measured by the actual genetic outcome of matings. For example, a mating system may be socially monogamous (male and female pairs spending most of their time together and cooperating), but genetic analysis may reveal that the male and/or the female had several mating partners.

> **KEY TERM** An EPC occurs when a member of a monogamous pair copulates with an animal outside the pair bond.

Birds are great examples of the potential for disparity between social and genetic mating systems. Most bird species are socially monogamous; a male and female spend much of their time together, they may collaborate in nest construction, and both birds often contribute to the care of the nestlings. Genetic testing reveals, however, that in many species of birds, the social father may not be the genetic father of the nestlings (Figure 11.16).

Nonetheless, despite the revolution in the understanding of mating systems led by DNA testing, the "how many males, how many females" approach to classifying mating systems is still valuable. Bonds associated with mating are at the core of animal social behavior, and the interplay between social and biological parentage is a key aspect of contemporary studies of mating systems. The following section outlines the different possibilities for male–female bonds and discusses the evolutionary dynamics that lead to each type of mating system.

373

Monogamy

Monogamy (Gr. *monos*, once; *gamos*, marriage)—the pairing of a single male with a single female—is common in birds but rare in most other animals

FIGURE 11.16
Plovers (left) are typically socially and genetically monogamous, with fewer than 10% of the chicks deriving from EPCs. In wild turkey (right), on the other hand, nearly half the nests contain chicks resulting from EPCs, suggesting a lack of genetic monogamy in this species. *Photos: Ben Pless (left) and Paul Bolstad, University of Minnesota, Bugwood. org (right).*

FIGURE 11.17

A pair of wood ducks, *Aix sponsa*, resting. The bright coloration of the male suggests that females of this species use this characteristic to choose their mates. The female's drab coloration makes her less conspicuous when she incubates eggs. In wood ducks, removal of the male from the pair during breeding season does not affect the ability of the widowed female to incubate the eggs, nor does it affect the survival of the eggs to hatching. However, the continued presence of the male makes production of a second brood more likely. This supports the hypothesis that the male may remain in the monogamous pairing because his chances of future reproduction are higher than if he leaves. *Manlove, C.A., Hepp, G.R., 1998. Effects of mate removal on incubation behavior and reproductive success of female wood ducks. Condor 100 (4), 688–693.*[48] *Photo: Michael Breed.*

FIGURE 11.18

Beavers probably engage in serial monogamy, changing mates during their lifetime. *Photo: Jeff Mitton.*

374

> **KEY TERM** Reproductive investment is the amount of time, energy, and nutrition a parent invests in its offspring.

> **KEY TERM** Bet hedging is when some investment is reserved and placed elsewhere. EPCs can be a form of bet hedging.

(Figure 11.17). Monogamy usually requires that animals be able to recognize their mate as an individual, a cognitive task that is not particularly easy. In theory, a monogamous pairing ensures that both of the mates will contribute to care of their offspring and to mutual defense. Monogamy may evolve when the cost of acquiring mates is very high, when females have the ability to restrict male behavior, or when offspring survival requires more intensive care than can be provided by a single animal.

The major cost of true monogamy is that the entire *reproductive investment* of an individual depends on the fitness of its mate. This puts a high premium on choosing an appropriate mate. Choosing an inappropriate mate could have catastrophic fitness costs for an animal. Thus, monogamy is usually more apparent, or social, than real, and monogamous animals commonly use two methods to *hedge their bets* on mate choice. First, they may actually engage in *serial monogamy*, bonding with a mate for one mating season, but choosing a different mate in a subsequent season (Figure 11.18). Second, many seemingly monogamous pairings are often subject to infidelities, or EPCs.[49] While over 90% of bird species are socially monogamous, genetic studies show that in most populations at least a few offspring in each generation result from matings with partners other than a pair member. In some species, EPCs produce over half of the offspring.

EPC and extrapair paternity (EPP) are best studied in birds, where they are surprisingly common in species that were presumed to be monogamous or polygynous (see below). From a biologist's perspective, EPCs can be observable behavior, while EPPs can generally only be measured by doing genetic testing. Ninety-five percent of birds are presumed monogamous, but across species the mean rate of EPP is 13% (range = 0–76%). EPCs are not restricted to monogamous species; they also occur in polygynous species when females mate with males other than the dominant male in the group (Figure 11.19).[51]

Given the frequency of EPCs, it should not be surprising that partners in a socially monogamous pairing might have evolved to attempt to enforce genetic monogamy. Among the possibilities are *mate guarding* (Figure 11.20), *copulatory plugs*,[52] prolonged copulation, aggressive behavior, and withholding parental care (or even infanticide) when paternity is uncertain.[53–56]

> **KEY TERM** In serial monogamy, an animal pair bonds for some period of time but then switches to a new mate.

> **KEY TERM** Mate guarding occurs when, after copulation, one member of the pair prevents the other from seeking additional mates, or prevents potential additional mates from approaching.

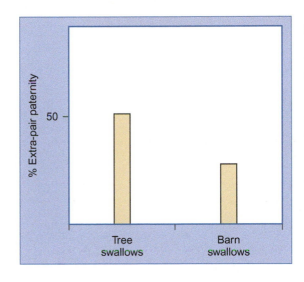

FIGURE 11.19
EPP is higher in tree swallows than in barn swallows. Variation among species in fidelity between socially monogamous partners can be due to a number of factors. If selective pressure is high for male parental involvement then EPP is usually lower, partly because males are spending more time caring for young and less time searching for new mates and partly because females may choose to mate only with males that exhibit signs of being good parents. *Kempenaers, B., Congden, B., Boag, P., Robertson, R.J., 1999. Extrapair paternity and egg hatchability in tree swallows: evidence for the genetic compatibility hypothesis? Behav. Ecol. 10, 304–311.*[50]

375

FIGURE 11.20
Prolonged copulation in arthropods is often a form of mate guarding. *Photos: Upper left, tiger beetles; upper right, robber flies; lower left, water striders (Jeff Mitton). Lower right, millipedes (Michael Breed).*

Polygyny

Polygyny (Gr. *poly*, many; *gyny*, female) is the association of one male with many females. The most common mating system in mammals, it is found in a few birds, insects, and other animals. The study of polygynous mating systems focuses on male–male competition prior to mating, which is sometimes called *precopulatory competition*. If a female mates with more than one male within a short time span, the male–male competition extends to competition among sperm. Along with copulatory plugs and postcopulatory mate guarding, this is *postcopulatory competition*. Sperm competition is an important aspect of mating systems, discussed in Section 11.9 later in this chapter.

> **KEY TERM** A copulatory plug is a secretion or object left by the male in the female's reproductive tract with the goal of preventing further matings by the female.

> **KEY TERM** A harem is a group of females associated with a single male.

> **KEY TERM** Resource defense polygyny occurs when male territoriality centers on a resource needed for breeding success. If multiple females join the male on the territory, then it is a polygynous mating system.

There are several types of polygyny; the two most common are as follows:

1. Groups of females centered around resources, or a male that holds those resources.
2. Membership in a *harem*, or a defended group of females. While the pattern of one male and many females is similar between these two types, they differ in very important ways.

FIGURE 11.21
Red-winged blackbirds are well-known examples of resource defense polygyny. The male defends a territory and the quality of that territory determines the size of his harem. *Photo: Michael Breed.*

376

Resource Defense Polygyny

Resource defense polygyny occurs when male territoriality centers on a resource needed for breeding success. In resource defense polygyny, male resource holding power is critical for male reproductive success. Sometimes discussions of territoriality mention *resource holding power* (RHP), the ability of an animal to retain a resource in the face of competition. An interesting point is that animals often are better able to hold onto a resource (higher RHP) that they already possess than they are to take a resource from another animal, even if their relative strength or size is equal. This means that males must be competitively equipped to dominate resources, and that females may sample and make choices among potential mates, using either direct measurements of male quality (size, plumage color, and so on) or measurements of territory quality. Because male movements are restricted by the actions of males on adjacent territories, any limitations on female movements (and choices) are imposed by competition among females for males. This competition can be intense, especially on better territories (Figure 11.21). A female may find that her potential reproduction is greater if she chooses a male on a high-quality territory, even if that male already has a mate, than it would be on a lesser territory, even if no other female is present on that territory. The *polygyny threshold model*[41,57,58] makes specific predictions about when females should choose to be a secondary mate in one territory rather than be the sole female in another. Note that in this form of polygyny, females exercise choice about males and resources.

OF SPECIAL INTEREST: MATING SYSTEM AND TERRITORIALITY IN THE YELLOWHEADED BLACKBIRD

Yellowheads (Figure 11.22), like their relatives, redwinged blackbirds, have a polygynous mating system in which a male territory holder may have more than one female nesting within his territorial boundaries.

Why might the male blackbird be so brightly colored? Here are three hypotheses:

1. The color may be the result of runaway sexual selection of a feature that attracts females.
2. The color may signal the male's good health to females. Unhealthy males may, due to poor nutrition, parasites, or disease, be unable to produce the bright colors. If the color is costly to produce, it may be a handicap.
3. The color may warn other males of the presence of a territorial male.

If the color attracts predators, it would also be a handicap.

How would you test these hypotheses?

FIGURE 11.22
The yellowheaded blackbird, *Xanthocepalus xanthocephalus*, perched near a marsh. *Photo: Jeff Mitton.*

377

Because male–male competition is central to resource defense polygyny, it is important to consider how competition among males occurs. Numerous studies have paired males and tested their fighting prowess.[59,60] Not surprisingly, generally larger males have an advantage in contests in neutral arenas; experience, or age, also can play an important role, as can the ability to search for widely distributed females, a form of scramble competition.[61–63] When contests are staged within a male's territory, though, the resident male often has a substantial edge in winning the encounters. Thus, the outcome of the resource defense polygyny may not be predictable in terms of variables that might influence non-territorial competition. This resident male effect can outweigh the effects of size, age, and experience (Figure 11.23). Because of this, the "best" male by measurable standards in a given area may not possess the richest territory. Overall, though, this mating system tends to favor larger, stronger, older males, and variance in mating success is skewed to these males. Social structure and mating system tend to be responsive to phylogenetic effects and to local variation in resource distribution.[65]

Harems, or Female Defense Polygyny

If females are clustered around a resource, then males may compete to monopolize that group of females. This type of polygyny puts the emphasis on male–male competition and minimizes the role of

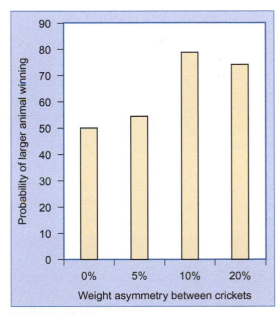

FIGURE 11.23
In these crickets, the greater the weight asymmetry between competing males, the greater the probability that the larger male would win the contest. *Hofmann, H.A., Schildberger, K., 2001. Assessment of strength and willingness to fight during aggressive encounters in crickets. Anim. Behav. 62, 337–348.*[64]

FIGURE 11.24
A zebra harem. *Photo: Mike Huffman.*

FIGURE 11.25
Harem polygyny in wild horses. *Photo: Allen Bridgman, South Carolina Department of Natural Resources, Bugwood.org.*

378

mate choice, either by males or females. In other cases, a band, or herd, of females is not associated with any resource but may have aggregated for other reasons (e.g., defense against predators). Regardless of the reason for the aggregation, the group itself becomes the focal point of competition among males. Because a female's movement is not limited by a male's hold on valuable resources and her evaluation of those resources or that male, direct male dominance of females themselves may come into play (Figures 11.24, 11.25).

Competition among females within a polygynous group is also potentially important and may be reflected by "pecking order" among the females. For instance, in domestic cows, priority of entrance through a gate or stable door is a reflection of dominance in the herd; while it is hard to imagine that getting into the stable first carries survival value, this is a reflection of other dynamics that may affect fertility, nutrition, and ultimately survival of a cow's calf. The best-known dominance hierarchies among females, such as those in flocks of domestic hens, are linear, meaning that each individual's rank can easily be determined by observation. In addition to hens, dominance hierarchies are known in various primates, in herds of ungulates, and in cockroaches. Because this sort of dominance is not as easily associated with the often conspicuous traits of competing males, this cause of variance in female mating success has probably not been sufficiently assessed. We discuss types of dominance relationships in Chapter 13.

> **KEY TERM** A lek is an aggregation of males, each seeking to attract a mate, that is not associated with a key resource.

FIGURE 11.26
A lekking male sage grouse. Males aggregate and females choose among the males. *Photo: Dave Menke, US Fish and Wildlife Service.*

Leks

A *lek* is an aggregation of males, each seeking to attract a mate, and each displaying fervently to do so. Leks are not associated with resources, such as food, water, or shelter. Conventional wisdom holds that males sacrifice their needs for food, water, and shelter by being near their competitors because the lek attracts more females than isolated males do.[66,67] Leks may be aggregations of males around a particularly attractive male (hotshots), or aggregations of males in a location in the landscape that females are likely to travel near (hotspots; Figure 11.26).

What do surplus males do in a polygynous mating system? If male reproductive success has high variance, then younger, weaker, and less-experienced males may be left without either mates or territories. Oftentimes these males herd together, forming bachelor groups. A bachelor herd is a group of adolescent or adult males that do not have harems. They travel together and gain benefits from mutual protection and food discovery. Such groups can also coalesce outside the mating season in species that form polygynous groups for only parts of the year. Membership in bachelor herds confers advantages in shared vigilance and group defense against predators and allows group members to take advantage of each other's capabilities in finding food. Sparring—male–male combat with less danger of injury—can set the stage for more intense competition during mating season (Figure 11.27).

FIGURE 11.27
A bachelor herd of walruses. *Photo: US Fish and Wildlife Service.*

Polyandry

Polyandry (Gr. *poly*, many; *andry*, male) is the reverse of polygyny—one female, many males (Figure 11.28). Because female choice of mates seems to drive many mating systems, polyandry, at least superficially, seems to be rare. There are a few cases of resource defense polyandry, in which the female is territorial and males coexist on her territory. This happens in American jacanas and spotted sandpipers. In the case of cooperative polyandry, seen in the Galapagos hawk, for instance, females have several mates and cooperate with them to raise a brood.

Polygynandry and Promiscuity

Polygynandry occurs when females mate with several males and the males care for the broods of several females. This stands in contrast to *promiscuity*, in which there are no pair bonds, and males and females in a population seem to mate randomly and with multiple partners. In general, the theory of mating systems suggests that promiscuity should be rare or absent in animals, as it is usually to the advantage of one or both of the sexes to choose their mates. Promiscuity might occur in species that live in completely unpredictable environments, that is, environments in which it is impossible to judge which mate might carry genes that would be good in the next generation. Another factor favoring promiscuity would be the inability of any animal to hold a territory and monopolize access to critical resources. In other words, promiscuity is expected when choice becomes unlikely or impossible.

Mere observation of seemingly random matings does not establish that a mating system is, in fact, genetically promiscuous. (Recall the distinction between the social appearances of a mating system and the actual genetic outcomes of that system.)

FIGURE 11.28
Honeybees are polyandrous; queens mate with 10–20 males. *Photo: Jessica Lawrence, Eurofins Agroscience Services, Bugwood.org.*

KEY TERM Polygynandry occurs when females mate with several males and males mate with several females.

KEY TERM Promiscuity occurs when there are no pair bonds, and males and females in a population seem to mate randomly.

379

Copulation can serve social functions other than reproduction; they include bonding, dominance, or even concealment of actual paternity of a female's offspring.[68,69] Genetic studies (usually using DNA fingerprinting or microsatellite marker techniques) establish true paternity of juveniles. If the norm for males in a population is to sire offspring with more than one female, and each of the males is more or less equally successful in fathering young, then the mating system is truly promiscuous. If there is substantial variation among males in their success in siring young, then even if mating frequencies suggest promiscuity, the system is probably actually polygynous or polygynandrous.[70]

In a study of collared pika, a North American mammal that lives at high elevation or high latitude and is distantly related to rabbits, Zgurski and Hik found that matings fit a polygynandrous model. Pika are relatively unique among mammals in that males and females are equally likely to disperse when they reach maturity and need to find a territory. This suggests that ecological pressures to hold territories are equal for males and females. In pika the main function of a territory is to allow collection of food, which is stored in "hay" piles for the long mountain winter. As territory is about individual maintenance, rather than using the resource to attract mates, other factors such as inbreeding avoidance become more important in mate choice.[71]

The snowshoe hare, *Lepus americanus*, while best known for its dramatic population cycles in Canada, offers a good example of a possibly promiscuous mating system. Snowshoe hare populations rise and fall on a 10-year cycle. While once thought to represent a classical predator–prey cycle, it is now recognized that hare population cycles are more complex, regulated not only by predators but also by food supply.[72] Snowshoe hares show no sign of being territorial, or of having any sort of complex social system, even at the peak of their population cycle. Going along with this lack of social complexity is a promiscuous mating system. Behavioral observations suggest that female and male snowshoe hares mate with many partners, and genetic analyses confirm that in many litters more than one father is represented. Some males, though, fathered more offspring by being the only father of some litters. Burton interprets these results to indicate that the mating system is not fully promiscuous, in that paternity is not entirely random.[73] Some sort of competition may occur among males. (Perhaps some males are more attractive to females and gain more matings.) The number of fathers represented in the offspring, and the frequency of litters representing more than one father, does make this mating system as close to true promiscuity as has been observed in any mammal.

DISCUSSION POINT: MONOGAMY

Now that we have completed our discussion of the types of mating systems, can you form some hypotheses about why social monogamy is the most common mating system among birds and polygyny is the most common mating system among mammals? Can you point to some basic difference between the biologies of birds and mammals that might account for this? Does sexual dimorphism seem more common in birds or mammals? When sexual dimorphism is present, what form does it typically take in birds? What about mammals? Returning to the original question about social monogamy versus polygyny, do you think this difference in social mating systems causes differences in sexual dimorphism between birds and mammals?

11.7 HORMONES AND SEXUAL BEHAVIOR

Chapter 2 reviewed the roles that hormones play in behavior and promised more detail about reproductive hormones in this chapter. Of all aspects of behavior, reproduction may be the most subject to hormonal influences. This section applies to mammals and birds. To begin with, luteinizing hormone (LH) and follicle-stimulating hormone (FSH) are both neuropeptides that are produced in the pituitary and that act on the ovaries and testicles. Behaviorally, LH is linked to sexual receptivity. A surge in LH occurs before ovulation, so it makes perfect sense that LH would prime female sexual behavior. In female rats, LH releases lordosis, a posture exhibited by an ovulating female in response to contact by a male. The male, in turn, mounts the female. FSH is less associated with behavioral responses. The gonads produce steroid hormones in response to LH and FSH from the pituitary—primarily testosterone by males and progesterone and estrogen by females, although all three hormones are produced by both sexes. Endocrinologically speaking, males are fairly uninteresting as testosterone levels increase with sexual maturity and then remain constant or cycle seasonally depending on the species. In females, estrogen provides feedback to the pituitary, so that LH and FSH production increases with maturity of the egg or eggs. This culminates with the LH peak, ovulation, and a return to low levels of estrogen. If the female becomes pregnant, progesterone levels increase; behaviorally and physiologically, progesterone is the hormone of pregnancy in mammals and of brooding in birds. This fits into a larger context of overall hormonal regulation, of course. For instance, stress hormones can interact with pair bonding in rodents such as prairie voles.[74]

Pituitary hormones that ultimately affect mating and parental behavior include LH and FSH, which activate the ovaries and testes of vertebrates, stimulating the production of estrogens and testosterone. Oxytocin and vasopressin, also both produced by the pituitary, play key roles in priming animals for pair bonding. Behaviorally, the vertebrate hormones, the estrogens and testosterone, have major impacts. These hormones affect the expression of mating behavior, parental behavior, and dominance and aggression.

The development of female sexual behavior in mammals is more complex than that of males. In most mammals, female sexual receptivity is linked to ovulation, so that females are primed and physiologically able to copulate when the probability of fertilization is high. In vertebrate development, the production of estrogen by the ovaries guides appropriate development of adult sexual morphology. Ovulation is stimulated by progesterone, a steroid secreted by the pituitary, and in many mammals progesterone is a prime trigger for female sexual receptivity. Once the female is pregnant, progesterone takes on an inhibitory role for sexual behavior, and after birth, hormones associated with milk production and suckling (particularly oxytocin and prolactin) further inhibit sexual behavior. In nonmammalian vertebrates, these hormones serve other functions.

DISCUSSION POINT: HORMONES AND GOVERNMENT

Until the last few years of the twentieth century, virtually no medical research was performed on women. In fact, women were not allowed in most clinical trials until after 1990. Researchers defended this practice by citing the observation that hormone levels in men changed more slowly than those in women; men were therefore thought to be less-complicated research subjects. In 1990, the Government Accountability Office issued a highly critical report, resulting in the formation of the Office of Research on Women's Health within NIH and a much more balanced approach to medical research.

Like LH and FSH, vasopressin and oxytocin are neuropeptides released into the circulation from the pituitary gland. Behaviorally, oxytocin has become known as the hormone of pair bonding and maternal behavior. Many of the adjectives associated with oxytocin (love, trust, and so on) are laden with human emotion and culture, but a well-known study of voles showed that one species, the prairie vole, has vasopressin and oxytocin receptors in the brain in locations that might associate monogamy (having a single mate) with dopamine receptors, a source of positive neural reinforcement. In the montane vole, which has multiple mates, this association with dopamine receptors and centers of positive neural reinforcement is lacking. Vasopressin may facilitate recognition of social partners, a key element of pair bonding.[75]

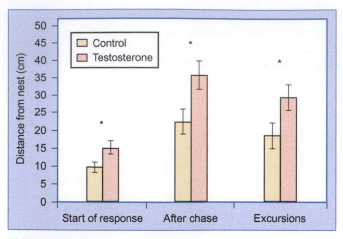

FIGURE 11.29
Typical responses of vertebrate males to testosterone implants. In this case, the study species is a fish, the rock-pool blenny. The pink bars represent individuals receiving supplemental testosterone. "Start of response" is the distance of the intruder from the nest when the territorial response starts, "after chase" is the distance the intruder is chased, and "excursions" compares the distance that treated and untreated males patrol from their nests. *Reprinted from Ros, A.F.H., Bruintjes, R., Santos, R.S., Canario, A.V.M., Oliveira, R.F., 2004. The role of androgens in the trade-off between territorial and parental behavior in the Azorean rock-pool blenny, Parablennius parvicornis. Hormones Behav. 46, 491–497.*[76]

382

11.8 HORMONES, TERRITORIALITY, AND AGGRESSION

The role of testosterone in regulating aggression and a general suite of male physical behavioral phenotypes is well known. Early awareness of this relationship is reflected in the agricultural practice of castrating male horses, cattle, swine, and chickens. The result is an animal that is both more manageable and, in the end, more edible. (In the absence of testosterone, the muscles are less well developed, more tender, and less sinewy.) Experience taught early domesticators of animals that castration was most effective if performed prior to the animal's attaining sexual maturity, and that once dominance and territoriality had been asserted in a male, castration was relatively ineffective in moderating these behaviors. This empirical knowledge of the testes as the seat of maleness led to the use of testes and testicular extracts to manipulate male traits (including treatment of impotence in humans), long before the chemical nature of the testicular factor was identified. The chemical structure of testosterone was discovered in the 1930s, and in 1939 the chemists Adolf Butenandt and Leopold Ruzicka received the Nobel Prize for synthesizing testosterone.

A long history of animal studies using testosterone injections or implants clearly links sexual behavior, aggression, and territoriality with this hormone (Figure 11.29). Note, though, that in some cases exposure to testosterone during maturation is sufficient to trigger irreversible adult behavioral patterns. While largely regarded as a male hormone, testosterone is produced in vertebrate ovaries as a derivative of estrogen and is important in modulating female sexual responses, as well as impacting some sexually selected traits in females.

Given current societal concerns about testosterone use as a performance-enhancing drug in humans, it is appropriate to ask what might be gleaned from knowledge about supplementation of natural testosterone in animal studies. It is clear from a broad range of animal studies that testosterone levels are associated with territoriality and aggression. Thus, "roid rage" has a real biological basis. In particular, rats treated with androgens respond much more aggressively to a provocation, such as a tail pinch, than do untreated controls. This effect in rats persists after the termination of the drug treatment. Human behavior, of course, is very much modulated by cultural effects, so care must be taken in extending conclusions from studies of nonhuman animals to humans.[77]

11.9 SPERM COMPETITION

Sperm competition is a form of male–male competition that warrants its own section in a textbook because it is indeed so basic and important.[78] Many[79] marine animals release their gametes into the ocean (this is called *spawning*), but this does not mean there is an absence of competition in mating, nor does it mean that the resulting zygotes are the results of random meetings of eggs and sperm. As with many topics in science, the absence of much evidence for a phenomenon–in this case, for competition among small objects in the vast sea–does not mean competition does not occur. Spawning is most common in aquatic or semiaquatic environments, where sperm can easily swim to eggs. In this system, animals that release the largest amounts of sperm with strong swimming abilities are most likely to sire young. Sperm competition was probably the first type of sexual competition to evolve, and as anisogamy evolved, sperm competition was central to the evolution of mating systems.

Sperm competition, coupled with external fertilization, is the simplest and, by far, the most common type of fertilization across the broad spectrum of the animal world. In many species, male and female never need meet; the female deposits her eggs, and later a male finds and fertilizes them. Even under these uncomplicated conditions, two processes make mating nonrandom. First, there is variation in the ability of males and females to find appropriate locations to deposit their gametes. Second, variations in sperm swimming speed and energy resources affect which sperm reaches the egg first.

> **KEY TERM** Sperm competition occurs among sperm after they are released from the male. The sperm from a single male may compete to reach an egg first; if the female has mated with more than one male, then sperm competition occurs among the males.

Sperm competition has been studied in many species of fish with external fertilization. Selection strongly favors all males that produce sperm with maximum speed and energy resources (ATP), but in many species there are still large enough differences among sperm from different males to affect fertilization probabilities. The amount of sperm, or sperm volume; the distance of sperm release from the eggs; and the timing of sperm release are also critical factors in sperm competition. The more sperm a male is able to produce, the higher the probability that one of his sperm will win the competition. Males that are able to get closer to eggs are generally more likely to be successful, and a few seconds' difference in sperm release can make all the difference in whether a male's sperm reaches the eggs first.

Generally, sperm volume is correlated with testicular size (Figure 11.30); larger testicles are necessary for production of larger amounts of sperm. A basic prediction in the biology of mating systems is that males of species with more intense sperm competition will have larger testes. The evolutionary responsiveness of testicular size to sperm competition is an important example of the effects of sexual selection on morphology and anatomy. Sometimes sperm limitation, which occurs when one male's ejaculate doesn't contain enough sperm to fertilize all a female's eggs, results in selection for polyandry, in which the female mates with more than one male.

Sperm may also cooperate. This seems odd in a situation that allows only one sperm to fertilize an egg, but isolated sperm are not efficient swimmers. An aggregate of a few sperm (5–10) swims together—working cooperatively, the sperm swim faster and more directly to the egg. When the egg is reached, competition takes over.[80]

In many cases external fertilization is combined with courtship; even though the eggs and sperm are released into the environment, the male and female first have the opportunity to assess the quality of their potential mate before committing their gametes. Bluegill sunfish (*Lepomis macrochirus*), a common fish of lakes in North America, exemplify how sperm competition can play a role in a more complex mating system that involves sophisticated behavioral interactions. Bluegill males can adopt a variety of mating strategies (Figure 11.31). Some, termed "parental" males, build nests in the bottom of the lake and court females, who spawn in the nest. Parental males then guard the eggs. "Sneakers" dash in and release sperm near the nests of parental males. And male "satellites" are able to come close to eggs because they mimic females. Sperm of all three types of males have similar speed and swimming ability, but even though sneaker males release their sperm farther from the eggs and later than parental males, they have an advantage over parental males.[81] This suggests that sneaker males produce higher-quality sperm. Satellites also have an advantage over parentals, but this could be explained

383

FIGURE 11.30
Large testicular size relative to body size may be an indication of an evolutionary history that includes sperm competition. Why might this male be displaying/exposing his scrotum?
Photo: Michael Breed.

FIGURE 11.31
A male bluegill sunfish and nest. This parental male is advertising the resources it has to support eggs. Sneaker males attempt to spawn with females that are approaching parental males. *Photo: Eric Engbretson, US Fish and Wildlife Service.*

384

by the fact that they can get closer to the eggs when they release their sperm.

Internal fertilization facilitates mating in terrestrial environments and, compared to external fertilization, allows for more control over gametes by both males and females; most terrestrial animals have evolved some form of internal fertilization. With internal fertilization, males do not need to respond so much to the effects of dilution and distance from the eggs on their gametes, and both sexes can exert a greater degree of choosiness over their mates. However, sperm competition remains an important mechanism in animals with internal fertilization. Sperm from a male will compete among themselves for fertilization opportunities, but any time a female mates more than once within a short time period, sperm competition among the sperm from the different males is likely.[82] In some cases, males even remove the sperm deposited by previous males.[83]

11.10 GOOD GENES MODELS FOR CHOOSING A MATE

As pointed out previously, most of mate choice seems to reflect the genetic quality of the potential mate. This is seen in considerations of the handicap principle and the idea that costly signals may be honest indications of a mate's quality. This section covers three other ways that animals might assess the genetic qualities of their mates. These models assume that mate choice is based on indicators of the mate's *good genes*. The possibility of mate choice based on good genes can be extended by considering characteristics that correlate with the viability of the resulting offspring. Ironically, strong mate choice based on a trait should render the trait noninformative; there will be no significant heritability for the trait if genetic variation associated with that trait is eliminated. This irony was noted on pages 367–368, along with the fact that despite this expectation, variation in sexually selected traits was often persistent. Table 11.1 shows causes of variation among animals in mating preferences. Note that all of these mechanisms preserve additive genetic variation.[19]

Inbreeding and Outbreeding

Laboratory experiments often show that inbreeding, that is, mating between closely related animals, results in lowered reproductive success. Inbreeding depression, evidenced by lowered survival rates in offspring of inbred matings, occurs across a wide range of taxa. Inbreeding exposes deleterious recessive genes by increasing the likelihood that both

> **KEY TERM** Good genes models hypothesize that mate choice relies on determining the potential of mates to pass strong genetic properties on to their offspring.

TABLE 11.1 Causes of Variation Among Animals in Mating Preferences
Genetic compatibility
Genetic complementarity
Heterozygosity
Mating facilitation
Variation among optimal offspring phenotypes
Phenotypic compatibility
Rare offspring advantage

of an animal's copies of an allele will be the same. In the ants, bees, and wasps, all members of the insect order Hymenoptera, inbreeding results in lethal genetic combinations. These observations suggest that animals should have mechanisms for avoiding mating with relatives. Inbreeding avoidance can be as simple as sex-biased dispersal: if males disperse from their birth area and females remain, then brother–sister matings will be unlikely. Learning the identities of littermates and parents early in development can also provide essential information for avoiding inbreeding. Sensory cues that allow genetic matching provide another way to facilitate inbreeding avoidance. If, for example, related animals smell alike, an animal may be able to avoid inbreeding by not mating with animals that smell similar to itself, or to its littermates. This type of social recognition is phenotype matching, which is discussed in more detail in Chapter 13. Females can also ensure against infertility resulting from inbred matings by mating with multiple males.

Of course, if an animal is performing well in a habitat, then selection would favor having genetically similar offspring. Sex has the potential for creating less-than-optimal genetic recombinations in this circumstance. Is inbreeding all bad under these conditions? The solution to this problem lies in optimal outbreeding, an idea developed by Patrick Bateson. By mating with animals that are related enough to share gene combinations, but not so related that inbreeding depression occurs, favorable genetic combinations can be maintained across generations (Figure 11.32).

If females exert choice among mates, does this really have an evolutionary effect on males? Moller and Alatalo presented a meta-analysis in which they tested the correlation between male traits and offspring survival.[85] The mean correlation coefficient was 0.122, which is significant, but explains only 1.5% of the variation in offspring survival. While small, the effect may be substantial over extended evolutionary timescales. (An aside: These authors also found a significant effect of publication year on the results of the original studies; this could be due to biases in the studies resulting from paradigm shifts in the field!)

Why not choose the same mate every time? If good genes are good, then why vary mating choice? The answer is simple and is the explanation at the heart of all biological diversity: Environments vary, and as the environment changes, so does the ideal genetic combination.

385

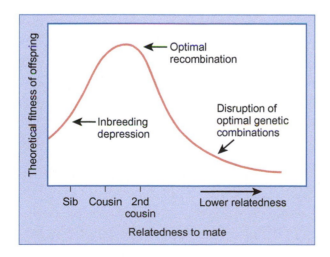

FIGURE 11.32
Patrick Bateson proposed that there was a level of outbreeding that would maximize fitness. He envisioned it as a compromise between the inbreeding depression caused by insufficient outbreeding and the disruption of optimal combinations of genes resulting from high levels of outbreeding. *Bateson, P., 1983. In: Bateson, P. (Ed.), Optimal outbreeding in mate choice, Cambridge University Press, Cambridge, pp. 257–276.*[84]

Immunocompetence and Immunocompatibility Hypotheses

The immune system relies on genetic information and precisely functioning mechanisms for antibody production. Not surprisingly, the genetic match between a male and female's immune systems can dramatically affect the survival and reproduction of their offspring. Immune systems rely on hypervariable genes that can be rapidly reshuffled from generation to generation to meet challenges posed by the evolution of disease organisms. This is the red-queen (see Figure 11.7) evolutionary arms race taken to the maximum. Hypervariable immune genes allow for much more rapid evolutionary responses than mutation could provide. Selection favors animals that confer maximal heterozygosity of immune system on their offspring. This means mate choice should favor non-matching or complementary immunotypes coming together.[86]

The major histocompatibility complex (MHC) loci play a key role in vertebrate mating systems and have been argued to be the "good genes" in mate choice in vertebrates.[87,88] Their primary function, of course, is one of mediating a vertebrate's immune response to diseases and parasites. MHC loci are hypervariable, meaning that a wide array of different alleles is possible at each locus. Many mammals detect MHC genotype by smell, giving them cues that can be used in behavioral decisions, including mate choice. MHC heterozygotes express twice the number of alleles as homozygotes and are hypothesized to have a broader range of resistance to disease. Because variability in MHC is a presumed advantage, heritability should be high. Honest signaling may be desirable or at least unavoidable. These are the ultimate good genes.

> **KEY TERM** Fluctuating asymmetry (FA) is the difference(s) within an organism from right to left. For example, an animal's right legs may be shorter than the left. The legs on each side are identical, genetically, so the length difference must be due to environment. Scientists have argued that more symmetrical animals have shown that they have better genetic combinations for their environment and, as a result, aren't so challenged in their development by the environment. Asymmetrical animals, following this argument, have genes that do not allow them to cope so well with the environment, resulting in developmental variation.

Hypothetically, pathogens evolve to evade common MHC genotypes, giving rare MHC genotypes a selective advantage. Thus, another prediction for female mate choice is that females should mate with males that are immunologically dissimilar. This can actually happen, as odors may correlate with immune system diversity.[89] Such choices would give their offspring high immune diversity and, in theory, make their offspring more disease-resistant (Figure 11.33).

Fluctuating Asymmetry

FA is the deviation from perfect bilateral symmetry caused by environmental stresses, developmental instability, and genetic problems during development.[20,91–94] It is thought that the more nearly symmetrical an organism is, the better it has been able to handle developmental stress and the more developmental stability it has, because environmental effects on development produce asymmetrical body form in animals. Females that favor symmetry may therefore be using an honest signal of "good genes." The relationship of this type of asymmetry to mating behavior is a controversial area of research in animal behavior. FA, then, may be another honest signal of good genes.

DISCUSSION POINT: SYMMETRY AND ATTRACTIVENESS

In humans, symmetrical men are more attractive to women, particularly during high fertility phases of the ovulatory cycle. Scents from symmetrical men are also more attractive to women. Conversely, men show no preferences among women based on either symmetry or scent. This is interpreted as support for good-genes preferences, with the scents being an "honest signal of phenotypic and genotypic quality of the human male." In men, there is a high correlation between symmetry and their lifetime number of sex

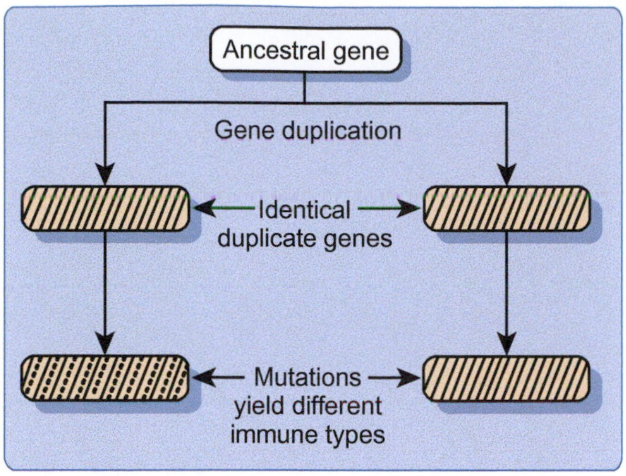

FIGURE 11.33

This figure alludes to two important principles in evolution. The first is that material for new genes can come from duplications of existing genes. If one of the duplicates retains the original function of the gene, then modifications of the other version can take on new roles. The second is that the immune system relies on genetic hyperdiversity to be able to deal with challenges from diseases and parasites (see the section on the red queen earlier in this chapter for more on this). The evolution of hyperdiversity at MHC loci can result from gene duplications, then mutations at the duplicated loci. *Adapted from Potts and Wakeland. Potts, W.K., Wakeland, E.K., 1993. Evolution of MHC genetic diversity: a tale of incest, pestilence, and sexual preference. Trends Genet. 9, 408–412.*[90]

partners. "One possible good-genes explanation is that symmetric men may have particularly rare major histocompatibility (MHC) genotypes making them attractive to many fertile women."[95] Find some photos of politicians, actors, and celebrities. Are these individuals symmetrical in their facial features? Do you think facial symmetry contributes to the kind of public success and attention that politicians, actors, and celebrities receive?

Parasites, Plumage, and Choosing a Good Mate

Another reason that animals may be unable to falsify signals of their fitness is that they cannot falsely alter their ability to cope with parasites.[23,96] Plumage color in some birds may be brightest in individuals with lower parasite loads and dullest in highly parasitized birds (Figure 11.34). If "good genes" are responsible for lower parasite loads, then bright plumage color reflects good genes.[97] This is termed the Hamilton-Zuk hypothesis and since its proposal in 1982 there have been many attempts to test it.[96] The level of acceptance of the Hamilton-Zuk hypothesis has varied over the years as empirical evidence has come in, with some studies providing supporting evidence and others finding no relationship between parasite load and signaling quality. A recent study of European green lizards revealed a negative effect of parasite load on coloration, while a meta-analysis of effects of parasites on bird plumage found less conclusive evidence for the Hamilton–Zuk hypothesis.[98,99]

11.11 FORCED COPULATIONS

Forced copulation is common reproductive behavior in ducks and geese but is much rarer in other types of animals.[100,101] The threat[102] of forced copulations may be a driving force in selecting for monogamy, at least in ducks and geese.[103] In this model, when the female chooses to pair bond with a male, she gains the

KEY TERM Forced copulation occurs when one animal subverts another's mating choice and uses physical means to force copulation on that animal.

FIGURE 11.34

This male mallard duck (*Anas platyrhynchos*) is an excellent example of how males of many bird species use bright colors to attract mates. Note the contrast of the vibrant green head, rich brown bib, and reddish-orange feet. The ability of males to produce these colors may be affected by their nutritional state and by whether or not they are infected by parasites. Males that are effective foragers with good nutrition or that have well-functioning immune systems may be more brightly colored than malnourished or infected males. Females who choose brightly colored males may be keying in on features that signal "good genes" in their prospective mate. Poorly fed or infected males, in theory, are unable to produce the bright colors of healthy plumage; coloration, then, becomes an honest signal that the male is carrying genes which make him successful in his environment. *Photo: Michael Breed.*

male's services in protecting her from forced copulations by other males. Females can also combat forced copulations by evolving methods of expelling or not using the resulting sperm.[104] The evolutionary dynamics surrounding forced copulations are complex; Clutton-Brock and Parker review the causes, costs, and modes of evading forced copulation.[105,106]

SUMMARY

Mating is a simple act with the most profound possible outcome: the production of the next generation. Mating is the point at which individual animals make a key evolutionary choice: how to combine their genes with the genes of another animal to give their offspring the greatest possible genetic advantage. Gametic investment determines sex roles; this is fundamental to defining maleness and femaleness. Many animal species produce two types of gametes: small, mobile sperm and large, nutrient-rich eggs. Males are defined as the producers of sperm; females are the producers of eggs. Mate choice and parental investment follow from these differences in investment in gametes.

Biologists have classified mating systems using observations of social interactions. Social monogamy is the pairing of a single male with a single female. Social polygyny is the pairing of one male—generally either a territory holder or a dominant male with a harem—and many females. Social polyandry, which is more rare, is one female associated with a number of males. Polygynandry and promiscuity result from more complex patterns involving multiple mates for both males and females. Animals may attempt to enforce monogamy; for example, mate guarding by males can limit mating opportunities for females.

Genetic characterization of mating systems leads to a far different picture. In many socially monogamous mating systems, organisms participate in EPCs that result in EPP. Sometimes the majority of the offspring of a socially monogamous pair are genetically fathered by other males. Females may have multiple mates during their lifetimes to increase the genetic diversity of their offspring or to hedge their bets against poor mate choices. Mating with more than one male may also ensure against male infertility. Males may mate with more than one female to produce more offspring than they might have with one mate.

Female gametic investment generates the prediction that females are often choosier than males when selecting mates. Females may prefer males that have "good genes," males who hold valuable resources, or males who are likely to produce sons that will have high chances of mating ("sexy sons"). The stereotype that males will be less choosy is sometimes not true; if a male's investment in his offspring is high or if he is unlikely to be able to obtain more mates in the future, then he, too, should be choosy.

Intermale competition plays a key role in obtaining mates for males of many species. Sexual selection resulting from this competition can result in large body size, enhanced musculature, and evolution of armament such as horns or antlers. Sperm competition among males extends intermale competition to the postcopulation interval. Interfemale competition can be as important as competition among males, and can result in selection for competitive abilities in females. Finally, forced copulations can subvert processes of choice. While they give males who would otherwise not mate opportunities to gain reproductive success, they are vigorously opposed by females and their mates.

STUDY QUESTIONS

1. Find a nonbiologist and try to explain to him or her why sex evolved and what the evolutionary function of sex is. Doing so will force you to organize your thoughts on this conceptually difficult question.
2. What are some attributes of females that increase the likelihood that they will be the caregiver for the offspring?
3. What is the role of dishonesty in courtship? Can you form a list of circumstances in which courtship signals should be honest? You may wish to refer to the discussion of honest signaling in Chapter 7 on communication as you answer this question.
4. What are the benefits to a female from social monogamy? As you think about this, focus on monogamy as a strategy the female might use to involve a male in parental care of their offspring. What are the disadvantages, from a female's point of view, of genetic monogamy? Can you explain some examples of how females may overcome these disadvantages?
5. What circumstances might favor the evolution of polyandry?
6. Investigators have argued that symmetry is a measure of "good genes" in an animal. A more detailed explanation is that symmetry reflects developmental stability and perhaps greater allelic diversity in the animal. If this is true, then animals should prefer to have symmetrical mates. How does this line of reasoning interact with thinking about environmental variability? Would you expect symmetry to be more important for mate choice in stable or unpredictable environments? Can you design an experiment to test your prediction?
7. Greek mythology states that the "eyes" on a peacock's tail are derived from Argus, whom Hermes killed after boring him to sleep with dull stories and insipid songs. Hera then cut the eyes off the corpse of her valued friend and scattered them on the tail of the bird now called the peacock. Can you give an explanation for the existence of these "eyes" that is more consistent with modern scientific thought?

Further Reading

Ahnesjo, I., 2010. Mate choice in males and females. In: Breed, M.D., Moore, J. (Eds.), Encyclopedia of Animal Behavior, vol. 2, Academic Press, Oxford, pp. 394–398.

Dixson, A.F., 2009. Sexual Selection and the Origins of Human Mating Systems. Oxford University Press, Oxford, pp. 232.

Grether, G.F., 2010. Sexual selection and speciation. In: Breed, M.D., Moore, J. (Eds.), Encyclopedia of Animal Behavior, vol. 3, Academic Press, Oxford, pp. 177–183.

Shuster, S.M., Wade, M.J., 2003. Mating Systems and Strategies. Princeton University Press, Princeton, NJ, pp. 520.

Thornhill, R., Alcock, J., 1983. The Evolution of Insect Mating Systems. Harvard University Press, Cambridge, MA, pp. 564.

Zahavi, A., Zahavi, A., 1997. The Handicap Principle: A Missing Piece of Darwin's Puzzle. Oxford University Press, Oxford, pp. 304.

Notes

1. Grammer, K., Fink, B., Moller, A.P., Thornhill, R., 2003. Darwinian aesthetics: sexual selection and the biology of beauty. Biol. Rev. 78, 385–407.
2. Mitchem, D.G., Purkey, A.M., Grebe, N.M., Carey, G., Garver-Apgar, C., Bates, T., 2014. Estimating the sex-specific effects of genes on facial attractiveness and sexual dimorphism. Behav. Genet. 44, 270–281.
3. Surbeck, M., Deschner, T., Schubert, G., Weltring, A., Hohmann, G., 2012. Mate competition, testosterone and intersexual relationships in bonobos, *Pan paniscus*. Anim. Behav. 83, 659–669.
4. Langergraber, K.E., Mitani, J.C., Watts, D.P., Vigilant, L., 2013. Male-female socio-spatial relationships and reproduction in wild chimpanzees. Behav. Ecol. Sociobiol. 67, 861–873.
5. Gorelick, R., Heng, H.H.Q., 2011. Sex reduces genetic variation: a multidisciplinary review. Evolution 65, 1088–1098.
6. Van Valen, L., 1973. A new evolutionary law. Evol. Theory 1, 1–30.
7. Fisher, R., 1930. The Genetical Theory of Natural Selection. New York, NY, Dover, 1957.
8. Trivers, R.L., Willard, D., 1973. Natural selection of parental ability to vary the sex ratio of offspring. Science 179, 90–92.
9. Lande, R., 1976. Models of speciation by sexual selection on polygenic traits. Proc. Natl. Acad. Sci. USA 78, 3721–3725.
10. Grether, G., 1996. Intrasexual competition alone favors a sexually dimorphic ornament in the rubyspot damselfly *Hetaerina americana*. Evolution 50, 1949–1957.
11. Hogg, J.T., Forbes, S.H., 1997. Mating in bighorn sheep: frequent male reproduction via a high-risk "unconventional" tactic. Behav. Ecol. Sociobiol. 41, 33–48.
12. Basolo, A.L., 1990. Female preference predates the evolution of the sword in swordtail fish. Science 250, 808–810.
13. Rosenthal, G.G., Evans, C.S., 1998. Female preference for swords in *Xiphophorus helleri* reflects a bias for large apparent size. Proc. Natl. Acad. Sci. USA 95, 4431–4436.
14. Coleman, S.W., Patricelli, G.L., Coyle, B., Sianil, J., Borgia, G., 2007. Female preferences drive the evolution of mimetic accuracy in male sexual displays. Biol. Lett. 3, 463–466.
15. Zahavi, A., 1975. Mate selection—selection for a handicap. J. Theor. Biol. 53 (1), 205–214; Zahavi, A., 1977. Cost of honesty (further remarks on handicap principle). J. Theor. Biol. 67 (3), 603–605.
16. Petrie, M., Halliday, T., 1994. Experimental and natural changes in the peacocks (*Pave cristatus*) train can affect mating success. Behav. Ecol. Sociobiol. 35 (3), 213–217.
    Petrie, M., Halliday, T., Sanders, C., 1991. Peahens prefer peacocks with elaborate trains. Anim. Behav. 41, 323–331.
    Petrie, M., Krupa, A., Burke, T., 1999. Peacocks lek with relatives even in the absence of social and environmental cues. Nature 401 (6749), 155–157.
17. Takahashi, M., Arita, H., Hiraiwa-Hasegawa, M., Hasegawa, T., 2008. Peahens do not prefer peacocks with more elaborate trains. Anim. Behav. 75, 1209–1219.
18. Loyau, A., Petrie, M., Saint Jalme, M., Sorci, G., 2008. Do peahens not prefer peacocks with more elaborate trains? Anim. Behav. 76, e5–e9.
19. Widemo, F., Saether, S.A., 1999. Beauty is in the eye of the beholder: causes and consequences of variation in mating preferences. Trends Ecol. Evol. 14, 26–31.
20. Manning, J.T., Hartley, M.A., 1991. Symmetry and ornamentation are correlated in the peacocks train. Anim. Behav. 42, 1020–1021.
21. Pomiankowski, A., Moller, A.P., 1995. A resolution of the lek paradox. Proc. Roy. Soc. Lond. B 260, 21–29.
22. Zuk, M., Kolloru, G.R., 1998. Exploitation of sexual signals by predators and parasitoids. Quart. Rev. Biol. 73, 415–438.

23. Zuk, M., Johnson, K., Thornhill, R., Ligon, J.D., 1990. Parasites and male ornaments in freeranging and captive red jungle fowl. Behaviour 114, 232–248; Zuk, M., Kolloru, G.R., 1998. Exploitation of sexual signals by predators and parasitoids. Quart. Rev. Biol. 73, 415–438.

24. Hall, M.L., 2000. The function of duetting in magpielarks: conflict, cooperation, or commitment? Anim. Behav. 60, 667–677; Hall, M.L., Magrath, R.D., 2000. Duetting and mateguarding in Australian magpielarks (*Grallina cyanoleuca*). Behav. Ecol. Sociobiol. 47 (3), 180–187.

25. McGraw, K.J., Mackillop, E.A., Dale, J., Hauber, M.E., 2002. Different colors reveal different information: how nutritional stress affects the expression of melanin- and structurally based ornamental plumage. J. Exp. Biol. 205, 3747–3755.

26. Jennions, M.D., Petrie, M., 1997. Variation in mate choice and mating preferences: a review of causes and consequences. Biol. Rev. 72, 283–327.

27. Gabor, C.R., Halliday, T.R., 1997. Sequential mate choice by multiply mating smooth newts: females become more choosy. Behav. Ecol. 8, 162–166.

28. Gowaty, P.A., Kim, Y.-K., Anderson, W.W., 2012. No evidence of sexual selection in a repetition of Bateman's classic study of *Drosophila melanogaster*. Proc. Natl. Acad. Sci. USA 109, 11740–11745.

29. Collet, J.M., Dean, R.F., Worley, K., Richardson, D.S., Pizzari, T., 2014. The measure and significance of Bateman's principles. Proc. R. Soc. Lond. B, Biol. Sci. 281, 20132973.

30. Sakuluk, S.K., 1984. Male crickets feed females to ensure complete sperm transfer. Science 223, 609–610.

31. Searcy, W.A., 1992. Song repertoire and mate choice in birds. Am. Zool. 32, 71–80.

32. Pruett-Jones, S., 1992. Independent versus nonindependent mate choice: do females copy each other? Am. Nat. 140, 1000–1009.

33. Alonzo, S., 2008. Female mate choice copying affects sexual selection in wild populations of the ocellated wrasse. Anim. Behav. 75, 1715–1723.

34. Schlupp, I., Ryan, M.J., 1997. Male sailfin mollies (*Poecilia latipinna*) copy the mate choice of other males. Behav. Ecol. 8, 104–107.

35. Tschirren, B., Postma, E., Rutstein, A.N., Griffith, S.C., 2012. When mothers make sons sexy: maternal effects contribute to the increased sexual attractiveness of extra-pair offspring. Proc. R. Soc. Lond. B, Biol. Sci. 279, 1233–1240.

36. Prokop, Z.M., Michalczyk, L., Drobniak, S.M., Herdegen, M., Radwan, J., 2012. Meta-analysis suggests choosy females get sexy sons more than "good genes." Evolution 66, 2665–2673.

37. Matysiokova, B., Remes, V., 2013. Faithful females receive more help: the extent of male parental care during incubation in relation to extra-pair paternity in songbirds. J. Evol. Biol. 26, 155–162.

38. Osorio-Beristain, H., Drummond, H., 2001. Male boobies expel eggs when paternity is in doubt. Behav. Ecol. 12, 16–21.

39. Magrath, M.J.L., Komdeur, J., 2003. Is male care compromised by additional mating opportunity? Trends Ecol. Evol. 18, 424–430.

40. Szykman, M., Engh, A.L., Van Horn, R.C., Funk, S.M., Scribner, K.T., Holekamp, K.E., 2001. Association patterns among male and female spotted hyenas (*Crocuta crocuta*) reflect male mate choice. J. Behav. Ecol. Sociobiol. 50, 231–238.

41. Orians, G.H., 1969. On the evolution of mating systems in birds and mammals. Am. Nat. 103, 589–603.

42. Emlen, S.T., Oring, L.W., 1977. Ecology, sexual selection and the evolution of mating systems. Science 197, 215–233.

43. Morris, M.R., Rios-Cardenas, O., Brewer, J., 2010. Variation in mating preference within a wild population influences the mating success of alternative mating strategies. Anim. Behav. 79, 673–678.

44. Petrie, M., Doums, D., Moller, A.P., 1998. The degree of extrapair paternity increases with genetic variability. Proc. Natl. Acad. Sci. USA 16, 9390–9395. Could be characterized as cryptic polyandry.

45. Schwagmeyer, P.L., Ketterson, E.D., 1999. Breeding synchrony and EPF rates: the key to a can of worms. Trends Ecol. Evol. 14, 47–48.

46. Stutchberry, B.J.M., 1998. Breeding synchrony best explains variation in extrapair mating system among avian species. Behav. Ecol. Sociobiol. 43, 221–222.

47. Westneat, D.F., Sherman, P.W., 1997. Density and extrapair fertilizations in birds: a comparative analysis. Behav. Ecol. Sociobiol. 41, 205–215.

48. Manlove, C.A., Hepp, G.R., 1998. Effects of mate removal on incubation behavior and reproductive success of female wood ducks. Condor 100 (4), 688–693.

49. Parker, G.A., Birkhead, T.R., 2013. Polyandry: the history of a revolution. Philos. Trans. R. Soc. Lond. B, Biol. Sci. 368, 20120335.

50. Kempenaers, B., Congden, B., Boag, P., Robertson, R.J., 1999. Extrapair paternity and egg hatchability in tree swallows: evidence for the genetic compatibility hypothesis? Behav. Ecol. 10, 304–311.

51. Webster, M.S., Pruett-Jones, S., Westneat, D.F., Arnold, S.J., 1995. Measuring the effects of pairing success, extra-pair copulations and mate quality on the opportunity for sexual selection. Evolution 49, 1147–1157.

52. Uhl, G., Kunz, K., Voecking, O., Lipke, E., 2014. A spider mating plug: origin and constraints of production. Biol. J. Linn. Soc. 113, 345–354.

53. Sparkes, T.C., Keogh, D.P., Pary, R.A., 1996. Energetic costs of mate guarding behavior in male stream-dwelling isopods. Oecologia 106, 166–171.

54. Emlen, D., 1997. Alternative reproductive tactics and male-dimorphism in the horned beetle *Onthophagus acuminatus* (Coleoptera: Scarabaeidae). Behav. Ecol. Sociobiol. 41, 335–341.

55. Alonzo, S.H., Warner, R.R., 2000. Allocation to mate guarding or increased sperm production in a Mediterranean wrasse. Am. Nat. 56 (3), 266–275.

56. Zamudio, K.R., Sinervo, E., 2000. Polygyny, mateguarding, and posthumous fertilization as alternative male mating strategies. Proc. Natl. Acad. Sci. USA 97 (26), 14427–14432.

57. Verner, J., Willson, M.F., 1966. The influence of habitats on mating systems of North American passerine birds. Ecology 47, 143–147.

58. Verner, J., 1964. Evolution of polygamy in the long-billed marsh wren. Evolution 18, 252–261.

59. Bronstein, P.M., 1985. Predictors of dominance in male *Betta splendens*. J. Comp. Psychol. 99, 47–55.

60. Petrulis, A., Weidner, M., Johnston, R.E., 2004. Recognition of competitors by male golden hamsters. Physiol. Behav. 81, 629–638.

61. Kovach, A.I., Powell, R.A., 2003. Effects of body size on male mating tactics and paternity in black bears. *Ursus americanus*. Can. J. Zool. 81, 1257–1268.

62. Duhrkopf, R.E., Hartberg, W.K., 1992. Differences in male mating response and female flight sounds in *Aedes aegypti* and *Ae. Albopictus* (Diptera: Culicidae). J. Med. Entomol. 29, 796–801.

63. Duvall, D., Schuett, G.W., 1997. Straight-line movement and competitive mate searching in prairie rattlesnakes, *Crotalus viridis* viridis. Anim. Behav. 54, 329–334.

64. Hofmann, H.A., Schildberger, K., 2001. Assessment of strength and willingness to fight during aggressive encounters in crickets. Anim. Behav. 62, 337–348.

65. Koenig, A., Scarry, C.J., Wheeler, B.C., Borries, C., 2013. Variation in grouping patterns, mating systems and social structure: what socio-ecological models attempt to explain. Philos. Trans. R. Soc. Lond. B, Biol. Sci. 368, 20120348.

66. Lank, D.B., Smith, C.M., Hanotte, O., Burke, T., Cooke, F., 1995. Genetic polymorphism for alternative mating behavior in lekking male ruff *Philomachus pugnax*. Nature 378 (6552), 59–62.

67. Aspbury, A.S., Gibson, R.M., 2004. Long-range visibility of greater sage grouse leks: a GIS-based analysis. Anim. Behav. 67, 1127–1132.

68. Drnevich, J.M., Papke, R.S., Rauser, C.L., Rutowski, R.L., 2001. Material benefits from multiple mating in female mealworm beetles (*Tenebrio molitor* L.). J. Insect. Behav. 14, 215–230.

69. Eakley, A.L., Houde, A.E., 2004. Possible role of female discrimination against 'redundant' males in the evolution of colour pattern polymorphism in guppies. Proc. Roy. Soc. Lon. B, Biol. Sci. 271, S299–S301.

70. Moscovice, L.R., Di Fiore, A., Crockford, C., Kitchen, D.M., Wittig, R., Seyfarth, R.M., et al., 2010. Hedging their bets? Male and female chacma baboons form friendships based on likelihood of paternity. Anim. Behav. 79, 1007–1015.

71. Zgurski, J.M., Hik, D.S., 2012. Polygynandry and even-sexed dispersal in a population of collared pikas, *Ochotona collaris*. Anim. Behav. 83, 1075–1082.

72. Krebs, C.J., Boutin, S., Boonstra, R., Sinclair, A.R.E., Smith, J.N.M., Dale, M.R.T., et al., 1995. Impact of food and predation on the snowshoe hare cycle. Science 269 (5227), 1112–1115.

73. Burton, C., 2002. Microsatellite analysis of multiple paternity and male reproductive success in the promiscuous snowshoe hare. Can. J. Zool. 80 (11), 1948–1956.

74. DeVries, A.C., DeVries, M.B., Taymans, S.E., Carter, C.S., 1996. The effects of stress on social preferences are sexually dimorphic in prairie voles. Proc. Natl. Acad. Sci. USA 93, 11980–11984.

75. Ross, H.E., Cole, C.D., Smith, Y., Neumann, I.D., Landraf, R., Murphy, A.Z., et al., 2009. Characterization of the oxytocin system regulating affiliative behavior in female prairie voles. Neuroscience 162, 892–903.

76. Ros, A.F.H., Bruintjes, R., Santos, R.S., Canario, A.V.M., Oliveira, R.F., 2004. The role of androgens in the trade-off between territorial and parental behavior in the Azorean rock-pool blenny, *Parablennius parvicornis*. Hormones Behav. 46, 491–497.

77. McGinnis, M.Y., Lumia, A.R., Breuer, M.E., Possidente, B., 2002. Physical provocation potentiates aggression in male rats receiving anabolic androgenic steroids. Horm. Behav. 41, 101–110.

78. Fisher, H.S., Hoekstra, H.E., 2010. Competition drives cooperation among closely related sperm of deer mice. Nature 463, 801–803.

79. Parker, G.A., Lehtonen, J., 2014. Gamete evolution and sperm numbers: sperm competition versus sperm limitation. Proc. R. Soc. Lond. B, Biol. Sci. 281, 20140836.

80. Fisher, H.S., Giomi, L., Hoekstra, H.E., Mahadevan, L., 2014. The dynamics of sperm cooperation in a competitive environment. Proc. R. Soc. Lond. B, Biol. Sci. 281, 20140296.

81. Stoltz, J.A., Neff, B.D., 2006. Sperm competition in a fish with external fertilization: the contribution of sperm number, speed and length. J. Evol. Biol. 19 (6), 1873–1881.

82. Corley, L.S., Cotton, S., McConnell, E., Chapman, T., Fowler, K., Pomiankowski, A., 2006. Highly variable sperm precedence in the stalk-eyed fly, *Teleopsis dalmanni*. BMC Evol. Biol. 6, 53.

83. Waage, J.K., 1979. Dual function of the damselfly penis: sperm removal and transfer. Science 203, 916–918.

84. Bateson, P., 1983. In: Bateson, P. (Ed.), Optimal outbreeding in mate choice, Cambridge University Press, Cambridge, pp. 257–276.

85. Moller, A.P., Alatalo, R.V., 1999. Good genes effects in sexual selection. Proc. Roy. Soc. Lond. B 266, 85–91.

86. Eizaguirre, C., Lenz, T.L., Kalbe, M., Milinski, M., 2012. Rapid and adaptive evolution of MHC genes under parasite selection in experimental vertebrate populations. Nat. Commun. 3, 621.

87. Penn, D.J., Potts, W.K., 1999. The evolution of mating preferences and the major histocompatibility complex genes. Am. Nat. 153, 145–164.

392

88. Yamazaki, K., Beauchamp, G.K., Singer, A., Bard, J., Boyse, E.A., 1999. Odortypes: their origin and composition. Proc. Natl. Acad. Sci. USA 96, 1522–1525.

89. Jordan, W.C., Burford, M.W., 1998. New perspectives on mate choice and the MHC. Heredity 81, 239–245.

90. Potts, W.K., Wakeland, E.K., 1993. Evolution of MHC genetic diversity: a tale of incest, pestilence, and sexual preference. Trends Genet. 9, 408–412.

91. Blanckenhorn, W.U., Reusch, T., Muhlhauser, C., 1998. Fluctuating asymmetry, body size, and sexual selection in the dung fly *Sepsis cynipsea*—testing the good genes assumptions and predictions. J. Evol. Biol. 11, 735–753.

92. Ligon, D.J., Kimball, R., MerolaZwartjes, M., 1998. Mate choice by female red junglefowl: the issues of multiple ornaments and fluctuating asymmetry. Anim. Behav. 55, 4150.

93. Morris, M.R., 1998. Female preference for trait symmetry in addition to trait size in swordtail fish. Proc. Roy. Soc. Lond. B 265, 907–911.

94. Oakes, E.J., Barnard, P., 1994. Fluctuating asymmetry and mate choice in paradise whydahs, *Vidua paradisaea*: an experimental manipulation. Anim. Behav. 48, 937–943.

95. Thornhill, R., Gangestad, S.W., 1999. The scent of symmetry: a human sex pheromone that signals fitness? Evol. Hum. Behav. 20, 175–201.

96. Hamilton, W.D., Zuk, M., 1982. Heritable true fitness and bright birds a role for parasites. Science 218 (4570), 384–387.

97. Buchholz, R., 1995. Female choice, parasite load and male ornamentation in wild turkeys. Anim. Behav. 50, 929–943.

98. Molnar, O., Bajer, K., Meszaros, B., Torok, J., Herczeg, G., 2013. Negative correlation between nuptial throat colour and blood parasite load in male European green lizards supports the Hamilton-Zuk hypothesis. Naturwissenschaften 100, 551–558.

99. Garamszegi, L.Z., Moller, A.P., 2012. The interspecific relationship between prevalence of blood parasites and sexual traits in birds when considering recent methodological advancements. Behav. Ecol. Sociobiol. 66, 107–119.

100. Mckinney, F., Derrickson, S.R., Mineau, P., 1983. Forced copulation in waterfowl. Behaviour 86, 250–294.

101. Seymour, N.R., 1990. Forced copulation in sympatric American black ducks and mallards in Nova Scotia. Can. J. Zool. 68 (8), 1691–1696.

102. Haig, D., 2014. Sexual selection: placentation, superfetation, and coercive copulation. Curr. Biol. 24, R805–R808.

103. Gowaty, P.A., Buschhaus, N., 1998. Ultimate causation of aggressive and forced copulation in birds: female resistance, the CODE hypothesis, and social monogamy. Am. Zool. 38 (1), 207–225.

104. Dunn, P.O., Afton, A.D., Gloutney, M.L., Alisauskas, R.T., 1999. Forced copulation results in few extrapair fertilizations in Ross's and lesser snow geese. Anim. Behav. 57, 1071–1081.

105. Clutton-Brock, T.H., Parker, G.A., 1995. Sexual coercion in animal societies. Anim. Behav. 49 (5), 1345–1365.

106. Shields, W.M., Shields, L.M., 1983. Forcible rape: an evolutionary perspective. Ethol. Sociobiol. 4, 115–136.

107. Tomkins, J.L., Simmons, L.W., 1998. Female choice and manipulations of forceps size and symmetry in the earwig *Forficula auricularia*. L. Anim. Behav. 56, 347–356.

Nesting, Parenting, and Territoriality

LEARNING OBJECTIVES

Studying this chapter should provide you with the knowledge to:

- Apply understanding of navigation and habitat preference to nest site selection.
- Analyze the complexities of animal architecture and appraise how animals use structure to manipulate microclimate in a nest, den, or burrow.
- Determine how parents control their investment in offspring and how these choices involve balances between current and future reproductive opportunities.

Animal Behavior. DOI: http://dx.doi.org/10.1016/B978-0-12-801532-2.00012-X

- Remember the origins of parent–offspring conflict over investment and the relationship of infanticide to paternal investment.
- Develop evolutionary explanations for how conflict over resources extends to interactions among siblings that in extreme cases can lead to siblicide.
- Connect the many causes of animal aggression and understand why aggression emerges from conflicts over resources, territories, or in the contexts of mating and predation.

12.1 INTRODUCTION

The main consequence of mating and reproduction is the production of young. This chapter covers topics that surround where to live and how to care for young. *Parental care* can be as simple as choosing a safe location for laying eggs, or as complicated as caring for young for months or years while they mature. For students of animal behavior, one of the most fascinating aspects of this material comes from the study of parent–offspring conflict and sibling–sibling (sib–sib) conflict, things that are well known to many people from personal experience (Figure 12.1).

> **KEY TERM** Parental care is behavior directed to feeding, protecting, and in some cases teaching offspring.

396

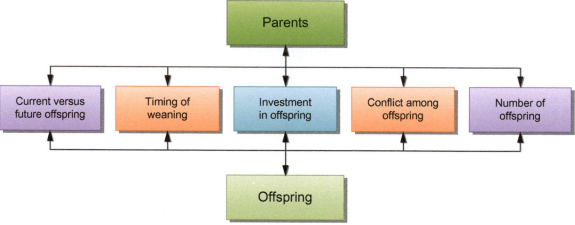

FIGURE 12.1

An overview of facets of parent–offspring interactions in which the evolutionary point of view of the parents and the offspring may differ. The balance tends to tip, for offspring, to favoring investment in themselves at the expense of sibs. Parents may restrict investment in any given offspring in order to distribute resources among members of a brood or to retain resources for future broods.

CASE STUDY: MEERKAT INSTRUCTION

Teaching survival skills to offspring can be a considerable investment. The meerkat is one of the few nonhuman animals in which the teaching of young has been demonstrated.[1] While it doubtlessly occurs in many species, it is difficult to demonstrate in the field, because it is difficult to distinguish from observational learning. (In observational learning, there is no attempt to teach; instead the learner simply observes.) Teaching young how to handle scorpions, a prey item that makes up approximately 5% of meerkat diets, is especially useful since some of the scorpion species consumed by meerkats can be lethal. Thornton and McAuliffe[2] set out to test the hypothesis that meerkats engaged in opportunistic teaching of young

(Continued)

CASE STUDY (CONTINUED)

meerkats about preying upon scorpions. Much of this study hinged on the fact that young meerkat calls change with age. By using age-specific playbacks of meerkat calls as controls, Thornton and McAuliffe showed that helper meerkats responded differently to meerkat young of different ages, and the younger the meerkat, the more help it received in preying upon scorpions. This help ranged from removal of stingers to nudging and retrieving escaped prey (Figure 12.2).

FIGURE 12.2
Adult meerkat with young. In meerkats the young observe older members of the social group, particularly their parents, and gain key information about diet and prey handling. *Photo: Katherine McAuliffe.*

Some important topics relating to parenting were covered in earlier chapters. For instance, the endocrine basis of parental care and parent–offspring bonding was discussed in Chapter 2 (in the section on hormones). Chapter 5 included the way offspring learn food preferences from parents, and Chapter 8 reviewed parental roles in orientation within the habitat and transmission of information about migratory routes in some species. Review these discussions when reading this chapter because much of what is presented here is a natural outgrowth of that material.

> **KEY TERM** A den or nest is a location that has been modified by the animal so that it is appropriate for rearing young. The modification can be minimal, as in clearing stray material from the area, or can be extensive, as exemplified by beaver construction of lodges and dams.

397

Often, parental care includes construction of a *den* or *nest*—a good starting place for this chapter. In many ways, this is "land use," from an animal perspective. The architecture of these structures is often fascinating and sometimes puts human construction attempts to shame. For generation after generation, without external blueprints, many animals construct nests that typify their species. Evolution shapes the nest or den architecture of species and gives animals the genetic information that shapes their construction strategies. Animal construction is a key element in keeping young safe from harm and in the right microclimate for development.

A nest is but one way that parents can care for their young. The choices about time and energy use that confront parents as they care for their offspring are striking. Do they invest more in current offspring or save energy for investment in future offspring? Such decisions naturally lead to conflict between parents and their offspring. The two parties may have strikingly different viewpoints about allocation of resources within a family; understanding this tension is part of the appeal of studying parent–offspring conflict. The same dynamics drive competition among sibs for parental investments, a competition that in extreme cases can be lethal. Infanticide is also covered in this chapter; while it most frequently involves males killing offspring (especially if the males are uncertain about paternity), it can also occur as a result of unequal investment in offspring during hard times.

A discussion of siblicide and infanticide naturally leads to a broader consideration of aggression and then to territoriality, the final two topics of this chapter. The journey from nesting, at the outset of this chapter, to territoriality, at the end, describes a circle that begins with a family in its nest and ends with that family's interactions with neighbors or with the family's offspring staking out its own claim to the future. Chapter 13 (on social behavior) expands this focus to extended families and larger social structures.

12.2 NESTS AND NESTING
Habitat Choice

Habitat choice in nesting is largely an issue of finding the appropriate microclimate for growth of the young and right level of protection from predators and parasites.[3] Behavioral strategies for finding appropriate habitats are covered in Chapter 8 on navigation. If appropriate nesting sites are scarce, then aggressive competition with other animals may be the first step in nesting.

Consider animals that nest in tree holes. Cavities in hollow trees, used as nests by many species, are the focus of classic examples of competition for nesting sites. Cavity-nesting birds, such as flickers, may compete among themselves and with other species for good nest sites. In addition, honeybees nest in hollow cavities, and some scientists hypothesize that their introduction into North America contributed to the decline and extinction of the Carolina Parakeet due to competition for nest sites.[4] In a habitat lacking adequate nesting sites, honeybees may end up nesting in marginal locations, such as hanging their comb from a tree branch; these marginal nests do not survive the winter.

Nest Construction and Animal Architecture

The simplest shelter is obtained by occupying a protected place, such as a cave, hollow tree, or crevice between rocks. The beauty of such accommodation is that they come labor-free; no time or energy is invested in making modifications to create a more habitable space. This section highlights some species with interesting and remarkable architectural and construction abilities (Figure 12.3).

FIGURE 12.3
Nests and nest construction in birds. (A) A cormorant swimming with nesting material; (B) cormorant nest; (C) black sandpiper nest; (D) rusty blackbird on nest; and (E) bald eagle in nest. *Photos: (A and B) Michael Breed; (C) US Fish and Wildlife Service; (D) Robin Corcoran, US Fish and Wildlife Service; (E) Donna Dewhurst, US Fish and Wildlife Service.*

Bat Roosts

Bat roosts are examples of the simplest type of nesting. Most bat species live in sheltered spots, such as hollow trees or caves, and they do not modify these homes (Figure 12.4). The drawback to such an arrangement is that the only control of *microclimate* comes from choosing a suitable spot and perhaps benefiting from heat and moisture generated by companions. Tent bats, on the other hand, choose large leaves for their nests, and modify these nests by cutting the leaves so that they fold over and protect the bats.

> **KEY TERM** Microclimate is the specific set of climatic conditions at a defined (micro) location. Conditions could include factors such as temperature, humidity, or light level.

FIGURE 12.4
Big brown bats in a roosting site. Most bats do not construct a nest, but they do seek a sheltered roosting site and huddle together for warmth and protection. *Photos: Rick Adams.*

Beaver Lodges

Beavers are among the most famous animal engineers. Beavers usually construct a dam and a lodge, but if trees are not around the stream or pond, the beavers simply dig a burrow in the bank; this construction flexibility gives them a broad range of habitat. For instance, prior to European colonization of North America, burrowing beavers were common in prairie ecosystems. In areas with trees, branches and logs from soft-wooded trees (e.g., poplars, aspens) are used to build the lodge and the dam. Mud may be used to fill cracks and cement the construction together. The beavers strip the bark from their construction materials and feed on the inner bark, as well as on nearby aquatic and terrestrial plants. The submerged entry to the lodge provides protection against predators (Figure 12.5). Beavers

399

FIGURE 12.5
A beaver lodge (diagram, top), beaver swimming in a pond (left), and a beaver lodge (right). The lodge provides a sheltered location for an extended family. *Photos: Jeff Mitton.*

are remarkable for their persistence in constructing dams across large spans of streams and rivers, and for their ability to maintain their dams even during flood conditions.

Prairie Dog Burrows

Prairie dogs re-engineer vast expanses of prairie; their towns can occupy up to 100 square miles. Above ground, they construct a mound of excavated soil, which provides a vantage point for observation of the surroundings (Figure 12.6). The mound also prevents rainwater from running into the burrow and flooding their living quarters. The area around the mound is cleared of grass, improving the view of approaching potential predators. Below ground, the tunnel leads to a chamber in which the young prairie dogs are reared, and is connected to other tunnels in the colony. The tunnel and chamber may be lined with grass, which acts as both insulation and a food store. The prairie dog's primary natural enemies are rattlesnakes, coyotes, and large birds of prey, such as eagles and hawks. The tunnel provides a defensible escape route that is effective against some of these predators, such as coyotes and birds. The young are vulnerable to snakes, and the adults work to prevent snakes from entering the chamber via the burrow.

FIGURE 12.6
A prairie dog at its burrow entrance. Note the pile of dirt that encircles the entrance. This keeps water from flowing into the burrow, and affords the prairie dogs a good vantage point for vigilance. *Photo: Michael Breed.*

400

Magpie Nests

The Black-billed Magpie is native to the American Northwest. Like many birds, it builds its nest of sticks. The nest appears to be a fairly sloppy construction, but is quite sturdy and holds up well against breezes that toss around the high branches where it is located. A major threat to magpies is predation of their eggs by crows, and magpies have responded by evolving to build stick roofs over their nest; unlike the traditional "cup" style of bird nest, magpie nests, at least in areas where crows occur, have specially designed architecture to thwart their foes (Figure 12.7).

Paperwasp Nests

Paperwasps, as well as yellowjackets and hornets, collect plant materials that they chew and glue together with saliva to form a nest with cells.[5] An egg is laid in each cell, and the growing larvae are fed by adults on the nest. Architecturally, small variations in building pattern can result in big differences in nest shape. In Figure 12.8, each new cell is begun near the base of the adjacent cell; this results in a flat comb. In some other species, the new cells are begun near the top of an older cell, resulting in a long, string-like nest that resembles a dead stick.

A

B

FIGURE 12.7
In habitats where crows are absent, magpies build open nests of sticks (A). When crows, which eat magpie eggs and nestlings, are members of the community, magpies build covered nests (B).

Termite Mounds

Most termites construct their mounds using a combination of feces, saliva, mud, and bits of chewed-up plant (Figure 12.9). The amazing thing about termites is that although each animal is only a few millimeters long, thousands of termites working collectively can build monumental nests that tower over a human. Such a large nest, occupied by so many animals, presents major architectural problems. It must support itself against collapse while providing enough porosity to allow air exchange: without this, carbon dioxide and water from the metabolism of the termites will accumulate in the nest, while oxygen will be depleted. The termites' architectural solution is to build passageways specifically for airflow

FIGURE 12.8
A paperwasp nest, *Polistes* sp. The nest is constructed of chewed plant material and hangs by a thin connection, or pedicel, from the substrate. The queen lays eggs in the cells within the comb, and the workers then feed the larvae. *Photo: Michael Breed.*

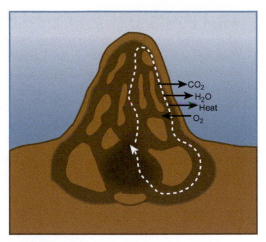

FIGURE 12.9
Airflow, temperature, and gas exchange in a termite mound. Oxygen comes into the nest with fresh air. As the termites metabolize, they use the oxygen and release carbon dioxide, water, and heat, which all leave the nest. The air circulation within the nest is driven by insolation and the position of the openings into the nest; it ensures that water and carbon dioxide do not build up, while adequate oxygen to support life is maintained. *Adapted from Turner, J.S., 2000. The Extended Organism. The Physiology of Animal-Built Structures. Harvard University Press, Cambridge, 234 pp.*[6]

and to rely on convective movement of air (hot air rising, cool air sinking) to maintain an ideal microclimate within the nest. Why build such a big mound? The mound protects the termites from voracious predators, such as aardvarks.

Sand Goby Nests

Like many fish, sand gobies, European saltwater fish that live on sandy bottoms of bays and estuaries, constructs nests of sand in which the eggs are laid and fertilized. The male constructs and defends the nest, which is an important protection against predators, such as shrimp. The presence of predators in the environment stimulates males to make better nests, using more sand, and large males are better at this kind of building than small males.[7] Females choose among nests when laying their eggs, preferring nests which have fewer eggs and sometimes choosing nests which are better protected[8-10] by having a deeper layer of sand. Female choice does not revolve entirely around the nest and at some points in time other male characteristics drive female choice. This case is interesting because of the presence of paternal care, the complexity of mate choice, and the responsiveness of the system to the presence of predators.

DISCUSSION POINT: TOOL USE, COGNITION, AND CONSTRUCTION

Tool use was one of the first behaviors to lead scientists to contemplate the cognitive lives of animals (see Chapter 6), and the distribution of this behavior across taxa is often thought to be limited by a taxon's cognitive ability. Mike Hansell and Graeme Ruxton[11] pose an alternate explanation for the fact that the great majority of animals do not seem to use tools. They suggest that most animals do not use tools because they don't need to use them; their anatomical adaptations are sufficient for the tasks they confront. Indeed, if tool use is viewed from the standpoint of construction—if the definition and emphasis on cognition are shifted—would we understand it better? How would you define tool use in the context of nesting, for instance?

12.3 PARENTAL INVESTMENT

There were hints about *parental investment* in Chapter 11, with the discussion of relative investment of males and females in offspring. Robert Trivers brought the idea of parental investment to the forefront of evolutionary thinking about reproduction in a 1972 paper that defined parental investment as the investment that a parent makes in an offspring that reduces the parent's future fitness.[12] It would be wrong to think of this as some species-specific number because the impact of such behavior on future fitness depends, in part, on a variety of factors (e.g., food abundance) that can change over time and space without regard to the parent–offspring interaction. Nonetheless, Trivers's concept provides a useful framework for understanding the allocation—and limits—of parental investment.[11]

This investment can take many forms. It can be as seemingly routine as provisioning gametes; the provisions can come from unexpected sources, such as nutrients associated with sperm transfer, or even cannibalistic mating. Indeed, when provisioning gametes is included, it is difficult to think of many species that do not invest in offspring. Parental investment does not stop with gamete provisioning, however. Once fertilized, eggs can be guarded and protected, either in the nest or inside the mother, and once the offspring are hatched/born, parents can continue to provision and/or guard them (Figure 12.10). Guarding of nests and territories where young are reared overlaps with mating behavior as well, because nest and territory quality may be used in mate assessment (see Chapter 11 on mating).[13]

OF SPECIAL INTEREST: DOLPHINS—THE COSTS AND BENEFITS OF PARENTAL CARE

Carrying young can be an energetically expensive proposition, as evidenced by the number of times you may have witnessed human parents "trade" that responsibility. It turns out to be no different for aquatic mammals. Cetacean mothers engage in "echelon swimming" with young, during which the calf swims very close to the mother's mid-lateral flank. This has considerable hydrodynamic benefits for the calf, which can then keep up with its mother, and it probably minimizes the risk of separation. However, it is not without cost for the mother, who swims at about 76% of the speed of her unencumbered mean maximum speed and whose distance per swimming stroke is reduced by 13%.[14,15]

Killer whales are the largest dolphins, and they often live in family groups; if they do so, they are called "resident" groups. In resident groups, both sons and daughters remain with their mothers, who may assist them in a variety of ways, from foraging to conflict with other killer whales. The sons mate with females in other *matrilines*, and in so doing, transmit their mother's genes to other groups. Male killer whale reproductive success increases with age, so long-lived sons are increasingly valuable contributors to their mother's fitness. Female killer whales may live for decades after they cease reproduction.

> **KEY TERM** A *matriline* is a group that traces its lineage through the mother, or the females in the group.

Recent killer whale census work in the Pacific Northwest suggests that the death of a mother killer whale, no matter what her age, has an adverse effect on the survival of her mature offspring, an effect that is particularly pronounced in her older (>30 years) male offspring.[16] In other words, the contributions of parental behavior to offspring fitness do not necessarily cease when the offspring matures, or even when the parent ceases reproduction.

While the nature of parental investment is more variable than is easily summarized, that variability is often the result of a combination of phylogenetic and ecological factors, and as such has significance for *life history traits*. Minimum investment per offspring, especially minimum nutritional investment, is most often seen in animals with high reproductive rates, whereas organisms that invest heavily in offspring have fewer of them. This is a logical association; if resources are limited, then it is impossible to invest a lot in each one of a hordesof offspring. This "choice" (often made over evolutionary time) is part of a "life history

403

FIGURE 12.10
Parents and offspring.
(A) Owl and chick; (B)
juvenile giraffe; (C) adult
giraffe; (D) blue-footed
boobie and chick;
(E) mallard and her
ducklings; and (F) plover
and chick. *Photos: (A)
Jeff Mitton; (B, C, and E)
Michael Breed; (D) Randy
Moore; (F) Ben Pless.*

strategy," that is, characteristics that determine the course of a life. Such characteristics include, but are not limited to, a variety of parameters that determine maturation, reproduction, and longevity. Life history theory places these investment extremes on a continuum. Low-investment, highly fecund organisms are said to be *r*-selected (the *r* comes from the term for reproductive ratio in population growth equations) and high-investment, less-fecund organisms are said to be *K*-selected (the *K* refers to carrying capacity in those equations).

KEY TERM Life history traits are a concept from ecology used to describe a species' or population's reproductive strategies. They include age at first reproduction, number of offspring in a clutch, number of clutches in an animal's lifetime, and so on. This concept is explored in more detail in many ecology textbooks.

Generally speaking, *r*-selected species with their low investment and high fecundity seem particularly well adapted to unpredictable environments that allow fast invasion, maturity, and reproduction. In contrast, *K*-selected species do well in more stable, highly competitive environments, where high investment in each offspring is necessary for that offspring to get a foothold in the habitat.

Clutch size—the number of young in a brood or litter—is another important life history trait. The number of young requiring care at any one time clearly has behavioral implications, but because clutch size is determined by evolutionary responses to long-term food availability and life history strategy, it is more a question of evolution and ecology than of behavior. Behavior becomes morbidly interesting when clutch size exceeds the ability of parents to feed the offspring, a situation addressed in the section on sibling conflict later in this chapter.

The preceding paragraphs paint with the broadest brush imaginable, but the continuum does provide a context for beginning to predict shifts in investment. Such shifts can occur even within a lifetime.[17] In species that experience senescence, older organisms may invest more in offspring because the probability of future reproduction is reduced, along with the negative impact of current investment on that future reproduction. Thus, reproductive behavior may change within a lifetime or may change across habitats. Life history theory is one way to understand those shifts and make predictions about when they might occur and what they might look like.

OF SPECIAL INTEREST: FOSSIL PARENTAL BEHAVIOR

Parental care is not a new behavior on the evolutionary scene, nor is all evidence of behavior gone when an animal dies. We can infer behavior from a variety of fossil structures (e.g., dentition is a clue to food, bone density to running speed, foot structure to habitat, footprints to social behavior); likewise, we can infer parental behavior from the existence of nests and careful egg placement. In fact, a recently discovered ostracod from the Silurian Period shows that ostracod parental care was probably practiced in much the same way 425 million years ago as it is today. An ostracod is a small crustacean that is shaped like a clam, with two valves forming its carapace, which almost entirely encloses its body.[18] In this group, as in some other crustaceans, a space within the carapace called the marsupium is used to brood eggs. This 425-million-year-old fossil had eggs, and possibly juveniles, in its marsupium, thus indicating that, much like today, Silurian ostracods directed parental care at their young.

The concepts associated with parental investment and life history strategies are so robust and pervasive that they have even been used to predict fossil parental behavior. For instance, plesiosaurs were Mesozoic marine reptiles, and one of them, in the species *Polycotylus latippinus*, died in the seas over what is now Kansas during the late Cretaceous period. She was pregnant, and her fossil remains include those of a single developing embryo that, at parturition, might have been almost half as long as its 4.7 m mother. Both the size and the absence of multiple embryos suggests that *P. latippinus* is more ecologically similar to extant cetaceans than to most current reptiles or to other fossil aquatic reptiles from the Mesozoic. Because of this, the working hypothesis put forward for this dinosaur group by scientists who have examined the *P. latippinus* fossil is an intriguing one: it may well be that these animals exhibited maternal care and social behavior similar to what we see in current-day cetaceans.[19]

12.4 PATTERNS OF PARENTAL CARE

Most behavioral research seems to have focused not so much on the existence of parental investment or the type of investment, but rather who invests and why. To begin, review the discussion about the evolution of sex in Chapter 11. For a simple case of uniparental care, the basic rules of anisogamy seem to apply: the sex that invests the most in the offspring and has the most to lose should provide the most care. Thus, eggs are more expensive to produce than sperm, so the initial working hypothesis is that females should provide parental care. Things are not always simple, however, and a variety of extenuating circumstances can cause modification of that hypothesis.

Even if females do invest, they should take steps to assure the identity of the target of their investment; any investment is a mistake in evolutionary terms if the female misidentifies her young and pours maternal investment into an unrelated animal.[20] As might be expected, there are recognition cues that help prevent this misidentification, and they are most highly developed in social animals for which the need to identify individual young is most pressing.[21,22] Tinbergen's classic work shows that in colonially nesting gulls, the ability to identify offspring develops simultaneously with the offspring's ability to leave the nest and move around. In contrast, because kittiwakes nest on cliff ledges, their young are under strong selection not to wander too far, and are unlikely to be confused with other young from other ledges.[23] As a result, recognition of the young by their parents even at the species level, has not evolved in kittiwakes.[24]

The ongoing evolutionary arms race between avian brood parasites, such as cuckoos, and their host species is another illustration of the importance of accurate assignation of parental care. This race can show up in very rapid changes in the appearance of parasite eggs.[25] If cuckoos can produce eggs that look like those of the host bird, they can obtain free parental care for their offspring. Host species respond by altering the appearance of their eggs, making this matching task a moving target. Thus, brood parasites employ aggressive mimicry in order to induce their hosts to misdirect parental care; the host cannot distinguish brood parasite eggs, and sometimes the hatchlings, from host offspring. Before cuckoos introduce their well-disguised eggs into nests, they must deal with the hosts' defensive response of mobbing. Cuckoos do this by employing Batesian mimicry;[26] cuckoos that exhibit a barred pattern on their underparts, similar to that of sparrowhawks, are less likely to be mobbed.[27]

For many birds, as well as for bottlenose dolphins,[28] auditory cues are involved in recognition and may develop into "signature" vocalizations that are used into adulthood for individual recognition[29] (see Chapter 13). Goats can recognize the vocalizations of their offspring for over a year after weaning, and this could be influential in social and reproductive behavior.[30] As might be expected, many terrestrial mammals seem more likely to use olfactory cues (see "Herding Behavior, Young Recognition, and the Old West"); these cues from young can affect maternal behavior by influencing hormonal state and even structures in the olfactory region of the brain (see Chapter 2).

405

BRINGING ANIMAL BEHAVIOR HOME: HERDING BEHAVIOR, YOUNG RECOGNITION, AND THE OLD WEST

In herds of ungulates (hooved mammals), switching of young can easily occur, and it is not surprising that mothers can easily recognize their offspring. For cattle, there is considerable lore supporting the idea that cows use the odors of their calves to make this identification.

For instance, calves often had difficulty keeping up with a cattle drive. One solution was to give them a ride in a wagon and to reunite them with their mothers at the end of the day. This passage from a book by J. Frank Dobie describes how nineteenth century cowboys handled this situation to ensure that mothers would take back their calves:

Charlie Goodnight said to me that when he began taking a "calf wagon" along, to pick up calves born on the trail, he had much difficulty in getting mother cows to accept their young after they were unloaded from the wagon in the evening. A half-dozen or twenty calves jostling together during the day would get their scents mixed, thus making each a stranger to its mother. Goodnight overcame the difficulty by putting over each calf a loose sack, so marked that it was used by the same calf day after day, being removed at evening; thus the scent proper to a calf would be retained for recognition by its mother.[31]

In some instances calves became orphaned. Here are two passages, from very different ranching situations, a century apart, which describe virtually the same procedure used to induce a cow to adopt an orphan calf. First, we have J. Frank Dobie reporting on nineteenth century cattle drives:

If a cow lost her calf and at the same time there was a calf that had lost its mother, the bloodless hide of the dead calf fastened loosely over the orphan would influence the cow to adopt it.[31]

In her account of life on a modern ranch, Judy Blunt describes a nearly identical method used to ensure the adoption of orphan calves: "He would live or die by morning. …If he died we'd peel his hide and make a jacket for one of the orphan calves in the barn, convince the cow it was hers."[32]

The similarities between these passages is striking, and emphasizes the importance of mother–calf recognition in the delivery of parental care.

Returning to the question of who should provide parental care, it only takes a short exposure to nature documentaries to realize that the "rule" of maternal care was meant to be broken in the natural world. There are obvious examples of male parental care; although female parental care is more common, male parental care is not rare. Why this deviation from the anisogamy-based prediction? For one thing, in many organisms, sperm are not the tiny, inexpensive propagules that work well for theory. Instead, they are produced in energetically costly packets (*spermatophores*). In other organisms, males make serious investments in territory holding and resource guarding. Once investment extends beyond the gamete, understanding which parent invests more can be a challenge.[33]

> **KEY TERM** A spermatophore is a structure that transfers sperm from males to females in some species. It may also contain nutrients that are valuable to the female. The suffix *-phore* is from a Greek word that means "to bear," or "bearing," and is common in scientific terminology.

In addition, certainty of paternity may influence the extent to which males are willing to provide care. In animals with internal fertilization, the female is more certain of her parentage than the male, whereas in externally fertilized organisms, males may be certain, especially if the fertilized eggs are maintained on the male's territory.[34,35] In addition, proximity to offspring can obviously offer opportunities for care, and a male that guards a territory may coincidentally guard eggs that he has fertilized that are within that territory.

Finally, many organisms exhibit biparental care. How does this evolve? An ecological, resource-based hypothesis seems to explain most cases. The reasoning begins with female uniparental care, which is the default case for parental care in general, based on the greater investment that females usually have in young. In a world where provisioning and guarding are not limiting reproductive success—that is, where females can rear offspring alone as successfully as they can with male assistance—the male is free to leave the female and her eggs or young and seek other reproductive opportunities for increasing his fitness.

Not all environments are so benign, and in some species, there are steep requirements for producing young that are likely to survive and reproduce themselves. These requirements can be environmentally imposed (scant resources and/or high predation pressure) and exacerbated by life history traits (young that are helpless for long periods of time or that mature at rates requiring intense provisioning). In these conditions, a male that leaves is likely to have no fitness because the female cannot do the task alone. Such conditions favor the evolution of monogamy and subsequent biparental care.[36–39]

For example, a comparison of paternal care in two congeneric species of hamsters, the Campbell's Russian dwarf hamster (*Phodopus campbelli*) and Siberian (*Phodopus sungorus*), revealed that the male Djungarian hamsters not only care for their young, but actually assist in the birth, whereas Siberian hamsters attend young much less often and do not assist at birth.[40] This is hypothesized to reflect the greater environmental challenges experienced by the Siberian hamsters, including thermoregulatory challenges.

Thus, evolutionary history, as well as ecological factors, influences both the degree of parental care and who performs it. Female mammals, for instance, gestate and then feed young. This makes male uniparental care almost impossible, although biparental care might be beneficial in some cases, often depending on ecological conditions. In addition, life history traits such as maturity at birth can strongly influence the opportunity for males to contribute significantly to offspring care. Altricial organisms (see page 45) are relatively immature at birth and require a great amount of parental care; in such animals, biparental care might increase fitness by increasing the survival probability of the offspring. In contrast, precocial animals are able to move and forage shortly after birth. These young need to be guarded from predators and they may need some time to learn survival skills, but they do not need to be given nutrients. Compared to altricial young, they require much less investment, so that biparental care may be less critical for precocial than for altricial young.

Parental Care When Fertilization Is External

In many aquatic species, the female lays her eggs prior to fertilization. The male then deposits his semen in the water, and both mother and father may have left by the time gametes have formed. In fish, some species simply scatter their eggs in moving water. Other species deposit their eggs on a surface. This last behavior opens the door for rudimentary parental care, as the female can then guard her eggs and young hatchlings (very small fish are called *fry*). Still other species of fish, such as Bettas, build nests of bubbles in which spawning takes place. A few species of fish keep their fry in their mouths to protect them; these are *mouth-brooders*, usually found within the family Cichlidae. Male participation in parental care is common in fish,[41] with males often constructing nests, guarding eggs or fry, or even participating in mouth brooding (Figure 12.11).

Amphibia—salamanders, frogs, and toads—also have external fertilization (Figure 12.12). Again, egg guarding is a possibility. In the well-known case of the midwife toad, the male

407

FIGURE 12.11
Spawning sockeye salmon. Salmon are anadromous fishes, that is, they spend part of their lives in freshwater, and part in the sea. Sockeyes are the only salmon to mature in lakes; these immature fish have "parr marks," vertical bars and spots that provide camouflage in a lake environment. All this changes when they move to the sea as adults; they lose the parr marks and while in the open sea, they have a light-colored underside, with a bluegreen dorsum (countershading!). As these "bluebacks" return to their freshwater breeding grounds, they turn bright red with a green head. As intriguing as it is, the name "sockeye" has nothing to do with the salmon's eye; it is probably derived from a Native American term. *Photo: US Fish and Wildlife Service.*

FIGURE 12.12
A frog egg mass. The gelatinous mass that surrounds the eggs may be rich in mucopolysaccharides and mucoproteins, and buffers the shell-less eggs from the external environment. *Photo: Michael Breed.*

takes over parental care, carrying the fertilized eggs on his body to protect them from predation. In salamanders, species living in streams lay larger eggs and are likely to guard their eggs and embryos, whereas pond-dwelling species' eggs are smaller and remain unguarded.[42]

Uniparental Caregiving

In terrestrial animals, internal fertilization is more common. In many, perhaps most, terrestrial animal species, the male and female do not maintain contact after mating. This leaves the female responsible for caring for the eggs. In many cases, this simply means finding an appropriate microclimate for egg development and then laying the fertilized eggs (Figure 12.13). For plant-feeding insects such as moths and butterflies, this typically involves ovipositing on a host plant; the emerging caterpillars then have a ready-made food source at hand. Cockroaches illustrate the range of possibilities for parental care in terrestrial animals. Many cockroaches lay their eggs in leathery masses called *oothecae*; the female cockroach may take some care in finding a hidden spot for her eggs, where she glues the egg mass in place. In some cockroach species, the ootheca is carried externally by the female until the eggs hatch. In still others, the ootheca is retracted into the body and protected in a brood pouch. In one cockroach genus, nutrients are transferred directly from the female to the embryos in the pouch, much like a placental mammal.

FIGURE 12.13
A cluster of insect eggs laid on a leaf. *Photo: Michael Breed.*

These same patterns play out in vertebrates with internal fertilization. Female reptiles may lay their eggs in a protected place; sea turtles do this when they bury their eggs on beaches. Some snake species, such as the milk snake, may guard their eggs, while others (the fer-de-lance, for instance) may give live birth.

Examples of male-only caregiving are rare enough to attract attention when discovered. In addition to the midwife toad, other well-known examples include seahorses, pipefish, and sea spiders.

Biparental Caregiving

Biparental caregiving is common in birds and is sometimes seen in reptiles and mammals (Figure 12.14). For the evolution of male parental behavior, the balance is fairly simple; the fitness gained by added survival of young due to the male's care is weighed against the fitness that might be gained from other potential matings. If male behavior can add to the survival of his young and he has little chance of mating with another female, then the balance favors male support of his offspring. In some cases, the prospect of the young being killed by another male (infanticide; see later section) is enough to keep a father present. Indeed, in some cases, the father's only role in care may be to ward off other males from the family group; this is a limited but crucial duty (Figure 12.15).

Intra- and Interspecific Adoptions

An odd and puzzling phenomenon is *adoption* of other females' young, within a species or across species lines. Adoptions within species, *intraspecific adoptions*, sometimes come about as a result of shared nests and communal brood care (see Chapter 13). *Interspecific adoptions* also occur: news stories report a cat that adopts and nurses puppies, a foal reared by a pig,

FIGURE 12.14
A pair of Canada geese with their young. Canada geese provide intense parental care during the first year of the gosling's life, and a goose defending a nest can be a dangerous animal. Although a young goose can fly by the time it is 7–10 weeks old, it remains with its parents until they complete the spring migration. *Photo: Michael Breed.*

and so on (Figure 12.16). There is no evolutionary advantage associated with adopting young from another species. Rather, these incidents are a testament to the power of maternal drive and to the impact of hormones in priming females, at certain times, to place their maternal response above other behavioral priorities.

> **KEY TERM** Adoption, also called *alloparenting*, involves giving parental care to a genetically unrelated animal.

409

12.5 HORMONES AND PARENTAL BEHAVIOR

While maternal behavior in vertebrates is strongly associated with prolactin, a peptide hormone produced in the anterior pituitary, parental care in male vertebrates is associated with declining levels of testosterone (See Chapter 2 for a review of endocrines and behavior). In female birds, prolactin stimulates nesting behavior and operates, via positive feedback, to promote the development of the brood patch. The brood patch is an area from which the female plucks the feathers so that she has direct skin contact with the eggs or hatchlings, improving her ability to warm the brood. In management of domestic poultry, such as chickens, females displaying nesting behavior and a brood patch due to high levels of prolactin are

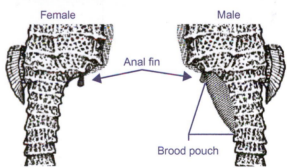

FIGURE 12.15
Male parental care in seahorses. The male protects the young in a brood pouch. *Photo: Rebekah D. Wallace, Bugwood.org.*

referred to as *broody*. If a farmer's object is to have the female produce more eggs, behavioral interventions (preventing hens from spending time in their nest, for example) are used to attempt to prevent broodiness and thus keep the birds laying. Drugs that block dopamine may also prevent broodiness. Prolactin also inhibits bird ovaries from developing more eggs and promotes foraging behavior. Prolactin-induced weight gain may be an important preparation for migration.

In mammals, prolactin has many similar effects. During pregnancy, prolactin levels increase, stimulating glandular development of the breasts. Prolactin primes maternal behavior and

FIGURE 12.16
An interspecific adoption. In this illustration, piglets have been adopted by a cat. Stories of cross-species adoptions illustrate the strength of parental responses in mammals.

modulates stress that is associated with pregnancy and maternal care. Repeated pregnancies reinforce better maternal care, partly because each pregnancy induces more prolactin receptors in the brain; the larger number of receptors then enable stronger maternal responses. The inhibitory effect of prolactin on ovulation reduces the likelihood that a nursing mammal will become pregnant; nursing can be seen as a natural contraceptive, and may be why women such as medieval royalty, who were concerned about fecundity, often employed wet nurses and did not nurse their own offspring. At least in humans, prolactin is also associated with sexual arousal and pleasure, due to its inhibition of dopamine secretion. The precise behavioral effects of prolactin and testosterone on parental behavior vary among species, but the general pattern is that prolactin enhances maternal behavior while testosterone inhibits or interferes with paternal behavior.

Finally, oxytocin is another hormone that is important in reproduction; it causes muscular contractions in the uterus associated with birth and milk-letdown in milk production, and is thought to be important in maternal bonding with young animals.

While care by male parents is less well studied that care by female parents, the role of the father in nest construction, provisioning, and in shaping the behavior of the developing young is often considerable.[43] A number of important evolutionary dynamics lead to situations in which males maximize their fitness by providing parental care to their young rather than seeking other mating opportunities.[44]

12.6 PARENTING AND CONFLICTS OF INTEREST

Family life, from an animal behavior perspective, might be seen as rooted in this assessment of investment (cost) versus fitness outcomes (benefit) that can often lead to conflict if the participants differ in these parameters. Not only is mating behavior and parenting heavily influenced by cost and benefit, but parent–offspring relationships and sibling relationships are thought to reflect such influences.

For behavioral biologists, one central question in parental care focuses on timing: when should it cease? Offspring and parents should differ in their assessment of the costs and benefits associated with the cessation of parental care; this is reflected in nature every time a recently fledged bird squawks loudly as it hops energetically after its (presumed) parent, which in turn studiously ignores the young bird's existence. From a fitness perspective, it is to the young bird's advantage to extract larger amounts of resources from the parent bird than benefits the parent bird. By restricting parental care once the young can meet most of its survival requirements, the parent can allocate those resources to its own future fitness. Trivers[12] addressed this issue, and it can be easily illustrated in graphical form (Figure 12.17). This is most readily seen in offspring that are nutritionally provisioned by parents. In

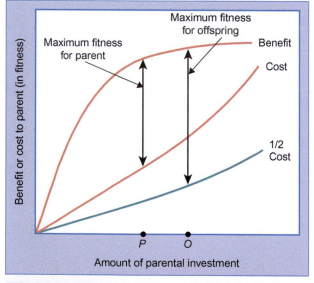

FIGURE 12.17
The upper line in this graph represents the benefit of parental care. The middle line expresses the cost of the care to the parent. The parent should stop giving care when the greatest benefit per unit cost has been achieved (point P on this graph). Because the young animal values itself twice as much as it values a sib, it views the cost of parental care as ½ the cost assessed by the parent; this is represented by the lower line. Given this discrepancy, the optimal investment from the young animal's point of view is higher (point O on the graph). The gap between the parent's and the young animal's assessment of the optimal investment is the region in which theory predicts parent–offspring conflict. *Adapted from Trivers, R.L., 1974. Parent–offspring conflict. Am. Zool. 14 , 249–264.*[12]

general, as the offspring gets older, it depends less on parental resources for its survival. At the same time, it requires more resources in general because it is getting larger and more active. At some point, there will be conflict between parent and offspring because the increasing demands of the offspring are not matched by increasing fitness benefits to the parent.

The manipulative potential of offspring demand for resources was convincingly demonstrated by Matthew Bell's study of banded mongoose food solicitation.[45] Banded mongoose breed communally, and each pup associates with an "escort" that cares for it. Pups begged more vigorously when resources had been withheld than when they were well provisioned, but they also begged more vigorously when cared for by escorts that were more likely to provide. Thus, female pups, which receive better care than males, begged more vigorously than males did given the same degree of food deprivation, and even more tellingly, pups with generous escorts begged more vigorously than pups with more stingy escorts. This raises the possibility that demand for resources among these offspring (and probably others) reflects both need and an assessment of the likelihood that the need will be met.

Similar reasoning can be applied to the distribution of resources among siblings by parents. Parents are equally related to all of their offspring, sharing 50% of their genetic material, and thus are expected to treat them equally when they all have equal expectations of survival. On the other hand, each offspring may be related, on average, to its siblings by 50% (some more, some less, given the whims of independent assortment of chromosomes), but it is related to itself by 100%. Thus, siblings of most species are not terribly excited by the prospect of sharing resources. In contrast, parents benefit greatly when siblings do cooperate.

There are times when it is in the parents' interest for some offspring to die, or at least become nonreproductive. Nonreproduction (e.g., sterile castes or helpers) is covered in Chapter 13, and is a hallmark of eusociality. Outright infanticide, however, differs in that it usually occurs more opportunistically and often results from either male–male competition (see Chapter 11 on mating) or offspring provisioning strategies.[46] The latter case is an extreme version of sibling rivalry, discussed later in this chapter.

12.7 BEGGING AND WEANING CONFLICT

Young birds and mammals (and even insects![47]) ask for food, sometimes almost continuously. It would seem to an observer that the food requests could exceed the actual need for food and likely exceed the parent's ability to deliver food. This section reviews the function of *begging* and the time when the parent determines it no longer needs to feed the young—*weaning*.

> **KEY TERM** Begging is signaling to induce feeding.

> **KEY TERM** Weaning is the process of removing parental nutritional support from a young animal. The young must replace parental support by developing their own foraging abilities.

Functions of Begging

From the chick's point of view, begging is a way of attracting food (Figure 12.18). Begging may reflect the actual needs of young for food (honest signaling), or it could reflect the young's point of view in parent–offspring conflict (dishonest signaling?). In most birds and mammals, there are multiple young in a litter or brood, so begging could be important in scramble competition among the offspring for food. In a recent study of redstarts, an honest signaling model was supported.[48] Hungrier nestlings moved to the front of the nest, which is the preferred location for parents to feed offspring, and there was no evidence of scramble competition (see Chapter 11). Studies of other species sometimes suggest scramble competition; begging is probably sensitive to evolutionary effects of food availability in habitat, with scramble competition likely in species that have an evolutionary history of food shortages.

FIGURE 12.18
Gaped-mouth begging in cormorants (A) and barn swallows (B). In some species, the color associated with the gape may communicate nestling condition to parents. *Photos: A, Michael Breed, B, Joanna Hubbard.*

FIGURE 12.19
Illustration of approach and rejection in parent–offspring interactions. Above, young animal solicits nursing. Below, parent pushes young away. Interactions following this pattern are called weaning conflict, and may occur when the parent's and offspring's interests do not coincide.

Begging: Parental Point of View

From the point of view of the parents, begging may help to ensure that they are feeding their offspring enough. Weakened or sick young that are unable to beg may not receive food, reducing "waste" in feeding young that are fated to die. A recent theoretical study argued that responding to begging reduces two risks: The first is wasting time with too-rapid returns to the nest; if nestlings are not begging intensively, then more time can be spent foraging. The second is overlooking nestlings by not responding to their begging.[49] A study in house sparrows supports the argument that parents should be responsive to begging.

Remember that if parents respond to begging, food will not necessarily be wasted. This is similar to filling the gas tank in a car when the tank is half full (as compared to waiting until it is empty); this does not waste gasoline. However, time may be wasted if filling is done too often, so an argument can be made that begging should be an honest signal of condition. Begging may also carry a cost if the noise and commotion attract predators, so there may be some counter-selection against continuous begging.

Begging by Cuckoo-Type Parasites

Sometimes brood parasites manage to get one of their eggs into the nest of a host. Once the parasite hatches, its begging is almost always more intense than nestlings of the host. The parasite begs whenever food is available (when the host parent arrives), instead of begging as a signal of need, and thus deprives the host's nestlings of food. In cowbirds, the nestling begs more intensely if host nestlings are large, suggesting an effect of competition on begging.[50]

Weaning Conflict

For birds and mammals, there comes a point in the development of a young animal at which it might feed itself, but it is likely to pursue being fed by its parents. This conflict derives from the young animal seeking a guaranteed supply of food while the parent may benefit from diverting food to other, less completely developed, offspring or to its own metabolism (to build fat resources for overwintering or to have another set of young; Figure 12.19).

OF SPECIAL INTEREST: THE CHEMISTRY OF PAIR BONDING

How are social bonds, particularly those of mating pairs, reinforced by neurotransmitters and hormones? A considerable number of urban legends tout the roles of dopamine and oxytocin in human–human bonding, but are these legends true and do dopamine and oxytocin rule pair bonding in animals in general?

Dopamine is a neurotransmitter found in the central nervous systems of nearly all animals. A simple molecule which is related to amino acids, it is synthesized from the amino acid tyrosine with the compound L-dopa being an intermediate in the synthetic pathway. Dopamine is key in the system that the brain uses to link the perception of reward with a memory. In other words, if something good happens, the reward at the level of the brain is dopamine release. Dopamine is a positive reinforcer for behavior which is pleasurable or which has a good outcome.

Oxytocin is a very different sort of compound. Produced by the hypothalamus, it is stored in the pituitary gland and secreted by that gland to regulate physiology and behavior. It is a peptide hormone, composed of nine amino acids. Oxytocin was first known as a regulatory hormone in birth in mammals, stimulating uterine contractions, and as a critical trigger of milk "let-down" during nursing. Oxytocin is now thought to have important roles in social recognition, maternal behavior, and pair bonding.

A little less publicized is the peptide hormone vasopressin. Very similar in structure to oxytocin, it has parallel effects to oxytocin in that high levels of vasopressin correlate with pair bonding. The main role of vasopressin is to regulate the concentration of salts in the blood by affecting whether water is reabsorbed or excreted by the kidneys, and its behavioral effects have only been discovered recently. Vasopressin's impacts on behavior are very similar to those of oxytocin, but pharmacological studies clearly show that each of these hormones can work separately to affect pair bonding.

In voles, small mouse-like mammals, oxytocin is present in high levels in a species which pair bonds, the prairie vole. Oxytocin binding to receptors is lower in the meadow vole, a nonmonogamous species. While this is the most well-developed example of the role of oxytocin in pair-bonding, comparative studies establish that this mechanism is probably a solid generality in birds and mammals. Zebra finches, for example, engage in stronger pair bonds when treated with drugs that mimic oxytocin and are less likely to bond when treated with drugs that block the effects of oxytocin.

Dopamine and oxytocin are also tied into the endogenous opiate system in the brain, which is best known for the role of endorphins in the perception of pleasure and in counteracting pain. Some scientists have even argued that the role of oxytocin is secondary to that of endorphins in reinforcing pair bonds. Much more interesting experimentation needs to be done to sort out how all of these regulators—dopamine, oxytocin, vasopressin, and endorphins—come together to shape pair bonding, and other levels of attachment.[51]

12.8 SIBLING CONFLICT

Returning to first principles, a parent shares the same proportion of its genes with each of its offspring (50%). As a result, parents might be expected to treat all offspring equally. That would be true if genetics were everything. However, the parent may discern that some offspring are more likely to survive and reproduce than others. In this case, the parent may create a situation in which siblings are in conflict for resources. Siblings may not all have an equal chance for many reasons, ranging from developmental problems to resource limitations that preclude investment in all offspring.

In many mammals, sibling conflict plays itself out in competition for access to the mother's nipples and for position among the nursing young. Some teats do, indeed, produce less milk than others (thus, the saying "sucking hind teat"); smaller, less-well-developed young will lose out in this competition. An animal that is the runt of the litter has little chance of surviving the competition.

But why would there be more offspring than the parents can support? Environments are unpredictable, and from the parental point of view, it is probably best to produce the number of newborns or hatchlings that could be supported in an extremely good year. The "extra" offspring also provide some level of insurance against the loss of one of the young to predation or disease. For birds, this is sometimes called the *insurance egg hypothesis*. If the insurance is not needed and food availability is constrained, then from the parent's point of view, it would be better for a few strong members of the litter to survive than to have them all survive as malnourished weaklings. In other words, it might be more profitable to invest in fewer offspring and raise them to fledging age than invest in the entire clutch only to see complete failure.

Parents generally do not kill "surplus" offspring. (Infanticide can occur, but for different reasons that are discussed later.) Instead, the issue is resolved either by death stemming from malnourishment, or in extreme cases by larger, older, members of the brood killing their sibs (siblicide). Siblicide has been particularly well studied in Laughing Kookaburras, boobies,[52] and egrets.[53] In these bird species, insurance eggs are laid. At least in boobies, high testosterone in the chicks is associated with fight behavior. Only under very favorable environmental conditions do all the chicks survive; usually, the last emerging chick is pecked to death.

The fate of parental investment is not always what it might seem. Male pipefish brood fertilized eggs in a special brood pouch for a few weeks. During this time, the number of embryos can decline. When that happens, who gets the nutrients? The father? The siblings? Male pipefish were allowed to mate with two types of females, one of which had radioactively labeled eggs. This meant that approximately one-half of the initial brood carried that label. No unlabeled half-sibs acquired the radioactive label, but it was found in the tissues of the brooding male. Thus, paternal pipefish practice a form of infanticide/cannibalism (see Section 12.9), acquiring nutrients from the female that initially provisions the embryos. Because pipefish do not form lengthy pairbonds, this raises a variety of questions about sexual conflict in this group of fish (see Chapter 11).[54]

414

OF SPECIAL INTEREST: ARE BABY BIRDS CUTE?

Baby American Coots look like bizarre Halloween party favors and, at least in terms of plumage and bare heads, do not resemble their more distinguished-looking parents (Figure 12.20). Lyon et al.[55] manipulated this spectacular plumage and showed that it did indeed influence parental provisioning. Chicks with bright plumage received more food, grew faster, and had higher survival rates than chicks that had the orange tips of their feathers removed. The bright plumage may trigger stronger parental responses.

FIGURE 12.20
Baby coots. *Photo: Böhringer Friedrich, Creative Commons.*

Why do parents not intervene? There is no parental benefit to intervention because the parents' fitness is best served by having fewer well-nourished young survive. Because of this, the idea that parents benefit if siblings cooperate (see above) needs some qualification. Sometimes that is not true; those times tend to occur when resources are limited, favoring rivalry that is taken to the extreme, with siblings competing to the death for those limited resources. Under such conditions, parents essentially allow sibling conflict to determine which offspring are most likely (and probably best equipped) to survive to reproductive age.

DISCUSSION POINT: INTRAFAMILY CONFLICT

Let's get real here. If you have sibs, did this discussion of parent–offspring conflict over investment and of sibling conflict prompt you to think about your own situation? Human parents often don't intervene in sib–sib conflict unless serious injury is threatened. Does that fit your experience? Does that level of parental noninvolvement fit with the evolutionary argument we've created?

12.9 INFANTICIDE

Infanticide by Males

Infanticide, killing of young animals, is common among mammals and has been observed in some birds. Members of the cat family (Felidae), including the domestic cat, are notorious for incidents of infanticide by roving males. Infanticide by males is also a feature of some primate societies. Infanticide by a male is almost always the result of paternity uncertainty—the male cannot be sure the young are his—or certainty that the young are not his.

One interesting example comes from Osorio-Beristain and Drummond, who found that male blue-footed boobies that were experimentally held so they could not return to their nest for several hours were likely to expel the eggs from their nest and remate with the female.[56] This is probably a response to opportunities the females had for extra-pair copulations while the male was gone. In primates, there are two mechanisms that at least hypothetically should prevent infanticide. The first is *concealed ovulation*, which together with promiscuous copulation may prevent infanticide because males do not want to risk killing young that are potentially theirs (chimpanzees). The second avenue to prevention of infanticide is pair-bonding; this creates an impression of paternity certainty but keeps group size small, as in gorillas, because the larger the group, the more difficult it is for a male to track the females' activities. As a result, risk of infanticide may be a constraint that limits social group size of not only primates, but possibly many other animals.

> **KEY TERM** Concealed ovulation occurs when a female is physiologically able to mate at times when she is not ovulating. Humans and a few other primates have concealed ovulation.

When Might Females Commit Infanticide?

Infanticide by females is predicted when resources for their offspring are limited; the female may then kill other females' infants. Columbian ground squirrel females kill other females' infants to increase the chances of their own infants maturing and entering the adult population.

The Bruce Effect

In at least some rodents, the presence of a new male causes a female to physiologically abort pregnancy; this is termed the *Bruce effect*. In mice, the Bruce effect is caused by odors from the new male, which allow the female to recognize him as new (different). Replacement of a male with a genetically identical (from the same clone) male does not cause the Bruce effect. The odors are associated with major histocompatibility complex differences, which are discussed in Chapters 11 on mating and 13 on social behavior. The Bruce effect makes the female available more quickly to mate with the new male, and the aborted young likely would have been victims of infanticide if the female had carried them to birth.

> **KEY TERM** The Bruce effect is physiological abortion induced by the presence of a new male. It is named for Hilda Bruce, who reported this phenomenon in 1959.[57]

415

12.10 AGGRESSION AND TERRITORIALITY

Aggression is one of those subjects in animal behavior that might be inserted in any one of a number of chapters. It is considered in this chapter simply because the majority of aggressive behaviors are related to reproductive and parenting concerns, broadly defined. Aggression occurs for a variety of reasons and is therefore frustratingly difficult to define and categorize. E. O. Wilson, in a 1973 book titled *Sociobiology* that inspired a new, vibrant field within animal behavior, attempted both definition and categorization; his approach is used here, with modification.[58] Other ethologists and popular writers have also published books on the behavioral biology of animal and human aggression, often stimulated by a desire to understand and explain why humans fight and sometimes kill one another. The titles of these books illustrate the depth of thought and concern about human aggression (examples are *On Aggression* by Konrad Lorenz; *The Territorial Imperative* by Robert Ardrey; and *Aggression and War: Their Biological and Social Bases*, edited by Groebel and Hinde). Humans have struggled to understand aggression among animals and to use that understanding to illuminate human conflict.

DISCUSSION POINT: IS AGGRESSION INEVITABLE?

One of the deepest question we can ask is whether aggression and war are inevitable for human societies. Is conflict a necessary outcome of the evolutionary origins of humanity, or can we use cognition and reasoning to allow human societies to overcome propensities for conflict? In other words, can moral and ethical sensibilities somehow separate humans from other animals? For that matter, do other animals compete in ways that have overtones that we fail to recognize as "ethical?" Indeed, given the often brazenly conspicuous nature of conflict and the subtleties of cooperation, have we underestimated the role of cooperation in nature (and overestimated aggression) simply because nonaggressive interactions do not grab our attention?

What Is Aggression?

Wilson noted that among humans aggression occurs when one individual surrenders something it owns or might have owned as a result of a physical act or a threat. When applied to animal behavior, the damage is calculated in terms of fitness. Broadly considered, conciliatory behavior and surrender can be part of this mix, and the term *agonism* is frequently used in animal behavior circles to include those actions and to diffuse some of the unpleasant connotations that might accompany "aggression." *Agonism* comes from a Greek word that refers to contenders in games, an origin that is reflected in more commonly used words such as *protagonist* and *antagonist*. Agonism usually refers to intraspecific interactions. For purposes here, the categories used in the following sections (loosely based on E. O. Wilson's categorization)[58] may illuminate this set of behaviors more than any definition. Above all, aggression—or agonism, or conflict—is not one thing.

Dominance Conflict

We will consider the type of interaction we call dominance conflict in more detail in Chapter 13, which deals with social behavior. Dominance interactions include aggression, threats of aggression, and submission. Oftentimes, aggression occurs when the subordinate animal finds it advantageous to remain within a group, even if it has lower access to resources in the group. However, aggression can occur in settings other than social groups… In these instances, dominance is exerted simply to withhold a resource from a competitor.

Sexual Conflict

Males can use force or threats to ensure that females remain with them or remain in their harems. For example, male bottlenose dolphins threaten, bite, tail-slap, body-slam, and otherwise "herd" females. They do this much more frequently to females that are not pregnant, which allows the inference that the motivation is indeed reproduction. Two or three males form alliances to herd the females and defend them from other males; they give chase when females attempt to escape.

Parent–Offspring Conflict

Parents may display mild forms of aggression toward young that are perhaps straying too far away or, at weaning time, remaining too close. In general, this is fitness-related behavior, either because it promotes the safety of the offspring (offspring fitness) or, at weaning time, assists the parent in shifting parental investment to future offspring (parental fitness). Parents also defend young against potential enemies with antipredator aggression (see below).

Predatory, Pain-, and Fear-Based Conflict

We do not consider interspecific predation as a purely aggressive act, although it is undeniably an act that results in a loss of fitness for one participant and a gain for the other. However, in the case of intraspecific predation, or cannibalism, especially cannibalism associated with territoriality, there is definite overlap with other forms of agonism. In addition, fear-related aggression is the unrecognized source of many—possibly most—human–animal conflicts, including conflict with domestic animals. This is yet another illustration of how the realities of nature can defy human categories.

Antipredatory Conflict

Antipredatory conflict is considered more fully in Chapter 10. It encompasses not only fighting back when pounced upon, but also (potential) prey-initiated attacks, such as mobbing. Animals at risk of being eaten have nothing to lose and use all of their available weaponry in the fight against a predator.

Territorial Conflict

Territorial conflict is related to the acquisition or maintenance of a *territory*. The distinguishing element of territoriality is agonism associated with a space or location.[59,60] Unlike any other space that an animal may occupy or use, a territory is defended. That defense is what distinguishes a territory from a *home range*, where an animal may spend the majority of its time, but does not defend. Male–male interactions are often either territorial or dominance, and can determine access to mates or can give males resources that are attractive to mates[61] (see Chapter 11 for more on this behavior). Territorial conflict also includes defense of young; mothers, in particular, may be very aggressive when they are guarding young.

> **KEY TERM** A territory is any defended space.

> **KEY TERM** A home range is the area in which an animal normally moves. A home range is not defended but may have a defended territory within it. Birds, for example, do not usually defend their feeding home ranges but are territorial around their nests.

417

DISCUSSION POINT: THE CASE OF THE "DEAR ENEMY"

The term "dear enemy phenomenon" has been coined to describe the frequent observation that territory holders are less defensive toward neighbors than toward strangers.[62] Territorial conflict may also create alliances between sometime-foes, and this is well-illustrated by fiddler crabs, which are notoriously territorial. A larger fiddler crab will come to the aid of its smaller neighbor—usually an enemy in boundary disputes—if and only if the potential invader is intermediate in size between the two adjacent territory holders.[63] Why might this alliance be advantageous?

What ties all of these categories of aggression together? In many cases the armaments evolved for use in one type of aggression work equally well in other contexts. Venomous snakes and predatory wasps bite or sting to subdue their prey, but they can also use these weapons to defend themselves. Bears are well armed as carnivores but use their teeth and claws in defense of their young. In vertebrates, the hormonal underpinnings of aggression may be the same across these categories (see Chapter 2).

We stress that while the types of aggression have some attributes in common, the widespread assumption that aggression in animals is monolithic in nature, a single behavior, is one of the biggest mistakes that can be made in animal management. There is no single cause for aggression, and because of that, there is no single solution for managing aggression in animals. An animal with offspring in the area may be much more aggressive than that same animal without offspring. A fearful animal may be extremely aggressive and give the appearance of dominance to the uninformed observer. A territory defender may be content, even meek, away from its territory. To be able to manage aggression in any species or individual, one MUST identify the motivation; in the absence of that knowledge, failure is very likely, and the results for the animal, especially a domestic animal, may be not only catastrophic, but sadly unnecessary.

BRINGING ANIMAL BEHAVIOR HOME: SHOULD A DOG BE "TOP-DOG"?

Here is an example for you to ponder as you consider animal conflict: many people believe that dog aggression is a "dominance" issue and that the true solution to such problems is to show the dog who is boss. It's true that some aggressive dogs insist on their dominance, but many more are thought to be fearful. Let's assume that you accept the dominance hypothesis and do everything possible to convince the dog that you are the boss, or "alpha," as the television shows like to say. What results might that have for a fearful dog? Will its fearfulness and fear-based aggression diminish as it is increasingly dominated?

BRINGING ANIMAL BEHAVIOR HOME: RITUALIZATION AND DOG BEHAVIOR

One way in which evolution tones down animal conflict is *ritualization*. In studies of aggression, this term is used when animals seem to apply symbolic value to their behavior. A dominant dog bares its teeth, threatening in a way that seems to symbolize its ability to bite. A submissive dog rolls on its back and exposes its vulnerable throat (Figure 12.21). Conflict among animals may be resolved by ritualized displays, without resort to physically damaging violence. It is nonetheless important to avoid the romantic fallacy that animals do not fight to injure or kill—they do, but ritualization has evolved because avoidance of fighting can benefit the fitness of both participants in a conflict. Human conflicts are similarly ritualized, with threat displays such as "spear-waving," which has evolved into "missile-waving," playing important roles in human territorial or resource conflicts. Note the ritualization of aggressive signals in this verse: "The shield of his mighty men is red; his soldiers are clothed in scarlet. The chariots come with flashing metal on the day he musters them; the cypress spears are brandished" (Nahum 2:3, The Bible, English Standard Version).

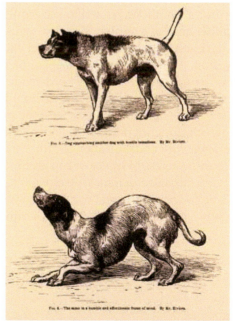

FIGURE 12.21
A dominant dog and a submissive dog. Note that the postures are distinctly, unmistakably different. *Adapted from Darwin, C., 1890. The Expression of the Emotions in Man and Animals. J. Murray, London, pp. 52–53.*[64]

KEY TERM In this context, ritualization means assigning symbolic value to signals of aggressive intent. Ritualized aggression may reduce the risk of injury and death in encounters.

418

Territoriality

The discussion of aggression, and territorial aggression in particular, leads us to a consideration of territoriality. A territory is a defended space, and that defense is what distinguishes it from any other use of space. Once again, the origins of our language inform its use, and the word *territory* is an excellent example. While the Latin *terra*, or earth, may have influenced this word, there is an intriguing, early link to a thirteenth-century French word that means "a place from which people are warned off" (*OED*, 2nd ed.) and that is also related to *terror*. The earliest use of *territory* in English has to do with land that is under the control of some town or ruler. All of this points strongly toward some essentials about the concept of territory: It is not just any space; it is a defended one. Nesting almost always creates a valuable resource for an animal, one that often requires territorial defense.

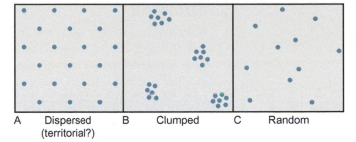

A   Dispersed   B   Clumped   C   Random
     (territorial?)

FIGURE 12.22
Animal distributions: (A) an even distribution that could reflect territoriality; (B) a clumped distribution; (C) a random distribution.

To test the hypothesis that a species is territorial, one needs to develop two lines of evidence. First, observations need to establish that the territory is defended. Second, the spatial distribution of the animals must be assessed and found to be nonrandom (Figure 12.22). Are the animals clumped so that they are grouped together in their environment, as in a flock or herd? Are they nonrandomly spaced so that they are evenly dispersed in the environment? In contrast to clumped or even spacing is a third alternative, the possibility that the animals are randomly spread through their environment. A truly random distribution is sometimes called an *ideal free distribution*.

An ideal free distribution means that the location of each animal is not influenced by either other animals or by environmental features. If a random numbers table is used to assign x and y coordinates to locations on a map, the result is an ideal free distribution. In contrast, throwing marbles into a box does not result in an ideal free distribution because the marbles' resting spots are influenced by their collisions with other marbles and with the walls. Statistical tests can be employed to test if animal distributions are random or nonrandom. Clearly, if an animal's distribution fits the ideal free model and is not influenced by biotic or abiotic features, the animal cannot be territorial; a territorial animal is very aware of the presence of other animals. Actual aggressive interactions may be difficult to observe, but if animals depart from an ideal free distribution, then biologists should entertain territoriality as one among several hypotheses that might explain that departure.

KEY TERM An ideal free distribution occurs when animals are located in the environment without being influenced by other animals or other environmental factors.

419

Both of these types of evidence—territory defense and nonrandomness of distribution—must be met to make an argument for territoriality. Birds may settle on a telephone wire at equally spaced intervals, but they do not fight over locations along the wire. Thus, birds on a wire are not territorial. In many animals, though, "personal space" within a social group is important, and animals will threaten or attack if a group member comes too close. Fights may break out among baboons in a troop, but the fights do not determine possession of specific locations or result in defended spaces; thus, within the troop the baboons are not territorial. Analyses of territoriality also need to take into account spatial scale. At a landscape level, bird species such as magpies may be concentrated—clumped—in subhabitats, such as cottonwoods along streams. But if the spatial scale is changed to consider only the cottonwood habitat, then nests may be evenly spaced within the subhabitat, which would suggest territoriality, especially if aggression is observed.

Territoriality is a driving force in animal behavior and almost always needs to be considered in behavioral field studies. In this chapter, the focus has been on the relationships among nesting, parenting, aggression, and territoriality. In Chapter 15, on conservation, this topic will come to the fore again because territorial behavior often determines the size of an area needed to preserve a species.

SUMMARY

Animals choose nest sites based on protection and microclimates. Manipulation of the nest site, through excavation and the addition of materials from the surrounding environment, feces, or glandular secretions, gives animals the capability for complex and elaborate architecture. Nest design incorporates protection from the elements, concealment and protection from predators, and manipulation of microclimate.

Parental investment choices involve balancing current and future reproductive opportunities, and allocating resources among young within a group of siblings. Parental decisions are made to maximize lifetime parental reproductive output, which sometimes results in less-than-optimal conditions from a young animal's point of view. Even infanticide can become a male's strategy for controlling his paternal investment. This conflict over resources includes interactions among siblings; the gain of resources by one sibling may result in diminished survivorship for others in the group. In extreme cases, this conflict can lead to siblicide.

Much of animal aggression emerges from these conflicts over investment and resources within families, from territorial conflicts between family groups, or in the context of mating and predation. Aggression among animals generally occurs as a result of conflicts over resources, and across the broad range of animals, conflicts appear to be much more likely to be resolved by some form of aggression than by negotiation. However, animal combat may evolve to be ritualized, thereby reducing the risk of injury to participants. Territoriality is defense of a location, often a nest. Finishing this chapter with a mention of territoriality as it relates to nests completes this circle and returns to parenting behavior.

STUDY QUESTIONS

1. Thinking about animal nests, are there unifying themes that make nest construction strategies similar among all animals? How is microclimate within nests regulated? What provisions for escape are included in nest design?
2. Consider the cuckoos that mimic host eggs and sparrowhawks (Section 12.4). How would you design experiments to test these mimicry hypotheses? Take this opportunity to review the discussion of mimicry in Chapter 10; what are some pitfalls associated with some tests of mimicry?
3. What are the causes of conflict over parental investment between parents and their offspring?
4. Why is it that parents often do not interfere in attacks among their offspring, even when an attack leads to the death of one of the young? What circumstances might favor interference on the parents' part?
5. When would you expect weaning conflict to occur? Would you expect weaning conflict to be more intense between a young mother, who might have litters in following years, and her offspring; or an older mother, who is near the end of her lifespan, and her offspring? Why?
6. We discussed the possible causes of aggression and the potentially tragic consequences of misidentifying them. How might you go about discerning the causes of aggression in an animal? What animal behavior tools would you bring to bear on this important question?

Further Reading

Ardrey, R., 1966. The Territorial Imperative. Atheneum, New York, NY, 390 pp.

Buntin, J.D., 2010. Parental behavior and hormones in non-mammalian vertebrates. In: Breed, M.D., Moore, J. (Eds.), Encyclopedia of Animal Behavior, vol. 2, Academic Press, Oxford, pp. 664–671.

Clutton-Brock, T.H., 1991. The Evolution of Parental Care. Princeton University Press, Princeton, NJ, 368 pp.

Groebel, J., Hinde, R.A. (Eds.), 1989. Aggression and War: Their Biological and Social Bases, Cambridge University Press, Cambridge, 237 pp.

Hansell, M.H., 2009. Built by Animals: The Natural History of Animal Architecture. Oxford University Press, Oxford, 280 pp.

Lorenz, K., 1969. On Aggression. Bantam Books, New York, NY, 306 pp.

van Schaik, C.P., van Noordwijk, M.A., 2010. Infanticide. In: Breed, M.D., Moore, J. (Eds.), Encyclopedia of Animal Behavior, vol. 2, Academic Press, Oxford, pp. 138–143.

Wynne-Edwards, K.E., 2010. Parental behavior and hormones in mammals. In: Breed, M.D., Moore, J. (Eds.), Encyclopedia of Animal Behavior, vol. 2, Academic Press, Oxford, pp. 657–663.

Notes

1. Storm, J.J., Lima, S.L., 2010. Mothers forewarn offspring about predators: a transgenerational maternal effect on behavior. Am. Nat. 175, 382–390.
2. Thornton, A., McAuliffe, K., 2006. Teaching in wild meerkats. Science 313, 227–229.;Tinbergen, N., 1953. The Herring Gull's World. Collins, London, 255 pp.
3. Farnsworth, G.I., Simmons, T.R., 1999. Factors affecting nesting success of wood thrushes in Great Smoky Mountains National Park. Auk 116, 1075–1082.
4. Seeley, T.D., 2003. Consensus building during nest-site selection in honey bee swarms: the expiration of dissent. Behav. Ecol. Sociobiol. 53, 417–424.
5. Downing, H.A., Jeanne, R.L., 1990. The regulation of complex building behaviour in the paper wasp, *Polistes fuscatus* (Insecta, Hymenoptera, Vespidae). Anim. Behav. 39, 105–124.
6. Turner, J.S., 2000. The Extended Organism. The Physiology of Animal-Built Structures. Harvard University Press, Cambridge, 234 pp.
7. Lehtonen, T.K., Lindstrom, K., Wong, B.B.M., 2013. Effect of egg predator on nest choice and nest construction in sand gobies. Anim. Behav. 86, 867–871.
8. Andren, M.N., Kvarnemo, C., 2014. Filial cannibalism in a nest-guarding fish: females prefer to spawn in nests with few eggs over many. Behav. Ecol. Sociobiol. 68, 1565–1576.
9. Lehtonen, T.K., Wong, B.B.M., 2009. Should females prefer males with elaborate nests? Behav. Ecol. 20, 1015–1019.
10. Lindstrom, K., Lehtonen, T.K., 2013. Mate sampling and choosiness in the sand goby. Proc. R. Soc. B Biol. Sci. 280, 20130983.
11. Houston, A.I., Szekely, T., McNamara, J.M., 2013. The parental investment models of Maynard Smith: a retrospective and prospective view. Anim. Behav. 86, 667–674.
12. Trivers, R.L., 1972. Parental investment and sexual selection. In: Campbell, B. (Ed.), Sexual Selection and the Descent of Man Aldine Transaction, Piscataway, NJ, pp. 139–179.
    Trivers, R.L., 1974. Parent–offspring conflict. Am. Zool. 14, 249–264.
13. Dulac, C., O'Connell, L.A., Wu, Z., 2014. Neural control of maternal and paternal behaviors. Science 345, 765–770.
14. Noren, S., 2008. Infant carrying behaviour in dolphins: costly parental care in an aquatic environment. Funct. Ecol. 22, 284–288.
15. Connor, R.C., Smolker, R.A., Richards, A.F., 1992. Two levels of alliance formation among male bottlenose dolphins (*Tursiops* sp.). Proc. Natl. Acad. Sci. USA 89, 987–990.
16. Foster, E.A., Franks, D.W., Mazzi, S., Darden, S.K., Balcomb, K.C., Ford, J.K.B., et al., 2012. Adaptive prolonged postreproductive life span in killer whales. Science 337, 1313.
17. Sarno, R., Franklin, W.L., 1999. Maternal expenditure in the polygynous and monomorphic guanaco: suckling behavior, reproductive effort, yearly variation, and influence on juvenile survival. Behav. Ecol. 10, 41–47.
18. Siveter, D.J., Siveter, D.J., Sutton, M.D., Briggs, D.E.G., 2007. Brood care in a Silurian ostracod. Proc. R. Soc. B 274, 465–469.
19. O'Keefe, F.R., Chiappe, L.M., 2011. Viviparity and K-selected life history in a mesozoic marine plesiosaur (Reptilia, Sauropterygia). Science 333, 870–873.
20. Yip, E.C., Rayor, L.S., 2014. Maternal care and subsocial behaviour in spiders. Biol. Rev. 89, 427–449.
21. Brown, C.R., 1984. Lay eggs in a neighbor's nest: benefit and cost of colonial nesting in swallows. Science 224, 518–519.
22. Lahti, D.C., Lahti, A.R., 2002. How precise is egg discrimination in weaverbirds? Anim. Behav. 63, 1135–1142.
23. Burger, J., 1974. Breeding adaptations of Franklin's gull (*Larus pipixcan*) to a marsh habitat. Anim. Behav. 22, 521–567.
24. Cullen, E., 1957. Adaptations in the kittiwake to cliff nesting. Ibis 99, 275–302.
25. Spottiswoode, C.N., Stevens, M., 2012. Host-parasite arms races and rapid changes in bird egg appearance. Am. Nat. 179, 633–648.
26. Gluckman, T.-L., Mundy, N., 2013. Cuckoos in raptors' clothing: barred plumage illuminates a fundamental principle of Batesian mimicry. Anim. Behav. 86, 1165–1181.
27. Welbergen, J.A., Davies, N.B., 2011. A parasite in wolf's clothing: hawk mimicry reduces mobbing of cuckoos by hosts. Behav. Ecol. 22, 574–579.
28. Mann, J., Smuts, B., 1999. Behavioral development in wild bottlenose dolphin newborns (*Tursiops* sp.). Behaviour 136, 529–566.

29. Chaiken, M., 1992. Individual recognition of nestling distress screams by European Starlings (*Sturnus vulgaris*). Behavior 120, 139–150.
30. Briefer, E.F., Padilla de la Torre, M., McElligott, A.G., 2012. Mother goats do not forget their kids' calls. Proc. R. Soc. B Biol. Sci. 279, 3749–3755.
31. Dobie, J.F., 1961. The Longhorns. Castle Books, New Jersey, p. 178.
32. Blunt, J., 2002. Breaking Clean. Vintage Books, New York, NY, p. 290.
33. Albo, M.J., Costa, F.G., 2010. Nuptial gift-giving behaviour and male mating effort in the neotropical spider *Paratrechalea ornata* (Trechaleidae). Anim. Behav. 79, 1031–1036.
34. Delehnty, D.J., Oring, L.W., 1993. Effect of clutch size on incubation persistence in male Wilson's Phalaropes (*Phalaropus tricolor*). Auk 110, 521–528.
35. Emlen, S.T., Wrege, P.H., 2004. Division of labour in parental care behaviour of a sex-role-reversed shorebird, the wattled jacana. Anim. Behav. 68, 847–855.
36. Thomas, J.A., Birney, E.C., 1979. Parental care and mating system of the prairie vole, *Microtus ochrogaster*. Behav. Ecol. Sociobiol. 5, 171–186.
37. Fillater, T.S., Breitwsch, R., 1997. Nestling provisioning by the extremely dichromatic Northern Cardinal. Wilson Bull. 109, 145–153.
38. Eggert, A.-K., Reinking, M., Muller, J.K., 1998. Parental care improves offspring survival and growth in burying beetles. Anim. Behav. 55, 97–107.
39. Spoon, T.R., Millam, J.R., Owings, D.H., 2006. The importance of mate behavioural compatibility in parenting and reproductive success by cockatiels, *Nymphicus hollandicus*. Anim. Behav. 71, 315–326.
40. Hume, J., Wynne-Edwards, K.E., 2006. Paternal responsiveness in biparental dwarf hamsters (*Phodopus campbelli*) does not require estradiol. Horm. Behav. 49, 538–544.
41. Blumer, L.S., 1979. Male parental care in the bony fishes. Q. Rev. Biol. 54, 149–161.
42. Nussbaum, R.A., 1987. Parental care and egg size in salamanders: an examination of the safe harbor hypothesis. Res. Popul. Ecol. 29, 27–44.
43. Stein, L.R., Bell, A.M., 2014. Paternal programming in sticklebacks. Anim. Behav. 95, 165–171.
44. Klug, H., Bonsall, M.B., Alonzo, S.H., 2013. Sex differences in life history drive evolutionary transitions among maternal, paternal, and bi-parental care. Ecol. Evol. 3, 792–806.
45. Bell, M., 2008. Strategic adjustment of begging effort by banded mongoose pups. Proc. R. Soc. B 275, 1313–1319.
46. O'Connor, R.J., 1978. Brood reduction in birds: Selection for fratricide, infanticide and suicide? Anim. Behav. 26, 79–96.
47. Smiseth, P.T., Andrews, C., Brown, E., Prentice, P., 2010. Chemical stimuli from parents trigger larval begging in burying beetles. Behav. Ecol. 21, 526–531.
48. Porkert, J., Spinka, M., 2006. Begging in Common Redstart Nestlings: scramble competition or signalling of need? Ethology 112 (4), 398–410.
49. Grodzinski, U., Lotem, A., 2007. The adaptive value of parental responsiveness to nestling begging. Proc. R. Soc. B 274, 2449–2456.
50. Rivers, J.W., 2007. Nestmate size, but not short-term need, influences the begging behavior of a generalist brood parasite. Behav. Ecol. 18, 222–230.
51. Arias-Carrión, O., Stamelou, M., Murillo-Rodríguez, E., Menéndez-González, M., Pöppel, E., 2010. Dopaminergic reward system: a short integrative review. Int. Arch. Med. 3, 24.
52. Ferree, E.D., Wikelski, M.C., Anderson, D.J., 2004. Hormonal correlates of siblicide in Nazca boobies: support for the Challenge Hypothesis. Horm. Behav. 46 (5), 655–662.
53. Mock, D.W., 1987. Siblicide, parent–offspring conflict, and unequal parental investment by egrets and herons. Behav. Ecol. Sociobiol. 20, 247–256.
54. Sagebakken, G., Ahnesjo, I., Mobley, K.B., Goncalves, I.B., Kvarnema, C., 2010. Brooding fathers, not siblings, take up nutrients from embryos. Proc. R. Soc. B 277, 971–977.
55. Lyon, B., Eadie, J.M., Hamilton, L.D., 1994. Parental choice selects for ornamental plumage in American coot chicks. Nature 371, 240–243.
56. Osorio-Beristain, M., Drummond, H., 2001. Male boobies expel eggs when paternity is in doubt. Behav. Ecol. 12, 16–21.
57. Bruce, H.M., 1959. An exteroceptive block to pregnancy in the mouse. Nature 184, 105.
58. Wilson, E.O., 1975. Sociobiology: The New Synthesis. Belknap Press of Harvard University Press, Cambridge, MA, 697 pp.
59. Hazlett, B.A., 1978. Shell exchanges in hermit crabs: aggression, negotiation or both? Anim. Behav. 26, 1278–1279.
60. Guerra, P.A., Pollack, G.S., 2010. Colonists and desperadoes: different fighting strategies in wing-dimorphic male Texas field crickets. Anim. Behav. 79, 2087–2093.
61. Dubois, N.S., Kennedy, E.D., Getty, T., 2006. Surplus nest boxes and the potential for polygyny affect clutch size and offspring sex ratio in house wrens. Proc. R. Soc. B 273, 1751–1757.
62. Fisher, J.B., 1954. Evolution and bird sociality. In: Huxley, J., Hardy, A.C., Ford, E.B. (Eds.), Evolution as a Process. Allen and Unwin, London, pp. 71–83, 307 pp.
63. Detto, T., Jennions, M.D., Backwell, P.R.Y., 2010. When and why do territorial conditions occur? Experimental evidence from a fiddler crab. Am. Nat. 175, E119–E125.
64. Darwin, C., 1890. The Expression of the Emotions in Man and Animals. J. Murray, London, pp. 52–53.

Social Behavior, Cooperation, and Kinship

423

LEARNING OBJECTIVES

Studying this chapter should provide you with the knowledge to:

• Use the concepts of cooperation and altruism to understand the evolution of group living and be able to measure the benefits for animals living in groups.

Animal Behavior. DOI: http://dx.doi.org/10.1016/B978-0-12-801532-2.00013-1

- Perceive the reasons for group living, including shared protection from predators, discovering and hunting food, opportunities for taking advantage of public information, and directing aid to others.
- Know that the simplest social groups are often families, which can be expanded by including multiple generations.
- Evaluate the importance of kinship structure in determining cooperative and competitive interactions.
- Understand the mutual benefits that accrue within social groups of nonkin.
- Distinguish among the evolutionary explanations for living in social groups and know that these are not mutually exclusive hypotheses.
- Apply evolutionary theories of cooperation to extreme altruism, or eusociality, and understand the remarkable mechanisms in eusocial animals for enhancing cooperation and colony efficiency.
- Integrate social recognition as a key to formation and maintenance of social groups.

13.1 INTRODUCTION

Some of the most ecologically successful species on the planet are social. What causes animals to group together? How is group membership determined? What are the benefits and costs of group living? Animals in groups can gain significant advantages over their solitary counterparts (Figure 13.1).[1]

Groups of predators can exploit prey that would be impossible for each predator to capture on its own (see Chapter 9). Group membership also provides protection from predators and other threats (see Chapter 10). Social dynamics keep family groups together, and social factors may eventually lead families to break up and individuals to disperse to establish independent life (see Chapter 12). Our study of groups includes many ways in which animals can come together and explores evolutionary models that explain group living.

Most evolutionary questions can best be analyzed using a cost–benefit approach; social behavior is no exception. The benefits of social living can be enormous; self-defense, foraging, and parental care may all be easier or more efficient in groups. To these benefits can be added potential efficiencies gained from dividing labor among members of the social group, so that individuals specialize in tasks. Group living also has significant costs: Groups may be more visible and therefore attract predators. Parasites and diseases may spread more

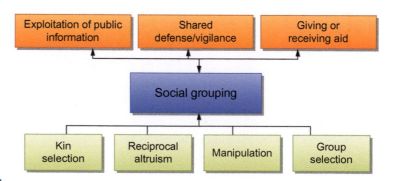

FIGURE 13.1
Animals come together in groups for a variety of reasons related to sharing or exploiting information, gaining protection, or giving and receiving aid. Even though animal groups are common, it is often difficult to ascertain the evolutionary advantage of grouping for each animal in the group. Possible explanations include kin selection, reciprocal altruism, manipulation, or group selection. One of the goals of this chapter is to develop ways in which these hypotheses for grouping can be tested.

easily within groups than among isolated animals. Finally, group members compete for food and perhaps for mates.

For group living to persist these costs and benefits must add up so that the fitness of the animals in the group is enhanced. For any one animal, the decision to join a group must be the result of a situation in which the benefits outweigh the costs, ultimately measured as fitness effects. Evolution can make this decision for an animal; selection favors behavior that has higher benefits than costs.

Species with sterile workers are termed *eusocial*. In these extreme cases, some animals within groups pay for group membership with a lifetime of sterility; they spend their lives working for others and gain reproductive benefits only if their relatives reproduce.

> **KEY TERM** Eusocial species live in colonies with overlapping generations in which the mother (queen) plays the reproductive role and the offspring are workers.

Social behavior is a big topic within animal behavior, so this coverage is divided into two chapters. This chapter focuses on the theories that aid in understanding the existence of social groups and then explores social recognition, a primary component of the glue that holds social groups together. Chapter 14 addresses how labor is divided within social groups and reveals fascinating examples of behavior that is both elaborate and highly social.

Sometimes animals come together in groups only to exploit social information. By observing other animals' choices for feeding locations, resting sites, and predator avoidance strategies, a creature can optimize its own choices without enduring the risks of gaining the information on its own. *Social information*, whether it is openly shared or kept private and exploited by eavesdropping, is a powerful driver for animals coming together in groups.

Direct cooperation is a step beyond information gathering as a motivation for group behavior. Cooperation includes intentional signaling of key events, such as giving alarm calls when a predator is present. Cooperation involves a cost for the donor animal, measurable in risk of lost direct fitness, and a potential fitness gain for the recipient. Donor behavior would never occur if the donor did not have a way to recover its fitness investment in the future. Theories to explain cooperation revolve around understanding how, over the long run, donor behavior is evolutionarily favored.

The theories that explain cooperation within groups are not mutually exclusive. For any instance of cooperation one or more of the following explanations may apply:

- Hamilton's *kin selection* model.[2] Kin selection plays a key role in conversations about *cooperation* and group living, and this theory deserves substantial emphasis.
- Group selection is an interesting set of ideas initially based on the thought that animals might behave "for the good of the species." We emphasize that modern ideas about group selection differ substantially from a "good of the species" approach.
- Equally important are models for *reciprocal altruism*—the trading of aid in the present for potential returned aid in the future.[3] This leads to complex transactional and social contract models for cooperation.
- Models for *selfish teamwork* and delayed competition are considered next; they can be powerful in explaining why animals with competing interests cooperate in the short term.
- Finally, the efforts of helpers can simply be stolen.

> **KEY TERM** Kin selection occurs when an animal behaves in ways that add to the fitness of its close relatives, with whom it shares genes.

> **KEY TERM** Cooperation is joint activity among animals to achieve a shared goal.

> **KEY TERM** In selfish teamwork, animals work together for some time so that the group as a whole benefits, but at a later date they compete for individual benefits.

The last section of the chapter explores the fascinating ability of animals to recognize group members and to discriminate group members from nongroup members. In many animals, evolution has led to social recognition that is intimately tied with immune recognition of self. For instance, insect and crustacean social groups have unexpected and intriguing ways of knowing who belongs in their group and who does not. The combination of theory and social recognition then leads into the amazing array of social life inhabiting Chapter 14.

Levels of Selection and Animal Behavior

We know that natural selection acts to increase the fitness of individuals and that evolution is a population-level phenomenon measured as change in gene frequencies over generations. But what unit does natural selection really act upon? The answer to this question is not as simple as might be thought. Does natural selection act on nucleotides, on genes, on cells within organisms, on organisms, or on the populations of which organisms are members? Some even use phrases like "for the good of the species," which assumes that natural selection can act on populations or even on species.

The issue of *levels of selection* has intrigued evolutionary biologists since Darwin's time. The basic argument of natural selection, that gene frequencies change from generation to generation based on which genes confer a reproductive advantage on the organism in which they are found, is key to evolutionary biology, but does selection act on genes, on organisms, or on groups? Earlier generations of biologists grappled with this question, and V. C. Wynne Edwards argued in a 1962 book that group selection, animals behaving for the good of the group, rather than for their individual fitness, explained many aspects of animal social behavior. Wynne-Edwards' views were never widely accepted and "for the good of the species" arguments are rarely seen in evolutionary biology.[4] Later Richard Dawkins made potent arguments in his 1976 book, *The Selfish Gene*, that selection acts at the level of the gene.

Certainly some genes can drive evolution—if variants of a particular gene have a strong selective advantage or disadvantage the strength of selection on that gene overwhelms the smaller fitness effects of other genes. A gene carried by an individual animal that results in a lethal developmental deformity acts essentially alone and the positive or negative effects of other genes in that animal's genome are evolutionarily irrelevant. However, absent such strong single-gene effects, many animal traits, including much of behavior, are shaped by the interaction of many genes, and it is reasonable to postulate that the net effects of all an animal's genes come together to shape its fitness. In other words, selection acts on the level of the organism rather than the isolated gene.

Does natural selection act on levels above that of the individual? Some contemporary biologists have focused on finding alternative explanations to kin selection for social evolution. Key to their argument is that selection can act on the levels above the individual. This concept is discussed in more detail later in this chapter, but it should be considered in this context as well. How might selection above the level of the organism play out? Imagine a group of animals exists in which one or more members behave detrimentally to themselves but to the benefit of other group members (these are *altruistic* groups). Other groups in the same population lack members who behave this way (these are *selfish* groups). If altruistic groups have a higher chance of surviving than the selfish groups because of the generous behavior of some group members then the genes for altruism will be evolutionarily favored. Of course, the animals carrying the altruistic gene need to reproduce; if their generosity extends to the point of self-sacrifice then the gene cannot be passed along.

The level of selection argument is fascinating when applied to the eusocial insects. Colonies of eusocial termites, ants, bees, and wasps survive or die as units. But for the main part, the workers in these colonies are sterile. So if selection is acting on the colony, is not the colony just a phenotypic extension of the queen's genes?

CASE STUDY: COOPERATIVE LOAD CARRYING

One of the most attention-getting stunts that some ants can do is cooperating to pick up a food item that is much larger than what could be carried by a single ant. The food is transported rapidly and directly to the nest, each ant seeming to know exactly how to contribute to the effort so there is no apparent confusion about the direction to carry the food. Australian weaver ants can carry very large food items, like birds or snakes, up the trunk of a tree, even though the food weighs thousands of times more than each ant.

This is much like a group of people picking up and carrying a heavy piece of furniture, except the ants cannot talk to one another about the direction to go or about how to shift the object to get around obstacles. How do they do this without obvious means of communication[5,6] (Figure 13.2)?

Likely each ant is able to orient its movements to the nest using external cues such as landmarks or a sun compass. They do not need to communicate about direction because they each carry the information with them. If there is disagreement among the ants—they are pulling in different directions—Helen McCreery has suggested the ants can resolve the impasse by some ants being more persistent (stubborn!) in their direction of pulling. If some ants give up while others persist, then the impasse is resolved and the food is moved in the direction chosen by the persistent ants. If the ants can update their assessment of whether they are moving in the correct direction by comparing their progress with landmark or compass information then they can correct their route if necessary.

We can only marvel at the accomplishments of these that animals with brains so different from humans that nonetheless

FIGURE 13.2
These ants have grasped food that is too large for any one of them to carry alone and they are working together to move it to their nest. In addition to being a fascinating example of cooperation, such behavior can provide the inspiration for the design of simple swarm robots that accomplish complex tasks. *Photo: Rachael Kaspar.*

achieve so much. Swarm robotics uses information from studies of ants to design programs for robots. This is a new and exciting field of robotics in which large numbers of small, simple, autonomous robots accomplish complex tasks when each robot is only given very simple rules to follow.[7,8]

427

13.2 ALTRUISM OR SELFISH INTERESTS?

Are animals generous or do they act out of self-interest? Altruism is a behavior performed without regard for self-interest. Biologists define a truly altruistic act as one that involves one animal giving aid to another with no opportunity for payback. Does this ever occur, or can apparently altruistic behavior ultimately be explained by self-interest? Animals often cooperate in foraging, defending themselves, and caring for young, but true altruism is entirely contrary to Darwinian thinking, which claims that everything an organism does is shaped by the drive to maximize its own fitness. Most cases of apparent altruism actually have roots in self-interest, so examples of truly altruistic aid-giving behavior are quite rare.[9,10] While altruism is clearly one possibility of explaining aid-giving behavior, the alternative hypotheses consider aid-giving and cooperation to be the results of self-interest.

In studies of cooperation and aid-giving behavior, the term *donor* signifies the animal that gives the aid. The *recipient* receives the aid and benefits. The donor usually incurs a cost for

giving the aid; it gives up food, puts itself at risk for predation, or delays its own reproduction as a result of aiding another. This is true whether the donor is motivated by altruism or self-interest. The cost to the donor translates into a benefit for the recipient, usually measurable in food obtained, protection from predation, or increased chances of reproduction. But is there also a benefit to the donor? In true altruism, the donor pays the cost without any consideration of potential benefits it might later receive. In contrast, any evolutionary explanation of aid-giving behavior must include benefits to both the donor and the recipient.

FIGURE 13.3
Mutual grooming in monkeys. Note the parallel in behavior with mutual grooming in chimpanzees shown in a later chapter (see Figure 14.8). *Photo: Pen-Yuan Hsing.*

DISCUSSION POINT: ALTRUISM AMONG SPECIES

Through a little searching on the Internet, you'll be able to find a few seemingly miraculous cases of true altruism between species. They include events such as dolphins saving drowning humans or dogs rescuing their human family members from fire or other catastrophes. It's interesting to think analytically about these events. Certainly, the potential for true altruism is there, but can you approach the event scientifically, forming alternative hypotheses that can be tested to explain such altruism? Are there hypotheses for this type of behavior that don't involve altruism? Might these events be more legend or wishful thinking than reality? When thinking about this, you might want to refer to the section on emotions in Chapter 6 on cognition.[11]

428

Sometimes the benefits to the donor are nearly immediate; mutual grooming among monkeys (Figure 13.3) exemplifies aid-giving with a rapid return. Benefits may also come to the donor at a later date. In most examples of cooperation among animals, the donor expects to receive benefits that at least balance the cost. These benefits may come directly to the donor during its lifetime, such as reciprocated protection from predators, or over a longer evolutionary span, due to increased propagation of the donor's genes in following generations.

This evolutionary trade-off can actually be quantified. For natural selection to favor cooperation and aid-giving, the costs of being a donor must be counterbalanced by current or future benefits to the donor. The very simple *cost–benefit equation*, $B > C$, describes the circumstances under which animals should become donors. The benefit (B) is the increase in the donor's genetic contribution to the next generation, and the cost (C) is any loss in contribution to the next generation that the donor incurs. A modification of this equation formulated by William D. Hamilton takes into account the genetic relationship between the donor and the recipient (see the box below).

KEY TERM A cost–benefit equation is a quantitative analysis of the balance between the cost of a behavior and the benefits gained from that behavior.

OF SPECIAL INTEREST: WILLIAM D. HAMILTON

Only a small handful of scientists contribute enough to a field to merit special mention in a textbook: William D. Hamilton is one of those select few in animal behavior (Figure 13.4). Hamilton contributed several major ideas to animal behavior, including three central concepts in this chapter: kin selection, cooperation, and selfish herds. Hamilton made other important contributions, notably the Hamilton–Zuk hypothesis that parasite load affects coloration and consequently mate choice. He also improved our understanding of dispersal and sex ratios.[12]

FIGURE 13.4
William D. Hamilton. Hamilton was a leading contributor to theory in animal behavior and evolutionary biology. Much of today's work on kin selection was stimulated by his theoretical contributions. He also developed the idea of the selfish herd and explained why sex ratios in some animals are very different from the 50:50 ratio that we normally expect.

Hamilton is credited with having more influence on evolutionary thought than any other individual biologist since Darwin. He made his points with elegant simplicity, and while some of those arguments are supported by mathematical models, the real genius of his work is that he developed principles that have survived nearly half a century of scrutiny. Hamilton, a humble and gentle-spirited person, received nearly every conceivable honor and award in his field during his lifetime. He died in 2000 of complications from malaria, an infection he contracted in Africa, where he was working on discovering the origin of HIV.

13.3 SCHOOLS, FLOCKS, HORDES, AND HERDS

Why live in a school, flock, or herd?[13-15] Groups of a few, a dozen, tens, hundreds, or even thousands of animals are a common phenomenon, one that is easily observed across major taxa ranging from flatworms to Crustacea, from insects to birds and mammals. Basic family groups (a female, perhaps a male, and their offspring) or slightly extended families (e.g., related females cooperatively rearing a brood) were covered in Chapter 12. Evolution, though, can draw together other kinds of family associations, such as schooling sibling tadpoles. People have crafted exotic descriptive names for animal groupings, but the terms *school*, *flock*, and *herd* capture the essential focus of this section.

Social Groups and Public Information

Animals rarely act in complete privacy, and an observable behavior becomes *public information*, meaning that other animals can exploit what they see (or smell or hear; see Chapter 7). An animal may find it advantageous to conceal some of its discoveries, such as food, the location of a food cache, the identity of its mate, or the location of its nest. The advantages of privacy and concealment are more difficult to obtain in social

> **KEY TERM** Public information is any feature that other animals can perceive. A wide range of attributes, such as signals, health status, mating condition, and size can become public information.

groups; an animal in a group makes much of its life "public information." On the other side of the coin, membership in a group gives an animal the ability to observe and gain leverage from information revealed by other animals.[16] The costs and benefits of revealing and gaining public information are part of the interplay that leads to the evolution of social groups.[17,18]

Within such groups, alarm signals can alert members to threats. Alarm signals are public information;[19] once the signal is given, any animal that can perceive the signal can take advantage of it, including the potential predator. These signals also can attract costly and unwelcome attention to the signaler, and under the rules of evolution, benefits such as improved survival of close kin are expected to at least balance that cost. Alternatively, alarm signals may cause a visually distracting commotion in the group, drawing attention away from the signaler. Such signals often have evolved so that it is difficult to locate their source, publicizing the information without revealing too much about the source. Many audible alarm signals are short and high-pitched calls which are particularly difficult to locate. Alarm signals serve as a great example of the balance between a need to create public information about the location of threat and the potential need to conceal the private information of the signaler's location.

Food Discovery

Following this line of reasoning about public information, there are two basic ways to find food: An animal can discover food for itself, or it can take advantage of food discovered by other animals. Taking advantage of public information within a group extends far beyond a situation in which more than one animal is attracted to the same food item; it includes group members' exploitation of the food discovery abilities of their group-mates.

A conspicuous example of food theft is the behavior of herring gulls on beaches. The first gull to discover a crab or other morsel is often not the gull that ends up enjoying the feast. Nearby gulls are always keenly interested in food discoveries by their companions, and a gull with a piece of food is a magnet for the attentions of its voracious flock-mates.

Similar behavior abounds in magpies, ravens, and crows, all members of the bird family Corvidae. Exploiting public information appears to draw crows or seagulls together (Figure 13.5); they stay close to other crows or seagulls to watch for food discovery. After one animal has found food, the others try to steal it. A bird that wants to forage in "private" may have a very difficult time evading observation because completely driving off the other birds is a fairly hopeless task. Thus, the group may contain unwilling participants.[20]

This grouping behavior can also involve multiple species. In fall and winter, mixed-species flocks of birds are common; in North America, house sparrows, house finches, and juncos

430

FIGURE 13.5
(A) A flock of seagulls competing for a food item; (B) pelicans, which often forage in flocks; and (C) a school of fish. These are examples of groups that might coalesce because individuals evolve to exploit public information. *Photos: Michael Breed (left and right), and Ben Pless (center).*

frequently occur together. Discovery of food by one bird in the flock inevitably leads to exploitation by the other members. Ultimately, the best foragers in the group may lose food; they cannot spend all their time and energy trying to drive away other animals, so groups formed around exploitation of other animals' foraging abilities are common.

Mutual Protection

In mutually protective groups, one possible advantage is shared vigilance. Perhaps all animals in the group benefit from having others in the group spend time watching for predators.[21] In groups of genetically related animals—a school, herd, or flock that is an extended family—mutual aid, including shared vigilance, can be explained by kin selection (see the section on this topic later in this chapter). Tadpoles often school in family groups, as do some species of trout and salmon. However, most schools, flocks, and herds consist of unrelated animals, requiring explanations other than kin selection—such as reciprocal altruism or *selfish herds* (see below)—for why the flocks come together.

> **KEY TERM** A selfish herd is one in which animals group together but are motivated by their own interests, rather than aid-giving.

In addition to gaining from shared watchfulness for predators, members of groups may benefit if predators are confused by the group's motion. Chasing a single animal can be much simpler than trying to select a prey item from a rapidly shifting group in which the animals' paths interweave. The flurry of motion when a flock of birds takes wing may make it much more difficult for the predator to catch any one animal. Related to predator confusion is predator saturation, a topic already discussed in some detail in Chapter 9.

Finally, sometimes the lead, or reproductively dominant, members of the group assume protective roles. In the herds of wild horses that thrived in the American West in the eighteenth and nineteenth centuries, the dominant stallion in the herd was also the chief protector of the mares and foals. Stallions actually hunt and kill mountain lions with their sharp front hooves if they catch the lion's scent (Figure 13.6). Bulls are similarly aggressive against interlopers in their herd's area. The balance between costs and benefits favors the dominant male that protects his reproductive investment in the herd. This type of protectiveness is comparable to the fierce defense that many animal mothers show for their young.

Selfish Herds

An animal can join a group for its own selfish reasons, seeking to push the risk of being a victim of predation onto other group members. A herd formed in this manner is aptly named a *selfish herd*; the name is self-explanatory. Selfish herds are groups of animals in which group membership confers protection on each animal in the group, but within the herd the animals compete for the position that maximizes the value gained from group membership.

FIGURE 13.6
Mountain lions. *Photo: US Fish and Wildlife Service.*

William D. Hamilton formalized the concept of the selfish herd in 1971. The idea is charming in its simplicity. For flocking, herding, or schooling animals group membership in any form may be more beneficial than remaining solitary, but nonetheless good and bad locations exist within a group. A bad location is one in which exposure to predators is high and access to public information about predators is hard to come by. A good location has low exposure to predators and high access to public information.

Generally speaking the good locations are at the center of the group, in positions with considerable social connectivity (see

431

Section 13.4 and with a good buffer of group-mates between the animal and any potential predator. Selfishness comes into play because not all positions within the flock are equally protected, so animals selfishly compete for position within the group.

For instance, the trailing fish in a school may be more vulnerable to predation; if the predator focuses on the last fish in the group, the others will scatter forward and to the sides. In this scenario, the scattering fish do not interfere with the predator's visual fix on the last fish, making the chase easier for the predator. In a selfish herd, animals should constantly jockey for position. Their constantly shifting positions reflect competition among group members for these favored positions. Models show that even if some animals spend more time being vigilant than the average member of the group they still gain an advantage as they are more likely to spot predators that pose a high risk to them as individuals.[22]

How Large Should a Group Be?

Another way of looking at joining a group is that it spreads the risk of being a victim across the group members. Thus for any given animal the risk is diluted. The importance of location within the group distinguishes a selfish herd from a group that depends primarily on dilution effects for defense.

What is the right number of animals in a group?[23] Begin by thinking about a herd, flock, or school that has come together for shared vigilance, and the value of mass movements confusing predators. If each member of the group has an equal chance of being caught by a predator, then the more the merrier—10 members means a 10% chance of being eaten, 100 members translates into a 1% chance, and so on. This is called a dilution effect, and in contrast to a selfish herd, group members may depend on group size, not location in the group, for relative safety.

Following the logic of the dilution effect, there might be no upper limit: A school might then include all the fish in the ocean, a herd should include all the ungulates on a continent. What might establish upper limits? Three factors seem most obvious; first, food supply will be limiting at some point, so the amount of food per animal establishes an upper group size. Second, larger groups may attract more predators. Third, the benefits of shared vigilance probably plateau after the group has a certain number of animals. The balance between the value of being in the group and the potential costs can result in an optimal group size.[24-26]

Another way of looking at group size is to consider reproduction of each group member related to the number of animals in the group.[27] Theoretically, animals should be willing to participate in a group as long as their reproduction is maximized. Figure 13.7 shows models for how group size relates to reproduction. If animals can assess their potential reproduction, then these models may give insight into whether a species may exhibit social behavior, and if it does, what group size might be.

Dominance Hierarchies in Social Groups

The best known *dominance hierarchies*, such as those in flocks of domestic hens, are linear, meaning that there is one individual (A) that dominates the others, there is a second individual (B) that dominates all others except A, there is a third individual that dominates all the others except A and B, and so on, so that if they were to be ordered by their ranks, relative to others in the group, they would form a straight line. A linear hierarchy is one in which each animal is dominated by higher-ranking group members and, in turn, dominates lower-ranking members. Other types of dominance hierarchies are known in various primates, in herds of ungulates, and in cockroaches.[28] Whether the group is composed of related or unrelated

KEY TERM A dominance hierarchy describes situations in which animals are physically or chemically dominant over other animals in their social group.

432

animals, dominance hierarchies can structure social interactions in groups.

High-ranking animals in dominance hierarchies may have preferred access to food, mating, grooming services of other group members, and perhaps to protected positions within the group (although when horses are on the move, mature mares often lead the way, and the stallion brings up the rear, a risky position from which he can defend the herd). Clearly, the benefits of group membership are not the same for all the animals when there is a dominance hierarchy (Figure 13.8); lower-ranking animals pay a much higher cost for group membership than the alpha animal does. A couple of dynamics help to keep groups together in these circumstances. First, the costs of not being a member (loss of access to food, higher predation risk) may be much higher than the costs associated with membership. Second, low-ranking animals may play a waiting strategy, buying a chance at rising to a higher rank later by paying the costs of low rank for the moment. In groups with multiple males, low-ranking animals, particularly males, may also have sneaker or satellite opportunities for mating (see Chapter 11 on mating). In addition, an animal's dominance with respect to one kind of resource may not mean it dominates all resources; the dominant animal at a watering hole may not be the dominant animal where some other resource is involved.

Generally, dominance hierarchies play out differently for males and females. Alpha, or dominant, males are much less likely to tolerate the presence of potentially reproductive males in the group than the presence of females. Holding the dominant position is very taxing, physically, for the male, and monopolizing reproduction in the group is the key benefit for the alpha male. Many females, on the other hand, can mate with one male, so they are more tolerant of each others' presence. This is not to say that dominance hierarchies do not exist among females; they do, and they play important roles in access to food and timing of reproduction in many species.

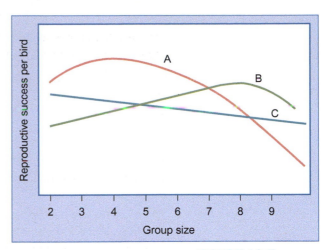

FIGURE 13.7
Group size as it relates to reproduction. In (A), reproduction per individual peaks in small groups, while in (B) larger groups are advantageous. In (C) grouping is disadvantageous at all group sizes. Ecological factors, such as effort needed to collect food for young animals or increased effectiveness of group defense by larger groups affect which of these models apply in a given species. *Adapted from Koenig, W.D., 1981. Reproductive success, group size, and the evolution of cooperative breeding in the acorn woodpecker. Am. Nat. 117, 421–443.*[27]

433

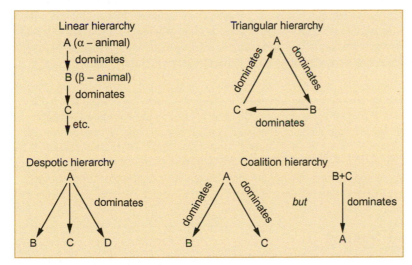

FIGURE 13.8
Common types of dominance hierarchies are diagrammed in this figure. In *linear hierarchies*, one "alpha" animal (A) dominates all other animals, a "beta" animal (B) dominates all animals except A, C dominates all animals except A and B, and so on. In a *despotic hierarchy*, one animal (A) dominates all other animals in the group (B, C, D). In a *triangular hierarchy*, A dominates B, B dominates C, and C dominates A. In a *coalition hierarchy*, A can dominate B and C, but if B and C cooperate, they can together dominate A.

One result of this difference between males and females in mammals is the formation of *bachelor herds*: outside the mating season, groups of males affiliate and gain benefits of group living, such as protection from predators. Bachelor herds break down with the onset of the mating season and intensified competition among males.

BRINGING ANIMAL BEHAVIOR HOME: DOGS AS MEMBERS OF HUMAN PACKS

Wolves, the wild ancestors of domestic dogs, live in packs led by an adult male and female; most of the pack members are the offspring of that pair. Much has been made about how a dog in a human household views itself as a pack member. Because early studies on wolves emphasized dominance relationships within packs, it seemed reasonable to interpret a dog's responses to the humans within its social group as a reflection of its dominance status in the group.

To a certain extent, the idea that dogs act like members of a human pack is true; dogs can be very territorial in protecting a house and the members of the household. Dogs respond to leadership within a human household and are more likely to follow direction from an older, deep-voiced male human than from a female or from younger group members. Unfortunately, many humans interpret this to mean that they should somehow establish their dominance over household dogs, and do so by physical means, such as hitting, kicking, and forced roll-overs, so-called "alpha rolls." These approaches may have disastrous consequences.

Early studies of behavior in wolf packs focused on captive packs that were composed of unrelated individuals who, because they were captive, were unable to disperse even if they had so desired. In these packs, agonistic behavior occurred more frequently than would be expected in a hierarchical social system. These studies formed the basis for much of the "wolf pack" view of domestic dogs. In the wild, however, members of wolf packs exhibit little aggression, although when it does exist, it can be lethal. In these packs, agonistic behavior does not conform to hierarchical expectations, but reflects a variety of attributes, including age, weight, and the nature of the contested resource.

One might then ask if wolves are good models for domestic dogs. Do dogs, perhaps by virtue of domestication, engage in social interactions that differ significantly from wolf social interactions? Studies of both feral dogs and neutered domestic dogs—animals that approximate the average family pet more than a wolf does—show that dogs differ substantially from wolves in their social behavior. Concepts like hierarchies and suppressed reproduction do not seem to apply to domestic dog interactions, and dogs do not seem to strive at every turn for dominance.

Instead, "dominance" in dogs may shift depending on motivation (how much do I value this contested resource?) and context (where have I seen this black dog before and did it mean trouble?). The Resource Holding Power model (see Chapter 11) seems to offer more promise for understanding dog social behavior than notions of dominance do. How fruitful this will be in the study of dog–human interactions remains to be seen, but many skilled dog trainers have used a variation on this theme. They know that their first task with a dog is to find an object that the dog will "work" for—that is, that the dog values. This is much more productive than a contest for "dominance."

Aggregated Nesting

Sometimes animals clump their nests together in a single location. Why they do this is one of the more intriguing mysteries in animal behavior. Clumped or aggregated nesting is common in mammals such as prairie dogs, birds such as swallows, and ground-nesting wasps and bees. Is it possible that the underside of a single, specific bridge is a perfect nesting habitat for swallows (see Chapter 14) while none of the bridges up- or downstream are suitable? What about large clusters of other types of nests (Figure 13.9)? Living in

FIGURE 13.9
Aggregations of animals and nests: (A) oropendolas, (B) cormorants, (C) seals, and (D) cliff swallows. *Photos: Ben Pless (seals), Michael Breed (oropendolas, cormorants, and cliff swallows).*

close proximity increases competition for food and enhances the chances for transmitting diseases and parasites. Clusters of nests might also attract predators. What possible benefits could outweigh these major disadvantages? Two that might immediately come to mind are collective defense and easy availability to mates, along with the previously discussed benefit of finding food by observing other animals foraging.

Beyond these three hypotheses to explain aggregated nests—defense, mates, and information sharing—there is no clear general answer to the question of benefits that outweigh the disadvantages of group living. It may be that microenvironmental variation is more important than humans understand, so a particular nesting site seems superior to the animal, but for reasons that are not obvious to us. (*Umwelt* remains important!)

It may also be that young prefer to nest where their parents nested; the success of the parents may predict success of the next generation in the same spot. This is called *philopatry*, and as generations pass and descendants accumulate, nesting populations may grow in a location, even though nesting together carries no particular advantage. Philopatry is a conservative choice of nesting site which relies on the prediction, made through evolutionary experience, that because the animal's parents survived and reproduced in a location, that spot remains an acceptable nesting site. Because the animal does not compare and assess other potential sites, philopatric choices are made based on limited information and may only be optimal if the cost of assessing other potential sites is high. The combination of philopatry and simple attraction to other animals of the same species can lead to large nesting aggregations occupying a few sites in areas which seem to have many acceptable nesting sites.[29]

KEY TERM Philopatry is the return, in a later breeding season, to the site where an animal was born.

OF SPECIAL INTEREST: LADYBIRD BEETLE AGGREGATIONS

One of life's mysteries is why some insects, such as ladybird beetles, overwinter in aggregations of hundreds or thousands of individuals (Figure 13.10). Every year, entomologists are flooded with phone calls and e-mails from curious members of the public who have found such a pile of beetles and want to know why the beetles aggregate. There's no clear answer to this question, but let's consider a couple of hypotheses for why they and other aggregating animals might do this:

1. The beetles may simply all be attracted to the same environmental feature, so there's no particular advantage or selective value in aggregating; they all just like the same place.
2. They may be attracted to each other; this would be an example of how public information provides a key element behind an interesting behavior.

From studies of the beetles, we know that they like to land on lighter-colored places, but that alone wouldn't explain how they aggregate.[30-32] In terms of why they aggregate, the two main ideas are that wintering as a group assists in thermoregulation as they insulate each other, and that the groups help them choose mates, which occurs in the spring before they leave the aggregation.

FIGURE 13.10
An aggregation of ladybird beetles. *Photo: Jeff Mitton.*

13.4 SOCIAL NETWORK ANALYSIS

Social network analysis is an exciting new frontier in animal social behavior.[33-35] Social network analysis draws on models of communication networks used in telecommunications. Each animal is considered a *node* in the network, and social relationships are diagrammed as links among nodes. A famous hypothesis related to network analysis is that all humans are separated by six degrees, or links (Figure 13.11). (See also the discussion of this concept in the context of information theory in Chapter 7.)

Several measures of social structure come out of network analysis. The more central animals in the group, in terms of social interactions, have more than the average number of links with other animals. Likewise, loss of a central animal is more disruptive to information flow and more damaging to network structure than the loss of a peripheral animal. Networks can be fairly open, with each animal having an equal probability of linkage to others, or they can be cliquish, with subgroups that are relatively isolated from the overall network.[36]

> **KEY TERM** Social network analysis portrays social relationships as connections in a web and analyzes the effects of the network on communication and social behavior.

> **KEY TERM** A node in a social network is a single animal that is connected with other animals.

Within a network, not all animals have the same amount of information. The locations of potential predators, food, and shelter may be better known to some individuals than others. Information can flow only between linked animals, so animals removed by several links from information centers are at much greater risk. Often the group leader is the animal with the most information. This leads to the prediction that animals should compete to have close linkages with leaders and that leaders might manipulate information flow within a network to maintain their status.[37]

In human social networks, people are very tuned to discovering which network members are better informed.[38] Slight social cues, such as location within the group, allow members to identify information centers. Groups in which leaders have been identified to group members prior to performing a task do better than groups in which leaders are not identified, provided that the leader has information that is useful in the task.

Network analysis has considerable potential for answering questions about how public information is passed around in animal groups, how social exclusivity impacts survivorship, and how disease is transmitted via social contacts.

13.5 EXPLAINING COOPERATION

There are six possible evolutionary explanations for aid-giving behavior and cooperation that are reviewed in this section:

1. Kin selection
2. Reproductive skew
3. Group selection
4. Social contract models
5. Delayed competition
6. Stolen aid/coercion

These concepts are not mutually exclusive; many behaviors can be explained equally well by more than one of the concepts.

Kin Selection

Kin selection occurs when an animal behaves in ways that add to the reproduction of its close relatives. In such an instance, the animal is a donor of aid to the close relative, even if there is a fitness cost to giving the aid, provided that the donor has a net fitness gain from the transaction. That fitness gain comes from added reproduction of the recipient of the aid.

When William D. Hamilton published his theory of kin selection in 1964, it reshaped how biologists think of animal behavior. His theory forced biologists to see family relationships as a critical element of animal behavior. This, in turn, led to the application of molecular technology to animal behavior studies in order to accurately determine family relationships. In kin selection, the benefits of cooperation and altruism are directed to close relatives. Natural selection then favors the spread of genes for these behaviors. Kin selection can function in extended families or in populations in which related individuals may come into contact after dispersal from their families. Whether or not animals in a group are related is central to investigations of animals living in groups. Groups of unrelated animals are almost certainly held together by individual cost–benefit decisions. While watching a flock of birds, for example, stop and ask what benefits each animal in the group might be gaining, in comparison to those associated with being alone. What costs might each bird incur by being in the group? Does it seem, based on casual observation, that the benefits outweigh the costs? What variables might be measured to gain a more scientific assessment of the cost–benefit equation for group living? The answer often lies in kinship.[39–42] Recently, however, kin selection as an explanation for social behavior has come under attack, a controversy that is covered in an "Of Special Interest" section below.

Kinship is a key element of most contemporary field studies of social behavior in animals. Kin selection requires and is based on genetic relatedness between the donor and the recipient of the altruistic act. Kin selection is the dominant explanation for the evolution of aid-giving behavior. The other explanations are interesting and important, but are restricted to more limited circumstances. As pointed out above, the explanations for the evolution

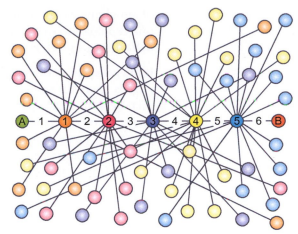

FIGURE 13.11
Six degrees of separation exist between animal A and animal B. Note that six comes from the number of links, not the number of individuals in the contact chain. For a message to reach animal B from animal A, it must pass through six links (five individuals).

437

of aid-giving behavior are not mutually exclusive and it would be misleading to argue that one explanation or another is the only possible mechanism for the evolution of aid-giving behavior in a given circumstance.

If the group is an extended family, or if the larger group can be subdivided into family groups, then kin selection can be an important dynamic in the costs and benefits of group membership. Kinship can be a powerful force in holding such groups together because all group members have a shared genetic interest in the survival of the group's offspring. Kinship is a part of the cost–benefit equation for some kinds of group living. However, even in groups of animals that have no close genetic relationships cost-benefit calculations can play an important role in shaping the evolution of behavior.

Careful consideration makes it clear that an animal has two ways of gaining fitness. The first, and more obvious, is to reproduce. The second is less obvious; an animal can gain fitness by aiding close relatives and adding to their reproduction. This is kin selection. Close relatives share genes, that is, have genes that are identical by descent. In fact, if a donor can act to increase a relative's reproduction enough, it may even benefit the donor to behave in ways that reduce its own reproduction. In the extreme case of the eusocial insects (termites, ants, bees, wasps, and a few other types of insect), workers give up their reproduction to assist their mother in raising her offspring.

Aid can be expressed in any currency, such as food or predation risk, as these ultimately translate into reproduction. What if the currency is restricted to that of reproduction, or fitness? If a person rescues a sibling from drowning, then she or he has protected that sibling's fitness. The rescuer has also gained fitness, because the genes shared by the rescuer and sibling will now be passed on to another generation, something that would not have happened if the sibling had drowned. This added reproduction of a close relative should count in an individual's fitness, but by how much?

The donor pays an evolutionary cost (C) for giving assistance. The cost is the risk involved in giving the aid times the donor's potential for future reproduction. If the donor has a 25% chance of dying when giving aid, and has the potential for producing offspring of its own in the future, then the cost to the donor is the 25% chance of dying times the potential future reproduction. The benefit (B) resulting from this aid is the added reproduction the recipient of the aid enjoys as a result of the donor's act. If the recipient was sure to die without aid, then all of its future reproduction is a benefit. In kin selection, we are also concerned about relatedness (r), the percentage of genes shared by the donor and recipient as a result of their common ancestry.

The basic principle of kin selection is captured in Hamilton's equation:

$$B/C > 1/r$$

this can be rearranged as:

$$rB > C$$

The elements of cost (C) and benefit (B) and relatedness (r) are as follows: The cost (C) is the lost potential fitness of the donor. The benefit (B) is the *added* fitness for the recipient due to the acts of the donor. The fundamental message of this equation is that aid-giving behavior by the donor should be favored in the course of evolution if the donor–recipient relatedness (r) times the added benefit to the recipient is greater than the cost to the donor.

Now apply the cost–benefit equation to the example of the drowning sibling:

C = the probability of the rescuer drowning in the rescue attempt
B = the subsequent reproduction of the sib
r = 0.50, the relatedness of rescuer and sib

Consider two scenarios. In both scenarios, each sibling will have two children in the future, if it survives. In the first, the risk of the rescuer drowning is small—say, 10%.

Substituting for rB yields $(0.5)(2) = 1$.
Substituting for C yields $(0.1)(2) = 0.2$.

Because $1 > 0.2$, Hamilton's inequality is met, and the plunge to rescue is worthwhile, evolutionarily speaking.

What if the risk of drowning is 0.9?

Doing the arithmetic reveals an alarming increase in C. Hamilton's inequality is not met, and if the world proceeds in accordance with Hamilton's Rule, then the equation predicts that the rescue would not be attempted.

Do animals make this cold-hearted calculation? Certainly not consciously, but evolutionary forces drive animals to behave in ways that are consistent with the equation. The rearranged form, $B/C > 1/r$, emphasizes the contrast between the benefit–cost ratio and the relationship between the two animals. To keep this straight, we refer to an animal's own reproduction as *classical fitness*; measures of fitness that include added reproduction of relatives due to aid-giving behavior are called *inclusive fitness*. Just to emphasize this point, inclusive fitness includes only the fitness benefits of the aid-giving behavior, not the entire reproduction of the relatives.

> **KEY TERM** Classical fitness is an animal's own reproductive output.

> **KEY TERM** Inclusive fitness is an animal's reproductive output, plus any added reproduction gained by relatives due to aid-giving by the first animal.

Ecological conditions can change the B/C ratio so that aid-giving is more or less likely; we discuss this principle in more detail later, but it also merits talking about in the context of kin selection. Suppose that an animal has virtually no chance of reproducing on its own. In this unhappy situation, there is practically no cost to aid-giving behavior; ecological conditions have already extracted the full price from the animal. If the cost is small, then small benefits, even from aiding more distantly related animals, will outweigh the costs. In studies of birds and mammals, often the low probability of successful dispersal of young animals weighs heavily in the likelihood that those animals will stay and help their parents, rather than attempting to start their own nest.

DISCUSSION POINT: KIN SELECTION

Kin selection theory predicts that you should be more likely to undertake a risky rescue, such as saving a drowning person from an icy pond, if that person is a close genetic relative. More creepy, given the choice between saving your child or your grandparent, theory predicts you should save your child because your child's potential future reproduction is higher. If forced to choose between your own child and someone else's, kin selection theory predicts that you would always choose your own. Do people actually behave as theory would predict? Why or why not? What is the role of culture in these decisions?

OF SPECIAL INTEREST: SOCIAL BEHAVIOR, ALTRUISM, AND KIN SELECTION: THE NEW CONTROVERSY

By the year 2000, most, but not all, experts on the biology of social behavior had come to accept kin selection as the most important explanatory hypothesis for a broad range of social behavior. Grouping into kinship units is the norm for many types of social affiliations, and strong evidence exists in many different animal species for the direction of helping behavior to relatives.

Despite the developing consensus around kin selection, a few biologists consistently expressed reservations about the hypothesis, its acceptance over other ideas, and the evidence supporting it. These included David Sloan Wilson, who had developed population models for the spread of altruistic genes under circumstances in which animals are helping nonrelatives. Famed sociobiologist Edward O. Wilson joined David Sloan Wilson in the critiques of kin selection and later collaborated with Martin Nowak in questioning the universal application of kin selection theory to understanding social behavior.[43] We invite you to read these papers and then to read counterarguments such as those put forward by Foster and colleagues and Crozier.[44–49] What do you conclude from this controversy?

Wilson and Wilson's models postulate that demes (sub-populations within a larger population) that contain animals with genes for helping will go extinct at lower rates than demes that lack the helping gene. In such models, the helping genes spread in the absence of kin selection. Special circumstances, such as small deme size, inexpensive helping, and relatively little movement among demes are required for these models to work. While the logic of the models for interdemic selection and the evolution of altruism is interesting, few data-based studies have uncovered animal populations that conform to the model. Like all of science, our understanding of social behavior is a work in progress, but at the moment, Hamiltonian thinking continues to carry the day.

Biologists have long known that under certain circumstances, delayed competition models can explain observed instances of cooperation, and that philopatry (discussed above) is a strong force in bringing animal groups together. To date, the current discussion has generated more heat than light but certainly should be provocative for classroom inquiry.

Reproductive Skew

Behavioral ecologists have developed a set of models, under the general rubric of "reproductive skew," that extend Hamilton's kin-selection models to events in hierarchical social groups characterized by differential reproduction. In such groups, the more highly ranked animals typically account for more than their share of the reproductive output of the social group.[50] In other words, reproductive skew occurs when animals within a social group differ in their reproductive potentials. In cases of skew within kin groups, skew can be viewed as a more modern formulation of kin selection. Skew also occurs when group members are unrelated. Reproductive skew theory attempts to be broad enough to explain aid-giving behavior under a very broad range of possible conditions.

In skew models, reproduction of a subordinate member of the group may actually be "allowed" as a kind of payment that induces the animal to remain in the group. This is called a "concession model of reproductive skew." In contrast, in "restraint" models, the subordinates control reproduction but can remain in the group only if the dominant allows it. The subordinates must "restrain" their reproduction, giving some reproductive rights to the dominant to be able to stay in the group. Finally, in a "tug-of-war" model of reproductive skew, group cohesion is driven by each individual's prospective maximum reproduction if it stays in the group, even though ultimately it would be impossible for all the members to maximize their reproduction. These models have primarily been tested in social insects, particularly paperwasps, in which a concession model seems to best describe worker behavior.[51] In this model, the continuing conflict, or tug of war, creates a situation in which the possible benefits of remaining in the group outweigh the potential reproductive benefits of leaving the group.[52]

All of this assumes that the aid-giver is making some kind of choice about its behavior. Instead, the choice has probably been made by natural selection over evolutionary time, and there is little room for an animal to assess its situation and respond. Older ideas in this field argued that if kin selection were driving the evolutionary process, then workers should cooperate in accepting the reproductives' role. If this were the case, then mechanisms that allow the reproductives to suppress the workers would not be expected. On the other hand, if the parental generation were coercing or stealing the labor of their offspring, then one would expect to find suppressive mechanisms, such as behavior dominance or pheromonal dominance. More recently, theorists have overlaid the concept of reproductive skew on this dichotomy, while at the same time recasting it slightly.[53]

Group Selection

Group selection is natural selection that acts at the level of the group, so individual survival depends on group survival. Group selection has an interesting history as a discredited theory that has come back to life in a very different form. In older literature or nonscientific representations of animal behavior, sometimes animals are said to behave "for the good of the species." For example, this is often given as a reason why nonhuman animals supposedly do not fight to the death. It is now clear that this is patently untrue: not only do animals never behave "for the good of the species," but there are many examples of animals of the same species fighting to the death. V. C. Wynne-Edwards attempted to argue for behavior that involved individual sacrifice for species-level benefit on scientific grounds;[54] his work in this area did not gain widespread acceptance at the time it was published and is still not generally accepted. Evolutionary thinking clearly posits that animals behave in ways that result in the passage of their genes to the next generation.[55]

There is another way of looking at group selection that has gained some credibility in recent years. In this type of group, or interdemic, selection, small subgroups within the larger population frequently go extinct. If a gene for altruism—behavior that helps the group to survive even if the behavior is disadvantageous for the carrier of the gene—arises in one of the subgroups, then that gene may be favored in evolution as subgroups without the gene will more rapidly go extinct. David Sloan Wilson, and more recently Edward O. Wilson, have championed this theory, but there are few convincing examples of altruism evolving by group selection in the field (Figure 13.12).[46] This is not surprising, given the strong negative

441

FIGURE 13.12
In this common model for group selection, a gene may arise for altruism in one of the groups. Individuals in the group do not reproduce as much because of the costs of the altruism, but the group has a higher probability of survival. In the illustration many of the other populations have gone extinct, even though reproduction was higher. This model is designed to explain the evolution of altruism via group selection. *Adapted from Wilson, D.S., Wilson, E.O., 2008. Evolution "for the good of the group". Am. Sci. 96, 380–389.*[46]

reaction to group selection as good-of-the-species thinking; new theoretical approaches may shift interest once again.[48] In sum, group selection might work if interdemic selection is high, but at this time this is viewed more as an interesting theory than a likely mechanism to explain the evolution of group living and helping within groups.

Social Contract Models for Cooperation

A social contract is an arrangement of trust, in which a donor gives something in return for a promise that the recipient will give back something of equal or greater value at a future date. A simple example of this is the Prisoner's Dilemma, discussed in detail in Chapter 7. Recall that in the Prisoner's Dilemma, both prisoners can cooperate to receive a small reward, or one can rat (defect) on the other and receive a large reward. If both defect, the result is catastrophic for both. Effectively, each promises to cooperate, forgoing the larger reward for defection in return for not being a victim of defection.

Put more broadly, aid given based on promises of future aid is *reciprocal altruism*. The prisoner's dilemma describes a simple, stable form of reciprocal altruism, which can persist in the absence of kinship. A large set of papers has been published on the theory of reciprocal altruism, and some arguments have been made that alarm calls could be an example of such behavior. The big question is enforcement of contract: How would animals be constrained by social conventions alone to reciprocate? Overall, the evidence is weak for reciprocal altruism outside of humans. Nonhuman animals seem to live much more "in the moment" than humans, and most, perhaps all, aid-giving behavior can be explained either by immediate returns or by kin selection. Recent overviews have argued that reciprocity is rare, perhaps absent, in nonhuman animals.[56–58]

Delayed Competition and Selfish Teamwork Models

In quite a few ant species, groups of unrelated queens establish a nest together. They lay eggs and their worker daughters build the nest, forage to feed more larvae, and tolerate each other's presence as the colony grows. The technical word for this behavior is *pleometrosis*. Sometimes these groups of queens are unrelated; in these cases, kin selection does not play a role in cooperation. The skew occurs in the production of the next generation of reproductives.[59]

Why participate in such an arrangement? The explanations are simple: if the queens are sisters, then kin selection can operate. If the queens are not sisters, then the persistence of such groups requires that the groups are much more likely to survive than individuals would be, and all of the members of the group start with a more or less equal chance of "winning." In this case, winning is the absence of reproductive skew (all queens reproduce equally) or the queen having reproductive skew in its favor. Working together in this case is "selfish teamwork" because each individual is in the group to promote its own reproduction. Part of what allows this to function is that the competition phase is delayed; the queens cooperate in colony growth by laying worker eggs and may later compete when the "real" reproduction involving the next generation of queens starts.

DISCUSSION POINT: SELFISH TEAMS AND THE *SURVIVOR* TV SHOW

Here's a quick summary of how the *Survivor* works, in case you've never watched. A group of 16–20 people are recruited to participate in an adventure television series. Participants cooperate, knowing cooperation is necessary for survival, but that only one will get a payoff at the end of the game. When the series begins, the participants are divided into two teams. In some seasons, the division is random; in other seasons, the division follows a theme, such as young versus old, male versus female, street-wise versus naïve. Participants are eliminated, by vote, one by one following team versus team competitions (each week the losing team votes out a member). When each team is down to a few members, the two teams are merged to form a final group; members of this final group then compete for individual advantage in the votes.

The striking result is that members of these randomly composed teams always cooperate and form alliances, and that the cooperation extends beyond the merger of the two teams. These team members have only superficial commonalities, but is there something about human psychology that causes us to see the value of working in teams, even if in the end there can be only one winner? Is this type of group behavior unique to humans, or do animals behave in similar ways?

Stolen Aid and Coerced Cooperation

The final model for aid-giving behavior is more insidious. What if the aid is coerced or stolen? This can clearly be the case in societies in which dominant animals have advantages over subordinates. Another major arena for this type of behavior is *parental manipulation*, in which parents can manipulate the reproductive choices of their offspring using nutritional dominance and social dominance to keep young animals in a pre-reproductive state. When combined with a lack of ecological choice, such as the lack of available territories, parental manipulation can be a powerful dynamic in promoting aid-giving behavior. In mammal societies in which dominant animals take advantage of their position, they manipulate social interactions so they gain food or shelter from subordinate animals.[57] This type of behavior has clear benefits to the dominant and clear costs to the subordinate.

The idea of parental manipulation was first developed in studies of primitively eusocial wasps and bees, in which the sterile workers are usually noticeably smaller than their mother, the queen.[60,61] They are smaller because the queen feeds them less than a potential reproductive will receive; this seems like a clear case of manipulation. Smaller workers then may be more easily physically dominated by their mother, and may also have lower reproductive potential if they were to attempt to start their own nests. This type of parental manipulation probably depends on workers deriving kin selection benefits from their aid-giving behavior;[45] it is an interesting model that needs further testing. Manipulation in parent–offspring relationships often occurs when food is at stake; refer to the discussions of begging and weaning conflict in Chapter 12 for more on this topic.

443

DISCUSSION POINT: CAN ALTRUISM EXIST?

At first glance, evolutionary theory predicts the complete absence of true altruism, and this unattractive idea might give you pause. You would not be alone. This leads to questions that we all should ask, and that inspire discussion extending far beyond the limits of this book: How should science affect personal behavior? Should science be constrained by its implications for that behavior? What are the roles of culture and learning in creating environments that favor altruistic behavior, and how subtle might those influences be? How well do we understand our own science as we create it? Note that this book is written from and consistent with a scientific worldview, a worldview that has served us remarkably well in many ways. Across time and space, humans have developed a variety of worldviews. Spend some time thinking about how worldviews create—and limit—perception.

13.6 EXTREME COOPERATION: EUSOCIALITY

Why in the world would an animal yield its reproductive potential to work as a nonreproductive, or sterile worker, in a colony? Put in evolutionary terms, it seems completely counterintuitive that selection would favor the loss of reproductive capacity. At the most extreme, nonreproductives, such as honeybee *workers*, can yield their lives in defense of their colony.

For Charles Darwin, the sterile workers found in termites, bees, ants, and wasps presented one of the largest barriers to his theories of evolution.

> **KEY TERM** A worker is a sterile or semisterile individual within a eusocial colony.

FIGURE 13.13
This stingless bee species, *Tetragonisca angustula*, is common in the neotropics. It builds large nests within hollow trees; the resinous structure pictured here contains the two entrances to this colony. *Photo: Michael Breed.*

Remember that species with sterile workers are called *eusocial*. Sterile workers of termites, ants, bees, and wasps represent extreme cooperation, in which an animal will give up some or all of its own reproductive potential to aid others. Recent discoveries add shrimp, aphids, thrips, and naked mole rats to the list of animals with sterile workers. Darwin, though, hinted at the solution: "This difficulty, though appearing insuperable, is lessened, or, as I believe, disappears, when it is remembered that selection may be applied to the family, as well as to the individual, and may thus gain the desired end."[62] The evolutionary underpinnings of extreme cooperation are examined next, followed by the diversity of animals that display aid-giving behavior at a surprising level (see Chapter 14).

Eusocial animals live in parent–offspring social groups in which the young are partially or fully sterile and aid their parents. In eusocial animals, the effort that the offspring might normally exert in parental care for their own young is redirected to care for siblings. These reproductively impaired animals are workers, and in many species the workers have very different physical characteristics than their reproductive parents. The definition of eusociality has been re-engineered in recent years to include animals such as thrips (small plant-eating insects), in which a semisterile generation cares for its own young. In fact, in primitively eusocial species, the founding generation of a colony is commonly relatively short-lived; the workers then take over, reproductively, for the founders. Primitively eusocial bees and wasps have workers that retain some reproductive potential. Eusociality differs from the helper-at-the-nest strategy discussed in Chapters 12 and 14 in the potential permanence of the reproductive state of the parents and their offspring.

In addition to having short-lived reproductives, primitively eusocial species, such as paperwasps, sweat bees, and bumblebees, have little physical differentiation between reproductives and workers and relatively small colonies. Highly eusocial species, such as honeybees, hornets, termites, and ants have workers that are well differentiated from the reproductives and tend to have large colonies (Figure 13.13). Highly eusocial species are ecologically dominant animals in nearly all terrestrial ecosystems.[63]

The big question here is why an animal appears to behave in ways that are extremely detrimental to itself for the apparent benefit of others. All of the explanations for the evolution of cooperation—kin selection, coercion and manipulation; reciprocal altruism; delayed competition; and reproductive skew theory—can be extended to provide as possible explanations for the evolution of eusociality.

As indicated, these hypotheses are not mutually exclusive, and it would be futile to look for a way to test among them. That said, this section focuses on kin selection explanations for the evolution of eusociality.

Diplodiploid Sex Determination

In most animals the receipt of sex-determining chromosomes from the parents determines the sex of offspring from a mating. In mammals, including humans, of course, the sex chromosomes have strikingly different appearance, so that one is called the X chromosome and the other the Y. XX animals are female, XY animals are male. In birds the situation is the opposite: Animals with the same sex chromosomes are male (ZZ), and those with WZ sex chromosomes are female. Other variations, such as X0, are found in some insects.

> **KEY TERM** In diplodiploid animals, both the male and the female are diploid, meaning that each has two sets of each chromosome, one received from its mother and one from its father.

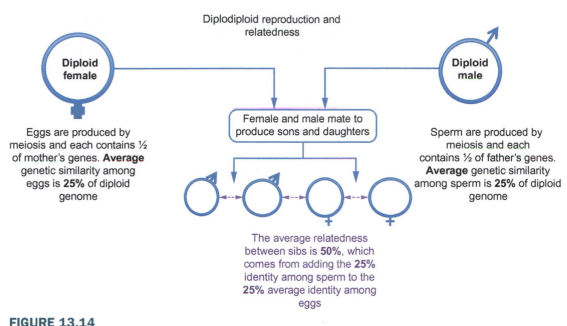

Diplodiploid reproduction and relatedness

Diploid female

Eggs are produced by meiosis and each contains ½ of mother's genes. **Average** genetic similarity among eggs is **25%** of diploid genome

Female and male mate to produce sons and daughters

Diploid male

Sperm are produced by meiosis and each contains ½ of father's genes. **Average** genetic similarity among sperm is **25%** of diploid genome

The average relatedness between sibs is **50%**, which comes from adding the **25%** identity among sperm to the **25%** average identity among eggs

FIGURE 13.14
Relatedness within a diplodiploid family. Mean relatedness between sibs is 0.5.

These modes of sex determination are called *diplodiploidy* because both males and females are diploid—they carry two sets of each chromosome. When sex is determined this way, the relationships among animals in a family are fairly straightforward. When unrelated parents mate, the resulting offspring, whether male or female, hold one-half parent-offspring ($r = 0.50$) of each of their parents' genes (Figure 13.14). While each offspring is related to a given parent by 50%, the amount of genetic material it shares with its sibling is variable, but on average, is 50%.

These proportions that represent average genetic identity are the r's used in Hamilton's equation. The second-generation offspring are related to their grandparents by one-fourth ($r = 0.25$), and so on. The relationships are symmetrical: gender does not change the relationship among individuals, either within a generation or between generations. Thus, in diplodiploid organisms, it is easy to calculate r among animals.

Haplodiploid Sex Determination

A less common mode of sex determination, *haplodiploidy*, has cropped up in some animals. In haplodiploid species the females have two sets of each of the chromosomes, but the males have only a single set. In other words, males are haploid and females are diploid. Males come from unfertilized eggs, so a female that has not mated still produces male offspring. The evolutionary history of this peculiar adaptation is murky, but haplodiploidy seems to have evolved most commonly in insect species that nest in bark and are exposed to *Wolbachia*, a bacteria that infects the ovaries of insects. In haplodiploid insects, the brothers are related to one another by one-half (0.5) because they develop from eggs that have only one set of each chromosome.

> **KEY TERM** In haplodiploid animals, one gender is haploid (one copy of each chromosome) and the other is diploid (two copies of each chromosome). In the Hymenoptera (bees, ants, and wasps), males are haploid, meaning they have no father; their single set of chromosomes comes from their mother.

Most importantly, a male haplodiploid insect produces sperm that are all identical, whereas a diplodiploid male produces sperm that vary genetically. In a haplodiploid male, there is only one copy of each chromosome, so meiosis and the accompanying recombination of genetic material cannot occur in males. Because a male's sperm are identical, the offspring of his matings all bear the identical genetic contribution from their father. In haplodiploids,

445

each sperm has 100% of the father's genes. Because males have only one chromosome, the father has only 50% of the genes of a female, that is, half of the "complete" diploid set of a female. The sperm in a haplodiploid contributes 50% of the genes to each offspring, just as in diplodiploids, but the key difference is that the sperm are all identical. Meanwhile, the contribution of the mother is the same in both haplodiploid and diplodiploid systems; each egg contains 50% of the mother's genes, and the eggs are on average 50% like each other. This yields the same proportions as diplodiploid gametes; the eggs contribute 50% of the genes to each offspring, and if the eggs come from the same mother, they are on average 50% alike (due to meiosis). So as in diplodiploid organisms, 0.5×0.5 ($50\% \times 50\%$) means that eggs are on average 25% alike ($r = 0.25$ when the mother's contribution alone is considered).

But what happens when the father's contribution is considered? All the sperm are genetically identical, so the average genetic similarity due to the father's contribution is $r = 50$. If we add the egg ($r = 0.25$, on average) to the sperm ($r = 0.50$), the average relatedness among sisters in haplodiploid species is $r = 0.75$! In haplodiploid species, sisters are more related to each other than to their mother or to their daughters.

In contrast, in diplodiploid organisms, the sons or daughters in monogamous families are, on average, 25% like their brothers and sisters due to genes from their mother's egg, and on average 25% like their brothers and sisters due to genes from their father's sperm. Summing the 25% average genetic similarity from the two parents shows that diplodiploid brothers and sisters are on average 50% alike genetically (Figure 13.15).

Haplodiploidy is not always necessary for the evolution of eusociality, but it seems to often prime the evolutionary pump. This conclusion is not without its detractors (e.g., Alpedrinha and colleagues[64]), but the majority of scientists working on eusocial insects at this point accept the idea that haplodiploidy is significant in the evolution of eusociality in those groups. Some species among sponge shrimp and naked mole rats as well as all termites have evolved eusociality but have diplodiploid sex determination. Some scientists argue that inbreeding, which raises the level of relatedness of animals within the inbred

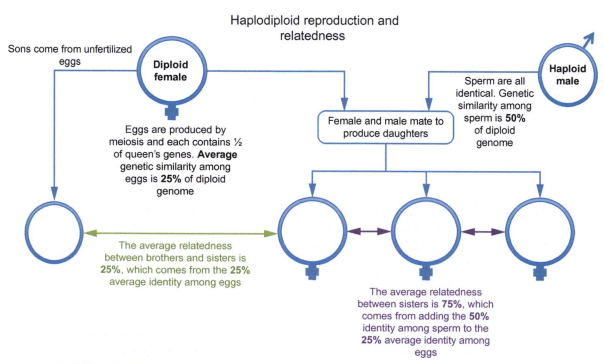

FIGURE 13.15
Relatednesses among relatives in a haplodiploid system. The mean relationship between full sisters is 0.75.

population, can create a set of circumstances that mirrors the genetic effects of haplodiploidy on relatedness. The most apparent common factor among sponge shrimp, termites, and naked mole rats is that they live in family groups in tunnel systems; this lifestyle constitutes a preadaptation for eusociality, a point that re-emerges in the next section.

This genetic reasoning about haplodiploidy and kin selection yields some intriguing predictions:

1. Haplodiploid eusocial colonies should consist of a mother (the queen) and her female offspring (the workers) working together to rear more daughters of the queen.
2. The queen should mate with only one male; if she mates with more than one male, the average relatedness among her daughters will be dramatically reduced.[65,66]
3. The genetic interests of the males do not favor working in the colony to help rear either sisters or nieces; males should not work.
4. There should be conflict between the queen and the workers over whose male offspring are reared. Females are more related to their sons than to either their brothers or their nephews.
5. If workers lay male eggs, then there should be conflict among the workers concerning which male eggs are actually reared. Because genetic relationships between workers and males favor first their own sons, then their brothers, and finally their nephews, workers may prevent each other from rearing worker-laid males; if this happens, then the queen wins the conflict over laying male eggs.

DISCUSSION POINT: E. O. WILSON

Beyond the argument over kin selection mentioned above you may wish to read more of E. O. Wilson's work, particularly his books. He is one of the most famous biologists of the twentieth century, having written numerous scientific and popular books on social insects, biological diversity, and the relationship between humans and nature. His book *Naturalist* should be required reading for those who want to be one (a naturalist). A famous, often repeated, quotation from Wilson captures his continued sense of wonder with nature: "Most children have a bug period. I never grew out of mine." Did you have a bug period? Have you grown out of it?

Preadaptations for Extreme Cooperation

Given these arguments about kin selection—and particularly with reference to haplodiploidy—is it reasonable to expect that all haplodiploid species will be eusocial and that many diplodiploid species might be eusocial as well? For eusociality to evolve, a number of pre-existing conditions need to be present. Most importantly, the species must have nesting and parental care with extended parent–offspring interactions. But nesting and parental care alone are apparently not sufficient for the evolution of eusociality; most bird species fit this requirement, yet eusociality is unknown among birds. Haplodiploidy, while not essential, is a major facilitator of eusocial evolution. The presence of parental abilities to physically, nutritionally, or chemically dominate their offspring can be preadaptations for the coercion of cooperation. Ecological pressure, in the form of predators, parasites, or seasonal unavailability can be powerful evolutionary factors that tip the balance so that young stay and help their parents. This last point sets the stage for further consideration of the effects of lack of choice on the evolution of helping.

13.7 LACK OF ECOLOGICAL CHOICE IN AID-GIVING DECISIONS

Ecological conditions that make dispersal risky are perhaps undervalued by scientists, but they are nonetheless important in aid-giving decisions. These conditions could range from soil that has become dry and difficult to excavate between the time the nest was founded

FIGURE 13.16
Honey bee guards defend their nest against intruders from other honey bee colonies. In this case a marked bee from another colony is being presented at the colony entrance, and a guard is poised to attack. *Photo: James Hanken.*

and the young are ready to disperse to a densely occupied habitat in which all the suitable territories are already occupied by nests. Overall, behavioral ecologists have grossly underestimated the effects of manipulation and ecological limitation of reproductive options on the evolution of cooperation. Ecological limitations enter the kin selection equation by reducing the cost of cooperation. (An animal with little reproductive options of its own stands to gain substantial inclusive fitness by working for a relative, but loses very little classical fitness by not testing its limited options.) The relative contributions of kin selection, ecological limitation, and manipulation to the evolution of extreme cooperation remain a very open question.

Anything that reduces the potential fitness of the offspring if they try to nest on their own increases the benefit they gain from helping. Similarly, parasites or predators that threaten either the parental nest or the nest that could be established by their young may create an evolutionary payoff for offspring that stay and protect the nest; group defense appears to have been an extremely important driving force in young staying in their parental nest. Factors that increase the benefit of helping in Hamilton's equation reduce the threshold required, in terms of genetic relationship between the aid-giver and the recipient, for natural selection to favor aid-giving.

Another often underestimated ecological factor in the evolution of extreme cooperation is the necessity for resource defense. In many primitively eusocial animals, the key element gained from the extreme altruists in the colony is improved colony defense. This is certainly the case in aphids, thrips, shrimp, and sweat bees, all of which face threats from parasites and predators, and all of which feature guards or soldiers in their social organization (Figure 13.16). Nesting tends to concentrate resources including food, young animals, and the adult colony members, and the nest itself can be an important resource. If the nest becomes an ecological magnet for animals that would take over the nest, steal the food, or eat the colony members, then selection for strong self-defense can easily lead to colony members—generally daughters in haplodiploid species—that specialize in defense.

13.8 SOCIAL RECOGNITION, KIN RECOGNITION, AND COOPERATION WITH CLOSE RELATIVES

How does an animal know where to direct its cooperative behavior? Cooperation often requires that an animal can identify the targets of its behavior. Take a step back and consider how animals might recognize each other. *Social recognition*, the ability to classify or individually recognize members of a social group, works on several levels:

1. *Species recognition* is the ability to discriminate between conspecifics and heterospecifics. This is particularly important in mating interactions but may play a role in other types of social behavior.
2. *Gender recognition* is the ability to discriminate male from female.
3. *Social group recognition* is the ability to discriminate members of one's own group (or social network) from members of other groups.
4. *Individual recognition* is the ability to identify other animals as individuals and to discriminate each individual in a social group from each other individual.[67,68]
5. *Kin recognition* is the ability to discriminate close kin from other animals in a population or to make distinctions among kin depending on their degree of relatedness.

Social recognition may evolve as a progression from the most general categorizations (same vs. different species, male vs. female) to successively narrow categories until finally individual recognition is achieved.[69,70] In many mammals, social behavior is largely structured around individual recognition, and kin recognition plays little, if any, role in shaping behavior. In all likelihood, the primacy of individual recognition over kin recognition is true for most birds and mammals. Dogs, for example, can clearly identify other dogs as individuals and use information from social encounters to shape future encounters; it is remarkable that dogs can generalize this ability to their interactions with humans, but, of course, they do.

There are two hugely important reasons for animals to recognize their kin. First, kin recognition is the glue that binds together the kin selection and reproductive skew explanations for cooperative behavior. For aid to be directed toward kin, animals must have ways of identifying kin or of discriminating kin from nonkin. Second, most animals avoid mating with close relatives because the genetic cost of inbreeding (inbreeding depression) is potentially very high. Kin are drawn together by helping dynamics but pushed apart by mating dynamics; in either case, a system for recognizing kin is essential for making correct decisions.[71–74]

What good, then, is kin recognition if it is just another relatively crude classification of animals in a social group? Kin recognition has the special property of being generalizable to previously unmet individuals; an animal that is unfamiliar can still be discriminated as kin or nonkin. This may be particularly important in social systems of animals with discrete litters; nonlittermate siblings can avoid mating only if there is a generalizable mechanism for discriminating kin. For animals that disperse from families in which the parents will have subsequent litters, the ability to classify others as kin or nonkin can be critical in preventing catastrophically inappropriate matings. Similarly aid-giving behavior, such as antipredator alarm calls, can be directed to closely related animals if kin recognition is used.

Most kin recognition relies on a simple mechanism called *phenotype matching*. In phenotype matching, an animal learns its own phenotypic characteristics or those of surrounding animals. This usually takes place when the animal is young and nearby animals are close kin—parents and siblings. The learned phenotypes can include odors, visual cues, vocalizations, or some combination of these. The learned characteristics are called a *template*, and whenever a new animal is met, its characteristics are compared with the template.

> **KEY TERM** Phenotype matching occurs when an animal learns its own phenotype, or the phenotype of animals around it, and then uses that information to classify previously unmet individuals.

449

Social Recognition in Clonal Invertebrates

Some sea anemone species engage in "wars." The anemones live in clones, with new members produced by budding of old members. When one clone comes into contact with another, the anemones use their stinging capability to "fight," and ultimately one clone comes to dominate an area. This is an example of how animals that are phylogenetically far removed from, say, mammals or birds can nonetheless show well-developed mechanisms of kin discrimination and use these mechanisms to exclude unrelated individuals from habitat patches. Highly polymorphic (hypervariable) genetic systems give members of clones distinctive signatures that differ from those of other clones.[75,76]

Social Recognition in Fish

Many fish species use phenotype matching in their social relationships.[77] Most studies have focused on inbreeding avoidance.[78–80] Salmonids (trout and salmon) may prefer the water occupied by sibs, even if they have had no prior contact with those sibs (Figure 13.17). Perhaps schooling is, as a generality, a nonrandom phenomenon involving sibs; much more research is needed to sort this out. Olsen and colleagues genotyped fish for major histocompatibility complex (MHC) loci and tested the fishes' water preferences, showing that water that had been exposed to sibling/MHC matches was more attractive to fish.[81] Mouthbreeding cichlids also use phenotype matching in kin recognition.[76,82]

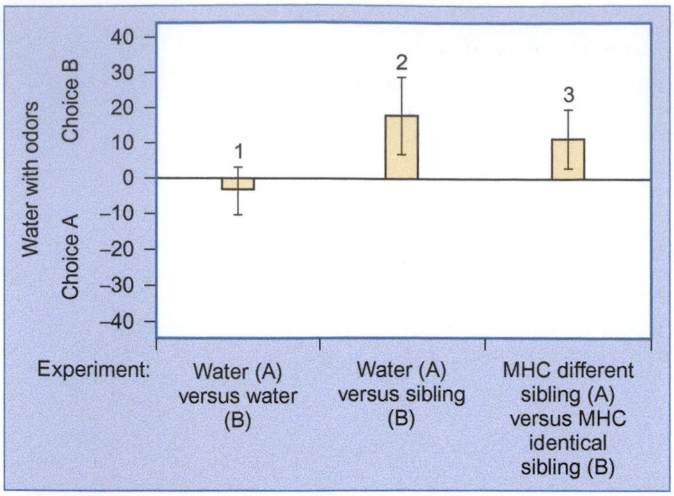

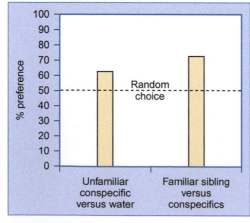

FIGURE 13.17

Preference for matching MHC genotypes in Arctic charr. Fish were allowed to choose between two water sources. The left column is a control, with untreated water for both choices. The middle column shows that the odor of sibling is preferred over untreated water, and the right column shows that siblings with similar MHC alleles to the test fish are preferred over a dissimilar sibling. *Adapted from Olsen, K.H., Grahn, M., Lohm, J., Langefors, A., 1998. MHC and kin discrimination in juvenile Arctic charr,* Salvelinus alpinus. *Anim. Behav. 56, 319–327.*[81]

Social Recognition in Amphibians

Amphibian kin recognition is very similar to that in fish. Many schooling frog and toad tadpoles are known to prefer to school with sibs and choose water that has been exposed to sibs, even if the sibs are not familiar to the individual doing the choosing (Figures 13.18 and 13.19). Much current interest in amphibian kin recognition is related to cannibalism (see p. 452).[84,85]

Social Recognition in Mammals

Mammalian social recognition systems seem to rely primarily on odors that reflect MHC variability, although some mammals (humans are a good example) use vocalizations and visual characteristics as well.[86,87] Most of the work on kin recognition in mammals has been done on rodents in which phenotype matching on MHC-related odors seems to be the rule, but there are also some interesting studies of primates.[88] Current studies on mammals focus largely on the role of the MHC[89] and on individual recognition as the primary method of social discrimination in carnivores, ungulates, and primates.[90]

DISCUSSION POINT: SOCIAL RECOGNITION

In a broad range of mammals, individual recognition has supplanted kin recognition as the key ingredient in social relationships. This is true of primates, hoofed animals such as horses, and carnivores such as dogs and cats. What are the evolutionary forces that favor individual recognition? How does this affect inbreeding avoidance, which remains an important factor in mate choice?

Social Recognition in Birds

Birds are unique in primarily using vocalizations rather than odor in kin recognition. It was thought for many years that birds were unlikely to have the song variability necessary for discrimination at the level needed to recognize kin, but Michael Beecher and his group led the way in showing that at least some species of birds can do exactly this.[91,92]

FIGURE 13.18

Preference of tadpoles for the odor in water of a conspecific versus untreated water (left column) and for the odor of a sibling versus a conspecific (right column). *Adapted from Waldman, B. 1985. Olfactory basis of kin recognition in toad tadpoles. J. Comp. Physiol. A. 156, 565–577.*[83]

FIGURE 13.19

Schooling tadpoles. *Photo: Jeff Mitton.*

Price and colleagues found that some of the wren's calls are learned from older male relatives and are specific to patrilineal families.[93] Palestis and Burger presented an interesting study that suggests a visual as well as an auditory component of recognition in common terns.[94] Individually distinctive calls are significant in social recognition in long-tailed tits, which breed in Europe and Asia.[95]

Social Recognition in Insects

Most studies of insect kin recognition have centered on one of two issues: avoidance of mating with close relatives and nestmate recognition in social insects.[96] Virtually all terrestrial organisms are covered with a mixture of hydrocarbons, and insects are no exception. The waxy or oily coating of hydrocarbons keeps water in, so the organism does not dessicate. Because the blend of hydrocarbons can vary without interfering with its antidessication function, it provides a perfect tool for identifying individuals or groups of animals. For instance, fruit flies, *Drosophila*, use cuticular hydrocarbons as signals in species recognition and in assortative mating.[97]

Nestmate discrimination in eusocial insects, including the honeybee, depends on odors that are blended among colony members, so that all colony members smell alike (Figure 13.20). In the honeybee, the wax comb in the nest plays an intermediary role; because all bees in the nest walk around on the same comb, they all smell the same.[98] In paperwasps the nest plays a similar intermediary role. Additionally, in at least some ants, the workers' postpharyngeal glands are critical in the development of a unified chemical recognition signal in the colony.[99]

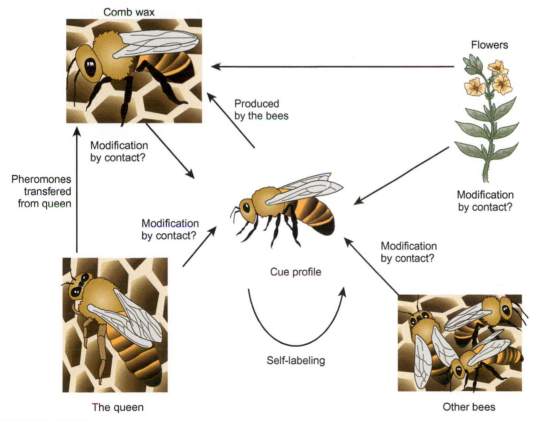

FIGURE 13.20
Nestmate recognition in honeybees. This figure illustrates proposed sources of a bee's chemical profile. A bee's cue profile could come from flowers (upper right), other bees in the colony (lower right), the queen (lower left), the comb in the colony (upper left), or the bee itself. Systematic testing of these possibilities shows that all the bees in a colony are labeled by walking on the comb. Fatty acids, which vary in concentration from colony to colony, are transferred from the comb to the bees. *Adapted from Breed, M.D., 1998. Recognition pheromones of the honey bee. Bioscience 48, 463–470.*[98]

FIGURE 13.21
Magpies on a warthog. The magpies are efficient at gleaning parasites and dead skin from the warthog, which tolerates their presence. *Photo: Michael Breed.*

Social Recognition and Cannibalism

Dining on relatives has a very real inclusive fitness cost, perhaps even higher than if they are also potential mates. Some salamanders (the tiger salamander) and toads (the spadefoot toad) have more than one tadpole morph; most tadpoles have normal, unappealing detritivorous amphibian feeding habits, but some are carnivores.[100] In the tiger salamanders, the frequency of the carnivore morph is dependent on the density of animals relative to the food supply, with the carnivore morph developing more frequently when food per animal is low.

13.9 SOCIAL SYMBIOSES

Symbioses are intimate relationships between pairs of species (Figure 13.21). One species lives in, on, or around the other so that the evolution of at least one of the species is affected. In *commensal relationships*, one species gains a benefit and the relationship is neutral for the other species. In *parasitism*, one species benefits to the other's detriment. In *mutualism*, both species benefit. The boundaries among all of these states and parasitism and predation can be a bit fuzzy; parasites generally do not cause the immediate death of their host, whereas predators kill their prey before consuming it. In addition, typical parasites are usually smaller than their hosts and may consume host resources—an indirect kind of robbery—rather than the host itself. Indeed, all of these categories are arbitrary; for instance, sometimes the "detriment" caused by a low level of parasite infection is immeasurably small and approaches commensalism. In other instances, one species of parasite can enhance resistance to another species; in that case, parasitism can rightly be called mutualism!

The focus here is on social parasitism. Brood parasitism in birds is covered in Chapter 12, and parasitism as a feeding strategy in Chapter 19. Social parasitism is yet a different type of exploitation. Social parasites take advantage of the benefits of the social group without paying any of the costs that come with membership. They use counterfeit signals or subvert the recognition system of the society to gain entrance and then take advantage of the food, shelter, or protection from predators that members of the group enjoy. Social parasites can come from the same species as the host—intraspecific social parasitism—or from another species—intraspecific social parasitism. These are insect examples, mostly ants, because they are the best-known cases of social parasitism.

Intraspecific Social Parasitism

Given the keys to the kingdom, in the form of appropriate cues for social recognition and behavior, why not use them? At least metaphorically, this is an interesting question for any member of a social species. Parasitizing a social group of the same species should be relatively easy because the host and parasite share the same signaling system, and likely the only barrier to be overcome is the host's social recognition system. Social wasp queens sometimes usurp an existing queen, taking over her nest. The larval workers, which are daughters of the first queen, then end up working for the second queen.[101] The barriers to this kind of behavior are relatively low, only requiring that the workers' labor can be harnessed by the second queen.

Interspecific Social Parasitism

SOCIAL PARASITISM BY CLOSELY RELATED SPECIES

Sometimes a social parasite is from a species that is closely related to its host. This relationship confers many, perhaps all, of the tools by which conspecifics can sometimes be parasites. Particularly interesting are two kinds of bees, *Psithyrus* and *Sphecodes*, which are parasites on other closely related bees. Both parasite genera exist only as reproductives (no workers) and lack structures for carrying pollen; to reproduce, these females must find a host nest. The *Psithyrus* species parasitizes bumblebees, which are always eusocial, by entering the nest before the first workers have emerged from their pupae. The *Psithyrus* queen kills the bumblebee queen and takes over the nest. The workers rear her eggs into larvae and ultimately to adults. The key to making this work is that the bumblebee workers treat the *Psithyrus* queen and her larvae as if they belonged in the colony. In *Sphecodes* the female sneaks into a nest and lays eggs that are reared by the resident bees (Figure 13.22). In some cases, the host species is social; in other cases, the host nest consists only of a reproductive female. (A host nest that consists of only a reproductive female is called a *solitary* species, as opposed to a eusocial one.)

Among ants there are numerous examples of species that, in the course of evolution, have lost their worker caste and depend on gaining entry into colonies of other species, where they lay eggs that are reared by the host species. In most cases, the parasite presents a "blank slate" in terms of social recognition cues and adopts the cues of the host colony as it gains acceptance. (See Chapter 12 for a discussion of nest parasitism in birds.)

> **KEY TERM** A symbiosis is a nonpredatory relationship between two species.

> **KEY TERM** A commensal relationship benefits one species and is neutral for the other.

> **KEY TERM** Parasitism is beneficial to one species and detrimental to the other.

> **KEY TERM** Mutualisms are beneficial to both species in an interspecific relationship.

> **KEY TERM** Myrmecophiles are guests in ant nests that take advantage of the food, shelter, and defense provided by the ants.

> **KEY TERM** Dulosis, or slave making, involves capture of a brood from one ant species and the use of that brood as the workforce in the other species' colony.

453

SLAVE MAKING IN ANTS

Another parasitic lifestyle that requires evasion of social recognition is *dulosis*, or *slave making*. Dulotic ants steal larvae and pupae from other colonies and bring them into their own colony, where they become part of the workforce. This requires integration of social recognition signals so the workers of the victimized species can work alongside the host species. Ants in the genus *Polyergus* are dulotic parasites on *Formica* ants; both genera are common in Europe and North America. A stream of medium-sized black or reddish ants carrying ant larvae and pupae to their nest is most likely an act of dulosis.

MYRMECOPHILES

Myrmecophiles are completely unrelated nonsocial species, such as beetles, butterflies, or mites that gain acceptance into an ant colony and benefit from being fed by their hosts. To enter and live unscathed in their host colony, they must be either blank slates or mimics in terms of the social recognition cues they present.

SUMMARY

Animals live in groups primarily for shared protection from predators, to cooperate in discovering and hunting food, and to take advantage of public information. The simplest social groups are often families, consisting of a mother with her

FIGURE 13.22
A *Sphecodes* bee. Photo: Jessica Lawrence, Eurofins Agroscience Services, Bugwood.org.

offspring and sometimes the father. One way of building larger social groups is by extending family groupings to include more offspring or multiple generations. Another way to obtain a larger group is to have unrelated individuals come together for the benefit of at least some group members.

Herds, flocks, and schools assemble for a variety of reasons, including protection, cooperation in foraging, and public information. Hamilton's theory of the geometry of a selfish herd explains why animals can come together in such groups even if they have no evolutionary interest in the survival of other members of the group. This reasoning explains groupings that include only animals of one species as well as groups that include two or more species.

For many groups the motivations for living together are more complicated and involve aid-giving among animals in the group. This chapter puts forward a series of ideas—group selection, kin selection transactional and social contract models, reproductive skew, delayed competition/selfish teamwork, and stolen aid—as explanations for living in groups. A key concept is that kinship structure—the web of genetic relationships among animals in such groups—is often important in determining cooperative and competitive interactions within groups. Sometimes groups consist of two or more species. Mutual benefits from the association, exploitation of public information, or parasitism underlie these types of social groups.

Extreme cooperation, or eusociality, in which some group members give up most or all of their own reproduction in order to maximize aid given to other group members is known in kin groups. Kin selection, parental manipulation, and ecological limitations on dispersal all play key roles in the evolution of eusociality. Asymmetric genetic relationships resulting from haplodiploid sex determination set the stage for eusociality; most but not all eusocial species are haplodiploid. Eusocial groups gain efficiency from dividing the labor among colony members so that the group has an advantage over solitary animals that might compete for the same resource.

Social recognition is key to formation and maintenance of many social groups. Social recognition can include recognition by classification, individual recognition, and kin recognition. Because kin groups often form the basis for social behavior, kin recognition, the ability to discriminate close kin from other animals in the population, is an important mechanism known in a variety of animals, from anemones to mammals.

STUDY QUESTIONS

1. How do kin selection and parental manipulation differ as explanations for the evolution of helping behavior?
2. What is phenotype matching? Design a simple experiment that would determine if a species is capable of phenotype matching.
3. Why are the workers in ant, bee, and wasp colonies almost always females?
4. Why might amphibians and fish school in sibling groups? What are some tests of this hypothesis?
5. What are the potential fitness costs and benefits to an individual who "helps at the nest"? Discuss how the various evolutionary models (kin selection, parental manipulation, reciprocal altruism, selfish teams) might aid in explaining helping behavior in young but potentially reproductive animals. How do you think these evolutionary models interact with proximate ecological forces, such as nest site availability and food availability?
6. "Selfish herds" could form in a variety of contexts, such as flocks or herds of migrating animals, feeding aggregations, and mating aggregations. But there may be other evolutionary dynamics that cause animals to group together. If you observe a herd, flock, or aggregation of animals, what alternative hypotheses for their grouping behavior should you consider? How could you test between the selfish herd hypothesis and the alternatives?

Further Reading

Axelrod, R., 1985. The Evolution of Cooperation. Basic Books, New York, NY, 241 pp.

Costa, J.T., 2010. Social evolution in 'other' insects and arachnids. In: Breed, M.D., Moore, J. (Eds.), Encyclopedia of Animal Behavior, vol. 3, Academic Press, Oxford, pp. 231–241.

Linksvayer, T.A., 2010. Subsociality and the evolution of eusociality. In: Breed, M.D., Moore, J. (Eds.), Encyclopedia of Animal Behavior, vol. 3, Academic Press, Oxford, pp. 358–362.

Queller, D.C., 2010. Kin selection and relatedness. In: Breed, M.D., Moore, J. (Eds.), Encyclopedia of Animal Behavior, vol. 2, Academic Press, Oxford, pp. 247–252.

Notes

1. Wilson, E.O., 1975. Sociobiology: The New Synthesis. Harvard University Press, Cambridge, MA.
2. Hamilton, W.D., 1964. The genetical evolution of social behaviour. I, II. J. Theor. Biol. 7, 1–52; Hamilton, W.D., 1967. Extraordinary sex ratios. Science 156, 477–488; Hamilton, W.D., 1971. Selection of selfish and altruistic behavior in some extreme models. In: Eisenberg, J.F., Sillon, W.S. (Eds.), Man and Beast: Comparative Social Behavior Smithsonian Institution Press, Washington, DC, pp. 59–91.
3. Trivers, R.L., 1971. The evolution of reciprocal altruism. Quart. Rev. Biol. 46, 35–57.
   Wilkerson, G.S., 1984. Reciprocal food sharing in the vampire bat. Nature 308, 181–184.
4. Williams, G.C. (Ed.), 1971. Group Selection Aldine-Atherton, Chicago, IL.
5. McCreery, H.F., Breed, M.D., 2014. Cooperative transport in ants: a review of proximate mechanisms. Insect. Soc. 61, 99–110.
6. Czaczkes, T.J., Ratnieks, F.L.W., 2013. Cooperative transport in ants (Hymenoptera: Formicidae) and elsewhere. Myrmecol. News 18, 1–11.
7. Sarker, O.F., Dahl Torbjorn, S., Arcaute, E., et al., 2014. Local interactions over global broadcasts for improved task allocation in self-organized multi-robot systems. Robot. Auton. Syst. 6, 1453–1462.
8. Rubenstein, M., Cornejo, A., Nagpal, R., 2014. Programmable self-assembly in a thousand-robot swarm. Science 345, 795–799.
9. Strassmann, J.E., Zhu, Y., Quellar, D.C., 2000. Altruism and social cheating in the social amoeba *Dictyostelium discoideum*. Nature 408, 965–967.
10. Clobert, J., McAdam, A.G., Corrigan, G., Alonzo, S., Lancaster, L., Chaine, A., et al., 2006. Self-recognition, color signals and cycles of greenbeard mutualism and altruism. Proc. Natl. Acad. Sci. USA 103, 7372–7377.
11. Wyatt, G.A.K., West, S.A., Gardner, A., 2013. Can natural selection favour altruism between species? J. Evol. Biol. 26, 1854–1865.
12. Axelrod, R., Hamilton, W.D., 1981. The evolution of cooperation. Science 211 (4489), 1390–1396.
    Hamilton, W.D., 1971. Selection of selfish and altruistic behavior in some extreme models. In: Eisenberg, J.F., Sillon, W.S. (Eds.), Man and Beast: Comparative Social Behavior Smithsonian Institution Press, Washington, DC, pp. 59–91.
    Hamilton, W.D., Zuk, M., 1982. Heritable true fitness and bright birds: a role for parasites? Science 218, 384–387.
    Hamilton, W.D., May, R.M., 1977. Dispersal in stable habitats. Nature 269, 578–581.
    Hamilton, W.D., 1964. The evolution of social behavior I. J. Theor. Biol. 7, 1–16.
    Hamilton, W.D., 1964. The evolution of social behavior II. J. Theor. Biol. 7, 17–52.
    Hamilton, W.D., 1967. Extraordinary sex ratios. Science 156, 477–488.
    Hamilton, W.D., 1971. Geometry for the selfish herd. J. Theor. Biol. 31, 295–311.; Hamilton, W.D., Zuk, M., 1982. Heritable true fitness and bright birds: a role for parasites? Science 218, 384–387.
    Hamilton, W.D., Axelrod, R., Tanese, R., 1990. Sexual reproduction as an adaptation to resist parasites (a review). Proc. Nat. Acad. Sci. USA 87 (9), 3566–3573.
13. Creel, S., Winnie Jr., J.A., 2005. Responses of elk herd size to fine-scale spatial and temporal variation in the risk of predation by wolves. Anim. Behav. 69, 1181–1189.
14. South, J.M., Pruett-Jones, S., 2000. Patterns of flock size, diet, and vigilance of naturalized Monk Parakeets in Hyde Park, Chicago. Condor 102 (4), 848–854.
15. Tringa-Tetanus Cresswell, W., 1994. Flocking is an effective anti-predation strategy in redshanks. Anim. Behav. 47 (2), 433–442.
16. Green, E., 1987. Individuals in an osprey colony discriminate between high and low quality information. Nature 329, 239–241.
17. Mann, O., Kiflawi, M., 2014. Social foraging with partial (public) information. J. Theor. Biol. 359, 112–119.
18. Cortes-Avizanda, A., Jovani, R.A., Donazar, J., et al., 2014. Bird sky networks: How do avian scavengers use social information to find carrion? Ecology 95, 1799–1808.
19. Charnov, E.L., Krebs, J.R., 1975. Evolution of alarm calls—altruism or manipulation. Am. Nat. 109, 107–112.
20. Van Horn, R.C., Engh, A.L., Scribner, K.R., Funk, S.M., Holekamp, K.E., 2004. Behavioural structuring of relatedness in the spotted hyena (*Crocuta crocuta*) suggests direct fitness benefits of clan-level cooperation. Mol. Ecol. 13, 449–458.
21. Lima, S.L., Bednekoff, P.A., 1999. Back to the basics of antipredatory vigilance: can nonvigilant animals detect attack? Anim. Behav. 58, 537–543.

455

22. van der Post, D.J., de Weerd, H., Verbrugge, R., et al., 2013. A novel mechanism for a survival advantage of vigilant individuals in groups. Am. Nat. 182, 682–688.

23. Janson, C.H., Goldsmith, M.L., 1995. Predicting group size in primates: foraging costs and predation risks. Behav. Ecol. 6, 326–336.

24. Blumstein, D.T., Evans, C.S., Daniel, J.C., 1999. An experimental study of behavioural group size effects in Tammar Wallabies, *Macropus eugenii*. Anim. Behav. 58, 351–360.

25. Grand, T.C., Dill, L.M., 1999. The effect of group size on the foraging behaviour of juvenile coho salmon: reduction of predation risk or increased competition? Anim. Behav. 58, 443–451.

26. Pollard, K.A., Blumstein, D.T., 2008. Time allocation and the evolution of group size. Anim. Behav. 76, 1683–1699.

27. Koenig, W.D., 1981. Reproductive success, group size, and the evolution of cooperative breeding in the acorn woodpecker. Am. Nat. 117, 421–443.

28. Bergman, T.J., Beehner, J.C., Cheney, C.L., Seyfarth, R.M., 2003. Hierarchical classification by rank and kinship in baboons. Science 302, 1234–1236.

29. O'Neil, S.T., Warren, J.M., Takekawa, J.Y., et al., 2014. Behavioural cues surpass habitat factors in explaining prebreeding resource selection by a migratory diving duck. Anim. Behav. 90, 21–29.

30. Honek, A., Martinkova, Z., Pekar, S., 2007. Aggregation characteristics of three species of Coccinellidae (Coleoptera) at hibernation sites. Eur. J. Entomol. 104, 51–56.

31. Nalepa, C.A., Kidd, K.A., Ahlstrom, K.R., 1996. Biology of *Harmonia axyridis* (Coleoptera: Coccinellidae) in winter aggregations. Ann. Entomol. Soc. Am. 89, 681–685.

32. Koch, R.L., 2003. The multicolored Asian lady beetle, *Harmonia axyridis*: a review of its biology, uses in biological control, and non-target impacts. J. Insect Sci. 3, 32.

33. Wey, T., Blumstein, D.T., Shen, W., Jordán, F., 2008. Social network analysis of animal behaviour: a promising tool for the study of sociality. Anim. Behav. 75, 333–344.

34. Croft, D.P., James, R., Krause, J., 2008. Exploring Animal Social Networks. Princeton University Press, Princeton, NJ, 208 pp.

35. Whitehead, H., 2008. Analyzing Animal Societies: Quantitative Methods for Vertebrate Social Analysis. University of Chicago Press, Chicago, IL, 320 pp.

36. Farine, D.R., Aplin, L.M., Garroway, C.J., et al., 2014. Collective decision making and social interaction rules in mixed-species flocks of songbirds. Anim. Behav. 95, 173–182.

37. Verdolin, J.L., Traud, A.L., Dunn, R.R., 2014. Key players and hierarchical organization of prairie dog social networks. Ecol. Complex. 19, 140–147.

38. Faria, J.J., Dyer, J.R.G., Tosh, C.R., Krause, J., 2010. Leadership and social information use in human crowds. Anim. Behav. 79, 895–901.

39. Bourke, A.F.G., 2014. Hamilton's rule and the causes of social evolution. Philos. Trans. R. Soc. B Biol. Sci. 369 article number 20130362.

40. Queller, D.C., 2011. Expanded social fitness and Hamilton's rule for kin, kith, and kind. Proc. Natl. Acad. Sci. USA 108, 10792–10799.

41. Costa, J.T., 2013. Hamiltonian inclusive fitness: a fitter fitness concept. Biol. Lett. 9 article number: UNSP 20130335.

42. Hatchwell, B.J., Gullett, P.R., Adams, M.J., 2014. Helping in cooperatively breeding long-tailed tits: a test of Hamilton's rule. Philos. Trans. R. Soc. B Biol. Sci. 369 article number: 20130565.

43. Allen, B., Nowak, M.A., Wilson, E.O., 2013. Limitations of inclusive fitness. Proc. Natl. Acad. Sci. USA 110, 20135–20139.

44. Foster, K.R., Wenseleers, T., Ratnieks, F.L.W., 2006. Kin selection is the key to altruism. Trends Ecol. Evol. 21, 57–60.

45. Crozier, R.H., 2008. Advanced eusociality, kin selection and male haploidy. Aust. J. Entomol. 47, 2–8.

46. Wilson, D.S., Wilson, E.O., 2008. Evolution "for the good of the group." Am. Sci. 96, 380–389.

47. Wilson, D.S., Wilson, E.O., 2007. Survival of the selfless. New Sci., 42–46.

48. Wilson, D.S., Wilson, E.O., 2007. Rethinking the theoretical foundation of sociobiology. Q. Rev. Biol. 82, 327–348.

49. Abbot, P., Abe, J., Alcock, J., et al., 2011. Inclusive fitness theory and eusociality. Nature 471, E1–E4.

50. Johnstone, R.A., 2000. Models of reproductive skew: a review and synthesis. Ethology 106, 5–26.

51. Reeve, H.K., Keller, L., 2001. Tests of reproductive-skew models in social insects. Annu. Rev. Entomol. 46, 347–385.

52. Scott, M.P., Lee, W.J., van der Reijden, E.D., 2007. The frequency and fitness consequences of communal breeding in a natural population of burying beetles: a test of reproductive skew. Ecol. Entomol. 32 (6), 651–661.

53. Young, C., Majolo, B., Schuelke, O., et al., 2014. Male social bonds and rank predict supporter selection in cooperative aggression in wild Barbary macaques. Anim. Behav. 95, 23–32.

54. Wynne-Edwards, V.C., 1962. Animal Dispersion in Relation to Social Behavior. Oliver & Boyd, London; Wynne-Edwards, V.C., 1986. Evolution Through Group Selection. Blackwell Scientific, Oxford, 340 pp.

55. Dawkins, R., 1976. The Selfish Gene. Oxford University Press, Oxford, 352 pp.

56. Hauser, M., McAuliffe, K., Blake, P.R., 2009. Evolving the ingredients for reciprocity and spite. Philos. Trans. R. Soc. Lond. B, Biol. Sci. 364, 3255–3266.

57. Clutton-Brock, T., 2009. Cooperation between non-kin in animal societies. Nature 462, 51–57.

58. Andre, J.-B., 2014. Mechanistic constraints and the unlikely evolution of reciprocal cooperation. J. Evol. Biol. 27, 784–795.

59. Bourke, A.F.G., Heinze, J., 1994. The ecology of communal breeding: the case of multiple-queen leptothoracine ants. Philos. Trans. R. Soc. (Lond) Series B 345, 359–372.

60. Liebig, J., Monnin, T., Turillazzii, S., 2005. Direct assessment of queen quality and lack of worker suppression in a paper wasp. Proc. R. Soc. B 272, 1339–1344.

61. Nonacs, P., 2006. Nepotism and brood reliability in the suppression of worker reproduction in the eusocial Hymenoptera. Biol. Lett. 2, 577–579.

62. Darwin, C., 1872. The origin of species by the means of natural selection. Chapter 8. Available online at <http://www.classicreader.com/book/107/59/>.

63. Wilson, E.O., 1971. The Insect Societies. Belknap Press of Harvard University Press, Cambridge, MA.

64. Alpedrinha, J., Gardner, A., West, S.A., 2014. Haplodiploidy and the evolution of eusociality: worker revolution. Am. Nat. 184, 303–317.

65. Cole, B.J., 1983. Multiple mating and the evolution of social behavior in the Hymenoptera. Behav. Ecol. Sociobiol. 12, 191–201.

66. Sherman, P.W., Seeley, T.D., Reeve, H.K., 1988. Parasites, pathogens and polyandry in social Hymenoptera. Am. Nat. 131, 602–610.

67. Sheehan, M.J., Straub, M., Tibbetts, E.A., 2014. Ethology. How does individual recognition evolve? Comparing responses to identity information in *Polistes* species with and without individual recognition. Ethology, 169–179.

68. Sheehan, M.J., Tibbetts, E.A., 2011. Specialized face learning is associated with individual recognition in paper wasps. Science 334, 1272–1275.

69. Tibbets, E.A., Dale, J., 2007. Individual recognition: it is good to be different. Tree 22, 529–537.

70. Breed, M.D., 2014. Kin and nestmate recognition: the influence of W.D. Hamilton on 50 years of research. Anim. Behav. 92, 271–279.

71. Lacy, R.C., Sherman, P.W., 1983. Kin recognition by phenotype matching. Am. Nat. 121, 489–512.

72. Solomon, N., 1991. Current indirect fitness benefits associated with philopatry in juvenile prairie voles. Behav. Ecol. Sociobiol. 29, 277–282.

73. Sun, L., Muller-Schwarze, D., 1997. Sibling recognition in the beaver: a field test for phenotype matching. Anim. Behav. 54, 493–502.

74. Mateo, J.M., 2003. Recognition in ground squirrels and other rodents. J. Mammal. 84, 1163–1181.

75. Hart, M.W., Grosberg, R.K., 1999. Kin interactions in a colonial hydrozoan (*Hydractinia symbiolongicarpus*): population structure on a mobile landscape. Evolution 53, 793–805.

76. Kamel, S.J., Grosberg, R.K., 2013. Kinship and the evolution of social behaviours in the sea. Biol. Lett. 9, 20130454.

77. Thuenken, T., Bakker, T.C.M., Baldauf, S.A., 2014. "Armpit effect" in an African cichlid fish: self-referent kin recognition in mating decisions of male *Pelvicachromis taeniatus*. Behav. Ecol. Sociobiol. 68, 99–104.

78. Gerlach, G., Lysiak, N., 2006. Kin recognition and inbreeding avoidance in zebrafish, *Danio rerio*, is based on phenotype matching. Anim. Behav. 71, 1371–1377.

79. Ward, A.J.W., Webster, M.M., Hart, P.J.B., 2007. Social recognition in wild fish populations. Proc. R. Soc. B Biol. Sci. 274, 1071–1077.

80. Hain, T., Neff, B., 2009. Promiscuity drives self-referent kin recognition. Curr. Biol. 16, 1807–1811.

81. Olsen, K.H., Grahn, M., Lohm, J., 2002. Influence of MHC on sibling discrimination in Arctic char, *Salvelinus alpinus* (L.). J. Chem. Ecol. 28, 783–795; Olsen, K.H., Grahn, M., Lohm, J., Langefors, A., 1998. MHC and kin discrimination in juvenile Arctic charr, *Salvelinus alpinus*. Anim. Behav. 56, 319–327.

82. Le Vin, A.L., Mable, B.K., Arnold, K.E., 2010. Kin recognition via phenotype matching in a cooperatively breeding cichlid, *Neolamprologus pulcher*. Anim. Behav. 79, 1109–1114.

83. Waldman, B., 1985. Olfactory basis of kin recognition in toad tadpoles. J. Comp. Physiol. A. 156, 565–577.

84. Blaustein, A.R., Waldman, B., 1992. Kin recognition in anuran amphibians. Anim. Behav. 44, 207–221.

85. Waldman, B., 1987. Mechanisms of kin recognition. J. Theor. Biol. 128, 159–185.

86. Heth, G., Todrank, J., Johnston, R.E., 1998. Kin recognition in golden hamsters: evidence for phenotype matching. Anim. Behav. 56, 409–417.

87. Ehman, K.D., Scott, M.E., 2001. Urinary odour preferences of MHC congenic female mice, *Mus domesticus*: implications for kin recognition and detection of parasitized males. Anim. Behav. 62, 781–789.

88. Yamazaki, K., Beauchamp, G.K., Curran, M., Bard, J., Boyse, E.A., 2000. Parent–progeny recognition as a function of MHC odortype identity. Proc. Natl. Acad. Sci. USA 97, 10500–10502.

89. Penn, D.J., Potts, W.K., 1999. The evolution of mating preferences and major histocompatibility complex genes. Am. Nat. 153, 145–164.

90. Holekamp, K.E., Boydston, E.E., Szykman, M., Graham, I., Nutt, K.J., Birch, S., et al., 1999. Vocal recognition in the spotted hyaena and its possible implications regarding the evolution of intelligence. Anim. Behav. 58, 383–395.

91. Beecher, M.D., 1990. The evolution of parent-offspring recognition in swallows. In: Dewsbury, D.A. (Ed.), Contemporary Issues in Comparative Psychology Sinauer, Sunderland, MA, pp. 360–380.

92. Medvin, M.B., Beecher, M.D., 1986. Parent–offspring recognition in the barn swallow (*Hirundo rustica*). Anim. Behav. 34, 1627–1639.

93. Price, J.L., 1999. Recognition of family-specific calls in stripe-backed wrens. Anim. Behav. 57, 483–492.

94. Palestis, B.G., Burger, J., 1999. Individual sibling recognition in experimental broods of common tern chicks. Anim. Behav. 58, 375–381.

95. Sharp, S.P., Hatchwell, B.J., 2005. Individuality in the contact calls of cooperatively breeding long-tailed tits (*Aegithalos caudatus*). Behaviour 142, 1559–1575.

96. Starks, P.T.B., Fischer, D.J., Watson, R.E., Melikian, G.L., Nath, S.D., 1998. Context-dependent nestmate discrimination in the paper wasp, *Polistes dominulus*: a critical test of the optimal acceptance threshold model. Anim. Behav. 56, 449–458.

97. Speiss, E.B., 1987. Discrimination among prospective mates in *Drosophila*. In: Fletcher, D.J.C., Michener, C.D. (Eds.), Kin Recognition in Animals John Wiley & Sons, Chichester, pp. 75–119.

98. Breed, M.D., 1998. Recognition pheromones of the honey bee. Bioscience 48, 463–470.

99. Soroker, V., Vienne, C., Hefetz, A., Nowbahari, E., 1994. The postpharyngeal gland as a gestalt organ for nestmate recognition in the ant *Cataglyphis niger*. Naturwissenschaften 81, 510–513.

100. Pfennig, D.W., Reeve, H.K., Sherman, P.W., 1993. Kin recognition and cannibalism in spadefoot toad tadpoles. Anim. Behav. 46, 87–94.

101. Klahn, J., 1988. Intraspecific comb usurpation in the social wasp *Polistes fuscatus*. Behav. Ecol. Sociobiol. 23, 1–8.

Comparative Social Behavior

LEARNING OBJECTIVES

Studying this chapter should provide you with the knowledge to:

- Apply your understanding of theories of social evolution developed in Chapter 13 to social systems of birds, mammals, and invertebrates.
- Comprehend division of labor in colonies of eusocial animals and the contribution of division of labor to the dominant ecological roles of eusocial animals.
- Compare the methods by which reproductive differences between queens and workers in eusocial insect colonies are maintained.

Animal Behavior. DOI: http://dx.doi.org/10.1016/B978-0-12-801532-2.00014-3

- Appraise the novel and highly effective defensive strategies of eusocial insects.
- Analyze the breadth of ecological roles and behavioral adaptations observed in eusocial animals.

14.1 INTRODUCTION

In this chapter prepare to delight in the broad sweep of social behavior, beginning with vertebrate societies and continuing to social invertebrates. The discussion builds on the simple groups such as herds, flocks, and schools discussed in Chapter 13; in this chapter, the focus is on more complex types of cooperation. It is easy to cloak social behavior in such a thick theoretical blanket that appreciation of its extraordinary nature all but disappears. This chapter highlights some outstanding examples of social behavior, such as swallows, lions, naked mole rats, honeybees, and leafcutter ants. Feel free to skip ahead to these fun pieces and then return to the theoretical layer—perhaps with renewed enthusiasm!

First consider that membership in a group has both benefits and costs (see Chapter 13 for more on this). Oftentimes a reduction in reproduction forms the major expense of group membership. Loss of reproductive value can be temporary, as when a subadult animal remains in its family group, or it can be permanent, as in workers in eusocial colonies (Figure 14.1). The benefits come from access to food, shelter, group defense, and sometimes from the possibility that the animal will "inherit" a resource owned by the group. Dominance hierarchies can play a key role in determining which animals gain access to reproduction as they mature; there is considerable social complexity beyond simple dominance relationships that makes the study of social groups interesting and challenging.[1] Much of the challenge of studying social behavior lies in assessing the costs and benefits of group membership.

This cost–benefit approach provides a common theme through social systems that otherwise might seem very different from one another. For instance, what could a worker army ant have in common with a lioness? The answer is that natural selection works with costs and benefits

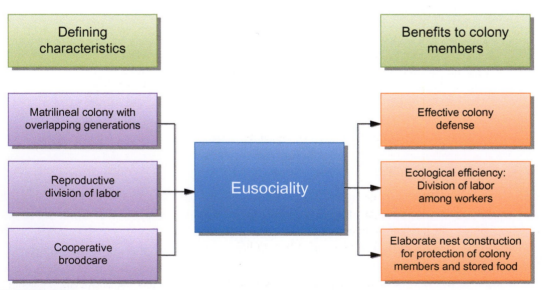

FIGURE 14.1
Eusociality, defined by the characteristics on the left, provides strong ecological benefits to colonies, so that eusocial organisms such as ants, wasps, bees, and termites, are among the most dominant animals in most terrestrial ecosystems.

in both species. The hypotheses presented in Chapter 13 apply to how social behavior evolves in any species, albeit with outcomes that can seem vastly different. An understanding of the costs and potential benefits of sociality in a species allows the development of hypotheses about the evolution of that species' social behavior.

Behavioral differences among group members are one of the major focal points in the current study of sociality. These differences are often subtle, require many hours of observation to document, and can be very difficult to interpret. Having a diversity of behavioral syndromes, or personalities, in a social group can result in better group functioning, as each animal can specialize in tasks to which it is suited.

Another theme that ties social groups together is ecological efficiency. Animals in social groups, working together effectively by coordinating their roles, have distinct advantages over nonsocial animals. Sometimes this efficiency is fairly obvious, as when a pack of wolves dines on a herbivore that would have been an impossible victim for a lone wolf. Other instances, such as very small leafcutter ants riding on leaves carried by larger ants, may initially be puzzling. The efficiency gained from this seemingly lazy behavior becomes apparent only when a parasitic fly arrives and the small ant drives it away.

CASE STUDY: LEAFCUTTER ANTS AND THEIR FUNGUS GARDENS

Leafcutter ants are herbivorous; their diet consists of pieces of leaf cut from living plants and carried back to their nest. As you probably know, a completely leaf-based diet presents nutritional challenges because it is low in nitrogen and high in undigestible cellulose. To be able to survive on this diet, leafcutter ants depend on symbiotic fungi; they feed their "fungus garden" the pieces of leaf and later eat parts of the fungus. Leafcutter colonies can be huge—more than a million workers—and have a highly elaborate system for division of labor.

Wilson's detailed analysis of the behavior of leafcutter ant workers shows how size and shape can be well matched with task.[2] Leafcutter ants build immense underground nests occupied by a single queen and hundreds of thousands or millions of workers. Leafcutters are agriculturalists;[3] they grow fungus gardens and live by eating the fungus. To support the fungi, they harvest leaves from plants, usually trees, in the area surrounding the colony. Processing the leaves involves a series of workers, starting with large-jawed, large-headed ants called *majors*, whose task is to cut semicircular chunks of leaf from the tree. From there, the leaf piece is passed to a medium-sized worker, who carries it, parasol-like, back to the nest (Figure 14.2). Within the nest, even smaller workers process the leaf into the fungus garden. Each task is performed by workers that are well matched to the demands of the task. In some ant species, the workers with the largest heads are soldiers, active in colony defense and territorial defense, while in leafcutting ants the large-headed workers use their jaws to cut leaves for the colony.[4]

How do leafcutter ants start their fungus garden? As you might guess, a randomly chosen fungus from the environment will not do. In fact, leafcutters are co-adapted with particular kinds of basidomycete (mushroom-producing) fungi. When a new queen

leaves to found a colony, she takes a bit of fungus and stores it in a pocket in her oral cavity. When she digs a new nest, she starts the fungus growing by feeding it glandular secretions; her first worker offspring then start collecting leaves to feed to the fungus. One of the challenges of fungus gardening is keeping unwanted fungi and microorganisms from invading the garden. Leafcutters have an interesting array of behavioral and antimicrobial adaptations to prevent disease and "weeds" from invading.[5]

461

FIGURE 14.2
Leafcutter ants cut pieces of leaves from living plants and carry them to their nest, where they are used to grow fungus gardens. Different-sized ants perform different tasks in this effort, forming the ant equivalent of a "bucket brigade" in their efforts to collect leaves.
Inset Photo: Scott Bauer, USDA-ARS.

On the other hand, *social conflict* over allocation of resources or access to reproduction within a group can easily outweigh efficiency gained from division of labor, communication, and protection from predators. Among the animals discussed in this chapter, wild horses are a particularly good example of groups that persist despite very costly conflict. The importance of this balance between conflict and cooperation always needs to be evaluated in analyses of social behavior.

The complexity and sophistication of the behavior captured in this chapter represent the culmination of all of the mechanisms and adaptations presented in this textbook. Although our coverage of humans in this chapter is fairly minimal, the behavioral principles developed here set the stage for thinking about how human social behavior fits into the larger picture of behavioral evolution and adaptation. As you proceed through this chapter, reflect on how behavioral genetics, learning, communication, cognition, mating systems, and parental care all coalesce when animals become social.

> **KEY TERM** Social conflict occurs when members of a social group don't benefit equally from an outcome.

14.2 VERTEBRATE SOCIAL SYSTEMS

This section highlights some examples of social behavior in birds and mammals. The evolutionary pathways to these two major vertebrate groups diverged early, and social evolution proceeded independently in the two groups. Recall from our discussion of mating behavior in Chapter 11 that most bird species are socially monogamous, whereas mammals are often polygynous. This difference alone appears to make large social groups based on extended families less common among birds than mammals.

Also important are the ecological effects of dispersal choices (Chapters 8 and 13) on bird and mammal sociality. If nesting sites or territories are in short supply, then young animals may face a choice between near certain failure in independent life and continued life with their family group. Added to this are the benefits gained by social groups in foraging and defense.

Three bird species—cliff swallows, acorn woodpeckers, and sociable weavers—and four mammal species—wild horses, hyenas, lions, and baboons—are highlighted here; they are a small, but representative, sample of the range of social behavior in birds and mammals.

FIGURE 14.3
Wild turkey (*Meleagris gallopavo*) males, such as the bird on the left in this photograph, display cooperatively in order to attract females. *Photo: Robert Burton, USFWS, Public domain.*

Males as Cooperators

It might be tempting to think of females as more likely to cooperate with one another in nesting, parental care, and other aspects of reproduction, but males also often participate in cooperative groups. Diaz-Munoz and his collaborators list nearly 90 vertebrate species, ranging from fish to mammals,[6] which are part of reproductively cooperating groups. Styles of cooperation include:

- Display (found in some birds and fish). This occurs when males displaying together are more likely to attract females than if males are isolated. Some lekking birds, such as wild turkeys (Figure 14.3), exemplify cooperative displays.
- Coalitions (most common form of cooperative reproduction in male mammals). Males form groups to cooperate in combat for resources against other groups of males. Lions are an example of this type of cooperation.
- Mutual tolerance and mate sharing (males share a mate and perhaps a territory but do not live in a social group). Striped hyenas seem to fit this model.
- Cooperative polyandry (males working together to rear young; most common form of cooperative reproduction in male birds). For instance, acorn woodpeckers live in groups

in which females may have more than one mate. In the most extreme form, such as tamarins, several males cooperate to aid a single female in raising her young.
- Cooperative parental care (e.g., subordinate males helping within a larger social group). In meerkats, for example, males that join social groups often take on this role.

In some instances, kin selection explains these types of cooperation, as when a group of related males forms a coalition. In other instances, males may work together in the short term in order to protect their individual long-term reproductive interests.

Swallows

Swallows are small insect-eating birds; of the several common species (barn, cave, tree, bank, and cliff), the focus here is on the barn swallow. The barn swallow is widely distributed in Europe and North America and the males are dark iridescent blue with a light-colored underside and a rusty colored band under their chin. Barn swallows construct their nests using mud; the cup-shaped structure protects a brood of two to seven eggs. They are socially monogamous, with both the male and the female contributing to care of the brood. Male ornaments, particularly their elongated tails, play a role in mate choice. Some extrapair copulations take place and, depending on the study population, up to 30–40% of the offspring may be sired by a male from outside the pair bond.

Among the bird species highlighted here, swallows are the least social, but they exemplify the level of sociality observed in many bird species. Barn swallows are opportunistically social; they often live in colonies ranging from a few to several dozen nests (Figure 14.4). This aggregated nesting (see Chapter 13) stems largely from mutual attraction to suitable nesting habitat and perhaps from some shared advantage in vigilance in watching for predators. Predators on barn swallows include hawks and owls, as well as mammals such as raccoons and weasels. Within a colony, nesting pairs are territorial, and there is no evidence of cooperative nest construction or brood care. Colony members must be able to identify their own young using calls so that their parental care is directed to the correct nestlings.

463

Sociable Weavers

Sociable weavers are small, relatively nondescript, tan-colored birds found in the deserts of southern Africa. Sociable weavers primarily eat insects, diet items that fluctuate considerably in abundance along with rainfall in the region. They are unique among birds in cooperatively constructing a single large nest that houses up to 100 nesting chambers.

FIGURE 14.4
Left: A barn swallow nest, showing an adult bird provisioning her young. Many of these nests may occur within the same barn. Right: A cliff swallow, which also nests in aggregations. What evolutionary forces cause animals to nest in aggregations? When this type of sociality is observed the question of benefit, such as increased defenses against predators, versus cost, such as easier passage of disease and parasites among animals, must be considered. *Photos: Matt Wilkins (left) and Ben Pless (right).*

FIGURE 14.5
Sociable weaver colony and birds. Dozens or hundreds of individual nests are within the colony, where birds have come together for thermoregulatory benefits. *Photo: Kenneth M. Gale.*

Cooperation extends to brood care. *Helpers* in colonies are young adults that lack a nest and have no young of their own; they forage and feed nestlings belonging to their parents. Helping may even extend to cooperative brood care for unrelated chicks. Up to nine such helpers can be associated with a single nest, so a colony may include several hundred adult birds, many more than two birds per nesting chamber.

While in many species of cooperatively breeding animals shared defense adds value to nesting together, sociable weavers show little sign of being able to defend their nests. They gather around falcons that approach, but do not mob them. Venomous snakes, such as cobras and boomslangs, are common in this habitat and are attracted more to larger colonies of sociable weavers, so that a snake may eat all of the young birds in a nesting chamber within the colony in a single meal.[7]

> **KEY TERM** A helper is an animal that can reproduce but instead remains with its family group and aids the group in foraging and defense, or plays other roles that benefit the group.

Desert habitats can be extremely hot during the day and equally cold at night, and the Kalahari is no exception. The large nest provides shelter from searing and freezing desert temperatures (Figure 14.5); the birds thermoregulate the structure so that the maximum temperature inside is 30°C and the minimum is 15°C. The main benefit of this cooperation probably comes from protection from cold temperatures; heat production is shared across all the birds in the colony, and the large nest provides insulation from the cold. The value of social behavior for this species, then, comes from shared thermoregulation of the nest rather than working together to defend the colony against predators.[8,9]

Acorn Woodpeckers

Acorn woodpeckers depend on large dead trees for their survival; Walter Koenig and his students[10] noted that the woodpeckers store acorns in their tree and use a larger cavity in the tree for their nest. Such trees can be a relatively rare resource in this bird's native habitat, the oak scrubland and forests of the arid southwest, including parts of Arizona and New Mexico, California, and much of Mexico. In addition to these large dead trees, acorn woodpeckers obviously require the presence of oak trees as a source of their primary food, acorns. Acorn abundance varies greatly from year to year. In a good year, oak trees "mast" with all the trees in a population, producing copious acorns; in a bad year, acorn production is low for all trees. Acorn woodpeckers store thousands of acorns, each in its own little niche carved into the surface of the trunk of their "granary" tree.

The value of the granary tree and the likelihood that a dispersing bird will fail as it seeks to find and establish a new granary are probably the primary pressures leading to social

FIGURE 14.6
Acorn woodpecker (left) and tree with cached acorns (right). The tree is a valuable resource that can be inherited by succeeding generations of birds. *Photos: Gary Kramer, US Fish and Wildlife Service (left); Tupper Ansel Blake, US Fish and Wildlife Service (right).*

behavior in the acorn woodpecker. A granary tree may have several breeding males and females, along with subadult helpers; as many as 10–15 birds, plus any nestlings present, can occupy the tree. Both males and females serve as helpers. All eggs are laid in the single nest cavity (Figure 14.6), and females may destroy each other's eggs. Reproduction is fairly evenly divided among female egg-layers but is typically dominated by one male from among the reproductive males.

Helpers could be viewed as waiting to fill a vacancy among the reproductives. However, unrelated birds from outside the group usually fill vacancies, even if a vacancy occurs when a helper is present that is the same sex as the vanished bird. This prevents inbreeding and also highlights the complexities of the benefits of helping. In this species, cooperation results in the rearing of nestlings that are related to the helper, supporting the idea that kin selection, combined with lack of ecological choice, are important dynamics in this system.

465

BRINGING ANIMAL BEHAVIOR HOME: WILD AND DOMESTIC HORSES

Domestic horses derive from smaller ancestor species found in the Americas. They are part of a larger evolutionary radiation of species in the genus *Equus* that includes zebras, donkeys, and asses. The species that gave rise to our domestic horses, *Equus caballus*, went extinct in the Americas between 12,000 and 10,000 years ago and was reintroduced as a domestic animal by Europeans in the 1500s. (The conservation of Przewalski's horse, an endangered subspecies of *E. caballus* found in Mongolia, is discussed in detail in Chapter 15.) Horses soon escaped and formed the foundation for wild horses, which were once numerous in North America.

Wild horses are now reduced to a smaller number of bands (Figure 14.7). They have a harem social system, in which a number of mares affiliate with a dominant stallion. In some larger bands, up to five stallions are present, and the band is then subdivided according to the stallion with which each mare is affiliated. While bands move within a familiar home range, they are not territorial, and more than one band may occupy the same general area. In general, grazing herbivores are not territorial, because their resource is widespread and difficult—perhaps even pointless—to defend.

Much of the motivation for membership in bands is reproductive; stallions and mares form long-term associations. For mares, one of the benefits of consorting with a stallion is protection from aggressive harassment by other stallions. Multistallion bands may be best explained by subordinate stallions grouping around a dominant stallion for potential future reproductive benefit, even though the cost of dominance interactions is high. Bands also benefit from public information about location of food and water resources, and from shared vigilance and defense against predators.[11]

FIGURE 14.7
Herds of wild horses like this are common in the western United States. Dominant stallions ardently defend their harem against other stallions and predators. *Photo: US Department of Interior, Bureau of Land Management.*

Domestic horse behavior reflects the fact that their ancestors were wild, and pastured domestic horses form bands and display combat among stallions. A stallion kept in a stall becomes formidable when the scent of a mare in season is in the air. Horses are highly social animals, and maintaining domestic horses in herds of mares and geldings is a key component of the welfare of these beautiful animals.

466

Hyenas

There are actually three species of hyena: brown, striped, and spotted (Figure 14.8). The spotted hyena has the most unusual social behavior of the three. (By the way, hyenas are not closely related to dogs; they are a distinct group within the carnivores.)

Spotted hyenas are highly effective and dangerous predators, and their social behavior has evolved in parallel with a shift from the scavenging habits of their relatives to predation. Perhaps because competition for meat from prey is so high, there is a premium on social aggressiveness in this species. Among newborns, competition for dominance status is quite important, and adult dominance status is determined by a combination of competition among sibs and their mother's status. Hyenas typically have either one or two cubs, and competition for milk between twin cubs can be intense. Female cubs have a competitive advantage over male cubs, being more likely to be dominant and thus more likely to survive.[12] Spotted hyena females have masculinized genitalia, with a large clitoris that looks like a penis. The canal of the clitoris serves for both copulation and, one might imagine, very painful birth. The masculization of the genitalia is associated with high testosterone levels during development; the testosterone is key to both behavior and morphology.

FIGURE 14.8
A pack of spotted hyenas at a kill. *Photo: Kay E. Holekamp.*

Spotted hyenas live in clans consisting of a half dozen to several dozen adult animals, plus their offspring, so that several generations can occur together in a clan. Social structure within the clan is driven by the females, which are organized into a linear dominance hierarchy. Clans fight for territory, and females remain in their clan for life; males can change clans, a behavior that reduces inbreeding. Females communally nurse the young of the clan.

The social structure of hyenas has been characterized as *fission–fusion*, in that small groups may at times break off from the larger clan (fission) and then later rejoin (fusion). This social structure is similar to that of African elephants. Fission and fusion represent a balance between the benefits of the social group (cooperative hunting, defense, and broodcare) and the possibility that food may be too scarce to support a large group.[13,14]

> **KEY TERM** Fission–fusion societies are large groups of animals that at times are grouped together (fusion) and at other times split into smaller groups (fission). Fusion may occur when resources are abundant and fission when resources are scarce.

Lions

A core group of up to 15 females, usually mothers, daughters, sisters, and nieces forms the foundation of a lion pride. They hunt together and cooperatively raise their young (Figure 14.9). One or a few males associate with the pride. The males participate in pride defense, but not in hunting. Because the females are a kin group, kin selection likely plays a role in the cohesiveness of the group and in cooperative behavior. Prides are territorial, and the ability to hold a feeding territory is key to the pride's survival. Some lionesses are more effective at hunting, group defense, and territoriality; these lionesses are termed *leaders*. *Laggards* are lionesses that share less equally in the tasks necessary for the pride's survival.

467

FIGURE 14.9
A pair of lions. A pride typically consists of a male and female and their female offspring.
Photo: Michael Breed.

A lion pride is a closed social group, with outside males and females being excluded by pride members. As male cubs approach adulthood, they are expelled from the pride. Prides undergo a fission–fusion process like that of hyenas, in which pride size adjusts according to environmental conditions.

Solo males have low survivorship due largely to their poor ability to obtain food, and males may form bachelor coalitions. These coalitions improve male survivorship, and ultimately a male coalition may take over a pride, driving the resident males away. Lionesses cooperate to defend their young against infanticide; cubs are particularly at risk when a new coalition of males takes over a pride.[15-18]

Lions are relatively unique among members of the cat family, the Felidae, in their habit of living in social groups. Other felids are generally solitary, only coming together as adults to mate. Domestic cats illustrate an interesting crossroads between solitary and social behavior. Domestic cats have, like lions, mutual tolerance and can live in large groups, but unlike lions they show no signs of social cooperation in hunting or group defense. As with dogs, domestication has heightened cat responses to bonding and affiliative behaviors, making them good household companions. Lions and domestic cats show how evolutionary flexibility can lead to social systems that are unlike those of these species' closest relatives.

Chimpanzees

Chimpanzee social behavior and ecology have been extensively studied, fascinating the public because of the close evolutionary relationship between chimpanzees and humans.[19,20] Chimpanzee social behavior varies among populations, and intriguing behavior, such as tool use in food collection, may be seen in some locations but not in others.

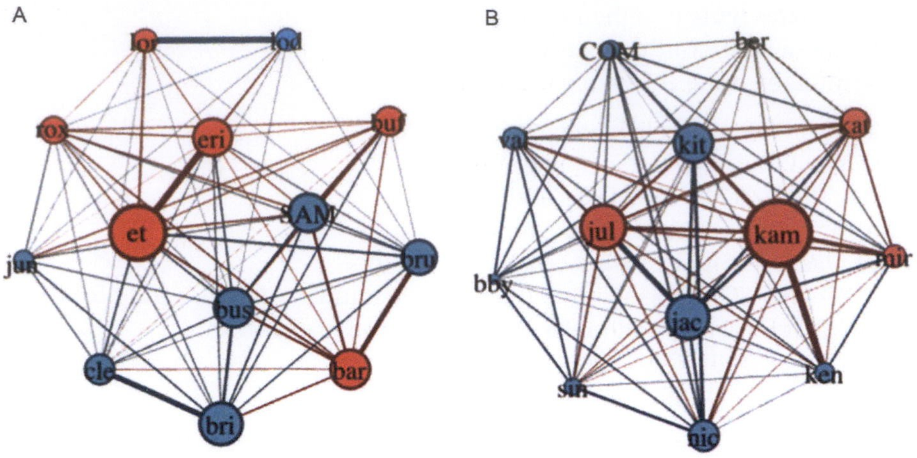

FIGURE 14.10

Social network diagrams of two chimpanzee social groups. Males are indicated by blue circles and females by red circles. The dominant male in each group is indicated by capital letters. The strength of the social connection between individuals is shown by the width of the line between the pair, and more connected animals are at the center of the network. The nature of a chimpanzee society depends on the personality of the dominant male, which has more of a central position β in network A. *Reprinted from Cronin, K.A., van Leeuwen, E.J.C., Vreeman, V., Haun, D.B.M., 2014. Population-level variability in the social climates of four chimpanzee societies. Evol. Human Behav. 35, 389–396.*[21]

FIGURE 14.11

Chimpanzees showing social grooming. Cooperative behaviors like this provide opportunities for reciprocity.
Photo: John Mitani.

The pattern of fission–fusion social groups that occurs in hyenas and lions remains a theme for chimpanzee social organization. Chimpanzees live in loosely organized social groups of 20 to well over 100 animals (Figure 14.10). Smaller subgroups derived from the community are called *parties*. Males form strong coalitions with other males and tend to stick with their subgroup through fission–fusion processes. Females are more often loners, often moving on their own or with their infants (Figure 14.11).

Male chimpanzees usually remain in their community, whereas females are likely to disperse to other communities. In this way, chimpanzees are unlike some of the other social systems in this text. Males within a community are likely to be related, whereas females are less likely to be related. Chimpanzee social behavior in large part reflects social relationships among animals in the group and is less structured by kinship than by histories of interactions between animals.[22,23] Males have a strong dominance hierarchy; mating is nonetheless promiscuous, and estrous females mate with many males. Dominance hierarchies are not so apparent among females, but female dominance status influences reproductive success.[24] Social network studies (Figure 14.10) show the tightly knit nature of chimpanzee groups. Given the focus on social relationships in chimpanzee communities, it is not surprising that investigators have observed reciprocity in trading for "political support" within the community.[25] Chimpanzees that have low political status or have violated reciprocity may suffer violent consequences.[26]

DISCUSSION POINT: BONOBOS

The bonobo is closely related to the chimpanzee but much less studied. You might want to do a little research on bonobo social systems and compare them with the better-studied chimpanzee. A good place to start is the 1998 article by Craig Stanford in *Current Anthropology*, "The Social Behavior of Chimpanzees and Bonobos: Empirical Evidence and Shifting Assumptions." Frans de Waal's book, *The Bonobo and the Atheist: In Search of Humanism Among the Primate*, W. W. Norton, 2013, is also a great starting spot for a discussion of comparative social systems in apes and morality.

Eusocial Mammals

Permanently sterile worker castes occur in only two species of mammal: the naked mole rat (Figure 14.12) and the Damaraland mole rat.[27-31] These burrowing rodents live in colonies of 50–100 individuals in southern Africa. The colony has one female reproductive, the queen, and a small number (up to three) of male reproductives. The queen, who can live 15 years or more, has a longer body than the workers, allowing her to support more or less continuous pregnancies. Internally, the queen's vertebrae are lengthened; the lumbar vertebrae grow during pregnancies so that the queen's repeated pregnancies result in elongation of her spine.[32] The rest of the colony members, daughters or sisters of the queen, serve as nonreproductive workers.

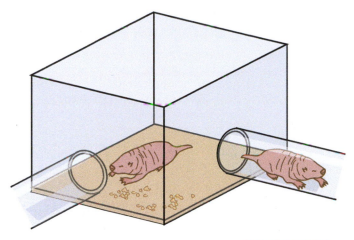

FIGURE 14.12
Naked mole rats are often maintained in zoo exhibits because of their unusual social structure. In addition to being eusocial, they have evolved to live in underground burrow systems, losing most of their hair and the functionality of their eyes in the process.

The queen uses physical aggression to establish her dominance in the colony, and the colonial dominance hierarchy is linear, with a clear ranking of the colony members. Queen removal experiments show that the loss of a queen is followed by intense competition, with the emerging dominant female assuming the reproductive role in the colony.[33] This might suggest that nonreproductive members of the society are simply repressed, but nonreproductive naked mole rats respond to oxytocin treatment with greater bonding and affiliative behaviors, suggesting that part of the underpinning of cooperation in this species is the hormonal framework for bonding shared by all mammals.[34]

Male workers have similarly suppressed reproductive capacities and, interestingly, can rapidly mature reproductively if removed from the social environment of the colony. Naked mole rats are decidedly atypical rodents, with the longest lifespan of any rodent and an inability to regulate body temperature, making them effectively cold-blooded. Worker mole rats are divided into two subsets: workers who work intensively through their lifespan and workers who work less frequently. These workers which work less have the physiological capacity to grow into queens, which they can do either by superseding the old queen in their colony or by dispersing to establish a new colony.

Two things set the stage for the emergence of eusociality in mole rats: colonial living and the high value of the tunnel system (and corresponding low probability of dispersal success for young). As in acorn woodpeckers with their tree, the tunnel system is an inherited high value resource and the importance of the tunnel system to the survival of the naked mole rats provides a strong impetus for social cooperation. Longevity clearly plays a key role in facilitating the existence of a complex society because rapid turnover of reproductives in the colony would disrupt both the social and genetic structures of the colony. Whether the unusual longevity of naked mole rats is a cause or a consequence of the evolution of eusociality is an interesting question.

Young adults in many bird and mammal species work as "helpers at the nest" prior to dispersing and establishing their own nests (or taking over the parental nest) (see also Chapter 12). Eusociality in mammals, such as the mole rats, can be seen as an extension of the ecological and evolutionary factors that lead to temporary helping in extended families in many mammals. The diplodiploidy of mammals shows that the asymmetry of relationships found in haplodiploid species is not essential for the evolution of eusociality.

In sum, social behavior in birds and mammals forms a continuum, beginning with minimally cooperative species through a large number of species with complex patterns of cooperation and at least temporary reproductive suppression, and reaching an extreme in some species of mole rat and the evolution of permanent reproductive suppression.

Naked mole rats are commonly displayed in zoos and are featured in a number of Internet videos. Visit a zoo display or watch some of the videos to get a better feel for the lifestyle of these animals. Now find some information on the lives of meerkats, another social mammal. Based on your comparisons of these species, what are the similarities and differences in their social lives? Can you hypothesize about key features (preadaptations; see Chapter 13) that may have favored eusociality in the naked mole rat but not in the meerkat?

14.3 INVERTEBRATE EUSOCIALITY: WORKERS AND THE DIVISION OF LABOR

Caste in eusocial insects can be defined in a couple of ways. The first deals with differences among a colony's *workers* in how they do their work. The second has to do with reproduction; caste in reproduction boils down to the female reproductive(s), or *queens*, the males (sometimes called *drones* in honeybees or *kings* in termites), and the sterile workers. This section focuses on *division of labor* among workers. Reproductive castes are covered in the next section of this chapter.

> **KEY TERM** Castes are differentiated morphological or behavioral groups within a society.

The theme of division of labor was captured in Shakespeare's play, Henry V:

> …for so work the honey-bees,
>
> …like merchants, venter trade abroad;
>
> Others, like soldiers, armed in their stings…
>
> The singing masons building roofs of gold,
>
> The civil citizens kneading up the honey,
>
> The poor mechanic porters crowding in
>
> Their heavy burdens at his narrow gate…
>
> Henry V, I.ii.188–202

Social Efficiency: Why Divide Labor?

Building a house is no easy matter. It requires mastery of carpentry, plumbing, electrical work, painting, laying tile, and so on. Each trade comes with a different set of skills, and some trades may require more physical strength than others to master. For one person to learn each skill and perhaps fail to become highly proficient at all of them can make the construction project slow and inefficient. If, on the other hand, skilled labor can be hired for each trade, the house can be built efficiently and rapidly.

> **KEY TERM** Division of labor occurs among members of the worker caste when they specialize in tasks.

The same logic can be applied to division of labor within social insect colonies. Colonies of specialists are more efficient than a collection of unspecialized individuals. How might social insect workers specialize? Some of the principal ways that workers can differ are listed in Table 14.1, and the balance of this section is organized around these types of differences among workers.

TABLE 14.1 Effects of Worker Differences on Social Efficiency in Insect Colonies

Type of Difference Among Workers	Potential Efficiency Gain	Examples
Status in dominance hierarchy	Each worker specializes in certain tasks, based on their status in the hierarchy	Found only in species with small colonies in which dominance hierarchies are effective—paper wasps and sweat bees
Size differences	Size/strength can be matched with task	Found in termites and ants, but not wasps and bees
Shape differences	Morphology matches task	Large-headed defenders in termites and ants
		Nozzle-headed termites
		Aphid and thrip soldiers
Age differences (also called *temporal division of labor* or *temporal polyethism*)	Can be flexible to match the number of workers performing a task with need. Younger workers can perform lower-risk tasks	Honeybees, some ants, paper wasps
Learning/experience	Can be a particular advantage for foragers, which may need to learn the landscape and how to return to their nest	Wasps, some ants, bees
Physiological development (glands ready for task)	Specific glands come into play for specialized tasks (e.g., honeybee wax glands for comb construction)	Almost all eusocial species

OF SPECIAL INTEREST: SOME CONVENIENT DEFINITIONS

In eusocial insects, the *colony* refers to the animals, and the *nest* is the physical structure in which the colony lives (Figure 14.13). Because colonies can sometimes change locations, and then construct a new nest, it is helpful to maintain this distinction in terms. Another specialized term is a *hive*, which is a honeybee nest.

FIGURE 14.13
Paper wasp nest and colony (left) showing the small size of this society and its ability to construct a nest by transporting materials to the nesting site. The harvester ant nest (right) is occupied by a much larger colony. Ants bring gravel to the surface for the mound, and also collect gravel from the surrounding area. *Photos: Michael Breed (left); Ave Bucy (right).*

Division of Labor Based on Dominance Hierarchies

In a smallish social group, perhaps up to 50 animals, a dominance hierarchy can be effective in establishing social roles. In the primitively eusocial sweat bee (*Lasioglossum zephyrum*), the queen establishes her dominance by nudging the other bees and by refusing to let them pass her in their tunnels. The second-ranking bee in the colony serves as a guard at the entrance of the nest (Figure 14.14). Lower-ranking bees specialize in either foraging or nest construction. In *Polistes* paper wasps, there is a similar correlation between dominance status and task performance.[35,36]

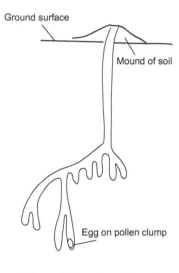

FIGURE 14.14

Sweat bees, and a diagram of a sweat bee nest in soil. The queen constructs a tunnel and rears some workers in the cells that extend from the tunnel. The workers, once they emerge, work to rear additional bees, including the next season's reproductives. *Photos: Russ Ottens, University of Georgia, Bugwood.org (left), Joseph Berger, Bugwood.org (right).*

472

OF SPECIAL INTEREST: THE UNIT OF SELECTION

What does selection actually act upon? Is it a single nucleotide in the DNA sequence? Or maybe it's the gene? Is it the chromosome? Cells within a multicellular organism? The individual animal? Richard Dawkins asks this question elegantly in his book *The Selfish Gene.* Here, we need to pause and consider the problem of the *unit of selection* in the specific context of eusocial invertebrates.

Up to this point, our discussion of animal behavior has focused on the individual as the unit of selection: evolutionary forces act on individuals and culminate in the level of reproductive success that individuals achieve. We emphasized the major problems with most group selection models in Chapter 13, and we argued that most seemingly altruistic behavior can be explained parsimoniously by individual natural selection acting on the donor animal.

What happens, though, when selection yields animals that are sterile, whose only option is to gain reproductive value through inclusive fitness by working with close relatives? It is easy to see that if an animal's reproductive value is invested in the success of the queen of the colony, and if that queen's success, in turn, depends on the survival of the colony, then selection is acting on the colony as a whole. Given that scenario, colonies which are able to organize their labor most effectively to exploit the environment will transmit the most genes to the next generation. Selection should favor organized, efficient colonies and punish chaotic, inefficient colonies. Colony-level selection is not group selection; it is selection on the queen's phenotype, which extends to include the activities of her workers.[37]

> **KEY TERM** The unit of selection is the level at which natural selection operates. This can be the individual worker or queen, the colony, or a local population of colonies.

Size and Shape Differences in Division of Labor

Some ant and termite species show amazing differences in size and shape among their workers. One intriguing question is why extreme caste differences appear in the two taxonomic groups with flightless workers—ants and termites—but do not occur in bee and wasp workers. Perhaps flight places evolutionary constraints on structural evolution in bee and wasp workers. This section explores how worker size and morphology interact in improving colony efficiency in eusocial insects.

Sometimes fairly subtle size differences are important in the division of labor. In some species of eusocial animals, all the workers have the same morphology, but differ in size, with a range of sizes occurring in the colony. One of the most feared animals in the New World tropics is the giant tropical ant. It is sometimes also called the *bala* (or bullet ant because being stung by one feels like being shot) or the *vienticuatro* (24, because its sting hurts for 24 h). Large workers—20 mm or more in length—are visible outside the nests, foraging, excavating, and when necessary, defending the colony. The very largest workers come to the nest entrance when the colony is disturbed. One might think that all the workers of this ant species are equally gigantic, but excavation of this ant's nests reveals that much smaller workers, down to 1.5 mm in length, toil inside the nest. When the nest is disturbed, it is these small workers that rescue the larvae.[38] This is a typical example of size-based division of labor, a commonly observed mechanism in ants.

E. O. Wilson envisioned an evolutionary scheme in which selection first favored colonies with workers of highly variable size.[39] Wilson argued that the evolution of a broad range of body sizes allowed body size to be matched to task; if the most appropriate animals engage each task, then the colony becomes more efficient and can outcompete colonies that lack this feature. This size variance would allow specialization of workers in size-appropriate tasks. Typically, small workers spend most of their lives inside the nest, caring for brood, cleaning the nest, and feeding other workers. Larger workers focus more on tasks that take them outside the colony, such as nest and territory defense and foraging. If selection favors colonies with greater variance in worker size, then over evolutionary time, 10-fold or even 100- or 1000-fold size differences between the smallest and largest workers in a colony become possible (Figure 14.15).

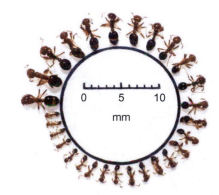

473

FIGURE 14.15
Worker size differences among fire ants, *Solenopsis invicta*. The extreme size difference between the largest and smallest workers is reflected in their roles in the colonies, with large workers more likely to defend the colony and collect food and the small workers more likely to work caring for the brood. *Photo: Sanford Porter.*

How does shape come into play? The answer lies in *allometric growth*—disproportionate growth of certain body parts as an animal gets larger. To understand this principle, think first of isometric growth. In *isometric growth*, body parts all grow at the same rate, so body proportions are the same in small and large individuals. (from Greek: "Iso" means "equal", "allo" means "different".) What if one body part, such as the head, grows at a much faster rate than the rest of the body? Then large animals will have disproportionately large heads (Figure 14.16). What good is a disproportionately large head? In insects, the head houses the muscles for their jaws (mandibles) and numerous glands, some of which can have important social functions. A large head can support bigger mandibles, larger mandibular muscles, and more glandular development. Allometric growth of the head, and sometimes the thorax, is the nearly universal route to extreme caste differentiation in eusocial insects.

What about the issue of the role of the tiniest of leafcutter workers, mentioned at the beginning of this chapter? Observers of leafcutter workers noticed that the smallest of workers (minims) "hitchhiked" on leaves; recall that a medium-sized worker carrying a leaf piece is often also burdened with one or two very small workers riding on the leaf. This observation was puzzling because the hitchhikers had no apparent function. Detailed tests showed that the hitchhikers

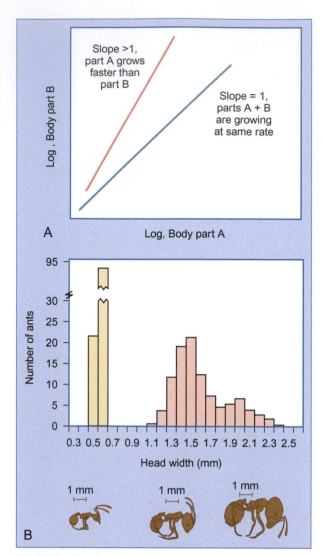

play a key role in warding off small parasitic flies (Phoridae) that attempt to lay eggs on the ant carrying the leaf. If the fly succeeds, its larva invades the head of the worker ant and ultimately kills the worker. The minims actively ward off the flies. The fact that much of the activity of the minims is at night, when the flies are absent, suggests that the minims have other functions, such as removing pathogens that might infect the colony's fungus garden from the surface of the leaves. Function and form are closely associated, and the minims play critical roles outside the nest, despite their small size.[40]

How do workers find the appropriate task? Natural selection on task preferences certainly plays a role, but the spatial positioning of workers is also an important part of the dynamic; a worker that is "programmed" to be outside the nest is likely to encounter a set of potential tasks that differs from the tasks encountered by workers who remain inside the nest.

Age Differences

In larger colonies, dominance hierarchies are less likely. The most commonly observed pattern of division of labor occurs when tasks are divided by age.[41–43] This type of division of labor was first documented in honeybees but also occurs in many ant species, and has been termed *temporal polyethism* (*temporal*=time-based, *polyethism*=many behaviors). In nearly all documented cases, the progression moves from young workers performing tasks inside the nest (construction, cleaning, caring for brood), to middle-aged workers busy on the periphery of the nest (guarding, ventilating, undertaking), and then to the oldest workers foraging outside the nest. Cornell University biologist Thomas Seeley built on earlier honeybee studies in Germany to reinforce this point;[41] simple programming of workers to prefer these locations (center of the nest, where brood is likely to occur; periphery of the nest, where nest defense is important; outside the nest, where foraging is paramount), combined with age-related division of tasks, can yield complex patterns of division of labor. Figure 14.17 shows how this plays out in honeybees and ants. The most elaborate systems of division of labor combine extreme morphological differences with temporal castes (Figure 14.18).

FIGURE 14.16

(A) Allometric growth relationships. The *x*-axis shows the size of one body part, such as the head; the *y*-axis shows another body part, such as the thorax. If the slope is equal to 1, growth is isometric. If the slope differs from 1, then growth is allometric. (B) A hypothetical size distribution of ants in a colony, with a group of small workers (tan bars on the left) and larger workers (pink bars). The drawings of these ants show allometry, with the head growing larger, proportionately, as body size increases.

Based on the observation that the oldest workers engage in the riskiest tasks, biologists have hypothesized that age equates with expendability for the society. In more formal terms, an animal's *marginal value* to the social group, or remaining work that it might do in a lifetime, falls as the animal approaches the end of its normal lifespan. Theory predicts that animals with little marginal value for the group should engage in the most dangerous occupations. As an aside, note that a different pattern seems to hold true for

KEY TERM The marginal value of life is the amount of work or reproduction expected for an individual of a given age.

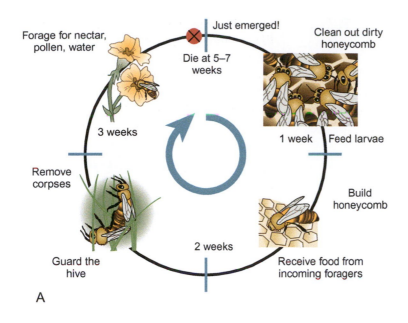

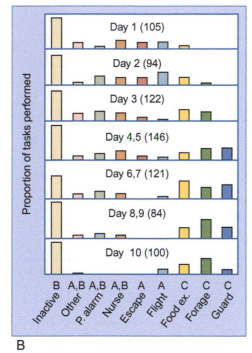

FIGURE 14.17

Age polyethism. (A) Sequence of events in the life of a worker honey bee. Workers progress from cleaning, to feeding larvae, to storing food, to guarding, and then foraging. (B) A similar analysis of ant behavior, with tasks listed on the *x*-axis, and each part of the plot representing a different age of ants. Young ants are more likely to nurse, and older ants guard and forage. *Adapted from Muscedere, M.L., Willey, T.A., Traniello, J.F.A., 2009. Age and task efficiency in the ant* Pheidole dentata: *young minor workers are not specialist nurses. Anim. Behav. 77, 911–918.*[42]

human societies, in which the oldest people, perhaps as a result of experience, are typically the most risk-aversive members of the society and are often highly valued as sources of wisdom and guidance. Yet another influence might be the fact that in many human societies, taking risks is perceived to confer status, which, in turn, may be age-related; status may be important for people who are more interested in attracting mates than in helping to care for an extended family.

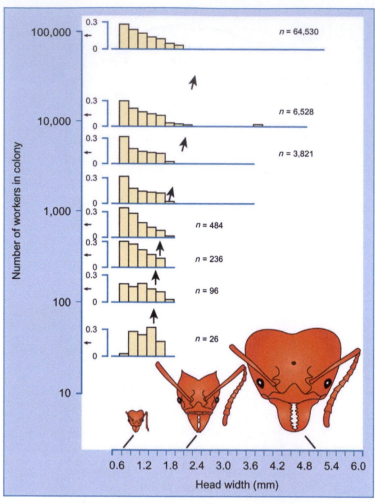

FIGURE 14.18

In this case, age and size combine to correlate with behavior. As a colony grows in size, from 10 to 100, and ultimately to 100,000 (y-axis in the drawing), the variability of worker size increases, and larger workers become more common. This allows more specialization in task performance by workers. *Adapted from Wilson, E.O., 1976. Behavioral discretization and the number of castes in an ant species. Behav. Ecol. Sociobiol. 1, 63–81.*[44]

DISCUSSION POINT: INSECT AND HUMAN SOCIETIES

Can we learn anything from insect societies that might improve the efficiency of human work? Generally, worker eusocial insects do not require instructions to complete their tasks; the information they need is genetic and is carried with them. Could humans function in a "boss-less" society? Why or why not?

Social Insects and Work Performance

Given this overview of division of labor in eusocial insect colonies, it is possible to think about two radically different models for organizing labor within a colony.

In the first scenario, work follows a masterplan, with strong centralized coordination of all of the activities in the colony. Information about colony state, including food use, amount of incoming food, repair and construction needs, threats requiring defense, and so on would flow to a central command post, from which directives would emerge to allocate labor as needed among the tasks. The term "queen" for the reproductive female in social insect colonies further invites and reinforces this image that a commander is an essential element of colony efficiency.

In the second scenario, each worker has a compact set of rules that it follows when deciding what work to do. These rules are partly geared to the need for a task to be done, such as "work on building if you encounter a partly built structure" or "forage if not enough food is coming into the colony." The are also partly geared to the worker's physiological readiness for a task, such as a honeybee worker having active wax glands, priming it to build comb. In this model each worker is fairly autonomous, just relying on the information it encounters and its level of priming for a task to determine its activities.

The second scenario is far closer to the truth. Division of labor is an *emergent property* of rules for engaging in action carried by each member of the colony. Each of these items is a key component of the mix that produces the division of labor:

1. Variance among workers in response threshold to tasks
2. Physiological readiness to perform a task
3. Feedback loops for information flow among workers
4. Spatial efficiency and grouping of tasks to minimize travel.

The first piece, variance among workers in response threshold, is particularly intriguing and is actually easy to grasp intuitively. Imagine a group of roommates each of whom has a favorite task: cleaning, cooking, shopping, laundry, and so on. If each performs their favorite task then all the work gets done and there's some level of equality of effort among the roommates. In social insect colonies some workers have a low threshold for doing one task (they're more likely to do it if it needs doing) and other workers have a high threshold for that task (they only do it if the task becomes a crisis). As in the human analogy, if there are at least some workers with a low threshold for each task, then all of the tasks get done.

> **KEY TERM** Social behavior can be an *emergent property* based on pre-existing behavior in solitary animals. Social behavior emerges when ecological or evolutionary factors bring members of a species together. For example, if in a pair of individuals one is more primed for nest construction and the other for foraging, division of labor between the two animals emerges as a result of being associated.

The second piece, physiological readiness, often correlates with age. In honeybees, the species in which division of labor has been studied most thoroughly, the hypopharyngeal glands in the worker's head produce food for the larvae and the wax glands on the underside of the abdomen produce wax for comb construction. These glands mature as they are needed and then regress when bees age beyond the required roles for the glands. If an experimenter constructs a honey bee colony from bees of just one age group, some bees will move ahead or fall behind the predicted tasks for their age group; this allows the colony to survive and demonstrates flexibility in performing tasks.

The third item addresses how workers gain needed information about the needs of their colony. They can obtain feedback from the condition of the nest or from other workers. A partially built structure might stimulate a worker to add material. The worker does not need a blueprint, it only needs to make its construction fit with the general rules for how nests are built by that species, information that can be genetically encoded. Workers may obtain feedback by the general level of activity in the colony or by how fast (or slow) food is being brought into the colony.

Finally, as mentioned earlier, age-related tasks are often grouped by location. This makes sense as travel can be metabolically costly and traffic jams from random flow of workers within the nest might also result in lost time. If workers perform a set of tasks related to broodcare, for example, their work will be spatially clustered around the brood.

Taken together, variance in response threshold, physiological readiness, feedback, and location can yield a very highly coordinated and efficient society with no need for direction or instructions from the top. It is intriguing that the lack of central organization is not a barrier to efficiency; the system relies on each element performing its tasks as if following a script. Natural selection favors more efficient colonies, so over evolutionary time, features like response thresholds evolve to enhance efficiency.

14.4 INVERTEBRATE EUSOCIALITY: QUEENS AND REPRODUCTION

Reproductive labor is divided (or skewed) among members of eusocial colonies. One or a few females lay the bulk of the eggs in the colony. By convention, these egg-layers are called *queens*, a term that can be a bit misleading because, as pointed out in "Social Insects and Work Performance", in most species of ant and social wasp the "queen" does little in the way of coordinating or directing activity within the colony.

The queen has evolved adaptations that have equipped it well for its jobs, with large ovaries contained in a very enlarged abdomen. In primitively eusocial species, such as sweat bees, bumblebees, and paper wasps, the queen does not differ morphologically from the workers, although she may be somewhat (5–10%) larger than her worker coterie. In primitively eusocial species, the queen is capable of establishing a nest on her own, without workers, and of rearing her first brood of workers.

In highly eusocial insects, queens are morphologically very different from their workers; for example, the honeybee queen lacks the hairs necessary for collecting pollen, she has relatively small eyes, and her sting is not barbed. All ants and termites are highly eusocial, with large differences between queens and workers (Figure 14.19). Even though ants and termites are far apart, evolutionarily, they share the trait of having queens (and in the case of termites, kings) that start their adult lives with wings. These winged reproductives fly during the mating and dispersal phase of their live cycle but then break off their wings once they establish their colony.

FIGURE 14.19
A fire ant queen with workers. The queen is well-adapted for her primary function of laying eggs. *Photo: Sanford Porter.*

In many species an evolutionary interplay continues between the queen and workers. This occurs even if substantial morphological differences exist so that the queens are specialized for reproduction and the workers are specialized for brood care, foraging, and colony defense. Workers usually retain small, but potentially functional ovaries; although they lack the apparatus for mating, ant, wasp, and bee workers can lay unfertilized (haploid) eggs that develop into males. Trivers and Hare pointed out that in a haplodiploid society, kin selection theory predicts different types of competition between queens and workers, depending on the sex of the young being reared.[45] Workers, which are female, should cooperate with queens in the production of female offspring. Assuming the queen is the mother of the workers, then the workers should prefer to assist the queen in producing their sisters, rather than producing their own daughters. But production of males is a different matter. Theory predicts that workers will benefit more, evolutionarily, from producing their own sons than from helping the queen to produce their brothers.

Trivers and Hare used this idea to develop an argument that if kin selection was responsible for the evolution of extreme altruism, then worker–queen conflict should occur over the laying of male eggs. In fact, in many species of ant, bee, and wasp, the ovaries of at least some workers develop to the point that they can lay male eggs. Queens and workers may respond to this competition by eating each other's male eggs.

Ratnieks and Visscher realized that this story is incomplete.[46] They focused on how a worker responds to another worker's reproductive efforts. A worker should prefer its own

478

sons, but given the choice between rearing a male egg laid by the queen (this egg is the worker's brother), and a male egg laid by a sister (this egg is the worker's nephew), the worker should prefer the queen's egg. They predicted that workers would monitor and control the reproductive efforts of their fellow workers by eating the worker-laid eggs. Such worker policing exists in a number of species, including many ants and honeybees.[47]

How Many Matings and How Many Queens?

Kin selection explanations of extreme altruism rest on the assumption that a queen mates no more than once, so her daughters are, indeed, supersisters. If a queen mates more than once, the genetic relatedness among workers in the colony is diluted (Figure 14.20), and the inclusive fitness benefits from working are reduced. However, selection could favor multiple mating by queens for several reasons, including the following: (1) genetic diversity among workers in the colony facilitates division of labor,[48] (2) genetic diversity among workers increases colony-level resistance to diseases and parasites, (3) multiple mating insures the queen against the costs of inbred mating (more matings, at least some of the matings won't be inbred), and (4) increased sperm supply allows the queen to lay more eggs over her lifetime. These possibilities create an interesting evolutionary balance among selective forces that impact the number of matings.

Workers in primitively eusocial species may have the evolutionary option of leaving the social group and reproducing solitarily, but workers in highly eusocial species lack this option (largely because they have lost the morphology necessary for mating). Based on this, researchers have predicted that multiple mating by queens would be more common in highly eusocial species than in primitively eusocial species. Once workers are "trapped" into that highly eusocial role, without the possibility of evolutionary reversal and escape from the worker role, selection can then act on the queens to gain the benefits of worker genetic diversity and insurance against inbreeding.

Kin selection theory also predicts that a colony should have only one queen; multiple queens dilute the genetic relatedness in the colony, even if the queens are sisters. In most primitively eusocial species, such as sweat bees, bumblebees, and paper wasps, colonies do, in fact, have

479

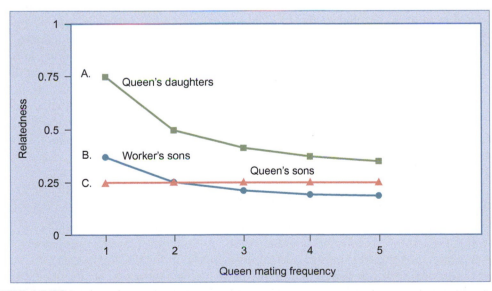

FIGURE 14.20
This illustration shows how the number of times a queen mates affects genetic relationships within an ant, bee, or wasp colony. The upper line shows the average relatedness among daughters of the queen; if she has mated only once, this is 0.75; added matings rapidly reduce this average. The same pattern applies to workers' sons, but not to the queen's sons, which are always related to each other by 0.25 (they have no father!).

only one primary reproductive female, and even if workers are capable of occasionally laying eggs, the workers have not mated, so their offspring are all males. (See the earlier section on worker policing for more on the topic of eggs laid by workers.)

In highly eusocial species, multiple queens (polygyny) are more common. One possible explanation for multiple queens is the delayed competition model discussed previously. In this case groups of queens must cooperate to found colonies; they later become competitors, and only one of the queens actually contributes reproductives to the next generation. The queens need not be genetically related because inclusive fitness has no role in their cooperation. In another model, related queens (generally sisters) remain together, and each contributes to colony reproduction. This is particularly advantageous in pioneering or "weedy" species that invade new habitats, such as Argentine ants.

Reproductive Suppression

Kin selection theory predicts that workers voluntarily give up their own reproduction in order to aid their mother. If this were strictly the case, conflict over reproduction would not occur in colonies, and there would be no need for the queen to actively suppress worker reproduction. As already discussed, in ants, bees, and wasps, the asymmetries in genetic relationships caused by the haplodiploid method of sex determination make conflict over production of males likely. Also, there is a distinct possibility that eusociality could be explained by coercion, rather than by kin selection.

The existence of dominance hierarchies in a broad range of primitively eusocial species (or, in the case of the naked mole rat, a highly eusocial species) suggests that reproductive suppression may not be an entirely voluntary act on the part of the workers. In some highly eusocial insects, there is a distinct possibility that queens engage in chemical suppression of the workers in the colony. Scientists have not yet sorted out the underlying evolutionary forces that lead to the presence of dominance hierarchies in so many eusocial species; division of labor, competition for production of males, and suppression of production of both males and females are the leading hypotheses.

Dominance Hierarchies

The dominance hierarchies of *Lasioglossum* sweat bees and *Polistes* paper wasps described in the discussion of division of labor also have reproductive implications, as studies of ovary development in workers consistently show that higher-ranking workers in the hierarchy have larger ovaries than lower-ranking workers.[49,50] Workers usually do not mate, so they lack the potential for laying female eggs, but ovarian development could be indicative of an ability to lay male eggs. Dominance hierarchies are also known in bumblebees and in some ant species that have small colonies.[51] Body size differences can facilitate the ability of a queen to dominate her workers; size confers an advantage in physical conflict. Thus, it is important to know that the workers of sweat bees, bumblebees, and paper wasps are all smaller than the queens and that, at least in sweat bees and bumblebees, smaller workers tend to have lower status in the hierarchy.

Chemical Suppression

Honeybee queens produce powerful pheromones that are distributed among the colony members. These pheromones could serve either to pharmacologically suppress worker reproduction or could simply act to inform workers of the queen's presence.[52] Pheromones from the queen as well as from the larvae in the colony may act together to suppress worker reproduction and to maintain worker behavior (Figure 14.21). In the older literature, the queen pheromone is called *mandibular pheromone* because some of its components were first identified from the mandibular glands of queens. *Retinue pheromone*

A

B

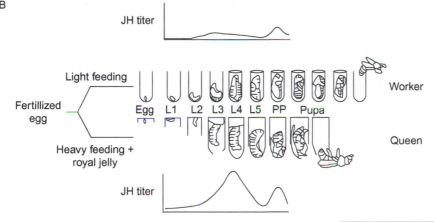

TRENDS in genetics

FIGURE 14.21

In the photo (A), workers from a retinue around a queen-like worker of the Cape honey bee; this is a normal response to queen pheromones in honey bees. The drawing (B) shows how feeding lower amounts of nutrients to a larva results in the development of a worker while a nutrition-rich diet including royal jelly results in a queen. Juvenile hormone (JH) is low in worker larvae, corresponding to the lack of development of their ovaries and is high in queen larvae. See Chapter 2 for more information about juvenile hormone and insect development. *Reprinted from Lattorff, H.M.G., Moritz, R.F.A., 2013. Genetic underpinnings of division of labor in the honeybee (Apis mellifera). Trends Genet. 29, 641–648.*[53]

may be more accurate because the chemical mix in the pheromone comes from several glands, and one of the behavioral effects of the pheromone is to attract a circle of bees (a "retinue") around the queen. Chemical communication of termite queen and king presence has also been suggested.[54,55] A mystery in ant behavioral biology involving the question of how the queen's reproductive dominance is maintained in the colony has recently been solved. Like honeybees and termites, in which chemical signals from the queen affect worker reproduction and activities, Oystaeyen and colleagues found that suppression of workers is due to chemicals produced on the surface of the queen ant.[56]

"Willing" Workers

In the vast majority of ant species and in the highly eusocial wasps, there is no evidence of either dominance or chemical reproductive suppression of the workers. Negative results like this are as tricky here as they are throughout science. Is the absence of evidence for behavioral or chemical dominance due to faulty experimental approaches? Is the investigator asking the right question? In this case, kin selection theory predicts that workers should willingly give up their reproductive "rights" to help their mothers rear their sisters, so the lack of evidence for reproductive suppression could support a kin selection hypothesis for eusociality.

> **KEY TERM** Retinue pheromone includes mandibular pheromone and other glandular products; the entire retinue pheromone is necessary for normal functioning in a honeybee colony.

14.5 INVERTEBRATE EUSOCIALITY: COLONY DEFENSE

Eusocial colonies face a variety of threats, but the common theme in studies of colony defense is that the lives of individual workers may be expended in efforts to protect the colony. This is not to say that workers are always expendable; instead, the claim is that the cost–benefit equation favors combat if the cost of the threat (and failing to oppose it) is greater than the cost of losing the workers needed to block the threat. Physical defenses usually include biting and in some cases stinging, along with many different types of chemical defenses.

While most ants and eusocial bees and wasps can sting (Figure 14.22), one large and ecologically important group of ants, the Formicinae, or *formic acid* ants, has lost, in the course of evolution, the ability to sting. Instead, these ants rely on an acid secretion for their defense. Another ecologically prominent group of eusocial insects, the stingless bees, has also lost the ability to sting, but these bees can readily defend their nests with biting and, in some cases, with chemical secretions. Because the sting is a modification of the egg-laying apparatus in the insect order Hymenoptera, it is found only in ants, bees, and wasps; other eusocial invertebrates, including termites, aphids, and shrimp, lack the ability to sting.

FIGURE 14.22
An ant stinger. Ants in the genus *Odontomachus* have very painful stings. *Photograph courtesy of Eli Sarnat.*

Often the major threats to a social insect colony are workers from other colonies of the same species, or workers from closely related species. Theft of brood or food is a continuing threat, and members of the same or closely related species are ideally adapted to use a colony's food or to employ workers as slaves. The key to preventing entry by larcenous workers is the ability to discriminate who belongs, and who does not belong, in the colony. In most species, odor differences among colonies play a key role in this recognition, but in some wasps, visual cues apparently serve this role.

Eusocial colonies have three types of resources that may need defending. The first is their brood, which can be a valuable source of food for other animals or a source of workers that are used as labor by raiding eusocial species (these are sometimes called *slave-makers*). Second, colonies may have stored food, such as honey, which also has high value for other species. Finally, colonies may have feeding territories that are subject to seizure by other colonies. Brood and stored food attract a wide range of species, while foraging territories are usually defended against conspecifics or other species with very similar ecological requirements. Because the value of brood and stored food is so high even relatively small eusocial insect colonies can attract vertebrate predators. Many eusocial insect species have defenses that are geared toward warding off vertebrates, a lesson that many humans learn in painful ways. Because there is a risk of complete loss of the colony and its future reproduction, colonies may make massive investments in defense.

In some social insects, the sting apparatus is barbed and remains in the victim after the sting is delivered; this intriguing defensive adaptation is called *sting autotomy* and is especially well known in honeybees. The honeybee stinger consists of a pair of barbed stylets, a muscular pump, and a venom sac. The muscles contract rhythmically after the sting is left in the skin of a potential predator, delivering a full dose of the potent toxins in bee venom. While the defensive work of the defending bee might appear to be done at that point, with the lost sting leaving a gaping wound and a damaged digestive tract, the bee might live a while longer and continue to be active, pestering and exhibiting defensive behavior by flying at and around the face of the target animal. This behavior intimidates because the animal has no way of knowing that the bee has already lost its sting. While all of the dozen or so species of honeybees have sting autotomy, only a few other social insects, all of them wasps, display this adaptation.

In summary, the sterile workers of ants, bees, and wasps have evolved an amazing range of defensive weapons. For instance, in one bizarre subfamily of termites, the Nasutitermitinae, the head of the soldier has been fashioned by evolution into a nozzle, from which the termite can spray a sticky glue that is effective against the termite's main enemy, ants.[57] Carpenter ants in Malaysia (*Camponotus cylindricus*) actually explode, releasing irritants and corrosive compounds, during territorial battles with other ant species.[58] One tropical stingless bee is known as the "fire bee" because its bite releases caustic chemicals that cause severe blistering and ultimate scarring of the skin. Ants in the subfamily Formicinae are well known for their secretion of formic acid as a defensive compound; in the course of evolution, these ants have forsaken a sting and rely instead largely on chemical secretions for their defense. And, of course, the honeybee and some species of wasp have evolved sting autotomy, in which the stinger is left in the victim, where it pumps venom after the bee or wasp has been brushed away. These defensive characteristics demonstrate the potential of evolution for yielding clever adaptations when an animal has no direct reproductive functions.

483

All of this defense can be called into action by *alarm pheromones*, which allow for rapid communication about threats to the colony. Most alarm pheromones of terrestrial animals are small molecules, which evaporate and spread rapidly in the air. When a honeybee stings, the stinger releases alarm pheromone, a complex chemical mixture in which isopentyl acetate (IPA) predominates. IPA smells like bananas to most people; it serves to alert other bees to the presence of a threat, such as a skunk, a bear, or a beekeeper. Motion and dark colors attract the alarmed bees, some of which sting the victim. The new stings release more alarm pheromone, so a cascade of positive feedback yields more and more stinging bees. The victim of the alarm response can terminate the stinging only by moving far enough away from the colony that new bees do not respond to the alarm pheromone. For European bees, 50–100 m is a safe distance from even the most alarmed colony. African bees may continue to have an alarm response as distant as a kilometer from the colony.

> **KEY TERM** Alarm pheromones are released by an animal when it is disturbed or injured; alarm pheromones function to alert other animals of impending danger.

DISCUSSION POINT: BENEFICIAL INSECTS AND STINGS

We depend on eusocial insects for pollination of crops and numerous other ecosystem services, yet many people are terrified of ants, bees, and wasps because of their stinging abilities. Of course, some people are indeed allergic to stings, but those people are far fewer than the number of people who fear insects even without allergic reaction. Are you entomophobic? Does the argument that eusocial insects provide essential ecosystem services outweigh your fear of insects? As a cure, we recommend the Insect Fear Film Festival—"scaring the general public with horrific films and horrific filmmaking since 1984"—an annual event at the University of Illinois, Urbana–Champaign.

14.6 EUSOCIAL INVERTEBRATES

This section highlights some interesting types of eusocial invertebrates. These examples point to family living, possession of a nest or shelter by the family group, and ecological pressures for group defense from predators or parasites, as the critical factors in the initial stages of the evolution of extreme cooperation. Note that strong colony defense is an important thread that ties these diverse kinds of animals together.

Eusocial Shrimp

Another case of eusociality is the snapping shrimp (Figure 14.23), which exhibits clear evidence for a soldier caste.[59] These shrimp have dramatically differing front claws. One, which is much larger and stronger than the other, can be cocked, much like a pistol, and "snapped" against itself to produce a jet of water that startles potential predators. The force of the snap is also sufficient to cause serious physical harm to other crustaceans. Some species of snapping shrimp, living in the canals of coral reef sponges, have a strong reproductive division of labor. While a colony may contain up to 300 individuals (average colony size is approximately 150 shrimp), only one shrimp can lay eggs. Genetic analyses have shown that the shrimp in the colony were likely all the offspring of this "queen" and a single male, but the "king" could not be identified.

Colony defense is a key element for snapping shrimp; shrimp engage in battles with noncolony members and use their snapping appendage to deliver lethal blows to intruders. Large male colony members are more active in colony defense than small shrimp, and they provide the bulk of the defensive effort. Members of the same colony are treated benignly, and noncolony members of the same species or of other species are attacked.[60] The snapping shrimp add another thread to the tapestry of eusociality, reinforcing the importance of living in kin groups, occupying a nest structure or cavity, and sharing colony defense in the evolution of eusociality.

Thrips and Aphids

One way to make a new discovery of extreme altruism—eusociality—is to generate a set of predictions using the characteristics that are known to set the evolutionary stage for eusociality. This is precisely what scientist Bernie Crespi did; he determined a set of criteria—haplodiploid species that live in family groups in a shelter or nest—and then set out to see if eusociality could be discovered in novel organisms.[61] He discovered it in an odd-looking

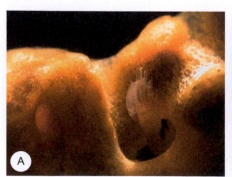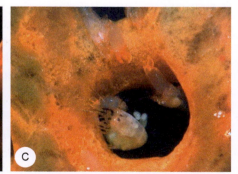

FIGURE 14.23

(A) This is an adult *Synalpheus regalis* shrimp at the surface of a sponge in a captive lab colony. Large individuals like this one typically are more active and aggressive than small ones and spend more time near the periphery of the sponge where intruders are likely to be encountered. (B) Typical posture adopted when two individuals encounter one another. Generally, they begin by tapping one another's antennae, which appears to yield information on identity at some level. (C) A queen and male, either her current mate or prospective mate (or both), in a captive lab colony. *Photos and captions courtesy of Emmett Duffy.*

insect called a *thrip*, an insect with wings that look like bird feathers (Figure 14.24). Thrips have another, less obvious, distinguishing feature; they are among the few insects other than Hymenoptera to have haplodiploid sex determination.[62] Haplodiploidy and living in a nest-like structure—in the case of thrips, a *gall* on a plant—give some thrips the necessary preadaptations for eusociality, which has been discovered in a half dozen species of Australian thrips.

On the surface, thrips (insect order Thysanoptera) seem like extraordinarily unlikely candidates for eusociality. These minute plant-feeders appear to have a limited behavioral repertoire and, if one expects complex communication as a part of eusocial life, the thrips are even more unlikely candidates. However, they have the two critical preadaptations for extreme altruism. They are haplodiploid, and some species of thrips live in galls formed by their host plants, probably in response to thrip secretions. The presence of a mother and her daughters living together in a protected structure (the gall) sets the stage for the evolution of extreme altruism. Add to this haplodiploidy, which tips the evolutionary balance in favor of aid-giving behavior, along with ecological pressure from species of thrips that are specialized to invade and take over the galls, and it becomes the perfect recipe for the evolution of extreme altruism.[63]

A colony of thrips begins with a single, large-winged (macropterous) female thrip starting a gall. She has previously mated, and she lays a batch of eggs inside her gall. The hatchlings, which in thrips look like miniature wingless versions of the adults, include both males and females. Some of these young develop into large-winged males and females, which may mate with females moving on to found new galls of their own. (Read more on the importance of brother–sister matings below.) Other young males and females develop very differently, with small (micropterous), useless wings. These are the extreme altruists; they work for the benefit of their mother and their sibs by defending the family's gall from invasion by socially parasitic species of thrip. The parasitic thrips attempt to exploit the work of the founding family by stealing the gall and the food and protection it provides.

Why might both males and females serve as soldiers when a central prediction of kin selection theory is that the worker castes of haplodiploid species should be female? The occurrence of male and female soldiers reflects high levels of inbreeding, which reduce the effects of the haplodiploid asymmetry in relatedness and increase the potential benefits to male soldiers.[64]

Females of some thrip species invade the stems of plants and then cause the plant to form a chamber, or gall, around them. The gall protects the female, and she feeds on plant cells on the inner surface of the gall. The female, or foundress, then lays a batch of eggs that all develop into a soldier caste with small wings and enlarged forelegs. The signals that cause the foundress's eggs to develop as soldiers, rather than the winged form, remain a mystery.

These soldiers are effective defenders of the gall against parasites and predators, and improve the foundress's chances for survival, but because they are unable to fly, the soldiers cannot disperse to start new galls. Soldiers are particularly adept at defending the gall against another genus of thrip, *Koptothrips*, which attempts to take over the gall for its own use.[65] Unlike many eusocial species, the soldiers retain the ability to reproduce. Brother and sister soldiers mate, and the female soldiers lay eggs inside the gall.

Thrip families are small: in *Kladothrips hamiltoni*, the number of soldiers is about 25 per gall. The next generation, which comes from eggs produced by the soldiers, is fully winged and capable of dispersing to establish new galls. Because the soldiers' matings are inbred, the winged dispersers coming from the soldiers are more related to the foundress than the

KEY TERM A gall is a growth of plant tissue stimulated by an insect. The insect lives within the gall and eats the plant tissue.

FIGURE 14.24
A eusocial thrip, *Kladothrips habrus*. Note the large front legs of the soldier on the right. *Photo: Bernie Crespi.*

485

FIGURE 14.25
Eusocial aphids, showing aphids feeding on a stem (left) and a colony with larger soldiers and smaller, winged, reproductives (right). *Photos: Patrick Abbott.*

foundress is to her own sons and daughters.[66] This supports a kin-selection argument for the evolution of soldiers in these thrips. The winged dispersers, most of which are female, overwinter, mate, and initiate new galls the next year.

The discovery of eusociality in the thrips shows that given the right set of preadaptations, extreme altruism can arise in surprising places. The eusocial thrips lack many of the intriguing adaptations for colony member communication and coordination found in termites, ants, bees, and wasps, yet the addition of the soldier generation greatly improves the chances that the foundress's genes will be passed to the next generation. With the thrips, evolutionary pressure from parasites and predators, combined with nesting in a confined space and haplodiploid sex determination, produces the perfect mix for the evolution of division of labor in colonies.

About the same time that Crespi uncovered eusociality in thrips, Aoki uncovered a sterile soldier caste in some colonial aphids. (Aphids are also small, plant-feeding insects.) Like thrips, a few species of aphids that inhabit galls are eusocial (Figure 14.25). The eusocial aphids are diplodiploid; the similarities between thrip and aphid colonies represent convergent evolution. Aphid soldiers defend their colony against possible predators, and can stab potential predators with their mouthparts (stylets).[67-71]

Termites

Termites probably evolved from colonial, wood-eating cockroaches, and in many ways their biology is still strongly linked to this evolutionary root. Nearly all termites use plant material as their primary dietary component. Depending on the species, termites may specialize on dead wood, grasses, or leaves. This food, while abundant in many ecosystems, is notably lacking in nutrients and is rich in near-indigestible cellulose. Termites make up for their nutrient-poor diet by employing intestinal symbionts to assist in digesting cellulose.

> **KEY TERM** Intestinal symbionts are microorganisms that aid in digestion and are found in the gut of many animals. In termites, flagellates or bacteria digest cellulose, making the carbon available to the termites.

Behavioral adaptations for the passage of the *intestinal symbionts* among workers are critical to termite social biology. These symbionts are a mixture of flagellated protozoa and bacteria in the ancestral termites and bacteria in the derived termites. The symbionts reside in the termite hindgut; at hatching, a termite infects itself with symbionts by feeding from the anus of another termite, repeating this process each time it molts and thus loses the hindgut contents. The need for anal–oral infection with symbionts may have been a major factor in the evolution of social cohesion of families in the wood-eating cockroach ancestors of termites.

Termites are among the master architects of the animal world, and their large mounds, essentially castles constructed of dirt, plant material, and feces, rise prominently from the

FIGURE 14.26
(A) Termite nest (*Nasutitermes*) in a tropical rainforest. (B) Termites from the nest pictured in (A) build covered walkways. When the covering is broken away, guards, which have spray nozzles and can shoot a gluey substance at ants, converge on the break. (C) A Formosan termite queen (*Reticulitermes*). (D) Formosan termite workers. *Photos: (A and B) Michael Breed; (C and D) USDA/ARS Formosan termite image gallery.*

tropical and subtropical savannas in Africa, Australia, and South America (Figure 14.26). Although not all termites are such expert builders (North American termites simply excavate cavities in wood or soil), this is such a noticeable feature of termite behavioral biology that it deserves special discussion. This elaborate construction raises two sets of questions: First, what is the functional significance of the architecture, and second, how can termites "know" how to construct these complex fortresses?

Termites, like all animals, have to contend with the physiological impacts of water, oxygen, carbon dioxide, temperature, and direct sunlight. They must also deal with potential predators, particularly ants, but also vertebrates. The nest serves to mitigate all of these problems. First, it protects termites from environmental extremes. Termite workers are typically soft-bodied and relatively unprotected from heat, dessication, and direct sun by their exoskeletons; the nest compensates for this vulnerability. Even North American termites construct coverings over their walkways when there is a risk of exposure to sun or predators.

Second, the nest helps solve problems that are created by life in an enclosed space (See Chapter 12). Animals produce heat, water, and carbon dioxide as byproducts of their metabolic activities. Metabolic heat and water can be put to use to help to maintain favorable conditions in the nest, but excesses of either need to be vented from the nest. In particular, carbon dioxide must not be allowed to build up in the nest because it can act as a metabolic poison, and instead, oxygen must be circulated into the nest. Because of this, large numbers of termites living in a closed space need a ventilation system that retains adequate heat and moisture, brings in necessary oxygen, and gives the option to vent carbon dioxide, heat, and moisture when needed.

The nest also serves a third function—that of protection against predators. The hardened exterior—termite feces are a great glue—protects the termites from all but the most determined vertebrates. By limiting the number and size of entrances and exits, concentrated soldier forces can repulse encroachments by ants.

Architecturally, it seems highly unlikely that termite workers individually have the cognitive ability to step back, look at the overall structure of the nest, and make decisions concerning what and how to build next. There is also no evidence for "foreman" or "supervisor" workers directing the building efforts of others. Instead, the ultimate construction, no matter how elaborate, arises from individual workers' responses to conditions at a given location and moment of time. This is seen most simply in the repair of covered walkways; if a covering is damaged, the damage or sunlight serves as a stimulus for workers to add construction material to the remaining walkway; if they follow the pattern established by the undamaged portions, the gap in the walkway is soon closed. Construction effort in response to a specific element of the existing structure is called *stigmergy*, and this explains much about how elaborate structures can arise from workers that use simple rules. If other simple rules for workers are added—plug up air flow if it is too cool or too dry, make an opening if carbon dioxide is too high, and so on—it becomes apparent how collective behavior yields complex buildings in the absence of either cognition or centralized direction. Computer models that are built on such simple rules can successfully mimic the construction behavior of termites.[72]

> **KEY TERM** Stigmergy is construction that follows a pattern already established by previous building. For instance, if a small deposit of dirt stimulates further deposits at the same spot, a pillar results.

Termite castes differ from those of ants, bees, and wasps in several dramatic ways. Termites are diplodiploid and, consistent with Hamilton's theory, both males and females serve as workers in the colony, following parallel developmental pathways (Figure 14.27). Termite colonies are headed by a queen and a king; these individuals are winged as young adults and mate following a nuptial flight. The mated pair sheds their wings and lives the balance of their lives within the nest; the king and queen mate again each time the queen runs low on sperm. Termites have incomplete metamorphosis; a termite hatchling, or nymph, looks like a miniature wingless adult. While the smallest hatchings cannot contribute much to colonial welfare, as an immature termite grows through several molt stages, its ability to work grows as well. Tasks within a termite colony include feeding and caring for the queen, excavation and construction of the nest, and colony defense.[74]

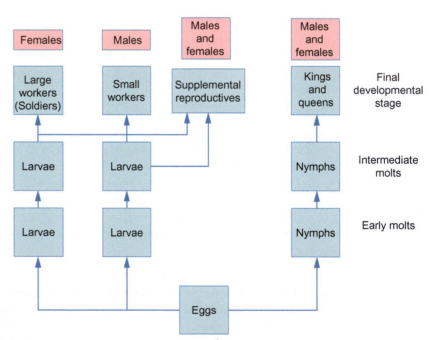

FIGURE 14.27

The typical developmental pathways of male and female termites. In termites both males and females are workers. In this illustration first molt termites are termed larvae. Males and females that will become the next generation's reproductives develop through nymphal stages to kings and queens. In some species one sex of the workers develops into larger individuals that have specialized foraging or defensive roles. In termites developing individuals fated to be workers are called larvae while developing queens and kings are called nymphs. In the illustration females are shown becoming soldiers. Either worker sex can also molt to become a supplemental (sometimes called ergatoid) reproductive.[73]

In most species, the largest females are major workers, whereas the largest males are soldiers, although in some cases these roles are reversed. Soldiers typically have enlarged heads and mandibles, and can use the mandibles as effective weapons in territorial combat with other termites or in battles with predators, which are often ants. The role of immatures as workers and the presence of both male and female workers make termite caste systems more complex than those of ants, bees, and wasps.

In addition to the nymphs that develop as workers, some male and female nymphs grow into the next generation's reproductives. Colonies, in response to environmental triggers such as seasonal changes, rainfall, and temperature, may release hundreds or thousands of winged reproductives on a single night. To ensure outbreeding, these flights of reproductives are usually synchronized among colonies, so there may be hundreds of thousands of winged termites flying at once. The winged females (potential queens) attract males with sex pheromones from a gland on their backs; once a male finds a female, he follows her as she runs on the ground, searching for a nest site. Very few of the termite pairs actually succeed in establishing nests; most become prey for ants, birds, rodents, and even humans. The mass release of winged termites is probably an example of predator saturation (see Chapter 9 on foraging); the winged termites are poorly defended, but so many are produced that the predators cannot possibly eat all of them. Once the nest is established, the queen's abdomen expands to contain her growing ovaries; in some species the abdomen is so large that the queen is immobilized. In very large termite societies, supplemental reproductives may develop; these females are intermediate in size between the queen and the workers and contribute by laying eggs that will develop into workers. Termite colonies of some species can become immense, containing hundreds of thousands or millions of workers and occupying underground cavities that span several hectares. (A hectare is 100 m on a side, about twice the length of an American football field.)

489

Ants

Ants are all highly eusocial and, in many ways, are the most diverse group of eusocial animals. This diversity is especially apparent in caste structure and in diet. Ant diets range from strictly carnivorous (army ants) to strictly herbivorous (leafcutter ants). The evolutionary ability of ants to exploit this wide range of diets has led to their ubiquity in terrestrial habitats.[75]

In castes, ants range from species such as the common wood ants, *Formica* sp., that show very little size variation in workers and no noticeable differences among workers in form, to species such as leafcutter ants, *Atta* sp., and members of the genus *Pheidole*, which have extreme physical worker castes (Figure 14.28). As discussed in the section on castes, caste differentiation allows task specialization, a key element in gaining efficiency for the social unit. The high level of eusociality in ants is established by the major morphological differences between ant queens and workers and the lack of reproductive potential (beyond laying male eggs) in workers of most ant species.

Ant queens exhibit two different patterns of nest founding, depending on the species. In some less-advanced species, the queen mates and then finds a nest site, to be used as a base for her foraging activities. Her first batch of eggs, which are all female, develop into workers, but to feed them she needs to actively forage and bring food back to the nest. This style of nest establishment is exemplified by the giant tropical ant, *Paraponera clavata*, in which the queens, during the initial stages of nest foundation, forage on insect prey. The obvious downsides to this strategy are that the queen must retain morphological adaptations for foraging and that she is exposed to predators while foraging.

Alternatively, ant queens may begin nests claustrally or semiclaustrally.[76] In claustral species, the queen relies solely on her metabolic resources, including digestion of her now-unnecessary wing

FIGURE 14.28

Ants: (A) Worker ants with brood. (B) Worker ants tending eggs. (C) An ant colony filled with plaster and then excavated (next to scientist Walter Tschinkel. (D) An ant tending aphids. *Photos: (A) USDA APHIS PPQ Archive, USDA APHIS PPQ, Bugwood.org; (B) Susan Ellis, Bugwood.org; (C) Charles F. Badland, with permission from Walter Tschinkel; (D) David Cappaert, Michigan State University, Bugwood.org.*

490

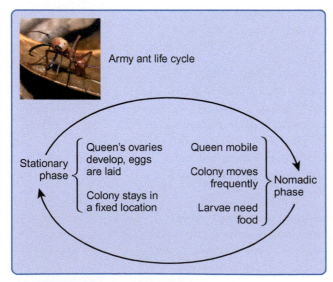

FIGURE 14.29

The life cycle of an army ant colony. Army ants alternate between a stationary phase, during which the queen lays eggs, and a nomadic phase, during which the colony moves frequently. Colony movement allows the ants to find new food resources to feed their actively growing larvae. Most army ant species specialize in preying on the broods of other ants. *Photo: California Academy of Sciences.*

muscles, to feed her first batch of larvae. Semiclaustral species rely largely on the queen's metabolic resources, but the queen occasionally forages. Claustral founding avoids the risk of exposure to predation, but the queen is limited by how much fat and muscle she can use as nutritional resources for her offspring. Semiclaustral founding, which is seen in leafcutter and harvester ants, bridges the gap between the two extremes, allowing the queen to collect some critical nutrients while she avoids the risks involved in full-time foraging.

Army Ants

Army ants are not the terror for humans portrayed in movies, but their biology is extraordinary and well worth discussing. Army ants lack a permanent nest; instead, colonies alternate between a stationary stage, during which the queen is enlarged and laying eggs, and a nomadic phase, during which the colony often moves in search of food for their voracious larvae (Figure 14.29). Armed with large mandibles and a painful sting, army ants are highly effective predators. Most army ants specialize on preying on other ants, but some conspicuous species eat a broad range of insects and even frogs, lizards, and nestling birds.

FIGURE 14.30
Paper wasps, *Polistes* spp. The queen on the relatively young colony in (A) is protecting her brood which will become workers when they mature. In (B) workers are present and the colony has grown. The cottony tops of the cells in the comb are the pupal cases of nearly mature wasps. *Photos: (A) Michael Breed, (B) Joan Strassman.*

Wasps

The painful stings of eusocial wasp workers get the attention of humans. Many, perhaps most, "bee stings" in late summer were actually delivered by a worker from a eusocial wasp colony. In North America, the eusocial wasps are easily divided into the *primitively eusocial* paper wasps (species in the genera *Polistes* and *Mischocyttarus*), and the *highly eusocial* yellowjackets and hornets. The colony cycle is annual, so new queens found nests in the spring; by mid- to late summer, from 20 to 100 or so workers populate the nest.

Paper wasps build a single comb using plant fibers (Figure 14.30). The paper wasp comb is open to the environment and, when full of larvae and pupae, is a tempting food item for rodents and birds, hence the strong defensive reaction of the workers. Primitively eusocial wasp colonies have a dominance hierarchy, in which the queen is quite physically aggressive to subordinates. Worker larvae may also receive inadequate nutrition for full reproductive development.

Yellowjackets and hornets construct paper nests consisting of a stack of combs that are covered by a spherical paper sheath (Figure 14.31). In highly eusocial wasps, the mechanism of queen dominance in the colony is not fully understood. Queens are much larger than workers, but as yet neither dominance nor pheromones have been implicated in regulation of these colonies, which can have hundreds or even thousands of workers. Yellowjackets and hornets have extremely aggressive and intimidating defensive responses to disturbances in the area around their nests.

The beginning stages of wasp colonies present an interesting evolutionary puzzle. Many wasp species found new nests with multiple queens. From a kin selection perspective, this seems counterintuitive: even if the queens are sisters, if one attains reproductive dominance, the reduction of the rest to the production

FIGURE 14.31
A yellowjacket nest. The colony is started by a single queen in the spring and grows to hundreds or thousands of highly defensive wasps by late summer. Yellowjackets and hornets, which have similar nests, are fiercely defensive and deliver painful stings. *Photo: Howard Ensign Evans, Colorado State University, Bugwood.org.*

491

> **KEY TERM** Primitively eusocial animals have relatively little physical differentiation between the queen and workers. Workers are morphologically capable of acting as queens.

> **KEY TERM** Highly eusocial animals have physical differentiation between the queen and the workers, so workers are morphologically incapable of functioning as queens.

of nieces may be a losing proposition, evolutionarily. How might multiple foundresses in wasp colonies be explained? Possibly, they are not really working but are waiting on the periphery for the chance to take over the nest, should the dominant foundress die. Or perhaps the foundresses work together to ensure the early success of the nest but then later compete intensely over reproductive rights. Finally, the option of founding a nest alone may be so unfavorable that ecological constraints force grouping of sisters (See Chapter 13). Work by Joan Strassmann of Washington University in St. Louis and other researchers has shown that foundresses are typically sisters and that kin selection is the best model to explain this cooperation.[77,78] However, ecological constraints are quite important, and under the right ecological conditions, nest founding by solitary wasps, rather than by groups, is favored.

Bees

Most of the world's 25,000 or so bee species are solitary, but the eusocial species get the most notice. All but a few bee species specialize on nectar and pollen diets; the evolutionary history of bees is inextricably interwoven with their association with flowers. Bees, both solitary and eusocial, are the most important pollinators in nearly every ecosystem. In North America, the primitively eusocial bumblebees are among the most commonly observed native bees. The highly eusocial honeybee, which was introduced to North America from Europe in the 1600s, is now the dominant pollinator in all but the highest elevation or latitude ecosystems.

Bumblebee queens leave their natal nest in the late summer or early fall and mate, usually with only one male. These queens then find a safe place, often underground, to spend the winter. In the spring, the queen establishes a nest, usually in an abandoned rodent burrow but sometimes in an old bird's nest or just in an underground cavity. The adopted nesting material from the rodent provides insulation and support for the bumblebee's nest, which is constructed of wax and cocoons spun by the larvae. In the early stages of colony life, the queen does all of the tasks in the nest; she constructs the nest, lays the eggs, forages for food, and defends the nest against predators and parasites. Her first batch of larvae, usually 4–8 or 10 in number, are all female and grow to be much smaller adults than the queen. These small adults, first seen in May or June, are workers and take over the work of the colony, except for egg laying. The queen maintains her reproductive dominance of the society by physical means, forming a dominance hierarchy in the nest once the workers are present. In warmer climates, additional female worker brood may be reared so that a bumblebee colony can grow to include several dozen workers. In late summer or early fall, the queen lays a mixed batch of male and female eggs; the female larvae hatching from these eggs are better fed than the males are and develop into large adults, which will complete the cycle by becoming the next summer's queens. The males are short-lived, serving only to mate with the queens. There are more than two dozen species of bumblebee in North America. The annual colony cycle, physical dominance hierarchy, overlap of queen and workers in their activities, and small colony size are all characteristics of primitive eusociality.

Honeybees are highly eusocial and have a perennial colony life cycle (Figure 14.32); the key to honeybee survival is the storage of large amounts of nectar (in the form of honey) and pollen so that the colony can survive the winter. Honeybee nests are constructed in cavities, such as hollow trees. Bees excrete the wax, which is shaped into extensive combs. The queen honeybee lacks the hairs needed to carry pollen and never forages; her roles in the society are restricted to egg-laying and secreting pheromones, which are spread through the colony. Honeybee queen pheromones inform the workers of the presence of the queen and may inhibit workers from laying eggs; if the queen is removed, the absence of her pheromones stimulates efforts by the workers to rear replacement queens and to lay haploid male eggs, something they can do in the absence of being mated.

All female eggs have the potential to become either workers or queens; the vast majority of eggs are reared as workers. Larvae receiving a carbohydrate-rich diet develop into workers;

FIGURE 14.32
(Upper left and lower right) Nest entrances of tropical stingless bees, showing guards. (Middle and lower left) Nests of the giant honey bee, *Apis dorsata*, in Southeast Asia. (Upper right) Honeybee comb, showing the repeated hexagonal structure of cells in which larvae are reared and food is stored. *Photos: Michael Breed.*

the workers are uniform in size, have hairs over their bodies for collecting pollen, have very small ovaries, and have barbed stings. Queen larvae receive a diet rich in lipid and protein (royal jelly, which is produced by glands in the workers' heads); this rich diet stimulates development of a larger, relatively hairless female with small eyes, an elongated abdomen, large ovaries, and no barbs on her sting. Queens are reared in the spring and leave their natal colony on mating flights; a queen mates between 10 and 20 times when she is young and stores the semen for use throughout her life, which can last 5 or 6 years. When a swarm begins, the old queen in a colony leaves with about half the workers; this swarm searches for a new nesting location. Meanwhile, the new queen then takes over reproduction in the nest in which she hatched. The clear physical differentiation of queens and workers and queen dominance via pheromones are characteristics of a highly eusocial society.

SUMMARY

This chapter builds on how the evolutionary forces of selfishness, reciprocity, and kin selection described in Chapter 13 can bring groups of animals together. The discussion of cooperation in birds started with aggregated nesting in swallows. In swallows, a distinct monogamous pair maintains each nest, and the aggregation results from widespread attraction of birds to (limited) good nesting sites, as well as possible benefits from shared vigilance against predators. Sociable weavers and acorn woodpeckers both nest cooperatively, with shared nest construction and maintenance and, particularly in acorn woodpeckers, cooperative rearing of nestlings. In both of these species, helpers at the nest also contribute to the survivorship of the group.

A similar progression occurs in mammals, ranging from the relatively independent lives of the members of a horse herd, which nevertheless gain antipredator benefits from group membership, to highly organized hyena and baboon societies. Among vertebrates, the permanently sterile workers in eusocial colonies of naked mole rat colonies represent the extreme of helping behavior and of differential roles within a society.

In invertebrates, there are numerous examples of eusociality. They include gall-occupying plant feeders, shrimp, termites, ants, wasps, and bees. Invertebrate eusocial societies typically feature division of labor in colonies that makes the social group ecologically efficient. Division of labor contributes much to the dominant ecological roles of these animals. Reproductive differences between queens and workers are maintained, highlighting evolutionary implications of worker reproductive suppression. Eusocial invertebrates hold substantial resources in their nest and become targets for predation. This has stimulated novel and highly effective defensive strategies. Stinging, biting, and chemical defenses are all present in eusocial invertebrates.

The breadth of ecological roles and behavioral adaptations in eusocial animals correlates with their ecological dominance. Termites are important consumers of cellulosic material. Bees pollinate plants and play key roles in the maintenance of many communities. Ants have a tremendous range of feeding niches and are among the most numerous animals in most terrestrial ecosystems. In sum, social behavior provides a set of remarkable adaptations for the ecological success of many animal species.

STUDY QUESTIONS

1. Our examples of bird social behavior represented a range from aggregated nests (swallows) to groups with cooperative brood care (sociable weavers and acorn woodpeckers). What common themes exist among these species? Are there key differences in their social behavior?
2. Group living has many advantages but can have the downside of focusing competition for food within the group. In the examples of bird and mammal social behavior, what balances have been struck between the value of living in a group and competition for food?
3. How has social competition affected female morphology and behavior in the spotted hyena?
4. Can you explain the differences between division of labor based on a dominance hierarchy and temporal division of labor?
5. Why are eusocial insects so successful ecologically?
6. Why do the eusocial insects have such strong defensive mechanisms? Give some examples of particularly effective defenses.

Further Reading

Breed, M.D., 2010. Honeybees. In: Breed, M.D., Moore, J. (Eds.), Encyclopedia of Animal Behavior, vol. 1, Academic Press, Oxford, pp. 89–96.

Brown, J.H., 2010. Swallow summer Encyclopedia of Animal Behavior, vol. 1. Academic Press, Oxford, pp. 548–552.

Dawkins, R., 1976. The Selfish Gene. Oxford University Press, New York, NY, pp. 368.

Fewell, J.H., 2010. Division of labor. In: Breed, M.D., Moore, J. (Eds.), Encyclopedia of Animal Behavior, vol. 1, Academic Press, Oxford, pp. 548–552.

Gadau, J., Fewell, J.H. (Eds.), 2009. Organization of Insect Societies: From Genome to Sociocomplexity, Harvard University Press, Cambridge, MA, pp. 640.

Holldobler, B., Wilson, E.O., 1990. The Ants. Belknap Press of Harvard University Press, Cambridge, MA, pp. 752.

Lubin, Y., 2010. Spiders: social evolution. In: Breed, M.D., Moore, J. (Eds.), Encyclopedia of Animal Behavior, vol. 2, Academic Press, Oxford, pp. 329–334.

Stanford, C.B., 1998. The social behavior of chimpanzees and bonobos: empirical evidence and shifting assumptions. Curr. Anthropol. 39 (4), 399–420.

Notes

1. Lea, A.J., Learn, N.H., Theus, M.J., Altmann, J., Alberts, S.C., 2014. Complex sources of variance in female dominance rank in a nepotistic society. Anim. Behav. 94, 87–99.
2. Wilson, E.O., 1975. Sociobiology: The New synthesis. Belknap Press of Harvard University Press, Cambridge, MA, pp. 697, 720.
3. Schultz, T.R., Brady, S.G., 2008. Major evolutionary transitions in ant agriculture. Proc. Natl. Acad. Sci. USA 105, 5435–5440.
4. Holldobler, B., Wilson, E.O., 2011. The Leafcutter Ants: Civilization by Instinct. W. W. Norton and Company, Inc., New York, NY, pp. 160.
5. Rodrigues, A., Bacci, M., Mueller, U.G., Ortiz, A., Pagnocca, F.C., 2008. Microfungal "weeds" in the leafcutter ant symbiosis. Microb. Ecol. 56, 604–614.
6. Díaz-Muñoz, S.L., DuValb, E.H., Krakauer, A.H., Lacey, E.A., 2014. Cooperating to compete: altruism, sexual selection and causes of male reproductive cooperation. Anim. Behav. 88, 67–78.
7. Spottiswoode, C.N., 2007. Phenotypic sorting in morphology and reproductive investment among sociable weaver colonies. Oecologia 154, 589–600.
8. Covas, R., Brown, C.R., Anderson, M.D., Brown, M.B., 2004. Juvenile and adult survival in a southern temperate colonial cooperative breeder. Auk 121, 1199–1207.
Covas, R., Doutrelant, C., du Plessis, M.A., 2004. Experimental evidence of a link between breeding conditions and the decision to breed or to help in a colonial cooperative breeder. Proc. R. Soc. Lond., B, Biol. Sci. 271, 827–832.
9. Doutrelant, C., Covas, R., 2007. Helping has signalling characteristics in a cooperatively breeding bird. Anim. Behav. 74, 739–747.
10. Koenig, W.D., Walters, E.L., Haydock, J., 2009. Helpers and egg investment in the cooperatively breeding acorn woodpecker: testing the concealed helper effects hypothesis. Behav. Ecol. Sociobiol. 63, 1659–1665.
Koenig, W.D., Mumme, R.L., 1987. Population Ecology of the Cooperatively Breeding Acorn Woodpecker. Princeton University Press., pp. 462.
Koenig, W.D., Haydock, J., Stanback, M.T., 1998. Reproductive roles in the cooperatively breeding acorn woodpecker: incest avoidance versus reproductive competition. Am. Nat. 151, 243–255.
Haydock, J., Koenig, W.D., 2002. Reproductive skew in the polygynandrous acorn woodpecker. Proc. Natl. Acad. Sci. USA 99, 7178–7183.
11. Linklater, W.L., Cameron, E.Z., 2000. Tests for cooperative behaviour between stallions. Anim. Behav. 60, 731–743.
Linklater, W.L., Cameron, E.Z., Minot, E.O., Stafford, K.J., 1999. Stallion harassment and the mating system of horses. Anim. Behav. 58, 295–306.
12. Benhaiem, S., Hofer, H., Kramer-Schadt, S., Brunner, E., East, M.L., 2012. Sibling rivalry: training effects, emergence of dominance and incomplete control. Proc. R. Soc. Lond. B, Biol. Sci. 279, 3727–3735.
13. Frank, L.G., 1986. Social organization of the spotted hyaena (Crocuta crocuta): I and II. Anim. Behav. 34, 1500–1527.
14. Holekamp, K.E., Sakai, S.T., Lundrigan, B.L., 2007. Social intelligence in the spotted hyena (Crocuta crocuta). Philos. Trans. R. Soc. Lond. B, Biol. Sci. 362, 523–538.
15. Grinnell, J., Packer, C., Pusey, A.E., 1995. Cooperation in male lions: kinship, reciprocity or mutualism? Anim. Behav. 49, 95–105.
16. Bygott, J.D., Bertram, B.C.R., Hanby, J.P., 1979. Male lions in large coalitions gain reproductive advantages. Nature 282, 839–841.
17. Packer, C., Pusey, A.E., Eberly, L., 2001. Egalitarianism in female African lions. Science 293 (5530), 690–693.
Packer, C., Heinsohn, R., 1996. Lioness leadership. Science 271 (5253), 1215–1216.
18. Spong, G., Creel, S., 2004. Effects of kinship on territorial conflicts among groups of lions, Panthera leo. Behav. Ecol. Sociobiol. 55, 325–331.
19. Goodall, J., 1986. The Chimpanzees of Gombe. Belknap Press, Cambridge, MA.
20. Nishida, T. (Ed.), 1990. The Chimpanzees of the Mahale Mountains University of Tokyo Press, Tokyo.
21. Cronin, K.A., van Leeuwen, E.J.C., Vreeman, V., Haun, D.B.M., 2014. Population-level variability in the social climates of four chimpanzee societies. Evol. Human Behav. 35, 389–396.
22. Langergraber, K., Mitani, J., Vigilant, L., 2007. The limited impact of kinship on cooperation in wild chimpanzees. Proc. Natl. Acad. Sci. USA 104, 7786–7790.
23. Muller, M., Mitani, J., 2005. Conflict and cooperation in wild chimpanzees. Adv. Study Behav. 35, 275–331.

24. Pusey, A., Williams, J., Goodall, J., 1997. The influence of dominance rank on the reproductive success of female chimpanzees. Science 277, 828–831.
25. Duffy, K., Wrangham, R., Silk, J., 2007. Male chimpanzees exchange political support for mating opportunities. Curr. Biol. 17, R586–R587.
26. Silk, J.B., 2007. Chimps don't just get mad, they get even. Proc. Natl. Acad. Sci. USA 104, 13537–13538.
27. Reeve, H.K., 1992. Queen activation of lazy workers in colonies of the eusocial naked mole-rat. Nature 358 (6382), 147–149.
28. O'Riain, M.J., Jarvis, J.U.M., Alexander, R., Buffenstein, R., Peeters, C., 2000. Morphological castes in a vertebrate. Proc. Natl. Acad. Sci. USA 97 (24), 13194–13197.
29. Bromham, L., Harvey, P.H., 1996. Behavioural ecology: naked mole-rats on the move. Curr. Biol. 6 (9), 1082–1083.
30. Cooney, R., Bennett, B.C., 2000. Inbreeding avoidance and reproductive skew in a cooperative mammal. Proc. R. Soc. Lond. B, Biol. Sci. 267, 801–806.
31. Scantlebury, M., Speakman, J.R., Oosthuizen, M.K., Roper, T.J., Bennett, N.C., 2006. Energetics reveals physiologically distinct castes in a eusocial mammal. Nature 440, 795–797.
32. Henry, E.C., Dengler-Crish, C.M., Catania, K.C., 2007. Growing out of a caste—reproduction and the making of the queen mole-rat. J. Exp. Biol. 210 (2), 261–268.
33. Clarke, F.M., Faulkes, C.G., 1997. Dominance and queen succession in captive colonies of the eusocial naked mole-rat, *Heterocephalus glaber*. Proc. R. Soc. Lond. B, Biol. Sci. 264 (1384), 993–1000.
34. Mooney, S.J., Douglas, N.R., Holmes, M.M., 2014. Peripheral administration of oxytocin increases social affiliation in the naked mole-rat (*Heterocephalus glaber*). Horm. Behav. 65, 380–385.
35. O'Donnell, S., 1995. Division of labor in postemergence colonies of the primitively eusocial wasp *Polistes instabilis de Saussure* (Hymenoptera, Vespidae). Insect. Soc. 42, 17–29.
36. Markiewicz, D.A., O'Donnell, S., 2001. Social dominance, task performance and nutrition: implications for reproduction in eusocial wasps. J. Comp. Physiol. A. Neuroethol. Sens. Neural. Behav. Physiol. 187, 327–333.
37. Dawkins, R., 1982. The Extended Phenotype. Oxford University Press, Oxford.
38. Breed, M.D., Harrison, J., 1988. Caste in the giant tropical ant, *Paraponera clavata*. J. Kans. Entomol. Soc. 61, 285–290.
39. Wilson, E.O., 1974. The Insect Societies. Belknap Press of Harvard University Press, Cambridge, MA, pp. 562.
40. Feener, D.H., Moss, K.A.G., 1990. Defense against parasites by hitchhikers in leaf-cutting ants: a quantitative assessment. Behav. Ecol. Sociobiol. 26, 17–29.
41. Seeley, T.D., 1982. Adaptive significance of the age polyethism schedule in honeybee colonies. Behav. Ecol. Sociobiol. 11, 287–293.
42. Muscedere, M.L., Willey, T.A., Traniello, J.F.A., 2009. Age and task efficiency in the ant *Pheidole dentata*: young minor workers are not specialist nurses. Anim. Behav. 77, 911–918.
43. Shorter, J.R., Tibbetts, E.A., 2009. The effect of juvenile hormone on temporal polyethism in the paper wasp *Polistes dominulus*. Insect. Soc. 56, 7–13.
44. Wilson, E.O., 1976. Behavioral discretization and the number of castes in an ant species. Behav. Ecol. Sociobiol. 1, 63–81.
45. Trivers, R.L., Hare, H., 1976. Haploidploidy and the evolution of the social insect. Science 191, 249–263.
46. Ratnieks, F.L.W., Visscher, P.K., 1989. Worker policing in the honey bee. Nature 342, 796–797.
47. Kaercher, M.H., Ratnieks, F.L.W., 2014. Killing and replacing queen-laid eggs: low cost of worker policing in the honeybee. Am. Nat. 184, 110–118.
48. Tarpy, D., Page, R.E., 2002. Sex determination and the evolution of polyandry in honey bees (*Apis mellifera*). Behav. Ecol. Sociobiol. 52, 143–150.
49. Kukuk, P.F., May, B., 1991. Colony dynamics in a primitively eusocial halictine bee *Lasioglossum (Dialictus) zephyrum* (Hymenoptera: Halictidae). Insect. Soc. 38, 171–188.
50. Pratte, M., 1997. Recognition and social dominance in *Polistes* wasps. J. Ethol. 15, 55–59.
51. van Doorn, A., Heringa, J., 1986. The ontogeny of a dominance hierarchy in colonies of the bumblebee, *Bombus terrestris* (Hymenoptera, Apidae). Insect. Soc. 33, 3–25.
52. Slessor, K.N., Winston, M.L., Le Conte, Y., 2005. Pheromone communication in the honeybee (*Apis mellifera* L.). J. Chem. Ecol. 31, 2731–2745.
53. Lattorff, H.M.G., Moritz, R.F.A., 2013. Genetic underpinnings of division of labor in the honeybee (*Apis mellifera*). Trends Genet. 29, 641–648.
54. Lüscher, M., 1960. Hormonal control of caste differentiation in termites. Ann. N. Y. Acad. Sci. 89, 549–563.
55. Korb, J., Weil, T., Hoffmann, K., Foster, K.R., Rehli, M., 2009. A gene necessary for reproductive suppression in termites. Science 324, 758.
56. Van Oystaeyen, A., Oliveira, R.C., Holman, L., et al., 2014. Conserved class of queen pheromones stops social insect workers from reproducing. Science 343, 287–290.
57. Prestwich, G.D., 1983. Chemical systematics of termite exocrine secretions. Annu. Rev. Ecol. Syst. 14, 287–311.
58. Jones, T.H., Clark, D.A., Edwards, A.A., Davidson, D.W., Spande, T.F., Snelling, R.R., 2004. The chemistry of exploding ants, *Camponotus* spp. (cylindricus complex). J. Chem. Ecol. 30, 1479–1492.
59. Duffy, J.E., 1996. Eusociality in a coral-reef shrimp. Nature 381, 512–513; Duffy, J.E., Morrison, C.L., Macdonald, K.S., 2002. Colony defense and behavioral differentiation in the eusocial shrimp *Synalpheus regalis*. Behav. Ecol. Sociobiol. 51 (5), 488–495.

Duffy, J.E., Morrison, C.L., Rios, R., 2000. Multiple origins of eusociality among sponge-dwelling shrimps (*Synalpheus*). Evolution 54 (2), 503–516.

60. Toth, E., Duffy, J.E., 2005. Coordinated group response to nest intruders in social shrimp. Biol. Lett. 1 (1), 49–52.

61. Crespi, B.J., 1992. Eusociality in Australian gall-thrips. Nature 359, 724–726.
    Crespi, B.J., 1994. Three conditions for the evolution of eusociality: are they sufficient? Insect. Soc. 41, 393–400.

62. Normark, B.B., Jordal, B.H., Farrell, B.D., 1999. Origin of a haplodiploid beetle lineage. Proc. R. Soc. Lond. B, Biol. Sci. 266 (1435), 2253–2259.

63. Chapman, T.W., Crespi, B.J., Kranz, B.D., Schwartz, M.P., 2000. High relatedness and inbreeding at the origin of eusociality in gall-inducing thrips. Proc. Natl. Acad. Sci. USA 97 (4), 1648–1650.
    Chapman, T.W., Kranz, B.D., Bejah, K.L., et al., 2002. The evolution of soldier reproduction in social thrips. Behav. Ecol. 13 (4), 519–525.

64. Bono, J.M., Crespi, B.J., 2008. Cofoundress relatedness and group productivity in colonies of social *Dunatothrips* (Insecta: Thysanoptera) on Australian Acacia. Behav. Ecol. Sociobiol. 62, 1489–1498.
    Bono, J.M., Crespi, B.J., 2006. Costs and benefits of joint colony founding in Australian Acacia thrips. Insect. Soc. 53, 489–495.

65. Perry, S.P., Chapman, T.W., Schwarz, M.P., Crespi, B.J., 2004. Proclivity and effectiveness in gall defence by soldiers in five species of gall-inducing thrips: benefits of morphological caste dimorphism in two species (*Kladothrips intermedius* and *K. habrus*). Behav. Ecol. Sociobiol. 56 (6), 602–610.

66. Kranz, B.D., Chapman, T.W., Crespi, B.J., Schwarz, M.P., 2001. Social biology and sex ratios in the gall-inducing thrips, *Oncothrips waterhousei* and *Oncothrips habrus*. Insect. Soc. 48 (4), 315–323.
    Kranz, B.D., Schwarz, M.P., Morris, D.C., Crespi, B.J., 2002. Life history of *Kladothrips ellobus* and *Oncothrips rodwayi*: insight into the origin and loss of soldiers in gall-inducing thrips. Ecol. Entomol. 27 (1), 49–57.
    Kranz, B.D., Schwarz, M.P., Mound, L.A., Crespi, B.J., 1999. Social biology and sex ratios of the eusocial gall-inducing thrips *Kladothrips hamiltoni*. Ecol. Entomol. 24 (4), 432–442.
    Kranz, B.D., Schwarz, M.P., Wills, T.E., Chapman, T.W., Morris, D.C., Crespi, B.J., 2001. A fully reproductive fighting morph in a soldier clade of gall-inducing thrips (*Oncothrips morrisi*). Behav. Ecol. Sociobiol. 50 (2), 151–161.

67. Ijichi, N., Shibao, H., Miura, T., Matsumoto, M., Fukatsu, T., et al., 2005. Comparative analysis of caste differentiation during embryogenesis of social aphids whose soldier castes evolved independently. Insect. Soc. 52 (2), 177–185.
    Fukatsu, T., Sariya, A, Shibao, H., 2005. Soldier caste with morphological and reproductive division in the aphid tribe *Nipponaphidini*. Insect. Soc. 52 (2), 132–138.

68. Aoki, S., 1991. Aphid soldiers discriminate between soldiers and non-soldiers, rather than between kin and nonkin, in *Ceratoglyphina bambusae*. Anim. Behav. 42, 865.
    Aoki, S., Imai, M., 2005. Factors affecting the proportion of sterile soldiers in growing aphid colonies. Popul. Ecol. 47 (2), 127–136.
    Aoki, S., Kurosu, U., 2004. How many soldiers are optimal for an aphid colony? J. Theor. Biol. 230 (3), 313–317.

69. Benton, T.G., Foster, W.A., 1992. Altruistic housekeeping in a social aphid. Proc. R. Soc. Lond. B, Biol. Sci. 247 (1320), 199–202.

70. Pike, N., Foster, W., 2004. Fortress repair in the social aphid species *Pemphigus spyrothecae*. Anim. Behav. 67, 909–914.

71. Fukatsu, T., Sarjiya, A., Shibao, H., 2005. Soldier caste with morphological and reproductive division in the aphid tribe *Nipponaphidini*. Insect. Soc. 52, 132–138.

72. Ladley, D., Bullock, S., 2005. The role of logistic constraints in termite construction of chambers and tunnels. J. Theor. Biol. 234 (4), 551–564.

73. Stewart, A.D., Zalucki, M.P., 2006. Developmental pathways in *Microcerotermes turneri* (Termitidae: Termitinae): biometric descriptors of worker caste and instar. Sociobiology 48, 727–740.

74. Miura, T., 2004. Proximate mechanisms and evolution of caste polyphenism in social insects: from sociality to genes. Ecol. Res. 19, 141–148.

75. Sendova-Franks, A.B., Hayward, R.K., Wulf, B., Klimek, T., James, R., Planqué, R., et al., 2010. Emergency networking: famine relief in ant colonies. Anim. Behav. 79, 473–485.

76. Reber, A., Meunier, J., Chapuisat, M., 2010. Flexible colony-founding strategies in a socially polymorphic ant. Anim. Behav. 79, 467–472.

77. Queller, D.C., Peters, J.M., Solís, C.R., Strassmann, J.E., 1997. Control of reproduction in social insect colonies: individual and collective relatedness preferences in the paper wasp, *Polistes annularis*. Behav. Ecol. Sociobiol. 40, 3–16.

78. Goodnight, K.F., Strassmann, J.E., Klingler, C.J., Queller, D.C., 1996. Single mating and its implications for kinship structure in a multiple-queen wasp. Ethol. Ecol. Evol. 8, 191–198.

497

Conservation and Behavior

499

LEARNING OBJECTIVES

Studying this chapter should provide you with the knowledge to:

- Know that conservation behavior offers a broad range of tools for understanding and possibly manipulating animal behavior to further conservation.

- Learn how maintenance of animals in captivity can serve the goals of preserving endangered species.

- Use animal behavior to conserve genetic resources and to manage captive populations with the goal of reintroducing animals to natural habitats.

Animal Behavior. DOI: http://dx.doi.org/10.1016/B978-0-12-801532-2.00015-5

- Understand that introduction of captive animals into natural habitats must be informed by knowledge of behavior.
- Recognize the key role that neophobia, the fear of new situations, has in protecting animals from risky situations and how neophobia can be manipulated to keep human and animal populations apart; the lack of neophobia, or a breakdown of neophobia due to habituation is central to the success of invasive animals and contributes to negative human–wildlife interactions.
- Gain a comprehensive understanding of how human behavior affects the behavior of wildlife; understand that activities such as hunting and feeding birds affect population sizes and competitive interactions among populations, and that carelessly making food available to wildlife brings animals into contact with human populations, often resulting in public pressure on wildlife.

15.1 INTRODUCTION: CONSERVATION AND THE FUTURE OF ANIMAL BEHAVIOR

All of the subfields of animal behavior converge when behavior is applied to conservation issues. Conservation behavior divides easily into the following five topics: species protection in natural habitats, extinctions and behavior, reserve design, captive breeding and reintroductions, and *human–wildlife interfaces*.

> **KEY TERM** The human–wildlife interface is a habitat in which humans are likely to come into close contact with members of a wildlife species.

This chapter reviews the important roles that animal behavior plays in each of these conservation priorities. We begin with conservation behavior in natural settings; then consider how behavior and extinction may be coupled in those settings; and then move to near-natural settings (reserves), and how understanding behavior is essential to their successful design. The behavior of animals in captive populations is considered along with reintroductions, because often animals are maintained in captivity with the goal of reintroducing them into the wild. Finally, we consider the least natural setting of all, and perhaps the most problematic—the one that develops when people join a neighborhood in which wild animals are the neighbors. In each of these settings, from natural habitats to suburbs, from reserves to captive breeding operations, knowledge of animal behavior is critical for success in conservation.[1–7]

How can an understanding of behavior be used to enhance the survival of threatened and endangered species (Figure 15.1)? How can that knowledge prevent species from becoming endangered, and what are the behavioral consequences of extinction for remaining community members? The protection of a species in its native habitat requires knowledge of its behavior, including diet choices, home range or territory size, and mating system. A successful reintroduction of a species into the wild requires knowledge of how these formerly captive animals can survive in a natural habitat:

- How will they know what to eat?
- How will they gain the knowledge to protect themselves from predators?
- Will they be able to find mates and rear young?
- Do they need information to migrate successfully?

The scientific discipline of conservation behavior provides the answers to these questions.

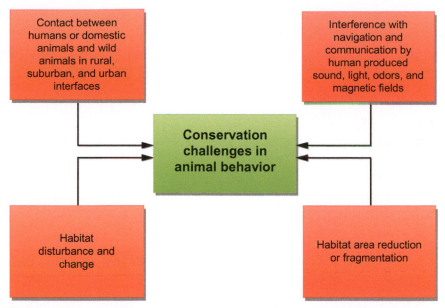

FIGURE 15.1
Much of what humans do impacts the welfare of animal populations. Human–wildlife interactions are complex and animal behaviorists have much to bring to conversations about conservation, habitat preservation, and protection of endangered species.

CASE STUDY: THE CALIFORNIA CONDOR

The California condor, the largest bird in North America, is an excellent example of what we need to know to be able to help animals survive (Figure 15.2). The California condor was 27 birds away from extinction in the 1980s; a successful captive breeding program in southern California zoos has allowed the release of more than 200 captive-reared condors. Disappointingly, very few chicks have fledged. In part, this failure results from the parents' seeming penchant for feeding nestlings indigestible junk from mountaintops, places that are frequented by both condors and people, thus giving new meaning to the term *junk food*—delicacies that include such snacks as bottle caps, plastic pipes, and even glass. These items—called *microtrash*—have one thing in common: they are small. People who would not leave a large piece of trash behind might carelessly drop a piece of microtrash. Researchers hypothesize that parent birds may be mistaking these things for bones, a normal condor chick diet item, but at this stage in the reintroduction, what the condor is doing is more important than what it might be thinking. In an effort to thwart the condor's fascination with microtrash, cleanup crews are scouring mountaintops (where condors like to visit and people like to leave trash), picking up all manner of small litter (http://www.fws.gov/cno/es/calcondor/condorthreats.cfm). The mortality of condor chicks has increased in the past 25 years, a reflection of both increasing use of wild places by people—and increasing trash left behind. Given the condor life history—maturity at 6 years, reproduction (one egg) every 2 years—and its ability to cover large areas of terrain, ignorance of condor behavior can be costly, both for the condor and for those who are trying to save it from extinction. In addition, recent studies have shown that the seeming success of bringing condors back from the brink may not continue without ongoing management.

FIGURE 15.2
The California condor. Plagued by microtrash and lead in its diet, this bird's numbers are slowly increasing, but only under intense management and captive release programs. *Photo: Noel Snyder/USFWS.*

This is because of another dietary threat: lead. Condors are long-lived birds, and unlike other carrion-eaters, have a distinct preference for large carcasses. These carcasses often contain lead shot, and lead to harmful levels of lead in condor blood. Thirty percent of free-flying condors have blood lead levels greater than 200 ng/mL, leading to subclinical health effects, and approximately 20% of the birds have over twice that level, causing clinical symptoms that may require intervention.[8] Finally, condors are not the only animals that sustain damage from microtrash. Microplastics (<5 mm long) are abundant in the sea, where they are not only eaten, but also contaminate gills.[9,10]

In addition, conservation behavior should impact the design of reserves that protect communities or ecosystems. Surprisingly, the importance of behavior is sometimes overlooked in the design of conservation plans. Essential elements of a reserve design include knowledge of territorial requirements and minimum population sizes for functional mating systems of *keystone species*. Dispersal patterns and how new animals are incorporated in the breeding population offer important insight into how genetic diversity can be maintained. Reserve designers also need to know how to assess the impacts of supplemental feeding, how to gauge the behavioral consequences of the absence of species in reserves that were present in undisturbed communities, and how to manage human–wildlife interactions in reserves.

> **KEY TERM** A keystone species is one that regulates the populations of other species via its actions as a predator or its effects on the ecosystem.

Perhaps the most noticeable lapses in reserve design and management come with underestimating the impact of human recreational activities on wildlife behavior. This impact can change wildlife behavior in ways that are highly detrimental to animal populations. With the growth of human populations near national parks, wildlife reserves, and other natural areas, behavioral conflicts between humans and wildlife are inevitable. Conservation behavior is as much about managing human behavior in relation to wildlife as it is about understanding and managing wildlife behavior, and behavioral biologists who address these conservation issues are far more essential than heretofore recognized.

In sum, the study and application of conservation behavior operates in many different arenas. Given the integrative nature of conservation behavior as a discipline, this is a rapidly evolving field. The goal of this chapter is to establish—undeniably—that animal behavior must play a central role in any credible discussion of the protection of species or ecosystems.

15.2 SPECIES PROTECTION IN NATURAL HABITATS

How is species protection status assessed and described? The answer to this question helps to direct this discussion of conservation behavior. Conservationists usually begin this process with a response to reports or surveys that show a species has a limited population size or has undergone rapid population declines. Some species are rare or limited in range simply as a reflection of their biology. *Endemics*, populations found in restricted geographic areas, often present particularly serious conservation challenges and are very susceptible to habitat destruction and climate change. For example, the Tahiti monarch is a bird that is endemic to Tahiti; fewer than 100 individuals of this endangered species persist in the wild. This illustrates the larger point that island animals are particularly at risk; for instance, many Hawaiian animals are threatened or endangered endemics.[11] An interesting contrast is that while two North American bird species, Carolina parakeets and passenger pigeons, have been pushed to extinction in the past 150 years and two others, ivory-billed woodpeckers and Bachman's warbler, may be extinct, more than 25 Hawaiian bird species have disappeared (http://hbs.bishopmuseum.org/endangered/ext-birds.html). The land area of Hawaii is about 0.2% that of the continental United States; endemism in a small land area makes Hawaii's species particularly vulnerable to extinction.

OF SPECIAL INTEREST: ISLAND BIOGEOGRAPHY, BEHAVIOR, AND CONSERVATION

The theory of *island biogeography*, first proposed by Robert MacArthur and E. O. Wilson, is another example of "academic" science that has profound implications for conservation, ranging from extinctions on islands to the design of the wildlife reserve—another type of "island."[12] MacArthur and Wilson focused on (1) how island size affects diversity (using what are called *species-area curves*; Figure 15.3); (2) how distance from the mainland and other islands affects dispersal to the island (Figure 15.4); (3) how persistence of a population on an island and species' *life history strategies*, which MacArthur and Wilson characterized along a continuum of *r and K selection*, interact with colonization

KEY TERM Island biogeography is a field of scientific inquiry, proposed by Robert MacArthur and E. O. Wilson, that integrates the study of the distribution of animals on islands with knowledge of the behavioral and ecological processes that determine that distribution.

KEY TERM A species–area curve depicts the relationship between island size and the number of species found on a given island.

KEY TERM A life history strategy is the combination of the age at first reproduction, the number of offspring per clutch, the number of clutches per year, and the number of years of reproduction for any one animal in a population. These factors together determine *r*, or the rate of population growth.

KEY TERM The terms *r* and *K* selection refer to a continuum of life history strategies, ranging from short-lived, colonizing species with very high reproductive rates (*r*-selected) to long-lived species that compete successfully in saturated environments and have low reproductive rates (*K*-selected) (see also Chapter 12, section 3).

503

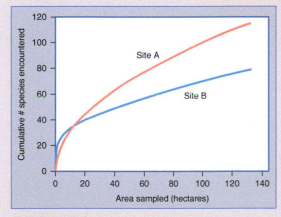

FIGURE 15.3
Example of a species–area curve. As the area sampled within a habitat increases (*x*-axis), the number of species increases as well (*y*-axis), but eventually nearly all of the species have been found, so further effort yields few new species. Some sites, such as A, require more sampling than others, such as B, to reach the point at which the curve flattens.

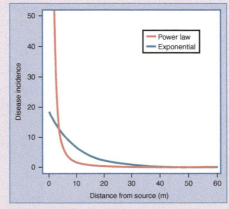

FIGURE 15.4
The effect of distance from a source population to a new habitat on dispersal. As distance increases (*x*-axis), the likelihood of successful dispersal diminishes rapidly.

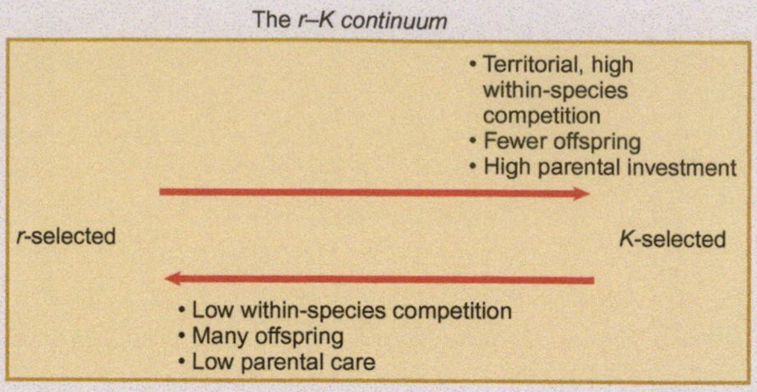

FIGURE 15.5

The *r–K* continuum of life history strategies; *r*-selected species are good dispersers, but often not behaviorally competitive in established communities.

and extinction (Figure 15.5). Many of the elements that make up the theory of island biogeography are behavioral; perhaps the most significant is dispersal (see Chapter 8). Because human activities fragment large tracts of habitat into smaller habitat islands, and because reserves are habitat islands, embedded in a "sea" of less welcoming places, this field of study has great relevance to conservation.[13]

Once a species is thought to be in trouble, more precise surveys are usually conducted. They can result in classification of a species or subspecies as being a conservation concern. Political arguments have surrounded the identification and classification process, and as a result, classifications by government agencies may differ from those of conservation organizations.

The listing of a species by the US Fish and Wildlife Service (USFWS) as endangered or threatened is a politicized process that may pit landowners against conservationists. This means that interpretation of USFWS listings needs to be tempered by knowledge of political issues. Presently, the International Union for the Conservation of Nature (IUCN) appears to provide the most credible evaluations; the IUCN has the advantage of taking a global, rather than national, view. The *IUCN Red List*, which is more politically independent and contains conservation assessments for many species, can be accessed online. Another good source of information about North American birds is the Audubon Society's watch list, which highlights birds with declining populations. Awareness of threatened or endangered birds and mammals tends to be much higher than that for other organisms; this leads to criticisms of listing processes as being biased to protect charismatic animals and consequently being not only incomplete, but also ill-considered with respect to animals that are less conspicuous but perhaps foundational in ecosystems. The IUCN Red Listing process is more inclusive than the USFWS process, in terms of including invertebrates and species that are less obvious to the public eye; nonetheless, the reality is that the conservation status of the vast majority of the planet's species is simply unknown.

KEY TERM A threatened species is one that is at risk of becoming endangered.

Three categories of species in trouble are particularly important in conservation—*threatened, endangered,* and *extinct in the wild.* Species that are extinct in the wild are divided into those that are completely extinct and those that are maintained only in captive populations. It is these captive species, whose last hope lies in the maintenance of captive populations and reintroductions in the wild, that have been the focus of the most intensive programs of behavioral research. Famous extinctions within the past two centuries include the dodo, a large nonflying bird that was endemic to Mauritius, and the passenger pigeon and Carolina parakeet in North America. The abundance of these two North American birds surpassed the descriptive abilities of most observers until early in the twentieth century; hunting was a major factor in their extinction. During that same time, no North American mammals have suffered complete extinction, although the American bison and the black-footed ferret came perilously close. The USFWS defines "endangered species" as ones that are in danger of extinction and "threatened" species as those that are at risk of becoming endangered. These definitions are necessarily nebulous, a factor that adds to contention in conservation discussions. In addition, the USFWS sometimes recognizes subspecies as threatened or endangered, even if other populations of the species are healthy.

> **KEY TERM** An endangered species is one that is at imminent risk of becoming extinct.

> **KEY TERM** A species that is extinct in the wild no longer exists in its natural habitat or in other noncaptive environments.

DISCUSSION POINT: CONSERVATION PRIORITIES

Is it better to focus conservation efforts and funding on identifying the potential for species loss and campaigning to save individual species, or to target habitat areas to save intact ecosystems? With classmates or friends, take turns defending each strategy.

Conservation efforts tend to focus on large and/or colorful birds and mammals because these species capture the interest of the public, and consequently government funding and private donations are available to support their conservation. In North America, many large mammals, such as mountain lions, wolves, and grizzly bears, are carnivores, and this sets up politically charged situations because such carnivores are regarded by private landowners as behavioral threats to livestock, pets, and perhaps to themselves. Animals do not recognize park or reserve boundaries as dispersal barriers. Populations in parks, such as wolves in the Greater Yellowstone Ecosystem inevitably serve as sources of dispersing animals into surrounding areas. Also, landowners may fear that legal protection of critical habitat will reduce the value of their land or prevent them from changing the use of the land.[14]

Animals maintained in parks or reserves can be reservoirs for disease that might be passed to livestock in surrounding areas. Bison and elk in the Greater Yellowstone Ecosystem may possibly harbor brucellosis, a bacterial disease that is devastating to livestock. Treating animals in wild populations for diseases such as brucellosis, rabies, or plague is difficult or impossible; when behavioral interactions between these populations and livestock, pets, or humans create risk for transmission of disease, landowner and public sentiment often turns against maintenance of the wild populations of animals. Many bats are the subject of serious conservation concerns; they are also incorrectly stereotyped as rabies-carrying vermin by much of the public, reducing the pressure for conservation of bats.

Unfortunately for some species that lack captive populations and are extinct in the wild, the only hope is that remnant undiscovered populations remain. This hope is not

OF SPECIAL INTEREST: BAT CONSERVATION

Bats that relied on forest elements, such as hollow trees, for roosting have suffered major impacts from deforestation. It may surprise you to learn that, like bluebirds and purple martins (discussed later in this chapter), bats can use nest boxes, and many people are involved in providing bat roosts by constructing nest boxes for bats (http://www.batcon.org/). Clearly, preserving natural roost sites is the highest priority, but when those have been removed, artificial roosts are an interesting conservation alternative[15] (Figure 15.6).

In tropical forests, bat species fill a wide variety of feeding niches; in addition to the familiar insectivores, bats serve as pollinators and dispersers of seed. Vampire bats, of course, are common in the tropics as well. Temperate zone bats are more typically insect eaters. Whatever their feeding niche, bats play important ecological roles that are disrupted by loss of nesting sites. Nest boxes in Costa Rican pastures support a range of bat species, an encouraging finding for bat conservation.[16] Artificial roosts also show promise in temperate zone areas.[17]

FIGURE 15.6
Bat boxes, designed to assist conservation of species such as the big brown bat, which suffers from habitat destruction. *Photo: Rick Adams.*

unfounded, as a few species thought to be extinct, such as the po'ouli honeycreeper (Hawaii), have been "rediscovered." A similar phenomenon occurred with the ivory-billed woodpecker. Until 2005, the last sightings of ivory-billed woodpeckers were in Louisiana in 1948 and in Cuba in 1988, and the species was thought to be extinct. Sightings reported in 2005 suggest that remnant populations of this species may exist in Arkansas. The search for this species continues, with behavioral traits (a distinctive call) giving biologists a major tool for recognizing the species. Another rediscovery is that of a golden-crowned manakin, spotted in Brazil in 2002 after a 45-year gap in sightings.[18] In contrast, the Golden Toad, a native to the mountains of Costa Rica, was never seen outside an area of about 4 square kilometers. This toad disappeared in 1989 and is presumed to be extinct, although the search continues for remnant populations. Understanding behavioral signs of species that are potentially extinct in the wild, such as scent marks, nests, food caches, flight patterns for birds spotted at a distance, and distinctive vocalizations, can play important roles in rediscoveries of species.

Other species, such as the Rodrigues fody, have recovered from extremely low population sizes. The continuing argument among conservationists is whether identifying and

OF SPECIAL INTEREST: THE PO'OULI HONEYCREEPER

What about species that are rare and elusive to begin with? The po'ouli honeycreeper was endemic to the Hawaiian island of Maui and was not known to scientists until 1973 (Figure 15.7). The range of this species was a 1300 hectare tract of forest. When discovered, this species' population was thought to be fewer than 150 birds; habitat destruction contributed to a decline, so that in 2000, only three birds, one male and two females, could be located in the wild. Each bird occupied a separate territory, and they did not seem likely to form a mating pair. This presented an extreme conservation dilemma; could the species be preserved based on such a limited population? Or was the species' fate already determined, making human intervention fruitless?

In a desperate effort to save the species, conservation biologists caught a female and translocated her to the male's territory. The female left the male's territory soon after she was moved and flew back to her previous territory. A male captured in an attempt to form a captive breeding population died, and the remaining two birds disappeared. Nevertheless, hope remains that this species will be rediscovered and that information gained from these experiences will facilitate a captive breeding program.

FIGURE 15.7
The po'ouli honeycreeper. *Photo: Paul E. Baker, USFWS.*

preserving (or reforesting) habitat areas is the best approach, or if the more extreme measures of translocation or captive breeding are more likely to succeed. This question is probably best resolved on a case-by-case basis, with the answer depending on the threats to the species and the likelihood of success of either approach. A glaring fault in the process is that decision making by governmental agencies can be excruciatingly slow, hampering emergency actions.

Many species that are extinct in the wild survive in populations in zoos or small reserves. Prior to the twentieth century, the American bison lived in vast herds in the United States, forming a key element of the plains ecosystem. Bison grazing, combined with fire and migratory locust swarms, prevented woody plants from invading grasslands. Bison were easy to kill with guns, and in the late nineteenth century, large numbers of bison were shot, often for their hides, but sometimes for no purpose other than the "pleasure" of the kill (Figure 15.8). Perhaps as many as 50,000,000 bison were killed, leaving a few small remnant populations on isolated ranches and Indian reservations; the National Park Service estimates that at the low point, the bison population numbered 600. A few ranchers and conservationists, including President Theodore (Teddy) Roosevelt, developed an interest in the preservation of this species, and captive and semicaptive breeding have generated a modern-day population of over 100,000. The American bison became federally protected in 1894; it is one of the first examples of using laws for conservation protection of a species.

FIGURE 15.8
A pile of bison skulls.
Public Domain.

FIGURE 15.9
American bison, a species that was re-established after near extinction. It is now abundant enough to be used as a source of lean meat. *Photo: US Fish and Wildlife Service.*

Behaviorally, bison are indeed wild animals—unpredictable, strong, and largely unmanageable; their wild behavior is part of their charisma. Probably much of the genetic variation of bison was lost in the population bottleneck of the late 1800s. An example of the remaining variation is the Wood Bison (Figure 15.9), found in northern Alberta, Canada; this subspecies is larger than Plains Bison found in the United States, with males weighing about a ton (900 kg). Plains and Wood subspecies of bison now live in several national parks in the United States and Canada, as well as on private ranches and reserves. One behavioral concern in the conservation of bison is their ability to hybridize with domestic cattle; beginning with the initial stages of the bison recovery, bison have been behaviorally managed in an effort to maintain the populations' genetic integrity. Another concern is that ranchers raising bison for slaughter might select for more behaviorally manageable (domesticated) stock.

The bison program for captive management of a much-reduced population and reintroduction into protected settings provides a model that conservationists have applied to a large number of species. (Six of these are discussed in detail in Section 15.5 on captive breeding and reintroduction.) The experience with bison proceeded largely through trial and error. It relied on possibly contradictory impulses to save the species, to provide stock for hunting expeditions, and to provide bison meat to food markets. A saving grace for the bison was its linkage to a romanticized image of the American West, embodied by larger-than-life figures such as Teddy Roosevelt and Buffalo Bill Cody. This theme of romanticizing emblematic species and making them targets of conservation efforts continues to this day; one role for conservation behaviorists is to emphasize the significance—and in some cases, the necessity—of less obvious and charismatic species in maintaining ecosystems.

Another way of thinking about conservation needs for species is to identify why a particular species is in trouble. Answers can include loss of critical habitat, reduction of population numbers below the minimum that prevents inbreeding and maintains genetic diversity, and emerging diseases or parasites. While one species may have more than one of these problems, each problem requires separate solutions. For example, black-footed ferrets, which are discussed later, suffered from a combination of reduction of populations below critical levels, susceptibility to canine distemper, and habitat loss. (The ferrets live in prairie dog towns.)

Scientists, conservation advocates, and politicians have not only quarreled over the processes for listing species as threatened or endangered, but they have also engaged in intense debate over whether conservation efforts should be focused on the preservation of significant endangered species, or on the preservation of critical ecosystems. The study of behavior brings much to either approach. As will become apparent, species conservation programs often succeed or fail based on how well, or poorly, they embrace an understanding of behavior. Knowledge of behavior is also central to reserve design because both the size of reserves and the importance of habitat components are dictated by the behavioral requirements of the animals in the community.

15.3 EXTINCTIONS AND BEHAVIOR

The loss of a species from a community can have far-reaching effects on the behavior of remaining species in that community. The extirpation of wolves and mountain lions from much of the United States, for example, has led to changed dietary habits and social behavior of the coyote. In the absence of both wolves and mountain lions, coyotes typically become the top mammalian predator in their communities. Coyote social groups become larger, and these larger social groups, by hunting cooperatively, can handle larger prey. Coyotes are also bolder and more commonly seen in daylight when wolves are absent. In the Greater Yellowstone Ecosystem, where wolves have been reintroduced, coyotes have been pushed back to the behavioral margin, subsisting on smaller prey and themselves risking predation from wolves.

More generally speaking, top predators are commonly among the most at-risk species in any community. The loss of top predators releases second-tier predators, allowing their populations to grow and exploit broader ranges of food. However, larger herbivores, which the second-tier predators cannot handle because of their size, may also experience a release of their populations, resulting in extensive loss of vegetation. In North America, when wolves and mountain lions are absent, the largest herbivores, such as deer (mule and white tailed), elk, and bison live virtually predator-free. How does this affect herbivore behavior? Hypothetically, herbivores without predators should be bolder, more willing to forage during daylight, and spend less time being vigilant. These effects have been demonstrated in elk, for example.[19]

Disruption of Migration: High Risks for Extinction

Long-distance migrants that depend on small patches of stopover habitats are very susceptible to habitat disruption. For many migratory birds, wetlands serve as important intermediate habitats; one of the compelling motivations for legal protection of wetlands is their key habitat role for migratory birds. Identification and protection of essential stopover points for birds other than waterfowl, such as warblers, is more challenging because of the warblers' small size, nocturnal migratory flights, and secretive habits. Conservation of the monarch butterfly, which migrates from the United States and Canada to the mountains of Mexico, has required international cooperation in attempts to prevent deforestation of wintering habitat in Mexico.

OF SPECIAL INTEREST: CONSERVATION MISTAKES

Sometimes big mistakes are made in well-intentioned conservation programs. An attempt to establish a new migratory population of Whooping Cranes is a disconcerting example of how not understanding bird behavior led to a failed conservation program. This species is highly endangered and is placed at greater risk because the cranes exist as a single population with only one overwintering site—Aransas National Wildlife Refuge on the Texas coast. The well-intentioned idea was simple: Whooping Crane eggs would be collected and placed into Sandhill Crane nests, a species that would accept and rear Whooping Cranes. Whooping Cranes would then, in theory, learn feeding and migratory habits from the Sandhills and establish a new population. Unfortunately, the program did not take into account the fact that birds typically imprint on the adults that care for their nest, and most importantly, that males then court females that look like their adopted mother. (See the section "Imprinting" in Chapter 5.) The young Whooping Cranes survived well, but failed to mate because they were imprinted on the Sandhills.

A more hopeful result emerges from a reintroduced population of (captive-bred) Whooping Cranes that summer in the vicinity of the Necedah National Wildlife Refuge (Wisconsin) and migrate to their wintering grounds in Florida's Chassahowitzka National Wildlife Refuge (Figure 15.10). For their first migration to Florida, the birds followed ultralight aircraft piloted by humans. After that experience, they were successful in migrating without the aircraft. Other birds certainly play a part; in post-ultralight migrations, young birds that travel with older birds are less likely than young birds without such guidance to deviate from the most direct route between ranges, but the use of ultralight guidance for the first migration is a successful intervention for helping the captive-bred birds survive in the wild.[20]

FIGURE 15.10
At the Necedah National Wildlife Refuge in Wisconsin, this crane caretaker, fresh from an animal behavior course the previous semester, is dressed for a good day's work in a hot marsh. She is testing an adoption protocol with sandhill cranes, and the costume she wears is meant to be crane-like; young cranes of any species should not be too comfortable around humans. *Photo: Marie Jones.*

Intentional Extinctions

Humans have often responded to perceived predation risk for livestock or competition for agricultural products with programs that attempt to eradicate entire species. Coyotes present an excellent example of a species that has been shot, trapped, and poisoned in an all-out effort at eradication. Many states offered bounties on coyotes; while coyotes were not exterminated, their populations were marginalized in many areas. Ranchers and agricultural economists failed to understand that the impacts of coyotes on livestock were potentially outweighed by the fact they limited herbivore populations.[21] Even now the relative balance of the impacts of coyotes on livestock and on potential competitors with livestock is not fully understood. The general result of the campaign against coyotes is probably an increase in rodent and rabbit (particularly jackrabbit) populations, which in turn directly compete for forage with cattle and sheep. A similar eradication program aimed at wolves nearly succeeded in driving them to extinction in North America.

Intentional eradication is not a modern idea. Some medieval Europeans believed that cats were associated with Satan, and they were systematically slaughtered, a decimation hypothesized to have contributed to the spread of the Black Plague. This disease, carried by rats and their fleas, killed approximately one-third of the human population of Europe and changed the course of history.

The populations of large predators have increased in some parts of the world, leading to conflicts between the goals of land and wildlife managers who are tasked with restoring ecological balance to these areas and the goals of livestock producers, who sometimes lose valuable stock to these predators and to zoonotic disease (disease acquired from other animals). Some ancient solutions are being revisited. From biblical times (Job 30:1) and before, dogs have helped to guard flocks. The use of protection dogs in farm animal husbandry declined during

FIGURE 15.11
Livestock protection dogs have been used by shepherds and other livestock managers for thousands of years. As predators are re-established in some areas, these dogs are becoming a viable part of an integrated approach to protection of livestock from predation and disease. *Photo: By Andy Fitzsimon from Brisbane, Australia (protector of the sheep) (CC-BY-SA-2.0 (http://creativecommons.org/licenses/by-sa/2.0)), via Wikimedia Commons.*

the nineteenth and twentieth centuries, a time when large predators were being exterminated, but interest in partnerships with these dogs is increasing. The type of dog and the way it works will vary depending on geography, livestock, farming practices, and climate, but in all cases, they work around the clock. By repelling both predators and some other wildlife, the livestock protection dogs not only prevent losses due to predation (a more desirable outcome than trapping or hunting the predator after the loss), but also keep potential sources of disease away from the flock. Some workers have suggested that the dogs may also serve as powerful aversive stimuli that could cause both predators and other wildlife to avoid guarded areas in the future. The long history of the use of these dogs in flocks and herds suggests that they can be effective; the time is ripe for more rigorous studies to understand the extent of their utility and how they can be integrated with other farming and ranching practices[22] (Figure 15.11).

> **KEY TERM** A zoonosis is a disease that is transmitted from an animal to another animal or to a human.

Extinction Due to Hunting and Fishing

Many animals are hunted for meat, as well as for the sale of pelts or feathers to the fashion trade. Historically, hunting has been a major cause of reduction of bird and mammal species; hunters have extirpated many local populations, although few of these species have been completely eliminated by hunting. In North America, beavers were once one of the most abundant species; hunting pressure greatly reduced beaver populations, which have only recently begun to recover in some parts of their previous range. Beavers are an interesting example because their comeback has generated vexing issues. Some behavioral attributes of beavers—their toppling of trees for food and construction, their creation of dams or blocking drains with tree branches, and their building of burrows in riverbanks—can cause land managers to view them as pests. Colonizing beavers are often euthanized or translocated away from urban and suburban habitats to preserve streamside plants or to protect dams or dikes built for flood control.

Animals for which hunting actually contributed to extinction include passenger pigeons and Carolina parakeets. Passenger pigeons were probably the most common North American birds prior to the arrival of Europeans on the continent. They numbered in the billions and were hunted for their meat. Carolina parakeets numbered in the millions and were hunted for their feathers. Both species were extinct by the early 1900s.

The greatest behavioral contributor to the extinction of the passenger pigeon was its apparent dependence on large, gregarious, flocks and the resulting ease with which they were hunted; small captive populations maintained after its extinction in the wild failed to reproduce and

FIGURE 15.12
The passenger pigeon. *Adapted from John James Audubon painting (passenger pigeons were common during Audubon's lifetime.*

ultimately dwindled to a single bird, "Martha," which died in the Cincinnati Zoo in 1914 (http://www.si.edu/Encyclopedia_SI/nmnh/passpig.htm; Figure 15.12). The last captive Carolina parakeet, "Incas," also died in Cincinnati, in 1918, and the last small wild flock was seen in Florida in 1920. Should this be taken as a cautionary note when the population sizes of other highly social species are reduced?

OF SPECIAL INTEREST: THE PASSENGER PIGEON

The passenger pigeon is yet another example of how behavior may be a strong influence in species conservation. In the seventeenth century, this species may have accounted for 25–40% of all birds in what was to become the United States (http://web4.audubon.org/bird/Boa/F29_G3a.html); nesting groups could occupy over 800 square miles. Cotton Mather described one flock's flight as stretching over a mile wide and taking several hours to pass overhead. The highly social habits of the birds—effective against most predators—became their undoing, as they were easy prey for professional hunters, who netted them at nesting sites or simply knocked young birds out of the nest and sold them to urban meat markets. There were no provisions for maintaining a breeding population. The species declined noticeably by 1860; nonetheless, in 1878 in one location alone, 50,000 birds were killed every day for almost 5 months. Within 300 years of Mather's observations, the passenger pigeon was extinct. Advances in genomic analyses have allowed the first estimates of passenger pigeon population size over time, and the results, when combined with a variety of ecological analyses, indicate that these birds, while capable of super abundance, also experienced serious population declines. Remarkable population fluctuations may have been precipitated by factors such as climate shifts or food shortages (e.g., bad acorn crops). Apparently, throughout its history, the species recovered from such fluctuations until they were combined with human exploitation.[23] Super abundance may be no guarantee against extinction if it is combined with severe population fluctuation or unfortunate timing on the part of eager human predators.

While the deforestation of the eastern United States would have doomed the passenger pigeon—there are no remaining forests that can accommodate flocks of over 130,000,000!—the professional hunter and the birds' inability to thrive in small groups were the immediate cause of its demise.

Invasive Species and Extinctions

Invasive species can exhibit ecologically dominant behavior, causing extinctions of indigenous species. The brown tree snake, native to the South Pacific, was accidentally introduced into Guam after World War II; prior to that introduction, only one species of snake existed on Guam, and that snake was a specialized resident of termite nests. In contrast, the brown tree snake is a voracious predator on birds, against which the birds of Guam have no evolved defenses. Ten of the 12 forest birds that were native to Guam are now extinct; because forest plants on Guam depended on these birds for pollination and seed dispersal, the effects of the snake reverberate through the ecosystem. The brown tree snake has proven unstoppable, a clear indicator of the extinction dangers that can come with species introductions (Figure 15.13).[24]

Successful invasive species often outcompete native species for critical niche components, such as food and nesting sites. Other events such as *habitat fragmentation* sometimes open opportunities for colonization by nonnative species, especially if native species do not adapt to modified conditions.[25]

Extinction threats posed by invasive species underscore the importance of understanding the behavioral predictors of invasiveness. Are there behavioral commonalities among brown tree snakes, fire ants, Argentine ants, cane toads, house sparrows, and Norway rats (Figure 15.14) that make them successful invaders of so many habitats? One trait is the ability to adapt easily to novel environments. Another is dispersal behavior that increases the probability that invasives will hitchhike on ships, airplanes, and automobiles. If invasives competitively overwhelm native species, then invasions can contribute to population collapses of native species and resulting community degradation. These two behavioral attributes—adaptability and a tendency to disperse—are at least in part due to a potentially dangerous lack of neophobia.

Neophobia, the fear of new objects or animals in the environment, can be a highly adaptive and critical element in animal's survival. Neophobia keeps animals from consuming potentially poisonous foods, from approaching predators, and from entering dangerous human-modified environments. In birds and mammals, feeding preferences are often shaped by juvenile experiences with food items; there is a temporal window during development when the young animals shape their diets. During this period, young need to be more open to novel foods simply because everything is new (Figure 15.15). In many species, the exposure of juveniles to new food is driven by parental diet or is carefully monitored by adults in a social group.[26-29] In contrast, *neophilia*, the desire for new or novel experiences, is rare in animals and usually limited to

FIGURE 15.13
A brown tree snake. As birds disappear on Guam, the snake is finding other prey. It can grow to over 8 ft in length. *Photo: Gordon H. Rodda, USFWS.*

> **KEY TERM** Habitat fragmentation is the breakdown, as the result of human activities, of large expanses of habitat into small habitat islands that are separated by human development.

513

FIGURE 15.14
The Norway rat, a common habitant of urban environments. Norway rats (*Rattus norvegicus*) originated in Asia, but they are distributed all over the globe, having traveled as stowaways wherever humans have gone. They are omnivorous and intelligent, and do well in social groups. Females reproduce throughout the year, and may produce as many as five litters. *Photo: Douglas E. Norris.*

> **KEY TERM** Neophobia is the fear of the new or unusual.

> **KEY TERM** Neophilia is an attraction to the new or unusual.

FIGURE 15.15
Young robin (left) and fox (right) exploring for food. *Photos: Left, Sean Varner, US Fish and Wildlife Service; Right, Lamar Gore, US Fish and Wildlife Service.*

514

FIGURE 15.16
Mechanical barriers such as this device designed to keep predators away from this bluebird box, can be effective in programs to support threatened or endangered species. *Photo: US Fish and Wildlife Service.*

periods of dispersal.[30] Domestic dogs tend to be neophilic in their choice of toys, a behavioral trait that may have been favored in the evolution of dogs as companions to humans.[31]

Summary: Extinctions and Conservation Behavior

Knowing about the behavior of imperiled animals can inform conservation decisions, such as protecting migratory stopover habitats or designing replacement habitats for migrating birds. Sometimes knowing the proximate behavioral causes of animal population declines, such as the predatory behavior of raccoons on bluebird nests, assists in the design of appropriate conservation plans; raccoons can be foiled by mechanical baffles that prevent their entry into a nest box (Figure 15.16). Biological invasions, and consequent extinctions, may be prevented if the dispersal behavior of the invader is appreciated. Although a fairly new subdiscipline of animal behavior, conservation behavior is already providing powerful tools for understanding and preventing animal extinctions.

15.4 RESERVE DESIGN

This section is necessarily far-ranging. The questions of how large reserves should be, how many reserves are needed to protect species against extinction, and how reserves should be managed are complex and far from resolved. Here the topic is divided into four issues that conservation behaviorists address:

1. Population genetics and reserve design
2. Fragmentation and edge effects in reserves
3. Special habitat requirements and reserve design
4. Human–wildlife interactions in reserves

In an ideal world, large tracts of land would be placed into absolute reserves, meaning that the species occupying the ecosystem would be free to survive through time in the absence of human interference. In the real world, reserves are limited in space and often influenced by human activities. Conservation behavior provides a major tool in the design of reserves that maximizes the conservation impact of those reserves.

Population Genetics and Reserve Design

A fundamental tenet of biology is that genetic diversity is the basis for adaptation—and as such, is the basis for survival of a population through evolutionary time. Environmental changes may favor new genetic combinations, certain genetic combinations may be better dispersers, and genetic diversity helps a population to avoid inbreeding effects. When population size decreases, the statistical chances of rare alleles surviving in the population also diminish. The reduction of a population to near extinction and then recovery of that population is called a *genetic bottleneck*. The analogy is simple; the population at its smallest point resembles the narrow neck of a bottle, in that not all of the previous genetic variation passes through.

> **KEY TERM** A genetic bottleneck is a low point in the number of animals in a population that results in decreased genetic diversity.

Genetic bottlenecks have been documented in a variety of species. Perhaps the two most famous examples are elephant seals and cheetahs. Elephant seals live on the Pacific coast of North America. They were hunted to very low numbers, about 50–100, and have since recovered and recolonized coastal areas. The cheetah in Africa and Asia is highly endangered at this time due to a combination of hunting pressure and competition with lions and hyenas; the genetic diversity of its wild populations is quite low. Censuses of known populations show that approximately 7,500 cheetahs remain in the wild, only half the number of cheetahs estimated in 1975, showing that this species is in serious decline (http://www.iucnredlist.org/details/219/0).[32]

There are at least three potential behavioral effects of genetic bottlenecks: lack of ability to express behavioral adaptations to environmental changes, expression of maladaptive behavior due to inbreeding effects, and possible disruption of the mating system due to lack of suitable mates. Presently, these effects are mostly hypothetical because they are difficult to measure in wild populations, particularly if those populations are small. Most captive populations of extinct-in-the-wild animals are undergoing a population bottleneck, but managers of these populations can plan pairings and use artificial insemination to conserve genetic resources.[33] This type of management is usually not feasible in a wild population.

How large does a wild population need to be to provide sustainability? This is a difficult question. Conservationists use a *population viability analysis* (PVA) that takes into account the age structure, rate of population growth, mating structure, and spatial distribution of the animals. Populations that fall below a critical number may, effectively, be extinct even if a few animals remain.[34,35]

African elephants provide a good example of viability analysis. With an average of 3.1 African elephants per square mile, the minimum reserve size for these elephants is 1000 square miles, or about half the size of the state of Delaware.[36] Over time, this elephant population would fluctuate between about 1000 animals and over 10,000, along with changes in ecological conditions, but it would be unlikely to go extinct. Approximately 2750 Rio Grande cutthroat trout are needed in a population to maintain viability and genetic diversity; this number yields about 500 breeding adults and a population with less than a 10% chance of going extinct over 100 years.[37] Finding space for populations this size of small animals, such as insects, is nowhere near as problematical as finding space for viable populations of large carnivores.[38]

> **KEY TERM** PVA takes into account the age structure, rate of population growth, mating structure, and spatial distribution of a species to determine the likelihood that the population will persist over time.

Given the lack of spaces large enough for conservation of many species, conservation biologists talk about maintaining *metapopulations*, which are networks of interlinked subpopulations. Good reserve design can preserve species by allowing dispersal of animals among subpopulations. This allows both retention of

515

the reservoir of underlying genetic variation and recolonization if local populations go extinct, a likely event if they are small. The behavioral biologist can provide knowledge of how dispersal affects the movement of genes among subpopulations of a species and how its mating system might maintain or reduce genetic variation.[39] For instance, given the mating system of the scimitar-horned oryx (discussed later), if the core reserve contains only one herd with one breeding male, interventions or dispersal interactions with other subpopulations will be necessary to retain genetic variation. Some tropical bird species have extremely limited dispersal capabilities, amplifying the isolating effects of living in forest fragments.[40] For conservation efforts to be successful, much more needs to be discovered about dispersal behavior and how animals living in metapopulations move among fragments in fragmented habitats.

If dispersal among subpopulations in metapopulations is cut off by distance, highways, or other factors, human-engineered *translocations* of animals among sites may be necessary. But how successful are translocations? Translocation, or moving an animal from one natural setting to another, is similar in concept to reintroduction, but usually involves moving wild adult animals to habitats that are already occupied by the species. Examples of commonly translocated animals include beavers, prairie dogs, and bears. Translocations of animals are costly and in many cases ineffective; in deer and gray squirrels, for which some data are available, more than 70% of the translocated animals die soon after they are moved. Relocated rattlesnakes suffer higher mortality than undisturbed conspecifics; over half of them return to the site where they were captured.

> **KEY TERM** Translocation is the movement of an animal from one location to another. This may be done to re-establish a population that is extinct, to supplement a small population, or to move a troublesome animal.

Oftentimes translocations are made to palliate public feelings even though the governmental agencies involved in the relocation know that the animals are likely to perish in their new habitat. The failure of relocated animals is typically due to stress from the translocation process, to inability of the animals to find territorial space in the habitat, or to their efforts to return to their original home ranges.[41] Stressful events in the translocation process include capture, transport, and release.[42] In some cases, the lack of documented success of translocations has led managers to favor euthanasia, or culling, despite strong public resistance to killing animals. In sum, considerable research needs to be done to develop methods of monitoring the results of translocation and to discover successful translocation techniques.

Fragmentation and Edge Effects in Reserves

Habitat edges differ in key ways from core habitats.[43,44] Under natural conditions, edges are called *ecotones*; the margin between a forested area and a grassland is a good example of an ecotone. Human modifications create a wide variety of edges, which are often much sharper than natural ecotones. Highways, parking lots, and grassy lawns all create well-defined edges with surrounding habitats. Some species do well in edges and, in fact, may be adapted for ecotonal life in undisturbed habitats, whereas others must have core habitat, away from edges, for nesting and foraging. Researchers determined habitat preferences of 31 bird species at a heath-woodland edge in southeastern Australia; of those species, 11 were conspicuous in edges but not limited to those areas, and three were "shy" of entering the ecotone.[45] The conservation effects of edges vary considerably among species; some species do better than they did before fragmentation and some do worse, but fragmentation always changes the community.[46,47]

Conservationists who are interested in managing metapopulations often ask if one large reserve is better or worse than a number of smaller reserves. An important factor in this is the relationship between reserve size and edge effects. Many small reserves have more edge than one large reserve, even if the overall areas are the same. Open-cup nesting birds have much lower nesting success in a fragmented rainforest habitat, where there is more edge relative to the size of the habitat,[48] than in larger tracts of forest. If the goal is to conserve a community that represents the core habitat, rather than ecotones, then enough core habitat needs to be preserved to maintain minimum viable populations of core habitat species. A minimum viable population is the number of animals required to maintain genetic diversity and avoid the effects of inbreeding, as well as to withstand normal fluctuations in population size without going extinct. Detailed research on community members and their responses to edges is necessary for a well-informed consideration of edge effects in reserve design.[49,50]

OF SPECIAL INTEREST: FRAGMENTED CULTURE?

Michael Huffman, a primatologist who studies chimpanzee behavior, told one of us about the hypothesis that young chimpanzees learned the medicinal use of leaf-swallowing from their mothers. All across the chimpanzees' range, the animals employ leaf-swallowing in response to gastrointestinal upset associated with nematode infection; the leaves are not taxonomically related, but they all have similar features, including a rough, "hairy" surface, perfect for "scrubbing" the parasites from the intestine. Dr. Huffman then wondered aloud about what the impact of the chimpanzees' fragmented habitat might be on the possible transmission of this or other information; after all, chimpanzees had once lived in a continuous swath of land stretching almost across Africa. Now, groups of chimpanzees, if they survive at all, do so in isolation from most other groups. Might this result in a depauperate chimpanzee culture, much as similar fragmentation has diminished the songs of some birds?[51]

Special Habitat Requirements and Reserve Design

Part of reserve management is maintenance or supplementation of specific habitat features to ensure survival of key species. Sunken ships can serve as artificial reefs in marine reserves, downed trees left in reservoirs serve the same function in freshwater systems, cut scrub and brush provide cover and habitat variation in grasslands and woodlands, and periodic managed fire maintains key grassland habitat features. Reserves in habitat fragments may not contain all the needed features for all species in the community, so active habitat management, aimed at meeting the behavioral requirements of the animals, has high importance.[52]

Competition for nest cavities is a good example of how reserve design can be informed by knowledge of behavior. For instance, hollow trees are a premium nesting site. Protected from the elements and usually easily defended, a cavity inside a hollow tree is attractive to a wide range of insects, birds, and mammals. Cavity occupants include honeybees, woodpeckers, squirrels, and bats. Human activities such as logging and burning of forests to clear land for agriculture can eliminate older, hollow trees in ecosystems.

In the southwestern United States, these dynamics have affected acorn woodpeckers, which depend on nesting trees for their survival. In addition, if an introduced species relies on tree cavities, it may exclude native species from those cavities; one alternative hypothesis for the extinction of the Carolina parakeet states that honeybees, introduced from Europe, outcompeted the parakeets for nesting cavities in trees. Understanding how competition affects cavity-nesting species and what cavity-nesters are looking for when they search for nests can be critical in conservation plan designs. Two excellent examples of public involvement in providing nest sites for cavity-nesters in North America are the purple martin, shown in Figure 15.17 (http://purplemartin.org/main/mgt. html) and the Eastern bluebird (http://www.sialis.org/).

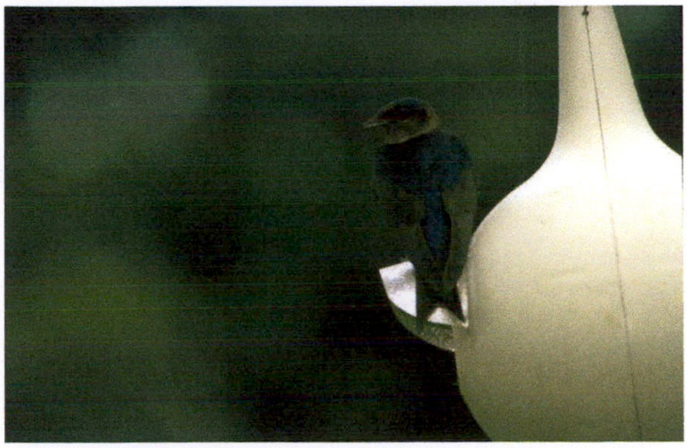

517

FIGURE 15.17
Purple martin at an artificial nest. Populations of this species are supported by a large public effort to provide nesting sites. *Photo: Thomas G. Barnes, US Fish and Wildlife Service.*

> ## OF SPECIAL INTEREST: SMALL RESERVES CAN HAVE BIG EFFECTS IF BEHAVIOR CHANGES
>
> The establishment of a reserve to protect a target species may have unintended consequences, some of them highly desirable. In marine systems, much like terrestrial ones, top predators may compete with humans for food (fish catch) and may be assumed to have ranges that are too large to benefit from the establishment of a reserve. The African Penguin (*Spheniscus demersus*) is one such predator, and its population has declined to 10% of its pre-twentieth-century size. The penguin not only covers hundreds of kilometers in its travels, so do its pelagic prey. How can this wide-ranging, declining predator be managed successfully?
>
> In January of 2009, purse-seine fishing was forbidden in a 20 km (radius) around a large penguin colony on St. Croix Island. Three months later, the penguins in that colony had reduced their foraging effort (including foraging path distance) by almost 30%, with a 43% energy saving. Before the closure, 75% of the penguins' foraging dives had occurred outside the closure area; after the closure, more than 70% of their dives were inside that area. In contrast, African Penguins in a neighboring colony 50 km away did not decrease effort or energy expenditure.[53]
>
> Marine protected areas (MPAs) lag behind similar terrestrial areas in terms of both area covered and distribution, with most MPAs occurring along coastlines. One obstacle is the perception that MPAs must be large to accommodate wide-ranging species. This study suggests that in at least some cases, small reserves can have large consequences. This may be particularly important in areas where political and economic opposition to reserves increases with their size.

Human–Wildlife Interactions in Reserves

Many parks, reserves, zoos, and open space systems depend on taxpayer support or paying visitors; the financial stability of conservation often depends on conservation co-existing with recreation or with opportunities for humans to observe and interact with animals. Indeed, in the United States, the national park system owes its very existence in large part to President Theodore Roosevelt's enthusiasm for hunting and, by extension, outdoor activities.

Tourists—either "ecotourists" or the regular kind of tourists—play a critical role in most conservation plans. Tourists bring much-needed cash to local economies; if they come solely to see a reserve, this provides a strong incentive to maintain the reserve. This is a stereotype of conservation in developing nations, but it is also true in North America, where tourism is the economic engine that maintains communities on the fringes of national parks. Unfortunately, in addition to cash, tourists bring air pollution, water pollution, and noise pollution, all potentially major disruptors of wildlife behavior. One basic element of reserve design is to incorporate mechanisms for the economic support of reserve protection. In many cases, this involves the simultaneous challenges of both attracting tourists and controlling the impacts of tourism to minimize stress on wildlife. Because many of the ways that humans affect animals in reserves are behavioral, animal behaviorists should be influential in designing reserves.

The obvious solution to the double-edged problem of visitor support and visitor impact is to provide adequate buffers between humans and animals in the reserve.[54] Unfortunately, for many species, the data that would allow scientists to determine this distance are not available. Behavioral data about sensitivity of animals to motion, noise and odors, their ability to habituate to these factors, and other human impacts on their behavioral environment are generally lacking. For example:

- How closely can a person approach a deer or elk before disrupting its foraging?
- Do these species habituate to human presence? If so, is that desirable?
- How do songbirds in the same environment differ in their behavioral responses to humans, if, in fact, they differ?

Off-road vehicle use—motorcycles, snowmobiles, ATVs, or trucks—brings noise and air pollution into core reserve areas. Vehicles crossing streams can cause serious damage to aquatic systems. Solutions such as restricting the use of motorized vehicles, providing adequate wastewater treatment, and enforcing noise regulations often are expensive or meet public resistance.

In the United States, much public land management emphasizes multiple uses, so wildlife shares space with recreational uses. The downsides of mixed use, in terms of behavioral disruption of wildlife, are obvious. In the United States, the public has not generally supported "absolute" reserves, where no human encroachment is allowed. Designated wilderness areas are regarded as recreational sites; and even if the general public is excluded, scientists often want to exempt themselves from exclusion. Scientific access to otherwise closed areas can be valuable, because it allows for monitoring of populations, but it can create an image of scientific elitism and increase public opposition to closures. In the developing world, immediate needs for food, shelter, and ways to generate cash can exert substantial pressure on reserves. Whether in the developed world or in developing nations, reserves are likely to experience recreational, hunting (including poaching), fishing, and perhaps logging and mining pressures.

In many parts of the world, reserves partly function as game parks for hunting and fishing. Often introduced species, such as pheasants and rainbow trout in North America, or native species, whose populations are augmented through translocations or supplemental feeding, are maintained to supply hunters and fishers. The coalition of hunting/fishing enthusiasts and conservationists that began in Teddy Roosevelt's time continues today with the conservation initiatives of groups such as Ducks Unlimited and Trout Unlimited. The bedfellows that inhabit the world of conservation can seem strange, but the economic and political power of hunters and fishers is considerable, and they are often a positive force in the establishment and maintenance of reserves.

519

Hunting typically targets herbivores, such as deer or elk, but sometimes carnivores are taken for trophies. Behavioral impacts of hunting include shifts by herds or flocks of animals to suboptimal habitat to avoid hunters, loss of dominant males (often hunted for the size of their horns or antlers) from social groups, and lowered food availability for carnivores.

Fishing usually targets carnivores, hence the efficacy of bait such as artificial flies, minnows, or worms; fishing tends to remove the top carnivores from a community, sometimes increasing competitive interactions among herbivores, whose populations can grow, limited only by food availability in the absence of carnivores. Removal of the largest fish also can affect mating systems because these fish are at their reproductive prime. Catch-and-release fishing, popular among fly-fishers and some bass fishers, mitigates some of the behavioral impacts of fishing.

DISCUSSION POINT: FISHING, HUNTING, AND CONSERVATION

Some of the most ardent enthusiasts for conservation and habitat preservation are hunters and fishers, who are represented by groups such as Ducks Unlimited and Trout Unlimited. Find out a bit more about these groups by looking at their websites. Do you support the marriage of hunting and fishing and conservation, or are you uncomfortable with this? Can hunting and fishing be used as effective population management tools? Many students of animal behavior view themselves as advocates for animals but are also very committed to conservation, so this is a question that can spark lively discussion.

Observing the community effects of the loss of top carnivores led Crooks and Soule to develop the *mesopredator release hypothesis*,[55] an important way to explain how the removal of top carnivores affects community structure. Top carnivores, through a combination

> **KEY TERM** The mesopredator release hypothesis postulates that the removal of top-level predators results in the release of population growth of the animals they were eating. Thus, mesopredators, which are frequently consumed by top predators, are ecologically released and flourish in the absence of top predators.

of direct consumption of middle-sized carnivores (mesopredators) and competition with them for food, control the population sizes of mesopredators. The Crooks and Soule study focused on the effects of coyotes, as a top carnivore, on smaller carnivores such as raccoons and domestic cats. These mesopredators are major threats to nesting songbirds; when coyotes are present, songbirds are protected because cat and raccoon populations are held in check. In the absence of coyotes, predation on songbirds increases, potentially resulting in a conservation threat to the songbirds.

Despite the clearcut need for spatial separation of humans and animal populations, it is surprising how often such measures are controversial or individuals attempt to evade regulations. A large body of literature addresses "the tragedy of the commons," that is, the fact that many people feel less responsibility about taking care of shared, common space than they do about caring for their private space.[56] We clearly have much to learn about how to keep people from loving reserve space so much that they "love it to death."

A Bike Path Runs Through it

Modern suburban development often leaves greenways—strips of vegetation that separate groups of houses, or houses from shopping areas. In older developments or urban areas, riparian areas along waterways serve similar functions as greenways. Conservationists often view greenways as conservation areas, even though uses of greenways are typically shared among conservation, flood control, and recreation. In suburban areas, open land may be viewed as ripe locations for soccer fields, and some members of the public have trouble distinguishing between the conservation value of a weedy-appearing open space and a field of planted and watered bluegrass. Greenways in urban areas often attract transients, a situation that can cause public officials to clear vegetation and take other measures to discourage homeless people from camping in greenways, actions that are often detrimental to conservation values. This section addresses the values and drawbacks of greenways as conservation areas, and suggests ways that conservation behaviorists can help to shape greenway management with an eye to maximizing conservation values.[57]

Riparian corridors are often the most "natural"-appearing habitat in urban areas. Because these strips of land along the sides of streams and rivers are prone to flooding, they tend not to be occupied by buildings. In recent years, the laudable environmental and recreational goal of creating bike paths has resulted in the encroachment of pathways into riparian corridors. These pathways create more edge and bring humans and their dogs into closer contact with wildlife that shelters or nests in the riparian areas. If, for example, a road runs parallel to but at some distance from a creek or river, conservation goals are better met if the bike path is built as close as possible to the road; this concentrates the disturbance away from the waterway and can help to minimize the behavioral impacts of recreation on animals. Unfortunately, bikers and joggers like the feeling of riding or running "in the woods" and often bike paths are built close to the waterway, even if this separates a wooded or shrubby riparian area from the creek or river. This design adds edges and is likely to lead to greater behavioral disruption of wildlife populations.[58–61]

In general, the density of trails (meters of trail per hectare of protected area) decreases as park size increases.[62] This means that small reserves are more likely to be designed in ways that put wildlife at risk of habitat disruption by human recreational activities. In aquatic and shoreline reserves, buffer zones from boat activity are important in protecting shorebirds and sea life, such as whales, manatees, and dolphins, from disturbance.[63,64]

520

BRINGING ANIMAL BEHAVIOR HOME: DOGS, TRAILS, AND CONSERVATION

Most dog owners enjoy taking their quadruped best friends for a walk along a scenic trail, preferably close to a creek. As much fun as that is, you may think differently about it after reading the following study. We have noted how creekside trails, while lovely, do not function well in the world of conservation. Do dogs make a difference?

To answer this question, Benjamin Lenth and colleagues used five indicators of wildlife activity — pellets, tracks, remote-triggered cameras, on-trail scat, and maps of prairie dog burrow location — to ask if the presence of dogs affected wildlife around trails.[65] They compared trails that prohibited dogs completely with trails that allowed off-leash dogs under "voice and sight" control. The trails were chosen to be as similar as possible with regard to ecological characteristics and visitation levels and were not within 300 m of roads or structures.

Lenth and coworkers found that wildlife activity was greatly altered by trails and even more altered in areas that permitted dogs on trails. For instance, in areas that prohibited dogs, mule deer were nonetheless less active within 50 m of trails. That boundary increased to 100 m in areas that were visited by dogs. Small mammals such as squirrels, rabbits, mice, and chipmunks reacted in similar ways, and bobcats tended to avoid areas with dogs. (Red foxes showed no aversion.) In areas where trails are within 100 m of each other or where their margins overlap considerably, it is possible that much of the area between trails is unsuitable for a wide range of animals. This situation is exacerbated by the disinclination of some people to remain on trails; the creation of "social" (unofficial) trails can generate a network of trails so close together that even in the absence of dogs, wildlife will find the area unattractive.

Of course, this study simply addressed the spatial aspects of wildlife reaction to trails and trail users. Depending on visitation patterns, the timing of wildlife activity may change in response to people and/or their pets. The sad news is that a natural area with a poorly designed trail is likely not all that natural after all.

Ecotourism: A Two-Edged Sword

Ecotourism can bring needed money to reserve areas, but tourists can disrupt the behavior of animals in a reserve. The most severe criticisms of ecotourists have been leveled at whale watchers, snorkelers at coral reefs, and various "adventure" activities, such as river rafting, all of which can bring human disturbances in close proximity to easily disturbed wildlife. Whale watching provides a good set of examples of the pros and cons of ecotourism. Some coastal communities rely on whale watching tours as a significant source of income, and whale watching can economically supplant the hunting of whales. Of course, not all whale watching takes place on a boat, and land-based whale watching is the least intrusive on whales. In contrast, motorized boats making close passes at foraging whales may be the most disruptive because noise pollution interferes with the ability of whales to echolocate and communicate.

> **KEY TERM** Ecotourism is activities in reserves, parks, or other relatively natural habitats that is rationalized by its educational benefits for the tourists and its economic benefits for local communities around the reserve or park.

DISCUSSION POINT: ECOTOURISM AND ADVENTURE TOURISM

Have you been on an outing that was billed as ecotourism but was really less "eco" and more "adventure" tourism? In retrospect, how big an ecological footprint did this activity have? How can you resolve the possible conflicts between economic needs of local communities for tourism dollars and the ecological impacts of ecotourism and adventure tourism?

Flight Initiation Distance: A Tool for Evaluating Human–Wildlife Interactions

Tarlow and Blumstein evaluated techniques for measuring how much stress human activity puts on wildlife.[66] They focused on birds, but the behavioral measures they used have broad applicability. Behavior plays a key role among the assessments they discussed; breeding success and flight initiation distance (FID) are particularly important (Figure 15.18). Managers assess breeding success by observing the number of nests originated, the number producing eggs, and the number of offspring that fledge. FID describes how close a bird can be approached before it flies away; analogous measures can be applied to fleeing mammals.[67] If FIDs are documented, then perimeters around sensitive nesting and feeding areas can be established and humans can be prevented from approaching an animal so closely that it is disturbed. Of course, animals may be disturbed and not flee—they may stop feeding and watch an intruder, for example—so perimeters can be established based on changes in vigilance as well as in actual flight. FIDs vary by nature of the disturbance (walking human, biking human, automobile noise), context, season, and population, so applying this principle is sometimes more difficult than it sounds, but overall this is an excellent tool for minimizing disturbance.

> **KEY TERM** FID describes how close an animal can be approached before it leaves.

522

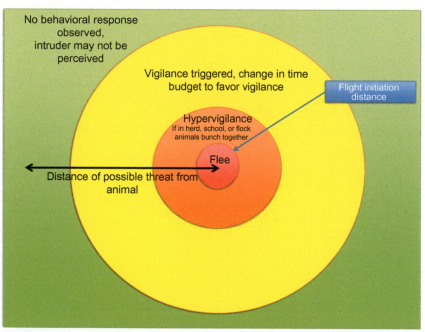

FIGURE 15.18

Perimeters at which disturbance of animals due to human activity can be assessed. Animals shift from being unaware of a potential threat, or at least not visibly responsive, in the green area, to time budget changes in the yellow area. These time budget changes are important because they affect feeding and other homeostatic activities and may also be correlated with stress-related physiological changes. In the orange zone the animals are hypervigilant, expressed as a complete shift in time budget to watching the perceived threat. The hypervigilant response can either de-escalate, with a return to more normal activities, or further escalate to flight. Not captured in the figure, but very important is flight distance—the distance the animal moves after being disturbed. Long flight distances may remove an animal from good foraging habitat, its nest, or from a sheltered spot.

15.5 CAPTIVE BREEDING PROGRAMS AND REINTRODUCTIONS

The basic concept of captive breeding and reintroduction is simple: Gather up enough animals of an endangered species, breed a reasonably large base population in protected conditions, and then release some of this base population into a protected wild or semiwild reserve.[68] This is Noah's idea, writ large, and for species on the brink of extinction, this may be their only hope.

A few such programs have been unqualified successes, but many have faltered because either captive breeding failed or the reintroduced animals did not survive. One of the more difficult cases, as detailed previously, involved a Hawaiian honeycreeper in which a translocation failed and a captive individual died. Mortality is typically high in reintroduced animals, and discovering the correct methods of reintroduction for any species at this time has a substantial trial-and-error component.

Many of the problems of reintroductions surround the ability of captive-reared animals to function in the wild. Some animals carry with them, in genetic form, much of the information they need to survive. Others, including most birds and mammals, rely on information learned from parents, sibs, and social group members, as well as on their own experiences, to shape the behaviors necessary for their survival.[69]

This section shows that managing imprinting, recognizing the importance of neophobia in shaping behavior, and learning about food preferences and predator avoidance can all play key roles in the success of a reintroduction program. It highlights six species, all of which were either endangered or extinct in the wild, that have been bred in captivity; most of them have successfully been reintroduced into the wild. These species are discussed in the order of difficulty they have presented in captive management and reintroduction; herbivores are probably easier to work with than carnivores, but all of these species have posed interesting behavioral questions and problems.

Scimitar-Horned Oryx

The scimitar-horned oryx (Figure 15.19) is a good model for discussing conservation of ungulates (hoofed mammals). There may have been as many as 1,000,000 oryx in North Africa as recently as 1900; the rapid extinction of this species in the wild, in many ways parallel to the extinction of the wild American buffalo, is shocking. Ungulates are capable of fairly rapid population growth, and in nature, populations are typically limited by predators and food resource availability, with major causes of decline being habitat destruction and hunting. Ungulates either graze on grass and herbaceous vegetation (cattle) or browse on leaves and buds of shrubs and low trees (deer, giraffes). They seem to easily adapt to any plant-based digestible diet, so feeding animals maintained in captivity is not challenging, and reintroduced animals do not seem to face problems with learning food preferences.

A fair amount is known about vigilance in ungulates, and shared vigilance in groups is probably an important part of the time budgets of these animals. Less is known about how ungulates learn about the identities of predators in their environments and whether reintroduced animals are more vulnerable to predators because they lack experience. Deer, for example, adjust their behavior seasonally to the presence of human hunters,[70] suggesting that learning of predators could be important. A key element of ungulate behavior is a harem-type mating system; in the scimitar-horned oryx, a single male leads a herd of 20–40 females and young. This reduces effective population size (N_e, a measure that takes into account the number of animals actually reproducing). If only a few males reproduce, even though they inseminate

523

many females, the N_e is low), which is accounted for in a genomic resource program organized by zoos for captive breeding of this species.

The Smithsonian National Zoological Park has used artificial insemination to overcome problems with behavioral incompatibility between pairs of scimitar-horned oryx. Progeny have successfully been reintroduced into the wild at Ben-Hedma National Park in Tunisia, where a population had grown to about 130 animals in 2005. The reintroduced population was initially maintained in a 10-hectare (a hectare is 100 m × 100 m) enclosure for the first 16 months; this allowed supplemental feeding with hay while the animals adapted to the natural vegetation. Supplemental foods were available to the animals following their release as well. The park, which is 2400 hectares, is fenced so the oryx are confined within the park; this space may be adequate for only one herd (one breeding male at a time). If this is the case, then clearly the reserve population, on its own, is not sustainable. However, this population and zoo animals may form the basis for future reintroductions at other sites; with management, a viable population could be maintained through translocations.

FIGURE 15.20
Przewalski's horses are bred in zoos, but a small wild population persists in Mongolia. *Photo: Michael Breed.*

524

Przewalski's Horse

Przewalski's horse, a native of Mongolia (Figure 15.20), at one point extinct in the wild, is the closest living relative of the domestic horse, and as a result, its greatest lure for behavioral biologists comes from the insights it gives into the roots of domestic horse behavior.[71] Sometimes regarded as a subspecies of the domestic horse and sometimes as a distinct species, Przewalski's horse has never been domesticated. Early in the 1900s, fewer than two dozen of these horses were captured; this small group of horses was the foundation stock for captive populations, which are a popular exhibit in many zoos, and for a population of reintroduced horses in Hustai National Park, Mongolia. While Przewalski's horses are not nearly as manageable as their domesticated relatives, knowledge of horse management has eliminated much of the guesswork often involved in captive management and reintroduction programs. The low genetic diversity of this species, due to the population bottleneck suffered in the early 1900s, has not had any visible negative effects on the species, and the major issue in the reintroduction has been development of a program to protect the population. Like the domestic horse, a single stallion moves with a harem of females over a relatively large home range. Home ranges of bands of Przewalski's horses can be overlapping, but a relatively large reserve is needed to maintain the number of bands required to preserve the species' remaining genetic diversity.

FIGURE 15.21
Black-footed ferrets are the target of a breeding and reintroduction program. Bold steps were taken when the only known wild population of black-footed ferrets was captured and used to found a captive breeding program, but reintroduction programs are promising. *Photo: Ryan Haggerty, US Fish and Wildlife Service.*

Black-Footed Ferret

Endemic to the western United States and specialized as a predator that lives only in prairie dog towns, the black-footed ferret (Figure 15.21) was thought to be extinct in the wild until 1981, when a Wyoming rancher's dog caught one and brought it to the rancher's house. From 1985 to 1987, all individuals in this wild population were caught, and they formed the basis for a captive breeding program, with the goal of re-establishing this species in the wild. The decline of black-footed ferrets paralleled the loss of prairie dog towns. Black-footed ferrets are highly susceptible to canine distemper, a vulnerability that may have contributed to their decline; this was one of the rationales for capturing the only known wild population. As with the scimitar-horned oryx, initial breeding failures in captivity led to use of artificial insemination to ensure reproduction. Reintroductions into the wild began in 1991 and continue to the present; to protect the long-term survival of the species and its genetic diversity, scientists hope to maintain several wild populations.

A number of behavioral issues had to be addressed during the black-footed ferret project; these issues included the ferrets' willingness to mate, given the available animals in the initial population, and the ferrets' ability to learn predatory strategies and food acceptability, as well as predator avoidance. Initial reintroductions failed because of a combination of disease and lack of skills that the ferrets needed to survive in the wild. Simulations of wild conditions for captive animals, such as exposing them to hunting opportunities within their captive confines and exposing them to cues of potential predators, have increased the success of reintroduction programs. Minimization of human interaction is also important for black-footed ferrets, as habituation to humans (such as in bottle feeding) can lead to diminished neophobia. A humorous, but telling, anecdote is that when ferrets were given a moving model of a predator in an attempt to teach them predator avoidance, they instead habituated to the model and started riding it around their enclosure. This suggests that even the best-laid plans for preparing animals for reintroduction to the wild can go awry, and that much trial and error is required to find strategies to prepare animals for reintroduction. Presently, the black-footed ferret reintroduction program is starting to succeed, with animals in several wild populations rearing young.

Golden Lion Tamarin

Because the golden lion tamarin (Figure 15.22) captive breeding project is largely based in zoos, it is one of the best-known breeding and release programs. Tamarins are small monkeys found in the Americas; the golden lion tamarin is native to the Atlantic Rainforest of Brazil, a unique ecosystem, separate and fairly distinct from the more recognized Amazonian Forest. Because the Atlantic Rainforest is close to Brazil's population centers, such as Rio de Janeiro, the survival of this forest and the animal species it shelters has been a major conservation concern. Between 1000 and 2000 golden lion tamarins live in the wild; about half of these are reintroduced captive animals. Another 500 or so live in zoos, and are maintained with a carefully managed breeding program to avoid inbreeding and maintain genetic resources.

The success of golden lion tamarin reintroductions has given behavioral biologists the opportunity to explore how to best ensure the survival of captive-bred animals in the wild.[72] Prerelease training of adult golden lion tamarins to handle foods they might find in the wild or to avoid certain predators is relatively ineffective, suggesting that prerelease training or reintroductions of young animals is preferable to prerelease training of adults. After reintroduction, golden lion tamarins' locomotory skills improved dramatically; after a few months, they fell less, climbed higher, and began to prefer natural habitat components. Foraging improved less dramatically, perhaps because the reintroduced animals received supplemental food. Animals with improved locomotory skills survived better and, not surprisingly, juvenile animals showed more behavioral improvement than adults. Animals that ranged freely in a seminatural environment prior to reintroduction and animals reintroduced directly from caged environments did not differ in survival.

The primary lesson from golden lion tamarin experience is that a combined effort of zoos and conservation agencies, combined with a charismatic animal that captures the public attention, can generate a very effective reintroduction program. The need for postrelease feeding and veterinary care of released tamarins is an expensive drawback for the program and brings into question just how "wild" the reintroduced animals are. The behavioral

FIGURE 15.22
Golden lion tamarin; this species has been the subject of an intensive breeding program in zoos and has been reintroduced in its native habitat in Brazil. *Photo: Michael Breed.*

lessons are that investment in pretraining should be carefully evaluated with follow-up field data on the performance and survival of animals. The difference between young and adult animals in development and adaptability may be generalized beyond tamarins and is an important point to consider in any captive breeding–reintroduction program.

FIGURE 15.23
A California condor in a breeding and release program. *Photo: Gary Kramer, USFWS.*

California Condor

One of the most controversial conservation decisions of the twentieth century was the capture of the few remaining California condors in the wild to create a core population for captive breeding (see the case study for this chapter). The California condor (Figure 15.23) is a large vulture, with a wingspan of nearly 3 m. By 1982, California condor populations had diminished to fewer than 30 birds. Captive breeding and release programs have increased the wild population of this species to about 200 individuals. Condors are highly susceptible to lead poisoning; because they scavenge on animal carcasses, they risk coming into contact with lead shot or bullet fragments. In response to this, California has banned the use of lead bullets in areas occupied by condors.

The major behavioral issues in the capture-and-release program have been concerned with imprinting of species recognition. Great care is used in the captive rearing program, through the use of feeding puppets, to imprint condor chicks on appropriate cues. Chicks are initially hand fed, using the puppets, and as they grow are visually imprinted on adult condors. Condor releases employ a double trap-door technique, in which a caged condor is given the opportunity to move into a second enclosure; when it does, a door closes behind it and a release door opens, so it can then fly, if motivated to do so. This process minimizes human contact and disturbance of the bird, but makes the timing of actual releases unpredictable. Appropriately imprinted condors, released using this technique, have succeeded in the wild, finding nesting sites, food, and mates. The wild population of condors is slowly growing but is still imperiled by the potential for lead poisoning, not to mention the microtrash described previously.

Peregrine Falcons

Peregrine falcons (Figure 15.24), like many birds of prey, suffered dramatic population declines due to the use of DDT as an insecticide in the middle part of the twentieth century. In a phenomenon known as *biomagnification*, the effects of DDT are compounded at the upper levels of food webs because the insecticide is not metabolized or excreted; it concentrates in the fatty tissues of animals. If, over time, a predator eats 1000 herbivores, then it ends up with 1000 times the individual herbivore's concentration of DDT in its system. A predator that eats predators ends up with hugely magnified concentrations of the poison in its system. In birds, DDT interferes with the calcification of eggshells; females lay thin-shelled eggs that are susceptible to breakage and lethal exposure of the embryo. While the insecticide saved millions of human lives before insects became resistant to it, the environmental cost was high. By the mid-1960s, peregrine falcons had disappeared from much of their range in the United States.

Under natural conditions, peregrine falcons nest on cliffs and ledges. They prey on birds and mammals on the ground, diving from great heights at high speeds (up to 200 mph), surprising the prey. The fairly narrowly specialized nesting habits of the peregrine falcon meant that they were rare even prior to the pesticide problems; thus, DDT diminished already-small populations.

Reintroductions of peregrine falcons, starting in 1980, have been spectacularly successful. A detailed report published by the Colorado Division of Wildlife gives insight into the

FIGURE 15.24
Peregrine falcons suffered greatly as the result of DDT accumulation in the food chain. Successful reintroductions have been made in both native habitats and urban environments, where they prey on pigeons. *Photos: US Fish and Wildlife Service.*

intricacies of the falcon recovery program.[73] A trait that favors captive breeding is the tendency of birds to replace eggs collected from a wild nest with a replacement clutch. (This is termed *recycling*.) Thus, eggs can be removed from the wild without eliminating reproduction of the wild population. Captive-reared chicks are gradually "hacked" (partially liberated) from hand feeding so that they can learn how to hunt on their own. Hacking is a technique developed by falconers over centuries (perhaps millennia), and the peregrine reintroduction programs benefited greatly from taking advantage of existing methods for handling and releasing birds. Perhaps the most interesting aspect of the reintroductions, from a behavioral standpoint, is that peregrines show unexpected adaptability in nesting and feeding habits, accepting ledges on bridges and tall buildings as nesting sites. In urban environments, they prey on pigeons, which are typically plentiful. However, many of the released peregrine falcons fail to survive hazards such as power lines and transmission towers, and others disperse from their release points. The peregrine falcon experience, like that with black-footed ferrets, points out how trial and error can be used to improve release programs and how important it is to release large numbers of animals to achieve success. (In the case of peregrines, 14–16 birds are needed to establish one wild pair.) As a result of the ban on DDT use, peregrine eggshell thickness has increased over the past 20 years, leading to increased reproductive success and expanding populations in both urban and natural cliff settings.

527

OF SPECIAL INTEREST: CAPTIVE BREEDING AND REINTRODUCTIONS

Here is a list of critical behavioral points for captive breeding and reintroduction programs. The role of animal behaviorists in captive breeding and release programs is to ensure that enough is known about the behavior of the species to maximize the chances that the program will succeed.

- Match important habitat characteristics of wild and captive environments.
- Mimic teaching that occurs between parents and offspring in the wild to shape food preferences and foraging strategies.
- Understand dispersal patterns and recognize that introduction into the wild may trigger long-distance dispersal.
- Minimize human contact and habituation to potential threats.
- Manipulate neophobia so that it does not impair foraging but provides needed protection from threats.
- Understand the role of imprinting in the species' biology and provide species recognition cues that will be necessary later in life for mating, parenting, and social behavior.
- Monitor and evaluate the fates of reintroduced animals so that the captive rearing program receives useful feedback.

Ethical Considerations

A scientific approach to animal behavior allows discovery of the knowledge that managers need to create appropriate living environments. With so many imperiled species, maintenance of captive populations with an eye to re-establishment of populations in the wild at a later time is becoming a critical strategy in maintaining biodiversity. Understanding and attending to behavioral indicators of *animal welfare*, such as repetitive movements or excessive self-grooming (see Chapter 4), are key aspects of this.

> **KEY TERM** Animal welfare involves the consideration of the effects of human activities and behavior on animal well-being.

DISCUSSION POINT: ETHICS AND CAPTIVE POPULATIONS

What are the ethics of maintaining animals in captive populations for conservation purposes? Is it better to be extinct than captive? Some extreme advocates of "animal rights" have argued that captivity is so cruel, or that captive conditions are so unnatural, that from the animals' point of view, extinction is a better option than captivity. Others say that captive-born animals may adapt easily to captive conditions. Once again, this points to the interface of innate tendencies, cognition, and umwelt.

Of course, welfare, in the sense of good health, is easier to measure than "happiness," a concept that eludes definition even for many humans. Anyone who has watched a puppy eagerly seize upon some horse droppings left in the middle of a trail knows that there are considerable, and possibly unfathomable, interspecific differences in what is apparently experienced as a source of delight.

To the extent that people can actually "interview" some of the great apes, through the use of a variety of communication tools, this gap may be bridged for at least a few species. This is a tantalizing possibility, but one that still suffers from great limitations. The great apes that have been taught to communicate with their human research partners may not be representative of animals, even apes, in general. For most species, it may be better to focus on stress indicators such as cortisol and the correlation of such indicators with specific behaviors to get an idea of how "happy" an animal is. Moreover, given the variation in perceived happiness among humans, even if a "happiness indicator" is discovered for some species, it may not be possible to devise an environment that produces happiness across a wide range of individuals. To add more complexity, living "naturally" may prove to be neither happy (stress-free) nor healthy for many animals.

A related question concerns the appropriateness of zoos for the task of managing populations for captive breeding. Before the prevalence of television and web-based information resources, zoos played critical roles in allowing the public to see the behavior of animals, albeit under highly modified circumstances. Zoos were the premier venues for public education and wildlife awareness, albeit at a high cost to animals. With video available of many animals and of almost every conceivable behavior, one might conclude that much of the informative value of zoos has been superseded. There is nonetheless something about the presence of a living animal that video does not replace, particularly for children. Scientific knowledge gives zookeepers strategies for behavioral assessments and habitat enrichment that reduce some of the harsher aspects of captivity,[74] but the main advantage of zoos remains the fact that the public funds them, while funding for captive breeding programs that do not include public displays is much more difficult to maintain. Of our primary examples, only the peregrine falcon program has not involved zoos as a major platform for captive breeding.

The human behavior that supports zoos also shapes conservation priorities. The public's greatest affinity is for vertebrates. This means that the conservation status of the vast majority of animal species, which are invertebrates, is unknown and that much of the hope for invertebrate conservation lies in reserve design, rather than captive rearing and release.

528

15.6 HUMAN–WILDLIFE INTERFACE IN THE SUBURBS

There is no sharp dividing line between a city and relatively undisturbed habitat; instead, densely built and populated cities give way to suburbs, with more space between houses, which then give way to suburbanized countryside, where farms or ranches have been subdivided in larger tracts (often 30–50 acres), each with a house. Mountainous areas are typically dotted with cabins and summer homes. This type of development means that, from the point of view of the wildlife, there is rarely an area that is clearly unpopulated by humans, and there is no clear boundary that demarks areas to be avoided. This section explores what happens when human settlement and wildlife habitats overlap, or when wildlife find their way into more dense human settlements.

FIGURE 15.25
A radiocollared elk moving with a herd of elk. In this species tracking a single animal gives insights into the movement of the entire herd. *Photo: Michael Breed.*

Surprisingly little is known about the movements of many animals in and around human settlements. For example, animals thought to be shy around humans and neophobic, such as mountain lions, have been revealed to frequently move through suburban yards at night. A very important set of tools for wildlife biologists and conservation behaviorists allows tracking animals as they move through their habitat (Figure 15.25). Radiocollars, collars with cellular telephone technology, RFID tags, and bar codes all have roles in helping biologists to understand animal movements. More knowledge about animal movements allows better planning for management of human/animal interactions.

OF SPECIAL INTEREST: ARE GREEN BUILDINGS ALL THAT GREEN?

The number of birds killed by collisions with buildings, especially windows in buildings, might be close to one billion per year in the United States alone. (The low estimate is 365 million.) This exceeds deaths resulting from automobile collisions by an order of magnitude and more[75] (see Chapter 8 for a discussion of automobiles as selective agents on swallows). It makes windows the second largest anthropogenic avian mortality source in the United States; only feral and pet cats kill more birds—four times as many. Often moving at top speed, sometimes on migration, birds crash into windows and fall to the ground, stunned at best, often dead or dying. In other cases, birds see the reflection of a "rival" (themselves) or good habitat (actually behind them) and fly into windows. (One of the authors of this book had a double-paned bedroom window eventually broken by a persistent territorial robin who would begin assaulting the rival from close range every day at the crack of dawn.)

The search for a solution is not as straightforward as it might seem. Some buildings that have received prestigious environmental awards from the US Green Building Council's Leadership in Energy and Environmental Design rating system have been particularly lethal; this is not surprising when one realizes that one hallmark of a "green" building is a large expanse of windows that allow illumination from natural light.

What can be done to make such windows more visible to birds? The answer to that question is a work in progress, although opaque parachute cords hung at intervals outside windows show promise. Conservationists who grapple with this problem are again stymied by avian *umwelt*. To what extent do birds see reflections of wave lengths that are invisible to humans? Indeed, forward vision of some birds, if it exists at all, is poor. After all, for as long as birds have flown, they have owned that air space above the height of a four story building (the space where many collisions occur); until humans began to build very tall buildings, there was nothing there to cause a collision.[76–78]

FIGURE 15.26
Black bears are frequent visitors to suburban habitats and may be attracted by garbage. Their presence is often unwelcome because people view them as threats. *Photo: Steve Hillebrand, US Fish and Wildlife Service.*

Garbage Dump and Other Feeding Effects

Some wildlife, particularly skunks, raccoons, and bears, is attracted to human garbage. In the early history of the US national parks, garbage dumps actually became wildlife viewing posts, particularly for bears; wildlife officials eventually realized that for bears, the behavioral effects of feeding at dumps, including dependency on human-supplied food and hazards to human observers from habituated bears, outweighed any educational advantage gained by the viewing public. This has led to the development of considerable technology to protect garbage from these bears (and also from skunks and raccoons); a surprising level of sophistication in design is needed to keep these clever and strong foragers from accessing garbage (Figure 15.26). It has been even more challenging to regulate human behavior in suburban settings, and dealing with bears that come into yards is a major preoccupation for wildlife managers in many states. Because bears can be hazardous to humans, the typical approach is to trap or tranquilize the bear and release it in a remote location. This approach is often ineffective, because once a bear has habituated to humans and has learned to take advantage of easy pickings in suburbs, it is highly likely to return, even from great distances. In Colorado, first offenders are marked before release, and second-time offenders are euthanized. This should be a strong incentive for humans to manage their garbage and other attractions.

In general, feeding pets outdoors creates behavioral issues with wildlife, attracting animals that may pose hazards, facilitating habituation, and often ending with tragic consequences for the animals. In Florida, pet food attracts alligators (obviously a serious hazard for both pets and humans) and cane toads, a nuisance invasive species. Animal behaviorists can help to educate the public about simple measures in managing pet food and garbage that maintain separation between people and wildlife. This approach supports conservation goals because it makes reserve management more compatible with human neighbors.

BRINGING ANIMAL BEHAVIOR HOME: BIRD FEEDERS

Garbage and dog food are obviously detrimental to wildlife, but what about bird feeders? Members of the general public and some biologists view feeding birds as providing a benefit to wildlife, as well as something that brings interest and entertainment into backyards (Figure 15.27). The bulk of bird food is seed. In the United States and the United Kingdom together, 500,000 metric tons of birdseed are put in feeders each year.[79] To place this amount into perspective, a gondola railcar carries a little over 100 metric tons; a coal train of average length, about 100 cars, carries around 10,000 metric tons of coal. The amount of birdseed provided annually in the United States and the United Kingdom is equal in weight to the loads of 45–50 coal trains. This huge supplement to natural food sources has contributed to population gains and range expansions of many seed-eating birds, such as the northern cardinal and tufted titmouse in the northeastern United States and the house finch in the southwest. The house finch was also introduced into the eastern United States as a pet. Birds supported by feeders have also delighted many cats. Seed generalists such as the cardinal, willing to consume a variety of seeds, benefit the most from backyard feeders. In urban environments, feeders support populations of house sparrows, allowing them to succeed in this manmade habitat. The ultimate results of bird feeding may include selection for birds that have behavioral attributes for success at feeders but would do less well in unsupplemented environments. Range expansions of seed generalists may also impact, through competition for food, populations of less-flexible

FIGURE 15.27
Birds at feeders. The upper feeder is intended to attract birds, and the lower feeder is for domestic animals and helps to support invasive species, such as these English sparrows. *Photos: Michael Breed.*

531

seed specialists; the house finch affects Cassin's and Purple finches in this way. Feeding birds is a huge, and interesting, experiment, and conservation behaviorists do not yet fully understand the consequences. Finally, even birds that do not feed at feeders may not prosper as a result of proximity to humans and their food. Crow nestlings in the suburbs show signs of nutrient deficiency compared to their rural counterparts. They are smaller and have lower levels of serum protein and calcium. There was no evidence of calorie restriction, but important nutrients may be limited in suburban nestling diets.[80]

Public Fear and Overcoming Zoophobia

Public fear of wild animals, or *zoophobia*, often leads to pressures for translocating or killing animals. People are particularly fearful of large mammalian carnivores, snakes, stinging insects, and spiders. Biologists are sometimes arrogantly dismissive of these fears, and it is hard for an educated scientist to have much sympathy for someone who has a great desire to live in a house "in the forest" but then wants to kill the animal residents of the forest—a forest without animals is just an empty set of trees. For conservation behaviorists to remain credible, though, they need to be sympathetic to members of the public who lose their dog or cat to a fox, coyote, or mountain lion. Public education needs to focus on managing the behavior of both pets and predators to keep domestic animals separated from potential predators; in the long run, this is a major element of enlisting public support for the presence of wildlife.

FIGURE 15.28
A rattlesnake in a defensive pose. *Photo: Jeff Mitton.*

Conservation behaviorists can help to build public sympathy for important species, such as venomous snakes, that are generally reviled. Timber and prairie rattlesnakes, once abundant throughout the eastern and plains states, have suffered tremendously from a kill-on-sight mentality. This has extended to nonvenomous snakes as well; many people cannot distinguish among snakes and therefore feel obliged to kill them all. Snakes are important in controlling rodent populations, and understanding their behavior helps humans to avoid snakebites (Figure 15.28).

Other species are less hated but create significant problems when interfacing with humans. Like their rural ranching and farming counterparts, suburban gardeners often suffer from herbivory by animals such as rabbits and deer on ornamental and vegetable plants. Many people hold an image of deer as "Bambi," and public resistance to management by culling deer herds comes into strong conflict with gardeners who would like to protect their plants. Landscapers turn to behavioral barriers such as high fences to exclude deer or fine mesh screens around plants to foil rabbits, but these barriers are unsightly and often ineffective. Considerable mythology surrounds the use of odorous repellents to drive herbivores away; among the solutions sprayed on vegetation are predator urine, including mixtures thought to mimic territorial markings and mixtures that attempt to make vegetation taste unpalatable to the herbivores. Some of these treatments, particularly the urine-based mixes, bring unpleasant odors to gardens. Hydrolyzed casein (milk protein) shows promise as a repellent[81]; when it is combined with latex, which makes the protein stick to the plants, deer substantially reduce their consumption of plant seedlings. Hydrolyzed casein also has shown good results in Australia, where marsupials present many of the same problems that deer cause in North America and Europe. Whether the use of repellents will ultimately have value as deer overcome their neophobia is open to question; the spectacle of crows roosting on a scarecrow spotlights the power of habituation in defeating efforts to repel wildlife.

Perceived direct threat of animals to humans or pets often turns human residents of areas against animal inhabitants. Threats can be predation: foxes, coyotes, and mountain lions all seem to enjoy snacking on household pets. Mountain lions, albeit rarely, can turn to humans as potential prey. Foxes are particularly adept at preying on cats, and coyotes may entice dogs into play or courtship prior to turning the tables and consuming the dogs. In the summer of 2008, a mountain lion entered a home in Colorado and captured a dog while its owners slept in the same room.

The conservation consequences of human/predator interactions often play out in calls for trapping or killing of predators; as developments encroach into wooded areas or farmland, homeowners who are intolerant of predators work to destroy the natural values that initially motivated them to live in semirural areas. Many homeowners in suburban southern California view coyotes as intolerable threats to their children and pets. Very little can be done to modify predator behavior once it is established, but humans can learn to take simple precautions to not attract predators. These include supervising pets and not leaving windows and doors open. Education of homeowners concerning predator behavior and ways to minimize risk to children and pets is an important conservation goal.

Diseases in wildlife populations are even more likely to stimulate harsh public responses. In addition to the potential for wildlife to carry diseases such as brucellosis that affect ranch animals, urban and suburban wildlife populations can bring serious disease risk to their human neighbors and their pets. Prairie dog colonies can harbor plague; pigeons in urban environments may carry a respiratory fungus, *Cryptococcosis*; many mammalian species have the potential to harbor rabies; and some wildlife (most notably foxes and raccoons in suburban environments) can carry canine distemper. Heated debates among homeowners, some of them conservationists who enjoy seeing wildlife in their neighborhood and others who are terrified of disease, are extremely difficult to resolve.

OF SPECIAL INTEREST: MARK TWAIN AND WILE E. COYOTE

One of the first non-native-American descriptions of the coyote was penned by the Prince of American Letters, Mark Twain. He had a grudging admiration for an animal that, over the intervening years, has proven to be an adaptable survivor (Figure 15.29). When Chuck Jones, a highly successful animation artist, encountered the passage below, he was inspired to create the unforgettable cartoon character—Wile E. Coyote.

> ...he was not a pretty creature or respectable either, for I got well acquainted with his race afterward, and can speak with confidence. The cayote is a long, slim, sick and sorry-looking skeleton, with a gray wolf-skin stretched over it, a tolerably bushy tail that forever sags down with a despairing expression of forsakenness and misery...He is always hungry. He is always poor, out of luck and friendless. The meanest creatures despise him, and even the fleas would desert him for a velocipede. He is so spiritless and cowardly that even while his exposed teeth are pretending a threat, the rest of his face is apologizing for it. And he is so homely!—so scrawny, and ribby, and coarse-haired, and pitiful. When he sees you he lifts his lip and lets a flash of his teeth out, and then turns a little out of the course he was pursuing, depresses his head a bit, and strikes a long, soft-footed trot through the sage-brush, glancing over his shoulder at you, from time to time, till he is about out of easy pistol range, and then he stops and takes a deliberate survey of you; he will trot fifty yards and stop again—another fifty and stop again; and finally the gray of his gliding body blends with the gray of the sage-brush, and he disappears...The cayote lives chiefly in the most desolate and forbidding desert, along with the lizard, the jackass-rabbit and the raven, and gets an uncertain and precarious living, and earns it. He seems to subsist almost wholly on the carcasses of oxen, mules and horses that have dropped out of emigrant trains and died, and upon windfalls of carrion, and occasional legacies of offal bequeathed to him by white men who have been opulent enough to have something better to butcher than condemned army bacon....He does not mind going a hundred miles to breakfast, and a hundred and fifty to dinner, because he is sure to have three or four days between meals, and he can just as well be traveling and looking at the scenery as lying around doing nothing and adding to the burdens of his parents.[82]

FIGURE 15.29
A coyote. This species has been reviled by ranchers but also revered for its cleverness and ability to elude eradication.
Photo: Jeff Mitton.

533

Maintaining Neophobia in Support of Conservation Goals

Social learning appears to help some species adapt to the presence of humans, cars, houses, and other features of the suburban landscape. Usually, this takes several generations of animals, as each new generation learns to be a little less fearful than previous generations. The persistence of coyotes in suburban settings is probably in part due to loss of neophobia, but a variety of animals, including deer, mountain lions, raccoons, foxes, and porcupines, have taken up residence in suburban settings. Among birds, the starlings, house (English) sparrows, house finches, robins, magpies, and crows all seem to easily adapt to suburban environments, overcoming any neophobia that might keep other species of birds in more natural settings. A yard in Denver, CO, may be frequented by Eastern fox squirrels, starlings, house sparrows, and house finches, none of which are native species to the area, but all of which adapt, behaviorally, to close co-existence with humans. Lack of neophobia may be an important part of such adaptability.

As mentioned previously, for many animals neophobia is a protective adaptation. Benign interactions with humans, particularly if they persist across several generations, seem to result in progressively bold wildlife. Feeding, including the availability of pet food and garbage, only accelerates this process of habituation. This does not matter much if the fed animal is a songbird but can have tragic results, for human and animal alike, if the fed animal is a raccoon, coyote, or bear. Obviously, eliminating human-provided food is an important goal.[83]

More methods of maintaining neophobia include use of loud noises; introduction of predator models, calls, or scents; and flashing lights. All of these have startle value, but if produced regularly and without physical consequence to the animal, habituation is the near-certain result. Deer respond to rifle hunting with shyness, not because they fear the sound of gunfire but because they are observant and see other animals injured or killed. Using a variety of aversive stimuli in random order is more effective than repeating a single stimulus. In species that are socially observant and can learn from the experiences of other animals, observing injuries or deaths is a powerful reinforcement of neophobia, but this approach is often unacceptable to the general public.

Developing techniques for encouraging neophobia is a current challenge for animal behaviorists. An increased understanding of how species of concern react to aversive stimuli will result in more effective manipulation of the human–wildlife interface, allowing people and wild animals to live in close proximity without unfortunate interactions.

SUMMARY: THE FUTURE AND CONSERVATION BEHAVIOR

Conservation behavior provides tools for understanding and possibly manipulating animal behavior to further conservation goals. Keeping animals in captivity can help to preserve endangered species, conserve genetic resources, or obtain populations of animals for reintroduction. To do this requires research on mating systems, territoriality, and foraging behavior. The introduction of captive animals into natural habitats must be informed by knowledge of dispersal behavior and of how animals obtain critical information for survival.

Neophobia, the fear of new situations, protects animals from risky situations and can be manipulated to keep human and animal populations apart. The lack of neophobia is a key element of success in invasive animals and contributes to negative human–wildlife interactions. Activities such as hunting, feeding birds, and leaving pet food outdoors have behavioral effects that often result in public pressure to relocate or kill wildlife. Human activities in conservation areas also affect wildlife behavior, and future design of reserves must take these effects into account.

This chapter, more than any other chapter in this book, highlights large and important gaps in current knowledge. There is some irony here, because conservation behavior is also the culmination of almost everything behavioral biologists have learned in the past few decades. From life history traits to foraging behavior, from observational learning to auditory signals and noise pollution, every aspect of animal behavior has the potential to help meet the challenge of conserving the natural world.

For conservation programs to be effective, far more knowledge is needed about habitat choice, mating systems, learning and imprinting, the functions of neophobia in animals' lives, as well as how to educate the public about interacting with wildlife. As the field of conservation behavior matures, it will play a key role in the protection of endangered species, reintroductions of animals that are extinct in the wild, and the success of reserve designs. Learning about animal behavior and its relationship to conservation is a critical field of biology for the twenty-first century. In fact, as human populations continue to invade and alter animal habitats, such understanding is more than an essential conservation tool—it is our ethical obligation.

STUDY QUESTIONS

1. More aggressive livestock protection dogs might be more successful at repelling large predators. Drawing on your knowledge of behavior genetics (Chapter 3), what might be some pitfalls associated with breeding for aggression? How might such a program conflict with other land uses?

2. Habitat fragmentation—the breakdown of a previously large area of habitat into small pieces by roads, buildings, and the like—has serious conservation implications. What effects might fragmentation have on mating systems? Think about all the different aspects of behavior that we've discussed—signaling, territoriality, types of mating systems—as you formulate your answer.

3. Captive rearing of animals for release in the wild has been successful with California condors and black-footed ferrets, but these efforts have been criticized because they take conservation dollars away from habitat protection and other priorities. Discuss the pros and cons of rearing animals in captivity for later release in natural habitats. If you were trying to predict the success of a captive rearing/release program, what do you think the most important behavioral factors would be? How would a behaviorist contribute to designing such a program?

4. You are given the job of designing a conservation plan for a large mammalian species. What would you need to know about the species' behavior before designing your plan? If money for research was limited, what information would you choose to gain first?

5. You are given a grant to study the conservation of an endangered animal species. Knowing only that the animal is rare and has a limited habitat, what, in your view, are the most important things to know about an animal's behavior? If your time and funds are limited, what would you want to discover first (about its behavior), and why?

6. You are an elected official and you have an opportunity to invest tax dollars in some land to be kept open and in a natural state, so that the wildlife species composition can be maintained. The public you serve has a variety of concerns about this—some want bike trails, some want horse trails, dog-owners want an off-leash space, and some constituents question this use of tax dollars when other needs are pressing. (You can imagine other public input!) Think carefully about these issues and their possible effects on the land in question and on future public support of natural areas. How do you respond, and why?

Further Reading

Angeloni, L., Crooks, K.R., Blumstein, D.T., 2010. Conservation and behavior: introduction. In: Breed, M.D., Moore, J. (Eds.), Encyclopedia of Animal Behavior, vol. 1, Academic Press, Oxford, pp. 377–381.

Benson-Amram, S., Weldele, M.L., Holekamp, K.E., 2012. A comparison of innovative problem-solving abilities between wild and captive spotted hyaenas, *Crocuta crocuta*. Anim. Behav. 85, 349–356.

Brumm, H., 2010. Anthropogenic noise: implications for conservation. In: Breed, M.D., Moore, J. (Eds.), Encyclopedia of Animal Behavior, vol. 1, Academic Press, Oxford, pp. 89–93.

Fuller, E., 2014. Lost Animals: Extinction and the Photographic Record. Princeton University Press, Princeton, NJ, 256 pp.

Simberloff, D., 2013. Invasive Species What Everyone Needs to Know. Oxford University Press, New York, NY.

Notes

1. Curio, E., 1996. Conservation needs ethology. Trends Ecol. Evol. 11, 260–263.
2. Angeloni, L., Schlaepfer, M.A., Lawler, J.J., Crooks, K.R., 2008. A reassessment of the interface between conservation and behaviour. Anim. Behav. 75, 731–737.
3. Caro, T., 2007. Behavior and conservation: a bridge too far? Trends Ecol. Evol. 22, 394–400.
4. Swaisgood, R.R., 2007. Current status and future directions of applied behavioral research for animal welfare and conservation. Appl. Anim. Behav. Sci. 102, 139–162.
5. Stamps, J.A., Swaisgood, R.R., 2007. Someplace like home: experience, habitat selection and conservation biology. Appl. Anim. Behav. Sci. 102, 392–409.
6. Kearney, M., Shine, R., Porter, W.P., 2009. The potential for behavioral thermoregulation to buffer "cold-blooded" animals against climate warming. Proc. Natl. Acad. Sci. USA 106, 3835–3840. http://dx.doi.org/10.1073/pnas.0808913106.
7. Mandel, J.T., Donlan, C.J., Armstrong, J., 2010. A derivative approach to endangered species conservation. Front. Ecol. Environ. 8, 44–49.
8. Finkelstein, M.E., Doak, D.F., George, D., Burnett, J., Brandt, J., Church, M., et al., 2012. Lead poisoning and the deceptive recovery of the critically endangered California condor. Proc. Natl. Acad. Sci. USA 109, 11449–11454.
9. Watts, A.J.R., Lewis, C., Goodhead, R.M., Beckett, S.J., Moger, J., Tyler, C.R., et al., 2014. Uptake and retention of microplastics by the shore crab *Carcinus maenas*. Environ. Sci. Technol. 48, 8823–8830.
10. Cozar, A., Echevarria, F., Gonzalez-Gordillo, J.I., Ingoien, X., Ubeda, B., Hernandez-Leon, S., et al., 2014. Plastic debris in the open ocean. Proc. Natl. Acad. Sci. USA 111, 10239–10244.
11. Vanderwerf, E.A., Rohrer, J.L., Smith, D.G., Burt, M.D., 2001. Current distribution and abundance of the O'ahu 'Elepaio. Wilson Bull. 113, 10–16.
12. MacArthur, R.H., Wilson, E.O., 1967. The Theory of Island Biogeography. Princeton University Press, Princeton, NJ, 203 pp.
13. Quammen, D., 1996. The Song of the Dodo: Island Biogeography in an Age of Extinctions. Scribner, New York, NY, 704 pp.
14. Simberloff, D., 1987. The spotted owl fracas: mixing academic, applied and political ecology. Ecology 68, 766–772.
15. Ritzi, C.M., Everson, B.L., Whittaker Jr., J.O., 2005. Use of bat boxes by a maternity colony of Indiana *Myotis* (*Myotis sodalist*). Northeastern Nat. 12, 217–220.
16. Kelm, D.H., Wiesner, K.R., von Helversen, O., 2008. Effects of artificial roosts for frugivorous bats on seed dispersal in a neotropical forest pasture mosaic. Conserv. Biol. 22, 733–741.
17. Whitaker, J.O., Sparks, D.W., Brack, V., 2006. Use of artificial roost structures by bats at the Indianapolis International Airport. Environ. Manage. 38, 28–36.
18. Groombridge, J., Massey, J.G., Bruch, J.C., Malcolm, T., Brosius, C.N., Okada, M.M., et al., 2004. An attempt to recover the Po'ouli by translocation and an appraisal of recovery strategy for bird species of extreme rarity. Biol. Conserv. 118, 365–375.
19. Liley, S., Creel, S., 2008. What best explains vigilance in elk: characteristics of prey, predators, or the environment? Behav. Ecol. 19, 245–254.
20. Mueller, T.R., O'Hara, B., Converse, S.J., Urbanek, R.P., Fagan, W.F., 2013. Social learning of migratory performance. Science 341, 999–1002.
21. Henke, S.E., Bryant, F.C., 1999. Effects of coyote removal on the faunal community in western Texas. J. Wildl. Manage. 63, 1066–1081.
22. Gehring, T.M., VerCauteren, D.C., Landry, J.-M., 2010. Livestock protection dogs in the 21st century: is an ancient tool relevant to modern conservation challenges? BioScience 60, 299–308.
23. Hung, C.-M., Shaner, P.-J.L., Zink, R.M., Liu, W.-C., Chu, T.-C., Huang, W.-S., et al., 2014. Drastic population fluctuations explain the rapid extinction of the passenger pigeon. Proc. Natl. Acad. Sci. USA 111, 10636–10641.
24. Henderson, R.W., 2004. Lesser Antillean snake faunas: distribution, ecology and conservation concerns. Oryx 38, 311–320.
25. Coulon, A., Fitzpatrick, J.W., Bowman, R., Lovette, I.J., 2010. Effects of habitat fragmentation on effective dispersal of Florida scrub-jays. Conserv. Biol. 24, 1080–1088.
26. Apfelbeck, B., Raess, M., 2008. Behavioural and hormonal effects of social isolation and neophobia in a gregarious bird species, the European starling (*Sturnus vulgaris*). Horm. Behav. 54, 435–441.
27. Echeverria, A.I., Vassallo, A.I., 2008. Novelty responses in a bird assemblage inhabiting an urban area. Ethology 114, 616–624.
28. Martin, L.B., Fitzgerald, L., 2005. A taste for novelty in invading house sparrows, *Passer domesticus*. Behav. Ecol. 16, 702–707.
29. Mettke-Hofmann, C., Winkler, H., Leisler, B., 2002. The significance of ecological factors for exploration and neophobia in parrots. Ethology 108, 249–272.
30. Aronsen, G., 2007. Animals behaving badly and the people who love/hate them. J. Mammal. Evol. 14, 71–73.
31. Kaulfuss, P., Mills, D.S., 2008. Neophilia in domestic dogs (*Canis familiaris*) and its implication for studies of dog cognition. Anim. Cogn. 11, 553–556.
32. Marker, L.L., Wilkerson, A.J.P., Sarno, R.J., Martenson, J., Breitenmoser-Wuersten, C., O'Brien, S.J., et al., 2008. Molecular genetic insights on cheetah (*Acinonyx jubatus*) ecology and conservation in Namibia. J. Hered. 99, 2–13.

33. Saura, M., Perez-Figueroa, A., Fernandez, J., Toro, M.A., Caballero, A., 2008. Preserving population allele frequencies in *ex situ* conservation programs. Conserv. Biol. 22, 1277–1287.

34. Agetsuma, N., 2007. Minimum area required for local populations of Japanese macaques estimated from the relationship between habitat area and population extinction. Int. J. Primatol. 28, 97–106.

35. Fujiwara, M., 2007. Extinction-effective population index: incorporating life-history variations in population viability analysis. Ecology 88, 2345–2353.

36. Peter, A., Lande, R., 1993. A population viability analysis for African elephant (*Loxodonta africana*): how big should reserves be? Conserv. Biol. 7, 602–610.

37. Cowley, D.E., 2008. Estimating required habitat size for fish conservation in streams. Aquat. Conserv. Mar. Freshw. Ecosyst. 18, 418–431.

38. Wiersma, Y.F., Nudds, T.D., 2009. Efficiency and effectiveness in representative reserve design in Canada: the contribution of existing protected areas. Biol. Conserv. 142, 1639–1646.

39. O'Brien, C.M., Crowther, M.S., Dickman, C.R., Keating, J., 2008. Metapopulation dynamics and threatened species management: why does the broad-toothed rat (*Mastacomys fuscus*) persist? Biol. Conserv. 141, 1962–1971.

40. Moore, R.P., Robinson, W.D., Lovette, I.J., Robinson, T.R., 2008. Experimental evidence for extreme dispersal limitation in tropical forest birds. Ecol. Lett. 11, 960–968.

41. Fischer, J., Lindenmayer, D.B., 2000. An assessment of the published results of animal relocations. Biol. Conserv. 96, 1–11.

42. Teixeira, C.P., de Azevedo, C.S., Mendl, M., Cipreste, C.F., Young, R.J., 2007. Revisiting translocation and reintroduction programmes: the importance of considering stress. Anim. Behav. 73, 1–13.

43. Hinam, H.L., Clair, C.C.S., 2008. High levels of habitat loss and fragmentation limit reproductive success by reducing home range size and provisioning rates of Northern saw-whet owls. Biol. Conserv. 141, 524–535.

44. Fuentes-Montemayor, E., Cuaron, A.D., Vazquez-Dominguez, E., Benitez-Malvido, J., Valenzuela-Galvan, D., Andresen, E., 2009. Living on the edge: roads and edge effects on small mammal populations. J. Anim. Ecol. 78, 857–865.

45. Baker, J., French, K., Whelan, R.J., 2002. The edge effect and ecotonal species: bird communities across a natural edge in southeastern Australia. Ecology 83, 3048–3059.

46. Burger, L.D., Burger, L.W., Faaborg, J., 1994. Effects of prairie fragmentation on predation on artificial nests. J. Wildl. Manage. 58, 249–254.

47. Baguette, M., Van Dyck, H., 2007. Landscape connectivity and animal behavior: functional grain as a key determinant for dispersal. Landscape Ecol. 22, 1117–1129.

48. Young, B.E., Sherry, T.W., Sigel, B.J., Woltmann, S., 2008. Nesting success of Costa Rican lowland rain forest birds in response to edge and isolation effects. Biotropica 40, 615–622.

49. Mbora, D.N.M., Meikle, D.B., 2004. Forest fragmentation and the distribution, abundance and conservation of the Tana River red colobus (*Procolobus rufomitratus*). Biol. Conserv. 118, 67–77.

50. Robinson, S.K., Thompson III, F.R., Donovan, T.M., Whitehead, D.R., Faaborg, J., 1995. Regional forest fragmentation and the nesting success of migratory birds. Science 267, 1987–1990.

51. Laiola, P., Tella, J.L., 2007. Erosion of animal cultures in fragmented landscapes. Front. Ecol. Environ. 5, 68–72.

52. Kazmaier, R.T., Hellgren, E.C., Ruthven III, D.C., 2001. Habitat selection by the Texas tortoise in a managed thornscrub ecosystem. J. Wildl. Manage. 65, 653–660.

53. Pichegru, L., Gremillet, D., Crawford, R.J.M., Ryan, P.G., 2010. Marine no-take zone rapidly benefit threatened penguin. Biol. Lett. 6, 498–501.

54. Smith, H., Samuels, A., Bradley, S., 2008. Reducing risky interactions between tourists and free-ranging dolphins (*Tursiops* sp.) in an artificial feeding program at Monkey Mia, Western Australia. Tourism Manage. 29, 994–1001.

55. Hardin, G., 1968. The tragedy of the commons. Science 162, 1243–1248.

56. Crooks, K.R., Soule, M.E., 1999. Mesopredator release and avifaunal extinctions in a fragmented system. Nature 400, 563.

57. Taylor, A.R., Knight, R.L., 2003. Wildlife responses to recreation and associated visitor perceptions. Ecol. Appl. 13, 951–963.

58. Fletcher, R.J., McKinney, S.T., Bock, C.E., 1999. Effects of recreational trails on wintering diurnal raptors along riparian corridors in a Colorado grassland. J. Raptor Res. 33, 233–239.

59. Keyser, A.J., 2002. Nest predation in fragmented forests: landscape matrix by distance from edge interactions. Wilson Bull. 114, 186.

60. Marzluff, J.M., Ewing, K., 2001. Restoration of fragmented landscapes for the conservation of birds: a general framework and specific recommendations for urbanizing landscapes. Restoration Ecol. 9, 280–292.

61. Mason, J., Moorman, C., Hess, G., Sinclair, K., 2007. Designing suburban greenways to provide habitat for forest-breeding birds. Landscape Urban Plan 80, 153–164.

62. McKinney, M.L., 2005. Scaling of park trail length and visitation with park area: conservation implications. Anim. Conserv. 8, 135–141.

63. Packard, J.M., Frohlich, R.K., Reynolds III, J.E., Wilcox, J.R., 1989. Manatee response to interruption of a thermal effluent. J. Wildl. Manage. 53, 692–700.

64. Rodgers, J.A., Schwikert, S.T., 2002. Buffer-zone distances to protect foraging and loafing waterbirds from disturbance by personal watercraft and outboard-powered boats. Conserv. Biol. 16, 216–224.

65. Lenth, B.E., Knight, R.L., Brennan, M.E., 2008. The effects of dogs on wildlife communities. Nat. Areas J. 28, 218–227.

537

66. Tarlow, E.M., Blumstein, D.T., 2007. Evaluating methods to quantify anthropogenic stressors on wild animals. Appl. Anim. Behav. Sci. 102, 429–451.

67. Stankowich, T., 2008. Ungulate flight responses to human disturbance: a review and meta-analysis. Biol. Conserv. 14, 2159–2173.

68. McDougall, P.T., Reale, D., Sol, D., Reader, S.M., 2006. Wildlife conservation and animal temperament: causes and consequences of evolutionary change for captive, reintroduced, and wild populations. Anim. Conserv. 9, 39–48.

69. Keith-Lucas, T., White, F.J., Keith-Lucas, L., Vick, L.G., 1999. Changes in behavior in free-ranging *Lemur catta* following release in a natural habitat. Am. J. Primatol. 47, 15–28.

70. Benhaiem, S., Delon, M., Lourtet, B., Cargnelutti, B., Aulagnier, S., Hewison, A.J.M., et al., 2008. Hunting increases vigilance levels in roe deer and modifies feeding site selection. Anim. Behav. 76, 611–618.

71. Souris, A.C., Kaczensky, P., Julliard, R., Walzer, C., 2007. Time budget-, behavioral synchrony- and body score development of a newly released Przewalski's horse group *Equus ferus przewalskii*, in the great Gobi B strictly protected area in SW Mongolia. Appl. Anim. Behav. Sci. 107, 307–321.

72. Stoinski, T.S., Beck, B.B., 2004. Changes in locomotor and foraging skills in captive-born, reintroduced golden lion tamarins (*Leontopithecus rosalia rosalia*). Am. J. Primatol. 62, 1–13.

73. Craig, G.R., Enderson, J.H., 2004. Peregrine Falcon Biology and Management in Colorado 1973–2001. Technical Publication., No. 43, Colorado Division of Wildlife. ISSN 0084-8883. http://cpw.state.co.us/documents/wildlifespecies/profiles/peregrine.pdf.

74. Maple, T.L., Finlay, T.W., 1989. Applied primatology in the modern zoo. Zoo Biol. 8, 101–116.

75. Brown, C.R., Brown, M.B., 2013. Where has all the road kill gone? Curr. Biol. 23, R233–R234.

76. Klem Jr., D., Saenger, P.G., 2013. Evaluating the effectiveness of select visual signals to prevent bird-window collisions. Wilson J. Ornithol. 125, 406–411.

77. Loss, S.R., Will, T., Loss, S.S., Marra, P.P., 2014. Bird-building collisions in the United States: estimates of annual mortality and species vulnerability. Condor Ornithol. Appl. 116, 8–23.

78. Ogden, L.E., 2014. Does green building come up short in considering biodiversity? BioScience 64, 83–89.

79. Robb, G.N., McDonald, R.A., Chamberlain, D.E., et al., 2008. Food for thought: supplementary feeding as a driver of ecological change in avian populations. Front. Ecol. Environ. 6, 476–484.

80. Heiss, R.S., Clark, A.B., McGowan, K.J., 2009. Growth and nutritional state of American Crow nestlings vary between urban and rural habitats. Ecol. Appl. 19, 829–839.

81. Kimball, B.A., Russell, J.H., DeGraan, J.P., Perry, K.R., 2008. Screening hydrolyzed casein as a deer repellent for reforestation applications. West. J. Appl. Forestry 23, 172–176.

82. Twain, M., 1872. Roughing It. Chapter 5, available online at: http://www.mtwain.com/Roughing_It/6.html.

83. Restani, M., Marzluff, J.M., Yates, R.E., 2001. Effects of anthropogenic food sources on movements, survivorship, and sociality of common ravens in the Arctic. Condor 103, 399–404.

541

543

CPI Antony Rowe
Chippenham, UK
2017-06-08 03:07